Lecture Notes in Artificial Intelligence 10234

Subseries of Lecture Notes in Computer Science

Jinho Kim · Kyuseok Shim
Longbing Cao · Jae-Gil Lee
Xuemin Lin · Yang-Sae Moon (Eds.)

Advances in Knowledge Discovery and Data Mining

21st Pacific-Asia Conference, PAKDD 2017
Jeju, South Korea, May 23–26, 2017
Proceedings, Part I

 Springer

Editors
Jinho Kim
Kangwon National University
Chuncheon
Korea (Republic of)

Jae-Gil Lee
KAIST
Daejeon
Korea (Republic of)

Kyuseok Shim
Seoul National University
Seoul
Korea (Republic of)

Xuemin Lin
University of New South Wales
Sydney, NSW
Australia

Longbing Cao ⓘ
University of Technology Sydney
Sydney, NSW
Australia

Yang-Sae Moon ⓘ
Kangwon National University
Chuncheon
Korea (Republic of)

ISSN 0302-9743   ISSN 1611-3349 (electronic)
Lecture Notes in Artificial Intelligence
ISBN 978-3-319-57453-0   ISBN 978-3-319-57454-7 (eBook)
DOI 10.1007/978-3-319-57454-7

Library of Congress Control Number: 2017938164

LNCS Sublibrary: SL7 – Artificial Intelligence

Printed on acid-free paper

This Springer imprint is published by Springer Nature
The registered company is Springer International Publishing AG
The registered company address is: Gewerbestrasse 11, 6330 Cham, Switzerland

PC Chairs' Preface

It is our great pleasure to introduce the proceedings of the 21st Pacific-Asia Conference on Knowledge Discovery and Data Mining (PAKDD 2017).

We received a record-breaking number of 458 submissions from 36 countries all over the world. This highest number of submissions is very encouraging because it reflects the improving status of PAKDD. To rigorously review the submissions, we conducted a double-blind review following the tradition of PAKDD and constructed the largest ever committee consisting of 38 Senior Program Committee (SPC) members and 196 Program Committee (PC) members. Each valid submission was reviewed by three PC members and meta-reviewed by one SPC member who also led the discussion with the PC members. We, the PC co-chairs, considered the recommendations from the SPC members and looked into each submission as well as its reviews to make the final decisions. Borderline papers were thoroughly discussed by us before final decisions were made.

As a result, 129 out of 458 papers were accepted, yielding an acceptance rate of 28.2%. Among them, 45 papers were selected as long-presentation papers, and 84 papers were selected as regular-presentation papers. Mining social networks or graph data was the most popular topic in the accepted papers. The review process was supported by the Microsoft CMT system. During the three main conference days, these 129 papers were presented in 23 research sessions. A long-presentation paper was given 25 minutes for presentation, and a regular-presentation paper was given 15 minutes for presentation. These two types of papers, however, are not distinguished in the proceedings.

We would like to thank all SPC members, PC members, and external reviewers for their hard work to provide us with thoughtful and comprehensive reviews and recommendations. Also, we would like to express our sincere thanks to Yang-Sae Moon for compiling all accepted papers and for working with the Springer team to produce the proceedings.

We hope that the readers of the proceedings find the content interesting and rewarding.

April 2017   Longbing Cao
                                                                          Jae-Gil Lee
                                                                          Xuemin Lin

General Chairs' Preface

Welcome to the proceedings of the 21st Pacific-Asia Conference on Knowledge Discovery and Data Mining. PAKDD has successfully brought together researchers and developers since 1997, with the purpose of identifying challenging problems facing the development of advanced knowledge discovery. After 14 years since PAKDD 2003 in Seoul, PAKDD was held again in Korea, during May 23–26, 2017, in Jeju Island.

We are very grateful to the many authors who submitted their work to the PAKDD 2017 technical program. The technical program was enhanced by three keynote speeches, delivered by Sang Cha from Seoul National University, Rakesh Agrawal from Data Insights Laboratories, and Dacheng Tao from the University of Sydney. In addition to the main technical program, the offerings of this conference were further enriched by three tutorials as well as four international workshops on leading-edge topics.

We would like to acknowledge the key contributions by Program Committee co-chairs, Longbing Cao, Jae-Gil Lee, and Xuemin Lin. We would like to extend our gratitude to the workshop co-chairs, U. Kang, Ee-Peng Lim, and Jeffrey Xu Yu; the tutorial co-chairs, Dongwon Lee, Yasushi Sakurai, and Hwanjo Yu; the contest co-chairs, Nitesh Chawla, Younghoon Kim, and Young-Koo Lee; the publicity co-chairs, Sang-Won Lee, Guoliang Li, Steven Whang, and Xiaofang Zhou; the registration co-chairs, Min-Soo Kim and Wookey Lee; the local Arrangements co-chairs, Joonho Kwon, Jun-Ki Min, Chan Jung Park, and Young-Ho Park; the Web chair, Ha-Joo Song; the finance co-chairs, Jaewoo Kang and Jaesoo Yoo; the treasury chair, Chulyun Kim; and the proceedings chair, Yang-Sae Moon. We would like to express our special thanks to our honorary chair, Kyu-Young Whang, for providing valuable advice on all aspects of the conference's organization.

We are grateful to our sponsors that include: platinum sponsors — Asian Office of Aerospace Research & Development/Air Force Office of Scientific Research, Mirhenge, Naver, NCSOFT, Seoul National University Big Data Institute and SK Holdings C&C; gold sponsors — KISTI (Korea Institute of Science and Technology Information); silver sponsors — Daumsoft, Douzone, HiBrainNet, Korea Data Agency, and SK Telecom; and publication sponsors — Springer for their generous and valuable support. We are also thankful to the PAKDD Steering Committee for its guidance and Best Paper Award, Student Travel Award, and Early Career Research Award sponsorship. In addition, we would like to express our gratitude to the KIISE Database Society of Korea for hosting this conference. Finally, we thank the student volunteers and everyone who helped us in organizing PAKDD 2017.

April 2017

Jinho Kim
Kyuseok Shim

Organization

Organizing Committee

Honorary Chair

Kyu-Young Whang   DGIST/KAIST, Korea

General Co-chairs

Jinho Kim   Kangwon National University, Korea
Kyuseok Shim   Seoul National University, Korea

Program Committee Co-chairs

Longbing Cao   University of Technology Sydney, Australia
Jae-Gil Lee   KAIST, Korea
Xuemin Lin   University of New South Wales, Australia

Tutorial Co-chairs

Dongwon Lee   Pennsylvania State University, USA
Yasushi Sakurai   Kumamoto University, Japan
Hwanjo Yu   POSTECH, Korea

Workshop Co-chairs

U Kang   Seoul National University, Korea
Ee-Peng Lim   Singapore Management University, Singapore
Jeffrey Xu Yu   The Chinese University of Hong Kong, SAR China

Publicity Co-chairs

Sang-Won Lee   Sungkyunkwan University, Korea
Guoliang Li   Tsinghua University, China
Steven Euijong Whang   Google Research, USA
Xiaofang Zhou   University of Queensland, Australia

Finance Co-chairs

Jaewoo Kang   Korea University, Korea
Jaesoo Yoo   Chungbuk National University, Korea

Treasury Chair

Chulyun Kim   Sookmyung Women's University, Korea

Proceedings Chair

Yang-Sae Moon   Kangwon National University, Korea

Contest Co-chairs

Nitesh Chawla   University of Notre Dame, USA
Younghoon Kim   Hanyang University, Korea
Young-Koo Lee   Kyung Hee University, Korea

Local Arrangements Co-chairs

Joonho Kwon   Pusan National University, Korea
Jun-Ki Min   Korea University of Technology and Education, Korea
Chan Jung Park   Jeju National University, Korea
Young-Ho Park   Sookmyung Women's University, Korea

Registration Co-chairs

Min-Soo Kim   DGIST, Korea
Wookey Lee   Inha University, Korea

Web Chair

Ha-Joo Song   Pukyong National University, Korea

Steering Committee

Co-chairs

Tu Bao Ho   Japan Advanced Institute of Science and Technology,
                              Japan
Ee-Peng Lim   Singapore Management University, Singapore

Treasurer

Graham Williams   Togaware, Australia (see also under Life Members)

Members

Tu Bao Ho   Japan Advanced Institute of Science and Technology,
                              Japan (Member since 2005, Co-chair 2012–2014,
                              Chair 2015–2017, Life Member since 2013)
Ee-Peng Lim   Singapore Management University, Singapore (Member
                              since 2006, Co-chair 2015–2017)
Thanaruk   Thammasat University, Thailand (Member since 2009)
  Theeramunkong
P. Krishna Reddy   International Institute of Information Technology,
                              Hyderabad (IIIT-H), India (Member since 2010)
Joshua Z. Huang   Shenzhen Institutes of Advanced Technology, Chinese
                              Academy of Sciences, China (Member since 2011)

Longbing Cao	University of Technology Sydney, Australia (Member since 2013)
Jian Pei	Simon Fraser University, Canada (Member since 2013)
Myra Spiliopoulou	Otto von Guericke University of Magdeburg, Germany (Member since 2013)
Vincent S. Tseng	National Cheng Kung University, Taiwan (Member since 2014)
Tru Hoang Cao	Ho Chi Minh City University of Technology, Vietnam (Member since 2015)
Gill Dobbie	University of Auckland, New Zealand (Member since 2016)

Life Members

Hiroshi Motoda	AFOSR/AOARD and Osaka University, Japan (Member since 1997, Co-chair 2001–2003, Chair 2004–2006, Life Member since 2006)
Rao Kotagiri	University of Melbourne, Australia (Member since 1997, Co-chair 2006–2008, Chair 2009–2011, Life Member since 2007, Treasury Co-sign since 2006)
Huan Liu	Arizona State University, USA (Member since 1998, Treasurer 1998–2000, Life Member since 2012)
Ning Zhong	Maebashi Institute of Technology, Japan (Member since 1999, Life Member since 2008)
Masaru Kitsuregawa	Tokyo University, Japan (Member since 2000, Life Member since 2008)
David Cheung	University of Hong Kong, SAR China (Member since 2001, Treasurer 2005–2006, Chair 2006–2008, Life Member since 2009)
Graham Williams	Australian National University, Australia (Member since 2001, Treasurer since 2006, Co-chair 2009–2011, Chair 2012–2014, Life Member since 2009)
Ming-Syan Chen	National Taiwan University, Taiwan, ROC (Member since 2002, Life Member since 2010)
Kyu-Young Whang	Korea Advanced Institute of Science and Technology, Korea (Member since 2003, Life Member since 2011)
Chengqi Zhang	University of Technology Sydney, Australia (Member since 2004, Life Member since 2012)
Zhi-Hua Zhou	Nanjing University, China (Member since 2007, Life Member since 2015)
Jaideep Srivastava	University of Minnesota, USA (Member since 2006, Life Member since 2015)
Takashi Washio	Institute of Scientific and Industrial Research, Osaka University, Japan (Member since 2008, Life Member since 2016)

Past Members

Hongjun Lu	Hong Kong University of Science and Technology, SAR China (Member 1997–2005)
Arbee L.P. Chen	National Chengchi University, Taiwan, ROC (Member 2002–2009)
Takao Terano	Tokyo Institute of Technology, Japan (Member 2000–2009)

Senior Program Committee

James Bailey	The University of Melbourne, Australia
Peter Christen	The Australian National University, Australia
Guozhu Dong	Wright State University, USA
Patrick Gallinari	LIP6, Université Pierre et Marie Curie, France
Joshua Huang	Shenzhen University, China
Seung-won Hwang	Yonsei University, Korea
George Karypis	University of Minnesota, USA
Latifur Khan	The University of Texas at Dallas, USA
Sang-Wook Kim	Hanyang University, Korea
Byung Suk Lee	University of Vermont, USA
Jiuyong Li	University of South Australia, Australia
Nikos Mamoulis	University of Hong Kong, SAR China
Wee Keong Ng	Nanyang Technological University, Singapore
Wen-Chih Peng	National Chiao Tung University, Taiwan
Vincenzo Piuri	Università degli Studi di Milano, Italy
Rajeev Raman	University of Leicester, UK
P. Krishna Reddy	International Institute of Information Technology, Hyderabad (IIIT-H), India
Dou Shen	Baidu, China
Masashi Sugiyama	RIKEN/The University of Tokyo, Japan
Kai Ming Ting	Federation University, Australia
Hanghang Tong	Arizona State University, USA
Vincent S. Tseng	National Chiao Tung University, Taiwan
Jianyong Wang	Tsinghua University, China
Wei Wang	University of California, Los Angeles, USA
Takashi Washio	Osaka University, Japan
Xindong Wu	University of Louisiana at Lafayette, USA
Xing Xie	Microsoft Research Asia, China
Hui Xiong	Rutgers University, USA
Yue Xu	Queensland University of Technology, Australia
Hayato Yamana	Waseda University, Japan
Jin Soung Yoo	Indiana University-Purdue University Fort Wayne, USA
Jeffrey Yu	The Chinese University of Hong Kong, SAR China
Osmar Zaiane	University of Alberta, Canada
Zhao Zhang	Soochow University, China

Yanchun Zhang	Victoria University, Australia
Yu Zheng	Microsoft Research Asia, China
Ning Zhong	Maebashi Institute of Technology, Japan
Xiaofang Zhou	The University of Queensland, Australia

Program Committee

Aijun An	York University, Canada
Enrique Muñoz Ballester	Università degli Studi di Milano, Italy
Gustavo Batista	University of Sao Paulo, Brazil
Johannes Blömer	Paderborn University, Germany
Kevin Bouchard	Université du Québec à Chicoutimi, Canada
Krisztian Buza	Rheinische Friedrich-Wilhelms-Universität Bonn, Germany
K. Selcuk Candan	Arizona State University, USA
Tru Hoang Cao	Ho Chi Minh City University of Technology, Vietnam
Wei Cao	HeFei University of Technology, China
Tanmoy Chakraborty	University of Maryland, College Park, USA
Jeffrey Chan	RMIT University, Australia
Chia-Hui Chang	National Central University, Taiwan
Muhammad Aamir Cheema	Monash University, Australia
Chun-Hao Chen	Tamkang University, Taiwan
Enhong Chen	University of Science and Technology of China, China
Shu-Ching Chen	Florida International University, USA
Ling Chen	University of Technology Sydney, Australia
Meng Chang Chen	Academia Sinica, Taiwan
Yi-Ping Phoebe Chen	La Trobe University, Australia
Songcan Chen	Nanjing University of Aeronautics and Astronautics, China
Zhiyuan Chen	University of Maryland Baltimore County, USA
Zheng Chen	Microsoft Research Asia, China
Silvia Chiusano	Politecnico di Torino, Italy
Jaegul Choo	Korea University, Korea
Kun-Ta Chuang	National Cheng Kung University, Taiwan
Bruno Cremilleux	Université de Caen, France
Alfredo Cuzzocrea	ICAR-CNR and University of Calabria, Italy
Xuan-Hong Dang	UC Santa Barbara, USA
Zhaohong Deng	Jiangnan University, China
Anne Denton	North Dakota State University, USA
Lipika Dey	Tata Consultancy Services, India
Bolin Ding	Microsoft Research, USA
Gillian Dobbie	The University of Auckland, New Zealand
Xiangjun Dong	Qilu University of Technology, China
Dejing Dou	University of Oregon, USA
Vladimir Estivill-Castro	Griffith University, Australia
Xuhui Fan	University of Technology Sydney, Australia

Philippe Fournier-Viger	Harbin Institute of Technology Shenzhen, China
Yanjie Fu	Missouri University of Science and Technology, USA
Jun Gao	Peking University, China
Yang Gao	Nanjing University, China
Junbin Gao	University of Sydney, Australia
Xiaoying Gao	Victoria University of Wellington, New Zealand
Angelo Genovese	Università degli Studi di Milano, Italy
Arnaud Giacometti	University François Rabelais of Tours, France
Lei Gu	Nanjing University of Post and Telecommunications, China
Yong Guan	Iowa State University, USA
Stephan Günnemann	Technical University of Munich, Germany
Sunil Gupta	Deakin University, Australia
Michael Hahsler	Southern Methodist University, USA
Choochart Haruechaiyasak	National Electronics and Computer Technology Center, Thailand
Tzung-Pei Hong	National University of Kaohsiung, Taiwan
Michael Houle	NII, Japan
Qingbo Hu	LinkedIn, USA
Liang Hu	Jilin University, China
Jen-Wei Huang	National Cheng Kung University, Taiwan
Nguyen Quoc Viet Hung	University of Queensland, Australia
Van-Nam Huynh	Japan Advanced Institute of Science and Technology, Japan
Yoshiharu Ishikawa	Nagoya University, Japan
Md Zahidul Islam	Charles Sturt University, Australia
Divyesh Jadav	IBM Almaden Research, USA
Meng Jiang	University of Illinois at Urbana-Champaign, USA
Toshihiro Kamishima	National Institute of Advanced Industrial Science and Technology, Japan
Murat Kantarcioglu	University of Texas at Dallas, USA
Hung-Yu Kao	National Cheng Kung University, Taiwan
Shanika Karunasekera	The University of Melbourne, Australia
Makoto Kato	Kyoto University, Japan
Yoshinobu Kawahara	Osaka University, Japan
Bum-Soo Kim	Korea University, Korea
Chulyun Kim	Sookmyung Women's University, Korea
Kyoung-Sook Kim	National Institute of Advanced Industrial Science and Technology, Japan
Yun Sing Koh	The University of Auckland, New Zealand
Irena Koprinska	University of Sydney, Australia
Sejeong Kwon	KAIST, Korea
Hady Lauw	Singapore Management University, Singapore
Ickjai Lee	James Cook University, Australia
Jongwuk Lee	Sungkyunkwan University, Korea
Ki-Hoon Lee	Kwangwoon University, Korea

Ki Yong Lee	Sookmyung Women's University, Korea
Kyumin Lee	Utah State University, USA
Yue-Shi Lee	Ming Chuan University, Taiwan
Sunhwan Lee	IBM Almaden Research Center, USA
Wang-Chien Lee	Pennsylvania State University, USA
SangKeun Lee	Korea University, Korea
Philippe Lenca	IMT Atlantique, France
Carson K. Leung	University of Manitoba, Canada
Zhenhui Li	Pennsylvania State University, USA
Dancheng Li	Northeastern University, China
Gang Li	Deakin University, Australia
Jianmin Li	Tsinghua University, China
Ming Li	Nanjing University, China
Sheng Li	Northeastern University, USA
Xiaoli Li	Institute for Infocomm Research, Singapore
Xuelong Li	Chinese Academy of Science, China
Yaliang Li	University at Buffalo, USA
Yidong Li	Beijing Jiaotong University, China
Zhixu Li	Soochow University, China
Hsuan-Tien Lin	National Taiwan University, Taiwan
Jerry Chun-Wei Lin	Harbin Institute of Technology Shenzhen, China
Jiajun Liu	Renmin University, China
Bin Liu	IBM T.J. Watson Research Center, USA
Wei Liu	University of Technology Sydney, Australia
Woong-Kee Loh	Gachon University, Korea
Shuai Ma	Beihang University, China
Giuseppe Manco	Università della Calabria, Italy
Florent Masseglia	Inria, France
Yasuko Matsubara	Kumamoto University, Japan
Xiangfu Meng	Liaoning Technical University, China
Jun-Ki Min	Korea University of Technology and Education, Korea
Nguyen Le Minh	JAIST, Japan
Yasuhiko Morimoto	Hiroshima University, Japan
Miyuki Nakano	Advanced Institute of Industrial Technology, Japan
Wilfred Ng	Hong Kong University of Science and Technology, SAR China
Ngoc-Thanh Nguyen	Wroclaw University of Technology, Poland
Xuan Vinh Nguyen	University of Melbourne, Australia
Kouzou Ohara	Aoyama Gakuin University, Japan
Salvatore Orlando	Ca' Foscari University of Venice, Italy
Satoshi Oyama	Hokkaido University, Japan
Jia-Yu Pan	Google, USA
Shirui Pan	University of Technology Sydney, Australia
Guansong Pang	University of Technology Sydney, Australia
Noseong Park	University of North Carolina, Charlotte, USA
Gabriella Pasi	University of Milano-Bicocca, Italy

Dhaval Patel	IBJ T.J. Watson Research Center, USA
Dinh Phung	Deakin University, Australia
Santu Rana	Deakin University, Australia
P. Krishna Reddy	International Institute of Information Technology Hyderabad, India
Chandan Reddy	Virginia Tech, USA
Patricia Riddle	University of Auckland, New Zealand
P.S. Sastry	Indian Institute of Science, Bangalore, India
Jiwon Seo	Ulsan National Institute of Science and Technology, Korea
Hong Shen	Adelaide University, Australia
Bin Shen	Zhejiang University, China
Chuan Shi	Beijing University of Posts and Telecommunications, China
Arnaud Soulet	François Rabelais University of Tours, France
Fabio Stella	University of Milano-Bicocca, Italy
Mahito Sugiyama	Osaka University, Japan
Yuqing Sun	Shangdong University, China
Yasuo Tabei	Tokyo Institute of Technology, Japan
Ichigaku Takigawa	Hokkaido University, Japan
Ming Tang	Chinese Academy of Sciences, China
David Taniar	Monash University, Australia
Xiaohui (Daniel) Tao	The University of Southern Queensland, Australia
Khoat Than	Hanoi University of Science and Technology, Vietnam
Hiroyuki Toda	NTT Cyber Solutions Laboratories, NTT Corporation, Japan
Ranga Raju Vatsavai	North Carolina University, USA
Zhangyang Wang	Texas A&M University, USA
Lizhen Wang	Yunnan University, China
Ruili Wang	Massey University (Albany Campus), New Zealand
Shoujin Wang	Advanced Analytics Institute, University of Technology Sydney, Australia
Jason Wang	New Jersey Institute of Technology, USA
Yang Wang	University of New South Wales, Australia
Wei Wang	University of New South Wales, Australia
Xin Wang	University of Calgary, Canada
Lijie Wen	Tsinghua University, China
Steven Euijong Whang	Google Research, USA
Joyce Jiyoung Whang	Sungkyunkwan University, Korea
Raymond Chi-Wing Wong	Hong Kong University of Science and Technology, SAR China
Brendon Woodford	University of Otago, New Zealand
Lin Wu	University of Queensland, Australia
Jia Wu	University of Technology Sydney, Australia
Xintao Wu	University of Arkansas, USA
Yuni Xia	Indiana University - Purdue University Indianapolis (IUPUI), USA

Guandong Xu	University of Technology Sydney, Australia
Congfu Xu	Zhejiang University, China
Bing Xue	Victoria University of Wellington, New Zealand
Takehiro Yamamoto	Kyoto University, Japan
Yusuke Yamamoto	Kyoto University, Japan
Jianye Yang	UNSW, Australia
Ming Yang	Nanjing Normal University, China
Shiyu Yang	UNSW, Australia
Min Yao	Zhejiang University, China
Ilyeop Yi	KAIST, Korea
Hongzhi Yin	University of Queensland, Australia
Ming Yin	Harvard University, USA
Yang Yu	Nanjing University, China
Long Yuan	UNSW, Australia
Xiaodong Yue	Shanghai University, China
Se-Young Yun	Los Alamos National Laboratory, USA
Yifeng Zeng	Teesside University, UK
Fan Zhang	University of Technology Sydney, Australia
Junping Zhang	Fudan University, China
Xiuzhen Zhang	RMIT University, Australia
Yating Zhang	Kyoto University, Japan
Ying Zhang	University of New South Wales, Australia
Du Zhang	Macau University of Science and Technology, SAR China
Min-Ling Zhang	Southeast University, China
Wenjie Zhang	University of New South Wales, Australia
Zhongfei Zhang	Binghamton University, USA
Peixiang Zhao	Florida State University, USA
Yong Zheng	Illinois Institute of Technology, USA
Shuigeng Zhou	Fudan University, China
Xiangmin Zhou	RMIT University, Australia
Chengzhang Zhu	National University of Defense Technology, China
Xingquan Zhu	Florida Atlantic University, USA
Arthur Zimek	University of Southern Denmark, Denmark

Additional Reviewers

Enzo Acerbi	Zhao Kang
Weiling Cai	Daehoon Kim
Minsoo Choy	Jungeun Kim
Thomas Devogele	Sundong Kim
Van Nguyen Do	Nicolas Labroche
Khan Chuong Duong	Trung Le
Laurent Etienne	Chengjun Li
Li Gao	Dominique Li
Viet Huynh	Wentao Li

Chun-Yi Liu
Jing Lv
Kiem-Hieu Nguyen
Oanh Nguyen
Thin Nguyen
Thuong Nguyen
Tu Nguyen
Vu Nguyen
Linshan Shen
Fengyi Song
Hwanjun Song
Gabriele Sottocornola

Linh Ngo Van
Yanran Wang
Yisen Wang
Kuoliang Wu
Hongyu Xu
Wanqi Yang
Xuesong Yang
Haichao Yu
Hanchao Yu
Chen Zhang
Chenwei Zhang
Yuhai Zhao

Sponsors

Platinum Sponsors

AFOSR/
AOARD

NAVER

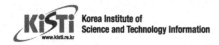
BIG DATA INSTITUTE
Seoul National University

SK holdings
C&C

Gold Sponsor

KiSTi Korea Institute of
www.kisti.re.kr Science and Technology Information

Silver Sponsors

Daumsoft
M I N I N G   M I N D S

DOUZONE

 HiBrain.Net

Kdata 한국데이터진흥원
Korea Data Agency

Publication Sponsor

Springer

Contents – Part I

Social Network and Graph Mining

Privacy-Preserving Mining and Security/Risk Applications

Spatio-Temporal and Sequential Data Mining

Contents – Part II

Feature Selection

Text and Opinion Mining

Clustering and Matrix Factorization

Classification and Deep Learning

Convolutional Bi-directional LSTM for Detecting Inappropriate Query Suggestions in Web Search

Harish Yenala[1(✉)], Manoj Chinnakotla[1,2], and Jay Goyal[2]

[1] IIIT Hyderbad, Hyderbad, India
harish.yenala@research.iiit.ac.in
[2] Microsoft, Hyderbad, India
{manojc,jaygoyal}@microsoft.com

Abstract. A web search query is considered *inappropriate* if it may cause anger, annoyance to certain users or exhibits lack of respect, rudeness, discourteousness towards certain individuals/communities or may be capable of inflicting harm to oneself or others. A search engine should regulate its query completion suggestions by detecting and filtering such queries as it may hurt the user sentiments or may lead to legal issues thereby tarnishing the brand image. Hence, *automatic detection and pruning* of such inappropriate queries from completions and related search suggestions is an important problem for most commercial search engines. The problem is rendered difficult due to unique challenges posed by search queries such as lack of sufficient context, natural language ambiguity and presence of spelling mistakes and variations.

In this paper, we propose a novel deep learning based technique for automatically identifying inappropriate query suggestions. We propose a novel deep learning architecture called *"Convolutional Bi-Directional LSTM (C-BiLSTM)"* which combines the strengths of both Convolution Neural Networks (CNN) and Bi-directional LSTMs (BLSTM). Given a query, C-BiLSTM uses a convolutional layer for extracting feature representations for each query word which is then fed as input to the BLSTM layer which captures the various sequential patterns in the entire query and outputs a richer representation encoding them. The query representation thus learnt passes through a deep fully connected network which predicts the target class. C-BiLSTM doesn't rely on hand-crafted features, is trained *end-end* as a single model, and effectively captures both local features as well as their global semantics. Evaluating C-BiLSTM on real-world search queries from a commercial search engine reveals that it significantly outperforms both pattern based and other hand-crafted feature based baselines. Moreover, C-BiLSTM also performs better than individual CNN, LSTM and BLSTM models trained for the same task.

Keywords: Query classification · Deep learning · Query auto suggest · Web search · CNN + Bi-directional LSTM · Supervised learning

---

H. Yenala—Work Done During Research Internship at Microsoft, India.

J. Kim et al. (Eds.): PAKDD 2017, Part I, LNAI 10234, pp. 3–16, 2017.
DOI: 10.1007/978-3-319-57454-7_1

1   Introduction

Web search engines have become indispensable tools for people seeking information on the web. Query Auto Completion (QAC) [1–3] feature improves the user search experience by providing the most relevant query completions matching the query prefix entered by the user. QAC has many advantages such as saving user time and keystrokes required to enter the query, avoiding spelling mistakes and formulation of better queries *etc* [4]. The candidates for query completion are usually selected from search query logs which record what other users have searched for [1][1]. During query time, those candidates which match the given query prefix are ranked based on various relevance signals such as time, location and user search history to finally arrive at the top k relevant completions to be displayed [4,5]. The value of k usually varies from 4–10 depending on each web search engine.

With more than a billion searches per day, search queries have become a mirror of the society reflecting the attitudes and biases of people. Hence, besides queries with a clean intent, search query logs also contain queries expressing violence, hate speech, racism, pornography, profanity and illegality. As a result, while retrieving potential completions from search logs, search engines may inadvertently suggest query completions which are *inappropriate* to the users.

Definition 1. *A search query is defined as* inappropriate *if its intent is any of the following - (a) rude or discourteous or exhibiting lack of respect towards certain individuals or group of individuals (b) to cause or capable of causing harm (to oneself or others) (c) related to an activity which is illegal as per the laws of the country or (d) has extreme violence.*

Figure 1 shows a few inappropriate query completions currently shown in some popular web search engines. For example, the query completions for *"christianity is"* are *christianity is fake, christianity is not a religion, christianity is a lie* and *christianity is the true religion.* Out of these, the first three suggestions are *inappropriate* since they are likely to offend and hurt the sentiments of christians. Similarly, the first query completion suggestion for *"angelina jolie is"* is

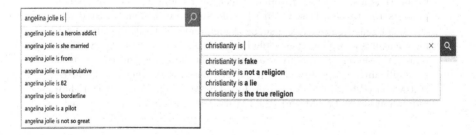

Fig. 1. Sample inappropriate query suggestions from popular web search engines.

---

[1] http://googleblog.blogspot.com/2004/12/ive-gotsuggestion.html.

angelina jolie is a heroin addict. This suggestion is inappropriate as it tries to characterize her entire personality with heroin addiction. *Inappropriate* queries are different from queries with an *adult* intent such as pornography and sexuality. Adult queries are much more coherent when compared to inappropriate class which includes a large number of sub-categories within it.

Although, users still have the right to search whatever they want, a search engine offering such inappropriate completions/suggestions inadvertently may - (a) be considered as endorsing those views thereby tarnishing the brand image or (b) damage the reputation of certain individuals or communities leading to legal complications or (c) help a person who is trying to harm oneself or others. In the past, there have been some instances[2] where search engines were dragged into legal tussles over such inappropriate suggestions. Hence, due to their potential to negatively influence their users and brand, it is imperative for search engines to *identify and filter or block* such inappropriate suggestions during QAC. Once the inappropriate queries are automatically identified, they can be pruned out during candidate generation phase thereby blocking their passage into the next modules of the QAC pipeline.

Automatic detection of inappropriate search queries is challenging due to lack of sufficient context and syntactic structure in web queries, presence of spelling mistakes and natural language ambiguity. For instance, a query like *"defecate on my face video"* may sound extremely offensive and hence inappropriate but it is the name of a famous song. Similarly, *"what to do when tweaking alone"* is a query where the term *tweaking* refers to the act of consuming meth - a drug which is illegal and hence the suggestion is inappropriate. A query like *"hore in bible"* has a spelling mistake where *hore* refers to *whore* which makes the query inappropriate. Previous approaches [6–9] have focused on identifying offensive language or flames in the messages posted on online or social networking forums such as twitter, facebook *etc.* They mainly rely on the presence of strong offensive keywords or phrases and grammatical expressions. They also explored the use of supervised machine learning based classification techniques, which require hand-crafted features, such as - Naive Bayes, SVMs, Random Forests *etc.* to automatically learn the target pattern with the help of labeled training data. However, such techniques are not suited well for inappropriate search query detection since the ambiguity is high, context is less and there is no linguistic structure to be exploited.

In this paper, we propose a novel deep learning based technique for automatic detection of inappropriate web search queries. We combine the strengths of both Convolutional Neural Networks (CNN) and Bi-directional LSTM (BLSTM) deep learning architectures and propose a novel architecture called *"Convolutional, Bi-Directional LSTM (C-BiLSTM)"* for automatic inappropriate query detection. Given a query, C-BiLSTM uses a convolutional layer for learning a feature representation for the query words which is then fed as input to the BLSTM which captures the sequential patterns and outputs a richer representation encoding those patterns. The query representation thus learnt passes through

---

[2] http://searchengineland.com/google-trouble-racist-autocomplete-suggestions-uk-184031.

a deep fully connected network which predicts the target class. C-BiLSTM does not require any hand-crafted feature engineering, is trained *end-end* as a single model and effectively captures both local features as well as their global semantics. Evaluating C-BiLSTM on real-world search queries and their prefixes from a popular commercial search engine reveals that it significantly outperforms both pattern based and other hand-crafted feature based baseline models in identification of inappropriate queries. Moreover, C-BiLSTM also performs better than individual CNN, LSTM and BLSTM models. In this context, the following are our main contributions:

- We introduce the research problem of automatic identification of inappropriate web search queries
- We propose a novel deep learning based approach called "Convolutional, Bi-Directional LSTM (C-BiLSTM)" for the above problem
- We evaluate the various techniques proposed so far, including standard deep learning techniques (such as CNN, LSTM and BLSTM), on a real-world search query and prefix dataset and compare their effectiveness for the task of inappropriate query detection

The rest of the paper is organized as follows: Sect. 2 discusses the related work in this area. Section 3 presents our contribution C-BiLSTM in greater detail. Section 4 discusses our experimental set-up. Section 5 presents our results and finally Sect. 6 concludes the paper.

2 Related Work

Vandersmissen *et al.* [6] applied Machine Learning (ML) techniques to automatically detect messages containing offensive language on Dutch social networking site Netlog. They report that a combination of SVM and word list based classifier performs best and observed their models perform badly with less context which is usually the case with search queries as well. Xiang *et al.* [7] used statistical topic modeling (LDA) [10] and lexicon based features to detect offensive tweets in a large scale twitter corpus. These try various models (SVMs, Logistic Regression (LR) and Random Forests (RF) with these features and report that LR performs best. It is pretty hard to extract topical features from search queries which are usually short and have less context. Razavi *et al.* [9] detect flames (offensive/abusive rants) from text messages using a multi-level classification approach. They use a curated list of 2700 words, phrases and expressions denoting various degrees of flames and then used them as features for a two-staged Naive Bayes classifier. Xu *et al.* [8] use grammatical relations and a curated offensive word list to identify and filter inappropriate/offensive language in online social forums. Chuklin *et al.* [11] automatically classify queries with adult intent into three categories - *black (adult intent), grey, white (clean intent)*. They use gradient boosted decision trees for classification.

Shen *et al.* [12] use a series of convolution and max-pooling layers to create a Convolution Latent Semantic Model (CLSM) aimed at learning low-dimensional

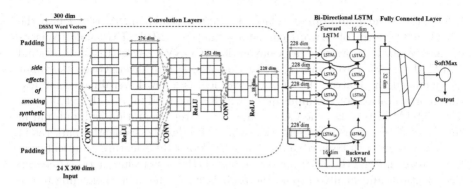

Fig. 2. Architecture of Convolutional Bi-Directional LSTM (C-BiLSTM) model.

semantic representations of search queries and web documents. CLSM uses character trigram and word n-gram features, is trained using click-through data. Results on a real-world data set showed better performance over state of the art methods such as DSSM [13].

Researchers have tried combining CNN and LSTM architectures in the context of other text mining problems such as Named Entity Recognition (NER) and Sentiment Analysis. Zhou et al. [14] combined CNN and LSTM architectures to create a hybrid model C-LSTM and apply it for sentiment analysis of movie reviews and question type classification. Although, the combination of CNN and LSTM is similar to our current model, there are some minor differences - (a) Through Convolutional layer, we're interested in learning a better representation for each input query word and hence we don't use max-pooling since it reduces the number of input words and (b) We use a bi-directional LSTM layer instead of LSTM layer since it can model both forward and backward dependencies and patterns in the query. Sainath et al. [15] also sequentially combine convolutional, LSTM and fully connected layers into a single architecture named CLDNN for the problem of speech recognition. Chiu et al. [16] combined Bi-directional LSTM and CNN models for NER. They augment the features learnt by the CNN with additional word features like capitalization and lexicon features for forming a complete feature vector. This complete feature vector then is then fed into Bi-directional LSTM layer for sequentially tagging the words with their NER tags. Unlike them, we only use the final outputs of forward and backward layers of the Bi-Directional LSTM since we're interested in query classification.

To the best of our knowledge, we are first one to introduce the research problem of detecting search queries with inappropriate/offensive intents and also to propose an end-end deep learning model for solving it.

3 C-BiLSTM for Inappropriate Query Detection

The architecture of our proposed C-BiLSTM model is shown in Fig. 2. C-BiLSTM takes an input search query and outputs the probability of the query belonging to the inappropriate class. The input search query is fed into the model in the form of a word embedding matrix. C-BiLSTM model consists of three sequential layers - (a) Convolution (CONV) Layer (b) Bi-directional LSTM (BLSTM) Layer and (c) Fully Connected (FC) layer. Given the input query embedding matrix, the CONV Layer learns a new lower-dimensional feature representation for the input query which is then fed into the BLSTM layer. The BLSTM layer takes the CONV layer query representation as input and in turn outputs a feature representation which encodes the sequential patterns in the query from both forward and reverse directions. This feature representation then passes through the FC layer, which models the various interactions between these features and finally outputs the probability of the query belonging to the inappropriate class. We share more details about each of the above layers below.

3.1 Input Query Embedding and Padding

For each word in the input query, we obtain its DSSM [13] word embedding in 300 dimensions and use it to form the input query matrix. We chose DSSM for representing the words since - (a) it has been specifically trained on web search queries and (b) it uses character grams as input and hence can also generate embeddings for unseen words such as spelling mistakes and other variations which are common in search queries. As the CONV layer requires fixed-length input, we pad each query with special symbols, indicating unknown words, at the beginning and at the end to ensure the length is equal to *maxlen* (in our case 24). We randomly initialize the DSSM word vectors for these padded unknown words from the uniform distribution $[-0.25, 0.25]$.

3.2 Learning Feature Representations Using Convolution Layer

Let $w \in \mathbb{R}^{MaxLen \times d}$ denote the entire query where $MaxLen$ is the length of the final padded query where $w_i \in \mathbb{R}^d$ be the DSSM word representation of the ith word of the input query in d dimensions. In our case, $MaxLen = 24$ and $d = 300$. Let $k \times l$ be the size of the 2-D convolution filter with weight $m \in \mathbb{R}^{k \times l}$ then each filter will produce a feature map $v \in \mathbb{R}^{MaxLen-k+1 \times d-l+1}$. We consider multiple such filters and if there are n filters then $C = [v_1, v_2, \ldots, v_n]$ will be the combined feature representation of these filters. After each convolution operation, we apply a non-linear transformation using a Rectified Linear Unit (ReLU) [17] as it simplifies back propagation and makes learning faster while avoiding saturation. In our case, we used four 3×25 filters. As shown in Fig. 2, we apply three successive steps of convolution and non-linearity to arrive at the final feature representation which has a dimension of 18×228.

3.3 Capturing Sequential Patterns with Bi-directional LSTM

Long-Short Term Memory (LSTMs) [18] are variants of Recurrent Neural Networks (RNN) [19,20] architectures which - (a) overcome the vanishing gradient

Table 1. Optimal hyper-parameter set for all models after tuning on validation set.

Parameter	CNN	LSTM	BLSTM	C-BiLSTM
Batch size	1000	1000	1000	1000
Max Len.	24	24	24	24
WordVecDim.	300	300	300	300
CNN depth	4	NA	NA	3
Filter size	2×20	NA	NA	3×25
Pooling	Max-pooling	NA	NA	NA
Non-linearity	ReLU	NA	NA	ReLU
LSTM cells	NA	40	40	32
Optimizer	Adagrad	Adagrad	Adagrad	Adagrad
Learning rate	0.01	0.05	0.05	0.05
Epsilon	1e-08	1e-08	1e-08	1e-08

problem of conventional RNNs and (b) have the ability to capture long-term dependencies present in a sequential pattern due to their gating mechanisms which control information flow. While LSTMs can only utilize previous contexts, Bi-Directional LSTMs (BLSTMs) overcome this limitation by examining the input sequence from both forward and backward directions and then combine the information from both ends to derive a single representation. This enables them to capture much richer sequential patterns from both directions and also helps in learning a much better feature representation for the input query.

The 18×228 feature representation from the previous convolution layer is fed as a sequence of 18 words, each with a 228 dimensional representation, to the BLSTM layer. The LSTM cells inside the forward and backward LSTM networks of BLSTM read the word representations in the forward and reverse orders and each of them output a 16 dimensional representation which is then combined to produce a 32 dimensional feature representation which encodes the various semantic patterns in the query.

The output of the BLSTM layer (32 dimensional feature vector) is given as input to a Fully Connected (FC) layer which models the interactions between these features. The final softmax node in the FC layer outputs the probability of the query belonging to the inappropriate class.

3.4   Model Training

We train the parameters of C-BiLSTM with an objective of maximizing their predication accuracy given the target labels in the training set. We randomly split the given dataset into train, validation and test sets. We trained the models using the training set and tuned the hyper-parameters using the validation set. The optimal hyper-parameter configuration thus found for various models is shown in Table 1. If t is the true label and o is the output of the network with the current weight configuration, we used Binary Cross Entropy (BCE) as the loss function which is calculated as follows:

Category	No. of. Queries	Sample Queries
Extreme Violence Self Harm Illegal Activity	1619	*woman beheaded video* *how many pills does it take to kill yourself* *growing marijuana indoors for beginners*
Race Religion Sexual Orientation Gender	2241	*new zealanders hate americans* *anti islam shirts* *gays are destroying this country* *butch clothing for women*
Other Offensive Celebrity	1124	*jokes about short people* *louie gohmert stupid quotes*
Clean	74057	*20 adjectives that describe chocolate* *what is the order of the planets*
Total	79041	

Fig. 3. Statistics of inappropriate categories in our evaluation dataset.

$$BCE(t, o) = -(t \cdot log(o) + (1 - t) \cdot log(1 - o))$$

We use Adagrad [21] as the optimization routine and the models were trained by minimizing the above loss function in a batch size of n which was tuned differently across models as given in Table 1.

4 Experimental Setup

In this section, we describe the details of our dataset, baseline approaches used for comparison and evaluation metrics.

4.1 Dataset Details

For evaluating this task, we created a dataset comprising of 79041 unique web search queries along with their class labels (inappropriate/clean) provided by human judges. The details of the dataset along with the statistics is shown in Fig. 3. These queries were sampled from the query suggestions being currently served through the auto-completion service of a well-known commercial search engine in the US market. The sampled queries were then judged by human judges through a crowd-sourcing platform. Each query had to be classified into one of two categories - (a) Clean (b) Inappropriate. In case of *Inappropriate*, they were also asked to mark one of the following three finer categories - (a) Violence/Illegal/Self-Harm (b) Race/Religion/Sexual/Gender and (c) Other/Profane. To avoid any individual bias, each query was judged by nine judges. An odd number of judges was chosen to easily decide the majority vote. The judges were given clear guidelines, along with sample query label pairs, on how to classify the queries. A query was finally labeled *inappropriate* if a majority of the judges labeled it as *inappropriate*. The inter-annotator agreement based on fleiss' kappa value [22] was found to be 0.632 which is substantial. As shown in Table 2, we randomly split the dataset into train, validation and test sets. The train set was used for training the models while the validation set was used for tuning the parameters of individual models and the test set for evaluation.

Table 2. Evaluation dataset label distribution across train, validation and test sets.

Label	Training	Validation	Test	Total
Inappropriate	4594	212	178	4984
Clean	65447	4788	3822	74057
Total	70041	5000	4000	79041

Sample Patterns	Sample Matching Queries	Sample Words in Category
Proper Noun AND *NegativePersonalityWord*	*hillary clinton is terrible* *marie harf is an idiot*	*NegativePersonalityWord:* terrible, idiot, moron, miser..
Proper Noun AND 'not'/'no' AND *PositivePersonalityWord*	*ellen degeneres is not funny*	*PositivePersonalityWord:* calm, affectionate, charming..
SelfHarmPrefix AND *SelfHarmSuffix*	*how can i commit suicide* *methods to kill myself*	*SelfHarmPrefix:* how can I, how should I, ways of... *SelfHarmSuffix:* hang myself, shoot myself, commit suicide...
Ethnicity/Religion AND *CommunityDislikeWord*	*americans hate black people* *muslims murdered christians*	*Ethnicity/Religion:* americans, jews, muslims.. *CommunityDislikeWord:* hate, disrespect, kill...
CoreOffensiveWord	*slut shaming quotes* *the bitch is back*	*CoreOffensiveWord:* fuck, asshole, bitch, slut..

Fig. 4. Sample patterns and keywords used in PKF baseline.

4.2 Baseline Approaches

We compare our approach with three baselines - (a) Pattern and Keyword based Filtering (PKF) and (b) Support Vector Machine (SVM) [23] based classifier and (c) Gradient Boosted Decision Trees (BDT) [24] based classifier. In PKF, based on manual query analysis, we curated a list of typical inappropriate query patterns and high-confidence inappropriate/offensive keywords. If a given query matches any of the inappropriate/offensive query patterns or contains one of the inappropriate/offensive keywords then it will be labeled as *inappropriate*. Some sample patterns and inappropriate/offensive keywords are shown in Fig. 4. SVM and BDT classifiers are supervised learning models and we implemented both of them using Scikit Learn [25]. We use the query words as features and the labeled training set for training. In order to test the efficacy of deep learning based word embeddings, we also implemented variants of SVM and BDT approaches called SVM-DSSM and BDT-DSSM where besides regular word features, DSSM embeddings of query words were also included as additional features. We optimized the performance of all models by tuning their parameters using the validation set. In the case of SVMs, we tuned the parameter C and also tried various kernels. We also handled class imbalance in all the models by appropriately setting the class weight parameter. The best performance was found with $C = 0.5$ and linear kernel. In case of BDT, the optimal parameter choice was found to be no. of trees $= 50$, max. depth $= 7$, min. samples split $= 350$ and min. samples leaf $= 10$. We use the standard classification evaluation metrics - Precision, Recall and F1 score [26].

Table 3. Final results of various models on test set. C-BiLSTM and BLSTM results are found to be statistically significant with $p < 0.005$

Model	Precision	Recall	F1 score
PKF	0.625	0.2142	0.3190
BDT	0.7926	0.2784	0.4120
BDT-DSSM	0.9474	0.3051	0.4615
SVM	0.8322	0.3593	0.5019
SVM-DSSM	0.9241	0.4101	0.5680
CNN	0.7148	0.8952	0.7949
LSTM	0.8862	0.7047	0.7850
BLSTM	0.8018	0.8285	0.8149
C-BiLSTM	0.9246	0.8251	**0.8720**

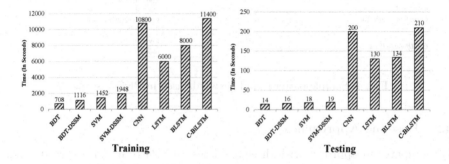

Fig. 5. Runtime performance of various models during train and test phases.

5 Results and Discussion

Table 3 presents the overall results of various techniques on the test set of the evaluation dataset. C-BiLSTM gives the best performance on F1 score which measures the overall quality of classification and also shows significant improvement over the baseline approaches which employ either pattern/keyword based or hand-crafted feature based identification techniques. It is also interesting to note that the other deep learning techniques such as CNN, LSTM and BLSTM also perform significantly better than the baseline approaches. The results show that DL approaches, with the help of automatically learnt feature representations, can perform better than competitively trained hand-crafted baselines. We can also observe that BDT and SVM perform better when we provided with DSSM word embedding features instead of word features alone. The results also prove that combining convolutional layer and BLSTM architectures in C-BiLSTM is better than individual CNN and BLSTM models and is especially helpful in improving precision as shown in the significant improvement of precision (more than 29% when compared to CNN). Although BDT baseline shows the highest

Query	True Label	C-BLSTM Label	Judgment Explanation	Comments
nigerplease.com	Inappropriate	Inappropriate	The word niger is an Inappropriate word	Since nigerplease.com is a single word, PKF, SVM, BDT models fail. Other deep learnt models also misclassify this query. C-BLSTM alone correctly classifies this one.
shake and bake meth instructions	Inappropriate	Inappropriate	"meth" is a drug which is illegal in US and some other parts of world	"meth" is a short form for Methamphetamine and hence PKF, SVM, BDT models fail. Other deep learnt models also misclassify this query. C-BLSTM alone correctly classifies this one.
a**monkey	Inappropriate	Inappropriate	It refers to the Inappropriate word - "assmonkey"	C-BLSTM perfectly classifies it. PKF, SVM, BDT fail because it includes "**". Other deep learnt models also fail in this case.
hore in the bible	Inappropriate	Inappropriate	It is a spell mistake of the word "whore" and is Inappropriate to christians	PKF, SVM, BDT fail because of not catching the spell mistake. Other deep learnt models also fail except for C-BLSTM.
marvin gaye if i should die tonight download	Clean	Clean	Not Inappropriate since it is a song download	PKF misclassifies it as "Inappropriate" due to presence of "die tonight" pattern. Remaining models classify correctly.
asshat in sign language	Inappropriate	Clean	An Inappropriate term in sign language	BDT perfectly classifies it. Remaining all models misclassify it.
why do asians speak the ching chong	Inappropriate	Clean	Ching chong is a pejorative term for chinese language	PKF classifies it correctly since "ching chong" is included in list of core Inappropriate words. Remaining classifiers fail.

Fig. 6. Qualitative analysis of C-BiLSTM results vis-a-vis other baseline approaches.

precision of 0.9474, the recall is poor (0.3051) which means it is precise only for a small class of inappropriate queries. Figure 5 shows a comparison of the relative running times taken by the various models during train and test phases.

5.1 Qualitative Analysis

Figure 6 presents a qualitative analysis of C-BiLSTM results vis-à-vis other baseline approaches using queries from the test set. Since, C-BiLSTM uses DSSM embeddings for query words, it understands the spelling mistakes and variations of a word. Due to this, it correctly classifies the queries - *"hore in the bible"* and *"a**monkey"* whereas the baseline approaches fail since the particular words may not have been observed in the training data or in the pattern dictionary. Similarly, C-BiLSTM correctly classifies the query *"nigerplease.com"* although the offensive word is fused with another word without any space. This shows that C-BiLSTM has effectively captured the character grams associated with the target label. However, there are still some queries where C-BiLSTM performs badly and needs to be improved. For example, it incorrectly classifies the query *"why do asians speak the ching chong"* which was perfectly classified by PKF due to the presence of the word *"ching chong"* in its inappropriate/offensive word list.

5.2 Query Autocompletion Filtering Task

In order to demonstrate the effectiveness of the proposed techniques for query autocompletion filtering task, we randomly selected 200 unique inappropriate/offensive queries from the pool of inappropriate queries excluding training set (i.e. from Validation (212) and Test (178)). From the above set, for each

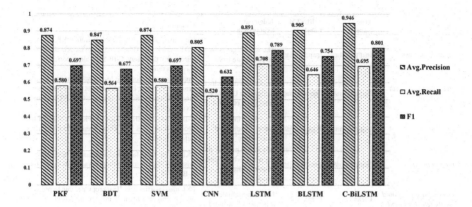

Fig. 7. Results on QPBAD dataset using top five query completion suggestions.

query, we generated a prefix of random length. Later, for each of these 200 query prefixes, we retrieved the current top 5 autocompletion suggestions offered by a popular commercial search engine and got them labeled from human judges using the same process described earlier in Sect. 4.1. We call this dataset of 200 query prefixes with 5 query suggestions each as the *"Query Prefix Based Auto-completions Dataset (QPBAD)"*. We ran all the models on QPBAD and report their average precision (at 5), average recall (at 5) and F1 scores across queries in Fig. 7. In tune with earlier observation, C-BiLSTM shows better performance in identifying inappropriate query completion suggestions based on query prefixes than the baseline and individual deep learning models.

6   Conclusions

We introduced the problem of automatically detecting inappropriate search queries and proposed a novel deep learning based technique called "Convolutional, Bi-Directional LSTM (C-BiLSTM)" for solving it. Given a query, C-BiLSTM uses a convolutional layer for extracting a sequence of high-level phrase representations from query words and then feeds it into a BLSTM which then captures the sequential patterns in the given query and learns an output representation for the entire query. The query representation thus learnt passes through a deep fully connected network which predicts the target class. C-BiLSTM does not require any hand-crafted feature engineering, is trained *end-end* as a single model, and effectively captures both local query features as well as its global semantics. Evaluation on real-world search queries and their prefixes from a large scale commercial search engine revealed that it significantly outperforms both pattern based and other hand-crafted feature based baselines. C-BiLSTM also proved to be better than other individual deep learning based architectures CNN, LSTM and BLSTM.

References

1. Bar-Yossef, Z., Kraus, N.: Context-sensitive query auto-completion. In: WWW 2011, pp. 107–116. ACM, New York (2011)
2. Shokouhi, M., Radinsky, K.: Time-sensitive query auto-completion. In: SIGIR 2012, pp. 601–610. ACM, New York (2012)
3. Whiting, S., Jose, J.M.: Recent and robust query auto-completion. In: WWW 2014 (2014)
4. Cai, F., de Rijke, M.: A survey of query auto completion in information retrieval. Found. Trends® Inf. Retrieval 10(4), 273–363 (2016)
5. Di Santo, G., McCreadie, R., Macdonald, C., Ounis, I.: Comparing approaches for query autocompletion. In: SIGIR 2015. ACM, New York (2015)
6. Vandersmissen, B., De Turck, F., Wauters, T.: Automated Detection of Offensive Language Behavior on Social Networking Sites, vol. xiv, 81 p. (2012)
7. Xiang, G., Fan, B., Wang, L., Hong, J., Rose, C.: Detecting offensive tweets via topical feature discovery over a large scale Twitter corpus, pp. 1980–1984 (2012)
8. Xu, Z., Zhu., S.: Filtering offensive language in online communities using grammatical relations. In: Proceedings of the Seventh Annual CEAS (2010)
9. Razavi, A.H., Inkpen, D., Uritsky, S., Matwin, S.: Offensive language detection using multi-level classification. In: Farzindar, A., Kešelj, V. (eds.) AI 2010. LNCS (LNAI), vol. 6085, pp. 16–27. Springer, Heidelberg (2010). doi:10.1007/978-3-642-13059-5_5
10. Blei, D.M., Ng, A.Y., Jordan, M.I.: Latent Dirichlet allocation. J. Mach. Learn. Res. 3, 993–1022 (2003)
11. Chuklin, A., Lavrentyeva, A.: Adult query classification for web search and recommendation. In: Search and Exploration of X-Rated Information (WSDM 2013)
12. Yelong, S., Xiaodong, H., Jianfeng, G., Li, D., Gregoire, M.: A latent semantic model with convolutional-pooling structure for information retrieval. In: CIKM, November 2014
13. Huang, P.S., He, X., Gao, J., Deng, L., Acero, A., Heck, L.: Learning deep structured semantic models for web search using clickthrough data. In: CIKM 2013 (2013)
14. Zhou, C., Sun, C., Liu, Z., Lau, F.C.M.: A C-LSTM neural network for text classification. CoRR abs/1511.08630 (2015)
15. Sainath, T.N., Senior, A.W., Vinyals, O., Sak, H.: Convolutional, long short-term memory, fully connected deep neural networks. US Patent App. 14/847,133, 7 April 2016
16. Chiu, J.P.C., Nichols, E.: Named entity recognition with bidirectional LTM-CNNs. CoRR abs/1511.08308 (2015)
17. Nair, V., Hinton, G.E.: Rectified linear units improve restricted Boltzmann machines. In: ICML 2010, 21–24 June 2010, Haifa, Israel, pp. 807–814 (2010)
18. Hochreiter, S., Schmidhuber, J.: Long short-term memory. Neural Comput. 9(8), 1735–1780 (1997)
19. Werbos, P.J.: Backpropagation through time: what it does and how to do it. Proc. IEEE 78(10), 1550–1560 (1990)
20. Rumelhart, D.E., Hinton, G.E., Williams, R.J.: Learning representations by backpropagating errors. Cogn. Model. 5(3), 1 (1988)
21. Duchi, J., Hazan, E., Singer, Y.: Adaptive subgradient methods for online learning and stochastic optimization. Technical report UCB/EECS-2010-24

22. Fleiss, J.L.: Measuring nominal scale agreement among many raters. Psychol. Bull. **76**(5), 378–382 (1971)
23. Cortes, C., Vapnik, V.: Support-vector networks. Mach. Learn. **20**(3) (1995)
24. Friedman, J., Hastie, T., Tibshirani, R.: The Elements of Statistical Learning. Springer Series in Statistics, vol. 1. Springer, Berlin (2001)
25. Pedregosa, F., Varoquaux, G., Gramfort, A., Michel, V., Thirion, B., Grisel, O., Blondel, M., Prettenhofer, P., Weiss, R., Dubourg, V., Vanderplas, J., Passos, A., Cournapeau, D., Brucher, M., Perrot, M., Duchesnay, E.: Scikit-learn: machine learning in Python. J. Mach. Learn. Res. **12**, 2825–2830 (2011)
26. Manning, C.D., Raghavan, P., Schütze, H.: Introduction to Information Retrieval. Cambridge University Press, New York (2008)

A Fast and Easy Regression Technique for k-NN Classification Without Using Negative Pairs

Yutaro Shigeto, Masashi Shimbo$^{(\boxtimes)}$, and Yuji Matsumoto

Nara Institute of Science and Technology, Ikoma, Nara, Japan
{yutaro-s,shimbo,matsu}@is.naist.jp

Abstract. This paper proposes an inexpensive way to learn an effective dissimilarity function to be used for k-nearest neighbor (k-NN) classification. Unlike Mahalanobis metric learning methods that map both query (unlabeled) objects and labeled objects to new coordinates by a single transformation, our method learns a transformation of labeled objects to new points in the feature space whereas query objects are kept in their original coordinates. This method has several advantages over existing distance metric learning methods: (i) In experiments with large document and image datasets, it achieves k-NN classification accuracy better than or at least comparable to the state-of-the-art metric learning methods. (ii) The transformation can be learned efficiently by solving a standard ridge regression problem. For document and image datasets, training is often more than two orders of magnitude faster than the fastest metric learning methods tested. This speed-up is also due to the fact that the proposed method eliminates the optimization over "negative" object pairs, i.e., objects whose class labels are different. (iii) The formulation has a theoretical justification in terms of reducing hubness in data.

1 Introduction

Let $\mathcal{X}$ be a feature space and $\mathcal{Y}$ be a set of class labels. The *k-nearest neighbor* (k-NN) classifier predicts the class label of an unknown query object $\mathbf{x} \in \mathcal{X}$ by its nearest neighbors. Given $\mathbf{x}$ and a set of labeled objects $\mathcal{D} = \{(\mathbf{x}_i, y_i)\}_{i=1}^n$ where $\mathbf{x}_i \in \mathcal{X}$ is the feature vector of the ith object and $y_i \in \mathcal{Y}$ its class label, the classifier first computes the distance between $\mathbf{x}$ and each labeled object $\mathbf{x}_i$. It then predicts the class label $\hat{y}$ of $\mathbf{x}$ by the majority among its k nearest labeled objects. When $k = 1$, the decision rule of the k-NN (1-NN) classifier is simply:

$$\hat{y} = \underset{y_i:(\mathbf{x}_i,y_i)\in\mathcal{D}}{\arg\min} \ f(\mathbf{x}, \mathbf{x}_i), \tag{1}$$

where function $f : \mathcal{X} \times \mathcal{X} \to \mathbb{R}$ is some distance/dissimilarity function.

Obviously, the choice of function f affects the accuracy of classification. Therefore, many researchers [2,11,13,15] have tackled *metric learning*, which is the task of learning a suitable distance function from data.

© Springer International Publishing AG 2017
J. Kim et al. (Eds.): PAKDD 2017, Part I, LNAI 10234, pp. 17–29, 2017.
DOI: 10.1007/978-3-319-57454-7_2

For Euclidean object space $\mathcal{X} = \mathbb{R}^d$, metric learning is usually formulated as the task of finding a Mahalanobis distance. In this formulation, the distance between two objects $\mathbf{x}, \mathbf{z} \in \mathbb{R}^d$ is defined by

$$f(\mathbf{x}, \mathbf{z}) = \sqrt{(\mathbf{x} - \mathbf{z})^{\mathrm{T}} \mathbf{M} (\mathbf{x} - \mathbf{z})}, \tag{2}$$

with some positive (semi)definite matrix $\mathbf{M}$. By defining matrix $\mathbf{L}$ by $\mathbf{M} = \mathbf{L}^{\mathrm{T}}\mathbf{L}$, we can write the distance in Eq. (2) as

$$f(\mathbf{x}, \mathbf{z}) = \|\mathbf{L}\mathbf{x} - \mathbf{L}\mathbf{z}\|. \tag{3}$$

This equation shows that learning Mahalanobis distance is equivalent to learning a suitable linear transformation $\mathbf{L}$.

In the context of k-NN classification, distance needs to be measured only between query (unlabeled) objects and labeled objects, as can be seen from Eq. (1); when distance $f(\mathbf{x}, \mathbf{z})$ is computed, the first object $\mathbf{x}$ is always a query object, and the second object $\mathbf{z}$ is always a labeled object $\mathbf{x}_i$. Moreover, function f need not be metric and can be any measure of dissimilarity; for instance, f being asymmetric is perfectly acceptable.

In this paper, we learn one such dissimilarity function. The idea is to compute a transformation of labeled objects to new points while unlabeled objects are kept at their original points. Thus, our objective is to find a suitable matrix $\mathbf{W}$ that defines a dissimilarity function

$$f(\mathbf{x}, \mathbf{z}) = \|\mathbf{x} - \mathbf{W}\mathbf{z}\|, \tag{4}$$

where $\mathbf{x}$ is a query object, and $\mathbf{z}$ is a labeled object.

Because the coordinates of query objects are fixed, our formulation might appear less flexible than Mahalanobis distance learning (Eq. (3)). However, as shown in a subsequent section, it gives a better k-NN classification accuracy than Mahalanobis distance learning methods on many datasets that feature high-dimensional space. Moreover, optimizing $\mathbf{W}$ in Eq. (4) is much easier and substantially (often more than two orders of magnitude) faster.

The effectiveness of the proposed approach has a theoretical foundation in terms of reducing *hubness* in data [7,9]. Recent studies have shown that the presence of hubs, which are a few objects that appear in the k-NNs of many objects, is an obstacle that can harm the performance of many vector space methods [7,8,10]. We show that metric learning is no exception, and the transformation of labeled objects restrains hubs from emerging. This approach is justified by a recent result of Shigeto et al. [9], who used regression to reduce hubness in zero-shot problems. In their work, the problem was cast as a task of cross-domain matching, whereas in this paper, we are concerned with improving the accuracy of k-NN classification in a single space.

Another notable feature of the proposed method is that it eliminates optimization over "negative" object pairs, i.e., objects belonging to difference classes. In other words, our method only attempts to make objects of the same class ("positive" object pairs) to be closer. Its objective function does not have any constraints or terms that promote negative object pairs to be apart from each other. Such constraints are indispensable in Mahalanobis metric learning to prevent trivial solutions $\mathbf{M} = \mathbf{O}$ in Eq. (2) or $\mathbf{L} = \mathbf{O}$ in Eq. (3), and metric learning typically optimizes over a large number of negative object pairs. Moreover, incorporating negative pairs results in a non-convex optimization problem with respect to matrix $\mathbf{L}$. Existing metric learning methods [2,5,11,13,15] hence resorts to optimizing $\mathbf{M} = \mathbf{L}^{\mathrm{T}}\mathbf{L}$ using computationally intensive methods such as semi-definite programming. By contrast, since we only transforms labeled objects, we need not worry about $\mathbf{W} = \mathbf{O}$ being the solution (see Eq. (4)), thus eliminating the need of negative pairs. This makes the solution easily obtained with ridge regression, which contributes to reduced computation time.

2   Related Work

We briefly review some of the metric learning methods, mostly those used in the experiments in Sect. 5. For comprehensive survey of the field, see [1,6].

A majority of the metric learning methods adopt Mahalanobis distance (Eq. (3)) as the distance function, and minimize the training loss under various constraints. As mentioned earlier, these methods do not make distinction between unlabeled (query) objects and labeled objects, in the sense that their coordinates are transformed by the same matrix, $\mathbf{L}$ in Eq. (3). Our approach differs from these methods in that it projects only the labeled objects to new coordinates.

There are various strategies for learning Mahalanobis distance. Xing et al. [13] formulated metric learning as a convex optimization problem, and demonstrated its effectiveness in clustering tasks. The *large-margin nearest neighbor* (LMNN) method [11] is probably the most popular of all metric learning methods. Its objective is to minimize distances between objects with the same label, and to penalize objects with different labels when they are closer than a certain distance. Hence objects from different classes are separated by a large margin. To make the problem convex, in Xing et al.'s method and LMNN, optimization is done over not $\mathbf{L}$ but $\mathbf{M} = \mathbf{L}^{\mathrm{T}}\mathbf{L}$, with semidefinite programming. Ying and Li [15] presented an eigenvalue optimization framework for learning Mahalanobis distance. Davis et al. [2] proposed *information-theoretic metric learning* (ITML). ITML minimizes the LogDet divergence subject to linear constraints. It thus requires no eigenvalue computation or semi-definite programming.

Although it has been shown that these methods work well in many applications, learning Mahalanobis distance typically incurs high computational cost. Indeed, as we show in an experiment (Sect. 5), these methods spend substantial time in optimizing $\mathbf{M}$, when applied to large datasets.

3   Proposed Method

In this section, we present our approach for improving the k-NN classification accuracy.

In nearly all metric learning methods, the objective function to be optimized involves a term that encourages objects of the same class to be placed closer. In the same vein, our method also optimizes the transformation matrix $\mathbf{W}$ in Eq. (4) by minimizing the distance between objects of the same class. However, in our formulation, the learned transformation $\mathbf{W}$ is only applied to labeled objects.

Our training procedure consists of two steps. We first make training object pairs for which the distance should be minimized. To this end, we follow Weinberger and Saul [11]: for each labeled object $\mathbf{x}_i \in \mathbb{R}^d$ in the training set, we define its "target" objects $\mathcal{T}_i$ to be the k objects in the training set that belong to the same class as $\mathbf{x}_i$ and are closest to $\mathbf{x}_i$ as measured by the original Euclidean distance (i.e., the one before training). We then find a matrix $\mathbf{W} \in \mathbb{R}^{d \times d}$ that moves objects in $\mathcal{T}_i$ towards $\mathbf{x}_i$, by solving the following optimization problem:

$$\min_{\mathbf{W}} \sum_{i=1}^{n} \sum_{\mathbf{z} \in \mathcal{T}_i} \|\mathbf{x}_i - \mathbf{W}\mathbf{z}\|^2 + \lambda \|\mathbf{W}\|_{\mathrm{F}}^2, \tag{5}$$

where $\lambda \geq 0$ is a hyperparameter for regularization and $\| \cdot \|_{\mathrm{F}}$ represents the Frobenius norm. Equation (5) is a familiar objective function of ridge regression, and we have the closed-form solution:

$$\mathbf{W} = \mathbf{X}\mathbf{J}\mathbf{X}^{\mathrm{T}}(\mathbf{X}\mathbf{X}^{\mathrm{T}} + \lambda\mathbf{I})^{-1}, \tag{6}$$

where $\mathbf{X} = [\mathbf{x}_1, \ldots, \mathbf{x}_n] \in \mathbb{R}^{d \times n}$, and $\mathbf{J} \in \{0,1\}^{n \times n}$ is an indicator matrix such that $[\mathbf{J}]_{i,j} = 1$ if $\mathbf{x}_j \in \mathcal{T}_i$ and 0 otherwise.

In the test phase, we first compute the image $\mathbf{x}_i' = \mathbf{W}\mathbf{x}_i$ of every labeled object $\mathbf{x}_i$ by the learned matrix $\mathbf{W}$. We then carry out k-NN classification by regarding $\mathcal{D}' = \{(\mathbf{x}_i', y_i)\}$ as the labeled objects in place of the original ones, $\mathcal{D} = \{(\mathbf{x}_i, y_i)\}$. In the case of 1-NN classification, for example, this amounts to using the dissimilarity function f given by Eq. (4) in the decision rule of Eq. (1), i.e.,

$$\hat{y} = \underset{y_i : (\mathbf{x}_i', y_i) \in \mathcal{D}'}{\arg\min} \|\mathbf{x} - \mathbf{x}_i'\|^2 = \underset{y_i : (\mathbf{x}_i, y_i) \in \mathcal{D}}{\arg\min} \|\mathbf{x} - \mathbf{W}\mathbf{x}_i\|^2. \tag{7}$$

4   Proposed Method Reduces Hubness

In this section, we argue that the proposed method is by design less susceptible to producing hubs [7] in the transformed labeled objects. This property is desirable, as hubs have been recognized as one of the major factors that harm the effectiveness of nearest neighbor methods.

4.1   Hubness Phenomenon

The hubness phenomenon [7] states that a small number of objects in the dataset, called *hubs*, may occur as the nearest neighbor of many objects. The presence of hubs will diminish the utility of nearest-neighbor methods, because the lists of nearest neighbors frequently contain the same hub objects regardless of the query.

Hubs occur in data because of an inherent bias present in Euclidean space, called *spatial centrality* [7]; i.e., objects closest to the mean of the data tend to be the nearest neighbors of many objects. This bias is known to grow stronger with the dimensionality of the space.

The following proposition by Shigeto et al. [9] quantifies the degree of spatial centrality as a function of the dimension of the space and the variance of data distribution, when the feature values of query and data follow zero-mean Gaussian distributions with different variances. Let $E_{\mathcal{X}}[\cdot]$ and $\mathrm{Var}_{\mathcal{X}}[\cdot]$ respectively denote the expectation and variance under a distribution $\mathcal{X}$.

Proposition 1 [9, Proposition 1]. *Let* $\mathbf{z}$ *be a* d-*dimensional random vector sampled i.i.d. from a normal distribution with zero means and diagonal covariance matrix* $s^2\mathbf{I}$*; i.e.,* $\mathbf{z} \sim \mathcal{Z}$*, where* $\mathcal{Z} = \mathcal{N}(\mathbf{0}, s^2\mathbf{I})$*. Further let* $\sigma = \sqrt{\mathrm{Var}_{\mathcal{Z}}[\|\mathbf{z}\|^2]}$ *be the standard deviation of the squared norm* $\|\mathbf{z}\|^2$*.*

Consider two fixed samples $\mathbf{z}_1$ *and* $\mathbf{z}_2$ *of random vector* $\mathbf{z}$*, such that the squared norms of* $\mathbf{z}_1$ *and* $\mathbf{z}_2$ *are* $\gamma\sigma$ *apart. In other words,*

$$\|\mathbf{z}_2\|^2 - \|\mathbf{z}_1\|^2 = \gamma\sigma.$$

Let $\mathbf{x}$ *be a point sampled from a distribution* $\mathcal{X}$ *with zero mean.*

Then, the expected difference between the squared distances from $\mathbf{x}$ *to* $\mathbf{z}_1$ *and* $\mathbf{z}_2$ *is given by*

$$\Delta = E_{\mathcal{X}}\left[\|\mathbf{x} - \mathbf{z}_2\|^2\right] - E_{\mathcal{X}}\left[\|\mathbf{x} - \mathbf{z}_1\|^2\right]$$
$$= \gamma s^2 \sqrt{2d}. \tag{8}$$

The quantity in Eq. (8) can be interpreted as the degree of the spatial centrality bias present in the data, which causes hub formation. If $\mathbf{z}_1$ is closer to the origin (data mean) than $\mathbf{z}_2$ is, $\Delta > 0$ because in this case $\gamma > 0$.

This implies that a query object $\mathbf{x}$ sampled from $\mathcal{X}$ is more likely to be closer to object $\mathbf{z}_1$ than to $\mathbf{z}_2$; i.e., given query object $\mathbf{x}$, $\mathbf{z}_1$ is more likely to become its nearest neighbor. Because this reasoning applies to any pair of objects $\mathbf{z}_1$ and $\mathbf{z}_2$ in the dataset, it can be concluded that the objects closest to the data mean is most likely to be a hub.

Further, the factor s^2 in Eq. (8) suggests that, for a fixed query distribution $\mathcal{X}$, the data distribution $\mathcal{Y}$ with smaller variance s is preferable to reduce spatial centrality, and hence hubness as well.

4.2   Hubness and the Proposed Method

Ridge regression reduces the variance of mapped feature values (observables) relative to that of target (response) variables; see, for example, Shigeto et al. [9, Proposition 2]. Thus, in our model of Eq. (5), the variance of the components of the mapped objects $\mathbf{Wz}$ tends to be smaller than that of $\mathbf{x}$. From the discussion on hubness in Sect. 4.1, reducing the variance of data objects (which correspond to the image $\mathbf{Wz}$ of the labeled objects $\mathbf{z}$ in the proposed method) relative to the query (unlabeled object $\mathbf{x}$) can reduce the spatial centrality. By combining these arguments, we expect that the proposed approach should alleviate the emergence of hubs, and, consequently, improve the accuracy of k-NN classification.

Note that we could think of a different regression problem in which query object $\mathbf{x}$, not labeled object $\mathbf{z}$, is mapped to new coordinates:

$$\min_{\mathbf{W}} \sum_{i=1}^{n} \sum_{\mathbf{z} \in \mathcal{T}_i} \|\mathbf{Wx}_i - \mathbf{z}\|^2 + \lambda \|\mathbf{W}\|_{\mathrm{F}}^2. \tag{9}$$

This would result in function f as follows:

$$f(\mathbf{x}, \mathbf{z}) = \|\mathbf{Wx} - \mathbf{z}\|^2. \tag{10}$$

However, this dissimilarity function is useless as it actually *promotes* hubness. The variance of the transformed query objects shrinks as a result of regression. Thus, in this model, the variance of the labeled objects is made larger than the transformed query objects, but this is not a desirable situation according to Proposition 1. We also verify this in one of the experiments in Sect. 5.

5 Experiments

We evaluate the proposed approach on various classification tasks. The objective of these experiments is to investigate whether the proposed approach can reduce the emergence of hubs, and improve the performance of k-NN classification. The performance is measured against several popular metric learning methods.

5.1 Experimental Setups

Dataset Description. Three types of datasets were used for our evaluation: UCI, document, and image datasets.

From the UCI machine learning datasets,[1] we chose balance-scale, glass, ionosphere, iris, and wine, as they are frequently used for evaluation in metric learning literature [2,5,11,15]. However, they are mostly toy problems, and their small feature dimensions, the numbers of labels and objects do not necessarily reflect real-world problems. We therefore used document and image datasets also for our evaluation.

For document and image classification, support vector machines are known to provide state-of-the-art accuracy. Notice, however, that our goal is not to design a state-of-the-art classifier. Rather, the main objective of this experiments is to evaluate the performances of the proposed method in comparison with metric learning methods, and to show its usefulness for k-NN classification.

---

[1] http://archive.ics.uci.edu/ml/.

Table 1. Dataset statistics. In document and image datasets, "original dim." indicates the number of raw dimensions before applying PCA.

(a) UCI datasets.

dataset	ionosphere	balance-scale	iris	wine	glass
#objects	351	625	150	178	214
#classes	2	3	3	3	6
dimension	34	4	4	13	9

(b) Document datasets.

dataset	RCV	News	Reuters	TDT
#objects	9625	18846	8213	10021
#classes	4	20	41	56
dimension	300	300	300	300
original dim.	29992	26214	18933	36771

(c) Image datasets.

dataset	AwA	CUB	SUN	aPY
#objects	30475	11788	14340	15339
#classes	50	200	717	32
dimension	300	300	300	300
original dim.	4096	4096	4096	4096

For document classification tasks, we used four publicly available document datasets: RCV1-v2 (RCV), 20 newsgroups (News), Reuters21578 (Reuters), and TDT2 (TDT).[2] In Reuters21578 and TDT2, we removed minority classes that hold less than 10 objects in the dataset. After this removal, Reuters21578 and TDT2 had 56 and 41 classes, respectively.

For image classification, we used the following image datasets: aPascal & aYahoo (aPY), Animals with Attributes (AwA), Caltech-UCSD Birds-200-2011 (CUB), and SUN Attribute.[3]

The computational cost of metric learning methods is heavily dependent on the dimension of the feature space. In our preliminary experiment, training of the metric learning methods (LMNN, ITML, and DML-eig; see below) did not complete in a reasonable time on document and image datasets. We therefore had to use principal component analysis to reduce the dimensionality of features to 300 for these datasets.[4]

The dataset statistics are summarized in Table 1.

All data (set of feature vectors) were centered before training. For the wine dataset, we further converted the features to z-scores, following the remark on the UCI website that a k-NN classifier achieved a high accuracy with this standardization.

---

[2] Datasets were downloaded from http://www.cad.zju.edu.cn/home/dengcai/Data/TextData.html.

[3] We used the publicly available features from https://zimingzhang.files.wordpress.com/2014/10/cnn-features1.key.

[4] We also conducted experiments where the dimensionality of features was set to 100. The results are not presented here due to lack of space, but the same trend was observed.

Each dataset was randomly split into training (70%) and test (30%) sets. Experiments were repeated on four different random splits, for which we report the average performance.

Compared Methods. We trained distance/dissimilarity functions using the following methods, and carried out k-NN classification on the datasets above.

- original metric: Euclidean distance in the original feature space, without any training. This is the baseline.
- LMNN: Large margin nearest neighbor classification [11]. This method has often been used in distance metric learning experiments as a baseline.
- ITML: Information theoretic metric learning [2].
- DML-eig: Distance metric learning with eigenvalue optimization [15].
- Move-labeled (proposed method): Learning to move labeled objects toward query objects. This is our proposed approach that optimizes the ridge regression objective of Eq. (5), and predicts the labels using Eq. (7).
- Move-query: Learning to move query (unlabeled) objects toward fixed labeled objects. This is the method discussed in Sect. 4.2 and optimizes Eq. (9). Like Move-labeled it is also based on ridge regression, but the roles of the input and response variables are exchanged. The resulting dissimilarity function of Eq. (10) is then used for k-NN classification.

Notice that Move-query was tested only to verify the claim made in Sect. 4.2; i.e., although both Move-labeled and Move-query are based on ridge regression, the proposed method, Move-labeled, is expected to perform well by reducing hubness, whereas Move-query is expected to do the contrary, by *promoting* hubness. It was therefore not meant as a competitive method.

LMNN, ITML, and DML-eig learn a Mahalanobis distance. For these methods, we used the publicly available MATLAB implementations provided by the respective authors[5]. We implemented the proposed method also in MATLAB[6], for fair evaluation of running time.

For LMNN, Move-labeled, and Move-query, the number of target objects for each training object was set to 1; i.e., for each object $\mathbf{x}_i$ in the training set, we made a training pair $(\mathbf{x}_i, \mathbf{z})$ whose distance should be minimized, where $\mathbf{z}$ is the object nearest to $\mathbf{x}_i$ among those with the same class label as $\mathbf{x}_i$ in the training set, with the distance measured by the original Euclidean metric. For the parameters of ITML on UCI datasets, the default values in the authors' implementation were used, and for document and image datasets, we followed Jain et al. [5]. For DML-eig, the default setting in the authors' code was used for obtaining pairwise constraints. We calibrated the parameter k of k-NN classification to be used at the test time and all other parameters (γ in ITML, μ in DML-eig, and λ in Move-labeled and Move-query) by cross validation on the training set.

---

[5] LMNN: https://bitbucket.org/mlcircus/lmnn/downloads,
   ITML: http://www.cs.utexas.edu/~pjain/itml/,
   DML-eig: http://www.albany.edu/~yy298919/software.html.
[6] This code will be made available at our homepage.

Evaluation Criteria. The methods were evaluated in three respects: (i) the accuracy of k-NN classification using the distance/dissimilarity measure learned by each method, (ii) training time, and (iii) the degree of hubness in the data with respect to the learned distance/dissimilarity.

Following the literature [4,7–10], we used the skewness of N_{10} distribution as the measure of hubness in the data. The N_{10} distribution is the empirical distribution of the number $N_{10}(i)$ of times each labeled object i is found in the 10-nearest neighbors of query (unlabeled) objects, and its skewness is defined as follows:

$$(N_{10} \text{ skewness}) = \frac{\sum_{i=1}^{n} (N_{10}(i) - \mathrm{E}\,[N_{10}])^3 / n}{\mathrm{Var}\,[N_{10}]^{\frac{3}{2}}}$$

where n is the total number of labeled objects, and $\mathrm{E}[N_{10}]$ and $\mathrm{Var}[N_{10}]$ are respectively the empirical mean and variance of $N_{10}(i)$ over n labeled objects. A large N_{10} skewness value indicates the existence of labeled objects that frequently appear in the 10-nearest neighbor lists of query objects, i.e., hubs.

5.2   Experimental Results and Discussion

Skewness. Table 2 shows the skewness of N_{10} distribution. For all datasets, we observe that the proposed approach (Move-labeled) reduced N_{10} skewness considerably compared with the original Euclidean distance, meaning that it effectively suppressed the emergence of hub objects. N_{10} skewness was reduced by metric learning methods (LMNN, ITML, and DML-eig) on many datasets, most notably by DML-eig. Also, as expected from the discussion of Sect. 4.2, Move-query increased N_{10} skewness except for the iris dataset.

Accuracy. Table 3 shows the classification accuracy. In most datasets, both the metric learning methods and the proposed method outperformed the original distance metric. The proposed method is comparable with, or slightly better than, the metric learning methods. Although Move-query optimized the minimizing distance between objects in same class (our proposed method also optimized such distance), the method obtained poor results even compared with the original Euclidean metric except for the iris datasets.

Note that, in UCI datasets, we observed that the proposed method did not work well, and even Move-query were competitive with others. This is an expected result, because the UCI datasets did not have much hubness even with the original metric (see Table 2a). Hubs tend to be emerge in high dimensional space [4,7,8], but all the UCI datasets have a small dimensionality (see Table 1a). Consequently, hub reduction/promotion methods did not affect the result significantly.

Table 2. Skewness of N_{10} distribution: a high skewness indicates the emergence of hubs (smaller is better). The bold figure indicates the best performer.

(a) UCI datasets.

method	ionosphere	balance-scale	iris	wine	glass
original metric	1.65	0.93	0.40	0.71	0.77
LMNN	1.05	0.63	0.39	0.61	0.74
ITML	0.96	0.79	**0.10**	0.43	0.70
DML-eig	**0.78**	0.66	0.41	**0.38**	**0.59**
Move-labeled (proposed)	1.04	**0.56**	0.32	0.55	**0.59**
Move-query	1.67	1.13	0.32	0.89	1.18

(b) Document datasets.

method	RCV	News	Reuters	TDT
original metric	13.35	21.93	7.61	4.89
LMNN	3.86	14.74	7.63	4.01
ITML	4.27	19.65	7.30	2.39
DML-eig	1.71	**1.45**	**3.05**	**1.34**
Move-labeled (proposed)	**1.14**	2.88	4.53	1.44
Move-query	21.57	33.36	17.49	6.71

(c) Image datasets.

method	AwA	CUB	SUN	aPY
original metric	2.49	2.38	2.52	2.80
LMNN	3.10	2.96	2.80	3.94
ITML	2.42	2.27	2.37	2.69
DML-eig	1.90	1.77	2.39	2.17
Move-labeled (proposed)	**1.24**	**0.97**	**1.02**	**1.23**
Move-query	7.81	7.83	7.48	11.65

Training Time. To investigate the computational cost, we measured the elapsed real time needed to train the proposed method and the metric learning methods.

Table 4 shows the average training time in document and image datasets. We observe that the proposed approach has a clear advantage in terms of training cost. It was faster than any metric learning methods compared. Indeed, on all datasets except RCV, it was more than two orders of magnitude faster than the fastest metric learning methods. This can be explained by the fact that the metric learning methods take burden of optimizing over Mahalanobis metric. To enforce the constraint that the matrix **M** in Eq. (2) should remain positive semi-definite, these methods pay high computational cost, e.g., to check the non-negativity of eigenvalues, at every training iteration. In contrast, the proposed approach has a closed-form solution; although this solution depends on matrix inverse, it needs to be calculated only once.

Table 3. Classification accuracy [%]: bold figures indicate the best performers for each dataset.

(a) UCI datasets.

method	ionosphere	balance-scale	iris	wine	glass
original metric	86.8	89.5	97.2	98.1	68.1
LMNN	**90.3**	90.0	96.7	98.1	67.7
ITML	87.7	89.5	**97.8**	**99.1**	65.0
DML-eig	87.7	**91.2**	96.7	98.6	66.5
Move-labeled (proposed)	89.6	89.5	97.2	98.6	**70.8**
Move-query	79.7	89.4	97.2	96.3	62.3

(b) Document datasets.

method	RCV	News	Reuters	TDT
original metric	92.1	76.9	89.5	96.1
LMNN	**94.7**	79.9	91.5	96.6
ITML	93.2	77.0	90.8	96.5
DML-eig	94.5	73.3	85.9	95.7
Move-labeled (proposed)	94.4	**81.6**	**91.6**	**96.7**
Move-query	89.1	70.0	85.9	95.4

(c) Image datasets.

method	AwA	CUB	SUN	aPY
original metric	83.2	51.6	26.2	82.2
LMNN	83.0	**54.7**	24.4	81.8
ITML	83.1	51.3	26.0	82.4
DML-eig	82.0	53.5	22.4	81.6
Move-labeled (proposed)	**84.1**	52.4	**28.3**	**83.4**
Move-query	79.2	43.3	14.6	78.7

Table 4. Training time [sec]: bold figures indicate the best performer for each dataset.

(a) Document datasets.

method	RCV	News	Reuters	TDT
LMNN	1713.0	1164.7	676.2	886.1
ITML	35.5	1512.5	124.1	169.0
DML-eig	762.2	6145.9	2710.4	2350.6
proposed	**6.0**	**7.0**	**4.6**	**16.1**

(b) Image datasets.

method	AwA	CUB	SUN	aPY
LMNN	1525.5	1098.2	15704.3	317.3
ITML	1536.3	577.6	1126.4	9211.2
DML-eig	2048.0	2084.7	2006.1	1787.1
proposed	**9.5**	**1.5**	**4.1**	**6.4**

6 Conclusion

In this paper, we have proposed a simple regression-based technique to improve k-NN classification accuracy. The results of our work can be summarized as follows:

- To improve k-NN classification accuracy, we proposed learning a transformation of labeled objects, without altering the coordinates of query (unlabeled) objects. This approach is justified from the perspective of reducing hubness in the labeled objects. Because our method is inherently designed to suppress hubness, it need not consider pairs of objects from different classes during training. The number of such pairs can be enormous and their use also renders the optimization problem non-convex, which is therefore a major obstacle to the scalability of metric learning methods.
- Our method deviates from the metric learning framework as the learned transformation $\mathbf{W}$ does not provide a proper metric. In k-NN classification, however, labeled objects can be interpreted not as mere points but rather a representation of the (non-linear) decision boundaries between classes. Our approach follows this interpretation and changes the decision boundaries, through $\mathbf{W}$, to improve classification accuracy.
- In the experiments of Sect. 5, our approach indeed improved the classification accuracy when the data was high-dimensional and hubs emerged. It outperformed metric learning methods on most document and image datasets, and comparable on the rest. However, its effect was not evident on the UCI datasets, in which hubness was not noticeable because of the low dimensionality of the data.
- The experiments showed that our approach was substantially faster than the compared metric learning methods. For large document and image datasets, the speed-up was more than two orders of magnitude over the fastest metric learning methods, although the classification accuracy was better or comparable.

We have focused on multi-class classification problems in this paper, but hubness is known to be harmful in other situations, such as clustering and semi-supervised classification in high-dimensional space [7]. We plan to extend our approach to deal with these situations. We will also extend our method to learn nonlinear metrics.

Another direction of future work is to investigate the effect of our approach on kernel machines. Metric learning has been shown to be an effective preprocessing for kernel machines [3,12,14], and we will pursue a similar line using our approach.

Acknowledgments. We thank anonymous reviewers for their comments and suggestions. This work was partially supported by JSPS Kakenhi Grant No. 15H02749.

References

1. Bellet, A., Habrard, A., Sebban, M.: A survey on metric learning for feature vectors and structured data. arXiv preprint arXiv:1306.6709 (2014)
2. Davis, J.V., Kulis, B., Jain, P., Sra, S., Dhillon, I.S.: Information-theoretic metric learning. In: ICML 2007, pp. 209–216 (2007)
3. Dhillon, P.S., Talukdar, P.P., Crammer, K.: Learning better data representation using inference-driven metric learning. In: ACL 2010, pp. 377–381 (2010)
4. Hara, K., Suzuki, I., Shimbo, M., Kobayashi, K., Fukumizu, K., Radovanović, M.: Localized centering: reducing hubness in large-sample data. In: AAAI 2015 (2015)
5. Jain, P., Kulis, B., Davis, J.V., Dhillon, I.S.: Metric and kernel learning using a linear transformation. JMLR **13**, 519–547 (2012)
6. Kulis, B.: Metric learning: a survey. Found. Trends Mach. Learn. **5**(4), 287–364 (2013)
7. Radovanović, M., Nanopoulos, A., Ivanović, M.: Hubs in space: popular nearest neighbors in high-dimensional data. JMLR **11**, 2487–2531 (2010)
8. Schnitzer, D., Flexer, A., Schedl, M., Widmer, G.: Local and global scaling reduce hubs in space. JMLR **13**, 2871–2902 (2012)
9. Shigeto, Y., Suzuki, I., Hara, K., Shimbo, M., Matsumoto, Y.: Ridge regression, hubness, and zero-shot learning. In: Appice, A., Rodrigues, P.P., Santos Costa, V., Soares, C., Gama, J., Jorge, A. (eds.) ECML PKDD 2015. LNCS (LNAI), vol. 9284, pp. 135–151. Springer, Cham (2015). doi:10.1007/978-3-319-23528-8_9
10. Suzuki, I., Hara, K., Shimbo, M., Saerens, M., Fukumizu, K.: Centering similarity measures to reduce hubs. In: EMNLP 2013, pp. 613–623 (2013)
11. Weinberger, K.Q., Saul, L.K.: Distance metric learning for large margin nearest neighbor classification. JMLR **10**, 207–244 (2009)
12. Weinberger, K.Q., Tesauro, G.: Metric learning for kernel regression. In: AISTATS 2007, pp. 608–615 (2007)
13. Xing, E.P., Ng, A.Y., Jordan, M.I., Russell, S.: Distance metric learning, with application to clustering with side-information. In: NIPS 2002, pp. 505–512 (2002)
14. Xu, Z., Weinberger, K.Q., Chapelle, O.: Distance metric learning for kernel machines. arXiv preprint arXiv:1208.3422 (2013)
15. Ying, Y., Li, P.: Distance metric learning with eigenvalue optimization. JMLR **13**, 1–26 (2012)

Deep Network Regularization via Bayesian Inference of Synaptic Connectivity

Harris Partaourides[(✉)] and Sotirios P. Chatzis

Department of Electrical Engineering, Computer Engineering and Informatics,
Cyprus University of Technology, Limassol, Cyprus
{c.partaourides,sotirios.chatzis}@cut.ac.cy

Abstract. Deep neural networks (DNNs) often require good regularizers to generalize well. Currently, state-of-the-art DNN regularization techniques consist in randomly dropping units and/or connections on each iteration of the training algorithm. Dropout and DropConnect are characteristic examples of such regularizers, that are widely popular among practitioners. However, a drawback of such approaches consists in the fact that their postulated probability of random unit/connection omission is a constant that must be heuristically selected based on the obtained performance in some validation set. To alleviate this burden, in this paper we regard the DNN regularization problem from a Bayesian inference perspective: We impose a sparsity-inducing prior over the network synaptic weights, where the sparsity is induced by a set of Bernoulli-distributed binary variables with Beta (hyper-)priors over their prior parameters. This way, we eventually allow for marginalizing over the DNN synaptic connectivity for output generation, thus giving rise to an effective, heuristics-free, network regularization scheme. We perform Bayesian inference for the resulting hierarchical model by means of an efficient Black-Box Variational inference scheme. We exhibit the advantages of our method over existing approaches by conducting an extensive experimental evaluation using benchmark datasets.

1 Introduction

In the last few years, the field of machine learning has experienced a new wave of innovation; this is due to the rise of a family of modeling techniques commonly referred to as deep neural networks (DNNs) [10]. DNNs constitute large-scale neural networks, that have successfully shown their great learning capacity in the context of diverse application areas. Since DNNs comprise a huge number of trainable parameters, it is key that appropriate techniques be employed to prevent them from overfitting. Indeed, it is now widely understood that one of the reasons behind the explosive success and popularity of DNNs consists in the availability of simple, effective, and efficient regularization techniques, developed in the last few years [10].

Dropout is a popular regularization technique for (dense-layer) DNNs [13]. In essence, it consists in randomly dropping different units of the network on

© Springer International Publishing AG 2017
J. Kim et al. (Eds.): PAKDD 2017, Part I, LNAI 10234, pp. 30–41, 2017.
DOI: 10.1007/978-3-319-57454-7_3

each iteration of the training algorithm. This way, only the parameters related to a subset of the network units are trained on each iteration; this ameliorates the associated network overfitting tendency, and it does so in a way that ensures that all network parameters are effectively trained. In a different vein, [14] proposed randomly dropping DNN synaptic connections, instead of network units (and all the associated parameters); they dub this approach DropConnect. As showed therein, such a regularization scheme yields better results than Dropout in several benchmark datasets, while offering provable bounds of computational complexity.

Despite these merits, one drawback of these regularization schemes can be traced to their very foundation and rationale: The postulated probability of random unit/connection omission (e.g., dropout rate) is a constant that must be heuristically selected; this is effected by evaluating the network's predictive performance under different selections of this probability, in some validation set, and retaining the best performing value. This drawback has recently motivated research on the theoretical properties of these techniques. Indeed, recent theoretical work at the intersection of deep learning and Bayesian statistics has shown that Dropout can be viewed as a simplified approximate Bayesian inference algorithm, and enjoys links with Gaussian process models under certain simplistic assumptions (e.g., [1,5]).

These recent results form the main motivation behind this paper. Specifically, the main question this work aims to address is the following: Can we devise an effective DNN regularization scheme, that marginalizes over all possible configurations of network synaptic connectivity (i.e., active synaptic connections), with the posterior over them being inferred from the data? To address this problem, in this paper, for the first time in the literature, we regard the DNN regularization problem from the following Bayesian inference perspective: We impose a sparsity-inducing prior over the network synaptic weights, where the sparsity is induced by a set of Bernoulli-distributed binary variables. Further, the parameters of the postulated Bernoulli-distributed binary variables are imposed appropriate Beta (hyper-)priors, which give rise to a full hierarchical Bayesian treatment for the proposed model.

Under this hierarchical Bayesian construction, we can derive appropriate posteriors over the postulated binary variables, which essentially function as indicators of whether some (possible) synaptic connection is retained or dropped from the network. Once these posteriors are obtained using some available training data, prediction can be performed by averaging (under a Bayesian inference sense) over multiple (posterior) samples of the network configuration. This inferential setup constitutes the main point of differentiation between our approach and DropConnect. For simplicity, and to facilitate reference, we dub our approach DropConnect++. We derive an efficient inference algorithm for our model by resorting to the Black-Box Variational Inference (BBVI) scheme [12].

The remainder of this paper is organized as follows: In Sect. 2, we provide a brief overview of the theoretical background of our approach. Specifically, we first briefly review DropConnect, which is the existing work closest related to our approach; subsequently, we review the inferential framework that will be used in the context of the proposed approach, namely BBVI. In Sect. 3, we introduce our approach, and derive its inference and prediction generation algorithms. Next, we perform an extensive experimental evaluation of our approach, and compare to popular (dense-layer) DNN regularization approaches, including Dropout and DropConnect. To this end, we consider a number of well-established benchmarks in the related literature. Finally, in the concluding section, we summarize our contribution and discuss our results.

2   Theoretical Background

2.1   DropConnect

As discussed in the Introduction, DropConnect is a generalization of Dropout under which each connection, rather than each unit, may be dropped with some heuristically selected probability. Hence, the rationale of DropConnect is similar to that of Dropout, since both introduce dynamic sparsity within the model. Their core difference consists in the fact that Dropout imposes sparsity on the output vectors of a (dense) layer, while DropConnect imposes sparsity on the synaptic weights W.

Note that this is not equivalent to setting W to be a fixed sparse matrix during training. Indeed, for a DropConnect layer, the output is given as [14]:

$$r = a((Z \circ W)v) \tag{1}$$

where $\circ$ is the elementwise product, $a(\cdot)$ is the adopted activation function, W is the matrix of synaptic weights, v is the layer input vector, and r is the layer output vector. Further, Z is a matrix of binary variables (indicators) encoding the connection information, with

$$[Z]_{i,j} \sim \text{Bernoulli}(p) \tag{2}$$

where p is a heuristically selected probability. Hence, DropConnect is a generalization of Dropout to the full connection structure of a layer [14].

Training of a DropConnect layer begins by selecting an example v, and drawing a mask matrix Z from a Bernoulli(p) distribution to mask out elements of both the weight matrix and the biases in the DropConnect layer. The parameters throughout the model can be updated via stochastic gradient descent (SGD), or some modern variant of it, by backpropagating gradients of the postulated loss function with respect to the parameters. To update the weight matrix W in a DropConnect layer, the mask is applied to the gradient to update only those elements that were active in the forward pass. Additionally, when passing gradients down, the masked weight matrix $Z \circ W$ is used.

2.2   BBVI

In general, Bayesian inference for a statistical model can be performed either exactly, by means of Markov Chain Monte Carlo (MCMC), or via approximate techniques. Variational inference is the most widely used approximate technique; it approximates the posterior with a simpler distribution, and fits that distribution so as to have minimum Kullback-Leibler (KL) divergence from the exact posterior [8]. This way, variational inference effectively converts the problem of approximating the posterior into an optimization problem.

One of the significant drawbacks of traditional variational inference consists in the fact that its objective entails posterior expectations which are tractable only in the case of conjugate postulated models. Hence, recent innovations in variational inference have attempted to allow for rendering it feasible even in cases of more complex, non-conjugate model formulations. Indeed, recently proposed solutions to this problem consist in using stochastic optimization, by forming noisy gradients with Monte Carlo (MC) approximation. In this context, a number of different techniques have been proposed so as to successfully reduce the unacceptably high variance of conventional MC estimators. BBVI [12] is one of these recently proposed alternatives, amenable to non-conjugate probabilistic models that entail both discrete and continuous latent variables.

Let us consider a probabilistic model $p(x, z)$ with observations x and latent variables z, as well as a sought variational family $q(z; \phi)$. BBVI optimizes an evidence lower bound (ELBO), with expression

$$\log p(x) \geq \mathcal{L}(\phi) = \mathbb{E}_{q(z;\phi)}[\log p(x, z) - \log q(z; \phi)] \tag{3}$$

This is performed by relying on the "log-derivative trick" [7,15] to obtain MC estimates of the gradient. Specifically, by application of the identities

$$\nabla_\phi q(z; \phi) = q(z; \phi)\nabla_\phi \log q(z; \phi) \tag{4}$$

$$\mathbb{E}_{q(z;\phi)}[\nabla_\phi \log q(z; \phi)] = 0 \tag{5}$$

the gradient of the ELBO (3) reads

$$\nabla_\phi \mathcal{L}(\phi) = \mathbb{E}_{q(z;\phi)}[f(z)] \tag{6}$$

where

$$f(z) = \nabla_\phi \log q(z; \phi)\,[\log p(x, z) - \log q(z; \phi)] \tag{7}$$

The so-obtained MC estimator, based on computing the posterior expectations $\mathbb{E}_{q(z;\phi)}[\cdot]$ via sampling from $q(z; \phi)$, only requires evaluating the log-joint distribution $\log p(x, z)$, the log-variational distribution $\log q(z; \phi)$, and the score function $\nabla_\phi \log q(z; \phi)$, which is easy for a large class of models. However, the resulting estimator may have high variance, especially if the variational approximation $q(z; \phi)$ is a poor fit to the actual posterior. In order to reduce the variance of the estimator, one common strategy in BBVI consists in the use of *control variates*.

A control variate is a random variable that is included in the estimator, preserving its expectation but reducing its variance. The most usual choice for control variates, which we adopt in this work, is the so-called weighted score function: Under this selection, the ELBO gradient becomes

$$\nabla_\phi \mathcal{L}(\phi) = \mathbb{E}_{q(z;\phi)}[f(z) - \varpi h(z)] \tag{8}$$

where the score function reads

$$h(z) = \nabla_\phi \log q(z; \phi) \tag{9}$$

while the weights ϖ yield the (optimized) expression [12]

$$\varpi = \frac{\mathrm{Cov}\left(f(z), h(z)\right)}{\mathrm{Var}\left(h(z)\right)} \tag{10}$$

On this basis, derivation of the sought variational posteriors is performed by utilizing the gradient expression (8) in the context of popular, off-the-shelf optimization algorithms, e.g. AdaM [9] and Adagrad [4].

3   Proposed Approach

The output expression of a DropConnect++ layer is fundamentally similar to conventional DropConnect, and is given by (1). However, DropConnect++ introduces an additional hierarchical set of assumptions regarding the matrix of binary (mask) variables $Z = [z_{ij}]_{i,j}$, which indicate whether a synaptic connection is inferred to be on or off.

Specifically, as usual in hierarchical graphical models, we assume that the random matrix Z is drawn from an appropriate prior; we postulate

$$p(Z|\Pi) = \prod_{i,j} p(z_{ij}|\pi_{ij}) = \prod_{i,j} \mathrm{Bernoulli}(z_{ij}|\pi_{ij}) \tag{11}$$

Subsequently, to *facilitate further regularization* for DropConnect++ layers under a Bayesian inferential perspective, the prior parameters $\pi_{ij} \triangleq p(z_{ij} = 1)$ are imposed their own (hyper-)prior. Specifically, we elect to impose a Beta hyper-prior, yielding

$$p(\pi_{ij}|\alpha, \beta) = \mathrm{Beta}(\pi_{ij}|\alpha, \beta), \ \forall i, j \tag{12}$$

Under this definition, to train a postulated DNN incorporating DropConnect++ layers, we need to resort to some sort of Bayesian inference technique. In this paper, we resort to BBVI, as we explain next.

3.1   Training DNNs with DropConnect++ Layers

Let us consider a DNN the observed training data of which constitute the set $\mathcal{D} = \{d_n\}_{n=1}^N$. In case of a generative modeling scheme, each example d_n is a single observation, say x_n, from the distribution we wish to model. On the other hand, in case of a discriminative modeling task, each example d_n is an input/output pair, for instance $d_n = (x_n, y_n)$. In both cases, conventional DNN training consists in optimizing a negative loss function, measuring the fit of the model to the training dataset $\mathcal{D}$. Such measures can be equivalently expressed in terms of a log-likelihood function $\log p(\mathcal{D})$; under this regard, DNN training effectively boils down to maximum-likelihood estimation [3,6].

The deviation of a DNN comprising DropConnect++ layers from this simple training scheme stems from obtaining appropriate *posterior* distributions over the *latent variables* of DropConnect++, namely the binary indicator matrices of synaptic connectivity, $\boldsymbol{Z}$, as well as the associated parameters with hyper-priors imposed over them, namely the matrices of (prior) parameters $\boldsymbol{\Pi}$. To this end, DropConnect++ postulates *separate* posteriors over each entry of the random matrices $\boldsymbol{Z}$, that correspond to each individual synapse, (i, j):

$$q(\boldsymbol{Z}) = \prod_{i,j} q(z_{ij}|\tilde{\pi}_{ij}), \text{ with}: \qquad q(z_{ij}|\tilde{\pi}_{ij}) = \text{Bernoulli}(z_{ij}|\tilde{\pi}_{ij}) \qquad (13)$$

Further, we consider that the matrices of prior parameters, $\boldsymbol{\Pi}$, yield a factorized (hyper-)posterior with Beta-distributed factors of the form

$$q(\pi_{ij}) = \text{Beta}(\pi_{ij}|\tilde{\alpha}_{ij}, \tilde{\beta}_{ij}) \qquad (14)$$

Our construction entails a conditional log-likelihood term, $\log p(\mathcal{D}|\boldsymbol{Z})$. This is similar to a conventional DNN, with the weight matrices $\boldsymbol{W}$ at each layer multiplied with the corresponding latent indicator (mask) matrices, $\boldsymbol{Z}$ (in analogy to DropConnect). The corresponding posterior expectation term, $\mathbb{E}_{q(\boldsymbol{Z})}[\log p(\mathcal{D}|\boldsymbol{Z})]$, constitutes part of the ELBO expression of our model. Unfortunately, this term is analytically intractable due to the entailed nonlinear dependencies on the indicator matrix $\boldsymbol{Z}$, which stem from the nonlinear activation function $a(\cdot)$. Following the previous discussion, we ameliorate this issue by resorting to an efficient approximation obtained by drawing MC samples. The so-obtained ELBO functional expression eventually becomes:

$$\mathcal{L}(\mathcal{D}) \approx - \sum_{i,j} \text{KL}\big[q(z_{ij}|\tilde{\pi}_{ij})||p(z_{ij}|\pi_{ij})\big] - \sum_{i,j} \text{KL}\big[q(\pi_{ij}|\tilde{\alpha}_{ij}, \tilde{\beta}_{ij})||p(\pi_{ij}|\alpha, \beta)\big]$$
$$+ \frac{1}{L} \sum_{l,n=1}^{L,N} \log p(\boldsymbol{d}_n|\boldsymbol{Z}^{(l)}) \qquad (15)$$

where L is the number of samples, $\boldsymbol{Z}^{(l)} = [z_{ij}^{(l)}]_{i,j}$ and $z_{ij}^{(l)} \sim \text{Bernoulli}(z_{ij}|\tilde{\pi}_{ij})$.

This concludes the formulation of the proposed inferential setup for a DNN that contains DropConnect++ layers. On this basis, inference is performed by resorting to BBVI, which proceeds as described previously. Denoting $\tilde{\boldsymbol{\pi}} = (\tilde{\pi}_{ij})_{i,j}, \tilde{\boldsymbol{\alpha}} = (\tilde{\alpha}_{ij})_{i,j}, \tilde{\boldsymbol{\beta}} = (\tilde{\beta}_{ij})_{i,j}$, the used ELBO gradient reads

$$\nabla_{\tilde{\boldsymbol{\pi}}, \tilde{\boldsymbol{\alpha}}, \tilde{\boldsymbol{\beta}}, \boldsymbol{W}} \mathcal{L}(\mathcal{D}) \approx \frac{1}{L} \sum_{l,n=1}^{L,N} \nabla_{\boldsymbol{W}} \log p(\boldsymbol{d}_n|\boldsymbol{Z}^{(l)}) - \sum_{i,j} \nabla_{\tilde{\boldsymbol{\pi}}, \tilde{\boldsymbol{\alpha}}, \tilde{\boldsymbol{\beta}}} \text{KL}\big[q(z_{ij}|\tilde{\pi}_{ij})||p(z_{ij}|\pi_{ij})\big]$$
$$- \sum_{i,j} \nabla_{\tilde{\boldsymbol{\alpha}}, \tilde{\boldsymbol{\beta}}} \text{KL}\big[q(\pi_{ij}|\tilde{\alpha}_{ij}, \tilde{\beta}_{ij})||p(\pi_{ij}|\alpha, \beta)\big] \qquad (16)$$
$$- \varpi \sum_{i,j} \nabla_{\tilde{\boldsymbol{\pi}}, \tilde{\boldsymbol{\alpha}}, \tilde{\boldsymbol{\beta}}} [\log q(z_{ij}|\tilde{\pi}_{ij}) + q(\pi_{ij}|\tilde{\alpha}_{ij}, \tilde{\beta}_{ij})]$$

where ϖ is defined in (10). As one can note, we do not perform Bayesian inference for the synaptic weight parameters $\boldsymbol{W}$. Instead, we obtain point-estimates, similar to conventional DropConnect.

3.2   Feedforward Computation in DNNs with DropConnect++ Layers

Computation of the output of a trained DNN with DropConnect++ layers, given some network input $\boldsymbol{x}_*$, requires that we come up with an appropriate solution to the problem of computing the posterior expectation of the DropConnect++ layers output, say $\boldsymbol{r}_*$.

Let us consider a DropConnect++ layer with input $\boldsymbol{v}_*$ (corresponding to a DNN input observation $\boldsymbol{x}_*$); we have

$$\boldsymbol{r}_* = \mathbb{E}_{q(\boldsymbol{Z})}[a((\boldsymbol{Z} \circ \boldsymbol{W})\boldsymbol{v}_*)] \tag{17}$$

This computation essentially consists in marginalizing out the layer synaptic connectivity structure, by appropriately utilizing the variational posterior distribution $q(\boldsymbol{Z})$, learned by means of BBVI, as discussed in the previous Section. Unfortunately, this posterior expectation cannot be computed analytically, due to the nonlinear activation function $a(\cdot)$.

This problem can be solved by approximating (17) via simple MC sampling:

$$\boldsymbol{r}_* \approx \frac{1}{L} \sum_{l=1}^{L} a((\boldsymbol{Z}^{(l)} \circ \boldsymbol{W})\boldsymbol{v}_*) \tag{18}$$

where the $\boldsymbol{Z}^{(l)}$ are drawn from $q(\boldsymbol{Z})$. However, an issue such an approach suffers from is the need to retain in memory large sample matrices $\{\boldsymbol{Z}^{(l)}\}_{l=1}^{L}$, that may comprise millions of entries, in cases of large-scale DNNs. To completely alleviate such computational efficiency issues, in this work we opt for an alternative approximation that reads

$$\boldsymbol{r}_* \approx a((\tilde{\boldsymbol{\Pi}} \circ \boldsymbol{W})\boldsymbol{v}_*) \tag{19}$$

where the matrix $\tilde{\boldsymbol{\Pi}} = [\tilde{\pi}_{ij}]_{i,j}$ is obtained from the model training algorithm, described previously. Note that such an approximation is similar to the solution adopted by Dropout [13], which undoubtedly constitutes the most popular DNN regularization technique to date. We shall examine how this solution compares to MC sampling in the experimental section of this work.

4   Experimental Evaluation

To empirically evaluate the performance of our approach, we consider a number of supervised learning experiments, using the CIFAR-10, CIFAR-100, SVHN, and NORB benchmarks. In all our experiments, the used datasets are normalized with local zero mean and unit variance; no other pre-processing is implemented in this work[1]. To obtain some comparative results, apart from our method we also evaluate in our experiments DNNs with similar architecture but: (i) application of no regularization technique; (ii) regularized via Dropout; and (iii) regularized via DropConnect.

---

[1] Hence, our experimental setup is *not completely identical* to that of related works, e.g. [14]; these employ more complex pre-processing for some datasets.

Table 1. Predictive accuracy (%) of the evaluated methods.

Method	CIFAR-10	CIFAR-100	SVHN	NORB
No regularization	74.47	41.96	90.53	90.55
Dropout	75.70	46.65	92.14	92.07
DropConnect	76.06	46.12	91.41	91.88
DropConnect++	76.54	47.01	91.99	93.75

Table 2. Computational complexity per iteration at training time ($L = 1$).

#Method	CIFAR-10	CIFAR-100	SVHN	NORB
No regularization	9 s	10 s	15 s	5 s
Dropout	9 s	10 s	15 s	5 s
DropConnect	9 s	10 s	15 s	5 s
DropConnect++	10 s	13 s	19 s	6 s

In all cases, we use Adagrad with minibatch size equal to 128. Adagrad's global stepsize is chosen from the set $\{0.005, 0.01, 0.05\}$, based on the network performance on the training set in the first few iterations[2]. The units of all the postulated DNNs comprise ReLU nonlinearities [11]. Initialization of the network parameters is performed via Glorot-style uniform initialization [6]. To account for the effects of random initialization on the observed performances, we repeat our experiments 50 times; we report the resulting mean accuracies, and run the Student's-t statistical significance test to examine the statistical significance of the reported performance differences.

Prediction generation using our method is performed by employing the efficient approximation (19). The alternative approach of relying on MC sampling to perform feedforward computation [Eq. (18)] is evaluated in Sect. 4.2. In all cases, we set the prior hyperparameters of DropConnect++ to $\alpha = \beta = 1$; this is a convenient selection which reflects that we have no preferred values for the priors π_{ij}. The Dropout and DropConnect rates are selected on the grounds of performance maximization, following the selection procedures reported in the related literature. Our source codes have been developed in Python, using the Theano[3] [2] and Lasagne[4] libraries. We run our experiments on an Intel Xeon 2.5 GHz Quad-Core server with 64 GB RAM and an NVIDIA Tesla K40 GPU.

---

[2] We have found that Adagrad allows for the best possible network regularization by drawing just one sample per minibatch; that is, we use $L = 1$ at training time. This alleviates the training costs of both DropConnect and DropConnect++. We train all networks for 100 epochs; we do not apply L2 weight decay.

[3] http://deeplearning.net/software/theano/.

[4] https://github.com/Lasagne/Lasagne.

CIFAR-10. The CIFAR-10 dataset consists of color images of size 32×32, that belong to 10 categories (airplanes, automobiles, birds, cats, deers, dogs, frogs, horses, ships, trucks). We perform our experiments using the available 50,000 training samples and 10,000 test samples. All the evaluated methods comprise a convolutional architecture with three layers, 32 feature maps in the first layer, 32 feature maps in the second layer, 64 feature maps in the third layer, a 5×5 filter size, and a max-pooling sublayer with a pool size of 3×3. These three layers are followed by a dense layer with 64 hidden units, regularized via Dropout, Drop-Connect, or DropConnect++. The resulting performance statistics (predictive accuracy) of the evaluated methods are depicted in the first column of Table 1. As we observe, our approach outperforms all the considered competitors.

CIFAR-100. The CIFAR-100 dataset consists of 50,000 training and 10,000 testing color images of size 32×32, that belong to 100 categories. We retain this split of the data into a training set and a test set in the context of our experiments. The trained DNN comprises three convolutional layers of same architecture as the ones adopted in the CIFAR-10 experiment, that are followed by a dense layer comprising 512 hidden units. As we show in Table 1, our approach outperforms all its competitors, yielding the best predictive performance. Note also that the DropConnect method, which is closely related to our approach, yields in this experiment worse results than Dropout.

SVHN. The Street View House Numbers (SVHN) dataset consists of 73,257 training and 26,032 test color images of size 32×32; these depict house numbers collected by Google Street View. We retain this split of the data into a training set and a test set in the context of our experiments, and adopt exactly the same DNN architecture as in the CIFAR-100 experiment. As we show in Table 1, our method improves over the related DropConnect method.

NORB. The NORB (small) dataset comprises a collection of stereo images of 3D models that belong to 6 classes (animal, human, plane, truck, car, blank). We downsample the images from 96×96 to 32×32, and perform training and testing using the provided dataset split. We train DNNs with architecture similar to the one adopted in the context of the SVHN and CIFAR-100 datasets. As we show in Table 1, our method outperforms all the considered competitors.

4.1 Computational Complexity

Another significant aspect that affects the efficacy of a regularization technique is its final computational costs, and how they compare to the competition. To allow for investigating this aspect, in Table 2 we illustrate the time needed to complete one iteration of the training algorithms of the evaluated networks in our implementation. As we observe, the training algorithm of our approach imposes an 11%–30% increase in the computational time per iteration, depending on

Table 3. Variation of the predictive accuracy (%) of the MC-driven approach (18) with the number of MC samples.

#Samples, L	CIFAR-10	CIFAR-100	SVHN	NORB
1	74.57	43.28	91.32	90.04
30	75.95	46.33	91.70	90.78
50	76.01	46.33	91.72	91.04
100	76.01	46.54	91.78	91.41
500	76.36	46.94	91.78	91.58

the sizes of the network and the dataset. Note though that DNN training is an offline procedure; hence, a relatively small increase in the required training time is reasonable, given the observed predictive performance gains.

On the other hand, when it comes to using a trained DNN for prediction generation (test time), we emphasize that the computational costs of our approach are exactly the same as in the case of Dropout. This is, indeed, the case due to our utilization of the approximation (19), which results in similar feedforward computations for DropConnect++ as in the case of Dropout.

4.2 Further Investigation

A first issue that requires deeper investigation concerns the statistical significance of the observed performance differences. Application of the Student's-t test on the obtained sets of performances of each method (after 50 experiment repetitions from different random starts) has shown that these differences are statistically significant among all relevant pairs of methods (i.e. DropConnect++ vs. DropConnect, DropConnect++ vs. DropOut, and DropConnect++ vs. no regularization); only exception is the SVHN dataset, where DropConnect++ and DropOut are shown to be of statistically comparable performance.

Further, in Table 3 we show how the predictive performance of DropConnect++ changes if we perform feedforward computation via MC sampling, as described in Eq. (18). As we observe, using only one MC sample results in rather poor performance; this changes fast as we increase the number of samples. However, it appears that even with a high number of drawn samples, the MC-driven approach (18) does not yield any performance improvement over the approximation (19), despite imposing considerable computational overheads.

Further, in Fig. 1(a) we illustrate predictive accuracy convergence; for demonstration purposes, we consider the experimental case of the CIFAR-10 benchmark. Our exhibition concerns both application of the approximate feedforward computation rule (19), as well as resorting to MC sampling. We observe a clear and consistent convergence pattern in both cases.

Finally, it is interesting to get a feeling of the values that take the inferred posterior probabilities, $\tilde{\pi}$, of synaptic connectivity. In Fig. 1(b), we illustrate the inferred values of $\tilde{\pi}$ for all the network synapses, in the case of the CIFAR-10 experiment. As we observe, out of the almost 300 K synapses, around 50 K take values less than 0.35, another 50 K take values greater than 0.6, while the rest 200 K take values approximately in the interval $[0.4, 0.6]$. This implies that, out of the total 300 K postulated synapses, almost half of them are most likely to be omitted during inference. Most significantly, this figure depicts that our approach *infers (in a data-driven fashion) which specific* synapses are most useful to the network (thus yielding relatively high values of $\tilde{\pi}_{ij}$), and which should rather be omitted. This is in contrast to existing approaches, which merely apply a *homogeneous, random* omission/retention rate on each layer.

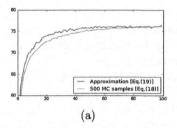
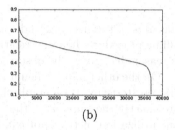

(a)   (b)

Fig. 1. (a) Accuracy convergence. (b) Inferred posterior probabilities, $\tilde{\pi}$.

5   Conclusions

In this paper, we examined whether there is a feasible way of performing DNN regularization by marginalizing over network synaptic connectivity in a Bayesian manner. Specifically, we sought to derive an appropriate posterior distribution over the network synaptic connectivity, inferred from the data. To this end, we imposed a sparsity-inducing prior over the network synaptic weights, where the sparsity is induced by a set of Bernoulli-distributed binary variables. Further, we imposed appropriate Beta (hyper-)priors over the parameters of the postulated Bernoulli-distributed binary variables. Under this hierarchical Bayesian construction, we obtained appropriate posteriors over the postulated binary variables, which indicate which synaptic connections are retained and which or dropped during inference. This was effected in an efficient and elegant fashion, by resorting to BBVI. We performed an extensive experimental evaluation, using several benchmark datasets. In most cases, our approach yielded a statistically significant performance improvement, for competitive computational costs.

Acknowledgment. We gratefully acknowledge the support of NVIDIA Corporation with the donation of one Tesla K40 GPU used for this research.

Appendix

$$\mathrm{KL}[q(z_{ij}|\tilde{\pi}_{ij})||p(z_{ij}|\pi_{ij})] = \tilde{\pi}_{ij}\log\tilde{\pi}_{ij} + (1 - \tilde{\pi}_{ij})\log(1 - \tilde{\pi}_{ij})$$
$$- \tilde{\pi}_{ij}\mathbb{E}_{q(\pi_{ij})}[\log\pi_{ij}] - (1 - \tilde{\pi}_{ij})\mathbb{E}_{q(\pi_{ij})}[\log(1 - \pi_{ij})] \tag{20}$$

$$\mathrm{KL}\big[q(\pi_{ij}|\tilde{\alpha}_{ij}, \tilde{\beta}_{ij})||p(\pi_{ij}|\alpha, \beta)\big] = \log\Gamma(\tilde{\alpha}_{ij} + \tilde{\beta}_{ij}) - \log\Gamma(\tilde{\alpha}_{ij}) - \log\Gamma(\tilde{\beta}_{ij})$$
$$+(\tilde{\alpha}_{ij} - \alpha)\mathbb{E}_{q(\pi_{ij})}[\log\pi_{ij}] + (\tilde{\beta}_{ij} - \beta)\mathbb{E}_{q(\pi_{ij})}[\log(1 - \pi_{ij})] \tag{21}$$

where:
$$\mathbb{E}_{q(\pi_{ij})}[\log\pi_{ij}] = \psi(\tilde{\alpha}_{ij}) - \psi(\tilde{\alpha}_{ij} + \tilde{\beta}_{ij}) \tag{22}$$

$$\mathbb{E}_{q(\pi_{ij})}[\log(1 - \pi_{ij})] = \psi(\tilde{\beta}_{ij}) - \psi(\tilde{\alpha}_{ij} + \tilde{\beta}_{ij}) \tag{23}$$

$\Gamma(\cdot)$ is the Gamma function, and $\psi(\cdot)$ is the Digamma function.

References

1. Baldi, P., Sadowski, P.: Understanding dropout. In: Proceedings of NIPS (2013)
2. Bastien, F., Lamblin, P., Pascanu, R., Bergstra, J., Goodfellow, I.J., Bergeron, A., Bouchard, N., Bengio, Y.: Theano: new features and speed improvements. In: Deep Learning and Unsupervised Feature Learning NIPS 2012 Workshop (2012)
3. Bengio, Y., Yao, L., Alain, G., Vincent, P.: Generalized denoising autoencoders as generative models. In: Proceedings of NIPS, pp. 899–907 (2013)
4. Duchi, J., Hazan, E., Singer, Y.: Adaptive subgradient methods for online learning and stochastic optimization. J. Mach. Learn. Res. **12**, 2121–2159 (2010)
5. Gal, Y., Ghahramani, Z.: Dropout as a Bayesian approximation: insights and applications. In: Deep Learning Workshop, ICML (2015)
6. Glorot, X., Bengio, Y.: Understanding the difficulty of training deep feedforward neural networks. In: Proceedings of AISTATS (2010)
7. Glynn, P.W.: Likelihood ratio gradient estimation for stochastic systems. Commun. ACM **33**(10), 75–84 (1990)
8. Jaakkola, T., Jordan, M.: Bayesian parameter estimation via variational methods. Stat. Comput. **10**, 25–37 (2000)
9. Kingma, D., Ba, J.: Adam: a method for stochastic optimization. In: Proceedings of ICLR (2015)
10. LeCun, Y., Bengio, Y., Hinton, G.: Deep learning. Nature **512**, 436–444 (2015)
11. Nair, V., Hinton, G.: Rectified linear units improve restricted Boltzmann machines. In: Proceedings of ICML (2010)
12. Ranganath, R., Gerrish, S., Blei, D.M.: Black box variational inference. In: Proceedings of AISTATS (2014)
13. Srivastava, N., Hinton, G.E., Krizhevsky, A., Sutskever, I., Salakhutdinov, R.R.: Dropout: a simple way to prevent neural networks from overfitting. J. Mach. Learn. Res. **15**(6), 1929–1958 (2014)
14. Wan, L., Zeiler, M., Zhang, S., LeCun, Y., Fergus, R.: Regularization of neural networks using DropConnect. In: Proceedings of ICML (2013)
15. Williams, R.J.: Simple statistical gradient-following algorithms for connectionist reinforcement learning. Mach. Learn. **8**(3–4), 229–256 (1992)

Adaptive One-Class Support Vector Machine for Damage Detection in Structural Health Monitoring

Ali Anaissi[1(✉)], Nguyen Lu Dang Khoa[1], Samir Mustapha[2], Mehrisadat Makki Alamdari[1], Ali Braytee[3], Yang Wang[1], and Fang Chen[1]

[1] Data61|CSIRO, 13 Garden Street, Eveleigh, NSW 2015, Australia
ali.anaissi@data61.csiro.au
[2] Department of Mechanical Engineering, American University of Beirut, Beirut, Lebanon
[3] Faculty of Engineering and IT, University of Technology Sydney, Sydney, Australia

Abstract. Machine learning algorithms have been employed extensively in the area of structural health monitoring to compare new measurements with baselines to detect any structural change. One-class support vector machine (OCSVM) with Gaussian kernel function is a promising machine learning method which can learn only from one class data and then classify any new query samples. However, generalization performance of OCSVM is profoundly influenced by its Gaussian model parameter σ. This paper proposes a new algorithm named Appropriate Distance to the Enclosing Surface (ADES) for tuning the Gaussian model parameter. The semantic idea of this algorithm is based on inspecting the spatial locations of the edge and interior samples, and their distances to the enclosing surface of OCSVM. The algorithm selects the optimal value of σ which generates a hyperplane that is maximally distant from the interior samples but close to the edge samples. The sets of interior and edge samples are identified using a hard margin linear support vector machine. The algorithm was successfully validated using sensing data collected from the Sydney Harbour Bridge, in addition to five public datasets. The designed ADES algorithm is an appropriate choice to identify the optimal value of σ for OCSVM especially in high dimensional datasets.

Keywords: Machine learning · Structural health monitoring · One-class support vector machine · Gaussian parameter selection · Anomaly detection

1   Introduction

Structural health monitoring (SHM) is an automated process to detect the damage in the structures using sensing data. It has earned a lot of interests in recent years and has attracted many researchers working in the area of machine learning [6,9,17]. With the advances in the sensing technology, it is becoming more

© Springer International Publishing AG 2017
J. Kim et al. (Eds.): PAKDD 2017, Part I, LNAI 10234, pp. 42–57, 2017.
DOI: 10.1007/978-3-319-57454-7_4

feasible to develop an approach for detection of structural damage based on the information gathered from the sensor networks mounted to the structure [5]. The focus now is to build a decision-making model that is able to detect damage on the structure using sensor data. This can be solved using a supervised learning approach such as a support vector machine (SVM) [2]. However, because of the lack of available data from the damaged state of the structure in most cases, this leads to the development of the OCSVM classification model [15]. The design of OCSVM is well suited this kind of problems where only observations from the positive (healthy) samples are required. Moreover, OCSVM has been extensively used in the area of SHM for detecting different types of anomalies [3,8,11].

The rational idea behind OCSVM is to map the data into a high dimensional feature space via a kernel function and then learn an optimal decision boundary that separates the training positive observations from the origin. Several kernel functions have been used in SVM such as Gaussian and polynomial kernels. However, the Gaussian kernel function defined in Eq. (1) has gained much more popularity in the area of machine learning and it has turned out to be an appropriate setting for OCSVM in order to generate a non-linear decision boundary.

$$K(x_i, x_j) = \exp(-\frac{\|x_i - x_j\|^2}{2\sigma^2}) \tag{1}$$

This kernel function is highly affected by a free critical parameter called the Gaussian kernel parameter denoted by σ which determines the width of the Gaussian kernel. This parameter has a great influence on the construction of a classification model for OCSVM as it controls how loosely or tightly the decision boundary fits the training data. To demonstrate the effect of the parameter σ on the decision boundary, we used a two-dimensional Banana-shaped data set. We applied the OCSVM on the dataset using different values of parameter σ, and then we plotted the resultant decision boundary of OCSVM for three different values of σ as shown in Fig. 1. Comparing the lower and upper bounds of σ, it can be clearly seen that the enclosing surface is very tight in Fig. 1a, while it is loose in Fig. 1b. The optimal one is shown in Fig. 1c as the decision boundary precisely describes the form of the data. At that point the issue is changed over into how to estimate the suitability of the decision boundary.

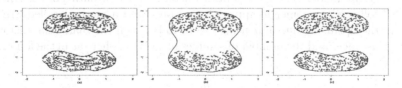

Fig. 1. Illustrations for enclosing surfaces at different values of σ. (a) $\sigma = 0.2$. (b) $\sigma = 1.3$. (c) $\sigma = 0.8$.

Several researchers have addressed the problem of selecting the proper value of σ [4,16,18]. However, they are not considered as appropriate methods to be applied on high dimensional datasets. Furthermore, tuning the Gaussian kernel for the OCSVM is still an open problem as stated by Tax and Duin [16] and Scholkopf *et al.* [15].

This paper addresses the problem of tuning the Gaussian kernel parameter σ in OCSVM to ensure the generalization performance of the constructed model to unseen high dimensional data. Following the geometrical approach, we proposed a Gaussian kernel parameter selection method which is implemented in two steps. The first step aims to select the edge and interior samples in the training dataset. The second step constructs OCSVM models at different settings of the parameter σ and then we measure the distances from the selected edge-interior samples to the enclosing surface of each OCSVM. Following these steps, we can select the optimal value of σ which provides the maximum difference in the average distances between the interior and edge samples to the enclosing surface. The algorithm was validated using a real high dimensional data collected using a network of accelerometers mounted underneath the deck on the Sydney Harbour Bridge in Australia.

The rest of this paper is organized as follows: Sect. 2 briefly presents some related work for tuning σ. The Gaussian kernel parameter selection method is provided in Sect. 3. Section 4 presents experimental results using different datasets. Section 5 presents some concluding remarks.

2   Related Work

Several methods have been developed for tuning the parameter σ in Gaussian kernel function. For instance, Evangelista *et al.* [4] followed a statistics-based approach to select the optimal value of σ using the variance and mean measures of the training dataset. This method is known as VM measure which aims to evaluate $\hat{\sigma}$ by computing the ratio of the variance and the mean of the lower (or upper) part of the kernel matrix using the following formula:

$$\hat{\sigma} = \max_{\sigma}(\frac{v}{\bar{m} + \xi}) \tag{2}$$

where v is the variance, m is the mean and ξ is a small value in order to avoid zero division. This method often generates a small value for $\hat{\sigma}$ which results in a very tight model that closely fits to a limited set of data points. Khazai *et al.* [7] followed a geometric approach and proposed a method called MD to estimate the optimal value of σ using the ratio between the maximum distance between instances and the number of samples inside the sphere as in the following equation:

$$\sigma^2 = \frac{d_{max}}{\sqrt{-\ln(\delta)}} \tag{3}$$

where the appropriate value for δ is calculated by:

$$\delta = \frac{1}{N(1 - \nu) + 1} \tag{4}$$

This method often produces a large value of σ which yields to construct a very simple and poor performance model especially when the training dataset has a small number of samples. In this case, the value of the dominator term in Eq. (3) ($\sqrt{-\ln(\delta)}$) becomes small. Xiao et al. [18] proposed a method known as MIES to select a suitable kernel parameter based on the spatial locations of the interior and edge samples. The critical requirement of this method is to find the edge-interior samples in order to calculate the optimal value of σ. The authors in [18] adopted the Border-Edge Pattern Selection (BEPS) method proposed by Li and Maguire [10] to select the edge-interior samples. This method performs well in selecting edge and interior samples when the data exists in a low dimensional space. However, it failed completely when it dealt with very high dimensional datasets where all the samples are selected as edge samples.

In this paper, we propose a method for tuning the Gaussian kernel parameter following a geometrical approach by introducing a new objective function and a new algorithm, inspired by [1], for finding the edge-interior samples of datasets exist in high dimensional space.

3   Gaussian Kernel Parameter Selection Method

The idea of the ADES method is based on the spatial locations of the edge and interior samples relative to the enclosing surface. The geometric locations of edge-interior samples with respect to the hyperplane plays a significant role in judging the appropriateness of the enclosing surface. In other words, the enclosing surface of OCSVM is very close to the interior samples when it has tightly fitted the data (as shown in Fig. 1a), and it is very far from the interior and edge samples when it is loose (as shown in Fig. 1b). However, the enclosing surface precisely fits the form of the data in Fig. 1c where the enclosing surface is far from the interior sample but at the same time is very close to the edge ones. This situation is turned up to be our objective in selecting the appropriateness of the enclosing surface. Therefore, we proposed a new objective function $f(\sigma_i)$ described in Eq. (5) to calculate the optimal value of $\hat{\sigma} = \underset{\sigma_i}{argmax}(f(\sigma_i))$.

$$f(\sigma_i) = mean(d_N(x_n)_{x_n \in \Omega_{IN}}) - mean(d_N(x_n)_{x_n \in \Omega_{ED}}) \qquad (5)$$

where Ω_{IN} and Ω_{ED}, respectively, represent the sets of interior and edge samples in the training positive data points, and d_N is the normalized distance from these samples to the hyperplane. This distance can be calculated using the following equation:

$$d_N(x_n) = \frac{d(x_n)}{1 - d_\pi} \qquad (6)$$

where d_π is the distance of a hyperplane to the origin described as $d_\pi = \frac{\rho}{\|w\|}$, and $d(x_n)$ is the distance of the sample x_n to the hyperplane obtained using the following equation:

$$d(x_n) = \frac{f(x_n)}{\|w\|} = \frac{\sum_{i=1}^{sv} \alpha_i k(x_i, x_n) - \rho.}{\sqrt{\sum_{ij}^{n} \alpha_i \alpha_j K(x_i, x_j)}} \tag{7}$$

where w is a perpendicular vector to the decision boundary, α are the Lagrange multipliers and ρ known as the bias term.

The aim of this objective function is to find the hyperplane that is maximally distant from the interior samples but not from the edge samples. In this new objective function, we use the average distances from the interior and edge samples to the hyperplane in order to reduce the effect of improper selection possibility of the interior and edge samples in high dimensional space datasets. The key point of this method now is how to identify the interior and edge samples in a given high dimensional dataset. Therefore, we propose a new method based on linear SVM to select the interior and edge samples in high dimensional space. This algorithm is described as follows: given a dataset of $x_i (i = 1, \ldots, n)$, the unit vector of each point x_i with its k closest points x_j is computed as follows:

$$v_j^k = \frac{x_j - x_i}{\|x_j - x_i\|} \tag{8}$$

Then we employ a hard margin linear SVM to separate v_j^k, the closest points to x_i, from the origin by solving the OCSVM optimization problem of the obtained unit vectors in Eq. (8). Once we get the optimal solution $\alpha_j, j = 1, \ldots, k$ and calculating ρ, we estimate the value of the decision function using,

$$f(v_j) = \sum_{i=1}^{sv} \alpha_i v_i . v_j - \rho. \tag{9}$$

The next step is to evaluate the optimization of the constructed linear OCSVM using the following equation

$$s_i = \frac{1}{k} \sum_{j=1}^{k} f(v_j) > 0 \tag{10}$$

where s_i represents the success accuracy rate of the model. If all the closest points v_j^k are successfully separated from the origin, then we count x_i as an edge sample. This approach may end up with a few number of edge samples. Therefore, we have used a threshold $1 - \gamma$ (γ is a small positive parameter), to control the number of edge samples by setting up a percentage of the acceptable success rate for each sample x_i to be an edge sample. For $\gamma = 0.05$, if 95% of the closest points to a sample x_i are successfully separated from the origin, then the sample x_i can be added to the edge sample set Ω_{ED}. We have also extended this method to select the interior samples based on the furthest neighbours of the edge samples. The assumption made is that the furthest neighbour samples to the edge sample should be added to the interior sample set Ω_{IN}. This method is presented in Algorithm 1.

Algorithm 1. Edge-interior samples selection method.

Input: A set of n positive samples $x = \{x_i\}_{i=1}^n$
For each sample x_i in x
    Find the k closest points to x_i: $x_j, j = 1, \ldots, k$.
    Calculate the unit vectors v_j^k of x_j according to (8).
    Separate v_j^k from the origin using a hard margin linear OCSVM.
    Calculate the decission values of v_j^k according to (9).
    Calculate s_i according to (10).
    If $s_i \geq 1 - \gamma$, then x_i and x_{jk} are added to Ω_{ED} and Ω_{IN}, respectively.

Output: Ω_{ED} and Ω_{IN}.

The algorithm starts with the entire set of positive samples. Two parameters are used in this algorithm; k, the number of the nearest neighbours which has been thoroughly studied by [10] and they set $k = 5 \ln n$, where γ take values in the range $[0, 0.1]$.

Once the edge and interior samples are identified, we start optimizing our objective function presented in Eq. 5. The complete proposed method of ADES is presented in Algorithm 2.

Algorithm 2. Gaussian kernel parameter selection method.

Input: A set of n positive samples $x = \{x_i\}_{i=1}^n$
  1. Obtain the sample sets Ω_{ED} and Ω_{IN} using Algorithm 1.
  2. Generate a candidate set σ_i $(i = 1, \ldots, m.)$ for parameter σ in the form of $[d_{min}, d_{max}]$.
  3. For each σ_i.
        Solve the optimization problem for OCSVM, that is, π_i (hyperplane).
        Calculate the normalized distances from samples in Ω_{ED} and Ω_{IN} to π_i according to (6).
        Calculate the objective function value $f(\sigma_i)$ according to (5).
  4. Select the biggest value $f(\sigma_i)$ as the optimal value $\hat{\sigma}$.

Output: the optimal kernel parameter $\hat{\sigma}$.

---

4   Experimental Results

Three experiments were carried to evaluate the performance of our proposed algorithm. We initially applied our method on two-dimensional toy datasets as it allows us to visually observe the performance of OCSVM by plotting its decision boundary. The performance was also tested on benchmark datasets that allows us to objectively compare our obtained results to the previously published ones. The final experiments were applied on datasets obtained from an actual structure, the Sydney Harbour Bridge, to demonstrate the ability of the proposed algorithm to detect damage in steel reinforced concrete jack arches.

4.1   Experiments on Artificial Toy Datasets

Three toys, {Round, Banana and Ring}-shaped, datasets were used in this section to visualize the performance of our method. The R package mlbench was used in order to generate these different geometric shaped datasets that vary in their characteristics. Figure 2 shows the selected edge and interior samples denoted by red "□" and green "✳", respectively. As it can be clearly observed that the proposed edge samples selection method has the ability to select the edge samples especially on the Ring-shaped dataset while the inner edge samples play a significant role in constructing the decision boundary.

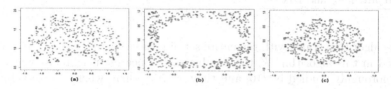

Fig. 2. Selected edge and interior samples: (a) Banana-shaped (b) Ring-shaped (c) Round-shaped. (Color figure online)

Figures 3 and 4 show the resultant decision boundary of each toy dataset. The enclosed surface of ADES method shown in each sub Figs. {3-4}(a) precisely fit the shape of each toy dataset without suffering from the overfitting nor the under-fitting problems. The MD and VM methods generate a loosely and tight decision boundary, respectively. The same results appeared with the Ring-shaped dataset (referring to Figs. 4(c) and (d)). The MIES method works well on the Banana-shaped dataset but failed in finding the optimal decision boundary in the Ring-shaped dataset. All the methods successfully enclosed the surface of the Round-shaped dataset with an optimal fitted decision boundary.

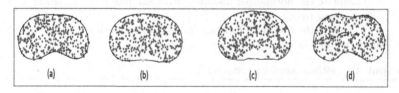

Fig. 3. Performance on the Banana-shaped dataset: (a) ADES (b) MIES (c) MD (d) VM.

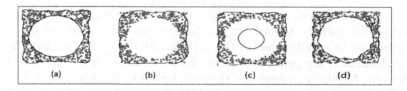

Fig. 4. Performance on the Ring-shaped dataset: (a) ADES (b) MIES(c) MD (d) VM.

4.2   Experiments on Benchmark Datasets

We further investigated the performance of our method using five publicly available datasets downloaded from the machine learning repository[1]. These datasets were previously used in previous related studies [18, 19]. The main characteristics of these datasets are summarized in Table 1. Each dataset has been pre-processed using the following procedure:

1. Label the class with a large number of samples as positive and the others as negative.
2. Randomly select 80% of the positive samples for training and 20% for testing in addition to the samples in the negative classes.
3. Normalize the training data to zero mean and unity variance.
4. Normalize the test data based on the mean and variance of the relating training dataset.

Table 1. Details of machine learning repository datasets.

Dataset	Dim	Positive	Negative
Breast	9	444	239
Heart	13	137	160
Survival	3	225	81
Diabetes	8	500	268
Sonar	60	111	97
Biomed	5	200	145

For each dataset, we generated 20 bootstrap samples from the training dataset to train OCSVM with σ parameter to be selected using Algorithm 2 and $\nu = 0.05$ for all tests. Once we construct the OCSVM model, we evaluate its classification performance on the test dataset and calculate the accuracy using g-mean metric defined as

$$g\text{-}mean = \sqrt{TPR^+ \times TNR^-} \qquad (11)$$

where TPR and TNR are the true positive rate and the true negative rate, respectively. Table 2 shows the classification performance comparison between ADES and the other methods described in Sect. 2. As shown in Table 2, the average g-mean of our method outperformed the other state-of-the-art methods on four datasets. ADES performed better than MIES algorithm on four datasets and generated a comparable result on the Biomed dataset. We can also notice that no results were reported in Table 2 for MIES method on the Sonar dataset. This is due to the fact that MIES algorithm does not work on high dimensional dataset where all the training data points are selected as edge samples.

---

[1] http://archive.ics.uci.edu/ml/datasets.html.

Table 2. Optimal values of σ selected by different methods for the benchmark datasets along with g-means, TPR and FPR.

Dataset	Method	σ	g-mean	TPR	FPR
Breast	ADES	11.78	**0.96**	0.93	0.03
	MD	9.19	0.94	0.92	0.03
	VM	0.40	0.54	0.29	0.00
	MIES	2.43	0.93	0.86	0.00
Heart	ADES	7.07	**0.73**	0.75	0.29
	MD	3.97	0.62	0.96	0.60
	VM	0.15	0.00	0.00	0.00
	MIES	2.40	0.69	0.75	0.35
Sonar	ADES	4.95	**0.76**	0.70	0.17
	MD	8.84	0.61	0.73	0.50
	VM	0.20	0.00	0.00	0.00
	MIES	—	—	—	—
Diabetes	ADES	5.56	**0.71**	0.64	0.22
	MD	5.87	0.40	0.92	0.83
	VM	0.21	0.68	0.55	0.15
	MIES	1.96	0.66	0.70	0.38
Biomed	ADES	3.22	0.83	0.82	0.16
	MD	4.52	0.78	0.79	0.22
	VM	0.30	0.00	0.00	0.00
	MIES	2.85	**0.85**	0.83	0.13

Further, it was observed from the results that the MD method achieves high classification accuracy on the positive samples represented by the value of the TPR, and low accuracy on the negative samples represented by the FPR measurement. This is what we anticipated discovering from the MD method based on the decision boundary resulted from the toy datasets. The same expectation with the VM method which achieves a high accuracy on the negative samples but a very low TPR. These findings also reflect what we have obtained using the toy datasets. Further, the VM method selects a very small value of σ when applied to the Heart and Sonar datasets. These small values lead to over-fit in the OCSVM model which completely failed to classify positive samples in the test datasets. This explains why we can see zero values for FPR and g-means. According to these results in Table 2, ADES has the capability to select the optimal value of σ without causing the OCSVM model neither to over-fit nor to under-fit. Moreover, ADES can work on high dimensional datasets while still being able to select the edge and interior samples which are crucial for the objective function presented in Eq. 5.

4.3   Case Studies in Structural Health Monitoring

This work is part of the efforts to apply SHM to the iconic Sydney Harbour Bridge. This section presents two case studies to illustrate how OCSVM using our proposed method for tuning sigma is capable to detect structural damage. The first case study was conducted using real datasets collected from the Sydney Harbour Bridge and the second case study is a reinforced concrete cantilever beam subjected to increasingly progressive crack which replicates one of the major structural components in the Sydney Harbour Bridge.

Case Study I: Sydney Harbour Bridge.

Experiments Setup and Data Collection. Our main experiments were conducted using structural vibration based datasets acquired from a network of accelerometers mounted on the Sydney Harbour Bridge. The bridge has 800 joints on the underside of the deck of the bus lane. However, only six joints were used in this study (named 1 to 6) as shown in Fig. 5. Within these six joints, only joint number four was known as a cracked joint [13,14]. Each joint was instrumented with a sensor node connected to three tri-axial accelerometers mounted on the left, middle and right side of the joint, as shown in Fig. 5. At each time a vehicle passes over the joint, defined as event, it causes vibrations which are recorded by the sensor node for a period of 1.6 s at a sampling rate of 375 Hz. An event is triggered when the acceleration value exceeds a pre-set threshold. Hence, 600 samples are recorded for each event. The data used in this study contains 36952 events as shown in Table 3 which were collected over a period of three months. For each reading of the tri-axial accelerometer (x,y,z), we calculated the magnitude of the three vectors and then the data of each event is normalized to have a zero mean and one standard variation. Since the accelerometer data is represented in the time domain, it is noteworthy to represent the generated data in the frequency domain using Fourier transform. The resultant six datasets (using the middle sensor of each joint) has 300 features which represent the frequencies of each event. All the events in the datasets (1, 2, 3, 5, and 6) are labeled positive (healthy events), where all the events in dataset 4 (joint 4) are labeled negative (damaged events). For each dataset, we randomly selected 70% of the positive events for training and 30% for testing in addition to the unhealthy events in dataset 4.

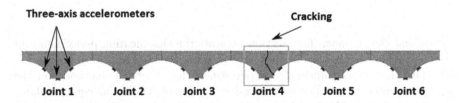

Fig. 5. Evaluated joints on the Sydney Harbour Bridge.

Table 3. Number of samples in each joint of the bridge dataset.

Dataset	Number of samples	Training	Test
Joint 1	6329	4430	1899
Joint 2	7237	5065	2172
Joint 3	4984	3488	1496
Joint 4	6886	0	6886
Joint 5	6715	4700	2015
Joint 6	4801	3360	1441

We trained the OCSVM for each joint (1, 2, 3, 5 and 6) using different values of the Gaussian parameter calculated using the three methods ADES, MD and VM. The MIES method was not used here because it does not work in high dimensional datasets.

Table 4. Optimal selection of the σ values based on different methods using the bridge datasets along with g-means, TPR and FPR.

Dataset	Method	σ	g-means	TPR	FPR
Joint 1	ADES	18.5	**0.986**	0.984	0.01
	MD	22.3	0.960	0.930	0.02
	VM	14.5	0.02	0.001	0.00
Joint 2	ADES	16.3	**0.978**	0.976	0.02
	MD	28.6	0.958	0.968	0.05
	VM	11.1	0.04	0.001	0.00
Joint 3	ADES	23.4	**0.983**	0.977	0.02
	MD	28.8	0.962	0.937	0.06
	VM	16.6	0.043	0.001	0.00
Joint 5	ADES	25.3	**0.951**	0.950	0.03
	MD	27.4	0.930	0.890	0.04
	VM	16.5	0.038	0.001	0.00
Joint 6	ADES	22.3	**0.969**	0.973	0.03
	MD	28.8	0.965	0.882	0.02
	VM	15.4	0.037	0.001	0.00

Results and Discussions. This section presents the classification performance of the OCSVM for each of the parameter selection methods. As shown in Table 4, the ADES method significantly outperformed the other two methods on the six experimented joints. The average g-mean value of ADES was equal to 0.971 compared to MD and VM, with their average values being 0.943 and 0.04, respectively. The VM method performed badly on the five joints due to the generation

Table 5. Average values of g-means, TPR and FPR over the five bridge datasets.

	ADES	MD	VM
TPR	0.961	0.921	0.001
FPR	0.022	0.038	0.00
g-means	**0.971**	0.943	0.04
p-value	—	0.01	2e−13

of a small value of σ which yields to over-fit the OCSVM model. It can be clearly noticed in the results presented in Table 4 where the FPR of the VM method was equal to zero which means that the model was able to fully predict the negative samples but completely failed in predicting the positive ones. With respect to the MD method, it is known from our discussion and experiments in Sect. 4.1 that MD often generates a large value of σ which yields to produce a loose decision boundary. As we expected, MD behaved similarly as it can be seen from Table 4 but with better results, since the number of samples was very large in these experiments which results in producing a small value of σ for the MD method. However, the values of FPR for the MD method are still consistently higher than ADES.

We further investigated the classification performance among the three methods by conducting a paired t-test (ADES vs MD and ADES vs VM) to determine whether the differences in the g-means between ADES and the two other methods are significant or not. The p-values were used in this case to judge the degree of the performance improvement. The paired t-test of ADES vs MD resulted in a p-value of 0.01 which indicates that the two methods do not have the same g-means values and they were significantly different. As shown in Table 5, the average g-means value of ADES is 0.971 compared to MD which has a mean value equal to 0.943. This indicates a statistical classification improvement of ADES over MD. The same t-test procedure was used to compare the classification performance of ADES and VM. The t-test generated a very small p-value of 2e−13 which indicates a very large difference between the two approaches. The average g-means value indicates that ADES significantly outperformed VM method and suggests not to consider VM method in the next experiments.

Case Study II: A Reinforced Concrete Jack Arch.

Experiments Setup and Data Collection. The second case study is a lab specimen which was replicated a jack arch from the Sydney Harbour Bridge. A reinforced concrete cantilever beam with an arch section was manufactured and tested as shown in Fig. 6 [12]. Ten accelerometers were mounted on the specimen to measure the vibration response resulting from impact hammer excitation. A data acquisition system was used to capture the impact force and the resultant acceleration time histories. An impact was applied on the top surface of the specimen just above the location of sensor A9. A total of 190 impact test responses were collected from the healthy condition. A crack was later introduced into the specimen in the location marked in Fig. 6 using a cutting saw. The crack is located between sensor locations A2 and A3 and it is progressively increasing towards sensor location A9. The length of the cut was increased gradually from 75 mm to 270 mm, and the depth of the cut was fixed to 50 mm. After introducing each damage case (four cases), a total of 190 impact tests were performed on the structure in the location prescribed earlier.

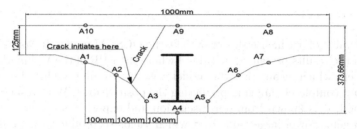

Fig. 6. Test specimen: intact structure with arrow indicating the cut location

Classification performance evaluation was carried out in the same way that was performed for the previous case study. The resultant 5 datasets has 950 samples separated into two main groups, Healthy (190 samples) and Damaged (760 samples). Each sample was measured for vibration responses resulted in a feature vectors with 8000 attributes representing the frequencies of each sample. The same scenario was applied here where the damaged cases were sub-grouped into 4 different damaged cases with 190 samples each.

Results and Discussions. In this section the classification results obtained using the OCSVM algorithm are presented for each of the parameter selection methods. Two sensors were used in this study, A1 and A4. As we mentioned in the above section, this dataset has four different levels of damage. The first level of damage, that is Damage Case 1, is very close to the healthy samples. This will allow us to thoroughly investigate the performance of the parameter selection methods considering the issues of under fitting and over-fitting. Table 6 shows the obtained results of each of the damage cases using sensors A1 and A4, respectively. Considering Damage Case 1 dataset, the results obtained by ADES are promising in comparison to MD. Although the samples in these dataset have a minor damage, ADES generated an optimal OCSVM hyperplane that was able to detect 80% of the damaged samples using sensors A1 and A4. 95% and 100% of the healthy samples were successfully classified using sensors A1 and A4, respectively. These results reveal the appropriateness of the generated enclosing surface of OCSVM using ADES. MD, on the other hand, detected only 43% of the damaged samples using A1 sensor, and 63% using A4 sensor. These results again reflect the general behavior of the MD method which often generates a loose OCSVM model. The results were improved with Damage Case 2 dataset where the severity of damage is not as close to the healthy samples. ADES also performs better than MD where the FPRs are 0.12 and 0.03 using A1 and A4, respectively. Both methods have similar performance on the two datasets, Damage Case 3 and Damage Case 4. MD performed well in these cases because the data points in these datasets were very far from the healthy samples.

Table 6. Optimal values of σ selected by different methods using specimen datasets (A1) along with g-means, TPR and FPR.

Dataset	Method	Sensor A1				Sensor A4			
		σ	g-means	TPR	FPR	σ	g-means	TPR	FPR
Damage case 1	ADES	42.2	**0.87**	0.95	0.20	41.8	**0.89**	1	0.20
	MD	57.8	0.64	0.95	0.57	56.1	0.79	1	0.37
Damage case 2	ADES	38.2	**0.93**	0.98	0.12	41.5	**0.98**	1	0.03
	MD	55.5	0.88	0.97	0.24	54.2	0.95	1	0.08
Damage case 3	ADES	42.7	**0.94**	0.89	0.00	37.3	**1**	1	0.00
	MD	53.3	0.94	0.89	0.00	51.8	1	1	0.00
Damage case 4	ADES	42.7	**0.94**	0.89	0.00	41.7	**1**	1	0.00
	MD	57.7	0.94	0.89	0.00	56.6	1	1	0.00

5   Conclusions

The capability of OCSVM as a warning system for damage detection in SHM highly depends upon the optimal value of σ. This paper has proposed a new algorithm called ADES to estimate the optimal value of σ from a geometric point of view. It follows the objective function that aims to select the optimal value of σ so that a generated hyperplane is maximally distant from the interior samples but at the same time close to the edge samples. In order to formulate this objective function, we developed a method to select the edge and interior samples which are crucial to the success of the objective function. The experimental results on the three 2-D toy datasets showed that the ADES algorithm generated optimal values of σ which resulted in an appropriate enclosing surface for OCSVM that precisely fitted the form of the three different shape datasets. Furthermore, the experiments on the five benchmark datasets demonstrated that ADES has the capability to work on high dimensional space datasets and capable of selecting the optimal values of σ for a trustworthy OCSVM model. We have also conducted our experiments on the bridge datasets to evaluate the performance of the OCSVM model for damage detection. Our ADES method performed well on these datasets and promising results were achieved. We obtained a better classification result on the five joint datasets with a low number of false alarms.

Overall, ADES algorithm for OCSVM classifier was superior in accuracy to VM, MD and MIES on the toy datasets, five publicly available learning datasets and Sydney Harbour Bridge datasets.

References

1. Bánhalmi, A., Kocsor, A., Busa-Fekete, R.: Counter-example generation-based one-class classification. In: Kok, J.N., Koronacki, J., Mantaras, R.L., Matwin, S., Mladenič, D., Skowron, A. (eds.) ECML 2007. LNCS (LNAI), vol. 4701, pp. 543–550. Springer, Heidelberg (2007). doi:10.1007/978-3-540-74958-5_51
2. Cortes, C., Vapnik, V.: Support vector machine. Mach. Learn. **20**(3), 273–297 (1995)
3. Das, S., Srivastava, A.N., Chattopadhyay, A.: Classification of damage signatures in composite plates using one-class SVMs. In: Aerospace Conference, pp. 1–19. IEEE (2007)
4. Evangelista, P.F., Embrechts, M.J., Szymanski, B.K.: Some properties of the gaussian kernel for one class learning. In: Sá, J.M., Alexandre, L.A., Duch, W., Mandic, D. (eds.) ICANN 2007. LNCS, vol. 4668, pp. 269–278. Springer, Heidelberg (2007). doi:10.1007/978-3-540-74690-4_28
5. Farrar, C.R., Worden, K.: An introduction to structural health monitoring. Philos. Trans. R. Soc. Lond. A: Math. Phys. Eng. Sci. **365**(1851), 303–315 (2007)
6. Farrar, C.R., Worden, K.: Structural Health Monitoring: A Machine Learning Perspective. Wiley, Hoboken (2012)
7. Khazai, S., Homayouni, S., Safari, A., Mojaradi, B.: Anomaly detection in hyperspectral images based on an adaptive support vector method. IEEE Geosci. Remote Sens. Lett. **8**(4), 646–650 (2011)
8. Khoa, N.L.D., Zhang, B., Wang, Y., Chen, F., Mustapha, S.: Robust dimensionality reduction and damage detection approaches in structural health monitoring. Struct. Health Monit. **13**, 406–417 (2014)
9. Khoa, N.L.D., Zhang, B., Wang, Y., Liu, W., Chen, F., Mustapha, S., Runcie, P.: On damage identification in civil structures using tensor analysis. In: Cao, T., Lim, E.-P., Zhou, Z.-H., Ho, T.-B., Cheung, D., Motoda, H. (eds.) PAKDD 2015. LNCS (LNAI), vol. 9077, pp. 459–471. Springer, Cham (2015). doi:10.1007/978-3-319-18038-0_36
10. Li, Y., Maguire, L.: Selecting critical patterns based on local geometrical and statistical information. IEEE Trans. Pattern Anal. Mach. Intell. **33**(6), 1189–1201 (2011)
11. Long, J., Buyukozturk, O.: Automated structural damage detection using one-class machine learning. In: Catbas, F.N. (ed.) Dynamics of Civil Structures, Volume 4. CPSEMS, pp. 117–128. Springer, Cham (2014). doi:10.1007/978-3-319-04546-7_14
12. Alamdari, M.M., Samali, B., Li, J., Kalhori, H., Mustapha, S.: Spectral-based damage identification in structures under ambient vibration. J. Comput. Civ. Eng. **30**, 04015062 (2015)
13. Mustapha, S., Hu, Y., Nguyen, K., Alamdari, M.M., Runcie, P., Dackermann, U., Nguyen, V.V., Li, J., Ye, L.: Pattern recognition based on time series analysis using vibration data for structural health monitoring in civil structures (2015)
14. Runcie, P., Mustapha, S., Rakotoarivelo, T.: Advances in structural health monitoring system architecture. In: Proceedings of the the fourth International Symposium on Life-Cycle Civil Engineering, IALCCE, vol. 14 (2014)
15. Schölkopf, B., Platt, J.C., Shawe-Taylor, J., Smola, A.J., Williamson, R.C.: Estimating the support of a high-dimensional distribution. Neural Comput. **13**(7), 1443–1471 (2001)
16. Tax, D.M.J., Duin, R.P.W.: Support vector data description. Mach. Learn. **54**(1), 45–66 (2004)

17. Worden, K., Manson, G.: The application of machine learning to structural health monitoring. Philos. Trans. R. Soc. Lond. A: Math. Phys. Eng. Sci. **365**(1851), 515–537 (2007)
18. Xiao, Y., Wang, H., Wenli, X.: Parameter selection of gaussian kernel for one-class SVM. IEEE Trans. Cybern. **45**(5), 941–953 (2015)
19. Zeng, M., Yang, Y., Cheng, J.: A generalized Gilbert algorithm and an improved MIES for one-class support vector machine. Knowl.-Based Syst. **90**, 211–223 (2015)

A Classification Model for Diverse and Noisy Labelers

Hao-En Sung[1], Cheng-Kuan Chen[1(✉)], Han Xiao[2], and Shou-De Lin[1]

[1] National Taiwan University, Taipei 10617, Taiwan
{b00902064,b98901048}@ntu.edu.tw, sdlin@csie.ntu.edu.tw
[2] Zalando, 10178 Berlin, Germany
han.xiao@zalando.de

Abstract. With the popularity of the Internet and crowdsourcing, it becomes easier to obtain labeled data for specific problems. Therefore, learning from data labeled by multiple annotators has become a common scenario these days. Since annotators have different expertise, labels acquired from them might not be perfectly accurate. This paper derives an optimization framework to solve this task through estimating the expertise of each annotator and the labeling difficulty for each instance. In addition, we introduce similarity metric to enable the propagation of annotations between instances.

Keywords: Noisy labeler · Crowdsourcing

1 Introduction

With the emerging of social networks and web services, it becomes popular to exploit crowdsourcing to obtain annotations of instances through online services such as *Amazon Mechanical Turk (AMT)*[1] for training a classification model. Although it is easy to obtain labels this way, those labels often come from imperfect labelers whose expertise toward the assigned task may vary significantly. Such noisy annotations can affect the performance of a traditional supervised machine learning model, which assumes all the training labels are reliable.

To address this issue, previous works [1–3] proposed a probabilistic framework to estimate the annotation quality from each annotator. One main disadvantage of such framework is that it fails to consider the feature-based similarities between instances. Moreover, they rely on certain predefined distribution to model annotator's expertise, which is often challenged in real scenario.

Instead of probabilistic framework, this paper proposes a novel optimization framework, which relaxes the predefined assumption of annotator's expertise toward instances. Our method further captures the similarity information shared among instances in feature space to yield a more effective solution. Our task can

---

H.-E. Sung and C.-K. Chen—denotes equal contribution.

[1] https://www.mturk.com/mturk/welcome.

© Springer International Publishing AG 2017
J. Kim et al. (Eds.): PAKDD 2017, Part I, LNAI 10234, pp. 58–69, 2017.
DOI: 10.1007/978-3-319-57454-7_5

be represented using Fig. 1. On the left side of this figure, we are obtaining a set of training instances i_1 to i_3, and each instances is labelled by every annotator from u_1 to u_3. The spirit of our proposed optimization framework is shown as the right figure of Fig. 1. It not only learns the annotator-instance confidence (dashed line) but also acquires the instance-instance relationship in feature space (solid line) to improve the performance.

The main contribution of this paper can be summarized as follows.

1. We introduce a novel framework that enables the propagation of annotations between instances. It relaxes the probabilistic distribution presumption on annotators' expertise as well as the independence assumption between annotations, which are usually required by probabilistic models.
2. Our model learns the latent variables to capture the expertise of each annotator and the labeling difficulty of each instance, which are essential information for most active learning frameworks.
3. We have conducted experiments on several datasets to verify our model.

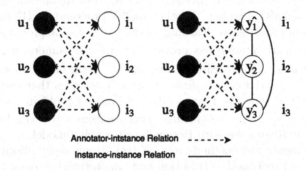

Fig. 1. Black nodes and white nodes represent annotators and instances, respectively. Dashed arrow that links one black node and one white node indicates the annotator-instance relationship; whereas, solid line that links two white nodes represents the instance-instance relationship.

The organization of this work is listed as follows. We first introduce the related works in Sect. 2. The formal problem definition and the derivation of our learning model are introduced in Sect. 3. In Sect. 4, we show the performance of our model on both simulated and real annotation datasets. We finally summarize this paper and propose future works in Sect. 5.

2 Related Work

There are mainly two kinds of scenarios for modeling multiple-annotator problems, and various algorithms solve either of them with different motivations. One of the prevalent frameworks tries to detect malicious annotators in order to remove or flip their responses; while the other models rank the expertise of each annotator and re-weight the annotation results based on the ranking.

For the first scenario, two-coin model for annotators was proposed in [2,4] to detect potential malicious annotators, which is also known as MAP-ML algorithm. A classification model with weighting matrix w is obtained during model learning and the label of each instance with feature $\boldsymbol{x}_i$ equips the probability $\sigma(w^T \boldsymbol{x}_i)$ of being true, where $\sigma(\cdot)$ is the sigmoid function. Each annotator flips a coin with bias α^i as sensitivity if the label is predicted true; whereas a coin with bias β^i as specificity is flipped if the label is predicted false. Under this framework, nasty annotations will be flipped automatically during model learning. In his later work [5], it further defines a criterion to evaluate spammer during learning process. These two works implicitly assume that the sensitivity and specificity of each annotator are independent from instances and they neglect the possibility that one annotator might equip varied levels of expertise toward different instances, which is often challenged in real world applications.

For the second scenario, [3] uses Gaussian Mixture Model combined with MAP-ML (known as GMM-MAPML) to evaluate annotator performances. Later works in [6,7] further define a specific threshold to eliminate low-quality annotations during model learning. Another model proposed in [1] and his later extended learning and active learning works [8–12] use a probabilistic model $p(y_i^{(u)}|\boldsymbol{x}_i, z_i)$ to learn annotations provided by different annotators, where z_i is the ground truth, $\boldsymbol{x}_i$ is the feature vectors, and $y_i^{(u)}$ is the label of instance i given by annotator u. Apart from the first scenario, it assumes that each annotator has varied levels of expertise toward different kinds of problems, which implies $p(y_i^{(u)}|\boldsymbol{x}_i, z_i) \neq p(y_i^{(u)}|z_i)$. The labeling expertise from u to i can be calculated through Logistic Regression with Bernoulli or Gaussian model.

The above-mentioned methods rely on two strong assumptions: annotator expertise follows predefined distribution and annotation processes are independent with one another. In our work, we relax these two assumptions and further integrate the similarity relationship between instances into our model. We also notice that some recent works [13,14] address the similar problem as ours. However, one main difference is that their work focus on active learning while ours is to design a new learning framework.

3 Methodology

To convey our idea, we formally define the problem in Sect. 3.1. In Sect. 3.2, we propose our learning model with detailed derivations.

3.1 Problem Definition

This paper mainly focus on the binary classification task, and leave multi-class one as our future work. We consider a dataset with n instances $\boldsymbol{X} = \{\boldsymbol{x}_1, ..., \boldsymbol{x}_n\}$, where $\boldsymbol{x}_i$ is a d dimensional feature vector for instance i, i.e. $\boldsymbol{x}_i \in \mathbb{R}^d$. Each instance i is assumed to be annotated by arbitrary number of annotators u with label $y_i^{(u)} \in \{0,1\}$, while we mainly follow the settings in [1] to consider full annotations in our following experiments.

The goal is to learn a model to predict each instance label by aggregating labels provided by all annotators toward all instances. Since the annotations are noisy, here we want to exploit the item similarity for a more robust model. For instance, if we want to predict the label $\hat{y}_1$ in Fig. 1, not only annotations toward instance 1 but also annotations toward instance 2 and 3 are taken into account, weighted by corresponding similarity between those items. The propagation of annotations weighted by similarity relationships is motivated by neighbor-based algorithm that similar instances are more likely to share similar annotations.

3.2   Learning Model

To model the annotator-instance relationship and the similarity relationship, we introduce two latent variables that will be jointly updated during optimization. One is the difficulty denoted as $dft \in \mathbb{R}$, which is used to model the labeling difficulty: the more difficult in labeling an instance, the higher it is. The other one is the expertise vector denoted as $\boldsymbol{expt} \in \mathbb{R}^d$, which is designed to convey the annotator expertise toward one instance. We model the annotator's expertise as a vector instead of a scalar because the annotator might have varied level of expertise toward different instances. We then define the task as minimizing an objective function f (Eq. 1), which consists of 4 components corresponding to 4 hypotheses as will be described later.

$$f = \alpha \cdot h_1 + \beta \cdot h_2 + \gamma \cdot h_3 + \delta \cdot h_4, \tag{1}$$

where α, β, γ, δ are hyperparameters chosen based on cross validation on each dataset. To simplify the notation, we denote sigmoid function as $S(\cdot)$. Intuitively, we also use $1 - S(dft)$ instead of $S(dft)$ to represent how easy it is to label such instance in following hypotheses.

Hypothesis 1 (h_1): similar instances should share the same annotation, unless they are difficult to be classified

To model the similarity relationship between instances, we compute the similarity score $R_{i,j}$ by euclidean distance in feature space and map them into $[0,1]$ by $e^{-|\mathbf{x}_i - \mathbf{x}_j|_2}$. The larger $R_{i,j}$ indicates i and j are more similar to each other. The prediction $\hat{y}$ is a real value, which is mapped into $[0,1]$ through $S(\cdot)$. Naively, we set 0.5 as the threshold for label 0 and 1. With the introduction of $R_{i,j}$, $\hat{y}$ and dft, we can write down our first hypothesis as follows.

$$h_1(dft_i, \hat{y}_i) := \sum_{i,j} R_{i,j} \cdot (S(\hat{y}_i) - S(\hat{y}_j))^2 \cdot ((1 - S(dft_i)) + (1 - S(dft_j))) \tag{2}$$

The equation shows that for any given pairs of prediction outcomes, if they are similar (i.e. large $R_{i,j}$), then their prediction shall less likely to be different, unless they are considered as instances that are not easy to be classified (representing by small latent variables $1 - S(dft_i)$). In other words, the predicted label of an instance j can be more easily propagated to another instance i if they are similar and assumed to be classified easier. There are actually multiple ways to represent the joint easiness measurement $1 - S(dft)$, while we find that simple summation is effective through the experiments.

Hypothesis 2 (h_2): the model shall trust labelers whose expertise matches the instance better

This hypothesis assumes the quality of annotation depends on how the expertise of annotators matches the instances to be announced. We assume a latent vector $\boldsymbol{expt}_u$ is used to represent the annotator's expertise, and its inner product with an instance shows the confidence of this annotator toward this specific instance. With these factors, we model the annotations as Eq. (3):

$$h_2(\boldsymbol{expt}, dft_i, \hat{y}_i) := \sum_{u,i} \left(S(\hat{y}_i) - y_i^{(u)} \right)^2 \cdot (S(\boldsymbol{expt}_u^\intercal \cdot \boldsymbol{x}_i) + (1 - S(dft_i))). \quad (3)$$

For any given pair of annotated label $y_i^{(u)}$ and predicted label $\hat{y}_i$, the model will favor one with higher annotator's confidence and lower instance difficulty by minimizing Eq. (3). Our model leverages the annotator's confidence toward each instance, and downplays the ones without sufficient confidence during learning.

Hypothesis 3 (h_3): instances are generally not difficult to be classified

Model with only terms h_1 and h_2 have the tendency to maximizes the instance difficulty, which will inevitably reduce the amount of information that can be used to make the final prediction $\hat{y}_i$. Thus, we add summation of reciprocal of $1 - S(dft)$ as regularization term to our model that encourages our model to reduce its belief to the difficulty of each instance.

$$h_3(dft_i) := \sum_i (1 - S(dft_i))^{-1} \quad (4)$$

Hypothesis 4 (h_4): each annotator's expertise vector should be smooth

To avoid overfitting, we need to constraint the annotator's expertise vector $\boldsymbol{expt}_u$ as a regularization term.

$$h_4(\boldsymbol{expt}) := \sum_u \|\boldsymbol{expt}_u\|_2^2 \quad (5)$$

Put everything together. The objective function to be minimized looks like:

$$f(\boldsymbol{expt}_u, dft_i, \hat{y}_i)$$

$$= \alpha \cdot \left(\sum_{i,j} R_{i,j} \cdot (S(\hat{y}_i) - S(\hat{y}_j))^2 \cdot ((1 - S(dft_i)) + (1 - S(dft_j))) \right.$$

$$+ \beta \cdot \left(\sum_{u,i} \left(S(\hat{y}_i) - y_i^{(u)} \right)^2 \cdot \left(S(\boldsymbol{expt}_u^{\mathsf{T}} \cdot \boldsymbol{x}_i) + (1 - S(dft_i)) \right) \right)$$

$$+ \gamma \cdot \left(\sum_i (1 - S(dft_i))^{-1} \right) + \delta \cdot \left(\sum_u \|\boldsymbol{expt}_u\|_2^2 \right) \tag{6}$$

We can infer annotations for each instance by jointly update the latent parameters to minimize the object. We apply gradient descent to get the local optima of $\boldsymbol{expt}_u$, dft_i, and $\hat{y}_i$. The update formulas are listed as follow:

– Update formula for annotator expertise

$$\frac{\partial f}{\partial \boldsymbol{expt}_u} = S(\boldsymbol{expt}_u^{\mathsf{T}} \cdot \boldsymbol{x}_i) \cdot (1 - S(\boldsymbol{expt}_u^{\mathsf{T}} \cdot \boldsymbol{x}_i))$$

$$\cdot \beta \cdot \left(\sum_i \left(S(\hat{y}_i) - y_i^{(u)} \right)^2 \cdot \boldsymbol{x}_i \right) + 2 \cdot \delta \cdot \boldsymbol{expt}_u. \tag{7}$$

– Update formula for instance difficulty

$$\frac{\partial f}{\partial dft_i} = -S(dft_i) \cdot (1 - S(dft_i)) \cdot \left[\alpha \cdot \left(\sum_{i,j} R_{i,j} \cdot (S(\hat{y}_i) - S(\hat{y}_j))^2 \right) \right.$$

$$\left. + \beta \cdot \left(\sum_{u,i} \left(S(\hat{y}_i) - y_i^{(u)} \right)^2 \right) - \gamma \cdot (1 - S(dft_i))^{-2} \right]. \tag{8}$$

– Update formula for predicted label

$$\frac{\partial f}{\partial \hat{y}_i} = S(\hat{y}_i) \cdot (1 - S(\hat{y}_i))$$

$$\cdot \left[2 \cdot \alpha \cdot \left(\sum_{i,j} R_{i,j} \cdot (S(\hat{y}_i) - S(\hat{y}_j)) \cdot ((1 - S(dft_i)) + (1 - S(dft_j))) \right) \right.$$

$$\left. + 2 \cdot \beta \cdot \left(\sum_{u,i} \left(S(\hat{y}_i) - y_i^{(u)} \right) \cdot (S(\boldsymbol{expt}_u^{\mathsf{T}} \cdot \boldsymbol{x}_i) + (1 - S(dft_i))) \right) \right] \tag{9}$$

4 Experiment

We compare the derived algorithm to the state-of-the-art learning model proposed in [1]. There are two learning models in Yan's work: **M.L-Bernoulli** and **M.L-Gaussian**. Since the former has better performance than the latter according to the original paper, we compare our results to **M.L-Bernoulli** only.

Similar to [1], we also compare our model with multiple baseline algorithms. Beside individual annotator models, where each annotator learns a logistic regression model disjointly from others, we also consider two majority-voting models learned with logistic regression — **L.R.-Majority** and **L.R.-Ensemble**. The former baseline, as in [1], takes the majority vote from all annotators as target labels while the later learns each annotator model separately in the first stage, and then combine learned models with a weighting matrix in the second stage.

We mainly perform our experiments on two kinds of dataset. One is simulated dataset that uses UCI datasets provided by [15], including UCI::Ionosphere, UCI::Cleveland, and UCI::Statlog. The experiment results are recorded in Sect. 4.1. The other one is a real dataset that uses Medical Text dataset cited in [16] with three different targets: Medical::Evidence, Medical::Focus and Medical::Polarity. The model performance is shown in Sect. 4.2. Finally, we demonstrate the contribution of each component of our model using UCI::Ionosphere in Sect. 4.3 and a summary of our experiment results in Sect. 4.4.

4.1 Simulated Datasets

For all UCI datsets, we follow similar setup in [1] with minor modifications to fit the scenario in real world better. The main procedure is summarized below.

1. Data preprocessing: including filling missing values, feature normalization and one-hot encoding.
2. Distribute instances into $K = 5$ clusters using K-Means algorithm. It tries to simmulate the instances into 5 different categories.
3. Assign $|U| = 5$ annotators to $K = 5$ clusters correspondingly. Each annotator is considered as an expert in its own cluster with higher labelling accuracy. Our simulation assumes the labelling accuracy of an expert to an instance is guilded by $0.6 + 0.4 \times e^{-\|x_i - c_k\|_2}$, where c_k is the center of cluster k. In other words, for one annotator, the labeling accuracy is closed to one in its own cluster, and can go down to 0.6 in the other clusters.

We consider fully-assigned annotations from each annotator to each instance for model learning and conduct experiments under 5-fold cross-validation. For evaluation, we use area-under-ROC curve (AUC) as the evaluation metric and report the average performance. We then repeat 5-fold experiments for $T = 30$ times to conduct *Wilcoxon Signed Rank Test* to examine whether the comparison is statistically significant under.

For all simulated experiments, we provide the ROC curve and calculate AUC as the performance metric, as shown in Figs. 2(b), 3(b) and 4(b). In addition, we provide an auxiliary cluster graph to show that our model can effectively locate difficult instances. To plot the cluster graph, we apply PCA to reduce high-dimension feature space to two-dimensional space then color each instance with its *dft* value: the lighter the color, the more difficult it is. Since each instance is annotated by five annotators during the experiment, they are represented as five centroids in K-Means. We would expect our model give higher difficulty value (i.e. lighter in color) to instances which are closer to boundary. The results are presented in Figs. 2(a), 3(a) and 4(a).

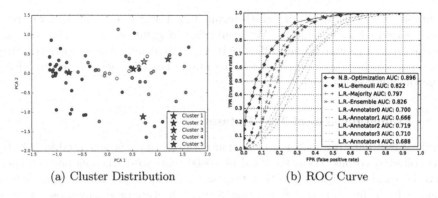

(a) Cluster Distribution

(b) ROC Curve

Fig. 2. UCI:: Ionosphere dataset

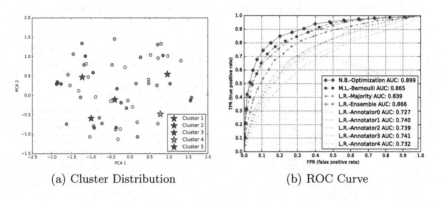

(a) Cluster Distribution

(b) ROC Curve

Fig. 3. UCI:: Cleveland dataset

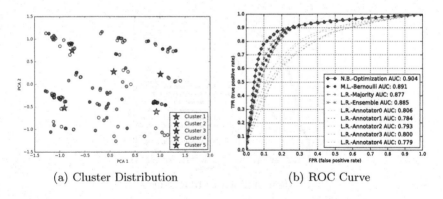

(a) Cluster Distribution

(b) ROC Curve

Fig. 4. UCI:: Statlog dataset

The results in Figs. 2(b), 3(b) and 4(b) show that our model (denoted as **N.B. optimization**) outperforms the state-of-the-art and other baselines. For the cluster distribution figures, we do find that the points near the cluster boundary are of lighter color, which means that the instance difficulty are correctly captured by our model. We would like to point out that being able to identify instances with higher difficulty is very important for tasks such as active learning, which implies our model as a suitable basis for active learning given multiple noisy annotators.

4.2   EvaluationMedical Text Dataset

Medical text data is annotated by real annotators and was first used in [16]. In the collected corpus, there are total of 10000 sentences; whereas one sentence may be consisted of multiple text fragments. Annotation process runs in two rounds. In the first round of annotation, 3 annotators are randomly chosen from 8 annotator-pool to label 10000 sentences. Later in the second round of annotation, randomly selected 1000 sentences are labeled by other 5 annotators.

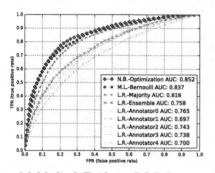

(a) Medical::Evidence ROC Curve

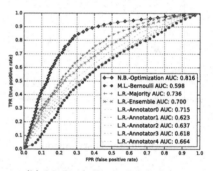

(b) Medical::Focus ROC Curve

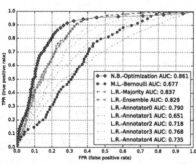

(c) Medical::Polarity ROC Curve

Fig. 5. Medical text dataset

Since there may be multiple labels for one text segment, Medical Text labeling is actually a multilabel-multiclass problem. Based on [16], available labels for each text fragment include focus (G for generic, M for methodology, S for

Table 1. h_1, h_2, are two hypotheses on annotators and instances we mentioned in Sect. 3.2. h_3, h_4 are two regularization terms we used to prevent model from overfitting.

Hypothesis combination	Area Under Curve (AUC)
$h_1 + h_2 + h_3 + h_4$	0.896
$h_1 + h_2 + h_3$	0.895
$h_1 + h_2 + h_4$	0.824
$h_1 + h_2$	0.847

Table 2. Hypothesis tests between Yan's and our algorithm are examined on 3 simulated datasets and a real dataset with three targets through AUC evaluation metric. Experiments on simulated datasets are repeated 30 times with 5-fold cross-validation; while experiments on real dataset are repeated 5 times with 5-fold cross-validation. P-value with * indicates siginificance.

	M.L.-Bernoulli	N.B.-Optimization	P-value
UCI:: Ionosphere	0.822 ± 0.033	$\mathbf{0.896 \pm 0.026}$	$< 0.00001^*$
UCI:: Cleveland	0.865 ± 0.027	$\mathbf{0.899 \pm 0.010}$	$< 0.00001^*$
UCI:: Statlog	0.891 ± 0.014	$\mathbf{0.901 \pm 0.009}$	0.000034^*
Medical:: Evidence	0.837 ± 0.005	$\mathbf{0.852 \pm 0.004}$	0.004065^*
Medical:: Focus	0.598 ± 0.029	$\mathbf{0.816 \pm 0.009}$	0.000034^*
Medical:: Polarity	0.677 ± 0.006	$\mathbf{0.861 \pm 0.008}$	$< 0.00001^*$

science), evidence (E0 means no evidence, E3 means direct evidence, while E1, E2 are in between), and polarity (P for positive, N for negative, ranging from N3 to N0, then from P0 to P3). In our experiment, we regard Medical::Evidence, Medical::Focus, and Medical::Polarity as different tasks, and transform each of them into a binary classification task. For Medical::Focus and Medical:Evidence, we follow the binarization process in [7]. For Medical:Polarity, the annotation contains N3, N2, N1, N0, P0, P1, P2, P3. We treat P0, P1, P2, P3 as 1 while others are 0. The results are presented in Fig. 5(a) through Fig. 5(c).

1. Select 1000 sentences that have been labeled from all 8 annotators.
2. Remove 309 sentences that are segmented differently by various annotators. 691 sentences remain.
3. Partition 691 sentences into 874 text fragments.
4. Apply stopword removal and rare term removal to get 848 text segments and 279 words as column features.
5. Calculate TF-IDF scores for each word in 279 column features as instance features and transform origin multi-class ground truths into binary ones according to different task targets.
6. Repeat 5-fold experiments for $T = 5$ times, and then put *Paired T-test* on averaged 5-fold results to judge the significance of performance improvement.

4.3   Component Importance of N.B.-Optimization

To examine which component has the most influence on the prediction quality, we use UCI::Ionosphere as the experimental dataset to evaluate some combinations of parameters, i.e. α, β, γ, and δ, and set some of them to 0. From Table 1, it is clear that h_1 and h_2 are both the crucial components to the model and the model become more robust with h_3. h_4 provides only marginal boost on the performance.

4.4   Experiment Summary

From the above experiment results, we can tell that our model is good at identifying instance difficulty and has great performance in both simulated and real dataset with four robust hypotheses.

5   Conclusion

Unlike the existence of ground truths in traditional supervised learning problems, perfect labels in crowdsourcing scenario are not guaranteed, as the labelers may equip varied levels of expertise toward different scope of knowledge. Thus, a model such as ours that can utilize information from highly-diversified and noisy data sources is highly demanded.

Acknowledgement. This material is based upon work supported by the Air Force Office of Scientific Research, Asian Office of Aerospace Research and Development (AOARD) under award number FA2386-15-1-4013, and Taiwan Ministry of Science and Technology (MOST) under grant number 105-2221-E-002-064-MY3.

References

1. Yan, Y., Rosales, R., Fung, G., Schmidt, M.W., Valadez, G.H., Bogoni, L., Moy, L., Dy, J.G.: Modeling annotator expertise: learning when everybody knows a bit of something. In: AISTATS, pp. 932–939 (2010)
2. Raykar, V.C., Yu, S., Zhao, L.H., Jerebko, A., Florin, C., Valadez, G.H., Bogoni, L., Moy, L.: Supervised learning from multiple experts: whom to trust when everyone lies a bit. In: Proceedings of the 26th Annual International Conference on Machine Learning, pp. 889–896. ACM (2009)
3. Zhang, P., Obradovic, Z.: Learning from inconsistent and unreliable annotators by a Gaussian mixture model and Bayesian information criterion. In: Gunopulos, D., Hofmann, T., Malerba, D., Vazirgiannis, M. (eds.) ECML PKDD 2011. LNCS (LNAI), vol. 6913, pp. 553–568. Springer, Heidelberg (2011). doi:10.1007/978-3-642-23808-6_36
4. Raykar, V.C., Yu, S., Zhao, L.H., Valadez, G.H., Florin, C., Bogoni, L., Moy, L.: Learning from crowds. J. Mach. Learn. Res. **11**(Apr), 1297–1322 (2010)
5. Raykar, V.C., Yu, S.: Eliminating spammers and ranking annotators for crowd-sourced labeling tasks. J. Mach. Learn. Res. **13**(Feb), 491–518 (2012)

6. Zhang, P., Obradovic, Z.: Integration of multiple annotators by aggregating experts and filtering novices. In: 2012 IEEE International Conference on Bioinformatics and Biomedicine (BIBM), pp. 1–6. IEEE (2012)
7. Zhang, P., Cao, W., Obradovic, Z.: Learning by aggregating experts and filtering novices: a solution to crowdsourcing problems in bioinformatics. BMC Bioinform. **14**(Suppl 12), S5 (2013)
8. Yan, Y., Fung, G.M., Rosales, R., Dy, J.G.: Active learning from crowds. In: Proceedings of the 28th International Conference on Machine Learning (ICML 2011), pp. 1161–1168 (2011)
9. Yan, Y., Rosales, R., Fung, G., Dy, J.: Modeling multiple annotator expertise in the semi-supervised learning scenario. arXiv preprint arXiv:1203.3529 (2012)
10. Yan, Y., Rosales, R., Fung, G., Farooq, F., Rao, B., Dy, J.G., Malvern, P.: Active learning from multiple knowledge sources. In: AISTATS, vol. 2, p. 6 (2012)
11. Yan, Y., Rosales, R., Fung, G., Dy, J.: Active learning from uncertain crowd annotations. In: 2014 52nd Annual Allerton Conference on Communication, Control, and Computing (Allerton), pp. 385–392. IEEE (2014)
12. Yan, Y., Rosales, R., Fung, G., Subramanian, R., Dy, J.: Learning from multiple annotators with varying expertise. Mach. Learn. **95**(3), 291–327 (2014)
13. Long, C., Hua, G.: Multi-class multi-annotator active learning with robust gaussian process for visual recognition. In: Proceedings of the IEEE International Conference on Computer Vision, pp. 2839–2847 (2015)
14. Rodrigues, F., Pereira, F., Ribeiro, B.: Learning from multiple annotators: distinguishing good from random labelers. Pattern Recogn. Lett. **34**(12), 1428–1436 (2013)
15. Lichman, M.: UCI machine learning repository (2013)
16. Rzhetsky, A., Shatkay, H., Wilbur, W.J.: How to get the most out of your curation effort. PLoS Comput. Biol. **5**(5), e1000391 (2009)

Automatic Discovery of Common and Idiosyncratic Latent Effects in Multilevel Regression

Sk Minhazul Islam[✉] and Arunvava Banerjee

Computer and Information Science and Engineering Department,
University of Florida, Gainesville, USA
smislam@cise.ufl.edu

Abstract. We present a flexible non-parametric generative model for multilevel regression that strikes an automatic balance between identifying common effects across groups while respecting their idiosyncrasies. The model is built using techniques that are now considered standard in the statistical parameter estimation literature, namely, Hierarchical Dirichlet processes (HDP) and Hierarchical Generalized Linear Models (HGLM), and therefore, we name it "Infinite Mixtures of Hierarchical Generalized Linear Models" (iHGLM). We demonstrate how the use of a HDP prior in local, groupwise GLM modeling of response-covariate densities allows iHGLM to capture latent similarities and differences within and across groups. We demonstrate iHGLM's superior accuracy in comparison to well known competing methods like Generalized Linear Mixed Model (GLMM), Regression Tree, Least Square Regression, Bayesian Linear Regression, Ordinary Dirichlet Process Regression, and several other regression models on several synthetic and real world datasets.

1 Introduction

Multilevel Regression is the method of choice for research design whenever response-covariate data is collected across multiple groups. When a common regressor is learned on the amalgamated data, the model fails to identify idiosyncratic effects for the responses across individual groups. Modeling separate groups via separate regressors results in a model that is devoid of common latent effects across the groups. Multilevel regression attempts to find a middle ground between these two extremes: a common regressor for the entire dataset versus separate regressors for the individual groups/levels. What this middle ground should be and how it may be inferred from the data is the subject matter of this paper.

The complexities that underlie the search for this middle ground are best motivated through examples. In Clinical Trials, for example, a group of people are prescribed either a new drug or a placebo to estimate the efficacy of the drug for the treatment of a certain disease. At a population level, this efficacy may be modeled using a single Normal or Poisson mixed model distribution with

© Springer International Publishing AG 2017
J. Kim et al. (Eds.): PAKDD 2017, Part I, LNAI 10234, pp. 70–82, 2017.
DOI: 10.1007/978-3-319-57454-7_6

mean set as a (linear or otherwise) function of the covariates of the individuals in the population. A closer inspection might however disclose potential factors that explain the efficacy results better. For example, there might be regularities at the subpopulation level—Caucasians as a whole might react differently to the drug than, say, Asians. Regularities might be found at a subpopulation level of a different kind—individuals with a particular genetic trait, whether they be Caucasians or Asians, might react similarly to the drug. Modeling the latent common and idiosyncratic effects of cross-cutting subpopulations is therefore an important problem to solve. Similar situations are found in the height imputation problem [15] for forest stands.

When the data is collected along a certain group structure, such as ethnic background in clinical trials, other group structures, such as genetic traits, naturally become subgroups within these groups. Identifying these latent subgroups through their similar responses, has, to our knowledge, not been satisfactorily accomplished. We present a framework here that fills this void.

We begin with a brief description of the weaknesses of Hierarchical Generalized Linear Models [8], the most popular multilevel regression technique. In regression theory, Generalized Linear Model (GLM), proposed in [13], brings erstwhile disparate techniques such as, Linear regression, Logistic regression, and Poisson regression, under a unified framework. Hierarchical Generalized Linear Model (HGLM), proposed in [8], extends GLM to grouped observations. HGLM is formally defined as:

$$f(y; \theta, \psi, v) = exp\left\{\frac{y\theta - b(\theta)}{a(\psi)} + c(y; \psi)\right\} \tag{1}$$

Here, ψ is a dispersion parameter and v is the random effect component. exp denotes the exponential family density. The mean response is $E[Y|\mathbf{X}] = b(\theta) = \mu = g^{-1}(X^T\beta + v)$, where g is the link function, $X^T\beta$ is the linear predictor and v is a strictly monotonic function of $u, \{v = v(u)\}$. Here, v signifies overdispersion. u has a prior distribution chosen appropriately.

It follows from above that in HGLM, the separate densities are characterized by two components. First, there is the fixed effect parameter, $(X^T\beta)$ of the density which includes the covariates X and its coefficients β. This remains the same across all the groups. Second, there is the random effect (v) which differs from group to group. Notwithstanding its effectiveness, the inherent assumptions in HGLM limit its applicability and need to be relaxed.

Firstly, the random effect (v) is not a function of the linear transformation of the covariates, $X^T\beta$. Therefore, this automatically assumes that the mean function and the variance of the outcomes in different groups depend neither on the covariate, X, nor on the coefficients. This makes the model suitable only for grouped data where properties of the outcomes in different groups vary independently of the covariates. Secondly, although the response-covariate pairs are grouped, two different pairs in the same group may come from different response-covariate densities (different latent effects within the same group). Alternatively, two pairs from two different groups may be generated from the same density (same latent effects in different groups). We need a robust model that can capture this hidden intra/inter clustering effect in grouped data. Thirdly, the covariate $(X^T \beta)$ is associated with the response-covariate density only through a linear function. Although we can introduce a non-linear function for the response at the output, it does not include the covariates. Finally, data may be heteroscedastic within individual groups, i.e., the variance of the response may be a function of the predictors within a group. The response variance however does not depend on the predictors in ordinary HGLM. Some later version [9] of HGLM pick heteroscedasticity between the groups (different variance for different groups), but not within a group.

In this article, we alleviate these shortcomings of HGLM by developing iHGLM, a Non-parametric Bayesian Mixture Model of the Hierarchical Generalized linear Model. The iHGLM framework is specified to all the models of HGLM, i.e. Normal, Poisson, Logistic, Inverse Gaussian, Probit, Exponential etc.

In iHGLM, we model outcomes in the same group via mixtures of local densities. This captures locally similar regression patterns, where each local regression is effectively a GLM. To force the density of the covariate, X, and its coefficients, β, to be shared among groups, we make the coefficients, β, and the covariates, X, for different groups be generated from the same prior atomic distribution. An atomic distribution places finite probabilities on a few outcomes from the sample space [19]. When the coefficients, β, and the covariates, X, are drawn from this atomic density, it enables the X and β in different groups to share densities. In this manner, in the Bayesian setting, the density of the random effect (v), as well as the density of fixed effect $(X^T \beta)$ are shared among groups. We obtain this prior atomic density for the fixed and random effects, while ensuring a large support, through a Hierarchical Dirichlet Process (HDP) prior [17].

From the HDP prior, our primary goal is to generate prior densities for u and $X^T \beta$ for each group. We draw a density G_0 from a Dirichlet Process $(DP(\gamma, H))$ [4]. In this case, γ is a scale parameter and the H (the base distribution) is basically the set of densities in the parameter space of random (v) and fixed effect $(X^T \beta)$. According to [16], this ensures that G_0 is atomic, yet with a broad support. Therefore, G_0 is a atomic density in the parameter space of u and $X^T \beta$ which puts finite probabilities on some discrete points which acts as its support. Then, for each group, we draw group specific densities G_j from $DP(\alpha, G_0)$. Since G_0 is already atomic, and according to [16] G_j is also atomic, the support of group specific densities G_js must share common points in their respective parameter space of fixed $(X^T \beta)$ and random effects (v). Now, these G_j's act as prior densities for the u and $X^T \beta$ for each group. Subsequently, both u and $X^T \beta$ are modeled through mixture of local densities which are shared among groups.

Although the mean function for each component (clusters within a group) in the mixture of response-covariate densities in a single group is linear, marginalizing out the local distribution creates a non-linear mean function. In addition, the variance of the responses vary among mixture components (clusters), thereby varying among covariates. The non-parametric model ensures that the data determines the number of mixture components (clusters) in specific groups and the nature of the local GLMs.

2   An Illustrative Example

We show a simple posterior predictive trajectory of the iHGLM Normal Model in a four-group synthetic dataset with a 1-D Covariate in Fig. 1. The "yellow" trajectory is the smoothed response posterior learned by the model. All the groups were created with four mixture components equally weighted. For the first group, responses were generated through four response-covariate densities with mean and standard deviation set as, $(1 + x, .5)$, $(1.75 + .5x, .8)$, $(1.15 + .8x, .2)$, $(2.40 + .3x, .4)$. For the 2nd group they were $(8.5 - x, 1.2)$, $(1.75 + .5x, .8)$, $(-18.25 + 4.5x, .1)$, $(1 + x, .5)$. For the 3rd, $(10.90 - .5x, .9)$, $(1.15 + .8x, .2)$, $(49.15 - 5.2x, 1.1)$, $(2.4 + x, .3)$, and for the 4th, $(3.55 + .2x, 1)$, $(10.90 - .5x, .9)$, $(-40.80 + 4.2x, .3)$, $(1.75 + .5x, .8)$. Observe that any two groups have at least one density in common. To capture this kind of multi-level data, a regression model is needed which captures sharing of latent densities between the groups. Also, every group must be modeled by a mixture of densities. The model must capture heteroscedasticity within groups where the variance of the responses depend upon the covariates in each group. The iHGLM normal model captures all of these hidden intra/inter-clustering effects between the groups as well as heteroscedasticity within the groups, as shown in Fig. 1.

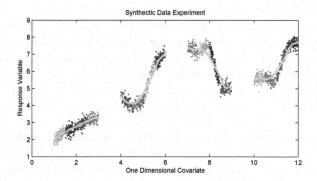

Fig. 1. The posterior trajectory of the synthetic dataset with 4 groups. Different colors represent different subgroups. (Color figure online)

3   Mathematical Background

3.1   Models Related to HGLM

After its introduction, Hierarchical Generalized Linear Model was extended to include structured dispersion [9] and models for spatio-temporal co-relation [10].

Generalized Linear Mixed Models (GLMMs) were proposed in [3]. The random effects in HGLM were specified by both mean and dispersion in [11]. Mixture of Linear Regression was proposed in [18]. Hierarchical Mixture of Regression was done in [7]. Varying co-efficient models were proposed in [5]. All of these models suffer the shortcomings of not picking up the latent inter/intra clustering effect as well as varying uncertainty with respect to covariates across groups, which the iHGLM inherently models.

3.2 Hierarchical Dirichlet Process and Chinese Restaurant Franchise

HDP defines a set of probability measures G_j, one for each group and a global random probability measure, G_0. G_0 is distributed as a Dirichlet Process with concentration parameter γ and base distribution H. A Dirichlet Process [4], DP (α_0, G_0) is defined as a probability measure over a sample space of probability measures, $G \sim DP(\alpha_0, G_0)$. According to the *Chinese Restaurant Process* or the *Polya urn scheme* [1,2], the density is given by, $(\theta_i|\theta_{1:(i-1)}) \sim \frac{1}{\alpha+i-1}\sum_{k=1}^{i-1}\delta_{\theta_k} + \frac{\alpha}{\alpha+i-1}G_0$. Here, θ_i is a draw from the *polya urn*.

$$G_0|\gamma, H \sim DP(\gamma, H), \qquad G_j|\alpha_0, G_0 \sim DP(\alpha, G_0),$$
$$\theta_{j,i}|G_j \sim G_j, \qquad x_{j,i}|\theta_{j,i} \sim F(\theta_{j,i}) \tag{2}$$

This proves the clustering/atomic property of DP. G_j, conditioned on G_0 follows a DP with parameters α_0 and G_0. HDP is used as a prior distribution for the grouped data. In HDP mixture, the latent variable $\theta_{j,i}$ is a draw from G_j and it parameterizes the density F of observed data $x_{j,i}$.

In the Chinese Restaurant Franchise (CRF), we have a finite number of restaurants (groups) with infinitely many tables (clusters) with shared dishes (parameter) among all restaurants. Let θ_{ji} be the customers, $\phi_{1:K}$ be the global dishes, Ψ_{jt} be the table-specific dishes and t_{ji} be the table index for the jth restaurant (Ψ_{jt}) and i^{th} customer (θ_{ji}). k_{jt} be the table menu index of the jth restaurant (Ψ_{jt}) and t^{th} table (ϕ_k). Furthermore, let $n_{jt.}$ and $n_{j.k}$ denote the number of customers in the tth table-j^{th} restaurant and j^{th} restaurant-k^{th} dish, respectively. Let m_{jk}, $m_{j.}$, $m_{.k}$ and $m_{..}$ denote the number of tables in the jth restaurant serving dish k, the number of tables in the jth restaurant serving any dishes, the number of tables serving dish k, and the total number of tables, respectively. Now, from Eq. (2), we have, $\theta_{ji}|\theta_{j1:j(i-1)}, \alpha_0, G_0 \sim \frac{\alpha_0}{\alpha_0+i-1}G_0 + \sum_{t=1}^{m_{j.}}\frac{n_{jt.}}{\alpha_0+i-1}\delta_{\Psi_{jt}}$. Integrating out G_0, we have, $\Psi_{jt}|\Psi_{11:j(t-1)}, \gamma, H \sim \frac{\gamma}{\gamma+m_{..}}H + \sum_{k=1}^{K}\frac{m_{.k}}{\gamma+m_{..}}\delta_{\phi_k}$ (Fig. 2).

4 iHGLM Model Formulation

4.1 Normal iHGLM Model

In Normal iHGLM, the generative model of the covariate-response pair is given by the following set of equations. Here, X_{ji} and Y_{ji} represent the ith continuous covariate-response pairs of the jth group. The distribution, $\{\mu_d, \lambda_{xd}\}$

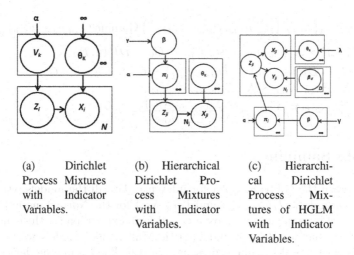

(a) Dirichlet Process Mixtures with Indicator Variables.

(b) Hierarchical Dirichlet Process Mixtures with Indicator Variables.

(c) Hierarchical Dirichlet Process Mixtures of HGLM with Indicator Variables.

Fig. 2. Plate notation of the iHGLM model.

(Normal-Gamma) is the prior distribution on covariates. The distribution, $\{\beta_d, \lambda_y\}$ (Normal-Gamma) is the prior distribution on the covariate coefficient β. Both the distributions are base distributions (H) of the first DP. The set $\{m_{xd0}, \beta_{xd0}, a_{xd0}, b_{xd0}\}$ and $\{m_{yd0}, \beta_{y0}, a_{y0}, b_{y0}\}$ constitute the hyperparameters for the covariates and covariate coefficients (β), respectively.

$$\{\mu_d, \lambda_{xd}\} \sim \mathcal{N}\left(\mu_d | m_{xd0}, (\beta_{xd0}\lambda_{xd})^{-1}\right) \text{Gamma}\left(\lambda_{xd} | a_{xd0}, b_{xd0}\right),$$
$$\{\beta_d, \lambda_y\} \sim \mathcal{N}\left(\beta_d | m_{yd0}, (\beta_{y0}\lambda_y)^{-1}\right) \text{Gamma}\left(\lambda_y | a_{y0}, b_{y0}\right),$$
$$G_0 \sim DP(\gamma, H), \quad G_j \sim DP(\alpha_0, G_0), \quad \{\mu_{kd}, \lambda_{xkd}^{-1}\} \sim G_j, \tag{3}$$
$$\{\beta_{kd}, \lambda_{yk}\} \sim G_j, \quad X_{jid} | \mu_{kd}, \lambda_{xkd} \sim \mathcal{N}\left(X_{jid} | \mu_{kd}, \lambda_{xkd}^{-1}\right),$$
$$Y_{ji} | X_{ji} \sim \mathcal{N}\left(Y_{ji} | \sum_{d=0}^{D} \beta_{kd} X_{jid}, \lambda_{yk}^{-1}\right)$$

4.2 Logistic Multinomial iHGLM Model

In the Logistic Multinomial iHGLM model, the continuous covariates are modeled by a Gaussian mixture (identically as the Normal model above) and a Multinomial Logistic framework is used for the categorical response (Number of Categories is P). Here, X_{ji} and Y_{ji} represent the ith continuous covariate and categorical response pair of the jth group. p is the index of the category. The distribution, $\{\mu_d, \lambda_{xd}\}$ (Normal-Gamma) is the prior distribution on the covariates. The distribution, $\{\beta_{pd}\}$ (Normal) is the prior distribution on the covariate coefficient β. Both the distributions are base distributions (H) of the first DP. The set $\{m_{xd0}, \beta_{xd0}, a_{xd0}, b_{xd0}\}$ and $\left\{m_{ypd0}, s_{ypd0}^2\right\}$ constitute the set of hyperparameters for the covariates and covariate coefficients (β), respectively. The complete model is as follows:

$$\{\mu_d, \lambda_{xd}\} \sim \mathcal{N}\left(\mu_d | m_{xd0}, (\beta_{xd0}, \lambda_{xd})^{-1}\right) \text{Gamma}\left(\lambda_{xd} | a_{xd0}, b_{xd0}\right),$$

$$\{\beta_{pd}\} \sim \mathcal{N}\left(\beta_d | m_{ypd0}, s_{ypd0}^2\right), \quad G_0 \sim DP\left(\gamma, H\right),$$

$$G_j \sim DP\left(\alpha_0, G_0\right), \quad \{\mu_{kd}, \lambda_{xkd}^{-1}\} \sim G_j, \quad \{\beta_{kpd}\} \sim G_j, \tag{4}$$

$$X_{jid} | \mu_{kd}, \lambda_{xkd} \sim \mathcal{N}\left(X_{jid} | \mu_{kd}, \lambda_{xkd}^{-1}\right),$$

$$\{Y_{ji} = p | X_{ji}\} \sim \frac{\exp\left(\sum_{d=0}^D \beta_{kpd} X_{jid}\right)}{\sum_{p=1}^P \exp\left(\sum_{d=0}^D \beta_{kpd} X_{jid}\right)}$$

5   Gibbs Sampling

We write down the Gibbs Sampler for inference. For all the models, we sample index t_{ji}, k_{jt} and ϕ_k ($\{\mu_{kd}, \lambda_{xkd}\}$ and $\{\beta_{kd}, \lambda_{yk}\}$ for the Normal model). As the Normal model is conjugate, we have a closed form expression for the conditional density of ϕ_k, but for Poisson and Logistic Multinomial models we have used Metropolis Hastings algorithm as presented in [12]. The Normal model's solution is given by the following,

$$\{\mu_{kd}, \lambda_{xkd}\} \sim \mathcal{N}\left(\mu_{kd} | m_{xkd}, (\beta_{xkd}, \lambda_{xkd})^{-1}\right)$$

$$\text{Gamma}\left(\lambda_{xkd} | a_{xkd}, b_{xkd}\right)$$

$$\{\beta_{kd}, \lambda_{yk}\} \sim \mathcal{N}\left(\beta_{kd} | m_{ykd}, (\beta_{yk}, \lambda_{yk})^{-1}\right) \tag{5}$$

$$\text{Gamma}\left(\lambda_{yk} | a_{yk}, b_{yk}\right)$$

Here,

$$m_{xkd} = \frac{\beta_{xd0} m_{xd0} + \sum_{z_{ji}=k} x_{ji}}{\beta_{xd0} + n_{j \cdot k}}$$

$$\beta_{xkd} = \beta_{xd0} + n_{j \cdot k} \quad a_{xkd} = a_{xd0} + n_{j \cdot k}/2$$

$$b_{xkd} = b_{xd0} + \frac{1}{2}\sum_{z_{ji}=k}\left(x_{jid} - \overline{x_{jid}}\right)^2 + \frac{\beta_{xd0} n_{n \cdot k}\left(\overline{x_{jid}} - m_{xd0}\right)}{2(\beta_{xd0} + n_{j \cdot k})} \tag{6}$$

$$m_{yk} = \left\{X^T X + (\beta_{y0}) I\right\}^{-1}\left\{X^T y + \beta_{y0} I m_{y0}\right\}$$

$$\beta_{y,k} = \left(X^T X + \beta_{y0} I\right) \quad a_{y,k} = a_{y0} + n_{j \cdot k}/2$$

$$b_{y,k} = b_{y0} + \frac{1}{2}\left\{y^T y + m_{y0}^T \beta_{y0} m_{y0} - m_{yk}^T \beta_{yk} m_{yk}\right\}$$

Again, the distribution of t_{ji} and k_{jt} is given below.

$$p\left(t_{ji} = t | t^{-ji}, k\right) \propto n_{jt \cdot}^{-ji} f_{k_{jt}}^{-x_{ji}, y_{ji}}\left(x_{ji}, y_{ji}\right) \text{ if t is used}$$

$$p\left(t_{ji} = t | t^{-ji}, k\right) \propto \alpha_0 p\left(x_{ji}, y_{ji} | t^{-ji}, k\right) \text{ if } t = t^{new} \tag{7}$$

If t^{new} is sampled, new sample of $k_{jt^{new}}$ is obtained from

$$p\left(k_{jt^{new}} = k\right) \propto m_{\cdot k}^{-ji} f_k^{-x_{ji}, y_{ji}}\left(x_{ji}, y_{ji}\right) \quad \text{if k is used}$$

$$p\left(k_{jt^{new}} = k\right) \propto \gamma f_{k^{new}}^{-x_{ji}, y_{ji}}\left(x_{ji}, y_{ji}\right) \quad \text{if } k = k^{new} \tag{8}$$

Sampling of k_{jt} is given by,

$$p\left(k_{jt^{new}} = k\right) \propto m_{\cdot k}^{-jt} f_k^{-x_{jt}, y_{jt}}\left(x_{jt}, y_{jt}\right) \quad \text{if k is used}$$

$$p\left(k_{jt^{new}} = k\right) \propto \gamma f_{k^{new}}^{-x_{jt}, y_{jt}}\left(x_{jt}, y_{jt}\right) \quad \text{if } k = k^{new} \tag{9}$$

Here, $p(x_{ji}, y_{ji})$, $f_k^{-x_{ji}, y_{ji}}(x_{ji}, y_{ji})$ and $f_{k^{new}}^{-x_{ji}, y_{ji}}(x_{ji}, y_{ji})$ is given by the following equations. For the Normal model, the integrals have close form solutions where it leads to a Student-t distribution. We solve other integrals by Monte Carlo integration.

$$
\begin{aligned}
p(x_{ji}, y_{ji}) &= \sum_{k=1}^{K} \frac{m_{\cdot k}}{m_{\cdot \cdot} + \gamma} f_k^{-x_{ji}, y_{ji}}(x_{ji}, y_{ji}) \\
&+ \frac{\gamma}{m_{\cdot \cdot} + \gamma} f_{k^{new}}^{-x_{ji}, y_{ji}}(x_{ji}, y_{ji}) \\
f_{k^{new}}^{-x_{ji}, y_{ji}}(x_{ji}) &= \int f(y_{ji}|x_{ji}, \phi) f(x_{ji}|\phi) h(\phi) \, d\phi, \\
f_k^{-x_{ji}, y_{ji}}(x_{ji}, y_{ji}) & \\
&= \int f(y_{ji}|x_{ji}, \phi_k) f(x_{ji}|\phi_k) h(\phi_k| - x_{ji}, y_{ji}) \, d\phi_k
\end{aligned}
\tag{10}
$$

6 Predictive Distribution

Finally, we derive the predictive distribution for a new response $(Y_{j(N+1)})$ given a new covariate $X_{j(N+1)}$ and the set of previous covariate-response pairs $\{D\}$. For prediction, we compute the expectation of $Y_{j(N+1)}$ given training data and $X_{j(N+1)}$ using M samples of $\psi_{j1:jT}$.

$$
\begin{aligned}
E[Y_{j(N+1)}|X_{j(N+1)}, D] &= E[E[Y_{j(N+1)}|X_{j(N+1)}, \psi_{j1:jT}]|D] \\
&= \frac{1}{M} \sum_{m=1}^{M} E[Y_{j(N+1)}|X_{j(N+1)}, \psi_{j1:jT}^m]
\end{aligned}
\tag{11}
$$

We now need to compute the likelihood of this expectation which is given in the following equation,

$$
\begin{aligned}
&E[Y_{j(N+1)}|X_{j(N+1)}, \psi_{jt} = \phi_{k_{jt}}] \propto (n_{jt \cdot}) E[Y_{j(N+1)}|X_{j(N+1)}, \psi_{jt} = \phi_{k_{jt}}] f_{k_{jt}}(x_{j(N+1)}), \\
&\text{if } t \text{ is used previously.} \\
&E[Y_{j(N+1)}|X_{j(N+1)}, \psi_{jt} = \phi_{k_{jt}}] \propto (\alpha_0 n_{jt \cdot}) E[Y_{j(N+1)}|X_{j(N+1)}, \psi_{jt} = \phi_{k_{jt}}] \, p(x_{j(N+1)}|t^{new}, k), \\
&\text{if } t = t^{new}.
\end{aligned}
\tag{12}
$$

Firstly, $p(x_{j(N+1)})$ is given by Eq. (11) with the y part omitted. A new sample of $k_{jt^{new}}$ (If t^{new} is sampled) is obtained from Eq. (9). A new sample of ϕ_k is obtained if $k = k^{new}$. After obtaining the specific table, ψ_{jt}, for $X_{j(N+1)}$ and corresponding ϕ_K, we compute the expectation $E[Y_{j(N+1)}|X_{j(N+1)}, \psi_{jt}]$. Averaging out successive expectations, we get the estimate of $Y_{j(N+1)}$ (Table 1).

Table 1. Algorithm: Gibbs sampler for iHGLM

1. **Initialize Generative Model Parameters in its State Space.**
Repeat
2. **Sample Model Parameters according to Eq. (6).**
3. **Sample t_{ji} according to Eq. (8).**
4. **Sample $k_{jt^{new}}$ according to Eq. (9), If Required.**
5. **Sample k_{jt} according to Eq. (10).**
until converged
6. **Evaluate $E[Y_{j(N+1)}]$ for a new covariate, $X_{j(N+1)}$, according to Eq. (12) and Eq. (13).**

7 Experimental Results

In all experiments, we collected samples from the predictive posterior via the Gibbs Sampler and compared the accuracy of the model against its competitor algorithms, including standard Normal GLMM, group specific Regression algorithms like Linear Regression(OLS), Random Forest, and Gaussian Process Regression [14].

7.1 Clinical Trial Problem Modeled by Poisson iHGLM

We explored a Clinical Trial problem [6] for testing whether a new anticonvulsant drug reduces a patient's rate of epileptic seizures. Patients were assigned the new drug or the placebo and the number of seizures were recorded over a six week period. A measurement was made before the trial as a baseline. The objective was to model the number of seizures, which being a count datum, is modeled using a Poisson distribution with a Log link. The covariates are: Treatment Center size (ordinal), number of weeks of treatment (ordinal), type of treatment–new drug or placebo (nominal) and gender (nominal). For ordinal covariates, we used a Normal-Gamma Mixture (Like the Normal model) as the Base Distribution. For nominal covariates, we used a Dirichlet prior Mixture as the Base Distribution (H). A Poisson distribution with log link was used for the count of seizures. Here, X_{ji} and Y_{ji} represent the ith continuous covariate and count response pair of the jth group. The distribution, $\{\mu_d, \lambda_{xd}\}$ (Normal-Gamma) is the prior distribution on the ordinal covariates. The distribution, $\{\beta_d\}$ (Normal) is the prior distribution on the covariate coefficient β. m is the index of the number of categories for the nominal covariate. p_{dm} is the probability of the m_{th} category of the d_{th} dimension. a_{dm0} is the hyper-parameter for the Dirichlet. Therefore, this becomes an infinite mixture of Dirichlet density. So, a draw G_0 is an infinite mixture of p_{dm}. Another draw G_j leads to an infinite collection of p_{dm} for groups separately, but this time the p_{dm}'s are shared among the groups because G_0 is atomic. After the draw of G_j, one of the mixture components, p_{kdm} gets picked for the j_{th} group and d_{th} dimension with k denoting mixture index. Then, covariate X_{jid} is drawn from a Categorical Distribution with parameters as p_{kdm}.

We found that most patient's number of seizures (they form the groups) comes from a single underlying cluster. This signifies that a majority of the patients across groups show the same response to the treatment. We obtained 10 clusters from 300 out of 565 patients (the remaining 265 were set aside for testing). Among them 8 clusters showed that the new drug reduces the number of epileptic seizures with increasing number of weeks of treatment while the remaining 2 clusters did not show any improvement. We also report the forecast error of the number of epileptic seizures of the remaining 265 patients in Table 4. Our recommendation for the usage of the new drug would be a cluster based solution. For a specific patient, if she falls in one of those clusters with decreasing trend in the number of seizures with time, we would recommend the new drug, and otherwise not. Out of 265 test case patients, 220 showed signs of improvements while 45 did not. Traditional Poisson GLMM cannot infer this findings since the densities are not shared at the patient group level. Moreover, only the Poisson iHGLM based prediction is formally equipped to recommend a patient cluster based solution for the new drug, whereas all traditional mixed models predict a global recommendation for all patients.

$$
\begin{aligned}
&\{\mu_d, \lambda_{xd}\} \sim \mathcal{N}\left(\mu_d | m_{xd0}, (\beta_{xd0}, \lambda_{xd})^{-1}\right) \text{Gamma}\left(\lambda_{xd} | a_{xd0}, b_{xd0}\right), \\
&\{\beta_d\} \sim \mathcal{N}\left(\beta_d | m_{yd0}, s_{yd0}^2\right), \quad p_{dm} \sim \text{Dir}\left(a_{dm0}\right), \\
&G_0 \sim DP\left(\gamma, H\right), \quad G_j \sim DP\left(\alpha_0, G_0\right), \quad \{\mu_{kd}, \lambda_{xkd}^{-1}\} \sim G_j, \\
&\{\beta_{kd}\} \sim G_j, \quad p_{kdm} \sim G_j, \quad X_{jid} \sim \text{categorical}\left(p_{kdm}\right), \\
&X_{jid} | \mu_{kd}, \lambda_{xkd} \sim \mathcal{N}\left(X_{jid} | \mu_{kd}, \lambda_{xkd}^{-1}\right) \\
&\{Y_{ji} | X_{ji}\} \sim Poisson\left(y_{ji} | \exp\left(\sum_{d=0}^{D} \beta_{kd} X_{jid}\right)\right)
\end{aligned}
\tag{13}
$$

7.2   Height Imputation Problem

We propose a new iHGLM based method for height imputation [15] based on height-diameter regression in forest stands. A forest stand is a community of trees uniform in composition, structure, age and size class distribution. Estimating volume and growth in forest stands is an important feature of forest inventory. Since there is generally a strong proportionality between diameter and other tree-attributes like past increment, forecasting height using diameter can proceed with limited loss of information. We processed data for five stands. The data incorporated in the model is through the logarithmic transformation $Y^{new} = \log\left(Y^{old} - 4.5\right)$ and inverse transformation $X^{new} = \left(1 + X^{old}\right)^{-1}$. We show the tree heights with respect to the diameters for each stand which clearly depicts the sharing of clusters among stands and different clusters within each stand. Also, different clusters within stands have different variability of growth, thereby modeling heteroscedasticity at the stand level. Roughly, there are 2 to 3 primary clusters in each stand totaling 5 primary clusters. The remaining clusters have very few trees (maximum 5) and represent outliers. We report the mean tree heights and also the variance of growth of the trees within each primary cluster in Table 3. We also report the forecast error of the trees of the testing set (20%) and compare against Normal GLMM, group specific OLS, Random Forest and Gaussian Process Regression.

7.3   Market Dynamics Experiment

In this experiment, instead of presenting a third example to demonstrate the efficacy of the model, we decided to demonstrate how the model could be used as an "exploratory" tool (as opposed to a classical "inference" tool) for analyzing the temporal dynamics of stocks from S&P 500 companies. This strength draws from the model's large support (i.e., hypothesis space). The companies belong to disparate market sectors such as, Technology (Microsoft, Apple, IBM and

Google), Finance (Goldman Sachs, JPMorgan, BOA and Wells-Fargo), Energy (XOM, PTR, Shell and CVX), Healthcare (JNJ, Novartis, Pfizer and MRK), Goods (GE, UTX, Boeing and MMM), and Services (WMT, AMZN, EBAY and HD). Using iHGLM Normal Model, we modeled each company's stock value at a given time point as a function of the values of the others at that time point (the remaining 23). Each stock of one particular sector(tech., finance, healthcare sector etc.) formed one group (e.g. tech. sector has 4 groups/stocks (IBM,MSFT,goog,aapl)) and a whole sector (tech., finance etc.) was modeled by one HGLM. Experiments were run over all such groupings. Past stock prices

Table 2. List of stocks with top 3 most significant stocks that influence each stock from all the sectors.

Time-Period	XOM	PTR	Shell	CVX	AAPL	MSFT	IBM	GOOG	BOA	JPM	WFC	GS
2009-14	PTR	XOM	PTR	XOM	IBM	GOOG	AAPL	AAPL	WFC	GS	GS	JPM
	CVX	CVX	XOM	SHELL	JPM	IBM	MMM	GS	GS	WFC	JPM	BOA
	GS	GOOG	HD	BOA	GOOG	GS	GOOG	MSFT	JPM	XOM	PFE	WFC
2007-09	HD	GS	MSFT	JNJ	BOA	GE	JPM	WFC	EBAY	MMM	GS	WMT
	PTR	PFE	XOM	IBM	WMT	GOOG	Shell	MMM	GE	AMZN	HD	GE
	JPM	CVX	MMM	HD	GS	JPM	NVS	UTX	MRK	CVX	PFE	GS
Time-Period	JNJ	NVS	PFE	MRK	GE	UTX	BA	MMM	WMT	AMZN	EBAY	HD
2000-14	MRK	JNJ	JNJ	NVS	BA	MMM	MMM	BA	AMZN	HD	HD	GOOG
	NVS	PFE	GS	JPM	MMM	GE	GE	AAPL	EBAY	EBAY	WMT	WMT
	GE	AAPL	MRK	JNJ	PTR	PFE	UTX	GE	GOOG	MSFT	MSFT	GS
2007-09	MSFT	BA	PTR	IBM	AXP	GS	JPM	WMT	HD	MMM	GS	WMT
	CVX	PFE	AAPL	CVX	P& G	BOA	MRK	PTR	GE	CVX	GE	GE
	GS	WFC	MMM	HD	GS	JPM	HD	WFC	MMM	HD	IBM	WFC

Table 3. MSE and MAE of the Algorithms for the Height Imputation Dataset and Means and Standard Deviation of the individual Clusters from many Stands. For Stand-1, the Main Clusters were C1,C2,C3, For S-2, these are C4,C5,C3, For S-3, they are C1,C4,C3, For S-4, these are C2,C3 and For S-5, they are C1,C4,C3.

Clusters	C1	C2	C3	C4	C5
Mean	.1317	.0692	.014	.0302	.0143
STD	.0087	.00086	.00049	.00038	.00015
iHGLM	GLMM	OLS	Rforest	GPR	CART
MAE (L1 Error)					
.0094	.0114	.01243	.01527	.01319	.0252
MSE (L2 Error)					
1.008e-2	9.8e-3	1.2e-2	4.2e-2	1.8e-2	3.4e-2

Table 4. MSE and MAE of the Algorithms for the Clinical Trial Dataset and Number of Patients in Clusters for Training and Testing Sets.

Patient Number in Clusters for Training Set								
Positive							Negative	
26	39	15	28	22	53	32	24	37 24
Patient Number in Clusters for Testing Set								
19	33	27	19	16	38	26	31	34 22
iHGLM	Poisson GLMM		Poisson Regression			CART		RForest
Mean Square Root Error(L2 Error)								
1.41	1.58		1.92			1.65		1.75
Mean Absolute Error Root Error(L1 Error)								
.94	1.34		1.51			1.23		1.62

were not included. We recorded the stocks having the most impact on the determination of the value of each stock. The impacts are by definition the magnitude of the weighted coefficients of the covariates (the stock values) in iHGLM. All the experiments were done on daily close out stock prices after the financial crisis (June-09 to March-14) and in the middle of the crisis (May-07 to June-09). Few trends were noteworthy.

Prices of a given set of Firstly, stocks from any given sector were impacted largely by the same stock (not necessarily from the same sector), with few stocks being influential overall. Secondly, the stocks having the most impact on a specific sector were largely the same. For example, Microsoft (tech. sector), is largely modeled by GOOG, IBM (tech), GS (Finance) after the crisis (in descending order of weights). However, during the crisis, the stocks showed no such trends. For example, Microsoft is impacted by GE, GOOG, JPM showing no sector wise trend. We report results for all the sectors/stocks in Table 2.

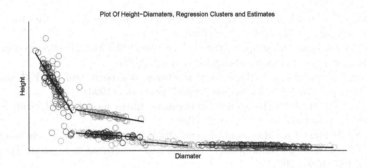

Fig. 3. Depiction of several clusters in the Height Imputation dataset for different stands which is shared by clusters. Every stand is shown with its own single color. (Color figure online)

8   Conclusion

In this paper, we have formulated an infinite mixtures of Hierarchical Generalized Linear Model (iHGLM), a flexible model for hierarchical regression. The model captures identical response-covariate densities in different groups as well as different densities in the same group. It also captures heteroscedasticity and overdispersion across groups. We proved posterior consistency of the iHGLM model, and experimentally evaluated it on a wide range of problems where traditional mixed effect models fail to capture structure in the grouped data. Although the Gibbs sampler turned out to be fairly accurate for the iHGLM models, developing a variational inference alternative would be an interesting topic for future research. Finally, the number of mixture components in each group depends on the scale factors γ and α (scale parameters of the DP) of the model, and at times grows large in specific groups. This occurs mostly when any group has a large number of data points compared to others. In most cases, beyond a few primary clusters, the remaining represent outliers. Although, careful tuning of scale parameters can mitigate these problems, a theoretical understanding of the dependence of the model on scale parameters could lead to better modeling and application (Fig. 3).

References

1. Antoniak, C.: Mixtures of dirichlet processes with applications to Bayesian non-parametric problems. Ann. Stat. **2**(6), 1152–1174 (1974)
2. Blackwell, D., MacQueen, J.B.: Ferguson distributions via polya urn schemes. Ann. Stat. **1**(2), 353–355 (1973)
3. Breslow, N.E., Clayton, D.G.: Approximate inference in generalized linear mixed models. J. Am. Stat. Assoc. **88**(421), 9–25 (1993)
4. Ferguson, T.: A Bayesian analysis of some nonparametric problems. Ann. Stat. **1**, 209–230 (1973)
5. Hastie, T., Tibshirani, R.: Varying-coefficient models. J. R. Stat. Soc. Ser. B (Methodological) **55**(4), 757–796 (1993)
6. http://www.ibm.com/support/knowledgecenter/SSLVMB_21.0.0/com.ibm.spss.statistics.cs/glmm_anticonvulsant_intro.htm. (2012)
7. Jordan, M., Jacobs, R.: Hierarchical mixtures of experts and the EM algorithm. International Joint Conference on Neural Networks (1993)
8. Lee, Y., Nelder, J.A.: Hierarchical generalized linear models. J. R. Stat. Soc. Ser. B (Methodological) **58**(4), 619–678 (1996)
9. Lee, Y., Nelder, J.A.: Hierarchical generalised linear models: a synthesis of generalised linear models, random-effect models and structured dispersions. Biometrika **88**(4), 987–1006 (2001)
10. Lee, Y., Nelder, J.A.: Modelling and analysing correlated non-normal data. Stat. Model. **1**(1), 3–16 (2001)
11. Lee, Y., Nelder, J.A.: Double hierarchical generalized linear models (with discussion). J. R. Stat. Soc.: Ser. C (Appl. Stat.) **55**(2), 139–185 (2006)
12. Neal, R.M.: Markov chain sampling methods for dirichlet process mixture models. J. Comput. Graph. Stat. **9**(2), 249–265 (2000)
13. Nelder, J.A., Wedderburn, R.W.M.: Generalized linear models. J. R. Stat. Soc. Ser. A (Gen.) **135**(3), 370–384 (1972)
14. Rasmussen, C., Williams, C.: Gaussian Processes for Machine Learning (Adaptive Computation and Machine Learning). MIT Press, Cambridge (2005)
15. Robinson, A.P., Wykoff, W.R.: Imputing missing height measures using a mixed-effects modeling strategy. Can. J. For. Res. **34**, 2492–2500 (2004)
16. Sethuraman, J.: A constructive definition of Dirichlet priors. Stat. Sin. **4**, 639–650 (1994)
17. Teh, Y.W., Jordan, M.I., Beal, M., Blei, D.: Hierarchical Dirichlet processes. J. Am. Stat. Assoc. **101**, 1566–1581 (2006)
18. Viele, K., Tong, B.: Modeling with mixtures of linear regressions. Stat. Comput. **12**(4), 315–330 (2002)
19. http://en.wikipedia.org/wiki/Atom_%28measure_theory%29

A Deep Neural Network for Pairwise Classification: Enabling Feature Conjunctions and Ensuring Symmetry

Kyohei Atarashi[1]([⊠]), Satoshi Oyama[1], Masahito Kurihara[1],
and Kazune Furudo[2]

[1] Graduate School of Information Science and Technology, Hokkaido University,
Sapporo, Hokkaido, Japan
{atarashi_k,oyama,kurihara}@complex.ist.hokudai.ac.jp
[2] NEC Solution Innovators, Ltd., Tokyo, Japan
f.kazune08031991@gmail.com

Abstract. Pairwise classification is a computational problem to determine whether a given ordered pair of objects satisfies a binary relation R which is specified implicitly by a set of training data used for 'learning' R. It is an important component for entity resolution, network link prediction, protein-protein interaction prediction, and so on. Although deep neural networks (DNNs) outperform other methods in many tasks and have thus attracted the attention of machine learning researchers, there have been few studies of applying a DNN to pairwise classification. Important properties of pairwise classification include using feature conjunctions across examples. Also, it is known that making the classifier invariant to the data order is an important property in applications with a symmetric relation R, including those applications mentioned above. We first show that a simple DNN with fully connected layers cannot satisfy these properties and then present a pairwise DNN satisfying these properties. As an example of pairwise classification, we use the author matching problem, which is the problem of determining whether two author names in different bibliographic data sources refer to the same person. We show that the method using our model outperforms methods using a support vector machine and simple DNNs.

1 Introduction

Pairwise classification is a computational problem to determine whether a given ordered pair of objects satisfies a binary relation R which is specified implicitly by a set of training data used for 'learning' R. It is an important component for entity resolution, network link prediction, and so on. The method commonly used for identifying two objects includes defining a manually tuned similarity. However, defining suitable similarities is difficult. Therefore, machine learning techniques for learning from labeled data have been used.

When typical machine learning methods are applied to pairwise classification using the "Learn a classifier directly" approach, the design of the feature

© Springer International Publishing AG 2017
J. Kim et al. (Eds.): PAKDD 2017, Part I, LNAI 10234, pp. 83–95, 2017.
DOI: 10.1007/978-3-319-57454-7_7

vector representing the pair of two objects is essential and important. Bilenko and Moony [3] represented a pair of objects by using a feature vector based on common features between the two objects and a support vector machine (SVM). This method is effective for a problem like citation matching, where two objects belonging to the same class have many common features. However, if the two objects have few common features, this method is not effective. An example problem is the "author matching problem" in which the task is to determine whether two author names in different bibliographic data sources refer to the same person. Oyama and Manning [14] proposed applying a kernel method to this problem that uses the conjunctions of not only the common features but also those of different features across the two objects and using an SVM. Their method outperforms Bilenko and Moony's method in the author matching problem.

Deep neural networks (DNNs) have begun attracting attention in the field of machine learning as they have better performance than existing methods (e.g., SVM) in image classification [13], speech recognition [10], and many other tasks. While there have been many studies of applying a DNN to datum-wise classification, there have been few studies of applying a DNN to pairwise classification. Tran et al. [17] used a DNN as a classifier for the author matching problem. Since they represented feature vectors representing pairs of objects by concatenating fixed similarities and distance metrics, their approach does not take advantage of a DNN's ability to obtain feature representations automatically. A method using a DNN should be able to outperform existing methods even in pairwise classification by using feature vectors that enable a DNN to effectively learn feature representations.

A straightforward way of creating a feature vector representing a pair of objects is to concatenate the feature vectors of the two objects. The resulting vector can be used as input to a simple DNN with fully connected layers. The fully connected layers should enable even a simple DNN to use feature conjunctions across the two objects. In many applications such as entity resolution, the classifier results for training and prediction should be invariant with respect to the order of the pair values (the symmetry property).

In this paper, we first show that determining whether two objects satisfy relation R while satisfying symmetry is difficult for a simple DNN with fully connected layers. Next, we present a DNN-based model, i.e., a pairwise deep neural network (pairwise DNN), that ensures symmetry and enables feature conjunctions across examples. Then we present experimental results demonstrating that the method using our model outperforms other methods in the author matching problem.

2 Pairwise Classification

2.1 Problem Formulation

Pairwise classification is the computational problem to determine whether a pair of objects, x^α and x^β, satisfies relation R. Therefore, our goal is to obtain classifier f:

$$f(\boldsymbol{x}^\alpha, \boldsymbol{x}^\beta) = \begin{cases} 1 \text{ (if } \boldsymbol{x}^\alpha \text{and } \boldsymbol{x}^\beta \text{satisfy } R) \\ -1 \text{ (otherwise).} \end{cases}$$

It is difficult to obtain accurate classifiers by using manually tuned similarities and thresholds. Therefore, machine learning methods in which learning is done from labeled data have been used. Many methods sample pair objects from the data, and then a person labels them in accordance with whether they satisfy relation R or not. Then these training examples are fed into classifier learning algorithms.

2.2 Symmetry

In many applications such as entity resolution, the classifier output should be invariant with respect to the order of the pair. Therefore, the classifier should satisfy symmetry; that is, $f(\boldsymbol{x}^\alpha, \boldsymbol{x}^\beta) = f(\boldsymbol{x}^\beta, \boldsymbol{x}^\alpha)$. Furthermore, the learning result should be invariant with respect to the order of the training data pairs.

3 Related Work

Machine learning approaches for determining the identity of two objects can be roughly grouped into three categories:

1. Learning a classifier directly.
2. Learning a similarity between two objects and deciding they are the same if the similarity exceeds a threshold.
3. Learning a distance between two objects and deciding they are the same if the distance is less than a threshold.

Then, toward lower approach, problem is more general and difficult. If the distance is obtained, the similarity can be obtained by reversing its sign. To learn the distance, it is necessary to satisfy the distance axiom. In many applications, there are cases in which only a classifier or a similarity suffice. In such cases, learning a classifier or a similarity is suitable according to Vapnik's principle [18]: "When solving a given problem, try to avoid solving a more general problem as an intermediate step". Furthermore, when learning a classifier directly, it is not necessary to set a threshold manually.

Bilenko and Moony [3] proposed a method for learning a classifier directly for the citation matching problem, in which a determination is made as to whether two citations in different bibliographic data sources refer to the same paper. They represent the original object as a bag-of-words feature vector $\boldsymbol{x}^\alpha = (x_1^\alpha, x_2^\alpha, \ldots, x_n^\alpha)^\mathrm{T}$ and a pair object consisting of $\boldsymbol{x}^\alpha = (x_1^\alpha, x_2^\alpha, \ldots, x_n^\alpha)^\mathrm{T}$ and $\boldsymbol{x}^\beta = (x_1^\beta, x_2^\beta, \ldots, x_n^\beta)^\mathrm{T}$ as feature vector $\boldsymbol{x}_\mathrm{Hadamard}$:

$$\hat{\boldsymbol{x}}_\mathrm{Hadamard} = (x_1^\alpha x_1^\beta, x_2^\alpha x_2^\beta, \ldots, x_n^\alpha x_n^\beta)^\mathrm{T}. \tag{1}$$

This feature vector is a Hadamard product of $\boldsymbol{x}^\alpha$ and $\boldsymbol{x}^\beta$. They labeled it y and used an SVM. The classifier satisfies the symmetry because the feature vectors representing the pair objects satisfy symmetry.

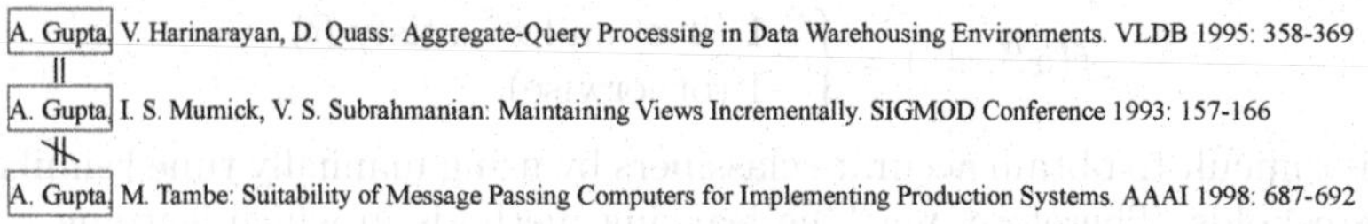

A. Gupta V. Harinarayan, D. Quass: Aggregate-Query Processing in Data Warehousing Environments. VLDB 1995: 358-369

A. Gupta I. S. Mumick, V. S. Subrahmanian: Maintaining Views Incrementally. SIGMOD Conference 1993: 157-166

A. Gupta M. Tambe: Suitability of Message Passing Computers for Implementing Production Systems. AAAI 1998: 687-692

Fig. 1. Matching authors

Bilenko and Mooney's method is effective for a problem like citation matching, in which two objects from the same class have many common features, but it is not effective if the two objects from the same class have few common features because it cannot distinguish between positive pairs and negative pairs. An example of this is the author matching problem, in which a determination is made as to whether two author names in different bibliographic data sources refer to the same person. In the illustrative example in Fig. 1, where A. Gupta is the abbreviated form of Ashish Gupta and Anoop Gupta, the first two records have no common words even though A. Gupta is the same person in both cases. One approach to such problems includes using not only common features but also conjunctions of different features across examples. Oyama and Manning [14] proposed using $\hat{x}_{\text{Cartesian}}$,

$$\hat{x}_{\text{Cartesian}} = (x_1^\alpha x_1^\beta, \ldots, x_1^\alpha x_n^\beta, x_2^\alpha x_1^\beta, \ldots, x_2^\alpha x_n^\beta, \ldots, x_n^\alpha x_1^\beta, \ldots, x_n^\alpha x_n^\beta)^{\text{T}}. \quad (2)$$

as a feature vector representing a pair of objects. This is the Cartesian product between x^α and x^β. However, the dimension of this feature vector is n^2, so doing this straightforwardly is computationally intensive. They thus represented a pair object by using feature vector $\hat{x}_{\text{concat}}$,

$$\hat{x}_{\text{concat}} = \begin{pmatrix} x^\alpha \\ x^\beta \end{pmatrix} = (x_1^\alpha, \ldots, x_n^\alpha, x_1^\beta, \ldots x_n^\beta)^{\text{T}}, \quad (3)$$

which concatenates original objects, and proposed using the following kernel:

$$K(\hat{x}_{\text{concat}}, \hat{z}_{\text{concat}}) = \langle \hat{x}_{\text{Cartesian}}, \hat{z}_{\text{Cartesian}} \rangle. \quad (4)$$

Using the SVM with this kernel enables classification on the Cartesian product space between two objects without high computational cost. Although this feature vector is not symmetrical, they showed that including a pair that is reversed with respect to the concatenation order or symmetrizing the kernel ensures classifier symmetry. A similar method has been used to predict protein-protein interactions [1].

As mentioned in the Introduction, DNNs are attracting attention in the field of machine learning as methods using them have better performance than existing methods in many tasks. Since DNNs can automatically obtain feature representations, a method using them should be able to outperform existing methods even in pairwise classification.

Also as mentioned in the Introduction, while there have been many studies of applying a DNN to datum-wise classification, there have been few studies of applying a DNN to pairwise classification, and most of these studies used learning of similarity or distance metric. Bromley et al. [5] proposed using a Siamese network for handwritten signature pair determination. In a Siamese network, two objects are input to a NN individually, and the similarity or distance between the outputs of the NN is output. Also, there have been several proposed methods based on a Siamese network for face verification [6,11,16]. Tran et al. [17] applied a DNN as a classifier to the author matching problem. Since they represented feature vectors representing pairs of objects by concatenating fixed similarities or distance metrics, this approach cannot utilize a DNN's ability to obtain representations automatically. A method using a DNN should be able to outperform existing methods even in pairwise classification by using feature vectors to effectively learn feature representations.

4   Problem of DNN in Pairwise Classification

Because only one object can be input to a DNN, it is necessary to design a feature vector representing a pair of objects. Given that a DNN can automatically obtain feature representations, the feature vector should self-sufficiently represent the information of the two objects. Hence, we first present a straightforward design of a feature vector representing a pair of objects that concatenates the feature vectors of the two objects, as given by Eq. (3). Because the input and hidden layers of a simple DNN are fully connected, the DNN should provide feature representation considering feature conjunctions across examples.

Since the feature vector represented by Eq. (3) does not satisfy symmetry, a symmetric DNN should be used when this feature vector is input. Bishop [4] classified the approaches to making a NN invariant into four approaches

1. The training set is augmented using replicas of the training patterns, transformed in accordance with the desired invariance.
2. A regularization term is added to the error function that penalizes changes in the model output when the input is transformed.
3. Invariance is built into the pre-processing by extracting features that are invariant under the required transformations.
4. Invariance is built into the structure of the NN.

In pairwise classification, the NN should be invariant with respect to the order of the data values in a pair; i.e., it should have symmetry. Approach 1, called data augmentation, incurs extra computational costs. Approach 2 cannot be used in pairwise classification because transformation that changes the data order is not continuous. Approach 3 is problematic because designing suitable features is difficult. Therefore, we took Approach 4. If the values of the hidden layer connected with the input layer satisfy symmetry, the NN output satisfies symmetry. Namely, let W be the weight matrix between the input and the hidden layer. If the following equation holds, the NN output satisfies symmetry.

$$W \begin{pmatrix} x^\alpha \\ x^\beta \end{pmatrix} = W \begin{pmatrix} x^\beta \\ x^\alpha \end{pmatrix} \tag{5}$$

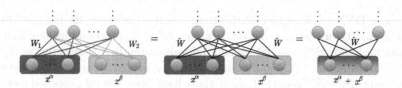

Fig. 2. Symmetric DNN

Let $\boldsymbol{W}_1$ be the weight matrix between the partial input layer for the first object and the entire hidden layer, $\boldsymbol{W}_2$ be the weight matrix between remaining input layer for the second object and the entire hidden layer. Since $\boldsymbol{W} = (\boldsymbol{W}_1, \boldsymbol{W}_2)$, Eq. (5) can be rewritten: $\boldsymbol{W}_1(\boldsymbol{x}^\alpha - \boldsymbol{x}^\beta) = \boldsymbol{W}_2(\boldsymbol{x}^\alpha - \boldsymbol{x}^\beta)$. For arbitrary $\boldsymbol{x}^\alpha$ and $\boldsymbol{x}^\beta$, this equation holds if $\boldsymbol{W}_1 = \boldsymbol{W}_2$. However, there is actually a problem here. If this equation holds, the following equation holds.

$$\boldsymbol{W}\boldsymbol{x}_{\text{concat}} = \hat{\boldsymbol{W}}(\boldsymbol{x}^\alpha + \boldsymbol{x}^\beta), \tag{6}$$

where $\hat{\boldsymbol{W}}$ is equal to $\boldsymbol{W}_1$ and $\boldsymbol{W}_2$. From Eq. (6), it follows that the addition of feature vectors of two objects is input to the DNN. However, the addition of feature vectors of two objects is not suitable for pairwise classification because the information about the features of the individual objects is lost. For example, in the author matching problem, it cannot be determined which words appear in which papers.

5   Proposed Method

5.1   Pairwise DNN

Our proposed pairwise DNN ensures symmetry without losing information about the features of individual objects and enables feature conjunctions across examples. In a pairwise DNN, two objects, $\boldsymbol{x}^\alpha$ and $\boldsymbol{x}^\beta$, are first mapped individually to $\boldsymbol{z}^\alpha = \hat{\boldsymbol{W}}\boldsymbol{x}^\alpha$ and $\boldsymbol{z}^\beta = \hat{\boldsymbol{W}}\boldsymbol{x}^\beta$, where $\hat{\boldsymbol{W}}$ is a common $m \times n$ weight matrix. Next, a Hadamard layer, which calculates the Hadamard product of $\boldsymbol{z}^\alpha$ and $\boldsymbol{z}^\beta$, is introduced, and $\hat{\boldsymbol{z}} = (z_1^\alpha z_1^\beta, z_2^\alpha z_2^\beta, \ldots, z_m^\alpha z_m^\beta)^{\text{T}}$ is calculated. Finally, $\hat{\boldsymbol{z}}$ is input to a simple fully connected DNN. $\hat{\boldsymbol{W}}$ is not a fixed parameter but a parameter that is learned when a simple fully connected DNN is trained by stochastic gradient descent (SGD) with backpropagation. Figure 3 provides an overview of a pairwise DNN.

5.2   Symmetry

Since a Hadamard product is commutative and $\hat{\boldsymbol{W}}$ is common between two objects, the output of a pairwise DNN satisfies symmetry. Let $\hat{\boldsymbol{w}}_i$ be the ith row vector of $\hat{\boldsymbol{W}}$ and w_{ij} be the jth element of $\hat{\boldsymbol{w}}_i$. Then $\hat{z}_i$, the ith element of $\hat{\boldsymbol{z}}$, can be written as

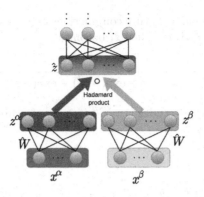

Fig. 3. Pairwise DNN

$$\hat{z}_i = \sum_{k=1}^{n} \sum_{l=1}^{n} \hat{w}_{ik} \hat{w}_{il} x_k^\alpha x_l^\beta. \tag{7}$$

In the elements of $\hat{\mathbf{z}}$, only $\hat{z}_i$ depends on $\hat{w}_{ij}$. The partial differential of $\hat{z}_i$ with respect to w_{ij} is given by

$$\frac{\partial \hat{z}_i}{\partial \hat{w}_{ij}} = \sum_{k=1}^{n} (\hat{w}_{ik}(x_j^\alpha x_k^\beta + x_k^\alpha x_j^\beta)). \tag{8}$$

Since this satisfies symmetry, the learning result satisfies symmetry. Therefore, a pairwise DNN satisfies symmetry.

A pairwise DNN can be considered to extract a feature $\hat{\mathbf{z}}$. However, since this feature extraction itself is also learned, invariance is built into the structure of a pairwise DNN. This is Approach 4 described above, making the NN invariant as classified by Bishop [4].

5.3 Feature Conjunctions Across Examples

In Eq. (7), which formulates $\hat{z}_i$, $x_k^\alpha x_l^\beta$ is the feature conjunction of the kth feature of the first object and lth is the feature of the second object. Therefore, each element of $\hat{z}_i$ can be considered to be the weighted summation of feature conjunctions across examples. Our approach is the same as Bilenko and Moony's with respect to using a vector expressed by a Hadamard product but different with respect to enabling feature conjunctions across examples by using the Hadamard product of a mapped vector. Like Oyama and Manning's approach, our approach using feature conjunctions should be effective when two objects belonging to the same class have few common features.

Even though the mapping of each object is linear, like a general NN, it is possible to use a non-linear activation function. However, if non-linear activation function $h(\cdot)$ is used, the ith element of $\hat{\mathbf{z}}$ becomes $\hat{z}_i = h(z_i^\alpha)h(z_i^\beta)$, so a pairwise

DNN does not always enable feature conjunctions across examples. For example, let $h(\cdot)$ be a sigmoid function. Then $\hat{z}_i$ becomes:

$$\hat{z}_i = \frac{1}{1 + \exp(-z_i^\alpha) + \exp(-z_i^\beta) + \exp(-z_i^\alpha - z_i^\beta)}.$$

Since $z_i^\alpha = \sum_{j=1}^{n} \hat{w}_{ij} x_j^\alpha$ and $z_i^\beta = \sum_{j=1}^{n} \hat{w}_{ij} x_j^\beta$, feature conjunctions do not appear in $\hat{z}_i$.

6   Experiment

6.1   Dataset and Overview

We experimentally evaluated our proposed method using the author matching problem on the DBLP dataset, which is a bibliography of computer science papers. We extracted 3,384 papers for which there were 729 unique author names. We only used papers for which the full name of author was given. Papers with the same author name were assumed to have been written by the same person. For the author matching problem, we abbreviated first names into initials and removed middle names. We used all words appearing in their titles, coauthor names and publication venues. Each original object was represented by a bag-of-words feature vector and the dimension of all vectors was 9,264.

There are two methods for creating a training set and a test set: (1) First, create a pair set from the original objects set. Next, split the pair set into a training set and a test set. (2) First, split the original objects set into a training objects set and a test objects set. Next, create a training set and a test set by making pairs from each objects set. With the method (1), the problem is easy because the objects constituting the test set are also in the training set. We thus used the method (2). If the authors specified by the objects of the pair are the same person, the pair was given a positive label, and if authors are not the same person, the pair was given a negative label. We only used pairs constituted by two papers with the same abbreviated author name.

Table 1. Size of dataset

	Fold 1	Fold 2
No. of papers	1,692	1,692
No. of pair objects	22,264	22,416

If the training set and test set are made using method (2), the number of negative pairs is much larger than that of positive pairs. We thus balanced the two sets by sampling the negative pairs, as described elsewhere [3]. The sizes of the datasets are shown in Table 1. Because author matching problem is binary classification problem, we evaluated accuracy, precision, and recall on the test set. Since which precision or recall is more important depends on situation, we evaluated the precision-recall curve and the area under the curve (AUC), as discussed elsewhere [2].

6.2 Experiment 1: Comparison of Proposed Method with Other Methods

In Experiment 1, we compared five methods:

Pairwise DNN Proposed method. The number of units in each layer was 1000, and the number of hidden layers behind the Hadamard layer was 3. The mapping of each objects was linear.

Concat DNN A simple fully connected DNN with feature vectors of pairs of objects represented by Eq. (3). The number of units in each layer was 1000, and the number of hidden layers was 4.

Concat DNN (aug) Same as Concat DNN except that feature vectors with the concatenation order reversed were included in the training set. Also, at the time of prediction, feature vectors with the concatenation order reversed were predicted, and the mean of the two outputs was used as the final output. In short, data augmentation was performed on the training and test sets.

Addition DNN A simple fully connected DNN with feature vectors of pairs of objects represented by the addition of two objects. This model is equivalent to a symmetric DNN, as illustrated in Fig. 2. The number of units in each layer was 1000, and the number of hidden layers was 4.

Pairwise SVM An SVM with the kernel proposed by Oyama and Manning [14].

In the DNN-based methods, the ReLU activation function [9] was used for the hidden layers, a sigmoid function was used for the output layer, and a loss function was used for cross entropy. Dropout [15] was used for all hidden layers and the Hadamard layer. All weight were initialized using Glorot's method [8], which uses a uniform distribution with the interval adjusted in accordance with the number of units.

We used SGD with a mini-batch of size 128 for 100 epochs. The Adam optimizer [12] was effective for **Pairwise DNN**. The AdaGrad optimizer [7] was effective for **Concat DNN, Concat DNN(aug)**, and **Addition DNN**. The default learning rate of Adam is 0.001, but we found that a learning rate several times higher produced better results. We thus set the learning rate to 0.002. We set it to 0.005 for the methods using AdaGrad. For hyper-parameter C of **Pairwise SVM**, we tested using 21 values, $[2^{-10}, 2^{-9}, \ldots, 2^{10}]$, and found that ten test values produced comparably good results. We thus set $C = 2^0 = 1$.

As shown in Table 2, **Pairwise DNN** had the highest recall, accuracy, and AUC. Although there is a trade-off relationship between precision and recall, the precision of **Pairwise DNN** was kept high. As shown in Fig. 4, **Pairwise DNN** and **Pairwise SVM**, both of which enable feature conjunctions across examples, retained high precision at higher recall levels, especially **Pairwise DNN**. The results of **Concat DNN(aug)** and **Concat DNN** indicated that data augmentation was ineffective in this experiment.

Table 2. Precision (P), recall (R), AUC, and accuracy (Acc) for five methods.

Method	P	R	AUC	Acc
Pairwise DNN	0.938	**0.717**	**0.924**	**0.835**
Concat DNN	0.881	0.600	0.869	0.760
Concat DNN (aug)	0.908	0.564	0.869	0.754
Addition DNN	0.932	0.413	0.839	0.692
Pairwise SVM	**0.968**	0.540	0.918	0.761

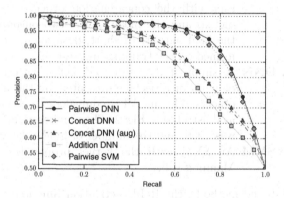

Fig. 4. Precision-recall curves for five methods.

6.3 Experiment 2: Evaluation of Proposed Method

In Experiment 2, we evaluated the proposed pairwise DNN. First, we changed the number of hidden layers behind the Hadamard layer, $[0, 1, 2, 3, 4, 5, 6]$. The results for **Pairwise DNN** in Experiment 1 (in which there were three hidden layers) were used as a baseline. The number of parameters in each model was made almost the same by adjusting the number of units. The other conditions were the same as for Experiment 1. As shown in Table 3, as the number of hidden layers was increased, recall and accuracy tended to increase while precision tended to decrease.

Table 3. Precision (P), recall (R), AUC, and accuracy (Acc) for seven models with different numbers of hidden layers (n = number of hidden layers).

Models	P	R	AUC	Acc
Layer 0	0.966	0.599	0.905	0.789
Layer 1	**0.971**	0.624	**0.926**	0.803
Layer 2	0.936	0.693	0.921	0.823
Layer 3	0.938	0.717	0.924	0.835
Layer 4	0.922	0.755	0.925	0.845
Layer 5	0.921	0.734	0.905	0.836
Layer 6	0.920	**0.757**	0.925	**0.846**

Table 4. Precision (P), recall (R), AUC, and accuracy (Acc) for four models whose mapping of each object was linear and non-linear.

Models	P	R	AUC	Acc
Linear	0.938	**0.717**	0.924	**0.835**
Sigmoid	0.913	0.453	0.799	0.705
Tanh	0.949	0.706	**0.925**	0.834
ReLU	**0.964**	0.569	0.898	0.774

Next, we compared four models whose mapping of each object was linear (**Linear**) and non-linear (**Sigmoid, Tanh,** and **ReLU**). The other conditions (e.g., the number of hidden layers) were the same as for **Pairwise DNN** in Experiment 1. As shown in Table 4, the results with **Sigmoid** and **ReLU** were significantly worse than with **Linear** except for precision. Clearly, a pairwise DNN does not enable feature conjunctions across examples when mapping each object with **Sigmoid** or **ReLU**. In contrast, the results with **Tanh** were nearly the same as those with **Linear**. Clearly, **Tanh** is almost linear for low absolute input values, and the weights are initialized using a uniform distribution with a mean of 0.

7   Conclusion and Future Work

Pairwise classification is an important component in entity resolution, network link prediction, protein-protein interaction prediction, and so on. We showed that there is a problem regarding symmetry and object information when a simple DNN is learned directly as a classifier with feature vectors represented by concatenating two objects. Our proposed DNN-based model, a pairwise DNN, satisfies symmetry without losing the information of each object and enables feature conjunctions across examples. Experimental results for the author matching problem showed that a pairwise DNN had higher recall, accuracy, and AUC. Experiments on a pairwise DNN clarified several properties.

This work focused on the author matching problem. Future work includes application of the pairwise DNN to other problems (e.g., face verification) and extension of the pairwise DNN model to convolutional neural networks, recursive neural networks, and so on.

References

1. Ben-Hur, A., Noble, W.S.: Kernel methods for predicting protein-protein interactions. Bioinformatics **21**(suppl 1), i38–i46 (2005)
2. Bilenko, M., Mooney, R.: On evaluation and training-set construction for duplicate detection. In: Proceedings of the KDD-2003 Workshop on Data Cleaning, Record Linkage, and Object Consolidation, pp. 7–12 (2003)
3. Bilenko, M., Mooney, R.J.: Adaptive duplicate detection using learnable string similarity measures. In: Proceedings of the 9th ACM SIGKDD International Conference on Knowledge Discovery and Data Mining, pp. 39–48 (2003)
4. Bishop, C.: Pattern Recognition and Machine Learning. Springer, New York (2006). pp. 261–267
5. Bromley, J., Bentz, J.W., Bottou, L., Guyon, I., LeCun, Y., Moore, C., Säckinger, E., Shah, R.: Signature verification using a "siamese" time delay neural network. In: Advances in Neural Information Processing Systems, pp. 737–744 (1993)
6. Chopra, S., Hadsell, R., LeCun, Y.: Learning a similarity metric discriminatively, with application to face verification. In: Proceedings of the IEEE Conference on Computer Vision and Pattern Recognition, vol. 1, pp. 539–546 (2005)
7. Duchi, J., Hazan, E., Singer, Y.: Adaptive subgradient methods for online learning and stochastic optimization. J. Mach. Learn. Res. **12**, 2121–2159 (2011)
8. Glorot, X., Bengio, Y.: Understanding the difficulty of training deep feedforward neural networks. In: Proceedings of the 13th International Conference on Artificial Intelligence and Statistics, vol. 9, pp. 249–256 (2010)
9. Glorot, X., Bordes, A., Bengio, Y.: Deep sparse rectifier neural networks. In: Proceedings of the 14th International Conference on Artificial Intelligence and Statistics, vol. 15, p. 275 (2011)
10. Hinton, G., Deng, L., Yu, D., Dahl, G.E., Mohamed, A.R., Jaitly, N., Senior, A., Vanhoucke, V., Nguyen, P., Sainath, T.N., et al.: Deep neural networks for acoustic modeling in speech recognition: the shared views of four research groups. IEEE Sig. Process. Mag. **29**(6), 82–97 (2012)
11. Hu, J., Lu, J., Tan, Y.P.: Discriminative deep metric learning for face verification in the wild. In: Proceedings of the IEEE Conference on Computer Vision and Pattern Recognition, pp. 1875–1882 (2014)
12. Kingma, D., Ba, J.: Adam: a method for stochastic optimization. arXiv preprint arXiv:1412.6980 (2014)
13. Krizhevsky, A., Sutskever, I., Hinton, G.E.: Imagenet classification with deep convolutional neural networks. In: Advances in Neural Information Processing Systems, pp. 1097–1105 (2012)
14. Oyama, S., Manning, C.D.: Using feature conjunctions across examples for learning pairwise classifiers. In: Boulicaut, J.-F., Esposito, F., Giannotti, F., Pedreschi, D. (eds.) ECML 2004. LNCS (LNAI), vol. 3201, pp. 322–333. Springer, Heidelberg (2004). doi:10.1007/978-3-540-30115-8_31
15. Srivastava, N., Hinton, G.E., Krizhevsky, A., Sutskever, I., Salakhutdinov, R.: Dropout: a simple way to prevent neural networks from overfitting. J. Mach. Learn. Res. **15**, 1929–1958 (2014)

16. Taigman, Y., Yang, M., Ranzato, M., Wolf, L.: Deepface: closing the gap to human-level performance in face verification. In: Proceedings of the IEEE Conference on Computer Vision and Pattern Recognition, pp. 1701–1708 (2014)
17. Tran, H.N., Huynh, T., Do, T.: Author name disambiguation by using deep neural network. In: Nguyen, N.T., Attachoo, B., Trawiński, B., Somboonviwat, K. (eds.) ACIIDS 2014. LNCS (LNAI), vol. 8397, pp. 123–132. Springer, Cham (2014). doi:10.1007/978-3-319-05476-6_13
18. Vapnik, V.: The Nature of Statistical Learning Theory. Springer, New York (2000). p. 30

Feature Ranking of Large, Robust, and Weighted Clustering Result

Mirka Saarela[(✉)], Joonas Hämäläinen, and Tommi Kärkkäinen

Department of Mathematical Information Technology,
University of Jyväskylä, P.O. Box 35 (Agora), 40014 Jyväskylä, Finland
{mirka.saarela,joonas.k.hamalainen,tommi.karkkainen}@jyu.fi

Abstract. A clustering result needs to be interpreted and evaluated for knowledge discovery. When clustered data represents a sample from a population with known sample-to-population alignment weights, both the clustering and the evaluation techniques need to take this into account. The purpose of this article is to advance the automatic knowledge discovery from a robust clustering result on the population level. For this purpose, we derive a novel ranking method by generalizing the computation of the Kruskal-Wallis H test statistic from sample to population level with two different approaches. Application of these enlargements to both the input variables used in clustering and to metadata provides automatic determination of variable ranking that can be used to explain and distinguish the groups of population. The ranking method is illustrated with an open data and then, applied to advance the educational knowledge discovery from large-scale international student assessment data, whose robust clustering into disjoint groups on three different levels of abstraction was performed in [19].

Keywords: Population analysis · Kruskal-Wallis test · Robust clustering · Educational knowledge discovery

1 Introduction

Various large-scale educational assessments, like the Programme for International Student Assessment (PISA), regularly collect large amount of data characterizing worldwide student populations to assess and compare arrangements and policies between different educational systems [16]. Although data originating from these assessments are of high quality and publicly available, there is surprisingly little research activity on the secondary analysis. This is due to the technical complexities within the different representations and transformations of data and the lack of methods that allow advanced analysis of these large datasets [18]. One example of the complication of analyzing PISA datasets are the weights. Through complex sampling designs only certain students of the studied population are selected for the assessment and weights are used to indicate the number of students in the population that a sampled student represents. This means that these weights must be taken into account in all steps of the

© Springer International Publishing AG 2017
J. Kim et al. (Eds.): PAKDD 2017, Part I, LNAI 10234, pp. 96–109, 2017.
DOI: 10.1007/978-3-319-57454-7_8

knowledge discovery to analyze the population instead of the collected sample (e.g., [14,20]).

The purpose of this paper is to advance the educational knowledge discovery from a robust, weighted clustering result. There exists various clustering methods and approaches, like e.g. density-based, probabilistic, grid-based, and spectral clustering [2], together with their comparisons and evaluations (e.g., [6]). Although hierarchical methods allow summarization and exploration of a given dataset through the visual dendrogram, the basic form of the technique is not scalable to large number of observations because of the pairwise distance matrix requirement [25]. Moreover, it is not clear how to take into account the weights in hierarchical clustering as presented, e.g., in PISA datasets. On the other hand, in [3] a robust (cf. [24]) prototype-based clustering algorithm was developed that can handle large datasets with high and unknown sparsity patterns (i.e., tens of percents of missing values). This paper continues the efforts of [19], where the weighted enlargement of the above-mentioned algorithm was applied to create prototypes for the PISA 2012 dataset on three different levels of abstraction, with different numbers of clusters of the student population. The dynamic numbers of clusters were based on the use of multiple cluster indices (e.g., [13]) suggesting the number of clusters, again taking into account the weights (see [19] for details).

One main advantage of crisp, prototype-based clustering result is the guarantee of globally separable subsets of data. The data division is completely determined by the disjoint labels, typically integers from 1 to K for K clusters, encoding the clustering result. This means that, in order to make an interpretation of the result, one can consider and compare data distributions of both the actual variables used in clustering as well as relevant metadata. Note that the use of a hierarchical clustering method with locally greedy aggregation could produce clusters of arbitrary shape in the data space, which could then be difficult or even impossible to interpret because of the overlapping variable distributions.

The results in [19] were obtained with a robust clustering method with (available data) spatial median as the cluster prototype, which is characterized by the Laplace density distribution. A feature selection approach for the robust EM-algorithm with Laplace mixture models was suggested in [5]. There the feature selection, similarly to the construction of classifiers [11], referred to ranking the given input features to select the most important ones for the clustering result. Here, our purpose is, similarly to the techniques proposed in [4,23], to assess the importance of variables with a given labeling. For this purpose, we apply the same method as in [5] where it was suggested that the feature ranking can be realized by Kruskal-Wallis (KW) statistical test. More precisely, the estimate of importance of a random variable with clustering provided labeling is supplied by the H statistics of the KW test [15], without need to compute the p-values and perform the actual statistical testing. To omit the hypothesis testing relaxes both the requirements of the KW test concerning the equal variances [15] and selection of appropriate distribution for the test statistics [21]. Moreover, because KW is a univariate method, it is easy to restrict the computation of the test statistic to the available values of a variable. This means utilizability with an arbitrary sparsity pattern.

Hence, one needs to generalize the KW H into the population level by using the weights. This is a difficult problem in statistics because of the reliance of KW on data ranking. After an extensive search for relevant literature and knowledge we were able to identify one related work generalizing KW [1], but not solving the problem at hand. The only article that was identified as fully relevant was [22], which suggested a very natural generalization of KW for *integer weights*: create univariate data to compute the KW test statistic, where each observation is copied as many times as the integer weight suggests. Clearly, we then precisely test the target population and not the sample. The purpose of this paper is to propose an approximate extension of this approach to real-valued weights, by utilizing the classical bootstrapping [8], and to compare this to an analytically derived novel heuristic formula. Both of these approaches are tested and evaluated with two different existing clustering results from [19], when ranking both actual input variables and selected set of metadata variables.

2   On PISA Data

The collected data of each PISA assessment, which since 2000 is conducted every three years, can be downloaded from the website[1] of the Organisation of Economical and Cultural Development (OECD). To select a reliable sample of the population, which in PISA are all 15-year-old students within the participating countries, the OECD applies a two-stage sampling design: First, schools attended by 15-year-old students are assigned to mutually exclusive groups based on explicit strata and schools from these groups are selected with probabilities proportional to their size. Then, students within those school are selected randomly with equal probability. The weight w_i assigned to each participating student i consists of the school base weight, the within-school base weight, and five adjustment factors, especially the one which compensates the non-participation of a sampled student [17]. Students that are sampled for the PISA test are asked to show their proficiencies in a cognitive test and answer a background questionnaire, which gathers information about demographics, activities, and attitudes of the students.

Table 1 details all PISA 2012 variables used in this study. The left-hand side of the table shows all the variables that in [19] were clustered on a population-level. The *ESCS* combines all information of the PISA background questionnaire that relate to the students' economic, social and cultural situation. The next five variables on the left-hand side of Table 1 are generally associated with the students' success in the PISA cognitive test, and the remaining nine variables relate directly to the students' mathematics performance, which was the main assessment area in PISA 2012. All of these 15 variables are so-called PISA scale indices that summarize many of the original questions in the students' background questionnaires by employing the Rasch model [17]. Since only a subset of all test item

---

[1] https://www.oecd.org/pisa/pisaproducts/.

Table 1. PISA variables used in this study with the original variables (i.e., the data that was used for clustering) on the left-hand side and metadata (i.e., additional PISA variables used to explain the clustering result) on the right-hand side.

PISA data used for clustering		PISA metadata	
Variable	ID	Variable	ID
Economic, social and cultural status	ESCS	ICT availability at home	ICTHOME
Sense of belonging	BELONG	ICT availability at school	ICTSCH
Attitude towards school: learning outcome	ATSCHL	ICT entertainment use	ENTUSE
Attitude towards school: learning activities	ATTLNACT	ICT use at home for school-related tasks	HOMSCH
Perseverance	PERSEV	Use of ICT at school	USESCH
Openness to problem solving	OPENPS	Use of ICT in math lessons	USEMATH
Self-responsibility for failing in math	FAILMAT	Positive attitudes towards computers	ICTATTPOS
Interest in mathematics	INTMAT	Positive attitudes towards computers	ICTATTPOS
Instrumental motivation to learn math	INSTMOT	Plausible values 1–5 in mathematics	PVMATH
Self-efficacy in mathematics	MATHEFF	Plausible values 1–5 in reading	PVREADING
Anxiety towards mathematics	ANXMAT	Plausible values 1–5 in science	PVSCIENCE
Self-concept in math	SCMAT		
Behaviour in math	MATBEH		
Intentions to use math	MATINTFC		
Subjective norms in math	SUBNORM		

are allocated to each student (this is called rotated design), around one third of the values for these 15 variables are missing.

On the right-hand side of Table 1, the meta-variables to be used in this study are listed. The first eight variables of general interest are all PISA scale indices that were computed to summarize the information obtained from the ICT questionnaire, which assessed the students' computing availability and familiarity as well as their attitudes towards computers. The next and last set of variables in Table 1 are the plausible values (PVs) for each assessment domain (mathematics, reading, and science). PISA does not provide individual test performance scores. Instead, to reliably assess the proficiencies of populations, five PVs for each assessment domain are estimated with Bayesian statistics and reported for each student. Note that we have allocated only one line in the table per assessment domain for the three sets of PVs but there are five single PVs vectors per assessment domain, i.e., 15 PVs altogether, that are used in the analysis.

The PVs are random draws from the Bayesian posterior distribution of a student's ability. In PISA, the prior distribution is a population model that is estimated with a latent regression model. This latent regression computes the average proficiencies of examinee subgroups given evidence about the distribution and associations of collateral variables in the data. In PISA 2012, these collateral variables included to the latent regression model were all available student-level information besides their performance in the cognitive test [17, p. 157]. That means, in particular, that also all variables listed in Table 1 except the 15 PVs themselves have been used to estimate the PVs, and therefore, the PVs cannot be seen totally independent of them. The likelihood of the success in test is a Rasch model, where the probability of success is a logistic function of the latent ability and some parameters (e.g. difficulties) of the test items. The obtained posterior distribution of a student's ability is specific for each student, since each student has different values of background variables and test results.

To sum up, student proficiencies in PISA are not directly observed. The PVs are estimates for group performance and only a selection of likely proficiencies for students that attained each score. Moreover, for the study at hand, it is important to note that all background information (i.e., all data that were clustered and all metadata except the PVs themselves) have been used in the latent regression model which contributes to the posterior distribution from which the PVs are drawn from.

3   Methods and Formulations

Let $\{x_i\}_{i=1}^N$ be a given, multidimensional dataset, where N observations $x_i \in \mathbb{R}^n$ are given. Assume further that a given set of positive, real-valued weights $\{w_i\}_{i=1}^N$ is also given. Moreover, assume that there is a set of missing values in $\{x_i\}$ with unknown sparsity pattern. To identify this pattern, define the projection vectors $p_i, i = 1, \ldots, N$, that capture the existing variable values:

$$(p_i)_j = \begin{cases} 1, \text{if } (x_i)_j \text{ exists}, \\ 0, \text{otherwise}. \end{cases} \tag{1}$$

3.1   Robust, Prototype-Based Clustering Method for Weighted Sparse Data

Let us briefly recapitulate the clustering method and the overall approach that was used hierarchically in [19], to produce three levels of disjoint clusters of PISA 2012 population with 2, 8, and 53 clusters, respectively.

The spatial median clustering algorithm, k-SpatMeds, proceeds similarly to any prototype-based method: first, an initial set of *complete* (i.e., no missing values) prototypes is created and second, these are refined by iteratively linking observations to the closest prototype whose value is then recomputed. The algorithm stops when there are no more changes in the linking. Mathematically, the score function that is locally minimized via the search procedure reads as follows:

$$\mathcal{J}_w = \sum_{j=1}^{K} \sum_{i=1}^{n_j} w_i \|\text{Diag}\{p_i\}(x_i - c_j)\|_2. \tag{2}$$

Here, Diag transforms a vector into a diagonal matrix. The latter sum is computed over the subset of data attached to the jth cluster. One observes from (2) that to take into account the first-order alignment of the sample data with the corresponding population is straightforward. Moreover, projection of the Euclidean distance between the observation and the prototype to available values creates an implicit (secondary) weighting that favors more complete observations over the sparser ones in cluster creation. Algorithmically, one still needs to check that the iterative refinement of the prototypes does not introduce missing values to them, because the resulting set of cluster prototypes $\{c_i\}_{i=1}^{K}$ should be complete to allow proper interpretation. The robustness of this algorithm as thoroughly described and tested in [3], refers to the tolerance of both missing values and noisy data. To this end, one can apply the k-SpatMeds algorithm hierarchically to refine a set of disjoint clusters further.

3.2   Construction of Test Statistic for Kruskal-Wallis with Weights

Next we describe two different approaches to estimate the test statistic H of the KW rank-test with real-valued weights. Because the KW test is univariate, we can restrict ourselves to univariate random variable.

Integer Approximation with Bootstrapping. Let $\{x_i, l_i\}_{i=1}^{N}$ be the pairs of a univariate observation $x_i \in \mathbb{R}$ and its cluster-indicating label $l_i \in \mathbb{N}$, where $1 \leq l_i \leq K$ for K denoting the number of clusters/groups. Let $n_k = |C_k| = \{i \in \mathbb{N} \mid l_i = k\}$ determine the size of cluster C_k. The original formula for the KW H is given by [15]

$$H = \frac{12}{N(N+1)} \sum_{k=1}^{K} \frac{s_k^2}{n_k} - 3(N+1), \tag{3}$$

where r_i denotes the *rank* of observation x_i in global sorting and $s_k = \sum_{i \in C_k} r_i$ the sum of ranks in cluster C_k. When there are equal values (ties) in data, one can compute the mean rank of equal observations and share this value among the ties.

As described, $w_i \in \mathbb{R}$ measures the amount of population that the ith observation represents. If all w_i's are integers, then in [22] it was proposed how to modify the basic KW test: rank a derived dataset representing the whole population, where each (available) observation is copied as many times as the weight suggests. This approach is referred from now on as *Integerweighted-KW, IW-KW*. Note that when such an enlarged data are ranked we end up with multiple ties whose mean ranks are then shared. In the following, we describe a novel approach how to approximate this integer-weighted KW using a bootstrapping technique.

Let w denote an arbitrary, real-valued weight. The proposed technique is, firstly, based on approximating w up to an accuracy of the first decimal place. This can be simply done as follows: determine the two integers $w_l = \lfloor w \rfloor$ and $w_h = \lceil w \rceil$ that provide lower and upper bound of w as integers. Let then $d = [10 * (w - w_l)]$ be the rounded integer that encapsulates the decimal place 1 of w. Vector v of ten integers, which is created by repeating w_l $10 - d$ times and w_h d times, provides an integer-approximating set of real-valued w in such a way that the mean of v is exactly the same as w up to the first decimal. For instance, for $w = 8.647$, $w_l = 8, w_h = 9$, and $d = 6$. And, for $v = [8\,8\,8\,8\,9\,9\,9\,9\,9\,9]$, we have $mean\{v\} = 8.6$. Similarly, in order to create an integer-approximation of w being accurate to the second decimal place, it is enough to just redefine $d = [100 * (w - w_l)]$. Proceeding with the example just given, the integer vector of size 100 with 65 nines and 35 eights would yield to $mean\{v\} = 8.65$. For the general procedure, the result of the just proposed integer approximation of all weights is stored in the matrix $\mathbf{W} \in \mathbb{N}^{N \times D}$, where D is 10 when approximating the first decimal place and 100 for the second decimal place, correspondingly.

Next we suggest to use the classical bootstrapping [8] to create a set of KW test statistics based on the IW-KW and W. Hence, we create a random sample of indices $\{1, \ldots, N\}$ with replacement, and for the resulting unique set of indices $\tilde{I}$, for the available values of $\{x_i\}_{i \in \tilde{I}}$, we apply IW-KW. When this is repeated D times for all the integer columns of W, we obtain D different samples of the bootstrap estimate of the KW H. To this end, similarly as with the derivation of W, we then simply take the mean of the D-vector to produce the final approximation of H for the real-valued weights.

Analytic Formula. Let $\bar{r}$ denote the global mean rank (equal to $\frac{1+N}{2}$) and $\bar{r}_k$ the mean rank of the observations in cluster C_k. An equivalent form of the original formula (3) for the KW test statistic H, as given in [9], reads as

$$H = (N - 1)\frac{\sum_{k=1}^{K} n_k(\bar{r}_k - \bar{r})^2}{\sum_{i=1}^{N}(r_i - \bar{r})^2}. \tag{4}$$

From this form, it is easy to derive an interpretation of the KW test statistic. With clusterwise $\bar{r}_k$ and global $\bar{r}$ mean ranks, the dividend presents sum of clusterwise variances multiplied by the size of the cluster whereas the divisor computes the global variance of ranks. Hence, when the weights represent the number of samples in the population, it is straightforward to derive an analogous formula to (4) in the population level. Hence, let $\bar{r}_w = \frac{\sum_{i=1}^{N} w_i r_i}{\sum_{i=1}^{N} w_i}$ be the weighted average rank and $(\bar{r}_w)_k$ the weighted average rank of cluster C_k. Then, we define

$$H_w = \frac{\sum_{k=1}^{K}(\sum_{i \in C_k} w_i)((\bar{r}_w)_k - \bar{r}_w)^2}{\sum_{i=1}^{N} w_i(r_i - \bar{r}_w)^2}. \tag{5}$$

Note that we have omitted the multiplier $(N - 1)$ from (4), which would be generalized into $(\sum_i w_i - 1)$ to represent the whole population. With PISA 2012 weights, which align the half a million students sample to the 24 million population, this means we do not include multiplication of H_w by over 24 million. Because the final ranking of variables, as suggested in [5], is based on sorting the H values of the variables in descending order, this omission does not change the result.

4   Evaluation

Implementation. We computed the KW rank-test H test statistics for real-value weighted data with two approaches, as described in Sect. 3. The bootstrapping with the IW-KW was tested with two different Ws. We will refer to the bootstrapping based method as Bootstrap KW. Further, Bootstrap KW with $D = 10$ refers to the one decimal place approximation of real-valued weights. Similarly, the two decimal place approximation is referred as Bootstrap KW with $D = 100$. In addition, the KW test statistics were computed directly from formula (5). In the following, this is shortly referred as Analytic KW. The two clustering results that are used in the experiments corresponded to 8 (*Labels 1*) and 53 (*Labels 2*) clusters from [19] in the second and third levels of refinement, respectively. The first result in [19] with the two clusters is excluded here, since the KW rank-test exactly generalizes the Mann-Whitney U-test for the two groups.

To speed up the computations, we implemented a parallel version of Bootstrap KW with Matlab PCT, SPMD blocks and message passing functions. The tests were run in Matlab 8.5.0 environment by using a cluster of 8 nodes. Each node consists of Intel Xeon CPU E7-8837 with 8 cores and 128 GB RAM. Each worker in the distributed computations corresponds to one of the 64 cores. Since Bootstrap KW computes the KW H values independently for each variable in a loop, those loop iterations can be easily parallelized with SPMD blocks. First, each worker reads one column of variable values from the data matrix and the corresponding sparsity indicator (1). Next, each worker computes the KW H values by utilizing its local data. Finally, results are aggregated and rankings for the variables based on the H values are formed. The number of workers is equal to the number of variables in all parallel runs.

The five individual PVs for mathematics, reading, and science, as given in Table 1, were first treated as independent variables, such that five H values were computed for them. The final value of the test statistic was then taken as the mean of these according to the recommended way of analysis in [17].

Results. To generally test the proposed approaches, we first used the Iris data from UCI machine-learning repository. For this, we created random integer weights in the range 5–25 and newly generated the data for each run. The KW H values for Analytic KW and Bootstrap KW $D = 100$ approaches gave the same variable ranking results in eight out of ten runs. After adding 5% zero-mean

Table 2. Rankings for full (original and metadata) variables for the different analysis approaches for both PISA clustering results.

Variable	Labels 1			Labels 2			Rank of rankings
	Analytic KW	Bootstrap KW		Analytic KW	Bootstrap KW		
		$D = 10$	$D = 100$		$D = 10$	$D = 100$	
ESCS	3	1	1	1	1	1	1
BELONG	11	13	13	9	13	13	12
ATSCHL	7	6	6	7	7	7	6
ATTLNACT	4	3	3	4	2	2	3
PERSEV	15	15	15	15	16	16	15
OPENPS	12	11	11	11	11	11	11
FAILMAT	20	18	18	17	18	18	19
INTMAT	1	2	2	3	3	3	2
INSTMOT	5	5	5	5	6	6	5
MATHEFF	9	9	9	10	12	12	9
ANXMAT	6	7	7	6	8	8	7
SCMAT	2	4	4	2	4	4	4
MATHBEH	14	14	14	12	9	9	13
MATINTFC	8	8	8	8	5	5	8
SUBNORM	13	10	10	13	10	10	10
ICTHOME	10	19	19	14	19	19	17
ICTSCH	25	24	24	25	25	25	25
ENTUSE	24	22	22	24	22	22	22
HOMSCH	22	21	21	23	21	21	21
USESCH	16	26	26	18	26	26	23
USEMATH	26	23	23	26	23	23	24
ICTATTPOS	21	20	20	21	20	20	20
ICTATTNEG	23	25	25	22	24	24	26
PVMATH	17	12	12	16	14	14	14
PVREADING	19	17	17	20	17	17	18
PVSCIENCE	18	16	16	19	15	15	16

uniformly distributed noise to make weights real-values, we obtained the same ranking order for the different approaches in nine out of ten runs. Moreover, similarly as in [7], features 4 and 3 were always selected as the important ones while features 1 and 2 were always last in the list. When we used the same data for each run the ranking order was always the same.

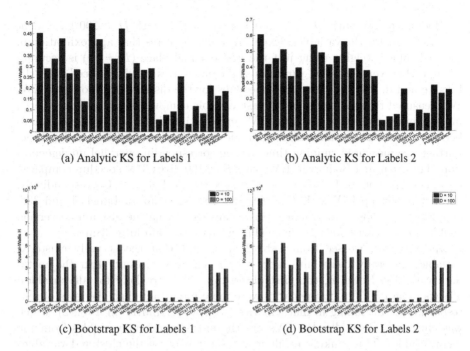

(a) Analytic KS for Labels 1   (b) Analytic KS for Labels 2

(c) Bootstrap KS for Labels 1   (d) Bootstrap KS for Labels 2

Fig. 1. KW H values for two clustering results for the combined (originally clustered and meta) PISA data determined with the analytic and the two bootstrap KW approaches.

Table 2 summarizes all ranking for the combined (originally clustered and meta) PISA data. In the table, the last column *rank of rankings* indicates for each variable the total rank, i.e. the rank of the sum of rankings of all methods on both labeling levels. KW H values for both clustering results are shown in Fig. 1. As can be seen from Table 2, variable rankings between the analytic and the bootstrapped results are highly similar with the exception that variable *USESCH* had a ranking difference 10 for Labels 1 and ranking difference 8 for Labels 2. In addition, variable *ICTHOME* had ranking difference 9 for Labels 1 and ranking difference 5 for Labels 2.

The Kendall's tau distance (see [10]) provides a way to compute distance between two ranking lists with an equal set of variables. The Kendall's tau distance is equal to the bubble sort algorithm steps to convert one list to the same order as the other one. If m is the number of elements in the list, then the maximum value for the Kendall's tau distance is $m(m-1)/2$ which is typically used to normalize this distance metric. Thus, the Kendall's tau distance is limited to an interval $[0, 1]$, where value 0 refers to the identical lists and value 1 to the case where one list is the reverse of the other list. The Kendall's tau distances between the Analytic KW and Bootstrap KW with $D = 100$ were 0.1015 for Labels 1 and 0.1138 for Labels 2. This concludes that, overall, the rankings are highly similar as measured by the Kendall's tau distance.

Bootstrap KW with $D = 10$ and Bootstrap KW with $D = 100$ gave identical rankings for the variables. Experimentally, it seems that approximation of the real-valued weights using just the first decimal place ($D = 10$) is accurate enough. However, for a few variables slight differences can be noticed from the Figs. 1c and d. We also computed speedups for the distributed Bootstrap KW. We measured running time for the first variable computations by using a serial implementation of the Bootstrap KW, and multiplied this with the total number of variables to get an estimate for the serial implementation running time. Further, we measured running time for the corresponding parallel implementation. Thus, parallel Bootstrap KW with $D = 100$ gives 34× speedup compared to sequential code for Labels 1 and 35× speedup for Labels 2. Correspondingly, parallel Bootstrap KW with $D = 10$ gives 28× speedup for Labels 1 and 33× speedup for Labels 2. In practice, this means that using the distributed version enables one to carry out the whole cluster analysis chain in realtime.

As expected, we see from Table 2 and Fig. 1 that the actually clustered variables generally contribute more to the clustering result than the metadata variables. However, this first observation does not hold for all variables: The metadata PVs in mathematics were more important than the level of self-responsibility for failing in mathematics (see row *FAILMAT* in Table 2), which was clustered. Generally, the PVs are the most important variables from the metavariables. This ranking result makes sense because the clustered variables are, as explained in Sect. 2, part of the posterior model from which the PVs were sampled. Moreover, most of the clustered variables are directly associated with the students' mathematics proficiencies. Hence, the PVs in mathematics should be important variables when explaining the clustering result and, thus, these observations support the validity of our results.

As can be seen in Table 2, the students' *ESCS* is the most important variable determining the different clusters. This was already assumed in [19] where the most distinguishing country clusters were those that showed different stages of development. Moreover, the students' *ESCS* is the single variable in the whole PISA data, which accounts for most of the variance in performance [16]. Therefore, it is reasonable to assume that the variable that explains the mathematics proficiency the most, is also the most important when variables associated with the mathematics performance, are clustered. The students' *ESCS* takes not only the highest parental education and occupation into account but also the students' home possessions. Therefore, the *ICTHOME*, which summarizes the home possessions in the ICT area, is partly associated with the students' *ESCS* [17, p. 132]. Hence, it seems reasonable that *ICTHOME* is next to the PVs one of the most important variables from the metadata (see Table 2).

To sum up, weighted enlargements with all approaches proposed in Sect. 3 successfully enabled ranking of input and metadata. Triangulation for both actual input and metadata by using two clustering results of a PISA dataset and two different algorithms/formulae showed very similar results for all methodological approaches and also for the two clustering results that were analyzed. Hence, it seems that the interpretation is not an artifact of the method used to analyze the data or only a result of the particular sample, but reflects genuine and overarching aspects of the data [12].

5   Discussion and Conclusions

Large-scale educational assessment data provide interesting and high quality resources for educational knowledge discovery. Although the data from these assessments are made available to the public a scarce pool of research outcomes exist that make use of those rich datasets because of the technical difficulties in them. Only one study [19] was identified, in which the whole PISA 2012 contextual data were clustered by taking the complexities of these data (especially the sparsity and the weights) into account. However, the work in [19] lacked a clear frame how to assess the importance of individual variables to interpret the clustering results.

In this study, we proposed weighted enlargements of the KW H test with different approaches, which as an independent statistical problem is not trivial. All approaches successfully enabled ranking of input and metadata. In particular, when applied to the two clustering results in [19], all approaches supported the finding that the students' *ESCS* is the most important variable determining the clusters—a fact that was also hypothesized in [19] but could not be statistically shown in there. Moreover, also the ranking of the other variables seem to support the interpretations made in [19].

The y-scales of Figs. 1c and d illustrate the very large size of the KW test statistic(s) H for a large population, which in our case is characterized by over 24 million students worldwide. Hence, even if the nonparametric KW test can be used for testing large samples [9], the actual hypothesis testing seems practically useless. We tested the computation of the p-values for the original sample, for both clustering results and for all data and metadata variables, and found in each case that the p-value was equal to zero up to six decimal places. Hence, the hypothesis test itself does not provide any useful information for educational knowledge discovery.

Based on the high similarity of the results of the different ranking approaches, we suggest the direct KW formula with weights to be used for quick evaluation of significance of a variable on the population level. If the weighted estimates are used to derive, e.g., confidence intervals for the test statistics and the resulting rankings, the bootstrap-based approach should be used. This approach is also better aligned to the existing literature [5, 8, 22]. To this end, we conclude that the proposed approach supports quantified educational knowledge discovery from PISA and similar large-scale educational datasets.

Acknowledgments. The authors would like to thank Ph.D. Salme Kärkkäinen for her kind and valuable suggestion to use bootstrapping.

References

1. Acar, E.F., Sun, L.: A generalized Kruskal-Wallis test incorporating group uncertainty with application to genetic association studies. Biometrics **69**(2), 427–435 (2013)

2. Aggarwal, C.C., Reddy, C.K.: Data Clustering: Algorithms and Applications. CRC Press, Boca Raton (2013)
3. Äyrämö, S.: Knowledge Mining Using Robust Clustering. Jyväskylä Studies in Computing, vol. 63. University of Jyväskylä, Jyväskylä (2006)
4. Ceccarelli, M., Maratea, A.: Assessing clustering reliability and features informativeness by random permutations. In: Apolloni, B., Howlett, R.J., Jain, L. (eds.) KES 2007. LNCS (LNAI), vol. 4694, pp. 878–885. Springer, Heidelberg (2007). doi:10.1007/978-3-540-74829-8_107
5. Cord, A., Ambroise, C., Cocquerez, J.P.: Feature selection in robust clustering based on Laplace mixture. Pattern Recogn. Lett. 27(6), 627–635 (2006)
6. Crabtree, D., Andreae, P., Gao, X.: QC4 - a clustering evaluation method. In: Zhou, Z.-H., Li, H., Yang, Q. (eds.) PAKDD 2007. LNCS (LNAI), vol. 4426, pp. 59–70. Springer, Heidelberg (2007). doi:10.1007/978-3-540-71701-0_9
7. Dash, M., Liu, H.: Feature selection for clustering. In: Terano, T., Liu, H., Chen, A.L.P. (eds.) PAKDD 2000. LNCS (LNAI), vol. 1805, pp. 110–121. Springer, Heidelberg (2000). doi:10.1007/3-540-45571-X_13
8. Efron, B.: Bootstrap methods: another look at the jackknife. Ann. Stat. 7, 1–26 (1979)
9. Elamir, E.A.: Kruskal-Wallis test: a graphical way. Int. J. Stat. Appl. 5(3), 113–119 (2015)
10. Fagin, R., Kumar, R., Sivakumar, D.: Comparing top K lists. In: Proceedings of the Fourteenth Annual ACM-SIAM Symposium on Discrete Algorithms, pp. 28–36. Society for Industrial and Applied Mathematics (2003)
11. Fung, P.C.G., Morstatter, F., Liu, H.: Feature selection strategy in text classification. In: Huang, J.Z., Cao, L., Srivastava, J. (eds.) PAKDD 2011. LNCS (LNAI), vol. 6634, pp. 26–37. Springer, Heidelberg (2011). doi:10.1007/978-3-642-20841-6_3
12. Gifi, A.: Nonlinear Multivariate Analysis. Wiley, Hoboken (1991)
13. Kim, Y., Lee, S.: A clustering validity assessment index. In: Whang, K.-Y., Jeon, J., Shim, K., Srivastava, J. (eds.) PAKDD 2003. LNCS (LNAI), vol. 2637, pp. 602–608. Springer, Heidelberg (2003). doi:10.1007/3-540-36175-8_60
14. Koskela, A.: Exploring the differences of Finnish students in PISA 2003 and 2012 using educational data mining. Jyväskylä Studies in Computing. University of Jyväskylä, Jyväskylä (2016)
15. Kruskal, W., Wallis, W.: Use of ranks in one-criterion variance analysis. J. Am. Stat. Assoc. 47(260), 583–621 (1952)
16. OECD: PISA 2012 Results: Excellence Through Equity: Giving Every Student the Chance to Succeed (Volume II). PISA, OECD Publishing (2013)
17. OECD: PISA 2012 Technical report. OECD Publishing (2014)
18. Rutkowski, L., Rutkowski, D.: Getting it "better": the importance of improving background questionnaires in international large-scale assessment. J. Curric. Stud. 42(3), 411–430 (2010)
19. Saarela, M., Kärkkäinen, T.: Do country stereotypes exist in PISA? A clustering approach for large, sparse, and weighted data. In: Proceedings of the 8th International Conference on Educational Data Mining, pp. 156–163 (2015)
20. Saarela, M., Kärkkäinen, T.: Weighted clustering of sparse educational data. In: Proceedings of the European Symposium on Artificial Neural Networks, Computational Intelligence and Machine Learning, pp. 337–342 (2015)
21. Spurrier, J.D.: On the null distribution of the Kruskal-Wallis statistic. Nonparametric Stat. 15(6), 685–691 (2003)

22. Tölgyesi, C., Bátori, Z., Erdős, L.: Using statistical tests on relative ecological indicator values to compare vegetation units-different approaches and weighting methods. Ecol. Ind. **36**, 441–446 (2014)
23. Verde, R., Lechevallier, Y., Chavent, M.: Symbolic clustering interpretation and visualization. Electron. J. Symbolic Data Anal. **1**(1), 1 (2003)
24. Yang, H., Zhao, D., Cao, L., Sun, F.: A precise and robust clustering approach using homophilic degrees of graph kernel. In: Bailey, J., Khan, L., Washio, T., Dobbie, G., Huang, J.Z., Wang, R. (eds.) PAKDD 2016. LNCS (LNAI), vol. 9652, pp. 257–270. Springer, Cham (2016). doi:10.1007/978-3-319-31750-2_21
25. Zaki, M.J., Meira Jr., W.: Data Mining and Analysis: Fundamental Concepts and Algorithms. Cambridge University Press, Cambridge (2014)

Purchase Signatures of Retail Customers

Clement Gautrais[1]([✉]), René Quiniou[2], Peggy Cellier[3], Thomas Guyet[4], and Alexandre Termier[1]

[1] University of Rennes 1, IRISA, Rennes, France
clement.gautrais@irisa.fr
[2] Inria Rennes, IRISA, Rennes, France
[3] INSA Rennes, IRISA, Rennes, France
[4] Agrocampus Ouest, IRISA, Rennes, France

Abstract. In the retail context, there is an increasing need for understanding individual customer behavior in order to personalize marketing actions. We propose the novel concept of *customer signature*, that identifies a set of important products that the customer refills regularly. Both the set of products and the refilling time periods give new insights on the customer behavior. Our approach is inspired by methods from the domain of sequence segmentation, thus benefiting from efficient exact and approximate algorithms. Experiments on a real massive retail dataset show the interest of the signatures for understanding individual customers.

1 Introduction

Retail, and more specifically understanding the behavior of supermarket customers, has been a strong motivation for data mining researchers since the early 1990s. Several methods have been developed in this field, such as mining frequent itemset [1], frequent sequential patterns [2] or more recently high utility itemsets [3]. These methods discover sets of products that are bought together in a large enough number of tickets, possibly with some extra information (e.g. sequencing, utility). They can be exploited to understand (large) *groups of customers*. However, with the success of loyalty programs and the increasing number of customers shopping at online grocery pick-up, a promising trend is "personalized marketing". This requires a fine grained understanding of the purchasing behavior of individual customers, in order to make relevant personalized suggestions.

In this context of personalized marketing, an important information is the "rhythm" of the individual customer. The main idea is to identify the set of products that the customer always wants to have stocked at home, and that she will thus buy on a more or less regular basis. The rhythm corresponds to the "refilling period". Extracting such information may help analysts to get insights about their customers in order to design personalized marketing campaigns. In practice, the problem is to discover from the set of the customer receipts, a set of products that are regularly purchased. A difficulty is that all the products that

© Springer International Publishing AG 2017
J. Kim et al. (Eds.): PAKDD 2017, Part I, LNAI 10234, pp. 110–121, 2017.
DOI: 10.1007/978-3-319-57454-7_9

the customer wants to have in stock are not likely to be bought at the same time: depending on depletion rates, renewing all such products will be distributed over several receipts.

Recently, Customer Relationship Management (CRM) has manifested interest in data mining [4], but with a focus on clustering techniques, for example to characterize segment profiles [5]. However, clustering cannot uncover the flexible time regularity of customers' purchases since time periods have to be fixed in advance. Most existing itemset mining algorithms [1,6] consider only sets of products that are bought on a single receipt, and cannot be used for this problem. Periodic patterns [7] can find regularities through a sequence of receipts however they extract patterns with strict temporal period. Mannila et al. [8] have proposed parallel episodes to extract temporal regularities but the approach requires fixed predefined equal size windows over the sequence of events, which lacks flexibility for the problem at hand. In [9], Casas-Garriga defined a method that also adapts the window size to the data, however it still requires a maximal time interval between two events.

In this paper, we define the problem of extracting *customer signature* through a sequence of receipts. A customer signature represents a maximal set of products that are bought regularly, possibly in several receipts and such that the regularity is not strict. We show that this problem can be formalized as a sequence segmentation problem. There is an important literature about sequence segmentation. We have adapted the formal setting provided in Bingham's survey [10]. The most significant adaptation lies in the notion of segment representatives that represent occurrences of a common set of products and on the distance from sequence elements to their related representatives. Roughly, we shift from a local error view to a global one (see Sect. 2 for more details).

The contributions of this paper are threefold. First, the signature mining problem is defined as a segmentation problem allowing to take advantage of the many algorithms that have been proposed in the sequence segmentation field (Sect. 3.1). Second, we have adapted and evaluated an algorithm based on dynamic programming for sequence segmentation [11] which gives an exact solution (Sect. 3.2). Third, a thorough experimental study on real massive supermarket data shows the interest of our approach (Sect. 4).

2   Background

This section provides the data mining vocabulary used in the sequel, and presents briefly the well-studied sequence segmentation problem.

In pattern mining, an *itemset* T is a set of literals called *items*. Let $\mathcal{I}$ be the set of all items. A *sequence* α is an ordered list of itemsets, denoted by $\alpha = \langle T_1, \ldots, T_m \rangle$. In the retail context, a receipt is an itemset (a set of purchased products) and a *customer purchase sequence* is a sequence of receipts identifying the products bought by a customer at each of her visits to a supermarket during the analysis period. For instance, $\langle (p_1, p_2)(p_3)(p_1)(p_4, p_2, p_3)(p_1) \rangle$ is a sequence of five receipts where four different products are bought one or several times.

A receipt may have an associated timestamp which indicates the purchase date. We assume that the timestamp of T_k is implicitly the index of T_k, *i.e.* k. By extension, the timestamp of any product of a receipt T_k is the timestamp associated with this receipt T_k. A *customer sequence database* SDB is a set of tuples (C_{id}, α) where C_{id} is a customer identifier and α is the sequence of her receipts.

Our proposal is grounded on sequence or time series segmentation which has received much attention in the literature. In [10], the segmentation problem is formulated as follows. Let $\alpha = \langle T_1, T_2, \ldots, T_n \rangle$ be a d-dimensional sequence where $T_i \in \mathbb{R}^d$. A k-segmentation S of α is a partition of α into k non-overlapping contiguous subsequences called *segments*, i.e. $S = \langle S_1, S_2, \ldots, S_k \rangle$ and $\forall i \in 1 \ldots k$, $S_i = \langle T_{b(i)}, \ldots, T_{b(i+1)-1} \rangle$, where $b(i)$ is the index of the first element of the i-th segment. A segmentation associates a *representative*, $\mu(S_i)$, with each segment by aggregating the values of the segment. Generally $\mu(S_i)$ is a single value such as mean or median, or a pair of values such as (min, max) or (mean, slope). This reduction results in a loss of information in the sequence representation which can be measured by the reconstruction error defined as:

$$E_p(\alpha, S) = \sum_{S_i \in S} \sum_{T \in \alpha} dist(T, \mu(S_i))^p$$

where $dist(T, \mu(s))$ represents the distance between the d-dimensional point T and the representative of the segment it belongs to. The p parameter refers to the Lp norm. In practice, the median ($p = 1$) or the mean ($p = 2$) usually serves as segment representatives. The *segmentation problem* consists in finding the segmentation that minimizes the reconstruction error:

$$S_{opt}(\alpha, k) = \arg \min_{S \in \mathscr{S}_{n,k}} E_p(\alpha, S)$$

where $\mathscr{S}_{n,k}$ represents the set of all k-segmentations of sequences of length n.

3   Mining Signatures

In this section, we present the signature mining problem in the sequence segmentation framework. Indeed, mining a signature from a customer purchase sequence α can be seen as segmenting α into k non-overlapping and non-empty segments that cover all receipts from α and such that every segment contains a common maximal subset of products, called the *customer signature*. The signature mining problem can thus be fitted to the segmentation problem providing the opportunity to use the many exact or approximate algorithms that have been proposed in the sequence segmentation field.

3.1   Mining Signatures with Sequence Segmentation

Section 2 introduced the problem of segmenting a sequence $\alpha = \langle T_1, T_2, \ldots, T_n \rangle$ into k segments. Let $\mathscr{S}_{n,k}$ denote the set of all k-segmentations of a sequence

α of length n and $S = \langle S_1, S_2, \ldots, S_k \rangle$ be an element of $\mathscr{S}_{n,k}$. Following the representation proposed in SPAM [12], a receipt can be represented by a bitmap[1] of dimension d such that if item i_j belongs to the receipt then the j-th bit of the bitmap is set to 1, otherwise the j-th bit is set to 0. The representative r_i of a segment is then defined as the set of items that belongs to at least one receipt in the segment, i.e. the union of the segment receipts. This union can be computed by a boolean disjunction on bitmaps: $r_i = \bigvee_{t \in S_i} t$. The k-signature of a purchase sequence is the set of items that are common to every segments from a segmentation of size k, so it corresponds to the intersection of the k segment representatives. It can be computed by a boolean intersection of the related bitmaps: $Sig_k(\alpha, S) = \bigwedge_{j=1}^{k} r_j$. As we intend to represent a customer purchase sequence by its signature, the reconstruction error is related to the loss of information in the signature. A simple way to estimate the error is to count the items that are not present in the signature, i.e. the number of bits equal to 0 in the bitmap:

$$E_k(\alpha, S) = |\mathcal{I}| - \|Sig_k(\alpha, S)\| = \|\overline{Sig_k(\alpha, S)}\|$$

where $\|X\|$ represents the number of bits equal to 1 in bitmap X and $\overline{X}$ represents the complement of X. The signature of a customer's purchase sequence T is the maximal signature for a segmentation of size k:

$$Sig_k(\alpha) = Sig_k(\alpha, S_{opt}(\alpha, k)), \quad \text{where } S_{opt}(\alpha, k) = \underset{S \in \mathscr{S}_{n,k}}{\arg\max} \|Sig_k(\alpha, S)\|$$

The segmentation size k is given a priori, either as an integer or as a percentage of n, the size of the input sequence (similar to support count and support [1]). The latter is called the *relative number of blocks* denoted by *RNB*.

3.2 Dynamic Programming for Signature Mining by Segmentation

Now, we present an algorithm for computing signatures by sequence segmentation based on Dynamic Programming (DP). This algorithm returns an optimal solution, i.e. a maximal signature.

Bingham [10] presents a formulation of DP for sequence segmentation. It is based on a table A of size $k \times n$ where k is the size of the segmentation and n is the number of itemsets (receipts) in the input sequence α. So, rows of A represent segments and columns represent itemsets of the input sequence. Let $\alpha[j, i]$ denote the subsequence of α starting at index j and ending at index i. A cell $A[s, i]$ of table A denotes the error of segmenting the sequence $\alpha[1, i]$ using s segments, formally defined by:

$$A[s, i] = \min_{2 \leq j \leq i} (A[s-1, j-1] + E(S_{opt}(\alpha[j, i], 1))) \tag{1}$$

---

[1] For the sake of simplicity, we focus here on a bitmap representation. To cope with memory consumption, a more efficient representation method, such as the dynamic bit vector (DBV) architecture, could be used [13].

where $E(S_{opt}(\alpha[j,i],1))$ is the minimum error that can be obtained for the sub-sequence $\alpha[j,i]$ when representing it as one segment. In our case, it is simply the number of items that are present in the segment receipts. In the signature mining problem, as presented in Sect. 3.1, the representative of a segment is not a numeric value but a set of items. In order to compute the reconstruction error, an A cell stores the bitmaps of the best signatures obtained so far. Since several signatures may exhibit the same reconstruction error value, an A cell contains a set of bitmaps. Intuitively, $A[s,i]$ is computed by considering, for all $j \in [s,i]$, the composition of a signature obtained for an $(s-1)$-segmentation of a subsequence $\alpha[1,j-1]$, stored in $A[s-1,j-1]$, and the signature of the new segment $Sig_1(\alpha[j,i])$.

Formally, it is defined by:

$$A[s,i] = \underset{s \le j \le i}{\mathrm{amaxN}}(\underset{Sig_{s-1} \in A[s-1,j-1]}{\mathrm{amaxN}} (Sig_{s-1} \wedge Sig_1(\alpha[j,i]))) \qquad (2)$$

where

$$\mathrm{amaxN}\, P(i) \equiv \underset{i}{\arg\max}\, |P(i)|$$
$$P(i)$$

($amaxN$ returns the maximal elements of a set with respect to norm $|.|$). The representation error associated with cell $A[s,i]$ is simply $E(P)$, which is identical for every bitmap $P \in A[s,i]$. Thus, all such signatures having a maximal size are stored in $A[s,i]$.

Table 1 displays the progressive segmentation by Dynamic Programming of sequence $\alpha = \langle (ab)(abc)(acd)(abd) \rangle$. The leftmost column gives the indices of segments. The bottom row gives the indices over sequence α as well as their associated itemset and bitmap. Other cells of Table 1 details the results of operations performed by DP formalized by Eqs. (1) and (2). m represents the error minimization operation of segmentation. Note, that in the signature mining we are looking for maximal signatures w.r.t. the number of items and thus, m is a max operator on signatures. For example, cell $[2,3]$ computes the best signature obtained by segmenting sub-sequence $\alpha[1,3]$ into 2 segments. There are 2 ways to segment $\alpha[1,3]$: $\alpha[1,2] - \alpha[3,3]$ and $\alpha[1,1] - \alpha[2,3]$. In the first case, the representative of $\alpha[3,3]$ (*i.e.* its associated bitmap) is composed with the

Table 1. DP segmentation table for sequence $\alpha = \langle (ab)(abc)(acd)(abd) \rangle$. To be read from bottom-left to top-right.

3			$m(\alpha[3,3]\circ A[2,2])$ $=m(1011\wedge1100)$ $=\{1000\}$	$m(\alpha[4,4]\circ A[2,3],\alpha[3,4]\circ A[2,2])$ $=m(1001\wedge1010,\ 1001\wedge1100,\ 1011\wedge1100\}$ $=\{1000\}$
2		$m(\alpha[2,2]\circ A[1,1])$ $=m(1110\wedge1100)$ $=\{1100\}$	$m(\alpha[3,3]\circ A[1,2],\alpha[2,3]\circ A[1,1])$ $=m(1011\wedge1110,\ 1111\wedge1100)$ $=\{1010,\ 1100\}$	$m(\alpha[4,4]\circ A[1,3],\alpha[3,4]\circ A[1,2],\alpha[2,4]\circ A[1,1])$ $=m(1001\wedge1111,\ 1011\wedge1110,\ 1001\wedge1111)$ $=\{1001,\ 1010,\ 1001\}$
1	$m(\alpha[1,1])$ $=\{1100\}$	$m(\alpha[1,2])$ $=\{1110\}$	$m(\alpha[1,3])=\{1111\}$	$m(\alpha[1,4])=\{1111\}$
	1: *(ab)* 1100	2: *(abc)* 1110	3: *(acd)* 1011	4: *(ad)* 1001

best signature obtained for sub-sequence $\alpha[1,2]$ given by $A[1,2]$. Actually, the composition operation (denoted by $\circ$ in the table), is simply a logical AND on bitmaps. In the second case, the representative of $\alpha[2,3]$ is composed with the best signature for sub-sequence $\alpha[1,1]$ given by $A[1,1]$. The representative of several sequence elements is simply a logical OR on their associated bitmaps. The best signature for the whole sequence α and a 3-segmentation is given by $A[3,4]$.

Algorithm 1. Dynamic Programming for segmentation-based signature extraction

---

    Input: $\alpha = \langle T_1, \ldots, T_n \rangle$: receipt sequence of length n, min_seg: the minimal
           segmentation size
    Result: Sig: signatures
1 $A[1,1] = T_1$;
2 /*Initialization of the first row of A*/
3 **for** $i = 2, n$ **do**
4 | $A[1,i] = T_i \vee A[1, i-1]$;
5 **end**
6 **for** $s = 2, min\_seg$ **do**
7 | **for** $i = s, n$ **do**
8 | | $Sig = \emptyset$;
9 | | **for** $j = s, i$ **do**
10 | | | $Sig = Sig \cup \{A[s-1, j-1] \wedge (\bigvee_{T \in \alpha[j,i]} T)\}$;
11 | | **end**
12 | | $A[s,i] = \arg\max_{p \in Sig} |p|$;
13 | **end**
14 **end**
15 $Sig = A[min\_seg, n]$;
16 **return** Sig

Algorithm 1 presents a DP algorithm for sequence segmentation and signature extraction. The first row of the DP table is initialized in lines 4–6. Then rows are added iteratively until reaching the min_seg threshold. To build A_k, for $k \in [2, min\_seg[$, we just have to add the row k to A_{k-1} (lines 9–13). Finally, $A[n, min\_seg]$ provides the best signatures and related min_seg-segmentations (line 16).

The dynamic programming algorithm has a complexity in $O(n^2 k)$ and computes the optimal solution, here the maximal signature and related segmentation.

4 Experiments

In this section, we compare signatures with some other data representation models for analyzing customer purchase regularity. We demonstrate that we are able to find new regularities, that can be used to answer practical questions, such as targeted marketing.

The experiments were performed on anonymized basket data provided by a major French retailer. They were collected from may 2012 to august 2014 (27 months) from customers owning a loyalty card. To remove occasional customers, whose data do not make sense for our experiments, only customers having more than 20 baskets during the period were kept. 149 942 distinct customers, worth 16.6 GB of data, remained. The resulting database contains 3,887,979 distinct items. The retailer also provided a taxonomy that relates items to sub-categories (item class). We ended up with a total of 3388 item categories. Such categories are used to get rid of minor items differences (*e.g.* packaging or brand).

4.1   Capturing Purchase Regularity

Mining methods that extract patterns while giving some insight of regularity, go from top-k item mining [14] to periodic pattern mining. Top-k items are the k most frequently bought items within all customer's baskets. However, top-k items do not provide an explicit information about purchase regularity. Yet, item frequency can be considered as a rough mean regularity. Periodic patterns [15] represent items that are purchased at a strict periodicity. However, some purchase delay could break the periodicity and prevent a pattern to be periodic. Signatures stand in the middle: they represent sets of items that are bought within a limited period of time and such items are bought together several times but under a non strict periodicity.

In the sequel, we compare signatures with top-k items and periodic patterns to exhibit some common and distinctive features.

Signatures vs Top-k Items. In this experiment, we compare the signature content with the top-k items for each customer. We compute signatures with a relative number of blocks of 0.15 (see Sect. 3.1). We try different values of k for the top-k items method, and compare all of them with the signature content in Fig. 1, on the left. Setting the value of k to the signature length for each customer is not possible in practive, as we do not know the signature length before hand. We therefore do not show experiments with this particualr value of k. More elaborate methods to adapt the k value to each customer, such as elbow methods [16], did not bring better results than the ones presented in Fig. 1-left. The Jaccard similarity between the signature and the top-k items of most customers is between 0.5 and 0.3. This means that top-k items and signature products overlap partially. When the k value is low, the number of top-k items is significantly lower than the number of items in the signature. This leads to a low Jaccard value, even though most of the top-k items are included the signature. A similar behavior is observed for large values of k, where the top-k contains more items than the signature, leading to a low Jaccard value. For values of k close to the mean signature length: between 5 and 10 items, the Jaccard goes higher, as both sets have a similar size. Overall, the signature overlaps partially with the top-k items, and the main source of difference comes from the fact that the number of items in the signature changes for each customer, whereas it is constant for all customers in the top-k computation. The number of items in

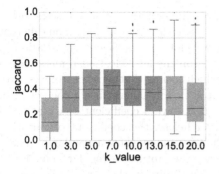
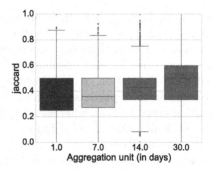

Fig. 1. On the left: Jaccard similarity between the signature and the top-k items, for different k values. This has been computed on 149 942 customers. On the right: Jaccard similarity between the signature and the longest periodic pattern, for different time scales. This was computed on 20 000 customers.

the signature could therefore be seen as a way to estimate a relevant value of k for the top-k items of a given customer. Another source of difference between signatures and top-k items is the fact that items that are very frequently bought during a short period of time do not appear in the signature, while they are more likely to appear in the top-k items.

Periodic Patterns Comparison. In this experiment, we compare the signature content with the periodic patterns for each customer. We used an algorithm that allows gaps between consecutive occurrences of periodic patterns [7]. As periodic patterns can only be found in a single time scale, a preprocessing step that aggregates the receipts on a given time scale (e.g., merge all receipts at the given granularity) is required. As we do not know in advance what is the relevant time scale for each customer, we are using 4 time scales to compute the periodic patterns: daily, weekly, bi-weekly and monthly purchases. For each time scale, we computed the Jaccard similarity between the longest periodic pattern and the signature. We chose the longest periodic pattern as the signature finds the longest regular pattern. If several longest periodic patterns are found, we take the one that has the largest Jaccard similarity with the signature. The results are presented in Fig. 1, on the right. In this figure, we can see that the Jaccard similarity between the signature and the longest periodic pattern is mostly between 0.3 and 0.45. This means that these two sets have common elements, but still differ. Further analysis showed that the longest periodic pattern is almost totally contained in the signature. This means that the signature is composed of most items from the longest periodic pattern. This periodic part of the signature represents between one third and one half of the total signature. The remaining part of the signature contains items that are not periodic but that are regularly bought. This highlights the flexibility of the signature, as it manages to capture periodic products, while also capturing non periodic regular purchases.

Signatures capture non periodic regularities because their segments can be of arbitrary length. More specifically, each customer signature segment can contain multiple baskets, and can therefore span on different time scales. On the other hand, periodic patterns have a fixed segment length and cannot span on different time scales. To illustrate this difference, we plot the coefficient of variation of the segment size for each customer in Fig. 2, on the left. Most customers have a coefficient of variation greater than 0.4, which means that most customers have variations in their purchase rhythms. Almost no customers have a coefficient of variation equal to zero, whereas all customers have a coefficient of variation of 0 for periodic patterns by definition. Nevertheless, the coefficient of variation remains mostly below 1, which means that customers show a regular purchase behavior. Therefore, the signature segment still captures a regular behavior of the customer. This shows that introducing flexibility in the period allows us to capture more regular products than existing methods (see Fig. 1-right), while capturing a regular behavior (see Fig. 2-left).

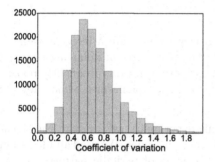 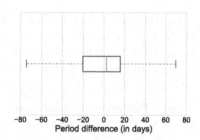

Fig. 2. On the left: distribution of the segment length coefficient of variation of 149 942 customers. On the right: difference between the signature period and the most similar periodic pattern period.

Because signatures are more flexible, their detected temporal regularity can be different than the one found by periodic patterns. To compare the period found by both methods, we compared the difference between the mean segment size of the signature, with the largest period of the periodic pattern that is the most similar with the signature (according to the Jaccard similarity). We choose the largest period of the periodic pattern, because signatures segments are as large as possible. This effect is due to the fact that segments have to cover the whole sequence. The results are presented in Fig. 2, on the right. We can see that most periods found by the signature are close to the period found by the most similar periodic pattern. While there can be some differences between both periods, these differences are usually contained within a reasonable time span.

To summarize, signatures are able to find regularly purchased products, whether they are periodically bought are not. The flexibility of the regularity definition of the signatures allows us to find these products without any pre-processing step. Moreover, signatures are able to find the underlying period of customers, that is consistent with the one found by periodic patterns. Signatures therefore find the time regularity of a customer, along with the regular products. This regularity cannot be totally captured by existing methods.

4.2   Insights from Signatures

As shown in the previous section, signatures group transactions into segments to find a set of regular products, that can not be totally found by existing methods. More specifically, let us consider the case of a real customer (named A). The store visits of customer A are represented in Fig. 3, on the top. For comparison, we also consider customer B whose store visits have a signature identical to her largest periodic pattern (shown in Fig. 3 on the bottom). The comparison of both Figures clearly shows that the customer A has no clear buying pattern, while the customer B has a clear buying pattern: she buys her groceries almost every Saturday. Nevertheless, by computing the signature on the customer A, we are able to detect her underlying period. Indeed, her signature contains 9 products:

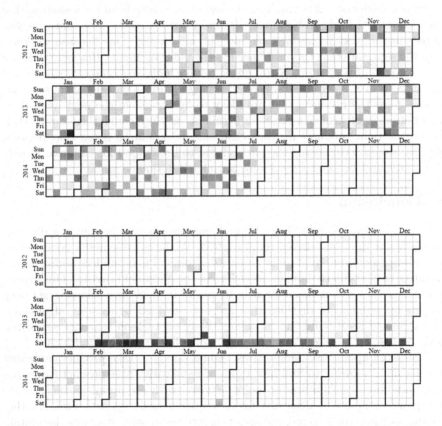

Fig. 3. Receipts of a non periodic customer (A), on the top, and periodic customer (B), on the bottom. Each green rectangle represents a visit to the store. The darker the green, the more products were bought during that visit. (Color figure online)

Biscuits, Hazelnut spread, cheese, frozen meat, pasta, cream, butter, ham and *chocolate powder*. Only some of them are bought during the same store visit, and these purchases usually spread over 4 transactions, for a segment length of 2 weeks on average. Among these products, some of them have a periodic buying pattern (*pasta, ham* and *hazelnut spread*), while the others are bought more sporadically. Nevertheless, this whole set of products has consistency and is related to meal and break food for children. The signature was therefore able to identify the purchase rhythm (both period and products) of a customer who had no clear buying pattern when using existing methods.

Signatures can also help marketers to answer the problem of finding the most appropriate time and products to give a coupon on, for a given customer. To achieve this targeted coupon policy, it would be interesting to be able to know what kind of products this customer is likely to buy in the next visits, to be able to give this customer targeted coupons. Thanks to the signature, we can provide the marketer with information about the time and content of next purchases. Indeed, if this customer has purchased *Biscuits, cheese, frozen meat, cream* and *butter* over 2 transactions in a week, we know from the signature that this customer is likely to be buying *Hazelnut spread, pasta, ham* and *chocolate powder* in the next 2 transactions over the next week. This because we know from the signature that this customer has the habit of buying *Biscuits, Hazelnut spread, cheese, frozen meat, pasta, cream, butter, ham* and *chocolate powder* in 4 transactions over 2 weeks. As we are observing a portion of a signature segment, we can guess the products that are likely to be bought in the next week. This information is of prime interest for retailers, as they could then target their ads on the right products for each customer. It should be noted that periodic patterns would have missed the part related to break food for children, as only *pasta, ham* and *hazelnut spread* were considered periodic.

5   Conclusion

Getting a better understanding of individual customers is becoming a differentiating factor in a data-driven retail context. We have presented a novel notion of *customer signature*, that gives for each customer a good understanding of the products most regularly bought, as well as of the household rhythm. Our experiments have shown that this approach, thanks to its flexibility, allows to get deep insights on purchasing rhythms that are not provided by existing algorithms. The approach itself builds up on a large body of work on sequence segmentation, taking advantage of years of research on efficient exact algorithms.

This work opens new perspectives. A first one is to take product categories into account, allowing to find new types of regularities over product categories or brands. From an application point of view, with our retail partner we are investigating the use of signatures for preventive actions against churn. Another exciting perspective is to test the use of signatures on other domains than retail. Thanks to the generality of the definitions, it can be easily applied on any sequence of itemsets where a segmentation is relevant. We performed preliminary experiments on datasets of labeled TV programs, with promising results:

while signatures with a high number of blocks detect regular daily programs, signatures with fewer segments but many items can detect relatively short span events (such as Roland-Garros tennis contest) for which TV channels devote many special programs, that are picked up by the signature.

References

1. Agrawal, R., Imieliński, T., Swami, A.: Mining association rules between sets of items in large databases. In: Proceedings of 17th International Conference on Management of Data, pp. 207–216 (1993)
2. Agrawal, R., Srikant, R.: Mining sequential patterns. In: Proceedings of 11th International Conference on Data Engineering, pp. 3–14 (1995)
3. Tseng, V.S., Shie, B.-E., Wu, C.-W., Yu, P.S.: Efficient algorithms for mining high utility itemsets from transactional databases. Trans. Knowl. Data Eng. **25**(8), 1772–1786 (2013)
4. Miguéis, V.L., Camanho, A.S., Falcão e Cunha, J.: Mining customer loyalty card programs: the improvement of service levels enabled by innovative segmentation and promotions design. In: Exploring Services Science, pp. 83–97 (2011)
5. Miguéis, V.L., Camanho, A.S., Falcão e Cunha, J.: Customer data mining for lifestyle segmentation. Expert Syst. Appl. **39**(10), 9359–9366 (2012)
6. Han, J., Pei, J., Yin, Y.: Mining frequent patterns without candidate generation. In: SIGMOD International Conference on Management of Data, pp. 1–12 (2000)
7. Cueva, P.L., Bertaux, A., Termier, A., Méhaut, J., Santana, M.: Debugging embedded multimedia application traces through periodic pattern mining. In: Proceedings of 12th International Conference on Embedded Software, pp. 13–22 (2012)
8. Mannila, H., Toivonen, H., Verkamo, A.I.: Discovery of frequent episodes in event sequences. Data Min. Knowl. Disc. **1**(3), 259–289 (1997)
9. Casas-Garriga, G.: Discovering unbounded episodes in sequential data. In: Proceedings 7th European Conference on Principles and Practice of Knowledge Discovery in Database, pp. 83–94 (2003)
10. Bingham, E.: Finding segmentations of sequences. In: Inductive Databases and Constraint-Based Data Mining, pp. 177–197 (2010)
11. Terzi, E., Tsaparas, P.: Efficient algorithms for sequence segmentation. In: Proceedings of SIAM Conference on Data Mining, pp. 314–325 (2006)
12. Ayres, J., Flannick, J., Gehrke, J., Yiu, T.: Sequential pattern mining using a bitmap representation. In: Proceedings 8th Conference on Knowledge Discovery and Data mining, pp. 429–435 (2002)
13. Vo, B., Hong, T.-P., Le, B.: DBV-miner: a dynamic bit-vector approach for fast mining frequent closed itemsets. Expert Syst. Appl. **39**(8), 7196–7206 (2012)
14. Han, J., Wang, J., Lu, Y., Tzvetkov, P.: Mining top-k frequent closed patterns without minimum support. In: Proceedings of International Conference on Data Mining (ICDM), pp. 211–218 (2002)
15. Ma, S., Hellerstein, J.L.: Mining partially periodic event patterns with unknown periods. In: Proceedings of 17th International Conference on Data Engineering, pp. 205–214 (2001)
16. Tibshirani, R., Walther, G., Hastie, T.: Estimating the number of clusters in a data set via the gap statistic. J. Roy. Stat. Soc.: Ser. B (Stat. Methodol.) **63**(2), 411–423 (2001)

Volatility Adaptive Classifier System

Ruolin Jia, Yun Sing Koh[⊠], and Gillian Dobbie

The University of Auckland, Auckland, New Zealand
rjia477@aucklanduni.ac.nz, {ykoh,gill}@cs.auckland.ac.nz

Abstract. A data stream's concept may evolve over time, which is known as the concept drift. Concept drifts affect the prediction accuracy of the learning model and are required to be handled to maintain the model quality. In most cases, there is a trade-off between maintaining prediction quality and learning efficiency. We present a novel framework known as the Volatility-Adaptive Classifier System (VACS) to balance the trade-off. The system contains an adaptive classifier and a non-adaptive classifier. The former can maintain a higher prediction quality but requires additional computational overhead, and the latter requires less computational overhead but its prediction quality may be susceptible to concept drifts. VACS automatically applies the adaptive classifier when the concept drifts are frequent, and switches to the non-adaptive classifier when drifts are infrequent. As a result, VACS can maintain a relatively low computational cost while still maintaining a high enough overall prediction quality. To the best of our knowledge, this is the first data stream mining framework that applies different learners to reduce the learning overheads.

Keywords: Data stream · Concept drift · Stream volatility

1 Introduction

Data streams are sequences of unbounded data arriving in real time. For example, electricity usage records produced by a power station, online tweets generated in a region, transactions recorded in a stock market can all be presented as data streams. Such real-world data are generated in order and are considered to be infinite. The task of data stream mining is to find valuable information from these unbounded streams of data. Data stream's properties raise various requirements when designing data stream algorithms. Instances in a stream can arrive very fast, allowing only limited time and memory for the algorithm to learn its underlying concepts. Moreover, a data stream may evolve over time such that the underlying concepts in a stream may change. Consequently, the learning model loses prediction accuracy over time. This is known as concept drifts. To maintain the quality of a learning model, stream learning algorithms are expected to detect changes and update their models to overcome these concept drifts. The frequency of concept drifts is known as stream volatility [6]. High volatility means a high frequency of concept drifts.

© Springer International Publishing AG 2017
J. Kim et al. (Eds.): PAKDD 2017, Part I, LNAI 10234, pp. 122–134, 2017.
DOI: 10.1007/978-3-319-57454-7_10

Some data stream learners can overcome a concept drift by adjusting their models to generalise the new concept and maintain a high prediction quality during the drift. These learning models can be classified as adaptive learners. Other data stream learners that cannot adjust their models are known as non-adaptive learners. Model adaptations come with a large computational cost. Thus, there is a trade-off between the model quality and the learning efficiency. It is also known that stream volatility may change in a stream over time [6]. For example, in stock market transactions, an anomalous event can result in an increasing number of concept drifts over a short period. One way to balance the trade-off between model quality and efficiency is to apply the model adaptation only when the stream volatility is high to maintain a stable prediction quality. When the volatility becomes low, we disable the model adaptation to save cost. We are addressing this problem by creating a new learning framework containing both adaptive and non-adaptive learners.

We designed a framework called Volatility Adaptive Classifier System (VACS). VACS has lower computational cost than the state-of-art adaptive learner while maintaining a similar prediction quality in a stream with volatility changes. VACS is composed of both adaptive and non-adaptive classifiers. VACS uses stream volatility [6] as the criterion to switch between classifiers. In particular, when the volatility is high, VACS applies the adaptive learner to maintain a better prediction quality. When volatility is low, it is deemed to be unnecessary to spend large overheads to handle infrequent concept drifts, so it switches to the non-adaptive classifier. As a result, VACS will maintain a sufficiently high prediction accuracy with relatively low overheads. Our contributions are as follows: (1) We proposed a Volatility Adaptive Classifier System (VACS), which is able to choose between the adaptive classifier and the non-adaptive classifier given different levels of stream volatility. (2) We show that the accuracy of VACS is comparable to state-of-the-art techniques, while maintaining low computational cost. To the best of our knowledge, this is the first data stream learning technique that uses stream volatility to adjust model adaptation behaviour to reduce computational cost while maintaining high model quality.

In the next section, we discuss the related work in the area. In Sect. 3, we illustrate how VACS works. In Sect. 4, we discuss the experimental results. Lastly, we conclude our paper in Sect. 5.

2   Related Work

There are various methods, derived from traditional learning methods, for mining data series including ensemble based methods [10] and neural network based methods [8]. In our research, we focus on tree based models: VFDT [4] and HAT-ADWIN [2] which have different expected properties.

VFDT is a decision tree classifier that is specifically designed for learning data streams. Because a data stream is infinite, VFDT splits an inner node using confident enough instances from a stream rather than seeing all instances. CVFDT [7] is a variant of VFDT. CVFDT maintains a sliding window storing recent instances learned, and it adjusts its tree model to be consistent with the instances in the window.

Hoeffding Adaptive Tree using ADWIN (HAT-ADWIN) was proposed by Bifet et al. [2]. This algorithm installs the drift detector ADWIN [1] on each node of the VFDT decision tree. The ADWIN drift detector monitors the attribute-class statistics on its host node. If a change in the attribute-class statistics at that node is detected, it starts to grow an alternative tree rooted at that node. When the alternative sub-tree has a better prediction accuracy, the current sub-tree is replaced by the alternative one.

We can categorise those two tree algorithms into two classes: adaptive learner and non-adaptive learner. VFDT is considered to be a non-adaptive learner. It does not have the ability to adjust its tree model to new concepts in an evolving data stream. Instead, VFDT with drift detector ADWIN (VFDT-ADWIN) can rebuild its tree model when a concept drift is detected. However, a severe prediction accuracy drop can be experienced during model rebuilding. In contrast, HAT-ADWIN and CVFDT are adaptive learners. Adaptive learners are able to partially update their models to fit the new concept in an evolving stream such that they can maintain a stable accuracy when encountering concept drifts. However, adaptive learners such as HAT-ADWIN have larger overhead than non-adaptive learners in terms of training time and memory. This is because adaptive algorithms need additional computation and storage to perform model adaptation.

Recurring Concept Drift (RCD) framework [5] is similar to our proposed system (VACS) in that they both use more than one learner to mine data streams. However, RCD is designed for improving the prediction quality, while VACS is designed to reduce learning overheads.

3   Volatility Adaptive Classifier Systems

We propose the Volatility Adaptive Classifier System (VACS). Intuitively, when mining a stream, VACS automatically applies the adaptive classifier in high volatility periods, and it switches to the non-adaptive classifier in low volatility periods. It aims to reduce learning overheads while maintaining high prediction quality. VACS is composed of several modules: Volatility Measurement Window, Double Reservoirs Classifier Selector, a drift detector and two component learners. Figure 1 presents an overview of VACS. In particular, we use VFDT with

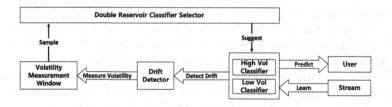

Fig. 1. VACS Overview

ADWIN (VFDT-ADWIN) in low volatility periods and we use HAT-ADWIN in high volatility periods. We use ADWIN as the drift detector.

3.1 Volatility Measurement Window

One task of VACS is to measure the volatility level of a stream such that it can switch classifiers based on different volatility levels. Huang et al. [6] calculate a stream's current volatility by calculating time intervals between each drift point in a buffer. In other words, their method measures the time differences among a fixed number of drifts. Small time differences denote high volatility while large differences denote low. Their method is appropriate to calculate relative volatility shift in a stream. However, it introduces a volatility measurement delay because it needs to wait for the next drift in order to calculate the new volatility level. The delay problem can be severe if the volatility drops from high to low. This is because the time difference between the next drift and the recent drift increases, and it needs to wait a longer period for the next drift to appear in order to update the measurement.

We develop a new method to measure the level of volatility using a sliding window with fixed size T. It contains indices of the most recent T instances learned from the stream. When a new instance's index is inserted into the window, the oldest instance's index is removed. The window maintains a value γ, which is the number of concept drifts detected in the most recent T instances. Then γ can represent the level of the current volatility. In the case when the level of volatility decreases, it does not need to wait until the next drift occurs. Instead, the window constantly updates γ over time. This new method mitigates the delay problem.

3.2 Double Reservoirs Classifier Selector

Double Reservoirs Classifier Selector (DRCS) is another module in VACS. DRCS uses the reservoir sampling technique [9] to sample and approximate both high and low levels of volatility of a stream while learning. We do not want to lose information about high volatility periods of a stream when sampling at low volatility. Similarly, we do not want to lose information about low volatility periods when sampling at high volatility. A single reservoir will not satisfy this requirement because reservoir sampling removes a random element when inserting a new element. DRCS separately samples the volatility levels from low and high volatility periods in a stream using two independent reservoirs.

In particular, DRCS has two functions: sampling and suggesting. The "sampling" function is called by VACS constantly when learning a stream. The sampling function of DRCS maintains two reservoirs named *High Reservoir* and *Low Reservoir*. Those two reservoirs sample each input γ value (volatility level) from the volatility measurement window using the reservoir sampling method [9]. The first γ is inserted into the low reservoir for initialisation. After initiation, when a new measured stream volatility γ arrives, it compares γ with the mean of the elements in the two reservoirs. If the value is lower than the mean, it stores this value into the *Low Reservoir*. If the value is greater than the mean, it stores the value into the *High Reservoir*. We specify a means' difference threshold λ. If the difference between two reservoirs' means is greater than λ, the DRCS is set to be active. VACS can only switch a classifier when DRCS is active. This setting can prevent two undesirable behaviours. Firstly, it prevents VACS from switching classifiers when there are rare volatility changes in the stream. Secondly, if there are volatility changes in the stream but these changes only appear in a later period, it prevents VACS from switching classifiers at the early stage in which a changing volatility has not been measured yet. Intuitively, λ is used to indicate the size of the volatility change that matters to the user. If there are volatility differences greater than this threshold in a stream, we can treat the stream as a volatility-changing stream and activate our system, otherwise, we use the single learner to handle the stream. λ is also related to the volatility measurement window size T. A larger T value can result in larger measured numbers of drifts in the window. Thus, larger γ (volatility level) can be obtained and inserted into reservoirs. So λ should increase with T. However, if λ is overly large, VACS may never be activated.

The second function of DRCS is a "suggesting" function. This is used when VACS queries DRCS. When VACS queries DRCS, it compares the most recent input γ with the mean of all other γ values in two reservoirs. If the recent γ is greater than the reservoirs' mean, it returns the high volatility classifier suggestion (adaptive learner). Otherwise, it returns the low volatility classifier suggestion (non-adaptive learner).

3.3   Pseudocode

In this section, we compose each module discussed in the previous sections into the complete Volatility Adaptive Classifier System (VACS). The pseudocode can be seen in Algorithm 1. The algorithm firstly initiates the Double Reservoirs Classifier Selector (DRCS) and a drift detector. By default, we use ADWIN for the drift detector. It also initiates the volatility measurement window counting concept drifts detected in the recent T instances. We provide two classifiers for VACS. One classifier is considered to be suitable in high volatility periods (adaptive learner) while the other is deemed to be appropriate in low volatility periods (non-adaptive learner). In our implementation, we use HAT-ADWIN as the adaptive learner and VFDT-ADWIN as the non-adaptive learner. In VACS, only one classifier is active at any time to perform the learning task. The user decides which classifier should be active at the start when no volatility level has been detected. When the algorithm starts, it takes each arriving instance from the stream and classifies it with the active classifier. If the classifier correctly classifies the instance, we input 0 into the drift detector. Otherwise, we input 1 into the drift detector. The drift detector is modified such that it signals a drift only if the prediction error is increasing. Next, it updates the volatility measurement window and γ (volatility level), which is the number of drifts detected in the recent T examples. If the number of classified instances since the last volatility level measurement reaches a user-specified count τ, the algorithm measures and inputs γ into DRCS and then queries DRCS. The reason for adding an interval τ between two consecutive γ measurements is because it is not likely to measure a change on γ if two measurements are close. Next, DRCS returns one classifier option from the two that are suggested to be used, best suited to the current level of volatility. If the suggested classifier is not consistent with the one active in VACS, it switches the current active classifier to the suggested one. It then re-initiates the new classifier. In the case of a decision tree, it resets the decision tree to a one-node tree without learning examples. When the user wants to make a classification with an instance with the unknown class, VACS will use the currently active classifier to make the prediction.

4   Results and Evaluation

We implemented VACS in the Massive Online Analysis (MOA) Framework [3]. In our experiments, we measure the performance of HAT-ADWIN, VFDT-ADWIN and VACS by evaluating total training time (Time), mean memory usage (Mem), maximal memory usage (Max Mem), the mean prediction accuracy (Acc) and the mean prediction accuracy when concept drifts occur (dAcc) on each algorithm. We compare the measurement results of those algorithms and contrast the differences among them.

Beyond those measurements, we also explore whether VACS switches between classifiers as expected. We introduce a new measurement called Percentage of Instances Classified by the Expected Classifier (PICEC). It denotes whether VACS applies the correct classifier accurately given a volatility level. Our synthetic datasets' volatility fluctuates between low and high volatility periods. VACS has two classifier options: high volatility classifier (adaptive) and low volatility classifier (non-adaptive). We obtain PICEC using the following calculation: we count the number of instances from the high volatility period in a stream classified by the high volatility classifier, and the number of instances from the low volatility period in a stream classified by the low volatility classifier. We divide the sum of these two numbers by the total number of instances in the streaming dataset.

Here we specify all parameters for VACS.

Each reservoir in DRCS has size 200, volatility measurement window size (T) is 300000, means' difference threshold (λ) is 15, the interval length between each volatility level measurement (τ) is 10000, the drift detector of VACS is ADWIN, the high volatility classifier is HAT-ADWIN, and the low volatility classifier is VFDT with ADWIN as the external drift detector (VFDT-ADWIN). The default starting classifier is VFDT-ADWIN. ADWIN uses the Hoeffding Bound whereby the δ value [1] is set to 0.002.

4.1   Experiments on Mutating Random Tree Generator Datasets

We show the evaluation results run on data generated by the mutating random tree generator. We intend to evaluate whether VACS has a reduction of

---

Algorithm 1. VACS: Volatility Adaptive Classifier System

---

    input   **:** S: Stream of examples
                 τ: Interval length between each volatility level measurement.
                 T: Size of sliding window for measuring current volatility level.
                 $HighVolClassifier$: The classifier used in high volatility period.
                 $LowVolClassifier$: The classifier used in low volatility period.
                 $StartingClassifier$: The classifier chosen at the start.
                 DT: Drift Detector.

1 **begin**
2 | Initiate drift detector DT;
3 | Initiate Double Reservoir Classifier Selector $DRCS$;
4 | Initiate Volatility measurement window W;
5 | Let $ActiveClassifier = StartingClassifier$;
6 | Let γ = Number of drifts detected when classifying the most T instances;
7 | **foreach** $example(x, y_k) \in S$ **do**
8 | | $ActiveClassifier$ classifies (x, y_k);
9 | | **if** $ActiveClassifier$ $correctly$ $classified$ (x, y_k) **then**
10 | | | Let $e = 0$;
11 | | **else**
12 | | | Let $e = 1$;
13 | | Input e into DT;
14 | | Update W and γ;
15 | | Let i = number of instances that has been classified;
16 | | **if** $i\%\tau = 0$ **then**
17 | | | Input γ in $DRCS$ (Call DRCS Sampling);
18 | | | Query $DRCS$ for the suggestion (Call DRCS Suggesting);
19 | | | Let $SuggestedClassifier$ = Suggested Classifier of $DRCS$;
20 | | | **if** $SuggestedClassifier$ is not null AND is not $CurrentClassifier$ **then**
21 | | | | **if** $SuggestedClassifier$ is $HighVolClassifier$ **then**
22 | | | | | Set $ActiveClassifier = HighVolClassifier$;
23 | | | | **else**
24 | | | | | Set $ActiveClassifier = LowVolClassifier$;
25 | | | | Re-initiate $ActiveClassifier$;
26 | | Train $ActiveClassifier$ with (x, y_k);
27 | **end for**
28 **end**

---

computational cost compared with the adaptive learner (HAT-ADWIN). We also inspect if VACS switches classifiers as expected.

We developed the mutating random tree generator based on the random tree generator presented in [4]. Our generator randomly chooses a branch of the tree and rebuilds it when we want to add a concept drift in the synthetic stream.

The synthetic data has 10 attributes and 2 classes. The maximal depth of the random tree is 5. We add 5% noise to the data. The synthetic data stream is made up of 28 blocks. Each block has 1 million instances. We have two types of blocks: high volatility blocks containing 50 concept drifts, and low volatility block containing 5 concept drifts. We interleave high volatility blocks and low volatility blocks such that the stream fluctuates between high volatility and low volatility over time. We generated three types of datasets. (1) balanced volatility periods streams: they contain equal numbers of high and low volatility blocks (2) majority of low volatility periods streams: they contain 8 high volatility blocks and 20 low volatility blocks (3) majority of high volatility periods stream: they contain 20 high volatility blocks and 8 low volatility blocks. For each pattern, we generate 20 stream samples with different random seeds, and we run experiments on each of them.

Experimental results of 20 sample streams are shown in Table 1. The bold font denotes the worst performance mean among the three algorithms. Generally, all experiments show that the prediction accuracy of VACS is close to the prediction accuracy of HAT-ADWIN in drifting periods. So it has similar stability to HAT-ADWIN when drifts occur. However, compared with HAT-ADWIN, VACS effectively saves on training time and memory usage. Results show that VACS has a better average prediction accuracy than VFDT-ADWIN and a slightly worse prediction accuracy than HAT-ADWIN. This is the expected result since

Table 1. Performance on mutating random tree generator datasets

Dataset	Balanced		Majority low vol		Majority high vol	
	Mean	Std dev.	Mean	Std dev.	Mean	Std dev.
VACS						
Acc %	84.86	0.28	86.71	0.456	83.43	0.38
dAcc %	81.2	0.41	81.85	0.49	80.9	0.55
Mem (B)	413178.93	30237.52	515455.88	48911.55	338391.59	22452.62
Max Mem (B)	2200813.6	540997.42	2372156	430484.32	2108248	480220.7
Time (s)	230.31	5.83	190.78	2.75	256.93	6.11
VFDT-ADWIN						
Acc %	**83.5**	0.24	**85.62**	0.5	**81.78**	0.4
dAcc %	**78.85**	0.37	**79.45**	0.6	**78.85**	0.37
Mem (B)	275669.66	29343.87	392995.25	55100	194866.07	24074.3
Max Mem (B)	2068718.4	519383.28	2274702	485691.92	1987230.4	469304.53
Time (s)	123.72	2.04	126.31	2.62	116.47	2.34
HAT-ADWIN						
Acc %	85.77	0.25	87.68	0.41	84.06	0.29
dAcc %	81.8	0.41	82.35	0.49	81.3	0.47
Mem (B)	**775754.48**	86571.52	**1115474.7**	158771.26	**520137.73**	77229.29
Max Mem (B)	**5139417.6**	1143583.34	**5754436**	1218915.82	**4763293.2**	959179.97
Time (se)	**364.69**	6.55	**392.04**	11.84	**333.41**	8.87

VACS's prediction accuracy is produced by the hybrid system composed of HAT-ADWIN and VFDT-ADWIN.

4.2 Evaluation of the Classifier Switch Quality

The objective of these experiments is to evaluate whether VACS applies appropriate learners as expected. We test VACS with the three different types of datasets from the mutating tree generator. The results are shown in Table 2. Results show decent PICECs for all types of data. PICECs for all datasets' experiments reach around 90%. The percentage of applying the low volatility learner and the high volatility learner are also consistent with the type of data. For example, in streams with a majority of low volatility periods, 72% of the instances are classified by the low volatility classifier (VFDT-ADWIN) and 28% of the instances are classified by the high volatility classifier (HAT-ADWIN).

Table 2. VACS behaviour on different streams

Dataset	Balanced		Majority low vol		Majority high vol	
	Mean	Std dev.	Mean	Std dev.	Mean	Std dev.
PICEC %	89	1.8	94	1.5	88	2.4
Low vol learner usage %	52	2.1	72	1.3	37	1.9
High vol learner usage %	48	2.1	28	1.3	63	1.9

4.3 Experiments on Different Volatility Measurement Window Size

The objective of these experiments is to inspect the influences of different Volatility Measurement Window Size T on the learner selection quality of VACS. In previous experiments, we use 300,000 as the default value for T. In this experiment, we vary T by both increasing and decreasing from its default value. We run VACS on 20 samples of the balanced volatility stream. The experimental results are shown in Table 3.

We can summarise that classifier selection quality of VACS is influenced by the volatility measurement window size T. Setting the value T to either too small or too large may cause low PICEC, which means a poor classifier selection quality. From our experiments, we found that T should be large enough such

Table 3. VACS with different volatility measurement window size

T	PICEC % (Mean)	PICEC % (Std dev.)
30000	50	0
100000	88	1.6
300000	89	1.8
3000000	29	4.3

that the measurement is not strongly influenced by the volatility fluctuation of randomness. Also, T should not be too large such that VACS can always measure one volatility level at a time in a stream with volatility changing between different levels. One suggestion for selecting T is to run a test on the stream before starting VACS. A user can start from a small T value and gradually increase T meanwhile evaluating whether measurements are susceptible to volatility fluctuations caused by noise. When VACS performs as expected, stop increasing T and use this value. We assume that the T value obtained in the pre-experiment period from a continuous stream is also appropriate in the upcoming period in that stream. This assumption holds true in all of our experiments.

4.4   Experiments on SEA Generator Datasets

We use SEA Generator as our second data generator from MOA. The SEA generator has 3 attributes and 2 classes by default. We generate the same three types of volatility-changing data streams as the previous experiments. The experimental results can be viewed in Table 4. The bold font denotes the worst performance metrics among three algorithms. Results show a similar conclusion for VACS reducing training time. VACS uses less training time than HAT-ADWIN.

Table 4. Performance on SEA generator datasets

Dataset	Balanced		Majority of low vol		Majority of high vol	
	Mean	Std dev.	Mean	Std dev.	Mean	Std dev.
VACS						
Acc %	92.14	0.08	92.62	0.07	91.69	0.19
dAcc %	90.85	0.37	90.9	0.31	90.75	0.44
Mem (B)	**173849.41**	3951.98	172856.88	2138.16	**163834.01**	2091.83
Max Mem (B)	572069.6	56281.64	576905.2	58845.55	532899.6	57047.26
Time (s)	127.33	4.33	105.41	2.39	143.82	3.3
VFDT-ADWIN						
Acc %	**92.09**	0.08	**92.52**	0.07	**91.67**	0.16
dAcc %	**90.85**	0.37	**90.8**	0.41	**90.65**	0.49
Mem (B)	56843.41	1871.21	63692.83	1191.53	41673.5	1123.55
Max Mem (B)	186323.2	15153.59	177403.6	9168.92	175032	5283.07
Time (s)	64.33	1.64	65.06	1.04	62.41	1.26
HAT-ADWIN						
Acc %	92.2	0.08	92.66	0.07	91.71	0.19
dAcc %	90.85	0.37	90.9	0.31	90.75	0.44
Mem (B)	161413.3	4963.52	**201208.75**	3811.19	112874.55	4215.42
Max Mem (B)	**666700**	71667.62	**688492.4**	88055.91	**637625.2**	67176.21
Time (s)	**188.22**	3.38	**209.84**	3.4	**171.86**	2.82

Table 5. VACS behaviour on different streams (SEA)

	Balanced		Majority of low vol		Majority of high vol	
	Mean	Std dev.	Mean	Std dev.	Mean	Std dev.
PICEC %	90	2.6	94	1.6	91	2.7
Low vol learner usage %	49	2.9	71	1.4	33	1.8
High vol learner usage %	51	2.9	29	1.4	67	1.8

Moreover, in streams composed of a majority of low volatility periods, VACS has the most effective reduction in time compared with HAT-ADWIN.

We also evaluate VACS's classifier switch quality in these datasets. We show the results in Table 5. VACS in all three types of datasets return around 90% values, which is similar to earlier experiments in mutating random tree datasets. This tells us the classifier switch quality of VACS is also decent on SEA generated datasets.

5   Experiments on Real-World Data

The aim of this set of experiments is to evaluate whether VACS can achieve the cost reduction compared with the adaptive learner on real-world datasets. We chose Poker Hand, Forest Cover Type, and Airlines real world datasets available on MOA, which contain volatility changes. After testing, we chose volatility measurement window size $T = 20,000$ and we set λ to 3 for VACS. Table 6

Table 6. Performance with real-world datasets

Dataset	Poker Hand	Forest Cover Type	Airline
VACS			
Acc %	72.0	82.14	64.98
Mem (B)	**124684.11**	**176295.35**	2642422.83
Max Mem (B)	**163392**	**357528**	10468520
Time (s)	4.29	11.62	10.14
VFDT-ADWIN			
Acc %	69.68	81.77	65.28
Mem (bytes)	28676.98	67016.01	1832277.51
Max Mem (B)	40192	131256	7855680
Time (s)	2.36	5.94	9.05
HAT-ADWIN			
Acc %	**66.42**	**81.42**	**63.37**
Mem (B)	23741.62	98633.23	**7024953.36**
Max Mem (B)	83600	304024	**13914984**
Time (s)	**5.12**	**14.37**	**15.82**

shows the experiment's results. We also plot the measured volatility level γ and the mean of double reservoirs in Fig. 2 to demonstrate the classifier switching behaviour of VACS. When γ is greater than the reservoirs' mean, VACS applies the high volatility classifier (HAT-ADWIN). When γ is lower than the mean, VACS uses the low volatility classifier (VFDT-ADWIN).

In all experiments, VACS achieves a training time reduction compared with HAT-ADWIN. In experiments with Poker Hand and Forest Cover Type, VACS has the highest prediction accuracy.

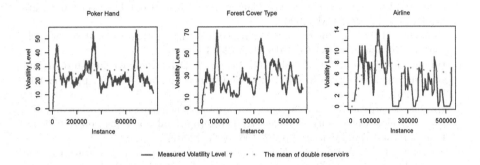

Fig. 2. Volatility measurements in the real world datasets

6   Conclusions and Future Work

We developed a system, called VACS, that can automatically choose the most suitable classifier between adaptive and non-adaptive algorithms in real time when mining a stream with changing volatility. The system applies the adaptive learner when the volatility is high and the non-adaptive learner when the volatility is low. It aims to reduce the learning costs while maintaining high enough prediction accuracy. We tested VACS on both synthetic and real-world data with changing volatility. In all of our experiments, VACS reduces the training time compared with the state-of-art adaptive learner, and VACS's prediction accuracy is also close or better than the adaptive learner. Through the experiments, we have shown that VACS is an effective approach to balance the trade-off between model prediction quality and efficiency.

One possible improvement is to enable VACS to adjust its volatility measurement window size T automatically when mining. The window is expected to shrink its size when the volatility changes notably and enlarge its size when the volatility is stable. When the volatility changes notably, it can remove the outdated elements by shrinking the window size such that it can react quickly to the changed volatility level. It can improve the accuracy of volatility measurement.

References

1. Bifet, A., Gavalda, R.: Learning from time-changing data with adaptive windowing. In: SDM, vol. 7, pp. 2007. SIAM (2007)
2. Bifet, A., Gavaldà, R.: Adaptive learning from evolving data streams. In: Adams, N.M., Robardet, C., Siebes, A., Boulicaut, J.-F. (eds.) IDA 2009. LNCS, vol. 5772, pp. 249–260. Springer, Heidelberg (2009). doi:10.1007/978-3-642-03915-7_22
3. Bifet, A., Holmes, G., Kirkby, R., Pfahringer, B.: MOA: massive online analysis. J. Mach. Learn. Res. **11**(May), 1601–1604 (2010)
4. Domingos, P., Hulten, G.: Mining high-speed data streams. In: Proceedings of the sixth ACM SIGKDD International Conference on Knowledge Discovery and Data Mining, pp. 71–80. ACM (2000)
5. Gonçalves, P.M., de Barros, R.S.M.: RCD: a recurring concept drift framework. Pattern Recogn. Lett. **34**(9), 1018–1025 (2013)
6. Huang, D.T.J., Koh, Y.S., Dobbie, G., Pears, R.: Detecting volatility shift in data streams. In: 2014 IEEE International Conference on Data Mining, pp. 863–868. IEEE (2014)
7. Hulten, G., Spencer, L., Domingos, P.: Mining time-changing data streams. In: Proceedings of the seventh ACM SIGKDD International Conference on Knowledge Discovery and Data Mining, pp. 97–106. ACM (2001)
8. Ng, H.T., Goh, W.B., Low, K.L.: Feature selection, perceptron learning, and a usability case study for text categorization. In: ACM SIGIR Forum, vol. 31, pp. 67–73. ACM (1997)
9. Vitter, J.S.: Random sampling with a reservoir. ACM Trans. Math. Softw. (TOMS) **11**(1), 37–57 (1985)
10. Wang, H., Fan, W., Yu, P.S., Han, J.: Mining concept-drifting data streams using ensemble classifiers. In: Proceedings of the ninth ACM SIGKDD international conference on Knowledge discovery and data mining, pp. 226–235. ACM (2003)

Distributed Representations for Words on Tables

Minoru Yoshida[✉], Kazuyuki Matsumoto, and Kenji Kita

Institute of Technology and Science, University of Tokushima,
2-1, Minami-josanjima, Tokushima 770-8506, Japan
{mino,matumoto,kita}@is.tokushima-u.ac.jp

Abstract. We consider a problem of word embedding for tables, and we obtain distributed representations for words found in tables. We propose a table word-embedding method, which considers both horizontal and vertical relations between cells to estimate appropriate word embedding for words in tables. We propose objective functions that make use of horizontal and vertical relations, both individually and jointly.

1 Introduction

Neural word embedding, also called distributed representation, has been studied extensively in recent years. It is a challenge to represent words by real-valued vectors, a method well known for its ability to obtain high-quality word similarity or provide answers to analogy questions [9].

Most previous neural word-embedding methods focus on obtaining distributed representations from sentences by modeling them as lists of words, learning parameters by predicting pivot words from their contexts (i.e., the words surrounding the pivot words).

In this paper, we focus on *tables* as a source of learning word embeddings. One principal property of table representation is that it is *two-dimensional*, in the sense that each cell in a table is related not only to the cells to the left and right of it, but also to the cells above and below it. See the example table shown in Table 1. Cell "25" is related not only to the cell "John Smith," indicating that the entity named "John Smith" is 25 years old, but also to the cell "age", which shows the attribute name for the value "25." Furthermore, cells in the same column, such as "25," "30," and "32" are related in the sense that they indicate values in the same class (i.e., person's age).

Table 1. An example table

Name	Age
John Smith	25
Taro Yamada	30
Hanako Tanaka	32

© Springer International Publishing AG 2017
J. Kim et al. (Eds.): PAKDD 2017, Part I, LNAI 10234, pp. 135–146, 2017.
DOI: 10.1007/978-3-319-57454-7_11

The above example suggests a need for approaches to embedding that differ from those used for sentences, which are essentially one-dimensional. We propose a method for obtaining word embeddings that is appropriate for representing words found on tables. Our *table word-embedding* method reflects the structure of tables, such as cells being in the same column having the same type, and cells being in the same row representing the same object.

In particular, we focus on the *attributes* shown in tables. The problem with embedding for attributes in tables is that there are two types of contexts for attributes, *the co-occurrence context*, which considers other co-occurring attributes as contexts, and *the value context*, which considers the values of the attributes as contexts. For example, word similarity using co-occurrence contexts finds the word "RAM" as a word similar to "CPU" because "CPU" and "RAM" co-occur frequently in personal computer specification tables. In contrast, word similarity using value contexts will find the attribute "event date" be similar to "birthday" because "birthday" and "event date" have similar values because both indicate some date. Word embedding for attributes in tables must locate words to reflect these two types of contexts. These are clearly different types of similarity compared to those of words in sentences, e.g., semantic contexts vs. syntactic contexts, or local similarity (i.e., contexts surrounding words in some small window) vs. document-level similarity (i.e., words co-occurring in the same document are considered to be similar, as in latent Dirichlet allocation, for example.) To deal with these issues, we define a new analogy task, which we call the "attribute analogy task", to measure how well each embedding reflects these similarities.

2   Related Work

Methods for obtaining neural word embeddings from sentences, including well-known word2vec [9] and GloVe [12] methods, have been studied extensively in recent years.

There have been also several research reports on using structured knowledge combined with sentences for learning distributed representations of words. In addition, there have been several efforts to obtain embeddings for hand crafted knowledge sources [2,8,16], and to combine these knowledge bases into word embeddings for sentences [1,11,15]. Tables also reflect a kind of knowledge structure, but the representation of that structure is implicit, and to make appropriate use of such structure requires new research efforts.

Tables have been receiving considerable attentions as knowledge sources in recent years owing to the increased number of available documents in the big data era. For example, [3] provided a system that can extract knowledge regarding attributes of objects from a large number of web tables. Pimplinkar and Sarawagi [13] proposed a system that accepts multiple queries describing multiple columns in HTML tables. Limaye et al. [7] presented a method for labeling each cell in HTML tables with labels from a knowledge base. Munoz et al. [10] proposed using Wikipedia tables as the source of triples. Our new table word-embedding methods and these table-based intelligent systems are complementary in that the new embedding provides a better way to represent tables, and better similarity calculations between tables to improve the quality of the output of table-related systems.

Recently, Yin et al. [17] proposed a question answering (QA) system that learn to extract information from database tables by modeling both query sentences and database table by distributed representations. Their work is aimed at task-oriented modeling of tables (i.e., to extract valid information from tables,) so different from ours in that our objective is to propose better models for words in tables that reflect attribute-value relations represented by two-dimensional layouts.

3   Data Sets and Notations

We used Japanese Wikipedia articles downloaded in 2013 as our corpus.[1]

We have a set of cells C, and each cell $c \in C$ is represented by one word $w(c)$. c has the *position* (x_c, y_c), where x_c is the horizontal position (i.e., the position of the column that contains c), and y_c is the vertical position (i.e., the position of the row that contain c).

We assume that the first row in each table represents attributes, and we perform different preprocessing operations for attribute and other cells. Many previous studies have attempted to find attribute positions in tables [5,18], but we cannot avoid estimation errors using such methods as preprocessing. However, most of the tables in our data set are row-wise[2], in the sense that cells in the same column have values of the same attribute (including the case in which attributes are omitted).

We call the cells in the first row *attribute cells*, and other rows *value cells*.[3]

For attribute cells (i.e., the cells in the first row), we extract the string in the cell as one word. For other cells, we use a morphological analyzer, Kuromoji, to break the string into a list of words, and obtain the last element of the list as the representative (head) word of the cell, because in many cases the strings found in table cells are noun phrases, which can be normalized by obtaining the last word (i.e., the head of a noun phrase is the last word in most cases).

We normalize each number using a significant digit approach. That is, we parse the number strings to obtain the numeric value (e.g., parse "123,000" to obtain the value 123 thousand), and retain only the significant digits (currently, the leftmost 2 digits) and discard the other digits, resulting in the string "120000" in this example.

---

[1] The total number of tables found in the corpus was 255,039.

[2] In our data set, 266 (93.7%) out of 284 randomly sampled tables were row-wise.

[3] In this research, we ignore tables that have no attribute names. Although this strategy can cause noise in the set of attribute vectors, the effects of such noise are small, because values are of many types and their frequency is relatively lower than that of the attributes.

Each cell $c \in C$ has a *horizontal context* H_c, which consists of the cells in the same row as the pivot cell, and a *vertical context* V_c, which consist of the cells in the same column.

Embedding for cells in the table must consider these two different kinds of contexts, in contrast to word embedding for sentences, as in word2vec, for example, which requires only one-dimensional contexts, i.e., the words to the left and right of the pivot word.

4   Proposed Method

In this section, we describe our algorithm for learning table word embeddings. Following the implementation of word2vec, which learns vectors using the skip-gram model with negative sampling (SGNS), we first introduce SGNS and then explain how to extend it to learn table word embeddings.

All of our methods define the objective function as the sum of the scores for every "pivot" ("target") cell.

4.1   SGNS

The objective function of this approach is the sum of the plausibility scores of predicting the "output word" w_O given the "input word" w_I where the input word is every word in the corpus and the output word is one of the context words of w_I. The score for the pair of w_I and w_O is defined as follows:

$$\log \sigma(v'_{w_O} \cdot v_{w_I}) + \sum_{k=1}^{K} \log \sigma(-v'_{w_k} \cdot v_{w_I})$$

where K is the number of negative samples, v_w and v'_w are two types of vectors called the "input" and "output" vectors, respectively, and σ is the sigmoid function. Negative samples w_k are selected randomly in each learning step.

SGNS maximize this function to obtain word vector representations v_w and v'_w.

4.2   Table Word Embedding

Our approach is a simple modification of SGNS as described above.

SGNS learns two vector representations, which are called "input" and "output" vectors. We use an approach similar to SGNS, but use four vectors in total, *attribute vectors* v_w^a, *attribute context vectors* v_w^{ac}, *value vectors* v_w^v, and *value context vectors* v_w^{vc} instead of output and input vectors.

These four vectors are defined as follows (see Fig. 1 for their positional relations).

Attribute and attribute context vectors: These are used for words in the cells in the attribute positions (i.e., the first row of the table). We use the skip-gram model in which the task is to predict words in the target cell represented by attribute vectors from the words in the other cells represented by attribute context vectors, resulting in maximizing the value of the dot product of v_w^a and v_z^{ac} where w is the word in the pivot cell and z is the word in the context cell.

Value and value context vectors: In value cells, value vectors v_w^v are predicted by value context vectors v_z^{vc}, resulting in maximizing the value of the dot product of v_w^v and v_z^{vc} where w is the word in the pivot cell and z is the word in the context cell.

The above explanations described the *horizontal relations* between cells. We can also consider another relation between attribute vectors and value vectors reflecting the vertical relation between attribute cells and cells in the same column as an attribute cell, resulting in maximizing the value of the dot product of v_w^a and $v_{z'}^v$ where z' is the vertical context word. This additional objective is intended to predict the attribute vector in the pivot cell from the value vector in the vertical context cell. Here we use the *vertical* relation of cells, assuming that *cells in the same column contain words in the same category* (see Table 1).

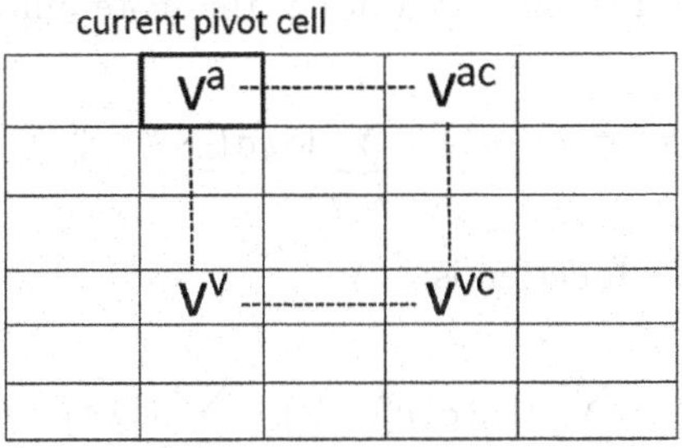

current pivot cell

Fig. 1. Relation of the vectors

For each cell, we take every pair of a pivot word w and its context word c, and vectors for w and c are selected from "value", "attribute context", "attribute", or "value context" according to the positions of the cell pair. More details are provided in the following sections. The vectors are learned to maximize the likelihood of the training samples. The definition of the likelihood function is also provided below.

4.3 Objective Functions

We now describe four types of objective functions. Each objective function uses its own combination of the vectors described above.

Horizontal Objective. In the horizontal approach, only two vectors v^a and v^{ac} and the relation between them are considered. The objective function in this approach is similar to that of SGNS[4].

For each word w in attribute cells (i.e., the cells in the first row) and its horizontal context word z, the score function is defined by:

$$\log \sigma(v_z^{ac} \cdot v_w^a) + \sum_{k=1}^{K} \log \sigma(-v_z^{ac} \cdot v_{w_k}^a)$$

where K is the number of negative samples, w_k is a randomly sampled word from a uniform distribution, and the global objective function for each table is:

$$l = \sum_{c \in C \wedge y_c = 1} \sum_{z \in H_c} \{\log \sigma(v_z^{ac} \cdot v_w^a) + \sum_{k=1}^{K} \log \sigma(-v_z^{ac} \cdot v_{w_k}^a)\}$$

Vertical Objective. In the vertical approach, only two vectors v^a and v^v and the relations between them are considered. The objective function in the vertical approach is similar to that in the horizontal approach, except that the words in the remaining cells in the same column as the pivot word are considered as contexts.

For each word w and its context word z, the score function is:

$$\log \sigma(v_z^v \cdot v_w^a) + \sum_{k=1}^{K} \log \sigma(-v_z^v \cdot v_{w_k}^a)$$

and the global objective function is:

$$l = \sum_{c \in C \wedge y_c = 1} \sum_{z \in V_c} \{\log \sigma(v_z^v \cdot v_w^a) + \sum_{k=1}^{K} \log \sigma(-v_z^v \cdot v_{w_k}^a)\}$$

Unified Objective I (CROSS). In the "cross" approach, we consider both horizontal and vertical contexts for each word in the attribute cells. This approach uses three types of vectors, v^a, v^{ac}, v^v, and the relations among them (i.e., the upper-left edges in Fig. 1.) For each such cell, we define score function taking words from both contexts.

$$l_{(z,w)} = \begin{cases} \log \sigma(v_z^v \cdot v_w^a) + \sum_{k=1}^{K} \log \sigma(-v_z^v \cdot v_{w_k}^a) & \text{if } z \text{ is from a vertical context,} \\ \log \sigma(v_z^{ac} \cdot v_w^a) + \sum_{k=1}^{K} \log \sigma(-v_z^{ac} \cdot v_{w_k}^a) & \text{if } z \text{ is from a horizontal context.} \end{cases}$$

---

[4] The original paper of word2vec derived this objective by maximizing the probability of a word appearing in the given contexts, but here we ignore these derivations and consider only the following objectives as merely the score function for the purpose of obtaining word-embedding vectors.

The final objective function for attribute vectors is thus the combination of the horizontal and vertical objectives, as follows:

$$l = \sum_{c \in C \wedge y_c = 1} [\sum_{z \in V_c} \{\log \sigma(v_z^v \cdot v_w^a) + \sum_{k=1}^{K} \log \sigma(-v_z^v \cdot v_{w_k}^a)\}$$

$$+ \sum_{z \in H_c} \{\log \sigma(v_z^{ac} \cdot v_w^a) + \sum_{k=1}^{K} \log \sigma(-v_z^{ac} \cdot v_{w_k}^a)\}])$$

Note that attribute vectors are combined with both attribute context vectors and value vectors in the objective function[5].

Unified Objective II (SQUARE). In the "square" approach, all the vectors v^a, v^{ac}, v^v, and v^{vc} are considered.

The final objective function for attribute vectors is thus the combination of the horizontal and vertical objectives considering all relations[6] (i.e., all edges shown in Fig. 1), as follows:

$$l = \sum_{c \in C \wedge y_c = 1} [\sum_{z \in V_c} \{\log \sigma(v_z^v \cdot v_w^a) + \sum_{k=1}^{K} \log \sigma(-v_z^v \cdot v_{w_k}^a)\}$$

$$+ \sum_{z \in V_c} \{\log \sigma(v_z^{vc} \cdot v_w^{ac}) + \sum_{k=1}^{K} \log \sigma(-v_z^{vc} \cdot v_{w_k}^{ac})\}$$

$$+ \sum_{z \in H_c} \{\log \sigma(v_z^{ac} \cdot v_w^a) + \sum_{k=1}^{K} \log \sigma(-v_z^{ac} \cdot v_{w_k}^a)\}])$$

$$+ \sum_{c \in C \wedge y_c \neq 1} \sum_{z \in H_c} \{\log \sigma(v_z^{vc} \cdot v_w^v) + \sum_{k=1}^{K} \log \sigma(-v_z^{vc} \cdot v_{w_k}^v)\}$$

4.4 Parameter Learning

We used the stochastic gradient descent (SGD) in the same manner as in the word2vec implementation [6], taking each pair of a word and its context as one training data and optimizing on it, iterating this process for every pair in the training data.

We used the same approach as the word2vec implementation, parallelized using the Hogwild approach [14], assigning each process a subset of tables in the corpus.

The number of iterations was set to 50.

---

[5] In addition, note that only two of these four terms are used for each (w, z) pair, rendering the SGD implementation for this model nearly the same as that of word2vec.

[6] Although two (the first and second) terms are used for the word w and its vertical context word c, we can differentiate each term independently because there are no vectors appearing both of the terms, thus we can use the iteration method similar to that of word2vec.

5 Experiments

To measure the appropriateness of the obtained vector representations for representing tables, we conducted two experiments: *a similar attribute ranking task* and *an attribute analogy task*. We compare our approach with simple distributional representation, which is so-called *vector space models*, which simply counts the number of words appearing in the table.

The dimension size was set to 50 and the negative sampling size, K, was set to 25.

5.1 Similar Attribute Ranking Task

As a standard evaluation criterion for word embeddings, we conducted a word similarity task designed for table word embedding. As mentioned previously, we used strings in the first row of the table as-is, without word segmentation or POS tagging. We normalized each word by removing all spaces and obtained a list of synonym sets in which all words in each set having the same normalized form. (For example, "n a m e" and "name" are regarded as synonyms and the list of synonyms contains the set that includes these two strings.) We discarded the strings that contained numeric characters, because such strings are noisy in most cases, as attributes such as mere numbers (e.g., "5") or simple variations of other attributes (e.g., "name3".)

For each word contained in these synonym sets, we created a word list ordered by cosine similarity values based on the vector representations for attribute vectors v^a.

We used *average precision* [4] for each list by regarding each synonym set as correct answers.

Assuming that we obtained a list of synonym candidates $\langle c_1, c_2, \ldots, c_n \rangle$ ranked on the basis of their similarity to the query, and given the synonym set $S = \{s_1, s_2, \ldots, \}$, the average precision of the resulting list is calculated as

$$\frac{1}{|S|} \sum_{1 \leq k \leq n} r_k \cdot precision(k),$$

where $precision(k)$ is the accuracy (i.e., ratio of correct answers [answers included in the gold standard list] to all answers) of the top k candidates, and r_k represents whether the k-th document is relevant (1) or not (0). (In other words, $r_k = 1$, if $c_k \in S$, and $r_k = 0$, otherwise.) This measure is the average of the precisions, where each precision is calculated for every correct answer x appearing in the output list as the precision for the set of all words above x in the list. The possible maximum value of the average precision is 1.0. We took the average (mean) of the average precisions over all the queries (the *mean average precision*) and used this value as the measure of total system performance.

For each set, we selected every string in the set as a query, regarding other strings as the answers, and calculated average precisions. This resulted in 2,869 queries. Final results were obtained by averaging the average precision values across all of these queries.

5.2   Attribute Analogy Task

As stated in the introduction, there are two types of attribute similarities, value and co-occurrence similarity. Here, we define another task, the "attribute analogy task" which is analogous to the analogy task for the evaluation of word embedding learning from sentences. Our assumption is that "if the attributes in the pair (a, b) co-occur in one table, and the attributes in another pair (c, d) co-occur in another table, where $a \simeq c$ and $b \simeq d$ according to value similarity, then $b - a \simeq d - b$, and we can guess the attribute d given (a, b, c) by searching for the attributes near $c + (b - a)$".

We constructed the test set for this task through the following steps.

1. We created a list of the top 1,000 most frequent schema from the tables in the data set. Here, the *schema* is the list of words in the first row in each table.

2. From these schema, we obtained similar word pair lists by calculating the similarity between every pair of words found in one or more schema. Here the similarity is calculated according to a simple vector space model in which the vector for each attribute is the bag of words found in the same column as the attribute. If the cosine similarity is greater than 0.5, then we regard the pair is similar.

3. For every similar word pairs $a \simeq b$ and $c \simeq d$, we checked whether the attribute pair (a, c) co-occur in one schema, and the attribute pair (b, d) also co-occurs in another schema. If this is true, then the question a:b - c:? whose answer is d, is added to the test set.

As a result, we obtained 15,248 analogy questions. For example, we obtained "Shimei (name)":"Chakunin (inauguration date)" - "Guest (guest name)": "Housoubi (broadcasting date)" as one tuple.

We again calculated the average precision for the list of words obtained by ordering the attributes by the similarity to the vector $c + b - a$. Here, the value is simply the precision when the output is the set of words from rank-1 to rank-n where n is the position of the correct answer in the output list.

Table 2 shows the results on both tasks. The combination of the two results is shown in the simple average and the harmonic mean (F-measure). The "baseline" is an approach that uses simple vector space models representing the contexts of each attribute word by the vector constructed by counting the number of words appearing in the same row or in the same column. We use three weighting methods, including the simple count, the pointwise mutual information (PMI), and the log of the counts (LOG). The label "mixed" means simply merging the horizontal and vertical contexts, which is equivalent to using the "horizontal" (or "vertical") method except that the contexts are a mixture of both ("horizontal" and "vertical").

The results clearly indicate that the vertical contexts are indispensable for this task because depending on only the horizontal contexts provided very poor results. In this task, the "vertical" method outperformed the "unified" method. We observed that the horizontal contexts had negative effects on ranking for

Table 2. Results: attribute similarity and attribute analogy task

Method	Similarity task	Analogy task	Average	F-measure
Full data				
Baseline (distributional vectors)				
Horizontal	2.46	13.04	7.75	9.72
Vertical	16.29	4.74	10.52	7.34
Mixed	12.96	18.04	15.5	15.08
Cross	13.10	**18.09**	15.60	15.20
Mixed (PMI)	16.08	10.28	13.18	12.54
Cross (PMI)	16.61	10.23	13.42	12.66
Mixed (LOG)	15.62	13.81	14.72	14.66
Cross (LOG)	13.60	15.89	14.75	14.66
Embedding (distributed vectors)				
Horizontal	1.47	12.22	6.85	2.62
Vertical	**20.25**	3.35	11.8	5.75
Mixed	14.65	15.59	15.12	15.11
Cross	13.51	16.52	15.02	14.86
Square	18.89	17.61	**18.25**	**18.23**

some queries, yielding high similarity scores for the attributes co-occurring with the query attribute (thus having high similarity in horizontal contexts) but low similarity in vertical contexts.

We observed that there are performance trade-offs between the similarity and analogy tasks. For example, using different configurations of weights for the baseline method showed that an increase in similarity task performance reduced the analogy task performance. We observed that the proposed embedding methods are slightly weaker than the baselines in that the F-measure of the embedding method was less than the baseline with the same configuration (e.g., "cross" in the embedding and "cross" in the baseline.) However, "square", which makes use of the property of embedding, by virtue of which it can modify vectors reflecting other vectors (i.e., to learn "attribute context vectors", the contexts of "attribute" vectors, with "value context vectors" as shown in Fig. 1) showed better performance (in F-measure) than the others.

Baseline showed good performance especially for the analogy task. We think the reason for this is that the test set contained a large number of contexts sufficient for a simple approach. Consequently, we also conducted another experiment, in which the data size for learning was reduced to one-tenth, with other conditions left unchanged.[7]

---

[7] Note that as a result, the size of the similarity and analogy task queries was reduced to 445 and 5,124, respectively.

Table 3 shows the results. In this setting, "square" slightly outperformed the baseline in the analogy task, and again showed the best performance in terms of the F-measure.

Table 3. Results: attribute similarity and attribute analogy task on the small data set

Method	Similarity task	Analogy task	Average	F-measure
Small data (1/10 Samples)				
Baseline (distributional vectors)				
Mixed	17.50	20.52	19.01	18.89
Cross	17.65	20.10	18.88	18.80
Embedding (distributed vectors)				
Cross	15.68	19.70	17.69	17.46
Square	**23.72**	**20.78**	**22.25**	**22.15**

6 Conclusions and Future Work

In this paper, we proposed a method for table word embedding that learns vector representations for words found in tables. We tested several approaches, one that considers horizontal relations as contexts, another that considers vertical relations as contexts, and another that considers both contexts. Experiments showed that our proposed method, especially the one considering vertical contexts, contributed to obtaining good representations of words in tables. Future work includes combining or comparing our table word embedding with word embeddings for sentences, such as sentences surrounding tables in Wikipedia articles.

Acknowledgement. This work was supported by JSPS KAKENHI Grant Numbers JP15K00309, JP15K00425, JP15K16077.

References

1. Bollegala, D., Alsuhaibani, M., Maehara, T., Kawarabayashi, K.I.: Joint word representation learning using a corpus and a semantic lexicon. In: Proceedings of AAAI 2016, pp. 2690–2696 (2016)
2. Bollegala, D., Maehara, T., Yoshida, Y., Kawarabayashi, K.I.: Learning word representations from relational graphs. In: Proceedings of AAAI 2015, pp. 2146–2152 (2015)
3. Cafarella, M.J., Halevy, A.Y., Wang, D.Z., Wu, E., Zhang, Y.: Webtables: exploring the power of tables on the web. Proc. VLDB Endowment **1**(1), 538–549 (2008)
4. Chakrabarti, S.: Mining the Web: Discovering Knowledge from Hypertext Data. Morgan-Kaufmann Publishers, Burlington (2002)
5. Embley, D., Hurst, M., Lopresti, D., Nagy, G.: Table-processing paradigms: a research survey. Int. J. Doc. Anal. Recogn. **8**(2), 66–86 (2006)

6. Ji, S., Satish, N., Li, S., Dubey, P.: Parallelizing word2vec in shared and distributed memory. CoRR abs/ 1604.04661 (2016)
7. Limaye, G., Sarawagi, S., Chakrabarti, S.: Annotating and searching web tables using entities, types and relationships. Proc. VLDB Endowment **3**(1), 1338–1347 (2010)
8. Lin, Y., Liu, Z., Sun, M., Liu, Y., Zhu, X.: Learning entity and relation embeddings for knowledge graph completion. In: Proceedings of AAAI 2015, pp. 2181–2187 (2015)
9. Mikolov, T., Sutskever, I., Chen, K., Corrado, G.S., Dean, J.: Distributed representations of words and phrases and their compositionality. In: Proceedings of NIPS 2013, pp. 3111–3119 (2013)
10. Munoz, E., Hogan, A., Mileo, A.: Triplifying Wikipedia's tables. In: Proceedings of the ISWC 2013 Workshop on Linked Data for Information Extraction (2013)
11. Neelakantan, A., Roth, B., McCallum, A.: Compositional vector space models for knowledge base completion. In: Proceedings of ACL 2015, pp. 156–166 (2015)
12. Pennington, J., Socher, R., Manning, C.D.: GloVe: global vectors for word representation. In: Proceedings of EMNLP 2014, pp. 1532–1543 (2014)
13. Pimplikar, R., Sarawagi, S.: Answering table queries on the web using column keywords. Proc. VLDB Endowment **5**(10), 908–919 (2012)
14. Recht, B., Re, C., Wright, S.J., Niu, F.: Hogwild: a lock-free approach to parallelizing stochastic gradient descent. In: Proceedings of NIPS 2011, pp. 693–701 (2011)
15. Toutanova, K., Chen, D., Pantel, P., Poon, H., Choudhury, P., Gamon, M.: Representing text for joint embedding of text and knowledge bases. In: Proceedings of EMNLP 2015, pp. 1499–1509 (2015)
16. Wang, Z., Zhang, J., Feng, J., Chen, Z.: Knowledge graph embedding by translating on hyperplanes. In: Proceedings of AAAI 2014, pp. 1112–1119 (2014)
17. Yin, P., Lu, Z., Li, H., Kao, B.: Neural enquirer: learning to query tables in natural language. In: Proceedings of IJCAI 2016, pp. 2308–2314 (2016)
18. Zanibbi, R., Blostein, D., Cordy, J.R.: A survey of table recognition. Int. J. Doc. Anal. Recogn. **7**(1), 1–16 (2004)

Link Prediction for Isolated Nodes in Heterogeneous Network by Topic-Based Co-clustering

Katsufumi Tomobe[1]([✉]), Masafumi Oyamada[2], and Shinji Nakadai[2]

[1] Department of Mechanical Engineering, Keio University, Kanagawa, Japan
katsufumi@keio.jp
[2] NEC Corporation, Kanagawa, Japan
m-oyamada@cq.jp.nec.com, s-nakadai@az.jp.nec.com

Abstract. This paper presents a new probabilistic generative model (PGM) that predicts links for isolated nodes in a heterogeneous network using textual data. In conventional PGMs, a link between two nodes is predicted on the basis of the nodes' other existing links. This method makes it difficult to predict links for isolated nodes, which happens when new items are recommended. In this study, we first naturally expand the relational topic model (RTM) to a heterogeneous network (Hetero-RTM). However, this simple extension degrades performance in a link prediction for existing nodes. We present a new model called the Grouped Hetero-RTM that has both latent topics and latent clusterings. Through intensive experiments that simulate real recommendation problems, the Grouped Hetero-RTM outperforms baseline methods at predicting links for isolated nodes. This model, furthermore, performs as effectively as the stochastic block model in the link prediction for existing nodes. We also find that the Grouped Hetero-RTM is effective for various textual data such as item reviews and movie descriptions.

Keywords: Relational model · Topic model · Link prediction · Recommendation

1 Introduction

Link prediction means to extract missing information in a network using existing links or nodes' attributes, and is applicable for various real problems. For example, the interactions between users and items are considered as links, and a recommendation or purchase forecasting can be transformed into a link prediction [4,5].

In link prediction, probabilistic generative models (PGM) are common approaches with many advantages: (1) they reveal community structures of nodes, (2) not only predict missing links but also identify possible spurious links, and (3) capture considerable roles of nodes in the network [14]. One of the famous PGMs, the stochastic block model (SBM), clusters nodes in accordance with existing links [19]. In a mixed membership stochastic block model

© Springer International Publishing AG 2017
J. Kim et al. (Eds.): PAKDD 2017, Part I, LNAI 10234, pp. 147–159, 2017.
DOI: 10.1007/978-3-319-57454-7_12

(MMSB), all nodes have distribution of the clusters, and overlapping clustering is taken into account [2]. However, since these approaches need existing links, they cannot be used to predict links of isolated nodes, such as for recommending new items.

The relational topic model (RTM) models documents and links for a homogeneous network [6]. On the other hand, real recommendation situations are basically heterogeneous textual and link data, for example, recommend movies to users on NetFlix with movie reviews or captions, propose videos on YouTube with comments, find web-sites on the Web, and predict which company will purchase a patent. In this article, we first naturally expand the RTM to a heterogeneous network. This model, called Hetero-RTM, predicts the purchases of new items by using only their textual data. This simple extension, however, degrades performance in link predictions for existing nodes. One reason for the degradation is that a latent value corresponds to both a link and a topic cluster.

We thus present a new model (Grouped Hetero-RTM) that has two separated latent values: link and topic cluster. Consequently, the Grouped Hetero-RTM performs as effectively as the SBM in link prediction for existing nodes. Furthermore, our models can also be used for topic labeling like a topic model [10]. This function helps model-users to understand the meaning of each item cluster and to label it with words.

Specifically, this article provides the following contributions:

- **Link prediction for isolated nodes.** As mentioned above, our models predict links of isolated nodes by using their textual data.

Proposed Link Prediction

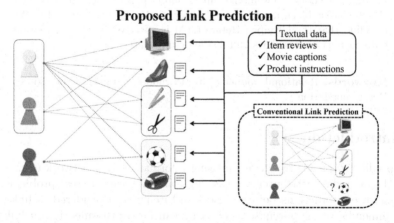

Fig. 1. Differences between proposed and conventional link predictions. Since conventional systems predict links by using existing links, they cannot predict links of isolated nodes, such as new items. Our models predict links by using nodes' textual data. Models also predict links that conventional models can predict. Red lines show the predicted links. (Color figure online)

- **No performance loss in link prediction for existing nodes.** In the link prediction for existing nodes, the Grouped Hetero-RTM performs as effectively as the SBM.
- **Labeling clusters.** Due to mixing with a topic model, item clusters are labeled with related words automatically (Fig. 1).

2   Related Works

Link prediction uses two types of features: graph based and collective local [13]. For the graph-based feature, purchase histories or ratings of items are converted into an interaction graph, and then the feature is analyzed by using various algorithms [9,21]. The SBM is one of the most widely used probabilistic relational models and predicts links using co-clustering of a heterogeneous network [1]. However, the SBM cannot predict any links of an isolated node correctly.

The collective local feature indicates items' attributes used for recommendation, and one of solutions for link prediction of an isolated node. In this study, we focus on textual data as a collective local feature of nodes. Topic models, such as the PLSA [12] and the LDA [3], are used to extract topics and make a topic-based clustering of the documents. In this model, documents have a topic distribution that generates a latent topic for each word. Simple topic models, however, ignore other rich information about the documents including their links, such as citations and hyperlinks.

Several methods that jointly model topics of documents and their links have emerged as extensions of the topic models. The Link-PLSA [7] and the Link-LDA [8] are simple extensions using links. The Pairwise-Link LDA [18] and the Link-PLSA-LDA [17] are developed, which are based on the concept that the simple extensions generate links from the same latent topic for word generations and fail to model the link generation.

Chang and Blei developed the RTM, which jointly models documents and their links [6]. Unlike the previous models, the RTM predicts links without any existing links using only their textual data. In case of paper citations, the model predicts citations only using their contents. The process of document generation is the same as that of the LDA; thus, a document has the same number of latent topics as words. A link is generated by the link probability by using the normalized topic distribution $\bar{z}_d = \frac{1}{N_d} \sum_n z_{d,n}$.

The RTM and its extensions, however, cannot be naturally used in a heterogenous network. For instances, while items have documents, it is difficult and unnatural to expand for users. In this study, we first transform and extend the RTM to capture the heterogeneous network as follows. We additionally developed a new model that combines the SBM and topic-based clustering.

3   Proposed Models

3.1   Probabilistic Generative Models

For modeling heterogeneous textual and link data, we developed two models: Hetero-RTM and Grouped Hetero-RTM. Hetero-RTM is a simple extension of the RTM, and Grouped Hetero-RTM is a new model that combines the SBM and topic-based clustering. In this section, we describe Bayesian models of our models. Nodes with and without textual data were defined as an item and a user, respectively. We assumed that $\mathcal{D} = \{(u_i, \boldsymbol{w}_j, y_{i,j})\}_{(i,j) \in \mathcal{I}}$ is a whole dataset, where u_i is a user i, $\boldsymbol{w}_j = \{w_{j,n}\}_{n=1}^{N_j}$ is a document of an item j, and $y_{i,j}$ is 1 or 0 if a user i and an item j are linked or not, respectively. N_j is the number of words in the document j, and $\mathcal{I}$ is a set of pairs of users and items. Graphical models of our two models are shown in Fig. 2.

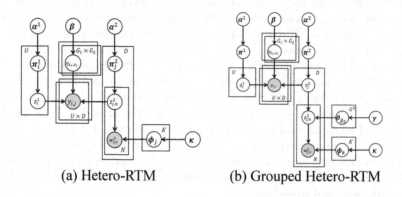

(a) Hetero-RTM   (b) Grouped Hetero-RTM

Fig. 2. Graphical model representations of the proposed models.

Hetero-RTM. We developed Hetero-RTM as a simple extension of the RTM. A topic is assigned from a multinomial distribution in each word n, $z_{j,n}^2 \sim$ Multi($\boldsymbol{\pi}_j^2$). A topic distribution in each document j is sampled from a Dirichlet prior, $\pi_j^2 \sim \text{Dir}(\boldsymbol{\alpha}_2)$. Then, a word is generated by the selected topic that has a multinomial distribution of words, $w_{j,n}^2 \sim \text{Multi}(\boldsymbol{\phi}_{z_{j,n}^2})$. The process of a user's cluster assignment is the same as above. A word distribution in each topic k is also sampled from a Dirichlet prior, $\phi_k \sim \text{Dir}(\boldsymbol{\kappa})$. However, the process to generate links is different from that of the RTM due to heterogeneous textual and link data. We use an averaged η, $\overline{\eta} = \frac{1}{N_j} \sum_{n=1}^{N_j} \eta_{z_i^1, z_{j,n}^2}$, as the Bernoulli parameter between i and j. Each η is sampled from a Beta prior like in the RTM, $\eta \sim \text{Beta}(\boldsymbol{\beta})$.

Grouped Hetero-RTM. In the Hetero-RTM, topics are shared in both relational and topic parts. However, a group structure in the link data may differ from that in the textual data. To take this into account, we developed the Grouped Hetero-RTM. A topic distribution is defined in each item cluster that is related to a group structure in the relational part, and a topic is sampled from this distribution, $z_{j,n}^2 \sim \text{Multi}(\boldsymbol{\theta}_{z_j^2})$. A item cluster is assigned for each document from the same multinomial distribution in a corpus like in the SBM, $z_j^2 \sim \text{Multi}(\boldsymbol{\pi}^2)$. Then, a word is generated from a word multinomial distribution in each topic.

Our final goal is to obtain the posterior distribution of the latent variables conditioned on the existing data and obtain statistics using the distribution. Unfortunately, this distribution is usually intractable. We used the Gibbs sampling, a Markov chain Monte Carlo sampling methods, to obtain an approximated posterior distribution.

3.2 Inference: Gibbs Sampling

Hetero-RTM. The full joint distribution of the Hetero-RTM is:

$$p(\boldsymbol{w}, \boldsymbol{z}^1, \boldsymbol{z}^2, \boldsymbol{y}, \boldsymbol{\pi}^1, \boldsymbol{\pi}^2, \boldsymbol{\phi}, \boldsymbol{\eta} | \boldsymbol{\alpha}^1, \boldsymbol{\alpha}^2, \boldsymbol{\kappa}, \boldsymbol{\beta})$$

$$= \prod_{i=1}^{U} p(z_i^1|\boldsymbol{\pi}_i^1)p(\boldsymbol{\pi}_i^1|\boldsymbol{\alpha}^1) \prod_{j=1}^{D} \prod_{n=1}^{N_j} p(w_{j,n}^2|\boldsymbol{\phi}_{z_{j,n}^2})p(z_{j,n}^2|\boldsymbol{\pi}_j^2)p(\boldsymbol{\phi}_{z_{j,n}^2}|\boldsymbol{\kappa})p(\boldsymbol{\pi}_j^2|\boldsymbol{\alpha}^2)$$

$$\prod_{(i,j)\in\mathcal{I}} p(y_{i,j}|\overline{\eta})p(\overline{\eta}|\boldsymbol{\beta}). \tag{1}$$

Algorithm 1. Gibbs sampling for Hetero-RTM

Inputs:
  $\{(\boldsymbol{w}_j, y_{i,j})\}_{(i,j)\in\mathcal{I}}$
Initialize:
  $\alpha_1, \alpha_2 \leftarrow 0.1, \kappa \leftarrow 0.01, \beta \leftarrow 1.0$
1: **for** $s = 1$ to S **do**
2:   **for** $j = 1$ to D **do**
3:     **for** $w = 1$ to W_d **do**
4:       Sample $z_{j,w}^2$ from (2).
5:     **end for**
6:     Sample π_j^2 from (4).
7:     **for** $(i,j) \in \mathcal{I}$ **do**
8:       Sample z_i^1 from (3).
9:     **end for**
10:     Sample π_i^1 from (6).
11:   **end for**
12:   Sample η from (7).
13:   **for** $k = 1$ to K **do**
14:     Sample ϕ from (5).
15:   **end for**
16: **end for**

From the full joint distribution and the graphical model, we derive conditional distributions of all parameters. The conditional distribution of latent topic $z_{j,n}^2$ is the LDA's conditional distribution of z multiplied by the link probability:

$$p(z_{j,n}^2 = k | w_{j,n}^2 = v, z_{\neg j,n}^1, w_{\neg j,n}^2, \phi, \pi^2, y, \eta)$$

$$\propto \frac{\pi_{j,k}^2 \phi_{k,v}}{\sum_{k'=1}^{K} \pi_{j,k'}^2 \phi_{k',v}} \prod_{(i,j)\in\mathcal{I}} \eta_{z_i^1,z_{j,n}^2}^{y_{i,j}} (1 - \eta_{z_i^1,z_{j,n}^2})^{1-y_{i,j}}, \tag{2}$$

$$p(z_i^1 = g_1 | \pi^1, y, \eta) \propto \pi_{i,g_1}^1 \prod_{(i,j)\in\mathcal{I}} \eta_{z_i^1,z_j^2}^{y_{i,j}} (1 - \eta_{z_i^1,z_{j,n}^2})^{1-y_{i,j}}, \tag{3}$$

where K is the number of topics. The conditional distributions of other parameters are as follows:

$$p(\pi_j^2 | z_{j,n}^2, \alpha^2) \propto \prod_{k'=1}^{K} (\pi_{j,k'}^2)^{n_{j,k'}+\alpha_k^2-1} \quad (n_{j,k'} = \sum_{n=1}^{N_j} \delta(z_{j,n}^2 = k')), \tag{4}$$

$$p(\phi_k | w_{j,n}^2, z_{j,n}^2, \kappa) \propto \prod_{v'=1}^{V} \phi_{k,v'}^{n_{v'}+\kappa_k-1} \quad (n_{v'} = \sum_{j=1}^{D}\sum_{n=1}^{N_j} \delta(w_{j,n}^2 = v', z_{j,n}^2 = k)), \tag{5}$$

$$p(\pi_i^1 | z_i^1, \alpha^1) \propto \prod_{g_1=1}^{G_1} (\pi_{i,g_1}^1)^{n_{i,g_1}+\alpha_{g_1}-1} \quad (n_{i,g_1} = \delta(z_i^1 = g_1)), \tag{6}$$

$$p(\eta_{g_1,k} | y, \eta, \beta) \propto \mathrm{Beta}(\beta_1', \beta_2') \quad (\beta_1' = \beta_1 + n_{g_1,k}^{y=1}, \beta_2' = \beta_2 + n_{g_1,k}^{y=0}), \tag{7}$$

where G_1 and V are the number of user clusters and words in the dictionary. $\alpha^1, \alpha^2, \kappa, and \beta$ are the hyperparameters of Dirichlet and Beta priors. $n_{g_1,k}^{y=1}$ and $n_{g_1,k}^{y=0}$ are the numbers of links and negative links between an user cluster g_1 and an item cluster k. Algorithm 1 summarizes the Gibbs sampling for the Hetero-RTM.

Grouped Hetero-RTM. The full joint distribution of the Grouped Hetero-RTM is as follows:

$$p(w, z_i^1, z^2, z'^2, y, \pi^1, \pi^2, \theta, \phi, \eta | \alpha_1, \alpha_2, \gamma, \kappa, \beta)$$

$$= \prod_{i=1}^{U} p(z_i^1 | \pi^1)p(\pi^1 | \alpha_1) \prod_{j=1}^{D} p(z_j^2 | \pi^2)p(\pi^2 | \alpha_2)$$

$$\prod_{n=1}^{N_j} p(w_{j,n}^2 | \phi_{z_{j,n}^2})p(z_{j,n}^2 | \theta_{z_j^2})p(\phi_{z_{j,n}^2} | \kappa)p(\theta_{z_j^2} | \gamma) \prod_{(i,j)\in\mathcal{I}} p(y_{i,j} | \eta_{z_i^1,z_j^2}) \prod_{g_1}^{G_1}\prod_{g_2}^{G_2} p(\eta_{g_1,g_2} | \beta), \tag{8}$$

where z'^2 is a vector of $z_{j,n}^2$. G_2 is introduced in the model and shows the number of item clusters. The assignment of a latent topic is simply estimated by following equation:

$$p(z_{j,n}^2 = k | w_{j,n}^2 = v, z_{\neg j,n}^2, w_{\neg j,n}^2, \phi, \theta) \propto \frac{\prod_{g_2'=1}^{G_2} \theta_{g_2',k}\phi_{k,v}}{\sum_{k'=1}^{K}\prod_{g_2'=1}^{G_2} \theta_{g_2',k'}\phi_{k',v}}. \tag{9}$$

Algorithm 2. Gibbs sampling for Grouped Hetero-RTM

Initialize:
  $\alpha_1, \alpha_2, \gamma \leftarrow 0.1, \kappa \leftarrow 0.01, \beta \leftarrow 1.0$
1: **for** $s = 1$ to S **do**
2:     **for** $i = 1$ to I **do**
3:         Sample z_j^2 from (11).
4:         **for** $w = 1$ to W_d **do**
5:             Sample $z_{j,w}^2$ from (9).
6:         **end for**
7:         **for** $(i,j) \in \mathcal{I}$ **do**
8:             Sample z_i^1 from (10).
9:         **end for**
10:    **end for**
11:    Sample π_1 from (14).
12:    Sample π_2 from (15).
13:    **for** $g_2 = 1$ to G_2 **do**
14:        Sample θ from (12).
15:    **end for**
16:    **for** $k = 1$ to K **do**
17:        Sample ϕ from (13).
18:    **end for**
19:    Sample η from (15).
20: **end for**

For the cluster assignment, we have to consider the network terms. The resulting conditional distributions for the assignment of a user and an item cluster are

$$p(z_i^1 = g_1 | y, \eta, \pi_1) \propto \pi_{g_1}^1 \prod_{(i,j) \in \mathcal{I}} \eta_{z_i^1, z_{j,n}^2}^{y_{i,j}} (1 - \eta_{z_i^1, z_{j,n}^2})^{1-y_{i,j}}, \tag{10}$$

$$p(z_j^2 = g_2 | y, \theta, \eta, \pi_2) \propto \pi_{g_2}^2 \prod_{n=1}^{N_j} \theta_{g_2, z_{j,n}^2} \prod_{(i,j) \in \mathcal{I}} \eta_{z_i^1, z_{j,n}^2}^{y_{i,j}} (1 - \eta_{z_i^1, z_{j,n}^2})^{1-y_{i,j}}. \tag{11}$$

Other parameters are sampled by the following equations:

$$p(\theta_{g_2} | z_j^2, z_{j,n}^2, \gamma) \propto \prod_{k'=1}^{K} \theta_{g_2, k'}^{n_{k'} + \gamma_k - 1} \quad (n_{k'} = \sum_{j=1}^{D} \sum_{n=1}^{N_j} \delta(z_{j,n}^2 = k', z_j^2 = g_2)), \tag{12}$$

$$p(\phi_k | z_{j,n}^2, w_{j,n}^2, \kappa) \propto \prod_{v'=1}^{V} \phi_{k,v'}^{n_{v'} + \kappa_k - 1} \quad (n_{v'} = \sum_{j=1}^{D} \sum_{n=1}^{N_j} \delta(w_{j,n}^2 = v', z_{j,n}^2 = k)), \tag{13}$$

$$p(\pi^1 | z_i^1, \alpha_1) \propto \prod_{g_1=1}^{G_1} (\pi_{g_1}^1)^{n_{g_1} + \alpha_{g_1} - 1} \quad (n_{g_1} = \sum_{j=1}^{D} \delta(z_i^1 = g_1)), \tag{14}$$

$$p(\pi^2 | z_j^2, \alpha_2) \propto \prod_{g_2=1}^{G_2} (\pi_{g_2}^2)^{n_{g_2} + \alpha_{g_2} - 1} \quad (n_{g_1} = \sum_{i=1}^{U} \delta(z_j^2 = g_2)), \tag{15}$$

$$p(\eta_{z_i^1, z_j^2} | \boldsymbol{y}, \boldsymbol{\eta}, \boldsymbol{\pi}_1) \propto \mathrm{Beta}(\beta_1', \beta_2') \quad (\beta_1' = \beta_1 + n_{g_1, g_2}^{y=1}, \beta_2' = \beta_2 + n_{g_1, g_2}^{y=0}). \tag{16}$$

Algorithm 2 summarizes the Gibbs sampling for the Grouped Hetero-RTM.

4 Experiments

4.1 Datasets

We evaluated our models on the following three different kinds of datasets: the MovieLens rating dataset [11] with the OMDb[1], and the Amazon review rating datasets [15] for Digital Music and Automotive. In the MovieLens dataset, we regarded ratings with more than three scores as links. The numbers of users, items, words (vocabulary), and links are listed in Table 1. Only 0.2% of negative links were randomly used for learning. We implemented all the approaches using Python (version 2.7.3) from scratch.

Table 1. Statistics of datasets used in our experiments.

	#users	#items	#vocabulary	#links
MovieLens & OMDb	943	1518	1265	79174
Amazon (digital music)	5541	1819	1101	39160
Amazon (automotive)	2928	1412	623	12904

For link predictions of new items, we separated items and link matrixes into training and test datasets. Our models used 80% of the items as a training dataset to learn the parameters and predicted links of 20% of the items as a test dataset. For a link predictions of existing items, we separated only links into training and test datasets with the same rates.

4.2 Baseline Method: KL Divergence-Based and Word2Vec-Based

We considered two baseline methods: KL divergence-based and Word2Vec-based.

Baseline 1 (KL divergence-based): Each user maintains a bag of words (BOW) of purchased items. The similarity between a new item and a user is defined by Kullback-Leibler (KL) divergence of multinomial distributions of words.

Baseline 2 (Word2Vec-based): Ozsoy assumes that items have a BOW and users also have a bag of items (locations) [20]. Word2Vec is a group of models to project words into a lower dimensional vector space via the neural network [16]. In this study, we also used an item-word space and a user-item space. When a new item is released, first we find the nearest conventional items in the sentence-word space and then recommend the new item to users who are near to the conventional items. The similarity was defined by using the cosine similarity between two vectors.

---

[1] http://www.omdbapi.com/.

4.3   Recommendation Using Link Prediction

All the models provide a way to predict a link using textual data, and we used the link prediction for recommendation. All links between users and items were ordered as follows:

Baseline1: KL divergence of word distributions between users and items, and recommended in ascending order.

Baseline2: (Step 1) In the item-word space, we picked out the item most similar to a new item using Word2Vec. (Step 2) In the user-item space, the new item was recommended to users who had high similarity with the item chosen in Step 1.

Hetero-RTM: The test documents were separated into two parts: D_{test1} and D_{test2}. D_{test1} was used for sampling and calculating a distribution of topics in each test document π^2_{test1}. Using π^2_{test1} and D_{test2}, we calculated the distribution of topics in each test document by using Eq. 2. Finally, new items were recommended in accordance with the link strength calculated by $p(z_i^1)\eta p(z_j^2)$.

Grouped Hetero-RTM: First, a distribution of topics was calculated by using Eq. 2. Then, the distribution of item cluster assignment $\overline{\theta}$ was calculated by using the following equation similarly to Eq. 11:

$$p(z_j^2 = g_2|\overline{\theta}, \pi_2) \propto \pi^2_{g_2} \prod_{n=1}^{N_j} \theta_{g_2, z_{j,n}^2} \tag{17}$$

The process of calculating the strength was the same as that of the Grouped Hetero-RTM.

5   Results

5.1   Performances of All Models

Figure 3(a) shows the Receiver Operator Characteristic (ROC) curves of our models and two baselines in the MovieLens & OMDb datasets. The proposed models dominate the baselines in all ROC space, which indicates that they outperform the baselines. Moreover, the proposed models always perform superiorly to the baselines in all thresholds. Figure 3(b) shows the ROC curve in the link prediction of existing items, and our models maintain high performances. Surprisingly, the Grouped Hetero-RTM suffers no performance loss. These results lead to the conclusion that the Grouped Hetero-RTM successfully predicts links of new items using their textual data without performance loss in predicting links of existing items.

We applied the same models to other two datasets. Table 2 summarizes the area under the curve (AUC) in all datasets. The boldfaced AUC values indicate the best performance in each dataset. The Grouped Hetero-RTM performs the best in all datasets, and the Hetero-RTM also predicts as much as the Grouped

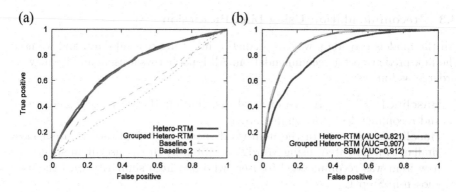

Fig. 3. ROC curves of each model for link prediction of (a) new items by textual data and (b) existing items in MovieLens & OMDb datasets.

Table 2. AUC values of link prediction of new items for various datasets.

	Baseline1	Baseline2	Hetero-RTM	Grouped Hetero-RTM
MovieLens & OMDb	0.482	0.503	**0.709**	**0.690**
Digital music	0.519	0.574	**0.609**	**0.625**
Automotive	0.528	0.536	0.557	**0.621**

Hetero-RTM. As textual data (vocabulary) decrease, new items cannot be significantly clustered and the prediction becomes harder. However, the Grouped Hetero-RTM maintains high performance in all datasets. In particular, only the Grouped Hetero-RTM predicts links in Automotive (which has a vocabulary below 650 words).

5.2   Comparison with Different Numbers of Clusterings

Next, we discuss effects of the multinomial parameters related to the number of clusters. The Grouped Hetero-RTM has one more parameter related to the number of clusters than the Hetero-RTM. Figure 4 shows the AUC values with different numbers of parameters in the Hetero-RTM and the Grouped Hetero-RTM. Both models maintain their performance with various numbers of parameters; however, the number of topics should be set below 15 for the Grouped Hetero-RTM. Although the LDA [3] and the SBM react simply for different numbers of parameters, the behaviors of the proposed models do not change monotonically. The behaviors are complex because (1) the model itself is a mixture of topic and relational models, and (2) the task (link prediction of new items) requires both abilities of topic and link based clustering.

5.3   How to Predict Links Using Textual Data

All above points make it clear that our models predict links of new items with high performance. Since our models are based on PGMs, they capture the group

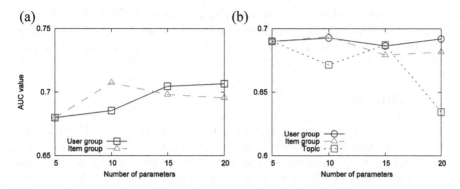

Fig. 4. AUC values of link prediction with different numbers of parameters in (a) Hetero-RTM and (b) Grouped Hetero-RTM. All AUC values are calculated in Movie-Lens & OMDb datasets. Squares, triangles and circles show the number of user clusters, item clusters and topics, respectively.

structure and role of the links in the network. In this subsection, we investigate the proposed links with high probability, particular those of the Hetero-RTM. The left part of Fig. 5 shows the parameter η that indicates the strength of relationships between the user and item clusters. As η becomes a high value, the probability of a link generation increases. The right part of Fig. 5 describes two highly recommended movies. Red and boldfaced words are ranked in the top 20 words of a multinomial distribution in item cluster 4. For instance, both

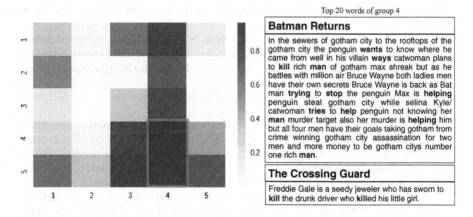

Fig. 5. Heat map of parameter η that shows strength of relationships between user and item clusters in Hetero-RTM. "Batman Returns" and "The Crossing Guard" are assigned as item cluster 4 with high probability. Red and boldfaced words are ranked in top 20 words of multinomial distribution in item cluster 4. These two movies were highly recommended to user clusters 4 or 5. These results were obtained in MovieLens & OMDb datasets. (Color figure online)

movies have a word ("kill") in their descriptions that highly contributes to their assignment to item cluster 4. From the parameter η, item cluster 4 has strong relationships with user clusters 4 and 5 ($\eta \simeq 1.0$). Therefore, the two movies are highly recommended to those who were weakly belong to user clusters 4 and 5.

6   Conclusion

In this paper, we focused on predicting links for isolated nodes. This link prediction can be used to recommend new items using their textual data (e.g. item reviews and product instructions). Our proposed models, the Hetero-RTM and the Grouped Hetero-RTM, are probabilistic generative models (PGMs) that treat heterogeneous textual and link data. We applied our models to three datasets and compared the results with those of baselines. The experimental results of the link prediction demonstrate that our models outperform baselines in all receiver operator characteristic (ROC) space. Furthermore, the Grouped Hetero-RTM predicts links of existing nodes as effectively as the stochastic block model (SBM). Overall, our models successfully predict purchases of new items without their purchase histories by using their textual data.

Acknowledgments. This work is supported in part by MEXT Grant-in-Aid for the Program for Leading Graduate Schools and Keio University Doctorate Student Grant-in-Aid Program.

References

1. Airoldi, E.M., Blei, D.M., Fienberg, S.E., Xing, E.P.: Mixed membership stochastic blockmodels. In: Advances in Neural Information Processing Systems, pp. 33–40 (2009)
2. Airoldi, E.M., Blei, D.M., Fienberg, S.E., Xing, E.P., Jaakkola, T.: Mixed membership stochastic block models for relational data with application to protein-protein interactions. In: Proceedings of the International Biometrics Society Annual Meeting, pp. 1–34 (2006)
3. Blei, D.M., Ng, A.Y., Jordan, M.I.: Latent Dirichlet allocation. J. Mach. Learn. Res. **3**, 993–1022 (2003)
4. Breese, J.S., Heckerman, D., Kadie, C.: Empirical analysis of predictive algorithms for collaborative filtering. In: Proceedings of 14th Conference Uncertainty in Artificial Intelligence (UAI 1998), pp. 43–52. Morgan Kaufmann Publishers Inc. (1998)
5. Cai, X., Bain, M., Krzywicki, A., Wobcke, W., Kim, Y.S., Compton, P., Mahidadia, A.: ProCF: probabilistic collaborative filtering for reciprocal recommendation. In: Pei, J., Tseng, V.S., Cao, L., Motoda, H., Xu, G. (eds.) PAKDD 2013. LNCS (LNAI), vol. 7819, pp. 1–12. Springer, Heidelberg (2013). doi:10.1007/978-3-642-37456-2_1
6. Chang, J., Blei, D.M.: Relational topic models for document networks. In: Proceedings of 12th International Conference on Artificial Intelligence and Statistics (AISTATS 2009), vol. 9, pp. 81–88 (2009)
7. Cohn, D., Hofmann, T.: The missing link - a probabilistic model of document content and hypertext connectivity. In: Advances in Neural Information Processing Systems, pp. 430–436 (2001)

8. Erosheva, E., Fienberg, S., Lafferty, J.: Mixed-membership models of scientific publications. Proc. Natl. Acad. Sci. **101**(Suppl. 1), 5220–5227 (2004)
9. Gori, M., Pucci, A., Roma, V., Siena, I.: ItemRank: a random-walk based scoring algorithm for recommender engines. IJCAI **7**, 2766–2771 (2007)
10. Griffiths, T.L., Steyvers, M.: Finding scientific topics. Proc. Natl. Acad. Sci. **101**(Suppl. 1), 5228–5235 (2004)
11. Herlocker, J.L., Konstan, J.A., Borchers, A., Riedl, J.: An algorithmic framework for performing collaborative filtering. In: Proceedings of the 22nd Annual International ACM SIGIR Conference on Research and Development in Information Retrieval, pp. 230–237. ACM (1999)
12. Hofmann, T.: Probabilistic latent semantic analysis. In: Proceedings of the 15th Conference on Uncertainty in Artificial Intelligence (UAI 1999), pp. 289–296. Morgan Kaufmann Publishers Inc. (1999)
13. Li, X., Chen, H.: Recommendation as link prediction in bipartite graphs: a graph kernel-based machine learning approach. Decis. Support Syst. **54**(2), 880–890 (2013)
14. Lü, L., Zhou, T.: Link prediction in complex networks: a survey. Physica A **390**(6), 1150–1170 (2011)
15. McAuley, J., Targett, C., Shi, Q., van den Hengel, A.: Image-based recommendations on styles and substitutes. In: Proceedings of the 38th Annual International ACM SIGIR Conference on Research and Development in Information Retrieval, pp. 43–52. ACM (2015)
16. Mikolov, T., Dean, J.: Distributed representations of words and phrases and their compositionality. In: Advances in Neural Information Processing Systems (2013)
17. Nallapati, R., Cohen, W.W.: Link-PLSA-LDA: a new unsupervised model for topics and influence of blogs. In: ICWSM 2008 (2008)
18. Nallapati, R.M., Ahmed, A., Xing, E.P., Cohen, W.W.: Joint latent topic models for text and citations. In: Proceedings of the 14th ACM SIGKDD International Conference on Knowledge Discovery Data Mining (ACM SIGKDD 2008), pp. 542–550. ACM (2008)
19. Nowicki, K., Snijders, T.A.B.: Estimation and prediction for stochastic blockstructures. J. Am. Stat. Assoc. **96**(455), 1077–1087 (2001)
20. Ozsoy, M.G.: From word embeddings to item recommendation (2016). arXiv preprint: arXiv:1601.01356
21. Salakhutdinov, R., Mnih, A.: Probabilistic matrix factorization. In: Proceedings of the 25th Neural Information Processing Systems (NIPS 2011), vol. 20, pp. 1–8 (2011)

Predicting Destinations from Partial Trajectories Using Recurrent Neural Network

Yuki Endo$^{(\boxtimes)}$, Kyosuke Nishida, Hiroyuki Toda, and Hiroshi Sawada

NTT Service Evolution Laboratories, NTT Corporation,
1-1 Hikarinooka, Yokosuka-shi, Kanagawa-ken 239-0847, Japan
endo-wop@hotmail.co.jp,
{nishida.kyosuke,toda.hiroyuki,sawada.hiroshi}@lab.ntt.co.jp

Abstract. Predicting a user's destinations from his or her partial movement trajectories is still a challenging problem. To this end, we employ recurrent neural networks (RNNs), which can consider long-term dependencies and avoid a data sparsity problem. This is because the RNNs store statistical weights for long-term transitions in location sequences unlike conventional Markov process-based methods that count the number of short-term transitions. However, how to apply the RNNs to the destination prediction is not straight-forward, and thus we propose an efficient and accurate method for this problem. Specifically, our method represents trajectories as discretized features in a grid space and feeds sequences of them to the RNN model, which estimates the transition probabilities in the next timestep. Using these one-step transition probabilities, the visiting probabilities for the destination candidates are efficiently estimated by simulating the movements of objects based on stochastic sampling with an RNN encoder-decoder framework. We evaluate the proposed method on two different real datasets, i.e., taxi and personal trajectories. The results demonstrate that our method can predict destinations more accurately than state-of-the-art methods.

1 Introduction

Mobile devices equipped with a GPS sensor enable us to easily collect location information on moving objects, known as *trajectories*. Predicting future destinations from their current trajectories is crucial for various location-based services such as personal navigation systems and ride sharing services. For example, it is more effective to deliver advertisements on sightseeing places around the destinations rather than advertisements on the current places.

To predict destinations from a partial trajectory, existing methods model the movement tendencies by using data-driven approaches and have achieved satisfactory results in some applications [7,13,14,19]. In particular, they are based on relatively low-order Markov processes, which count the number of transitions of short sequences in historical trajectories to alleviate a data sparsity problem. However, the ability to learn *long-term dependencies* between a destination and long sequences towards the destination is important for accurate

© Springer International Publishing AG 2017
J. Kim et al. (Eds.): PAKDD 2017, Part I, LNAI 10234, pp. 160–172, 2017.
DOI: 10.1007/978-3-319-57454-7_13

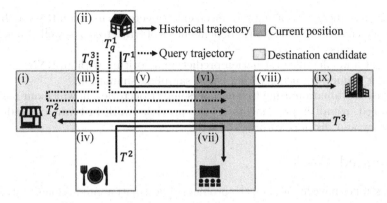

Fig. 1. Our destination prediction problem. Given a query trajectory, our method predicts top-k destinations on the basis of only historical trajectories. E.g., given T_q^1, it iterates one-step ahead predictions, starting from the cell (vi), to derive multi-step ahead predictions such as the cell (ix).

prediction because individuals move through the same area with different contexts (**challenge 1**). Figure 1 shows an example. We use a grid space because such discretized information is useful for modeling changes in trajectories. Given query trajectories T_q^1, T_q^2, and T_q^3, the task is to predict likely destinations, on the basis of historical trajectories T^1, T^2, and T^3. For example, given T_q^1, cell (ix) is the location that the user is most likely to visit. To compute this, Markov processes first calculate the transition probabilities from the current location (vi) to the subsequent locations (v), (vii), and (viii) for each query trajectory. If we assume a first-order Markov process for T_q^1, we can compute the transition probabilities $P(\mathrm{v}|T_q^1) = P(\mathrm{vii}|T_q^1) = P(\mathrm{viii}|T_q^1) = 0.33$ by counting the same transitions in the historical trajectories. Meanwhile, if we assume a second-order one, $P(\mathrm{vii}|T_q^1) = P(\mathrm{viii}|T_q^1) = 0.5$, and a fourth-order one can narrow down the candidates, that is, $P(\mathrm{viii}|T_q^1) = 1$. However, high-order Markov processes cause *the data sparsity problem* (**challenge 2**). For example, when a user departs another place, like in T_q^2, or takes a detour, like in T_q^3, $P(\mathrm{vii}|T_q^2)$ and $P(\mathrm{vii}|T_q^3)$ are not calculable in the fourth-order one. These two challenges lie in modeling the movements of specific individuals and also unknown individuals such as taxi users.

To overcome these two challenges, we exploit a recurrent neural network (RNN) [12], which can store sequential information in hidden layers. In order to model the transition information consisting of location points by using an RNN model, we represent the sequence of locations in a discretized grid space and let the model learn transitions from one cell to the next in each timestep. Different from the other count-based models, the RNN embeds sequences of sparse representations of cell locations into dense vectors consisting of statistical weights for the sequence of locations. The RNN can also handle variable lengths of trajectories without a strict built-in limit. These characteristics are useful for avoiding the data sparsity problem.

Specifically, the contributions in this paper are summarized as follows:

- Location sequence modeling for movement trajectories using an RNN architecture, which can learn long-term dependencies of trajectories as well as alleviate the data sparsity problem.
- Efficient and effective destination prediction algorithm with an RNN encoder-decoder framework using voting-based sampling simulation.
- Extensive evaluation using taxi and personal trajectories, in which our method produced overall improvements on the previous work in terms of predictive accuracy and distance error.

2   Related Work

Several studies have focused on the problem of destination prediction using external information such as trip time distribution [6–8] and road conditions [19]. Although external information is often useful to improve predictive accuracy, it is costly to obtain.

As a method that takes only trajectory data, Xue et al. [13,14] proposed Sub-Trajectory Synthesis (SubSyn) algorithm. This algorithm first estimates every transition probability in a grid space by using sub-trajectories based on low-order Markov process. Given a query trajectory, the algorithm then estimates visiting probabilities for destination candidates by using the transition probabilities from the starting and current locations to the candidates. Although their method efficiently alleviates the data sparsity problem, modeling transitions based on low-order Markov processes is not sufficient in terms of predictive accuracy for trajectories with diverse and long movements.

Brébisson et al. [1] formulated the destination prediction as a regression problem and solved it by using a multilayer perceptron (MLP). Their method assumes that the location of a destination can be represented as a linearly weighted combination of popular destination clusters, and the weights are computed using the MLP. Because the number of dimensions of the input of the MLP must be fixed, the oldest five points and newest five points in each trajectory are fed to the MLP. While their sequence-to-point prediction can minimize the overall distance error in dense areas of the training data, accurately predicting destinations is difficult especially in areas where trajectories are sparse. Furthermore, unlike the SubSyn algorithm [13,14] and ours, theirs is designed for predicting a single destination; it does not output multiple top k results. Although they also tried to use RNNs based on this method but it did not outperformed their MLP-based method.

Recently, another RNN-based model [9] was proposed and achieved satisfactory results in the task of the next location prediction based on check-in history. However, their model is not suitable for the multi-step prediction based on trajectory data. That is, while a next destination can be directly computed by considering a single transition for check-in data, transitions between multiple location points on the routes from a current location to a destination must be considered for trajectory data. Our focus in this paper is to efficiently predict destinations from trajectory data, which are collected at shorter intervals and enable earlier prediction than check-in data.

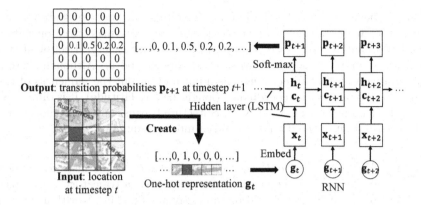

Fig. 2. Modeling transitions in trajectories using the RNN.

3 Method

Given a query trajectory $T_q = \{g_1, g_2, ..., g_c\}$ from starting time $t = 1$ to current time $t = c$, our goal is to accurately predict the probabilities $\mathbf{P} \in \mathbb{R}^{|C|}$ of visiting destination candidates $d \in C$. The trajectories are represented in a grid space, where an index g is assigned to each cell. The data available for solving this problem are only training data D consisting of historical trajectories $T = \{g_1, g_2, ..., g_{e_T}\}$ from starting time $t = 1$ to arrival time $t = e_T$, which are obtained from either unknown or specific individuals.

Note that our focus is to predict not a single destination but the visiting probabilities of multiple destinations. This is because the probabilities are useful in diverse applications. For example, location-based services can deliver multiple advertisements to users according to probability values. Additionally, car navigation systems allow users to efficiently select their destinations from candidates.

The procedure of our method is split into two phases: learning and prediction. In the learning phase, it uses historical trajectories to generate destination candidates $d \in C$ and estimates the model parameter θ of an RNN for the destination prediction. In the prediction phase, our method computes the visiting probabilities $\mathbf{P}$ from a query trajectory T_q by using the learned RNN model with θ. Section 3.1 describes our RNN model, and Sect. 3.2 describes our prediction algorithm based on the learned model.

3.1 Model

Architecture. Figure 2 illustrates our RNN architecture. The RNN takes as input a vector $\mathbf{g}_t$, which is a one-hot representation computed from a location index g_t at timestep t. The one-hot vector $\mathbf{g}_t$ is then embedded into a relatively low-dimensional space to obtain semantic representations of a location. Next, the embedded features are fed to the hidden layers consisting of long short-term memory (LSTM) units [3,5], which can memorize long-term sequences with variable sequence lengths. The LSTM units also take two previous state vectors, hidden state $\mathbf{h}_{t-1}$ and cell state $\mathbf{c}_{t-1}$. Finally, the soft-max layer outputs a transition probability $\mathbf{p}_{t+1} = P(g_{t+1}|g_t, g_{t-1}, ..., g_1)$ for each grid cell while the LSTM units compute the next two state vectors $\mathbf{h}_t$ and $\mathbf{c}_t$. Although there is an alternative way that directly uses as input and output the sequences of two scalar values of raw spatial coordinates (longitude and latitude), it is difficult to train such a simple model in the RNN architecture because any prior information on the distribution of the data is not taken into account [1].

Learning. We first generate destination candidates C. Given historical trajectories $T \in D$, we consider the last location point of each trajectory T as a past destination (e.g., taxi drop-off locations and stay points). The algorithm extracts the index d on the basis of the past destination for each trajectory and uses the set of the indices as the destination candidates C.

Next, we learn the model parameters of the RNN by maximizing the conditional likelihood over the set of all historical trajectories as:

$$\hat{\theta} = \arg \max_{\theta} \sum_{T \in D} \prod_{t=1}^{e_T - 1} P(g_{t+1}|g_t, g_{t-1}, ..., g_1; \theta). \tag{1}$$

This is equivalent to minimizing the cross-entropy loss between the output probability distributions $\mathbf{p}_t$ and the one-hot representations $\mathbf{g}_t$ at $t = 2, 3, ..., t_{e_T}$. To optimize θ based on Eq. (1), we use truncated back propagation through time (BPTT) [20] based on mini-batch AdaDelta [15], which is more efficient than vanilla stochastic gradient decent (SGD) or full-batch optimization. In our experiments, we set the length of truncated BPTT and size of a mini-batch to 20 and 100, respectively. We also clip the norm of the gradients (normalized by mini-batch size) at 5 to deal with exploding gradients [11].

3.2   Destination Prediction

Using the RNN model with θ, our method predicts visiting probabilities $\mathbf{P}$ for C from an input query trajectory T_q. Specifically, we exploit the RNN encoder-decoder framework. First, we compute hidden states $\mathbf{h}_c$ and $\mathbf{c}_c$ at the current timestep $t = c$ by using the RNN encoder with the LSTM units, which takes as input the sequence of one-hot representations $\mathbf{g}_1, ..., \mathbf{g}_c$ of T_q and recurrently updates the state vectors. The visiting probabilities $\mathbf{P}$ are then computed from the current hidden states $\mathbf{h}_c$ by using the RNN decoder with the soft-max layer. However, the soft-max layer only estimates transition probabilities $P(g_{c+1}|T_q)$ on the next timestep because the RNN decoder is based on sequence-to-sequence modeling. In most cases, more than one timestep is taken to get to destinations from current locations, and thus we need additional operations to obtain $\mathbf{P}$.

A naïve solution is to directly calculate transition probabilities for all transition patterns and integrate them. Given the maximum step size M for detours, which is computed from training data, the visiting probability for a destination

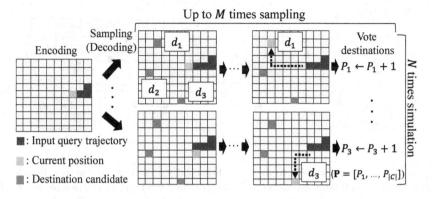

Fig. 3. Sampling simulation for destination prediction using the RNN encoder-decoder.

d of $\mathbf{P}$ is obtained as $P(g_{c+1} = d|T_q) + P(g_{c+2} = d|T_q) + \ldots + P(g_{c+M} = d|T_q)$, where $P(g_{c+m}|T_q)$ for variable m and any d is defined as:

$$\sum_{\forall g_{c+m-1}} \cdots \sum_{\forall g_{c+1}} \prod_{t=c}^{c+m-1} P(g_{t+1}|g_t, g_{t-1}, \ldots, g_{c+1}, T_q; \theta). \qquad (2)$$

Compared with low-order Markov processes, the RNN decoder takes a large amount of time to compute $P(g_{c+m}|T_q)$ because RNN depends on longer sequences. For example, to compute $P(g_{c+M}|T_q)$ as exact as possible, we need to iterate RNN decoding $|G|^{M-1}$ times; this is impossible because $|G|$ and M are usually a few thousand and a few dozen, respectively. If we assume that users move to adjacent cells in a single transition, the computational complexity reduces to $O(8^{M-1})$, but the prediction still takes a long time.

Instead, analogous to what is done in Monte Carlo methods and word sequence generation [4], our algorithm efficiently estimates the visiting probabilities for each destination candidate. As shown in Fig. 3, the algorithm simulates the movement of objects by stochastically sampling a position at the next timestep according to the transition probabilities obtained by the RNN decoder. Specifically, the algorithm first samples a position according to $P(g_{c+1}|T_q)$ and then samples a position according to $P(g_{c+2}|g_{c+1}, T_q)$. This process is repeated up to M times or until the sampling reaches one of the destination candidates. If one of the destination candidates is reached, a vote is cast on the element of $\mathbf{P}$ corresponding to the destination. After $\mathbf{P}$ is initialized to a zero vector, this sampling simulation is iterated N times. Finally, by normalizing $\mathbf{P}$, we can estimate the visiting probability distribution for each destination candidate. This procedure can be easily parallelized for each simulation step.

Considering Spatial Proximity. In our sampling simulation, there is no constraint on the distance of a single transition in a discretized grid space. This results in the algorithm predicting wrong destinations far away from a true destination because a simulation sample may suddenly jump to a distant cell.

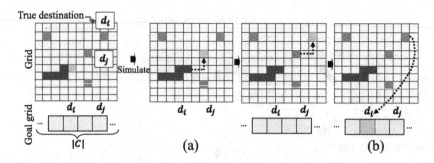

Fig. 4. Virtual goal grid cells to distinguish arrival states from goal states. If indices of destination candidates are defined on the common grid space, the sample stops at a false destination candidate on the routes to a true destination (a). If indices of goal grid cells are defined, the sample does not stop at the false destination candidates and jumps to the goal grid cell of a true destination from a nearby cell (b).

To consider distance information, we control the transition probabilities on the basis of the spatial proximity of the cells. In practice, we update the transition probabilities $\mathbf{p}_t$ estimated using the RNN decoder and obtain new transition probabilities $\mathbf{p}'_t$ so that a sample does not easily jump to a distant cell:

$$\mathbf{p}'_t = \frac{\mathbf{p}_t \circ \mathbf{s}_{g_{t-1}}}{|\mathbf{p}_t \circ \mathbf{s}_{g_{t-1}}|}, \tag{3}$$

$$\mathbf{s}_g = [\exp(-\frac{\text{Dist}(g, 1)}{\sigma^2}), ..., \exp(-\frac{\text{Dist}(g, |G|)}{\sigma^2})], \tag{4}$$

where $\circ$, $\text{Dist}(\cdot, \cdot)$, and $|G|$ denote element-wise multiplication, distance (in meters) between two cells, and the number of all grid cells, respectively. σ^2 denotes the variance of the distribution. The smaller σ^2 is, the harder it is for the sampling to jump to a distant cell. We set σ^2 to 200 m based on the size of a cell (i.e., 150 m × 150 m as explained in the Sect. 4.1). σ^2 can also be automatically determined from historical trajectories.

Distinguishing Arrival States from Moving States. If indices of destination candidates are defined on the common grid space, the sampling procedure stops when a sample reaches any of the other false destinations on the route to the true destination (Fig. 4(a)). This fact may prevent the sample from getting to a true destination far away from its current place. As shown in Fig. 4(b), we solve this problem by assigning destination candidates with indices of virtual goal grid cells to distinguish arrival states from moving states. The goal grid cells are connected with other common grid cells and indicate whether the sample arrives at a destination or moves through it.

4   Experiments

To validate the effectiveness of our method, we compared our method with the SubSyn algorithm [14], which is a state-of-the-art parameter-free algorithm that uses low-order Markov process modeling, and the MLP-based algorithm [1].

Table 1. Geographical ranges and statistics of the datasets.

Dataset	TST	GL
Latitude	[41.131571, 41.162477]	[39.68, 40.19]
Longitude	[−8.613876, −8.565437]	[116.05, 116.72]
# of trajectories	154,616	8,390 (in 46 users)
Avg. distance (m)	2127.4 ± 1130.0	5258.1 ± 8172.7
Avg. cell length	14.8 ± 7.0	13.3 ± 13.8

4.1 Datasets

We used a taxi service trajectory (TST) dataset [10], which contains trajectories for hundreds of taxis. Each trajectory includes sequences of latitude and longitude from a boarding location to a drop-off location. Additionally, we used a GeoLife dataset [16–18], which is publicly available dataset of personal trajectories in Beijing. We assumed that a change point of a transportation mode was a stay point and considered a segment with the same transportation mode as a single trajectory from an origin to a destination. We omitted users that had less than ten trajectories from the dataset. For our method and the SubSyn algorithm, cell size and a grid space need to be defined. We used 150 m × 150 m cells and a grid space consisting of a part of the full dataset to reduce the high computational costs of conducting various experimental conditions. Table 1 summarizes statistics of these datasets.

4.2 Experimental Settings

Evaluation Measures. We used *Accuracy@k* and *Distance@k* as the evaluation measures. *Accuracy@k* indicates the ratio of destinations that are accurately predicted in a cell to all query trajectories T_q. On the other hand, *Distance@k* indicates the average distance error between the true destinations and the predicted destinations for all query trajectories T_q. We computed these measures for the top k destinations based on destination visiting probabilities **P** and used the best values in top k.

Evaluation Methods. We sorted the trajectories in each dataset in ascending order of time and used the first 70% for training and the remaining 30% for test. For the TST dataset, we generated a single model for all of the trajectories to evaluate our method for unknown individuals. Meanwhile, for the GL dataset, we generated multiple models for multiple users to evaluate our method for specific individuals. In this case, we computed *Accuracy@k* and *Distance@k* for each user and averaged them. As a query trajectory T_q, we used the older $\alpha\%$ location points in each test trajectory and took the last location point to be the ground-truth destination.

4.3 Results and Analysis

In this section, we answer the research questions that correspond to the two challenges, explained in Sect. 1, by analyzing our experimental results.

RQ1. Can the proposed method make improvements to the destination prediction by learning long-term dependencies?

Figure 5 shows *Accuracy@k* and *Distance@k* for each dataset. As can be seen in the results for the TST dataset, our method outperformed SubSyn in terms of both *Accuracy* and *Distance* when using the same input length α. In particular, the degree of *Accuracy* improvement was remarkable when larger values of α were used (i.e., query trajectories have longer sequences). Moreover, Ours50% outperformed SubSyn70% that used a longer input. MLP, which formulates the destination prediction as a regression problem, minimizes the overall *Distance* in the training dataset; on the other hand, our method optimizes the probabilities of visiting destination cells, i.e., the *Accuracy* measure. Therefore, our method performed significantly better than MLP in terms of *Accuracy@1*. In contrast, MLP significantly outperformed our method in *Distance@1* when $\alpha = 30\%$, but ours yielded comparable or slightly poor results to MLP when $\alpha = 50\%$ and performed slightly better than MLP when $\alpha = 70\%$. These results indicate the capability of our method for handling long-term dependencies.

Our method worked much better than the other methods on the GL dataset, especially for larger values of α. One reason for the improvements is that our method can capture personal routines based on long sequences of trajectories. Meanwhile, larger values of α worsened the predictive accuracy and distance error of SubSyn in contrast to the results of Ours and MLP. In the GL dataset, there are several frequent origins (e.g., home), and the current location of a query trajectory will be distant (near) from such origins when α is large (small). That is, larger values of α seemed to decrease the number of training transitions between the current and destination locations although SubSyn needs the transition probabilities between current and destination locations.

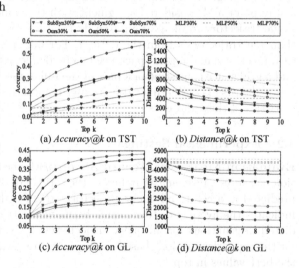

(a) *Accuracy@k* on TST

(b) *Distance@k* on TST

(c) *Accuracy@k* on GL

(d) *Distance@k* on GL

Fig. 5. Overall performance of destination prediction.

RQ2. Can the proposed method alleviate the data sparsity problem?

Figure 6 shows performance for each method with different training data sizes on the TST dataset. Despite modeling longer sequences, our method often worked well even when the number of historical trajectories for learning was small. This is because our RNN stored the statistical weights of variable-length trajectories instead of counting fixed-length transitions. Additionally, as shown in Table 1 and Fig. 5, although the GL dataset had a smaller number of trajectories than the TST dataset, the improvements of our method over the other methods on it were larger than on the TST dataset. In this case, MLP performed worst because it faced the data sparsity problem when determining the

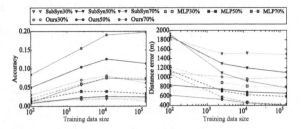

Fig. 6. *Accuracy*@1 (left) and *Distance*@1 (right) with different training data size in the TST dataset.

density-based destination clusters and when learning the sequence-to-point relation. Although SubSyn performed slightly better than MLP thanks to their sub-trajectory approach, Ours performed best because our sequence-to-sequence RNN model can cope with a small dataset more effectively.

Computational Time. Figure 7(a) shows the time needed to make a prediction for a query trajectory. We parallelized the sampling simulation using 10 processes on the CPU. In this figure, computational times linearly increase depending on the number of sampling simulations N, except when $N =$ 10 because most time was

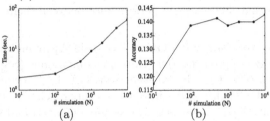

Fig. 7. (a) Prediction time and (b) *Accuracy*@1 depending on the number of simulation N.

taken for overhead costs of the parallelization. In particular, the prediction takes only a few seconds when N is a few hundred. Figure 7(b) shows *Accuracy* depending on N in 2,500 trajectories sampled from the TST dataset. As can be seen, while a large N tends to give better performance, the gain in accuracy begins to level off after $N = 100$. As for the learning time, we took a few days worth of data from the TST dataset, and took about a half hours worth for one user of the GL dataset. Although the learning time exceeded the prediction time, this does not matter much in real application services because the model can be learned in advance.

Parameter Sensitivity. We evaluated the sensitivity of our method to variations of the cell size using the GL dataset as shown in Table 2. For the grid-based methods (Ours and SubSyn), the cell size is important to determine the granularity of predicted destinations. The larger cell, the more difficult it is to identify the true location of a destination in a cell. That is, the large cell increases *Accuracy* while it decreases the resolution of predicted locations, which affects *Distance*. Nevertheless, ours outperformed the existing methods for every cell size in terms

Table 2. Performance of each method with different cell sizes when $\alpha = 50\%$.

Cell width	5000 m	1200 m	150 m	40 m
MLP (Accuracy@1)	0.466	0.268	0.101	0.012
SubSyn (Accuracy@1)	0.506	0.310	0.103	N/A
Ours (Accuracy@1)	**0.618**	**0.408**	**0.147**	**0.029**
MLP (Distance@1)	4448	4448	4448	4448
SubSyn (Distance@1)	4376	4247	4346	N/A
Ours (Distance@1)	**3427**	**2401**	**2380**	**2647**

of both metrics. For MLP, *Distance* did not depend on the cell size because it directly predicts a location of a specific destination. When the cell size was small (the number of cells was large), SubSyn could not predict destinations because its orders of computational complexity and memory space were $O(|G|^{3.5})$ and $O(|G|^{2.5})$, where $|G|$ is the number of cells.

For the network structure, we evaluated our method with other parameters (128–1024 LSTM units and 1–3 RNN layers). In the results, there were no significant differences in performance between them.

Comparison with Other Possible Approaches. We also validated our individual approaches (i.e., sampling simulations (SS), spatial proximity (SP), and goal cells (GC) described in Sect. 3.2). Table 3 compares the performance of the methods with and without these approaches using 2,500 trajectories sampled from the TST dataset. The results demonstrated that each of these approaches improved the accuracy and distance metrics.

Table 3. Performance of ours and other possible approaches when $\alpha = 50\%$.

Method	Ours	Ours w/o SS	Ours w/o SP	Ours w/o GC
Accuracy@1	**0.141**	0.051	0.115	0.050
Distance@1	**701**	1200	939	770

5   Conclusion

We proposed a method of predicting destinations from partial trajectories. Our method represents trajectories as sequences of one-hot representations on a grid space and explicitly learns transitions of objects by using an RNN model that stores statistical weights of sequential information in its hidden layers. This enables us (i) to model long-term dependencies while (ii) avoiding the data sparsity problem. Additionally, our method efficiently predicts destinations using a stochastically sampling simulation based on the RNN encoder-decoder framework. We conducted evaluation experiments using two different datasets and demonstrated that our method was effective for both unknown individuals and specific individuals.

Limitations and Future Work. The computational time for learning the RNN model linearly increases with the number of grid cells $|G|$, and thus it seems to be difficult to target a huge area with fine grids. Although we did not use finer grids with $150\,\text{m} \times 150\,\text{m}$ cells in our experiments, our method performed in large areas such as downtown in Beijing. A simple solution to expand a target area would be to divide a single RNN model that covers a huge area into multiple RNN models that cover a small area.

Another challenge is to incorporate time information into our method. Time information is helpful for predicting user activities; e.g., a user goes to an office in the morning and a bar in the evening. Multi-view models [2] for handling such information in trajectories would thus be an interesting topic of future work.

References

1. de Brébisson, A., Simon, É., Auvolat, A., Vincent, P., Bengio, Y.: Artificial neural networks applied to taxi destination prediction (2015). CoRR, abs/1508.00021
2. Elkahky, A.M., Song, Y., He, X.: A multi-view deep learning approach for cross domain user modeling in recommendation systems. In: WWW, pp. 278–288 (2015)
3. Gers, F., Schmidhuber, J., Cummins, F.: Learning to forget: continual prediction with LSTM. In: ICANN (2), pp. 850–855 (1999)
4. Graves, A.: Generating sequences with recurrent neural networks (2013). CoRR, abs/1308.0850
5. Hochreiter, S., Schmidhuber, J.: Long short-term memory. Neural Comput. **9**(8), 1735–1780 (1997)
6. Horvitz, E., Krumm, J.: Some help on the way: opportunistic routing under uncertainty. In: UbiComp, pp. 371–380 (2012)
7. Krumm, J., Horvitz, E.: Predestination: inferring destinations from partial trajectories. In: Dourish, P., Friday, A. (eds.) UbiComp 2006. LNCS, vol. 4206, pp. 243–260. Springer, Heidelberg (2006). doi:10.1007/11853565_15
8. Krumm, J., Horvitz, E.: Predestination: where do you want to go today? Computer **40**(4), 105–107 (2007)
9. Liu, Q., Wu, S., Wang, L., Tan, T.: Predicting the next location: a recurrent model with spatial and temporal contexts. In: AAAI, pp. 194–200 (2016)
10. Moreira-Matias, L., Gama, J., Ferreira, M., Mendes-Moreira, J., Damas, L.: Predicting taxi-passenger demand using streaming data. IEEE Trans. Intell. Transp. Syst. **14**(3), 1393–1402 (2013)
11. Pascanu, R., Mikolov, T., Bengio, Y.: On the difficulty of training recurrent neural networks. ICML **28**, 1310–1318 (2013)
12. Rumelhart, D.E., Hinton, G.E., Williams, R.J.: Learning representations by back-propagating errors. In: Neurocomputing: Foundations of Research, pp. 696–699 (1988)
13. Xue, A.Y., Qi, J., Xie, X., Zhang, R., Huang, J., Li, Y.: Solving the data sparsity problem in destination prediction. VLDB J. **24**(2), 219–243 (2015)
14. Xue, A.Y., Zhang, R., Zheng, Y., Xie, X., Huang, J., Xu, Z.: Destination prediction by sub-trajectory synthesis and privacy protection against such prediction. In: ICDE, pp. 254–265 (2013)
15. Zeiler, M.D.: ADADELTA: an adaptive learning rate method (2012). CoRR, abs/1212.5701

16. Zheng, Y., Li, Q., Chen, Y., Xie, X., Ma, W.: Understanding mobility based on GPS data. In: UbiComp 2008, pp. 312–321 (2008)
17. Zheng, Y., Xie, X., Ma, W.: GeoLife: a collaborative social networking service among user, location and trajectory. IEEE Data Eng. Bull. **33**(2), 32–39 (2010)
18. Zheng, Y., Zhang, L., Xie, X., Ma, W.: Mining interesting locations and travel sequences from GPS trajectories. In: WWW, pp. 791–800 (2009)
19. Ziebart, B.D., Maas, A.L., Dey, A.K., Bagnell, J.A.: Navigate like a cabbie: probabilistic reasoning from observed context-aware behavior. In: UbiComp, pp. 322–331 (2008)
20. Zipser, D.: Subgrouping reduces complexity and speeds up learning in recurrent networks. In: Advances in Neural Information Processing Systems 2, pp. 638–641. Morgan Kaufmann Publishers Inc. (1990)

Preventing Inadvertent Information Disclosures via Automatic Security Policies

Tanya Goyal, Sanket Mehta$^{(\boxtimes)}$, and Balaji Vasan Srinivasan

BigData Experience Lab, Adobe Research, Bangalore, India
{tagoyal,samehta,balsrini}@adobe.com

Abstract. Enterprises constantly share and exchange digital documents with sensitive information both within the organization and with external partners/customers. With the increase in digital data sharing, data breaches have also increased significantly resulting in sensitive information being accessed by unintended recipients. To protect documents against such unauthorized access, the documents are assigned a security policy which is a set of users and information about their access permissions on the document. With the surge in the volume of digital documents, manual assignment of security policies is infeasible and error prone calling for an automatic policy assignment. In this paper, we propose an algorithm that analyzes the sensitive information and historic access permissions to identify content-access correspondence via a novel multi-label classifier formulation. The classifier thus modeled is capable of recommending policies/access permissions for any new document. Comparisons with existing approaches in this space shows superior performance with the proposed framework across several evaluation criteria.

Keywords: Digital Rights Management · Extreme Multi-label Learning

1 Introduction

Enterprises constantly share and exchange digital documents containing sensitive information - internally with employees, externally with partners and with customers. With the increase in data sharing, the incidents of data breaches have also increased significantly [2]. A *data breach* is an incident in which sensitive, protected or confidential data, such as personally identifiable information (PII), personal health information (PHI), trade secrets and enterprise financial information is maliciously or inadvertently viewed, stolen or used by unauthorized entities. Until recently, the most common concept of a data breach embodied only malicious attackers hacking into enterprise networks to steal sensitive information. Data Loss Prevention (DLP) e.g. McAfee, Symantec are a class of software that detect and prevent such data breaches by malicious attackers by continuously monitoring sensitive data and providing appropriate encryption based on the sensitivity of the data.

However, according to a study by Johnson [14] in 2008, inadvertent disclosure of sensitive information represents one of the largest classes of security breaches

© Springer International Publishing AG 2017
J. Kim et al. (Eds.): PAKDD 2017, Part I, LNAI 10234, pp. 173–185, 2017.
DOI: 10.1007/978-3-319-57454-7_14

exceeding even the number of malicious data attacks. The consequences of such inadvertent data disclosures for enterprises could be severe: compromising customer's privacy, losing market share or damaging intellectual property. DLP software cannot tackle the inadvertent data disclosures since these occur outside the enterprise boundary. This has led to the development of Digital Rights Management (DRM) solutions e.g. Microsoft which are designed to protect sensitive data outside the enterprise boundary as well by ensuring only intended recipients can access the shared sensitive information regardless of their location.

The data being shared in DRM often vary in the degree of sensitive information. A *security policy* is applied to protect it against unauthorized access. A security policy is a collection of information that includes the confidentiality settings and a list of authorized users corresponding to the confidentiality settings. The confidentiality setting in the policy determines how the recipient can use the shared data, for example whether recipients can *print, copy,* or *edit text* in a protected document is dictated by the confidentiality settings corresponding to those recipients.

There are two major pitfalls with existing DRM solutions in identifying data breaches. First, the data breach identification in DRM is heavily based on keyword matching with a manually curated dictionary, thus limiting their capabilities severely [13]. Further, the policies in DRMs are manually assigned, which can lead to an error prone process [20]. With the increase in online document transactions, such a manual process is incapable of scaling to the enterprise document volumes, calling for an automated approach to protect against unauthorized access of information.

In this paper, we present an algorithm that analyzes the sensitive information and suggests appropriate access to the documents thus mitigating the risk from inadvertent disclosures. Our algorithm analyzes the historic information and extracts the semantics of the underlying content. The access permissions (that constitutes the document's DRM policy) associated with the information is then analyzed to identify content-access correspondence via a multi-label classifier formulation. The classifier thus modeled is capable of recommending policies/access permissions for any new document.

This paper is organized as follows. In Sect. 2, we describe existing literature in the light of our problem. In Sect. 3, we formulate the DRM policy modeling in a multi-label classification framework [17] and adapt the cost function to simultaneously optimize for precision and recall to suit the needs of the policy modeling. In Sect. 4, we compare the proposed framework against several alternative frameworks along with state-of-the-art security modeling systems showing the viability and superior performances of the proposed framework. Section 5 concludes the paper.

2   Related Work

While automatic recommendation of security/DRM policies for documents is less studied, one direction of explorations that is close to our problem is the prediction of email recipients based on its content. Carvalho and Cohen [8] have proposed an algorithm to generate a ranked list of intended recipients of an email via several algorithmic formulations including an expert search framework and a multi-class framework. Graus et al. [12] address the same problem via a graphical model framework of already composed email messages. Carvalho and Cohen [7] and Liu et al. [16] model the problem of identifying unintended recipients of emails as an outlier detection problem and a binary classification problem respectively based on the textual analysis of the email content overlayed with a correlation based on social network analysis. Zilberman et al. [22] propose an algorithm that extracts topics for all recipients and approves recipients for an email based on its topics and the common topics between the sender and the recipient. All these works focus just on identifying who are the right recipients of an e-mail from a curated list of recipients. However, in the context of DRMs, it is required not only to identify the 'recipients' of a content but also to determine the access permissions of the identified recipients on the content.

Evaluating the *sensitive* nature of information in a content is another direction of exploration that is related to our work. Existing DLP systems use regular expressions [6] and keyword matching to identify PII and other confidential information. Cumby and Ghani [9] present a semi automatic method to identify and redact private content from documents. However, they do not factor in the intended recipients of these documents while making such decisions. Hart et al. [13] propose a binary text classification algorithm to categorize a document as sensitive or non-sensitive. Geng et al. [11] use association mining between different types of PIIs to predict PIIs in emails. All these works are purely based on the document content, without any emphasis on the intended recipients of the content, which is a key factor in our problem.

To the best of our knowledge, there are no prior works that addresses the problem of suggesting appropriate DRM policy by a joint modeling between the document content and its intended recipients (along with their access permissions) which is the novel contribution of our work here.

3   Method

Consider an intra-organizational document repository that comprises of a set of documents with appropriate DRM policies attached to each of them. Let D denote the set of documents in the system, U be the set of all authorized users in the system and ACL be the set of access rights (permissions) in the system. For example, 'read', 'read-modify', 'read-modify-delete' are potential access rights for documents. Each document $d \in D$ is assigned a security policy p which describes the access rights of the authorized users for the document. A security policy is represented as a set of $(user, permission)$ pairs, where each pair defines an authorized $user \in U$ and its corresponding $permission \in ACL$, i.e., $p = \{(user_i, permission_j) | 1 \leq i \leq |U|, 1 \leq j \leq |ACL|\}$. Let P represent the set of all such policies available in the system. Further, we denote L to be the set of all $(user, permission)$ pairs, thus, $|L| = |U| \times |ACL|$.

Given such a repository with documents and their respective security policies, we propose an algorithm to suggest a set of $(user, permission)$ pairs for any new document based on its textual content. More formally, for a new document d, the algorithm aims to provide a ranked list $L' = [(user_1, permission_1), (user_2, permission_2), \cdots, (user_k, permission_k)]$ as the policy to protect the document from unauthorized access.

Feature Extraction: For a given document, we first extract features that capture the personal/sensitive information present in the document. These features include mentions of individuals and organizations, locations, dates, references to currency/money, phone number, social security number and email addresses. We use the Stanford Named Entity Tagger [3] to identify individual and organization names, locations, dates and currency. We further define regular expressions to identify phone numbers, social security numbers and email addresses in the document text. The feature set is further expanded to include individual words found in the document (after removing stop words and stemming the root words) using a "bag of words" representation. TFIDF (Term-Frequency-Inverse-Document-Frequency) [15] is extracted based on the bag-of-words representation for every document d.

With increasing size of document repository, the dimensions of the feature space also increases due to introduction of new words. We therefore reduce the dimensionality by removing features that might be irrelevant to the policy modeling to improve both the overall accuracy and scalability of the model. We calculate the information gain [10] of each feature f in the feature set as,

$$IG(f) = \sum_{l \in L} \sum_{f' \in \{f, \bar{f}\}} P(f', l).log \frac{P(f', l)}{P(f')P(l)} \tag{1}$$

where, l is a $(user, permission)$ pair. Information Gain measures the amount of information in bits obtained for $(user, permission)$ pair prediction by knowing the presence or absence of a feature, f. We retain the top $k\%$ (70% in our experiments later) informative features for our modeling.

Policy Modeling: Our framework to model security policies aims to suggest $(user, permission)$ pairs for any new document. A brute-force way to formulate this could be as a multi-class classification task where each class represents a security policy. A classifier $g : R^d \longrightarrow P$ can be learned based on the training set $\{(x_i, p_i) \mid 1 \leq i \leq |D|\}$, where $x_i \in R^d$ is the set of features for document d_i, and $p_i \in P$ is the corresponding security policy. For a new document, the above classifier can provide a probability of the existing policies $p_i \in P$ being suitable for the given document which can be used to decide on the final security policy for the document. However, modeling the problem as a multi-class framework problem captures neither the relation between the various security policies nor the overlap between security policies in terms of common users and permissions. Moreover, such a multi-class framework is also incapable of suggesting new policies outside what already exists in the repository and hence can be restrictive.

The above shortcomings can be addressed by formulating the problem in a multi-label classification framework, considering each unique $(user, permission)$ pair as a separate label on the document. In this framework, a function $f :$ $R^d \longrightarrow 2^L$ is learned from the training set $\{(x_i, y_i) \mid 1 \leq i \leq |D|\}$, where x_i is defined as before and $Y = \{y_1, y_2, ..., y_{|L|}\}$ denotes the label space. For a new document d_i, the multi-label classifier f predicts the set of labels, i.e., identifies the most relevant $(user, permission)$ pairs that must be assigned to a document as its policy. Unlike the multi-class framework, such a formulation is capable of recommending any set of $(user, permission)$ pairs in the repository.

There are several alternatives for multi-label classification. One class of algorithms transforms the multi-label classification problem to a binary classification, either by training binary classifier for each label in data-set (binary relevance method) [18] or by training binary classifier for multiple label subsets in data-set [21]. Another class of algorithms adapt traditional classification frameworks to the multi-label task [18]. However, both these methods are highly sensitive to the distribution of labels and do not perform well when the labels have very few training examples. Also, as these methods rely on independent models for each label/label sets, prediction cost increases with increasing number of labels.

In light of these drawbacks, a recent exploration in this space is centered on **Fast Extreme Multi-label Learning (FastXML)** which deals with large number of labels having skewed label distributions. FastXML learns a hierarchy over the feature space by recursively partitioning a parent's feature space between its children. The partitioning at each node is done by optimizing a ranking loss function, i.e., normalized Discounted Cumulative Gain (nDCG). Agrawal et al. [5,17] observe that in the case of XML problems, only a small number of labels are active in each partition of feature space, thus improving the modeling capacity.

However, the ranking-loss function in FastXML [17] only includes a reward for a high recall (correctly predicting all the relevant labels) without accounting for the precision (reducing incorrectly predicted labels) in the prediction. For modeling security policies, it is also important to penalize wrongly predicted $(user, permission)$ pairs, as that means that an ineligible user has been given permission to sensitive information. Given a ranking r of $(user, permission)$ pairs and the ground truth vector y_i, the discounted cumulative gain $L_{DCG}(r, y_i)$ is modified as,

$$L_{DCG@k}(r, y_i) = \sum_{l=1}^{k} \frac{(y_{r_l}) + (y_{r_l} - 1)}{log(1 + l)}, \tag{2}$$

where, y_{r_l} is the binary ground truth for the lth label according to ranking r (as defined in [17]), i.e. it has the value 1 if the lth label is attached to document d_i. We add $(y_{r_l} - 1)$ term to the definition of L_{DCG} [17] to introduce a $-1/log(1+l)$ term for each wrongly predicted $(user, permission)$ pair in the top-k labels. This ensures that apart from positive labels predicted with high ranks being rewarded, highly ranked negative labels are also penalized. Algorithm 1 outlines the steps required to obtain the hierarchy.

Algorithm 1. FastXML: Grow Tree

Require: $\{x_i, y_i\}_{i=1}^N$
1: $N_{root} \leftarrow$ new node
2: **if** no. of $(user, permission)$ pairs active in $N_{root} < MaxLeaf$ **then**
3:     Create N_{root} into a leaf node
4: **else**
5:     Learn linear separator w
6:     $n^+ = \{x_i | w^T x_i > 0\}$
7:     $n^- = \{x_i | w^T x_i < 0\}$
8:     $N_{root}(\text{linear\_separator}) = w$
9:     $N_{root}(\text{left\_child}) = \text{GrowTree}(\{x_i, y_i\}_{i \in n^+})$
10:     $N_{root}(\text{right\_child}) = \text{GrowTree}(\{x_i, y_i\}_{i \in n^-})$
11: **end if**
12: **return** N_{root}

Finally, the FastXML algorithm learns a linear separator w at each node of the tree, which divides the feature space into the positive and negative partition respectively, by minimizing a ranking loss function given by,

$$min||w||_1 + \sum_i log(1 + \exp(-\delta_i w^T x_i)) - \sum_i (1 + \delta_i) L_{nDCG}(r^+, y_i)$$

$$- \sum_i (1 - \delta_i) L_{nDCG}(r^-, y_i) \qquad (3)$$

where, $w \in R^d, \delta \in \{-1, +1\}, r^+, r^-$ is the ranked list of labels in the positive and negative partitions respectively. The labels are ranked in decreasing order of the number of documents in the partition that they are assigned to.

Prediction: To suggest $(user, permission)$ pairs for a new document, first a feature representation (x_i) of the new document d_i is extracted as before. The algorithm starts with the root node of each tree in the model and traverses down the tree till it reaches a leaf node. For traversal at each node, it calculates the value of the term $w^T x$ where w is the linear separator at that node. Since the linear separator at each node divides the feature space into two parts, depending on the sign of $w^T x$, the document d_i is passed down to the left child node (if $w^T x < 0$) or the right child node (if $w^T x > 0$) till it reaches a leaf node. Each leaf node contains a subset of points from the training data. The algorithm returns the ranked list of the top k $(user, permission)$ pairs active in the leaf nodes of all trees, where the rank is defined as:

$$r(x) = rank(\frac{1}{T} \sum_{t=1}^{|T|} P_t^{leaf}(x_i)) \qquad (4)$$

where T is the number of trees, $P_t^{leaf}(x_i) \propto \sum_{S_t^{leaf}(x_i)} y_i$ and $S_t^{leaf}(x_i)$ are the label distributions of the set of points in the leaf node that x_i reaches in tree t.

4   Experimentation and Results

The problem of automatic DRM policy assignment can be framed as a multi-class classification problem, with each policy being considered as a separate label. However, we had articulated the shortcomings of such a formulation and overcame these with our multi-label framework. In our experiments, we first compare our formulation with a multi-class framework and show the superior performance of our formulation to test our hypothesis. We also test our algorithm against several multi-label frameworks to show the viability of the proposed algorithm. Finally, we compare the performance of the proposed framework against existing security modeling in the context of email protection.

Dataset: All our experiments were performed on the Wikileaks Cablegate [4] data which includes diplomatic cables sent by the US State Department to its consulates, embassies, and diplomats around the world. Each cable is marked with a group of recipients to whom it was addressed, and a classification scale denoting the security level of the document. The classification scale on each document was one of *Unclassified, Limited Official Use, Confidential* or *Secret.*

In order to replicate an intra-organization document collection, we considered only the unique documents sent by Department of State for our experiments. Documents with insufficient textual content were discarded. The document's classification scale was used as the access permission given to its corresponding recipients. The set of all such (recipient, classification level) combinations for a document yielded the document's policy for our experiments. The filtered data-set contained 11,760 unique documents, 842 unique policies, 114 unique users/recipients and 452 (*user, permission*) pairs across all documents. This included 3,301 unclassified, 3,318 limited official use, 3,676 confidential and 1,465 secret documents. Table 1(a) provides additional details about the data-set, at the user level and the (*user, permission*) level.

Table 1. Wikileaks cablegate dataset

(a) WikiLeaks

(b) (*user, permission*)   distribution.

Label Description	User	User-Permission
Unique Labels	114	452
avg. labels/doc	5.11	5.11
avg. docs/label	185.01	47.18
min doc/label	49	1
max doc/label	1093	371

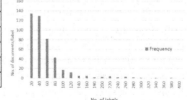

The average cardinality of labels, i.e., (*user, permission*) pairs, for documents is 5.11, which indicate that the security policies on each document

constitutes a set of 5 $(user, permission)$ pairs on an average. The average
number of documents for each $(user, permission)$ pair is 47. However, the
$(user, permission)$ pair distribution is highly skewed as seen in Table 1(b) and
ranges between a maximum 371 documents with the same user-permission pair
to a minimum of 1 document for a user-permission pair.

For all our experiments, the FastXML was built with 50 trees and the maximum number of instances allowed in a leaf node (MaxLeaf) was set at 10. The
number of labels k in a leaf node whose probability scores are retained was set
to 10.

4.1   Comparison with Multi-class Frameworks

Here, we compare our proposed method against two multi-class formulation to
check the viability of our formulation over the multi-class formulation. The first
framework is a **frequency based approach**, where all candidate policies are
ranked in decreasing order according to the number of times they were assigned
to any document in the training set. The top k policies, thus ranked, constitute
the set of policies suggested to the user. The second framework is a **1-vs-All
approach** where a separate SVM [19] is trained for each policy. For any new
document, the scores from each of the corresponding classifiers decides the policy ranking. To obtain a ranking of the policies for our proposed approach, we
obtain a binary vector representation $(\{0, 1\}^{|L|})$ of the algorithmically suggested
$(user, permission)$ pairs with value 1 for entries corresponding to the top k pairs.
Also, we represent all existing security policies present in the system into equivalent binary representations. Then, we use *cosine similarity* metric to identify
the nearest security policies and rank them accordingly.

To evaluate the performance of different approaches, we define the
$Accuracy@k$ which measures the probability of the actual policy being in the
top k predicted policies,

$$Accuracy@k = \frac{1}{|test\_set|} \sum_{doc \in test\_set} \mathbb{1}_{p \in f_k(doc)}, \tag{5}$$

where p is the actual policy of document doc and $f_k(doc)$ is the set of top-k
policies predicted by the algorithm f for the document doc.

Because of the training data requirements of an SVM model, we evaluated
only on those policies that had at least n corresponding documents and report
the accuracy for various values of n. Note that the proposed approach does not
suffer from this limitation since it does not try to generate individual models for
each $(user, permission)$ pair.

Table 2 shows the performance of the proposed framework against the multi-
class baselines. The results of our experiments indicate the superiority of our
proposed multi-label approach in comparison to the multi-class baselines. In
particular, the proposed approach performs significantly better in identifying
the relevant policy with smaller values of n. For instance, the $Accuracy@10$ for
our proposed approach is almost 6% higher than that achieved by 1-vs-All for

Table 2. Comparison against multi-class approaches

Dataset		Freq based		1-vs-All		Proposed approach	
n	#policies	Acc@5	Acc@10	Acc@5	Acc@10	Acc@5	Acc@10
20	95	0.205	0.323	0.565	0.656	**0.581**	**0.682**
30	73	0.226	0.356	0.601	0.691	**0.664**	**0.758**
40	54	0.258	0.406	0.634	0.728	**0.686**	**0.789**
50	43	0.278	0.442	0.685	0.771	**0.713**	**0.791**

the data-set with $n = 30$. This indicates that our algorithm is relatively agnostic to lower number of documents per policy in the training data. As the number of documents per label increases, the performance of the 1-vs-All approach is at par with the proposed approach - indicating that a multi class framework requires a lot of training data per policy to perform on par with the proposed framework.

Real world data-sets often contain numerous policies that might have very few supporting documents to train on as reflected in the Wikileaks data-set as well where only 95 out of 842 policies have been applied to more than 20 documents. Thus, any algorithm that aims to suggest policies by training on such a data-set needs to be robust enough to provide reasonably accurate predictions with lesser documents per policy.

4.2 Comparison with Multi-label Approaches

Next, we compare our approach against multi-label frameworks, 1-vs-All and Rakel [21]. For both these approaches, each $(user, permission)$ pair is considered as a separate label, similar to the proposed algorithm. In the 1-vs-All framework, a separate linear SVM classifier is trained for each unique $(user, permission)$ pair different from the multi-class framework where the modeling was done at the policy level. For Rakel, linear SVM classifiers are built for random ensembles of labels. For any new document, the final score for any label (user-permission pair in our context) is obtained by taking the cumulative score of all ensembles that the label is a part of.

In multi-label frameworks, output is generated at a lower granularity - (user, permission) pairs instead of the entire policy. We therefore use a different set of metrics to evaluate the algorithms in this framework. Since the candidate set of labels is typically large and only a small fraction of those labels is attached to any document, the algorithm needs to be measured on both its ability to correctly predict the highly ranked labels (precision), as well as the ability to retrieve all relevant labels (recall). Here, we define these metrics in the context of our problem. The *Precision@k* counts the correct pairs in the top k predicted pairs, whereas the *Recall@k* counts the fraction of the actual pairs that the algorithm is able to predict in the top k pairs. More formally,

$$Precision@K(y',y) = \frac{1}{k} \sum_{i \in rank_k(y')} y_i, \text{ and } Recall@K(y',y) = \frac{\sum_{i \in rank_k(y')} y_i}{\sum_{i \in |y|} y_i}.$$

(6)

where y' is the vector of predicted $(user, permission)$ pair vector, y is the ground truth permission set. In our evaluations, we report the average $Precision@k$ and $Recall@k$ across all documents in the test set.

We first evaluate our algorithm's prediction capability at the user level by retaining only the unique users from the k predicted $(user, permission)$ pairs for each document. These are compared against the actual users that have been given some permission by the policy. Modeling at the user level provides a measure of how correctly the set of users who should have permission to a given document is identified. High precision in these cases suggests that fewer irrelevant users are being given access to a certain document whereas high recall suggests that most users that must be given some level of access to the document have been identified. Tables 3 and 4 summarize the results at the user level. Table 4 shows that the algorithm is able to surface 85% of all users if it returns a list of the top 25 labels. Thus, the administrator has to only go through this much condensed list instead of the entire list of 114 users in the system.

Table 3. Precision for different values of k evaluated at user and user-permission level

	Precision@1	Precision@2	Precision@3	Precision@4	Precision@5
User level					
1 vs All	0.619	0.557	0.496	0.447	0.408
Rakel	0.508	0.478	0.441	0.4	0.365
Proposed algorithm	**0.653**	**0.609**	**0.548**	**0.502**	**0.463**
User-permission level					
1 vs All	0.34	0.32	0.3	0.286	0.272
Rakel	0.256	0.257	0.259	0.26	0.25
Proposed algorithm	**0.524**	**0.490**	**0.454**	**0.420**	**0.387**

Table 4. Recall for different values of k evaluated at user and user-permission level

	Recall@5	Recall@10	Recall@15	Recall@20	Recall@25
User level					
1 vs All	0.502	0.652	0.729	0.767	0.791
Rakel	0.486	0.64	0.674	0.705	0.723
Proposed algorithm	**0.521**	**0.688**	**0.767**	**0.815**	**0.848**
User-permission level					
1 vs All	0.29	0.416	0.483	0.524	0.553
Rakel	0.283	0.42	0.493	0.54	0.57
Proposed algorithm	**0.466**	**0.614**	**0.698**	**0.753**	**0.788**

Next, we evaluate the performance of the algorithms at the user-permission level. As previously elaborated, each policy assigns a unique permission to a user. That is, a policy cannot contain multiple $(user, permission)$ pairs with the same user. However, in multi-label classification, the predicted list of labels may contain multiple labels/$(user, permission)$ pairs corresponding to the same user. We leverage the nature of our problem to solve this. Since the algorithm aims to suggest policies to safeguard against data leak, we hypothesize that a stricter policy is preferable to a more lenient one which may provide extraneous rights to some users. Hence, if a particular user is a part of multiple $(user, permission)$ pairs finally predicted, we make use of the inherent hierarchy in the permissions and assign the strictest permission to that user. For instance, if both (u_i, p_m) and (u_i, p_n) are in the top k predicted labels, we compare the permissions p_m and p_n and retain the stricter permission. In *ACL* permissions, for example, a *Read* permission is stricter than a *Modify* permission, as the latter gives more rights to the user. Tables 3 and 4 shows the results at the $user, permission$ level. Table 4 shows that the algorithm is able to surface 78% of all $user, permission$ pairs if it returns a list of the top 25 labels. This is greatly reduced from the 452 total $user, permission$ pairs that exist in the system.

The proposed algorithm performs the best among all compared frameworks. One reason for this is the ability of our model to handle the skewed distribution of $(user, permissions)$. There exists some $user, permission$ pairs that have been assigned to a very few documents as reflected in the Wikileaks dataset and our proposed algorithm is robust in such scenarios. Traditional methods like 1-vs-All and Rakel, which depend on training independent models for each label/label sets do not perform well in policy modeling with very few training examples.

4.3   Evaluation on Enron Email Dataset

While none of the existing literature addresses the exact problem of assigning security policies for documents, a related exploration is the task of predicting recipients for an email based on its content, e.g. [8]. Due to the robustness of our framework, our approach is capable of handling these tasks as well and here we evaluate the performance of our algorithm for the aforementioned task on the Enron Email Dataset [1] and compare it against the best performing approach in [8]. Our experiments are run on emails sent by 30 Enron users, selected

Table 5. Results on the Enron dataset

	Precision@1	Precision@2	Precision@3	Precision@4	Precision@5
Carvalho and Cohen [8]	0.606	0.440	0.352	0.301	0.264
Proposed algorithm	**0.892**	**0.550**	**0.415**	**0.341**	**0.293**
	Recall@1	Recall@2	Recall@3	Recall@4	Recall@5
Carvalho and Cohen [8]	0.485	0.640	0.711	0.761	0.789
Proposed algorithm	**0.747**	**0.809**	**0.835**	**0.850**	**0.861**

based on the volume of emails sent, after discarding emails having insufficient textual content. Table 5 provides a summary of the results. The proposed approach clearly outperforms the baseline in [8] thus proving the robustness of the proposed formulation.

5 Conclusion

In this work, we have addressed the problem of automatically recommending appropriate access permissions for users by deciding the appropriate DRM security policies for the document. We have proposed an algorithm that analyzes the sensitive information in the document and determines the right policies for it. Experiments on a real world dataset against several alternate frameworks establish the viability of the proposed approach. Comparison with existing baseline modeling approaches further establishes the superiority of the proposed approach across several evaluation criteria.

References

1. Enron email dataset webpage. http://www.cs.cmu.edu/~enron/
2. The history of data breaches. https://digitalguardian.com/blog/history-data-breaches. Accessed 22 Oct 2016
3. Stanford Named Entity Recognizer (NER). http://nlp.stanford.edu/software/CRF-NER.shtml. Accessed 26 Oct 2016
4. Wikileaks cablegate data. https://wikileaks.org/plusd/
5. Agrawal, R., Gupta, A., Prabhu, Y., Varma, M.: Multi-label learning with millions of labels: recommending advertiser bid phrases for web pages. In: Proceedings of the 22nd International Conference on World Wide Web, pp. 13–24. ACM (2013)
6. Aura, T., Kuhn, T.A., Roe, M.: Scanning electronic documents for personally identifiable information. In: Proceedings of the 5th ACM Workshop on Privacy in Electronic Society, pp. 41–50. ACM (2006)
7. Carvalho, V.R., Cohen, W.W.: Preventing information leaks in email. In: International Conference on Data Mining (SDM). SIAM (2007)
8. Carvalho, V.R., Cohen, W.W.: Ranking users for intelligent message addressing. In: Macdonald, C., Ounis, I., Plachouras, V., Ruthven, I., White, R.W. (eds.) ECIR 2008. LNCS, vol. 4956, pp. 321–333. Springer, Heidelberg (2008). doi:10.1007/978-3-540-78646-7_30
9. Cumby, C.M., Ghani, R.: A machine learning based system for semi-automatically redacting documents. In: IAAI (2011)
10. Gabrilovich, E., Markovitch, S.: Text categorization with many redundant features: using aggressive feature selection to make SVMs competitive with C4. 5. In: Proceedings of the Twenty-First International Conference on Machine Learning, p. 41. ACM (2004)
11. Geng, L., Korba, L., Wang, X., Wang, Y., Liu, H., You, Y.: Using data mining methods to predict personally identifiable information in emails. In: Tang, C., Ling, C.X., Zhou, X., Cercone, N.J., Li, X. (eds.) ADMA 2008. LNCS (LNAI), vol. 5139, pp. 272–281. Springer, Heidelberg (2008). doi:10.1007/978-3-540-88192-6_26

12. Graus, D., Van Dijk, D., Tsagkias, M., Weerkamp, W., De Rijke, M.: Recipient recommendation in enterprises using communication graphs and email content. In: Proceedings of the 37th International ACM SIGIR Conference on Research and Development in Information Retrieval, pp. 1079–1082. ACM (2014)

13. Hart, M., Manadhata, P., Johnson, R.: Text classification for data loss prevention. In: Fischer-Hübner, S., Hopper, N. (eds.) PETS 2011. LNCS, vol. 6794, pp. 18–37. Springer, Heidelberg (2011). doi:10.1007/978-3-642-22263-4_2

14. Johnson, M.E.: Information risk of inadvertent disclosure: an analysis of file-sharing risk in the financial supply chain. J. Manag. Inf. Syst. **25**(2), 97–124 (2008)

15. Leopold, E., Kindermann, J.: Text categorization with support vector machines. How to represent texts in input space? Mach. Learn. **46**(1–3), 423–444 (2002)

16. Liu, T., Pu, Y., Shi, J., Li, Q., Chen, X.: Towards misdirected email detection for preventing information leakage. In: 2014 IEEE Symposium on Computers and Communications (ISCC), pp. 1–6. IEEE (2014)

17. Prabhu, Y., Varma, M.: FastXML: a fast, accurate and stable tree-classifier for extreme multi-label learning. In: Proceedings of the 20th ACM SIGKDD International Conference on Knowledge Discovery and Data Mining, pp. 263–272. ACM (2014)

18. Sorower, M.S.: A Literature Survey on Algorithms for Multi-label Learning. Oregon State University, Corvallis (2010)

19. Suykens, J.A., Vandewalle, J.: Least squares support vector machine classifiers. Neural Process. Lett. **9**(3), 293–300 (1999)

20. Verizon RISK Team: 2014 data breach investigations report (2014)

21. Tsoumakas, G., Vlahavas, I.: Random k-labelsets: an ensemble method for multilabel classification. In: Kok, J.N., Koronacki, J., Mantaras, R.L., Matwin, S., Mladenič, D., Skowron, A. (eds.) ECML 2007. LNCS (LNAI), vol. 4701, pp. 406–417. Springer, Heidelberg (2007). doi:10.1007/978-3-540-74958-5_38

22. Zilberman, P., Dolev, S., Katz, G., Elovici, Y., Shabtai, A.: Analyzing group communication for preventing data leakage via email. In: 2011 IEEE International Conference on Intelligence and Security Informatics (ISI), pp. 37–41. IEEE (2011)

Personalized Deep Learning for Tag Recommendation

Hanh T.H. Nguyen$^{(\boxtimes)}$, Martin Wistuba, Josif Grabocka,
Lucas Rego Drumond, and Lars Schmidt-Thieme

Information Systems and Machine Learning Lab, University of Hildesheim,
Universitätsplatz 1, 31141 Hildesheim, Germany
{nthhanh,wistuba,josif,ldrumond,schmidt-thieme}@ismll.de

Abstract. Social media services deploy tag recommendation systems to facilitate the process of tagging objects which depends on the information of both the user's preferences and the tagged object. However, most image tag recommender systems do not consider the additional information provided by the uploaded image but rely only on textual information, or make use of simple low-level image features. In this paper, we propose a personalized deep learning approach for the image tag recommendation that considers the user's preferences, as well as visual information. We employ Convolutional Neural Networks (CNNs), which already provide excellent performance for image classification and recognition, to obtain visual features from images in a supervised way. We provide empirical evidence that features selected in this fashion improve the capability of tag recommender systems, compared to the current state of the art that is using hand-crafted visual features, or is solely based on the tagging history information. The proposed method yields up to at least two percent accuracy improvement in two real world datasets, namely NUS-WIDE and Flickr-PTR.

Keywords: Image tagging · Convolutional Neural Networks · Personalized tag recommendation

1 Introduction

Tags assigned freely by users can be used to support users organizing or searching resources of social media systems [1]. However, a considerable number of shared resources has few or no tags because of the time-consuming aspect of the tagging task. For example, between February 2004 to June 2007, around 64% of Flickr uploaded photos had 1 to 3 tags and around 20% had no tags [20]. To encourage users annotating their resources, tag recommendation systems are used to facilitate the tagging task by suggesting relevant tags for them. These systems can be personalized systems that recommend different tags depending on the users' preferences, or non-personalized ones that omit the users' interests. Because the tags represent the user's view to his resource, the recommended tag list for a user is practically a personalized list containing his "favorite" keywords.

© Springer International Publishing AG 2017
J. Kim et al. (Eds.): PAKDD 2017, Part I, LNAI 10234, pp. 186–197, 2017.
DOI: 10.1007/978-3-319-57454-7_15

The personalized models can be based on the relation between users, items and tags, or otherwise on the correlation information of tags [4,16,19].

The personalized approaches are not efficient for new images with no historical information. As Sigurbjörnsson and Van Zwol mentioned [20], people usually choose the words related to the contents or contexts such as location or time to annotate images. The visual information can be considered to be used in the personalized recommendation models in order to enhance the prediction quality. The recommended tags of a personalized content-aware tag recommendation express personal and content-aware characteristics as in Fig. 1.

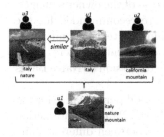

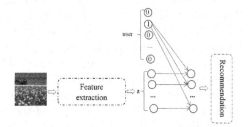

Fig. 1. The tags recommended for *u1* contain his favorite word *italy*, a word *mountain* related to the content of the image and a word *nature* from *u3* being similar to *u1*.

Fig. 2. The architecture of CNN-PerMLP

In this work, we show how a deep learning approach can be adapted to solve a personalized image tag recommendation. For a personalized problem, the features used in a deep learning model have to include the information of a user and an associated image. We propose a new way to add the user's information into the CNN models. A new layer that captures the interaction between users and visual features plays a bridged role between a CNN image feature extractor and a multilayer perceptron as in Fig. 2. In addition, we adapt the Bayesian Personalized Ranking optimization [18] in a different way to apply for the model.

Empirically, our experiments obtained in two real datasets, namely NUS-WIDE and Flickr-PTR, show that the proposed model outperforms the state-of-the-art personalized tag recommendation models, which are purely based on tagging history, up to at least four percent. The experiments also indicate the stronger support of the supervised features to increase the prediction quality up to at least two percent compared to low-level features.

2   Related Work

A large number of tag recommendation approaches focus on various features of objects, such as the contents of media objects, the relation between users and images, or the objects' contexts. The neighbor voting model [10] assembles the votes of similar images to retrieve the relevant tags. The collective knowledge approach [20] recommends the correlated tags with the user-provided tags based on the co-occurrence metric. The metric is also used for the personalized tag recommendation [4]. The model predicts the relevant tags for users based on the global and personal correlated scores of tags.

The correlated scores of tags achieved from different contexts are aggregated to look for the relevant tags [15]. The contexts include the information of the whole system, the social contacts of a user and the attending groups. In another approach, both the content and context information are used to find the neighbors of a given image from the historical tagging collection of the owner's images. The most frequent tags selected from its neighbors are recommended for the image [14]. In the model proposed by Chen and Shin [2], textual and social features that are extracted from tags, titles, contents, comments or users' social activities are combined to represent tags. Then, logistic regression or Naïve Bayes is employed as the recommender.

Factorization models are widely applied and show a good performance for tag recommendation. One of the state-of-the-art models is the Pairwise Interaction Tensor Factorization (PITF). It models all interactions between different pairs of users, items and tags, and accumulates all pairwise scores to the tags' scores [19]. Factorization Machine (FM) [16] is an approach that takes advantage of feature engineering and factorization. It can be applied to solve different tasks, such as regression, classification or ranking.

Tag recommendation based on the visual information of items only can be viewed as a multilabel classification or an image annotation task. A Convolutional Neural Network (CNN), a strong model for image classification and recognition [8,9,22], is applied to solve image annotation [5,23]. The approach can learn the predictor by optimizing different losses, such as pairwise, or Weighted Approximate Ranking (WARP), either to deal with the ranking problem [5], or to predict labels from arbitrary trained objects [23].

Because the factorization models merely depend on the relation between users, images and tags, they perform worse when predict new images. Our proposed model relied on users and visual features of images overcomes the limitation of recommending tags for new images.

The image annotation models do not contain the user's information so they work poorly in a personalized scenario. The proposed model has a personalized layer that captures the user-aware features so that the deep learning model can be adapted into a personalized tag recommendation.

3   The Proposed Model

3.1   Problem Formulation

The personalized tag recommender suggests a ranked list of relevant tags to a user annotating a specific image. The set of tag assignments $\mathcal{A}$ can be represented as a combination of users U, images I and tags T. It is denoted as $\mathcal{A} = (a_{u,i,t}) \in$

$\mathbb{R}^{|U| \times |I| \times |T|}$ [11] where $a_{u,i,t} = 1$ if the user u assigns the tag t to the image i, or otherwise $a_{u,i,t} = 0$. The observed tagging set is defined as $S := \{(u,i,t)|a_{u,i,t} \in \mathcal{A} \wedge a_{u,i,t} = 1\}$. The set of relevant tags of a user-image tuple (u,i) is denoted as $T_{u,i} := \{t \in T \mid (u,i,t) \in S\}$. Let $P_S := \{(u,i) \mid \exists t \in T : (u,i,t) \in S\}$ be all observed posts [17].

In addition, the collection of all RGB squared images is defined as R. The visual features of the i-th image R_i is a vector $z_i \in \mathbb{R}^m$. In this paper, we crop each image into Q patches to enhance the value of extracted features so we can define the collection of images $R = \{R_{i,q} | R_{i,q} \in \mathbb{R}^{d \times d \times 3} \wedge i \in I \wedge q \in Q\}$.

The scoring function of the recommendation model computes the scores of tags for a given post $p_{u,i}$ which are used to rank tags. The score of a tag to a given post is represented as $\hat{y}(u,i,t) : U \times I \times T \to \mathbb{R}$. If the score $\hat{y}_{u,i,t_a}$ is larger than the score $\hat{y}_{u,i,t_b}$, the tag t_a is more relevant to the post $p_{u,i}$ than the tag t_b. The tag recommendation model is expected providing a top-K tag list $\hat{T}_{u,i}$ that is ranked in descending order of tags' scores for a post $p_{u,i}$.

$$\hat{T}_{u,i} := \underset{t \in T, |\hat{T}_{u,i}| = K}{\operatorname{argmax}} \hat{y}(u,i,t) \tag{1}$$

3.2 Personalized Content-Aware Tag Recommendation

The architecture of the proposed model called **CNN-PerMLP** based on the relation between the user and the visual features of the given image is illustrated in Fig. 2. The supervised visual features are achieved by passing a patch q of the image i through the CNN feature extractor.

To personalize the visual features, a proposed specific layer called the personalized fully-connected layer is obtained following the extractor. The layer captures the interaction between the user and each visual feature to generate the latent features for the post $p_{u,i}$.

A neural network is deployed as a predictor to compute the relevant probabilities of tags. The network receives the user-image features as the input and its outputs are used to derive a ranking of recommended tags.

In this paper, we divide images into several patches and the final scores of tags are the average scores computed from different patches. If the score of a tag to a given post $p_{u,i}$ and a patch q is represented as $\hat{y}' : U \times R \times T \to \mathbb{R}$, the final tag's score is

$$\hat{y}(u,i,t) = \underset{R_{i,q}, q \in Q}{\operatorname{avg}} \hat{y}'(u, R_{i,q}, t) \tag{2}$$

Convolution Neural Networks. The CNN is obtained to represent high-level abstraction of image features. One or more convolutional layers are employed to generate several feature maps by moving kernel windows smoothly across images. The k-th feature map of a given layer is denoted as τ^k, the weights and the biases of the filters for τ^k are $W^k \in \mathbb{R}^{p_1} \times \mathbb{R}^{p_2} \times \mathbb{R}^{p_2}$ and b_k where p_1 is the number of the previous layer's feature maps and p_2 is the dimension of kernel windows. The element at the position (i,j) of τ^k is acquired as

$$\tau_{ij}^k = \varphi\left(b_k + \sum_{a=1}^{p_1}(W_a^k * \xi^a)_{ij}\right) \tag{3}$$

with $*$ being the convolution operator, ξ^a being the a-th feature map of the previous layer and φ being the activation function. The subsampling layer, which pools a rectangular block of the previous layer to generate an element for the current feature map, follows the convolutional layer. If the max pooling operator is used, the element at the position (i, j) of the k-th feature map τ^k is denoted as follows

$$\tau_{ij}^k = \max_{a,b}(\xi^k)_{a,b} \tag{4}$$

where a and b are the positions of the element associating to the pooled block. The output of the CNN feature extractor is a dense feature vector representing the image. We define the extraction process of the patch q of the i-th image by

$$z_i^q = f_{cnn}(R_i^q) : \mathbb{R}^{d \times d \times 3} \to \mathbb{R}^m \tag{5}$$

Personalized Fully-Connected Layer. The extracted features from the CNN only contain the information of the image i. To personalize visual features of an image, the user's information has to be added or combined with these features. For this reason, a layer stood between the feature extractor and the predictor is employed to generate the user-aware features that are used as the input of the predictor.

If the model uses only the user's id as the personalized information, the user's features u are described as a sparse vector represented $\kappa_u := \{0, 1\}^{|U|}$. Both the visual feature vector z_i^q and the sparse vector κ_u are the input of this layer. The layer is responsible to capture the interaction between the user and to each visual feature. If the output of this layer is denoted by $\psi \in \mathbb{R}^m$, it is obtained as follows

$$\psi_j(u, z_i^q) = \varphi(b_j + w_j^{per} \cdot (z_i^q)_j + V_j \kappa_u) \tag{6}$$

where $w^{per} \in \mathbb{R}^m$ is the weights of the visual features and $V \in \mathbb{R}^{m \times |U|}$ is the weights of the user features. As in the convolutional layer, the elementwise activation function φ is used after combining the weighting visual feature and the user's features.

Multilayer Perceptron as the Predictor. To compute the scores of the tags, a multilayer perceptron is adopted as a predictor and its input is the output of the personalized fully-connected layer ψ. The output of the network is the relevant scores of tags associated with the post (u, i) and the patch q of the image i. Because the network in the proposed model has one hidden layer, we denote the neural network score function as follows

$$\hat{y}'(u, R_i^q, t_j) = \varphi\left(w_j^{out} \cdot \varphi(W^{hidden}\psi + b^{hidden}) + b_j^{out}\right) \tag{7}$$

where W^{hidden} and b^{hidden} are the weights and the biases of the hidden layer; $w_j^{out} \in W^{out}$ and b^{out} are the weights and the biases of the output layer.

3.3 Optimization

We adapt the Bayesian Personalized Ranking (BPR) optimization criterion [18] in a different way so that the algorithm can be applied to learn the deep learning personalized image tag recommendation.

Algorithm 1. Learning BPR

1: **Input:** P_S, S, R, N, α
2: **Output:** Θ

3: Initialize $\Theta \leftarrow \mathcal{N}(0, 0.1)$
4: **repeat**
5:    Pick $(u, i) \in P_{S_{train}}$ and $R_i^q \in R$ randomly
6:    Get $T_{u,i}^+ := \{t \in T \mid (u, i, t) \in S\}$
7:    Pick $T_{u,i}^- := \{t \in T \mid (u, i, t) \notin S\}$ randomly where $\mid T_{u,i}^- \mid = N$
8:    Compute $z_i^q = f_{cnn}(R_i^q)$ and $\psi(u, z_i^q)$
9:    **for** $t \in 1, \ldots, |T|$ **do**
10:       **if** $t \in T_{u,i}^+ \vee t \in T_{u,i}^-$ **then**
11:          Compute $\hat{y}'(u, R_i^q, t)$
12:       **end if**
13:    **end for**
14:    Update $\Theta \leftarrow \Theta + \alpha \left(\frac{\partial \, \mathrm{BPR}(u,i)}{\Theta} \right)$
15: **until** convergence
16: **return** Θ

The optimization based on BPR finds the model's parameters that maximize the difference between the relevant and irrelevant tags. In addition, the stochastic gradient descent applied for BPR is in respect of the quadruple (u, i, t^+, t^-); i.e., for each $(u, i, t^+) \in S_{train}$ and an unobserved tag of $p_{u,i}$ drawn at random, the loss is computed and is used to update the model's parameters. The aforementioned BPR is not efficient to be used to learn the proposed model. The BPR criterion with respect to the posts is proposed to use and it is defined as

$$\mathrm{BPR}(u, R_i^q) := \frac{1}{\mid T_{u,i}^+ \mid \mid T_{u,i}^- \mid} \sum_{t^+ \in T_{u,i}^+, t^- \in T_{u,i}^-} \ln \sigma(\hat{y}'(u, R_i^q, t^+, t^-)) \tag{8}$$

where $T_{u,i}^+ := \{t \in T \mid (u, i, t) \in S_{train}\}$ is the set of tags selected by the user u for the image i. The rest of tags is the unobserved tag set denoted as $T_{u,i}^- := \{t \in T \mid (u, i) \in P_{S_{train}} \wedge (u, i, t) \notin S_{train}\}$. The function $\sigma(x)$ is described as $\sigma(x) = \frac{1}{1+e^{-x}}$. The difference between the score of relevant tags and irrelevant tags is defined as $\hat{y}'(u, R_i^q, t^+, t^-) = \hat{y}'(u, R_i^q, t^+) - \hat{y}'(u, R_i^q, t^-)$.

The learning model's parameters process is described in Algorithm 1. For each random post, a random patch of the associated image is chosen to extract the visual features. An irrelevant set having N tags is selected at random from the unobserved tags of the post. The system computes the scores of all relevant tags and the drawn irrelevant tags. From Eq. (8), the gradient of BPR with respect to the model parameters is obtained as follows:

$$\frac{\partial \text{BPR}}{\partial \Theta} = \Omega \sum_{t^+ \in T_{u,i}^+} \sum_{t^- \in T_{u,i}^-} \Psi_{t^+,t^-} \frac{\partial \hat{y}'(u, R_i^q, t^+, t^-)}{\partial \Theta} \tag{9}$$

where

$$\Omega = \frac{1}{|T_{u,i}^+||T_{u,i}^-|} \qquad \Psi_{t^+,t^-} = \frac{e^{-\hat{y}'(u,R_i^q,t^+,t^-)}}{1 + e^{-\hat{y}'(u,R_i^q,t^+,t^-)}}$$

To learn the model, the gradients $\frac{\partial \hat{y}'(u,R_i^q,t^+)}{\partial \Theta}$ and $\frac{\partial \hat{y}'(u,R_i^q,t^-)}{\partial \Theta}$ have to be computed. Depending on the weights in the different layers, one or both the gradients are computed. For example, if the parameter θ_j depends on the relevant tags t_j^+, Eq. (9) becomes

$$\frac{\partial \text{BPR}}{\partial \theta_j} = \frac{\partial \text{BPR}}{\partial \hat{y}'(u,i,t_j^+)} \times \frac{\partial \hat{y}'(u,R_i^q,t_j^+)}{\partial \theta_j} = \frac{\partial \hat{y}'(u,R_i^q,t_j^+)}{\partial \theta_j} \cdot \Omega \sum_{t^- \in T_{u,i}^-} \Psi_{t_j^+,t^-} \tag{10}$$

To find the gradients of the CNN parameters, the derivatives with respect to the visual features are propagated backward to CNN. From the Eqs. (6) and (9), the derivatives are computed as

$$\frac{\partial \text{BPR}}{\partial (z_i^q)_j} = \Omega \sum_{t^+ \in T_{u,i}^+} \sum_{t^- \in T_{u,i}^-} \Psi_{t^+,t^-} \frac{\partial \hat{y}'(u,R_i^q,t^+,t^-)}{\partial \psi_j} \cdot w_j^{per} \tag{11}$$

4   Evaluation

In the evaluation, we performed experiments addressing the impact of supervised visual features and the personal factor on the tag recommendation process.

4.1   Dataset

We obtained experiments on subsets of the publicly available multilabel dataset NUS-WIDE [3] that contains 269,648 images and Flickr-PTR [12] that was created by crawling around 2 million Flickr images. We preprocessed the NUS-WIDE dataset as follows: keeping available images annotating by the 100 most popular tags, sampling 1.000 users, refining to get 10-core dataset referring to users and tags where each user or tag occurs at least in 10 posts [6] and removing tags assigning more than 50% of images by one user to avoid the case that users tag all their images by the same words. Similarly, the Flickr-PTR dataset is preprocessed by mapping all tags to WordNet [13], refining dataset to get the 40-core regarding to users and 400-core to tags dataset, sampling 500 users and removing tags assigning more than 50% of images by a user.

Table 1. Dataset characteristics

Dataset	Users $|U|$	Images $|I|$	Tags $|T|$	Triples $|S|$	Posts $|P_S|$	Training posts $|P_{S_{train}}|$	Test posts $|P_{S_{test}}|$
NUS-WIDE	1000	27.662	100	81.263	27.858	25.858	2.000
Flickr-PTR	323	29.095	133	94.387	29.096	23.402	5.694

Table 2. Layer characteristics of the convolutional architectures

Layer	NUS-WIDE	Flickr-PTR
The 1st ConvL	$6 \times 6 \times 3$ (stride: 3)	$5 \times 5 \times 3$ (stride: 2)
The 1st MaxPoolL	2×2	2×2
The 2nd ConvL	$6 \times 6 \times 10$ (stride: 2)	$5 \times 5 \times 10$ (stride: 1)
The 2nd MaxPoolL	2×2	3×3
The 3rd ConvL	$2 \times 2 \times 30$ (stride: 1)	$3 \times 3 \times 30$ (stride: 1)

We adapted leave-one-post-out [11] for users to split the dataset. For each user, 20% of Flickr-PTR posts and 2 NUS-WIDE posts are randomly picked and put into the test sets. These subdivided dataset can be described with respect to users, images, tags, triples and posts as in Table 1. Images crawled by Flickr API[1] were cropped from the aspect ratio retained 75×75 for NUS-WIDE or 50×50 for Flickr-PTR into 3 pieces at 3 positions top-left, center and bottom-right to be used as the input patches for training and predicting.

4.2 Experimental Setup

The architectures used for both datasets contain 3 convolutional layers (*ConvL*) alternated with 2 max-pooling layers (*MaxPoolL*). *ConvL*s in these architectures have the same number of kernels that are 10 for the first, 30 for the second and 128 for the third. Because of the difference of the image size, the dimensions of convolutional kernels and pooling blocks in these architectures are different shown in Table 2. The hidden layers of the predictor have the dimension 128 for both architectures and the rectifier function $max(0, x)$ is used as the activation function. The evaluation metric used in this paper is F1-measure in top K tag lists [19].

$$\text{F1@K} = \frac{2 \cdot \text{Prec@K} \cdot \text{Recall@K}}{\text{Prec@K} + \text{Recall@K}} \qquad (12)$$

where

$$\text{Prec@K} = \underset{(u,i) \in S_{test}}{\text{avg}} \frac{|\hat{T}_{u,i} \cap T_{u,i}|}{K} \qquad \text{Recall@K} = \underset{(u,i) \in S_{test}}{\text{avg}} \frac{|\hat{T}_{u,i} \cap T_{u,i}|}{|T_{u,i}|}$$

---

[1] https://www.flickr.com/services/api/.

$$\hat{T}_{u,i} = \text{Top}(u, i, K) = \underset{t \in T, |\hat{T}_{u,i}| = K}{\text{argmax}} \ \hat{y}(u, i, t)$$

The grid search mechanism was used to find the best learning rate α among the range $\{0.001, 0.0001, 0.00001\}$ for all $ConvLs$ and $\{0.01, 0.0001, 0.0001\}$ for all fully-connected layers, the best L2-regularization λ from the range $\lambda \in \{0.0, 0.0001, 0.00001\}$ while the momentum value μ was fixed to 0.9. The 64-dimension color histogram (CH) and 225-dimension block-wise color moments (CM55) provided by NUS-WIDE's authors [3] and the 64-dimension color histogram (CH) of Flickr-PTR images are used for comparison.

The proposed model **CNN-PerMLP** is compared to following personalized tag recommendation methods that use only the users' preference information and do not consider the visual features: Pairwise Interaction Tensor Factorization (**PITF**) [19], Factorization Machine (**FM**) [16], most popular tags by users (**MP-u**) [6].

It is also compared to the non-personalized models including most popular tags (**MP**) [6], the multilabel neural networks (BP-MLLs) [24] that have low-level visual features as the input (**CH-BPMLL, CM55-BPMLL**), **CNNs** obtained for image annotation which optimizes the pairwise ranking loss to learn the parameters as the loss used by Zhang and Zhou [24]. The reimplemented CNN is similar to the proposed model of Gong et al. [5] with respect to optimizing the loss under the ROC curve (AUC).

The adjusted models (**CH-PerMLP** and **CM55-PerMLP**) of the proposed model using low-level features were obtained for the comparison. We used the Tagrec framework [7] to learn MP and MP-u, and the Mulan library [21] to learn CH-BPMLL and CM55-BPMLL.

4.3   Results

As shown in Fig. 3, the non-personalized models cannot capture the user's interests and they just recommend tags related to the content. The prediction quality of these models is lower than that of the personalized model. However, the model with CNN supervised features captures more information than the models using low-level features, leading to a boosted performance around 2%.

The claim that visual features improve the prediction quality is more serious in Fig. 4. In the test having most new images, the weights associated with these images are not learned in the training process. So the prediction of the personalized content-ignored models like FM and PITF solely depends on users and their results are clearly comparable to the prediction of MP-u. The personalized content-aware models work better than them in this case and recommended tags rely on both users and visual image information. The visual features help increasing the prediction quality around 4%. The supervised features also prove their strength in the recommendation quality compared to the low-level features. The performance is improved around 2 to 3% as a result of using the learned visual features.

Examples in Table 3 show that the proposed model can predict both personal tags and content tags compared to MP-u that purely predicts personal tags and

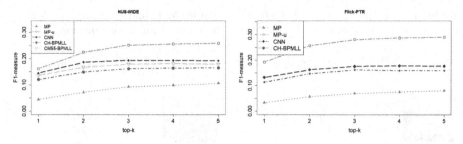

Fig. 3. The results of the non-personalized models are not as good as the personalized models but the model using supervised features outperforms the model using low-level features.

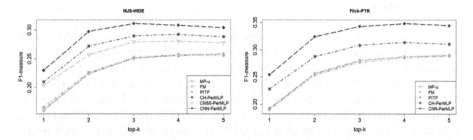

Fig. 4. CNN-PerMLP outperforms the personalized models based purely on tagging history and the personalized models using low-level visual features.

Table 3. Examples of top 5 recommended tags of CNN-PerMLP, CNN and MP-u

Image	Ground truth	CNN-PerMLP	CNN	MP-u	Image	Ground truth	CNN-PerMLP	CNN	MP-u
	flower	red	red	flowers		green	landscape	green	beautiful
	red	white	woman	white		grass	green	bravo	park
	orange	flowers	girl	orange		landscape	sky	blue	landscape
	white	orange	white	pink		park	park	nature	color
		flower	people	flower			grass	flowers	animal

CNN recommending content tags. As a result, the CNN-PerMLP suggests more relevant tags to the image. For example, in the first photo, the recommender can catch personal tags as "flowers", "flower", "orange" and content tags as "white" or "red". Through Tables 4 and 5, CNN-PerMLP works well in the case that people use their frequent tags or tags related to the image's content to annotate a new image. However, the prediction quality of the model is poor if the users assign tags that are new and do not relate to the content.

Table 4. Example having the highest accuracy of top 5 recommended tags

Image	Ground truth	Prediction
	lake	sunset
	sunset	water
	water	lake
	blue	blue
	sun	sun

Table 5. Example having the lowest accuracy of top 5 recommended tags

Image	Ground truth	Prediction
	green	sea
		beach
		sunset
		clouds
		ocean

5 Conclusion

In this paper, we propose a deep learning model using supervised visual features for personalized image tag recommendation. The experiments show that the proposed method has advantages over the state-of-the-art personalized tag recommendation purely based on tagging history, like PITF or FM in the narrow folksonomy scenarios. Moreover, the learnable features strongly influence the recommendation quality compared to the low-level features. The information of users used in the proposed models is plainly the users' id and it does not really represent the characteristic of the users, such as favorite words or favorite images. In the future, we plan to investigate how to use the textual features of users, in combination with the visual features of images to enhance the recommendation quality.

References

1. Ames, M., Naaman, M.: Why we tag: motivations for annotation in mobile and online media. In: Proceedings of the SIGCHI Conference on Human Factors in Computing Systems, pp. 971–980 (2007)
2. Chen, X., Shin, H.: Tag recommendation by machine learning with textual and social features. J. Intell. Inf. Syst. **40**, 261–282 (2013)
3. Chua, T.S., Tang, J., Hong, R., Li, H., Luo, Z., Zheng, Y.: NUS-WIDE: a real-world web image database from National University of Singapore. In: Proceedings of the ACM International Conference on Image and Video Retrieval, p. 48 (2009)
4. Garg, N., Weber, I.: Personalized, interactive tag recommendation for flickr. In: Proceedings of the 2008 ACM Conference on Recommender Systems, pp. 67–74 (2008)
5. Gong, Y., Jia, Y., Leung, T., Toshev, A., Ioffe, S.: Deep convolutional ranking for multilabel image annotation (2013). arXiv preprint: arXiv:1312.4894
6. Jäschke, R., Marinho, L., Hotho, A., Schmidt-Thieme, L., Stumme, G.: Tag recommendations in folksonomies. In: Kok, J.N., Koronacki, J., de Mantaras, R.L., Matwin, S., Mladenič, D., Skowron, A. (eds.) PKDD 2007. LNCS (LNAI), vol. 4702, pp. 506–514. Springer, Heidelberg (2007). doi:10.1007/978-3-540-74976-9_52
7. Kowald, D., Lacic, E., Trattner, C.: TagRec: towards a standardized tag recommender benchmarking framework. In: Proceedings of the 25th ACM Conference on Hypertext and Social Media, pp. 305–307 (2014)

8. Krizhevsky, A., Sutskever, I., Hinton, G.E.: ImageNet classification with deep convolutional neural networks. In: Advances in Neural Information Processing Systems, pp. 1097–1105 (2012)

9. LeCun, Y., Bottou, L., Bengio, Y., Haffner, P.: Gradient-based learning applied to document recognition. Proc. IEEE **86**, 2278–2324 (1998)

10. Li, X., Snoek, C.G., Worring, M.: Learning tag relevance by neighbor voting for social image retrieval. In: Proceedings of the 1st ACM International Conference on Multimedia Information Retrieval, pp. 180–187 (2008)

11. Marinho, L.B., Hotho, A., Jäschke, R., Nanopoulos, A., Rendle, S., Schmidt-Thieme, L., Stumme, G., Symeonidis, P.: Recommender Systems for Social Tagging Systems. Springer Science & Business Media, New York (2012)

12. McParlane, P.J., Moshfeghi, Y., Jose, J.M.: Collections for automatic image annotation and photo tag recommendation. In: Gurrin, C., Hopfgartner, F., Hurst, W., Johansen, H., Lee, H., O'Connor, N. (eds.) MMM 2014. LNCS, vol. 8325, pp. 133–145. Springer, Cham (2014). doi:10.1007/978-3-319-04114-8_12

13. Miller, G.A.: WordNet: a lexical database for English. Commun. ACM **38**, 39–41 (1995)

14. Qian, X., Liu, X., Zheng, C., Du, Y., Hou, X.: Tagging photos using users' vocabularies. Neurocomputing **111**, 144–153 (2013)

15. Rae, A., Sigurbjörnsson, B., van Zwol, R.: Improving tag recommendation using social networks. In: Adaptivity, Personalization and Fusion of Heterogeneous Information, pp. 92–99 (2010)

16. Rendle, S.: Factorization machines. In: 2010 IEEE 10th International Conference on Data Mining (ICDM), pp. 995–1000 (2010)

17. Rendle, S., Balby Marinho, L., Nanopoulos, A., Schmidt-Thieme, L.: Learning optimal ranking with tensor factorization for tag recommendation. In: Proceedings of the 15th ACM SIGKDD International Conference on Knowledge Discovery and Data Mining, pp. 727–736 (2009)

18. Rendle, S., Freudenthaler, C., Gantner, Z., Schmidt-Thieme, L.: BPR: Bayesian personalized ranking from implicit feedback. In: Proceedings of the Twenty-Fifth Conference on Uncertainty in Artificial Intelligence, pp. 452–461 (2009)

19. Rendle, S., Schmidt-Thieme, L.: Pairwise interaction tensor factorization for personalized tag recommendation. In: Proceedings of the Third ACM International Conference on Web Search and Data Mining, pp. 81–90 (2010)

20. Sigurbjörnsson, B., Van Zwol, R.: Flickr tag recommendation based on collective knowledge. In: Proceedings of the 17th International Conference on World Wide Web, pp. 327–336 (2008)

21. Tsoumakas, G., Katakis, I., Vlahavas, I.: Mining multi-label data. In: Maimon, O., Rokach, L. (eds.) Data Mining and Knowledge Discovery Handbook, pp. 667–685. Springer, New York (2009)

22. Wan, L., Zeiler, M., Zhang, S., Cun, Y.L., Fergus, R.: Regularization of neural networks using DropConnect. In: Proceedings of the 30th International Conference on Machine Learning (ICML-2013), pp. 1058–1066 (2013)

23. Wei, Y., Xia, W., Huang, J., Ni, B., Dong, J., Zhao, Y., Yan, S.: CNN: single-label to multi-label (2014). arXiv preprint: arXiv:1406.5726

24. Zhang, M.L., Zhou, Z.H.: Multilabel neural networks with applications to functional genomics and text categorization. IEEE Trans. Knowl. Data Eng. **18**(10), 1338–1351 (2006)

Information-Theoretic Non-redundant Subspace Clustering

Nina Hubig[1](✉) and Claudia Plant[2]

[1] Technical University of Munich, Bavaria, Germany
nina.hubig@tum.de
[2] University of Vienna, Vienna, Austria

Abstract. A comprehensive understanding of complex data requires multiple different views. Subspace clustering methods open up multiple interesting views since they support data objects to be assigned to different clusters in different subspaces. Conventional subspace clustering methods yield many redundant clusters or control redundancy by difficult to set parameters. In this paper, we employ concepts from information theory to naturally trade-off the two major properties of a subspace cluster: The quality of a cluster and its redundancy with respect to the other clusters. Our novel algorithm NORD (for NOn-ReDundant) efficiently discovers the truly relevant clusters in complex data sets without requiring any kind of threshold on their redundancy. NORD also exploits the concept of microclusters to support the detection of arbitrarily-shaped clusters. Our comprehensive experimental evaluation shows the effectiveness and efficiency of NORD on both synthetic and real-world data sets and provides a meaningful visualization of both the quality and the degree of the redundancy of the clustering result on first glance.

1 Introduction

Clustering is a powerful data exploration tool capable of uncovering previously unknown patterns in data. Subspace clustering is an extension of traditional clustering. It is based on the observation that different clusters, i.e., groups of mutually similar data objects, may exist in different subspaces of a data set. Research on subspace clustering has attracted a lot of attention, since in a plethora of applications domains, it is natural that objects are clustered flexibly in different subspaces. Typical applications for subspace clustering include:

- life science data, like genes under different experimental conditions [4] or in identification and classification of diseases [7]
- security and privacy in recommended systems [17],
- computer vision, e.g. image/motion/video segmentation [5,6] as well as image representation and compression [16].

How can we define the *quality of a clustering*? Our solution relies on the information-theoretic idea of linking data mining to data compression. Any kind of non-random patterns can be exploited to compress data. The stronger the

© Springer International Publishing AG 2017
J. Kim et al. (Eds.): PAKDD 2017, Part I, LNAI 10234, pp. 198–209, 2017.
DOI: 10.1007/978-3-319-57454-7_16

patterns and the better an algorithm succeeds in detecting them, the better is the compression rate. We define the quality of a cluster as its contribution to the overall compression rate.

The challenge is to distinguish interesting novel information from undesired redundancy. We approach this challenge by our novel clustering method NORD (for NOn-ReDundant) via information theoretic measures that automatically balances the novelty and quality of a subspace cluster. The information we gain by a subspace cluster on our data is two-fold: we learn which attributes span the subspace of the cluster and we learn which objects belong to that cluster. Information-theoretic measures allow to quantify the amount of novel information provided by each cluster as well as its redundancy to the remaining clusters. We combine this idea with the quality measure to obtain a novel information-theoretic optimization goal for subspace clustering, that automatically balances both aspects.

Contributions

1. **Non-redundant Subspace Clustering by Balancing Quality and Information:** Besides optimizing the *cluster quality* we consider maximizing the *amount of non-redundant novel information in terms of cluster and subspace identification* as a second major optimization goal for subspace clustering. NORD is the first approach relying on information theory to make both aspects measurable in a comparable way such that they can be integrated into a common objective function.
2. **Complex Clustering easy depicted:** Measuring both properties in bits opens up the opportunity to plot the clusters in the space of quality and novelty. This plot gives a summary on the overall cluster structure.
3. **Efficient and Automatic Clustering:** Besides the parameters required for redundancy control the comparison methods require further parameters specific to the underlying clustering method, e.g., density thresholds. We also avoid such difficult to estimate parameters relying on the Minimum Description Length Principle.

2   How to Balance Quality and Novelty

Notations

Definition 1 (Database). *Let $DB \subset R^d$ be a finite collection of d-dimensional vectors. We call $n := |DB|$ the database size, d the dimensionality of DB, and we call the set $\overline{S} := (1, ..., d)$ the set of all dimensions-, or full space.*

Definition 2 (A Subspace Cluster). *A subspace cluster C is defined as a pair $C = (O \subseteq DB, S \subseteq \overline{S})$, where O is a set of data points existing in a cluster C of subspace S. In the following, if multiple clusters are considered, we use $C.O$ and $C.S$ to reference the set of objects and set of dimensions of C, respectively.*

Definition 3 (Subspace Clustering). *A subspace clustering $C = \{C_1, ..., C_k\}$ is a set of k subspace clusters such that two clusters sharing the same subspace do not contain the same objects, formally:*

$$\forall C_i, C_j \in C, C_i \neq C_j : C_i.S = C_j.S \Rightarrow C_i.O \cap C_j.O = \emptyset.$$

Thus, subspace clustering sharing the same subspace are assumed to have disjointed objects [1,2]. We note that different subspace clusters are allowed to share, or all, dimensions in their respective subspace, and subspace clusters having non-identical subspaces may share common objects.

2.1   Making Cluster Information Measurable

The information provided by a subspace cluster on the data is two-fold: information about which objects are contained in the cluster and information about which dimensions span the corresponding subspace, more specifically:

Definition 4 (Information of a Subspace Cluster). *The amount of information $H(C)$ of a subspace cluster $C = (S, O)$ is defined as:*

$$H(C) = H(O) + H(S) \tag{1}$$

$H(O)$ denotes the entropy of the cluster labels, i.e.

$$H(O) = -\frac{|O|}{|DB|} \cdot \log_2 \frac{|O|}{|DB|} - \frac{|DB \backslash O|}{|DB|} \cdot \log_2 \frac{|DB \backslash O|}{|DB|},$$

and $H(S)$ is defined analogously for the dimensions S of C. We note that $H(C)$ is a measure which quantifies, in bits, the novel information provided by cluster C on the data set $\mathcal{D}$. As a subspace cluster represents a subset of objects and dimensions of the data space, $H(C)$ is the sum of the entropies of object- and dimension-assignments.

The information of multiple subspace clusters can be highly redundant. High redundancy can be due to the fact that multiple clusters are composed by similar subsets of the objects and/or reside in similar subspaces of the data space. It is therefore interesting to quantify how much non-redundant information a single cluster contributes to the overall clustering result.

Definition 5 (Non-redundant Information). *The amount of non-redundant information of two subspace clusters $C_1 = (O_1, S_1)$ and $C_2 = (O_2, S_2)$ is defined as:*

$$NR(C_1, C_2) = VI(O_1, O_2) + VI(S_1, S_2), \tag{2}$$

where VI denotes the Variation of Information, i.e. $VI(C_1, C_2) = H(C_1) + H(C_2) - 2I(C_1, C_2)$ with $I(C_1, C_2)$ is the mutual information of both clusters $I(C_1, C_2) = \frac{|O_1 \cap O_2|}{|DB|} \cdot \log_2 \frac{|O_1 \cap O_2|}{|DB|} \cdot \frac{|O_1|}{|DB|} \cdot \frac{|O_2|}{|DB|}.$

NR is a metric, since it is the sum of two metrics. In [8] the authors used VI as similarity measure for comparing the results of different approaches to classical partitioning clustering. We use the basic concept of VI to quantify the similarity of the information which two subspace clusters from the result set of our method provide on the data. A NR of zero among C_1 and C_2 implies that $VI(O_1, O_2)$ and $VI(S_1, S_2)$ are zero, i.e. both clusters are composed of independent subsets of objects and dimensions of the data.

2.2   Encoding Quality

We follow an information-theoretic perspective [14] to quantify the quality of a subspace cluster. To measure the quality of a subspace cluster, we regard a subspace cluster as a pattern which we can exploit to compress the data. The more information (in bits) we can save by having identified some subspace cluster C in the data, the higher its quality. Intuitively, the aim of this section is to describe the points of a cluster by a probability distribution. The better this function fits the data, the more of the data is explained by this function. The remaining, unexplained, error needs to be encoded separately from the parameters of the PDF (probability density function) to achieve a lossless compression. We first define how we can encode the set of objects O belonging to some subspace cluster $C = (O, S)$.

We model the data of a subspace cluster $C = (O, S)$ by a probability density function pdf_C. In a nutshell, the coding cost of a subspace cluster C corresponds to the deviation of the subspace clusters' points from its expectation. Formally,

Definition 6 (Coding cost of a Subspace Cluster). *Let $pdf_C = No(\mu_C, \sigma_C)$ be a probability density function. The coding costs of a subspace cluster $C = (O, S)$ consist of two parts, data costs dc and parameter costs pc, i.e. $cost_{pre}(C) = dc(C) + pc(C)$ with*

$$dc(C) = \sum_{o \in C} \log_2 \frac{1}{p(o, pdf_C(\pi_S(o)))} \ \text{and}$$
$$pc(C) = \frac{|params|}{2} \cdot \log_2 |O|, \tag{3}$$

where $\pi_S(o)$ is the projection of object o to subspace S, params is the set of parameters of pdf_C, $p(o, pdf_C(o))$ is a function that returns the probability that the distance between o and a random point x sampled from pdf_C is greater than the distance between o and the expectation $E(pdf_C)$ of pdf_C, formally:

$$p(o, pdf_C) = \int_{x \in O} pdf_C(x) \cdot I(dist(o, x) > dist(E(pdf_C), o)),$$
$$E(pdf_C) = \int_{x \in O} pdf_C(x) \cdot x. \tag{4}$$

The data cost dc is provided by the negative log-likelihood of all associated points w.r.t. the PDF. The parameters of the PDF are determined after projecting the objects into the subspace S of the cluster. The better the cluster model fits the data, the smaller is this term. By using the dual logarithm, we obtain the coding length in bits. In $pc(C)$ the first term represents the costs required to encode the

$|params|$ model parameters of the pdf, in case of a spherical Gaussian PDF we have $|p| = 2 \cdot |O|$. Following central results from information theory [8,13], we use $\frac{1}{2} \cdot log_2 |C|$ to represent a parameter.

In the remainder of this paper we use a spherical Gaussian PDF requiring $|p| = 2 \cdot |S|$ parameters (mean, variance) but our coding scheme can be extended to support different distributions and rotated clusters.

Definition 7 (Baseline Coding Cost). *The baseline coding cost for some unclustered data set U is as follows:*

$$db(U) = \sum_{u \in U} log_2(\frac{1}{pdf_{DB}(u)}). \tag{5}$$

The baseline is unimodal with parameters estimated from the complete data set DB and therefore denoted by PDF_{DB}.

It is useful to compare the baseline coding cost of the full data set DB in full-dimensional space $\mathcal{A}$ which models all data as one unimodal PDF in comparison to the coding costs of any intermediate or final subspace clustering result. Subspace clustering only makes sense if we can improve on the baseline coding cost, otherwise we have evidence that our data does not contain any subspace clusters which can be represented by our cluster model. It is also useful to consider the baseline coding cost of a subspace cluster to define its quality in a comparable way:

Definition 8 (Quality of a Subspace Cluster). *The quality of a subspace cluster $C = (S, O)$ is provided by the savings in bit over the baseline:*

$$Q(C) = db(U) - cost_{pre}(C). \tag{6}$$

To facilitate the comparison of the quality of different clusters, we often also consider the normalized quality $\hat{Q}(C) = \frac{Q(C)}{|O|}$ which represents the average number of bits saved per object over the baseline.

2.3   Balancing Quality and Information

Having formalized quality and information, we exploit the Minimum Description Length (MDL) [14] for coping with two challenges: We not only aim at automatically selecting the number and dimensionality for the subspace clusters, but also aim at balancing information and quality of the clusters. In other words, the second aspect means that the better the quality of a subspace cluster, the more redundancy we allow w.r.t. the remaining clusters in terms of object assignment and subspace identification. When a cluster provides a greater quantity of novel information, our algorithm allows a lower quantity of quality.

Definition 9 (Encoding of a Subspace Cluster). *The full description length of a subspace cluster C extends the coding costs $cost_{pre}(C)$ in Definition 6 for* *extra parameter costs pc describing the cluster object assignments and the cluster subspace assignments.*

$$cost(C) = cost_{pre}(C) + |O| \cdot \log_2 \frac{n}{|O|} + |S| \cdot \log_2 \frac{d}{|S|}. \tag{7}$$

This coding scheme naturally balances quality and information: For a result consisting of many high quality but very redundant clusters, the data costs are low but the parameter costs are unnecessarily high, due to many similar subspaces and clusters. In his case the MDL score would be far from its optimum. Also the result with random non-sense clusters has excessively high data costs due to low cluster quality. In MDL it scores poorly as well. However, this basic coding scheme encodes each cluster separately and does not consider any dependencies among clusters. Since we already know that dependencies exist, because objects and dimensions can be assigned to multiple clusters, we should also exploit this information for more compact coding.

Last, we define the overall optimization goal for our subspace clustering which corresponds to minimizing the overall cost of the sorted cluster sequence:

Definition 10 (Overall Optimization Goal). *Our optimization goal is to minimize the coding cost of the overall subspace clustering result, i.e.*

$$\min \sum_{C_i \in \mathcal{C}} cost(C_i | C_j <_s C_i).$$

3   NORDs Algorithmic Procedure

After formalizing our quality functions and how to optimize the compression, we will now explain the algorithm of our approach in detail. Our algorithm is a greedy bottom-up approach divided into two phases: first, the initialization phase for creating the microclusters and second a recursive refinement step in which the initial quality of the clustering is improved by (a) combining these microclusters and by (b) removing redundancy by choosing and merging the most similar clusters first.

3.1   Initialization Phase

Our initialization phase – although quite simple – contains one of the main concepts used in our algorithm and provides the algorithms' flexibility to find arbitrarily-oriented clusters: creating so called *microclusters* in each single dimension similar to the one in [15]. A microcluster $M(\sigma, \mu)$ is a gaussian cluster generated by K-Means of variance σ and mean μ with as few data points as possible but no less than a given minimum m. Up to this point, the quality of each dimension of the data set consisting of very small microclusters is very low. Now, raising the quality in this first step considers only the quality *per dimension*, merging the microclusters to larger ones in each dimension.

Algorithm 1. Non-redundant Algorithm NORD

Require: Numeric, high dimensional data set DB
Ensure: Overlapping label (optional)
 1: Initialization;
 2: **for** all one dimensional microcluster M **do**
 3:     Generate the coding cost CC as base cost bc cf. Sect. 2.3
 4: **end for**
 5: Apply quality step
 6: Re-generate CC
 7: **while** CC decreases AND $CC > 0$ **do**
 8:     **for** every two (one dimensional) cluster **do**
 9:         Calculate symmetric VI-Matrix $VI(C_i, C_j)$ cf. Sect. 2.1
10:     **end for**
11:     Rank the minimal VI entries
12:     Select (next) minimum entry from VI
13:     Merge the clusters with highest redundancy
14:     Re-calculate CC cf. Sect. 2.3
15: **end while**
16: Compute visualization
17: **return** non-redundant Clustering $\mathcal{C}$.

3.2 Merging Phase

After merging each single dimension to its highest possible quality outcome, the dimensions are combined to higher dimensional subspaces. For this process we need a smart suggestion on how to search through every subspace not to end up in exponential possibilities. This smart suggestion is provided by the information-theoretical concept of Variation of Information (VI). With VI we are enabled to measure novel content/information compared to other clustering results. Those cluster pairs, providing the least novel information in terms of VI are the ones which are very similar to each other, thus not very interesting on their own. These clusters are the ones suggested by VI to be merged with each other. But a suggestion is not a proof that this would be a correct decision. The real decision whether it is a good choice to merge two - often multi-dimensional - clusters *globally* is done by our quality function. If the suggestions holds, which means in general the costs are decreased, the algorithm goes on recursively, if not, the merging is rolled back and other suggestions are tried until the algorithm finishes in a local cost optimum. The overall procedure is described in Algorithm 1.

3.3 Complexity Analysis

The single steps of our heuristic approach are efficient: Our initialization step where the creation of microclusters takes place is determined by the runtime of our partitioning clusterer which is $O(nkdi)$, with k being the number of micro-clusters chosen and i the number of iterations. Then, raising the quality of these one dimensional clusters needs to create the cost matrix M_{cc} for each combination of clusterlabel l, for which the MDL for every cluster combination is calculated. The quality function from the MDL is linear in the number of objects O by exploiting the gaussian entropy, the overall runtime of this step is $O(l^2 O)$. For the last step, the heuristic search via a matrix M_{VI} that holds the variation of information (VI) is also quadratic in the number of matrix entries. The VI itself needs a linear calculation over the number of clusterlabel l for the entropy H and a quadratic calculation for the mutual information I for the number of dimensions d. Therefore the overall procedure for creating a smart suggestion from the VI, costs $O(M_{VI}^2 l d^2)$.

4   Experimental Evaluation of NORD

In this section, we compare NORD with recent competitor approaches for non-redundant subspace clustering paradigms like RESCU [10], INSCY [2] and STATPC [9]. For a fair comparison we implemented NORD in JAVA and used the evaluation frameworks OpenSubspace [11] and OutRules [12], WEKA extensions, where all mentioned competitor methods are implemented in JAVA as well. All runtime experiments were done on the same machine, an Intel Core Quad with 3 GHz and 6 GB main memory.

We compare all synthetic and real world data sets with four algorithms, which three of them are specifically created to tackle the problem of too many redundant subspace clusters in the final result.

– Curler [15] is an approach for finding non-linear correlation clusters without considering redundancy in their results. We chose it as baseline as the algorithmic procedure resembles NORD, because the concept of microclusters is applied fo find arbitrary clusters. Both algorithms use microclusters as an divide approach to create small gaussian mixture models to find non-linear clusters, Curler by applying the EM algorithm and NORD by applying kMeans. Both create some visualization for arbitrary clusterings. But in Curler microclusters are applied to find the different orientation of the neighbors, an information that is necessary for their objective function, where in NORD the orientation of the microclusters in space does not matter. Their objective function does not consider redundancy in subspaces. The visualization differs in the sense that Curlers NNC plot plots the microclusters and their co-sharing level (the number of tuples overlaping), whereas NORD plots the final clustering result without any microclusters. Besides, Curler needs five input parameter to work properly. By applying EM for the microclusters some level of approximation is considered.

– INSCY [2] is the most related competitor to our approach in terms of redundancy reduction and cluster definition we could find in literature so far. Both aim at finding all shapes of cluster no matter in which dimensions, so called arbitrary clusters. As for most newer clustering algorithms INSCY allows overlapping objects and overlapping dimensions. It needs seven parameters to be set by the user to work properly.

– RESCU [10] is a complex non-redundant subspace clusterer using a grid to find combined clusters. With such cluster definition it does not find all arbitrary

cluster shapes. RESCU applies even eight parameters to process redundancy accurately.
- STATPC [9] is an approximative non-redundant statistical method which defines cluster as hyper-rectangle. It needs three input parameters given by the user.

With these methods being the state-of-the-art for redundancy removal in clustering, we can say that so far correct parametrization was crucial to process redundancy adequately.

4.1   Synthetic Data

For each synthetic dataset, we generate a number *clusterlabel* clusters. For each cluster $C = (O, S)$, we randomly select a set $\subset S$ of dimensions randomly, by choosing each dimension with a probability of 0.5. We generate the same number of $|O| = \frac{n}{clusterlabel}$ of objects for each cluster. These points follow a uniform distribution in $[0,1]$ in all dimensions except when in subspace S. In each dimension in S, the points in O follow a gaussian distribution, using a uniformly chosen value in $[0,1]$ as mean, and using a variance scaling from *variance* on average. Average variance means the average of all variances of each cluster per dimension.

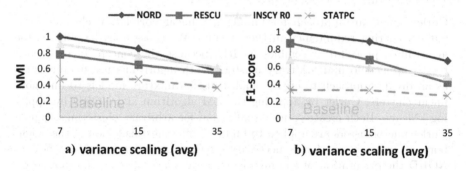

Fig. 1. Quality evaluation for scaling the variances (average variance shown on x-axis).

Scaling the variance allows us to experimentally evaluate the level of overlap between different clusters and how the algorithm reacts to it. The variances are given in Fig. 1 as the average variance of two clusters. For example, if the plot shows a variance of 35, then this combines a higher variance of 50 and a lower variance 20. Figure 1 shows clearly, that for all methods the quality decreases with a higher overlap. Depending on how redundancy is implemented and removed in the different approaches raises the quality. STATPC as approximate method has the most problems with the mutually overlapping data points that come with varying the variance of the cluster while NORD achieves the highest results in NMI as well as in F1-score. The baseline, Curler, is far off from these results due to no redundancy reduction.

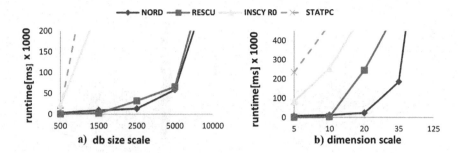

Fig. 2. Scalability runtime experiments for dimension size and db size.

For scalability we show runtime experiments for db size (number of data points) and dimension size as well as quality evaluation for scaling the variances of our synthetic data sets. Figure 2 shows the results for all comparison methods. While NORD scales similar with RESCU for the db size, it outperforms all other algorithms when scaling the dimensions of the data set. Besides the level of redundancy removal also seems to affect the runtime performance. STATPC has the worst approximative redundancy removal scales compared to the fully non-redundant comparison methods. CURLERs results on efficiency are omitted as it does not tack non-redundancy.

4.2   Real World Data

For real world data sets we apply two data sets. The data set *wages* is publicly available on a webpage[1] and the *genes* data set belongs to the Spellman gene expression data available at the MINE projects webpage[2]. If possible we applied the exact same parameter settings for the competitors that were given by their authors[3].

The "wages" data that we derived from UCI Machine Learning Repository [3] consist of a random sample of 534 persons from the Current Population Survey (CPS). This social studies goal was to determine the impact of gender and other attributes like years of education, work experience and age on wage. The study provides information on wages and other characteristics of the workers, including sex, number of years of education, years of work experience, occupational status, region of residence and union membership. From all attributes only wage, age, work experience and year of education were numeric and thus relevant. The goal of the study was to determine (i) correlations between wage and characteristics of the workers, and (ii) whether there is a gender gap in wages. Our clustering algorithm NORD is able to find 9 meaningful clusters on this data set which are depicted in Fig. 3(a). The cluster with the highest quality shows the strong

---

[1] http://lib.stat.cmu.edu/datasets/CPS_85_Wages.

[2] http://www.exploredata.net/Downloads/Gene-Expression-Data-Set.

[3] http://dme.rwth-aachen.de/de/OpenSubspace/RESCU.

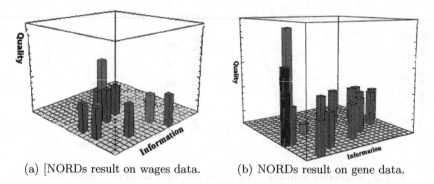

(a) [NORDs result on wages data.   (b) NORDs result on gene data.

Fig. 3. The information content of two subspace clusters. The overlapping part of cluster 1 and cluster 2 is redundant, the other parts are novel information provided by this one cluster. Together, novelty and redundancy form the full information of a cluster.

correlation between work experience and age. The three close cluster combine wage to age and experience. Years of education had no apparent correlation with wage. RESCU had a very similar result with 10 clusters. STATPC scored well with 15 clusters but INSCY did not manage to gain a high quality score with F1-score of 0.27.

The Spellman gene expression data does study the mitotic cell cycle of yeast genes. This data set is also quite well known in our discipline as a means of identifying functionally related genes using cluster analysis. In this relatively large data set (nearly 5000 data points), NORD finds 15 meaningful clusters depicted in Fig. 3(b). Clearly, the two clusters with highest quality are also relatively similar to one another, both contain 13 dimensions but are both important in quality. RESCU could not process this dataset (and logically also not the metabolic data with nearly 10,000 samples). Even after we running the data on a machine with 48 GB RAM, the JVM got an out of memory error. For INSCY we needed to modify this data set somewhat, because it works with positive values only.

5   Conclusion

In this paper, we introduced a novel approach to subspace clustering that balances the quality of a clustering with the novel information gained. As the first information-theoretic algorithm that is applied to the topic of non-redundant subspace clustering. NORD can be considered parameter-free in the sense, that no sensitive input parameter are necessary to gain a highly valuable result. We showed clearly in our experiments that the heuristic search method is relatively fast compared to other state-of-the-art algorithms and achieves a high quality even without the need for parameters. Last but not least we proposed a visualization of the clustering result of our algorithm, that intuitively shows the relationship between quality and novelty. To conclude, we feel that our proposed solution is able to yield more useful subspace clustering results: Instead of yielding a potentially overwhelmingly large set of high-quality clusters, that might be highly redundant, our solutions narrows down the space of interesting clusters to the most representative clusters.

References

1. Agrawal, R., Gehrke, J., Gunopulos, D., Raghavan, P.: Automatic subspace clustering of high dimensional data for data mining applications. In SIGMOD Conference, pp. 94–105 (1998)
2. Assent, I., Krieger, R., Müller, E., Seidl, T.: INSCY: indexing subspace clusters with in-process-removal of redundancy. In: ICDM Conference, pp. 719–724 (2008)
3. Bache, K., Lichman, M.: UCI machine learning repository (2013)
4. Baumgartner, C., Plant, C., Kailing, K., Kriegel, H.-P., Kröger, P.: Subspace selection for clustering high-dimensional data. In: ICDM, pp. 11–18 (2004)
5. Costeira, J., Kanade, T.: A multibody factorization method for independently moving objects. Int. J. Comput. Vis. **29**(3), 159–179 (1998)
6. Kanatani, K.: Motion segmentation by subspace separation and model selection. In: IEEE ICCV, vol. 2, pp. 586–591 (2001)
7. Kannan, R., Vempala, S.: Spectral algorithms. Found. Trends Theor. Comput. Sci. **4**(3&4), 157–288 (2009)
8. Meila, M.: Comparing clusterings: an axiomatic view. In: ICML, pp. 577–584 (2005)
9. Moise, G., Sander, J.: Finding non-redundant, statistically significant regions in high dimensional data: a novel approach to projected and subspace clustering. In: KDD Conference, pp. 533–541 (2008)
10. Müller, E., Assent, I., Günnemann, S., Krieger, R., Seidl, T.: Relevant subspace clustering: mining the most interesting non-redundant concepts in high dimensional data. In: ICDM, pp. 377–386 (2009)
11. Müller, E., Günnemann, S., Assent, I., Seidl, T.: Evaluating clustering in subspace projections of high dimensional data. PVLDB **2**(1), 1270–1281 (2009)
12. Müller, E., Keller, F., Blanc, S., Böhm, K.: *OutRules*: a framework for outlier descriptions in multiple context spaces. In: Flach, P.A., De Bie, T., Cristianini, N. (eds.) ECML PKDD 2012. LNCS (LNAI), vol. 7524, pp. 828–832. Springer, Heidelberg (2012). doi:10.1007/978-3-642-33486-3_57
13. Rissanen, J.: An introduction to the MDL principle. Technical report, Helsinki Institute for Information Technology (2005)
14. Rissanen, J.: Information and Complexity in Statistical Modeling. Springer, New York (2007)
15. Tung, A.K., Xu, X., Ooi, B.C.: CURLER: finding and visualizing nonlinear correlation clusters. In: SIGMOD, pp. 467–478 (2005)
16. Yang, A.Y., Wright, J., Ma, Y., Sastry, S.S.: Unsupervised segmentation of natural images via lossy data compression. Comput. Vis. Image Underst. **110**(2), 212–225 (2008)
17. Zhang, A., Fawaz, N., Ioannidis, S., Montanari, A.: Guess who rated this movie: identifying users through subspace clustering (2012). CoRR, abs/1208.1544

Cost Matters: A New Example-Dependent Cost-Sensitive Logistic Regression Model

Nikou Günnemann$^{(\boxtimes)}$ and Jürgen Pfeffer

Technical University of Munich, Munich, Germany
{Nikou.Guennemann,Juergen.Pfeffer}@tum.de

Abstract. Connectivity and automation are evermore part of today's cars. To provide automation, many gauges are integrated in cars to collect physical readings. In the automobile industry, the gathered multiple datasets can be used to predict whether a car repair is needed soon. This information gives drivers and retailers helpful information to take action early. However, prediction in real use cases shows new challenges: misclassified instances have not equal but different costs. For example, incurred costs for not predicting a necessarily needed tire change are usually higher than predicting a tire change even though the car could still drive thousands of kilometers. To tackle this problem, we introduce a new *example-dependent cost sensitive prediction model* extending the well-established idea of logistic regression. Our model allows different costs of misclassified instances and obtains prediction results leading to overall less cost. Our method consistently outperforms the state-of-the-art in example-dependent cost-sensitive logistic regression on various datasets. Applying our methods to vehicle data from a large European car manufacturer, we show cost savings of about 10%.

1 Introduction

Automation has become of prime importance to improve the quality of our life. An example from the vehicle industry, where predictive maintenance [17] looms large, is to predict whether the tires of a car need to be changed soon. Goals are (i) providing customers services with less latency for tire change, and (ii) forecasting tire delivery in correct number for all customers with a tire change need.

Given historical data, such potential 'malfunctions' (required tire change) can be predicted based on binary classification algorithms like logistic regression, support vector machines, ARIMA models or neural networks etc. [2–5]. Such approaches, however, often do not meet the real world use cases since intuitively they try to minimize the so called zero-one loss with the assumption that all misclassified instances have equal cost. Meaning correct classifications lead to a cost of zero and misclassification gets a cost of one [6].

In many applications, however, the costs for misclassified instances might vary significantly from one instance to the other. Predicting tire change is ranked among these applications. The cost associated with, e.g., an incorrect early tire

© Springer International Publishing AG 2017
J. Kim et al. (Eds.): PAKDD 2017, Part I, LNAI 10234, pp. 210–222, 2017.
DOI: 10.1007/978-3-319-57454-7_17

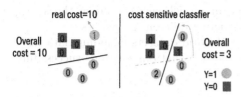

	True Pos. $(y_i = 1)$	True Neg. $(y_i = 0)$
Predicted Pos. $(\hat{y}_i = 1)$	0	c_i^{FP}
Predicted Neg. $(\hat{y}_i = 0)$	c_i^{FN}	0

Fig. 1. Left: result of a cost-insensitive classifier; right: result of a cost-sensitive classifier with smaller overall cost

Fig. 2. Cost matrix for cost-sensitive classification

change prediction is smaller than the expected cost for not predicting an imminent tire change at all, which could even cause customer dissatisfaction. Also, instances where the tire change has been predicted too early might show different costs. In some cases the tires might actually be used for further 20,000 km compared to only 1,000 km. Hence, in practical applications different misclassified instances can cause different costs. Since standard binary classifiers are not suited for such scenarios, *example-dependent cost-sensitive classification* [7] has been introduced, considering the different costs of instances during learning.

Technically, to distinguish between different misclassified instances, a predefined cost value can be assigned to each instance in the dataset. Figure 2 shows the cost for an instance according to their actual class vs. their predicted class. When the instance is a false positive, we have cost of c_i^{FP}, if it is a false negative, the cost is c_i^{FN}. If the instance is correctly classified (i.e. true positive or true negative), we assign a cost of zero. Note that each instance i might get different cost (indicated by the index i). Having assigned costs to each instance i, a possible way to estimate the overall misclassification cost is

$$Cost = \sum_{i}^{m} y_i \cdot (1 - \hat{y}_i) \cdot c_i^{FN} + (1 - y_i) \cdot \hat{y}_i \cdot c_i^{FP} \tag{1}$$

where $y_i \in \{0, 1\}$ is the observed and $\hat{y}_i \in \{0, 1\}$ the predicted label of instance i, with m as the number of overall instances [3]. Accordingly, instead of considering each instance equally, our goal is to train a classifier that takes the overall misclassification cost into account. The benefit of such an approach is shown in Fig. 1. On the left, the cost-insensitive classifier misclassified only one instance but its associated cost is very high (10). On the right, the potential result of a cost-sensitive classifier is shown. Although now two instances are classified false, they have only very low cost of 1 and 2; the previously misclassified instance with cost of 10 is classified correctly. Thus, the overall misclassification cost is only 3. Based on this motivation, in this work, we focus on the well established classification model of logistic regression – and we extend it by the principle of example-dependent cost sensitive learning. The contributions of this paper are:

- We propose an enhanced binary classification model that includes the individual costs of instances while fitting a model to the training set – thus, costs are not simply used in a post-processing step but during training.

- We propose four different variants of our model each extending the sound principle of logistic regression but considering different properties.
- We perform experiments on multiple real-world datasets including data from a leading European car manufacturer showing that our methods successfully lower misclassification costs.

2 Background

In a binary classification problem, input vectors $X = \{x_1, \ldots, x_m\}$ with $x_i \in \mathbb{R}^d$, and class labels $Y = \{y_1, \ldots, y_m\}$ with $y_i \in \{0, 1\}$ are given. Here, x_i is the i-th instance described by d features, and $y_i = 1$ represents the class of instances with a certain issue (tire change required) and $y_i = 0$ the opposite. In our scenario, each instance i is associate with a certain predefined cost c_i.[1] The higher the cost, the worse is a potential misclassification of this instance. Our goal is to fit a model on observed data X to predict Y denoted by $\hat{Y}$ at best. More precisely, our aim is to find a model that leads to small overall misclassification cost.

2.1 Logistic Regression and Important Properties

Logistic regression treats the binary classification problem from a probabilistic perspective. Given the instance x, the probability of the occurrence of an issue (i.e. $y = 1$) is denoted by $p(y = 1 \mid x)$, and $p(y = 0 \mid x) = 1 - p(y = 1 \mid x)$ respectively for $y = 0$. Here, $p(y = 1 \mid x)$ is defined as the sigmoid function, known as logit:

$$p(y = 1 \mid x) = f(g(x, \beta)) = \frac{1}{1 + e^{-g(x,\beta)}} \qquad (2)$$

where $0 \leq f(g(x, \beta)) \leq 1$ and $g(x, \beta) = \beta_0 + \sum_{j=1}^{m} \beta_j \cdot x_j$ is a linear expression of Eq. 2 including the explanatory features and the regression coefficients β. Considering the sigmoid equation, the question is how to estimate β in $g(x, \beta)$ to make $f(g(x, \beta)) = \hat{y}$ close to y? To formalize this, and assuming that the m samples in the data are independent, we can write $p(Y|X; \beta)$ as a product, leading to the following overall Likelihood function:

$$L(Y, X, \beta) = \prod_{i=1}^{m} f(g(x_i, \beta))^{y_i} \cdot (1 - f(g(x_i, \beta)))^{1-y_i} \qquad (3)$$

The β in logistic regression can be obtained by maximizing Eq. 3, i.e. it corresponds to the maximum likelihood estimate. Obviously, instead of maximizing $L(Y, X, \beta)$, we can equivalently minimize the negative log likelihood given by

$$l(Y, X, \beta) = \sum_{i=1}^{m} y_i \cdot (- \log f(g(x_i, \beta))) + (1 - y_i) \cdot (- \log(1 - f(g(x_i, \beta)))) \quad (4)$$

---

[1] Note that we do not have to explicitly distinguish between c_i^{FP} and c_i^{FN}. If $y_i = 0$, then $c_i^{FP} = c_i$, if $y_i = 1$, then $c_i^{FN} = c_i$. For a single instance, c_i^{FP} and c_i^{FN} can never occur together.

which is the logistic loss function. Figure 3 shows the logistic loss function for $y_i = 1$ (e.g. tire change)[2]: $y_i \cdot (-\log f(g(x_i, \beta)))$. Clearly, if $f(g(x_i, \beta)) = 1$ the prediction is correct and we have zero loss. For $f \to 0$, in contrast, the loss will increase. Thus, minimizing the loss means lowering the prediction error.

Loss of Correctly Classified Instances: In logistic regression the return values of the sigmoid function are between 0 and 1. Therefore, we have no deterministic decision which samples are classified correctly and which are classified wrong. To turn the predicted probabilities into binary responses, a threshold is used. Based on the probabilistic view and as default in literature, we choose 0.5 as threshold. That is, if $f(g(x_i, \beta)) \geq 0.5$, the predicted class is 1, otherwise 0. The resulting observation is that *even for correctly classified instances the logistic loss is not zero*. This becomes obvious in Fig. 3: e.g. an instance with $f(g(x_i, \beta)) = 0.9$ has a loss of 0.05 even if it is correctly classified.

Assuming the correctly classified instances get a probability $f(g(x_i, \beta))$ uniformly random between 0.5 and 1, then the average loss of a correctly classified instance is proportional to $T_{log} := \int_{0.5}^{1} y_i \cdot (-\log f(g(x_i, \beta))) \, \mathrm{d}f \approx 0.15$.[3] Here, T_{log} can also be illustrated as the area under the 'logistic loss'-curve as shown in Fig. 3. Likewise, the incorrectly classified instances get an average loss proportional to $F_{log} := \int_{0}^{0.5} y_i \cdot (-\log f(g(x_i, \beta))) \, \mathrm{d}f \approx 0.85$ where F_{log} represents the area from 0 to 0.5. Also note that $T_{log} + F_{log} = 1$. That is, the average loss assigned to an instance (independent if correctly or incorrectly classified) is 1. Obviously $F_{log} > T_{log}$, which means that a correct prediction actually leads to smaller loss. However, the two loss terms F_{log} and T_{log} are constant and identical for each instance. That is, the standard logistic loss function does not distinguish between different losses caused by different instances with different costs.

3   Example Dependent Cost-Sensitive Logistic Regression

The above discussion leads to the core motivation of our paper: How can we adapt the logistic loss function in a sound way, so that different samples having different costs are treated differently? How can we define a loss function to make sure that instances with higher costs are more likely to be predicted correctly?

General Framework. To answer these questions, we adapt the standard logistic loss function to a cost sensitive one in four different ways. The general framework we explore in these versions is to minimize the loss function $l(Y, X, \beta)$ defined as

$$\sum_{i=1}^{m} a_i \cdot y_i \cdot (-\log f(g(x_i, \beta))^{b_i}) + a_i \cdot (1 - y_i) \cdot (-\log(1 - f(g(x_i, \beta)))^{b_i}) \quad (5)$$

---

[2] The case $y_i = 0$ is equivalent; only mirrored. W.l.o.g. we consider in the following only $y_i = 1$.

[3] More precise, the average loss for correctly classified instances would be $2 \cdot T_{log}$.

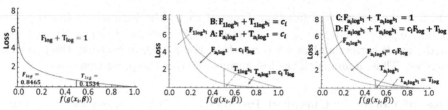

Fig. 3. Loss function of standard logistic regression for $y = 1$.

Fig. 4. Loss function for A & B. The loss ratio for B is smaller. (Color figure online)

Fig. 5. Loss function for variant C & D. Both variants control the loss ratio.

where a_i and b_i depend on c_i. That is, $a_i = a(c_i)$ and $b_i = b(c_i)$ based on functions $a : \mathbb{R}^+ \rightarrow \mathbb{R}^+$ and $b : \mathbb{R}^+ \rightarrow \mathbb{R}^+$. As shown next, given different choices of a and b, we realize different properties. For convenience, let us already introduce the following notation: $F_{a_i log^{b_i}} := \int_0^{0.5} a_i \cdot (y_i \cdot (-\log f(g(x_i, \beta)))^{b_i}) \, df$ represents the average loss for misclassified instances and $T_{a_i log^{b_i}} := \int_{0.5}^{1} a_i \cdot (y_i \cdot (-\log f(g(x_i, \beta)))^{b_i}) \, df$ is the average loss for correctly classified instances.

Variant A: Weighting the Logistic Loss Function. The first, simplest way is to weight the logistic loss function depending on the cost value c_i by setting the area under the curve – representing the average loss – equal to the cost.

$$\int_0^1 a_i \cdot (y_i \cdot (-\log f(g(x_i, \beta)))) \, df \overset{!}{=} c_i \tag{6}$$

This way, instances with a higher cost will get a higher average loss value. Since in standard logistic regression the area is 1, it obviously holds that the weight factor a_i needs to be equal to c_i. The variable b_i is equal to 1. The purple curve in Fig. 4 shows the plot for $c_i = 3$. Clearly, by weighting the loss function, we oversample the instances proportional to their costs, i.e. an instance with cost 2 is basically considered twice. But this solution has one drawback: By weighting the loss, not only the misclassification loss $F_{a_i log^1}$ but also the 'correct' loss $T_{a_i log^1}$ will be higher. *Correctly classified instances are penalized by this version, too.* In particular, the ratio between $F_{a_i log^1}$ and $T_{a_i log^1}$ does *not* change. Thus, for every instance a misclassification has always $\frac{F_{log}}{T_{log}} \approx 5.5$ higher loss than a correct classification. Thus, this solution might not well represent the intuition that the cost of *misclassification* will be higher.

Variant B: Logistic Loss Function to the Power of b. To avoid penalizing correctly classified instances, we exchange the weighting in Eq. 6 by an exponentiation of the logistic loss function to the power of b. That is, we increase the average loss from 1 to c_i by using the term b_i and keeping $a_i = 1$:

$$\int_0^1 y_i \cdot (-\log f(g(x_i, \beta)))^{b_i} \, df \overset{!}{=} c_i \tag{7}$$

Since Eq. 7 is equal to $\Gamma(b_i + 1)$, the solution for b_i given a specific c_i is equal to

$$b_i = \Gamma^{-1}(c_i) - 1 \tag{8}$$

Γ^{-1} is the inverse of the Gamma function Γ which can be computed numerically.

Figure 4 shows the corresponding loss function by the blue curve with $c_i = 3$. While the loss area $T_{1\log^{b_i}}$ is pressed downwards, the loss area of instances in $F_{1\log^{b_i}}$ wins on more importance since instances with high costs are more important to be classified correct. Thus, not only the average loss increases for these instances but also the ratio between $F_{1\log^{b_i}}$ and $T_{1\log^{b_i}}$. A potential drawback is that the ratio $F_{1\log^{b_i}}/T_{1\log^{b_i}}$ is not controlled explicitly.

Variant C: Controlling the Ratio - I. We aim to control the ratio between the loss area of $F_{a_i\log^{b_i}}$ and $T_{a_i\log^{b_i}}$. That is, for an instance with cost c_i we want to ensure $\frac{F_{a_i\log^{b_i}}}{T_{a_i\log^{b_i}}} \overset{!}{=} \frac{F_{\log}}{T_{\log}} \cdot c_i$. The ratio between the loss of false and correct classification is c_i times higher than for an instance with cost 1. Simultaneously, the average loss of the instances should be independent of c_i. The motivation is that in average each instance is equally important, but for some of them the *misclassification* should be penalized stronger. That is, the area under the curve has to be equal to 1, meaning $F_{a_i\log^{b_i}} + T_{a_i\log^{b_i}} \overset{!}{=} 1$. This constraint implies that

$$a_i = \frac{1}{\Gamma(b_i + 1)} \tag{9}$$

The value of $b_i > 0$ can be computed numerically by solving (see Appendix)

$$\frac{\Gamma(b_i + 1)}{\Gamma(b_i + 1, 0.6931)} = 1 + \frac{T_{\log}}{c_i \cdot F_{\log}} \tag{10}$$

where $\Gamma(s, r)$ is the incomplete gamma function. The effect of this variant is shown in Fig. 5 as variant C, again for $c_i = 3$. Here, we have $F \approx 0.94$ and $T \approx 0.057$. Thus, the ratio is c_i times higher than in the standard case. Still the average loss is identical (i.e. equal to 1).

Variant D: Controlling the Ratio - II. In version C, we kept the average loss at 1 but increased the ratio between false and correct classification; thus, the area $T_{a_i\log^{b_i}}$ needs to decrease. Accordingly, instances with a high costs will not only have higher *misclassification loss* but also lower *correct classification loss* compared to instances with low costs – which again, might not be intended since the costs for correct classification is constant. Therefore, we introduce our last version which (i) directly controls the ratio, and (ii) ensures that the correct classification loss stays constant. This idea can be transformed to $\frac{F_{a_i\log^{b_i}}}{T_{a_i\log^{b_i}}} \overset{!}{=} \frac{F_{\log}}{T_{\log}} \cdot c_i$ and $T_{a_i\log^{b_i}} \overset{!}{=} T_{\log}$. Solving this, we obtain for b_i the identical solution as in variant C; only the weighting a_i changes to

$$a_i = \frac{T_{\log}}{\Gamma(b_i + 1) - \Gamma(b_i + 1, 0.6931)} \tag{11}$$

variant	a_i	b_i	avg. loss	ratio F/T	T
LR	1	1	1	constant	T_{log}
A	c_i	1	c_i	constant	$c_i \cdot T_{log}$
B	1	Eq. 8	c_i	adaptive	adaptive
C	Eq. 9	Eq. 10	1	$\propto c_i$	adaptive
D	Eq. 11	Eq. 10	$T_{log} + c_i \cdot F_{log}$	$\propto c_i$	T_{log}

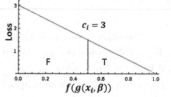

Fig. 6. Proposed variants and their properties   **Fig. 7.** Loss function used in [7].

Indeed, what we observe is that for this variant we have $F_{a_i log^{b_i}} = c_i \cdot F_{log}$. Thus, we only increase the average loss for the misclassified instances by a factor of c_i. This effect is shown in Fig. 5 as variant D. As one can also observe, the area $F_{a_i log^{b_i}}$ in variant D is equal to the area $F_{a_i log^{b_i}}$ of variant A; in both variants the area increases by a factor of c_i compared to standard logistic regression. While in variant A, however, also the area $T_{a_i log^{b_i}}$ increases, it stays constant in variant D. Thus variant D better captures the increased costs for *misclassification*.

Summary and Algorithmic Solution. Figure 6 summarizes our different variants. While variants A and B focus on increasing the average loss according to the costs, variants C and D focus on increasing the fraction between false and correct classification loss.

Our final goal is to find the parameter β that minimizes the loss function in Eq. 5: $\beta^* = \arg\min_\beta l(Y, X, \beta)$. For this purpose we exploit a gradient descent search. Starting from a random solution, we iteratively follow the steepest descent direction: $\beta^{t+1} \leftarrow \beta^t - \alpha\nabla l(\beta)$ where α is the learning rate.

4   Related Work

Various research papers are published with focus on cost sensitivity [1,7–13]. Often, the main objective is predicting potential customers with financial obligation based on their existing financial experience. While [11–14] use constant costs for misclassified instances, the authors in [9] propose a Bayes minimum risk classifier including the financial costs of credit card fraud detection in order to have a cost sensitive detection system. Another interesting approach is introduced in [15], by presenting a taxonomy of cost-sensitive decision tree algorithm using the class-dependent cost. An extension of [15] with focus on example-dependent cost for decision trees is published by [16].

The only method similar to ours is [7], which proposes an example-dependent cost sensitive logistic regression. Here the loss function of logistic regression is changed to a cost sensitive one by integrating the cost as a factor into its calculation. A drawback of [7] is that the loss function is no longer a logarithmic function but linear. That is, for the case that correct classification has 0 cost, [7] uses $\frac{1}{m}\sum_i^m y_i(1 - f(g(x_i, \beta)))c_i + (1 - y_i)f(g(x_i, \beta))c_i$. Thus, the loss decreases linearly: starting from c_i to 0 (see Fig. 7). Using a linear loss function causes weak differentiation between false and correctly classified instances. The two areas marked by F and T in Fig. 7 show this problem. As we will see in our experimental analysis, this principle will often perform worse than our technique.

5   Experimental Analysis

In this section we compare our four variants A–D with standard logistic regression LR and the competition model proposed in [7]. For this purpose, we test our designed models on the basis of three different datasets: (i) a vehicle dataset from a large European car manufacturer for predicting tire change service, (ii) the dataset breast cancer[4] to predict whether a patient is affected by breast cancer or not, and (iii) data from the 2011 Kaggle competition Give Me Some Credit[5] to predict whether a customer will experience financial distress in the next two years.

Our main goal is to achieve low overall misclassification cost (see. Eq. 1). Thus, a technique is successful if it obtains the lowest overall cost. As an evaluation measure we compute the savings of our techniques w.r.t. logistic regression $savings = \frac{Cost_{LR} - Cost_x}{Cost_{LR}}$ where $Cost_{LR}$ is the obtained misclassification cost (Eq. 1) based on the result of logistic regression and $Cost_x$ the cost based on the result of the technique x. In each scenario we used $\frac{2}{3}$ of the data for training our models and $\frac{1}{3}$ to evaluate them.

5.1   Tire Change Service

The vehicle dataset is a binary classification dataset containing 1,800 instances, each with 40 features. The features are indirectly influenced by tire wear and which, thus, indicate a resultant tire change. An example of such features could be acceleration.[6] Important to mention is that features resulting through, e.g. a sensor which directly measures the tire tread to asses a tire wear are not considered here. The target variable is whether a vehicle needs a tire change: yes = 1 or no = 0. Instances requiring a tire change account to $\sim 15\%$ amount of the whole data. Each instance is assigned with a cost; the higher the cost value, the more urgently a tire change is needed. The degree of urgency was determined by the domain experts.

Figure 8 shows the results. Here, the threshold to cast the predicted probabilities to binary responses is set to 0.5. Generally all of our four versions obtain lower overall misclassification cost than traditional logistic regression. But the best savings are achieved by B, C, and D with a win of around 10%. Applying the competing variant from [7] shows even higher overall cost than our variants. As discussed, this is caused by the used non-discriminative loss function in their logistic regression model.

While a threshold of 0.5 is from a probabilistic view the correct one, it might, however, not lead to smallest misclassification cost. Thus, in Fig. 9 we report the results for each method when individually using the threshold that leads to lowermost misclassification cost. These 'optimal' thresholds are given in Table 1. Note that in practice such a tuning is not possible since we a-priori do not know the true class of an instance.

---

[4] https://goo.gl/U2Uwz2.

[5] http://www.kaggle.com/c/GiveMeSomeCredit/.

[6] Due to nondisclosure agreements we unfortunately can not provide more details on the dataset. The two other datasets studied in this work are publicly available.

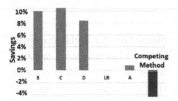
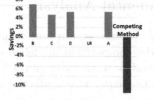

Fig. 8. Savings of the methods when threshold is 0.5. Our techniques significantly outperform logistic regression.

Fig. 9. Savings of the methods with variable threshold with minimal misclassification cost. Our techniques perform better

The results show that even in a scenario of complete knowledge our techniques significantly outperform logistic regression. Example dependent cost improves the ability to make less failure than the standard process. Also, the results from Table 1 are very close to $t = 0.5$. Thus, applying our models for $t = 0.5$ without tuning the threshold in practice would still obtain very good performance.

Table 1. Variable thresholds leading to minimal costs in vehicle datasets

LR	A	B	C	D	Competing
0.4542	0.4509	0.4950	0.5060	0.4775	0.5001

5.2 Breast Cancer

The Wisconsin breast cancer dataset is a binary classification dataset, e.g., available in scikit-learn. The total number of instances is 569, each with 30 attributes. 212 ($\sim 37.2\%$) of records are malignant denoted by 1 and 357 ($\sim 62.8\%$) records are benign presented by 0. Since the instances are not presented by different costs we randomly assigned each instance a cost between 1 to 5 to guaranty the fairness. To obtain reliable results, we generated $n = 10$ such datasets.

Table 2 shows the number of correctly classified instances and false classified ones (on avg. on the test data). Also, the average corresponding misclassification cost of the false classified instances are presented by the column *cost*. The left part of the table presents results for $t = 0.5$, the right part shows the results for the 'optimal'/tuned thresholds.

For $t = 0.5$, versions B and D return the lowest overall cost with a small number of false classified malignant instances. Surprisingly, not only the cost is lower in our variants, but also the classification accuracy increases. The same behavior can be seen for the variable threshold. In comparison to the competing variants LR and [7] our results are much better.

Figures 10 and 11 show the relative savings of the techniques w.r.t. logistic regression. Since we applied the algorithms 10 times based on different randomly assigned costs, different savings are observed. In Fig. 10, the bars show the mean savings for $t = 0.5$ achieved by each algorithm; the black lines represent the standard deviation over the 10 runs. In average, D as well as B save at most whereas the model in [7] cause even more loss than standard LR. *As shown by*

Table 2. Wisconsin breast cancer dataset. Left: threshold $t = 0.5$; right: variable thresholds with overall minimum misclassification cost.

Method, t	Incorrect	Correct	Avg. cost	Method, t	Incorrect	Correct	Avg. cost
LR, 0.5	≈ 17	≈ 171	52.61	0.39	≈ 14	≈ 174	38.07
A, 0.5	≈ 13	≈ 175	40.29	0.52	≈ 11	≈ 177	36.23
B, 0.5	≈ 12	≈ 176	**36.96**	0.4857	≈ 11	≈ 177	**33.37**
C, 0.5	≈ 14	≈ 174	43.72	0.496	≈ 13	≈ 175	38.61
D, 0.5	≈ 12	≈ 176	**37.21**	0.496	≈ 11	≈ 177	**34.02**
Competing, 0.5	≈ 87	≈ 101	282.8	0.499	≈ 38	≈ 150	115.49

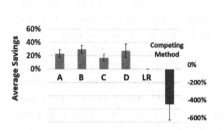

Fig. 10. Average savings when threshold $t = 0.5$. Average over 10 runs.

Fig. 11. Average savings when selecting variable t with minimal misclassification cost (see. Table 2). Average over 10 runs.

the black lines, these results are significant. While all our 4 versions return very similar good results, [7] shows bad performance and strong fluctuation. A similar behavior can be considered in Fig. 11 for variable thresholds.

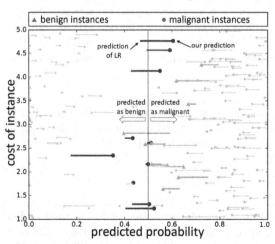

Fig. 12. Comparison between LR and D. The predicted probabilities of D (big end of each line) reflect better the true labels. Triangles should be on the left; circles on right. (Color figure online)

Finally, Fig. 12 shows why our techniques (here variant D) perform better than LR, that is we show the difference in their prediction. Each line in Fig. 12 represents one instance from the dataset. The start of each line indicates the predicted probability based on LR, while the end (shown by a circle/triangle) the probability assigned by our model. In an optimal classification, all triangles (green) should be on the left of the threshold line ($t = 0.5$); all circles (blue) on the right. We *grayed out* all instances which are correctly classified by both techniques;

thus, the *colored lines* show the interesting differences between LR and model D. As we can see, our method pushes more instances to the correct side of the line, i.e. classifies them correctly. Specifically, we also observe these changes for instances with high costs, thus, leading to overall lower misclassification cost.

5.3   Credit Datasets

The Kaggle Credit dataset is a bi-class dataset containing 112,915 credit borrowers as instances. Each instance has 10 features with a proportion of 6.74% positive examples. Some features are, e.g., monthly income or monthly debt. Our goal is to perform credit score prediction using our different versions of logistic regression. For assigning instances different costs, we took the same costs proportion as in [7]. Table 3 shows the corresponding results. In contrast to the first two experiments, the best models for Kaggle Credit data, are A and [7]. For $t = 0.5$, indeed version A has the best accuracy of around 93.0% but it saves slightly less than [7]. In contrast, using the optimal threshold, version A wins more on savings than [7] but the number of false classified instances is higher. This means that in model A, primarily instances with low costs are classified incorrectly. The model D performs good w.r.t. the savings and shows in both cases a low number of false classified instances.

Table 3. Results on credit dataset. Left: threshold $t = 0.5$; right: variable t with overall minimum misclassification cost.

Method, t	Incorrect	Correct	Cost	Savings	t	Incorrect	Correct	Cost	Savings
LR, 0.5	2470	34792	8260.35	-	0.39	2573	34689	8056.65	-
A, 0.5	**2444**	34818	7972.32	**3%**	0.38	**2586**	34676	7732.88	**4%**
B, 0.5	2474	34788	8248.05	0.1%	0.36	2483	34779	8056.65	−2%
C, 0.5	2471	34791	8261.57	−0.01%	0.40	2547	34715	8225.40	−2%
D, 0.5	2466	34796	8152.35	1%	0.33	2480	34782	7893.33	2%
Comp., 0.5	**2695**	34567	7823.53	**5%**	0.503	**2497**	34765	7766.05	**3%**

In summary, considering all datasets together, model D has consistently ranked among the best competing methods. Based on our model description (Sect. 3) it also very naturally captures example-dependent cost.

6   Conclusion

In this paper we have presented four different extensions of logistic regression to a cost sensitive one, each using a different loss functions having different properties. We evaluated the impact of each model based on two different public datasets as well as on a vehicle dataset to predict tire change. Our results confirm that a cost sensitive model not only classifies instances with higher importance better but can also improve the accuracy of classical logistic regression. For our use case on tire change service, we obtained significant savings of 10%.

A Appendix

$$\frac{F_{a_i log^{b_i}}}{F_{a_i log^{b_i}}} = \frac{a_i \Gamma(b_i + 1, 0.6931)}{a_i(\Gamma(b_i + 1) - \Gamma(b_i + 1, 0.6931))} \overset{!}{=} c_i \cdot \frac{F_{log}}{T_{log}}$$

$$\Leftrightarrow \Gamma(b_i + 1, 0.6931) \overset{!}{=} c_i \cdot \frac{F_{log}}{T_{log}} \cdot \Gamma(b_i + 1) - c_i \cdot \frac{F_{log}}{T_{log}} \Gamma(b_i + 1, 0.6931)$$

$$\Leftrightarrow \frac{\Gamma(b_i + 1)}{\Gamma(b_i + 1, 0.6931)} \overset{!}{=} \frac{1 + c_i \cdot \frac{F_{log}}{T_{log}}}{c_i \cdot \frac{F_{log}}{T_{log}}} = 1 + \frac{T_{log}}{c_i \cdot F_{log}}$$

References

1. Zadrozny, B., et al.: Cost-sensitive learning by cost-proportionate example weighting. In: ICDM, pp. 435–442 (2003)
2. Günnemann, N., et al.: Robust multivariate autoregression for anomaly detection in dynamic product ratings. In: WWW, pp. 361–372 (2014)
3. Murphy, K.P.: Machine Learning: A Probabilistic Perspective. MIT Press, Cambridge (2012)
4. Quinlan, J.R.: Induction of decision trees. Mach. Learn. **1**, 81–106 (1986)
5. Haykin, S.: A comprehensive foundation. Neural Netw. **2**, 41 (2004)
6. Weiss, G.M.: Learning with rare cases and small disjuncts. In: ICML, pp. 558–565 (1995)
7. Bahnsen, A.C., et al.: Example-dependent cost-sensitive logistic regression for credit scoring. In: ICMLA, pp. 263–269 (2014)
8. Anderson, R.: The Credit Scoring Toolkit: Theory and Practice for Retail Credit Risk Management and Decision Automation. Oxford University Press, Oxford (2007)
9. Bahnsen, A.C., et al.: Cost sensitive credit card fraud detection using Bayes minimum risk. In: ICMLA, pp. 333–338 (2013)
10. Bahnsen, A.C., et al.: Improving credit card fraud detection with calibrated probabilities. In: SIAM, pp. 677–685 (2014)
11. Alejo, R., García, V., Marqués, A.I., Sánchez, J.S., Antonio-Velázquez, J.A.: Making accurate credit risk predictions with cost-sensitive MLP neural networks. In: Casillas, J., Martínez-López, F., Vicari, R., De la Prieta, F. (eds.) Management Intelligent Systems. AISC, vol. 220, pp. 1–8. Springer, Heidelberg (2013). doi:10.1007/978-3-319-00569-0_1
12. Beling, P., et al.: Optimal scoring cutoff policies and efficient frontiers. J. Oper. Res. Soc. **56**(9), 1016–1029 (2005)
13. Oliver, R.M., et al.: Optimal score cutoffs and pricing in regulatory capital in retail credit portfolios. University of Southampton (2009)
14. Verbraken, T., et al.: Development and application of consumer credit scoring models using profit-based classification measures. Eur. J. Oper. Res. **238**(2), 505–513 (2014)

15. Lomax, S., et al.: A survey of cost-sensitive decision tree induction algorithms. CSUR **45**(2), 16 (2013)
16. Bahnsen, A.C., et al.: Ensemble of example-dependent cost-sensitive decision trees (2015). arXiv preprint arXiv:1505.04637
17. Mobley, R.K.: An Introduction to Predictive Maintenance. Butterworth-Heinemann, Oxford (2002)

Social Network and Graph Mining

Beyond Assortativity: Proclivity Index for Attributed Networks (PRONE)

Reihaneh Rabbany[(✉)], Dhivya Eswaran, Artur W. Dubrawski,
and Christos Faloutsos

School of Computer Science, Carnegie Mellon University, Pittsburgh, PA, USA
{rrabbany,deswaran,awd,christos}@andrew.cmu.edu

Abstract. If Alice is majoring in Computer Science, can we guess the major of her friend Bob? Even harder, can we determine Bob's age or sexual orientation? Attributed graphs are ubiquitous, occurring in a wide variety of domains; yet there is limited literature on the study of the interplay between the attributes associated to nodes and edges connecting them. Our work bridges this gap by addressing the following questions: Given the network structure, (i) which attributes and (ii) which pairs of attributes show correlation? Prior work has focused on the first part, under the name of *assortativity* (closely related to *homophily*). In this paper, we propose PRONE, the first measure to handle pairs of attributes (e.g., major and age). The proposed PRONE is (a) *thorough*, handling both homophily and heterophily (b) *general*, quantifying correlation of a single attribute or a pair of attributes (c) *consistent*, yielding a zero score in the absence of any structural correlation. Furthermore, PRONE can be computed fast in time linear in the network size and is highly useful, with applications in data imputation, marketing, personalization and privacy protection.

Keywords: Attributed networks · Homophily · Heterophily · Assortativity

1 Introduction

Suppose we know that Alice is majoring in Computer Science. To what extent can we comment on the major of her friend Bob? How accurately can we predict his age or sexual orientation? At a broader level, given the structure of a network and some attributes (e.g., major, age) on the nodes, how can we find out (a) which attributes (b) which pairs of attributes show correlation?

Attributed networks are ubiquitous, occurring in a number of domains. For instance in social networks, where nodes represent people, and edges indicate friendships, the attributes may include interests/demographics of individuals. Similarly in citation networks, where papers (nodes) cite each other (edges), each paper also incorporates information regarding the venue or keywords (attributes). However, despite the prevalence of attributed graphs, the vast

© Springer International Publishing AG 2017
J. Kim et al. (Eds.): PAKDD 2017, Part I, LNAI 10234, pp. 225–237, 2017.
DOI: 10.1007/978-3-319-57454-7_18

majority of network science has dealt solely with the graph structure/topology [4,5] ignoring the attributes.

Studies focusing on the interplay between the network structure and attributes are fairly recent [9,14,15]. For example, in a typical social network, the similarity of individuals motivates them to form relations (social selection) and in turn the individuals may themselves be affected by their relations (a.k.a. social influence) [14]. This *assortative mixing* and peer influence results in a homophily pattern observed in many real world networks [17], where neighboring nodes exhibit similar characteristics/attributes. Several works use this observation to cluster data [6], build realistic generative models [2,12] and accurate prediction models [1,11]. There are still fewer studies that try to understand assortativity in networks: by quantifying the correlation of nodal attributes and the structure in a static network [17,18], or by investigating the interplay of social selection and influence over time [9].

Table 1. All variants of PRONE are **thorough**, **general** and **consistent** in contrast to the baseline *assortativity* measures.

	(i) self-proclivity			(ii) cross-proclivity	
	Given: node's color; Guess: neighbor's colors			Given: node's color; Guess: neighbor's shape	
	i.1	i.2	i.3	ii.1	ii.2
	homophily	heterophily	random	correlation	no correlation
Q [16]	0.75	-0.25	-0.25	✗	✗
r [17]	**1.0**	-0.33	-0.33	✗	✗
PRONE$_1$	**1.0**	**1.0**	0.21	0.67	**0.0**
PRONE$_2$	**1.0**	**1.0**	0.11	0.5	**0.0**
PRONE$_3$	**1.0**	**1.0**	0.05	0.33	**0.0**

Assortativity, as a measure for structural correlation of a single attribute, presents a major drawback that it can capture homophily mixing pattern (*i.e.*, when nodes of same attribute value link together) only. This is demonstrated in Table 1. Assortativity (*r*-index) gives a full score of 1 to perfect homophily (i.1); but is unable to distinguish between perfect heterophily (i.2) and randomness (i.3). Further, it cannot characterize or distinguish the mixing patterns involving a pair of attributes (e.g., ii.1 where there is correlation between color and shape based on structure, and ii.2. where shape and color are independent).

The goal of this work is the formal characterization of the *proclivity* of attributed networks, *i.e.*, *the inclination or predisposition of nodes with a certain value for an attribute to connect to nodes with a certain other value for the same (self-proclivity) or a different attribute (cross-proclivity).* The problem we address in this work can be informally stated as:

Informal Problem 1. *Given: an attributed network $\mathcal{G}$, two different attributes a_1, a_2*
To measure:

- *Self-proclivity which captures how predictable neighbors' attribute values for a_1 are, given a node's value for a_1.*
- *Cross-proclivity which captures how predictable neighbors' attribute values for a_2 are, given a node's value for a_1 or vice versa.*

We propose PRONE (PROclivity index for attributed NEtworks) for quantifying both self- and cross-*proclivity* in attributed networks, by drawing upon the clustering validation literature. In place of the confusion matrix (a.k.a. contingency table) which is used to measure the agreements between two groupings of datapoints, we propose to consider the *mixing matrix* (which will be introduced in Sect. 3) of attributes. PRONE has the following desirable properties:

✓ **Thoroughness:** ability to capture homophily and heterophily
✓ **Generality:** applicability in characterizing both self- and cross-proclivity
✓ **Consistency:** quantification of the absence of correlation as zero
✓ **Scalability:** linear running time with respect to the number of edges

PRONE will help with numerous settings, including:

- *data imputation:* what attributes should we use to guess a missing attribute of Alice, given the attributes of her friends
- *marketing:* for ad placement and enhancing e-shopping experience
- *personalization:* for early depression-detection from online networks [7,8]
- *anonymization/privacy:* which attributes, or pairs of attributes can reveal sensitive information about Alice and thus should be masked

The outline of the paper is as follows. In Sect. 2, we review related work and present the assortativity indices proposed in literature. Section 3 formally introduces our proposed metric PRONE and Sect. 4 establishes its theoretical properties. After presenting the results upon applying PRONE to Facebook attribute networks in Sect. 5, we finally conclude in Sect. 6.

2 Related Work and Background

In this section, we briefly review the prior work for attributed graphs and present more background on the two assortativity measures proposed in the literature which we will use as our baseline for comparison.

2.1   Related Work

We will group related work under the following four categories: (i) measures for attribute correlation [16–18] (ii) dynamic patterns in attributed graphs [9,12] (iii) models for attributed networks [13,19] (iv) link prediction and inference [10,11,14,21,24].

The correlation of attributes with the structure of the network was first studied in [17], in which the assortative mixing of a single attribute was quantified through r-index. To the same end, Q-modularity is proposed [16] based on the surprise in encountering edges connecting attributes of the same value. For vector attributes, assortativity is extended by considering average similarities of connected nodes (e.g., using euclidean or cosine similarity) [18]. There is little work beyond this on quantifying structural correlation of attributes.

On the other hand, several studies try to better understand the dynamics of homophily [9,12]. For example, a clear feedback effect between social influence and selection in the network of Wikipedia editors has been discovered in [9], where they observe a sharp increase in the average cosine similarity of users right before they interact for the first time followed by a steady increase in their similarity. In a related study, patterns of attributes in Google+ network have been investigated [12] by modeling it as a social-attribute network (SAN), which simply augments the graph by adding nodes which correspond to attribute values and connects them to the individuals who have those attributes. Multiplicative attributes graph model [13] is proposed for attributed networks using a link-affinity matrix, where they assume that the attributes are binary and are independent. To incorporate the attribute correlations into this model, [19] an accept-reject sampling framework was used to filter the edges generated from the underlying model and selectively accept those that match the desired correlations.

Since nodal similarities and social interactions are two tangled factors which affect the evolution of networks [9], models which incorporate the correlation between attributes and relations better predict links and infer attributes, as confirmed by many recent studies [10,11,14,24]. A large body of predictive models extract topological features from the network and combine them with the nodal features to achieve better classification [23] while others directly utilize the generative graph models to jointly predict links and infer attributes [10,11].

We are interested in the more fundamental question of quantifying structural correlations of a single attribute (more general than assortative mixing) or a pair of attributes and thus our work falls into group (i). We will review our only competitors – r-index [17] and Q-modularity [16] in the following section.

2.2   Background

r-index: Given an attributed network, r-index for assortativity constructs the $k \times k$ normalized mixing matrix E whose $(i, j)^{th}$ entry, e_{ij}, determines the fraction of edges connecting nodes with attribute value i to nodes with value j. This matrix can be then summarized by an assortativity coefficient [17] defined as:

$$r = \frac{\sum_i e_{ii} - \sum_i e_{i.}e_{.i}}{1 - \sum_i e_{i.}e_{.i}} = \frac{Tr[e] - ||e^2||}{1 - ||e^2||} \qquad (1)$$

where $e_{i.} = \sum_j e_{ij}$, $e_{.i} = \sum_j e_{ji}$, and $\sum_{ij} e_{ij} = 1$. Here, $r = 1$ shows perfect assortative mixing and $r = 0$ when there is no assortative mixing.

Q-Modularity: An alternate characterization of assortativity is to measure how unexpected the edges between the nodes with the same attribute value are compared to *random*. Here, *random* refers to the distribution of edges at random after fixing the degree distribution of the nodes. Mathematically,

$$Q = \sum_i e_{ii} - e_{i.}^2 = \text{Tr}[e] - ||e^2|| \qquad (2)$$

Observation: We can see that Q-modularity (Eq. 2) is equivalent to the numerator of r-index (Eq. 1). In fact, the normalized Q proposed for measuring the assortativity in [16] is equivalent to Eq. 1 (since the maximum value of $\text{Tr}[e]$ is 1).

3   Proposed Method: PRONE

Consider the $k \times r$ mixing matrix E for two categorical/nominal attributes a_1 and a_2, with respectively k and r distinct values (cardinality). More precisely, elements of E denote *the number of* edges connecting nodes with the corresponding attributes, *i.e.*, e_{ij} represents the number of edges that connect a node that possesses i^{th} value of a_1 ($v_i^{a_1}$) to a node that has the jth value of attribute a_2 ($v_j^{a_2}$). The resulting mixing matrix (and its marginals) is summarized in the following table and form the basis of our PRONE index for measuring the structural correlation between a_1 and a_2 (Table 2).

Table 2. Mixing matrix of two categorical attributes, a_1 and a_2

	$v_1^{a_2}$ $v_2^{a_2}$ $\ldots$ $v_r^{a_2}$	marginal sums
$v_1^{a_1}$	e_{11} e_{12} $\ldots$ e_{1r}	$e_{1.}$
$v_2^{a_1}$	e_{21} e_{22} $\ldots$ e_{2r}	$e_{2.}$
$\vdots$	$\vdots$ $\vdots$ $\ddots$ $\vdots$	$\vdots$
$v_k^{a_1}$	e_{k1} e_{k2} $\ldots$ e_{kr}	$e_{k.}$
marginal sums	$e_{.1}$ $e_{.2}$ $\ldots$ $e_{.r}$	$e_{..}$

Here, we have $e_{i.} = \sum_j e_{ij}$, $e_{.i} = \sum_j e_{ji}$, and $e_{..} = \sum_i \sum_j e_{ij}$. This mixing matrix is analogous to the confusion matrix or contingency table of two clusterings if we assume distinct values of each attribute are class labels for a grouping based on that attribute. Hence, we can quantify the divergence in this matrix as [20]:

$$D_f = \frac{\sum_i [f(e_{i.}) - \sum_j f(e_{ij})] + \sum_j [f(e_{.j}) - \sum_i f(e_{ij})]}{\sum_i f(e_{i.}) + \sum_j f(e_{.j}) - 2\sum_i \sum_j f(\frac{e_{.j}e_{i.}}{e_{..}})} \tag{3}$$

The numerator aggregates the per-row and per-column divergences of this matrix, while the denominator normalizes this quantity using the maximum divergence value when the marginals are fixed. The correlation or agreement of the two attributes a_1 and a_2 is then obtained from $1 - D_f$. We consider three specific derivations of this measure using $f(x) = x \log x$, $f(x) = x^2$, $f(x) = x^3$; the first two correspond to the two most commonly used clustering agreement indexes: respectively Normalized Mutual Information (NMI), and Adjusted Rand Index (ARI). Specifically, if we normalize E so that $e_{..} = 1$, the PRONE$_l$ ($f(x) = x \log x$) and PRONE$_2$ ($f(x) = x^2$) derivations are simplified as:

$$\text{PRONE}_l = \frac{\sum_j e_{.j} \log(e_{.j}) + \sum_i e_{i.} \log(e_{i.}) - \sum_{ij} e_{ij} \log(e_{ij})}{\frac{1}{2}[\sum_j e_{.j} \log(e_{.j}) + \sum_i e_{i.} \log(e_{i.})]} \tag{4}$$

$$\text{PRONE}_2 = \frac{\sum_{ij} e_{ij}^2 - (\sum_i e_{i.}^2)(\sum_j e_{.j}^2)}{\frac{1}{2}[\sum_i e_{i.}^2 + \sum_j e_{.j}^2] - (\sum_i e_{i.}^2)(\sum_j e_{.j}^2)} \tag{5}$$

4   Theoretical Properties

4.1   Thoroughness

PRONE considers all combinations of attribute values when measuring proclivity. Therefore, it can capture all proclivity patterns inherent in the data including homophily and heterophily; whereas the original assortativity index only considers the matched attribute values abd hence can only capture homophily. In particular, PRONE can capture any mixing patterns between the nodes which are regularly link together, *i.e.*, of two given but not necessarily the same attributes.

For instance, in the test case of (i.2) in Table 1, PRONE detects perfect proclivity as red nodes always connect to yellow nodes, and light blue nodes always link to dark blue nodes. This is not captured by the assortativity index (Q or its normalized version r) which only measures the links between nodes of the same color (homophily) and neglects the off-diagonal elements in the mixing matrix E. These indices in fact have the exact same value for (i.2) and (i.3), even though (i.3) has random color assignments and hence zero proclivity. PRONE, however, returns the maximum value 1 for the perfect proclivity in (i.2) and is close to zero for the random case.

Lemma 1. (r-index is not thorough). *R-index does not capture perfect heterophily, especially when the number of attribute values is high.*

Proof. We prove this by giving a counter example. Consider a graph with a single attribute which takes values $\{1, 2, \ldots, 2k\}$ and shows perfect heterophily in the following manner: Nodes with value i (for $i = 1, \ldots, k$) are connected only to nodes with value $k + i$. If p_i is the fraction of total edges connecting nodes of attribute value i with nodes of attribute value $k + i$, the $k \times k$ normalized mixing matrix E is given by $e_{i,k+i} = e_{k+i,i} = p_i/2$ and $e_{ij} = 0$ otherwise. Note that the leading diagonal elements are zero and the row/column sums are $e_{i.} = e_{.i} = e_{k+i.} = e_{.k+i} = p_i/2$ for $i = 1, \ldots, k$. Using these, the r-index (Eq. 1) may be calculated as

$$r = \frac{0 - 0.5 \sum_{i=1}^{k} p_i^2}{1 - 0.5 \sum_{i=1}^{k} p_i^2}$$

Taking $p_i = 1/k$, $r = \frac{-1}{2k-1}$. The maximum negative assortativity of -1 is attained only when $k = 2$. As k is increased, the value approaches zero (randomness). Thus r-index fails to capture *perfect heterophily*, particularly for large k.   ∎

Lemma 2 (Heterophily and self-proclivity). *Perfect heterophily leads to a perfect self-proclivity score of 1, for any choice of f.*

Proof. Let an attribute assume values $1, \ldots, k$ and let π be a permutation of the values such that $\pi_i \neq i$ and $\pi_i = j \iff \pi_j = i$. Let the probability of edge between i and j be $p_i = p_j$ if $j = \pi_i$ and 0 otherwise. Also, let $\sum_i p_i = 1$.

The row/column marginals are $e_{i.} = e_{.i} = p_i$ while $\sum_i \sum_j f(e_{ij}) = \sum_i f(e_{i\pi_i}) = \sum_i f(p_i)$. From Eq. 3,

$$\text{PRONE} = 1 - \frac{\sum_i f(p_i) + \sum_j f(p_j) - 2\sum_i f(p_i)}{\sum_i f(p_i) + \sum_j f(p_j) - 2\sum_i \sum_j f(p_i p_j)} = 1$$

∎

4.2   Generality

Equation 3 and its PRONE derivations including Eqs. 4 and 5 do not impose any assumptions on the mixing matrix $E_{r \times k}$ and hence can be applied to general cases. On the other hand, the definition of previous measures for assortativity in Eqs. 1 and 2 require E to be a square matrix $(r = k)$ and hence cannot be extended to measure cross-proclivity of two attributes which have different cardinalities.

For instance, in the test case of (ii.1) in Table 1, we see a mixing pattern between color and shape: *i.e.*, red and yellow circles mix together while light and dark blue squares link to each other. The assortativity measure Q and its normalized version, r, cannot be applied in this case, as the diagonal is not defined for the 4×2 mixing matrix. PRONE, on the other hand, is able quantify this non-square mixing matrix, since it is defined based on average divergence/dispersion in the rows and columns of E. We can see that all variations of PRONE correctly detect a high correlation between shape and color for this case, whereas they return the baseline of 0.0 for the random case of (ii.2) where there is no such correlation.

Lemma 3 (r-index and PRONE). *Squashing the off-diagonal elements in formula of PRONE$_x$ yields r-index.*

Proof. Let E be the normalized mixing matrix with $e_{..} = \sum_i e_{i.} = \sum_j e_{.j} = 1$.
Using $f(x) = x$, we have

$$\text{PRONE}_x = 1 - \frac{\sum_i e_{i.} + \sum_j e_{.j} - 2\sum_i \sum_j e_{ij}}{\sum_i e_{i.} + \sum_j e_{.j} - 2\sum_i \sum_j e_{i.}e_{.j}} = 1 - \frac{1 - \sum_i \sum_j e_{ij}}{1 - \sum_i \sum_j e_{i.}e_{.j}}$$

Squashing the off-diagonal products to 0 using the indicator function $\mathbb{I}(i = j)$,
we get

$$1 - \frac{1 - \sum_i \sum_j \mathbb{I}(i = j)e_{ij}}{1 - \sum_i \sum_j \mathbb{I}(i = j)e_{i.}e_{.j}} = 1 - \frac{1 - \text{Tr}[e]}{1 - \sum_i e_{i.}e_{.i}}$$

which is the expression for r-index. ∎

4.3 Consistency

PRONE is expected to return zero when there is no structural correlation in the
network. This is a known desired property for the clustering validation indexes.
ARI, in particular, is called Adjusted Rand Index for the very same reason that
it returns a constant baseline of zero for agreements by chance. This complies
with the ~ 0 correlations we observed for random color assignments in the two
test cases of (i.3) and (ii.2) of Table 1.

Lemma 4 (Consistency of PRONE). *For any choice of f, PRONE is consistent (adjusted for chance), i.e., if values for a nodal attribute are drawn from a categorical distribution ignoring the network structure, its self-proclivity is zero in expectation.*

Proof. Let the multinomial distribution from which the values for attributes a_1
and a_2 are drawn be parameterized by $p_1, \ldots, p_k$ and $q_1, \ldots, q_r$ where k and
r are the cardinalities of categorical attributes a_1 and a_2 respectively. Here,
$\sum_i p_i = \sum_j q_j = 1$. In the absence of structural correlation of attributes, the
expected fraction of edges that connect nodes of attribute values $a_1 = i$ and
$a_2 = j$ is $p_i q_j$, which is the expected entry e_{ij} in the normalized mixing matrix
E. The expected marginal of row i (or column j) in E is $\sum_j p_i q_j = p_i$ (or q_j).
Thus, in expectation,

$$\text{PRONE}_f = 1 - \frac{\sum_i f(p_i) + \sum_j f(q_j) - 2\sum_i \sum_j f(p_i q_j)}{\sum_i f(p_i) + \sum_j f(p_j) - 2\sum_i \sum_j f(p_i q_j)} = 0$$

which proves the consistency of PRONE. ∎

4.4 Scalability

PRONE has the same computational complexity as the previous measures Q and
r, which is the cost of building the mixing matrix E. E can be computed by a

single pass over all edges in the graph and hence PRONE is linear in order of number of edges.

In more detail, if we assume m is the total number of edges in the network and k represents the maximum cardinality of attributes, PRONE can be computed in $O(m+k^2)$ time. This matches the computational order for the previous measures, $O(m+k)$, as $k \ll m$ (the number of edges in a graph is typically much larger than the cardinality of a nodal attribute).

Here, we also empirically measure the computation time of PRONE for networks of varying sizes to show the scalability of the PRONE. In particular, we a generate network of size m, and assign nodes a single attribute with cardinality k, i.e., we assign to each node u, a value in $\{1, \ldots, k\}$ chosen uniformly at random. Figure 1 plots the computational time in seconds as the number of edges grows. The observed linear trend confirms our claim.

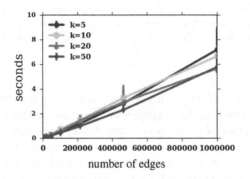

Fig. 1. Scalability of PRONE on networks generated using Barabási and Albert [3] model with 1K nodes and ~10K edges. The attribute cardinality was varied in $\{5, 10, 20, 100\}$ and the results were averaged over 10 runs.

Choice of f in Practice

Although the above properties are valid for arbitrary choice of f, we recommend choosing f to be a superadditive function[1] satisfying $f(x) \geq 0 \forall x \in [0, 1]$ and $f(1) = 1$ for the proclivity scores to be bounded in $[0, 1]$ [20].

5 Empirical Studies Using Real World Data

Here, we study the PRONE in Facebook friendship network of 100 US collages available in a.k.a. *Facebook 100 dataset* [22]. In networks of this dataset, each user has six categorical attributes: (1) *gender* (male/female), (2) *status* (faculty/student/etc.), (3) *major*, (4) *second major/minor* (high missing values), (4) *dormitory* of residence, (5) class *year* and (6) *high school*. Figure 2 shows

---

[1] f is superadditive $\iff f(x + y) \geq f(x) + f(y)$.

one sample network of this dataset which has 6386 nodes and 217662 friendships edges. The same network is plotted with six different color codings of the nodes, *i.e.*, one plot per attribute in which nodes are colored based on their value for that particular attribute.

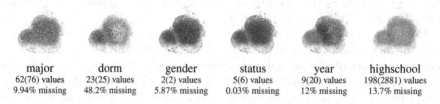

major	dorm	gender	status	year	highschool
62(76) values	23(25) values	2(2) values	5(6) values	9(20) values	198(2881) values
9.94% missing	48.2% missing	5.87% missing	0.03% missing	12% missing	13.7% missing

Fig. 2. An example Facebook friendship network, where nodes are colored based on their corresponding attribute value (missing values are white, and non-frequent values are gray). For attribute *status* and *year*, we visually observe some correlation between the color of the nodes and their locations, whereas the locations are derived from a layout algorithm that looks only at the connectivity between the nodes. (Color figure online)

In Fig. 2, locations of nodes are derived from a network visualization algorithm which only looks at the topology or structure of the graph and tries to place nodes together as cohesive groups. Depending on the layout algorithm used, we can visually observe some of the correlations between attributes (colors) and the structure. In particular, with this example layout, the self proclivityof *year* might be obvious. PRONE provides a fast and quantitative way to detect both the obvious and the hidden structural correlations in such a dataset.

We can see the values of PRONE for Facebook dataset in Fig. 2 reported in Table 3. The diagonal of this matrix show the self-proclivity values for the corresponding attributes, and the off-diagonal values provide the cross-proclivity measurements between the corresponding pairs of attributes.

Table 3 reports the results using PRONE$_2$; we observe a similar trend using the PRONE$_l$ and PRONE$_3$ variations. These are reported in Table 4. The choice of PRONE, *i.e.*, the generative function used in Eq. 3, depends on the application at hand.

We observe similar patterns over different samples in the Facebook 100 dataset. Here, for example, we report the proclivity for another sample, *i.e.*, Rice31 network from this collection which has 4087 nodes and 184828 edges.

Discussion: From the PRONE scores, we infer that the *dormitory* is significantly correlated with friendship as it has a high self-proclivity. This is also the case for *status* (faculty or student) and *year*. What this means is the following: Given Smith's *dormitory* (or *status* or *year*) attribute value, we can predict the *dorm* (or *status* or *year*, respectively) value of his friends. On the other hand, *highschool* and *minor* show zero self-proclivity and the same cannot be said of them. Also,

Table 3. Proclivity of attributes for the Facebook dataset in Fig. 2 using PRONE$_2$. The diagonal and off-diagonal entries represent the self-proclivity and the cross-proclivity values respectively. Nodes with missing values were removed before the computation.

PRONE$_2$	Major	Gender	Year	Status	Dorm	Highschool	Minor
Major	0.01	0.00	0.00	0.00	0.00	0.00	0.00
Gender	0.00	0.00	−0.00	−0.00	−0.00	0.00	−0.00
Year	0.00	−0.00	**0.22**	**0.03**	**0.04**	0.00	0.00
Status	0.00	−0.00	**0.03**	**0.27**	**0.02**	0.00	0.00
Dorm	0.00	−0.00	**0.04**	**0.02**	**0.11**	0.00	0.00
Highschool	0.00	0.00	0.00	0.00	0.00	0.00	0.00
Minor	0.00	−0.00	0.00	0.00	0.00	0.00	0.00

Table 4. Proclivity of attributes for the Facebook dataset in Fig. 2 using PRONE$_l$ and PRONE$_3$. These tables provide alternative measurements to Table 3.

	PRONE$_l$							PRONE$_3$						
	Major	Gender	Year	Status	Dorm	Highschool	Minor	Major	Gender	Year	Status	Dorm	Highschool	Minor
Major	0.01	0.00	0.01	0.00	0.01	**0.05**	0.01	0.00	0.00	0.00	0.00	0.00	0.00	0.00
Gender	0.00	0.00	0.00	0.00	0.00	0.00	0.00	0.00	0.00	0.00	0.00	0.00	0.00	0.00
Year	0.01	0.00	**0.25**	0.07	**0.07**	0.07	0.01	0.00	0.00	**0.15**	0.01	0.02	0.00	0.00
Status	0.00	0.00	**0.07**	0.09	0.02	0.02	0.00	0.00	0.00	0.01	**0.29**	0.00	0.00	0.00
Dorm	0.01	0.00	**0.07**	0.02	**0.16**	**0.10**	0.02	0.00	0.00	0.02	0.00	**0.05**	0.00	0.00
Highschool	**0.05**	0.00	**0.07**	0.02	**0.10**	**0.31**	**0.07**	0.00	0.00	0.00	0.00	0.00	0.00	0.00
Minor	0.01	0.00	0.01	0.00	0.02	**0.07**	0.01	0.00	0.00	0.00	0.00	0.00	0.00	0.00

Table 5. Proclivity of attributes for Rice31 dataset using different derivations of PRONE.

	PRONE$_l$							PRONE$_3$						
	major	gender	year	status	dorm	highschool	minor	major	gender	year	status	dorm	highschool	minor
major	0.02	0.00	0.01	0.00	0.01	0.03	0.01	0.00	0.00	0.00	0.00	0.00	0.00	0.00
gender	0.00	0.00	0.00	0.00	0.00	0.00	0.00	0.00	0.00	0.00	0.00	0.00	0.00	0.00
year	0.01	0.00	**0.17**	0.08	0.00	**0.05**	0.01	0.00	0.00	**0.07**	0.02	0.00	0.00	0.00
status	0.00	0.00	0.08	**0.11**	0.00	0.02	0.00	0.00	0.00	0.02	**0.30**	0.00	0.00	0.00
dorm	0.01	0.00	0.00	0.00	**0.25**	**0.09**	0.01	0.00	0.00	0.00	0.00	**0.15**	0.00	0.00
highschool	0.03	0.00	**0.05**	0.02	**0.09**	0.21	**0.05**	0.00	0.00	0.00	0.00	0.00	0.00	0.00
minor	0.01	0.00	0.01	0.00	0.01	**0.05**	0.01	0.00	0.00	0.00	0.00	0.00	0.00	0.00

	PRONE$_2$						
	major	gender	year	status	dorm	highschool	minor
major	0.01	0.00	0.00	0.00	0.00	0.00	0.00
gender	0.00	0.00	0.00	0.00	0.00	0.00	0.00
year	0.00	0.00	**0.13**	**0.05**	0.00	0.00	0.00
status	0.00	0.00	**0.05**	**0.26**	0.00	0.00	0.00
dorm	0.00	0.00	0.00	0.00	**0.24**	0.00	0.00
highschool	0.00	0.00	0.00	0.00	0.00	0.01	0.00
minor	0.00	0.00	0.00	0.00	0.00	0.00	0.01

we uncover a surprising pattern that attribute values for *year* and *dorm* show correlation given the friendship network, based on their cross-proclivity of 0.04. Thus, given Smith's *dorm*, it may be possible to predict Smith's friends' *year* values, an inference which is otherwise not possible, from just visualization.

In sum, PRONE is (i) novel and is the first to characterize pairwise attribute correlations given the structure; (ii) is fast to compute and scales linearly with network size; (iii) is effective and discovers interesting correlation patterns when applied to real world graphs. These together make PRONE extremely useful in practice – with applications in anonymizing networks, marketing, data imputation and many more (Table 5).

6   Conclusion

In this paper, we proposed PRONE to measure the self- and cross-proclivity patterns and quantify the correlation of a single attribute or a pair of attributes with the network structure. Our proposed PRONE has the following desirable characteristics:

✓ **Thoroughness:** PRONE can capture the full range of mixing patterns in networks, including homophily and heterophily (Lemma 2).
✓ **Generality:** PRONE can capture both self-proclivity (mixing patterns of a single attribute) and cross-proclivity (mixing patterns of any pair of attributes) (Lemma 3).
✓ **Consistency:** In the absence of structural correlation of nodal attributes, PRONE consistently returns a value of zero in expectation (Lemma 4).
✓ **Scalability:** PRONE can quantify the mixing patterns, a.k.a. structural correlation, in $\mathcal{O}(m)$ time where m is the number of edges in the network and is fast, processing million-scale graphs in a few seconds.

PRONE is also highly useful, with applications in (i) *data imputation* to guess the values of missing attributes of nodes, (ii) *marketing* for ad-placement, (iii) *personalization* for early depression detection and (iv) *privacy protection* and anonymization of social network.

References

1. Akcora, C.G., Carminati, B., Ferrari, E.: User similarities on social networks. Soc. Netw. Anal. Min. 1–21 (2013)
2. Akoglu, L., Faloutsos, C.: RTG: a recursive realistic graph generator using random typing. Data Min. Knowl. Disc. **19**(2), 194–209 (2009)
3. Albert, R., Barabási, A.L.: Statistical mechanics of complex networks. Rev. Mod. Phys. **74**(1), 47 (2002)
4. Bianconi, G.: Interdisciplinary and physics challenges of network theory. EPL (Europhys. Lett.) **111**(5), 56001 (2015)
5. Boccaletti, S., Latora, V., Moreno, Y., Chavez, M., Hwang, D.U.: Complex networks: structure and dynamics. Phys. Rep. **424**(4), 175–308 (2006)
6. Bothorel, C., Cruz, J.D., Magnani, M., Micenkova, B.: Clustering attributed graphs: models, measures and methods. Netw. Sci. **3**(03), 408–444 (2015)
7. Choudhury, M.D., Counts, S., Horvitz, E., Hoff, A.: Characterizing and predicting postpartum depression from shared Facebook data. In: CSCW (2014)

8. Colombo, G., Burnap, P., Hodorog, A., Scourfield, J.: Analysing the connectivity and communication of suicidal users on Twitter. Comput. Commun. **73**, 291–300 (2016)
9. Crandall, D., Cosley, D., Huttenlocher, D., Kleinberg, J., Suri, S.: Feedback effects between similarity and social influence in online communities. In: Proceedings of 14th ACM SIGKDD International Conference on Knowledge Discovery and Data Mining, pp. 160–168 (2008)
10. Gong, N.Z., Talwalkar, A., Mackey, L., Huang, L., Shin, E.C.R., Stefanov, E., Shi, E.R., Song, D.: Joint link prediction and attribute inference using a social-attribute network. ACM Trans. Intell. Syst. Technol. (TIST) **5**(2), 27 (2014)
11. Gong, N.Z., Talwalkar, A., Mackey, L., Huang, L., Shin, E.C.R., Stefanov, E., Song, D., et al.: Jointly predicting links and inferring attributes using a social-attribute network (SAN) (2011). arXiv preprint arXiv:1112.3265
12. Gong, N.Z., Xu, W., Huang, L., Mittal, P., Stefanov, E., Sekar, V., Song, D.: Evolution of social-attribute networks: measurements, modeling, and implications using Google+. In: Proceedings of 2012 ACM Conference on Internet Measurement Conference, pp. 131–144 (2012)
13. Kim, M., Leskovec, J.: Multiplicative attribute graph model of real-world networks. In: Kumar, R., Sivakumar, D. (eds.) WAW 2010. LNCS, vol. 6516, pp. 62–73. Springer, Heidelberg (2010). doi:10.1007/978-3-642-18009-5_7
14. La Fond, T., Neville, J.: Randomization tests for distinguishing social influence and homophily effects. In: Proceedings of 19th International Conference on World Wide Web, pp. 601–610. ACM (2010)
15. Lewis, K., Gonzalez, M., Kaufman, J.: Social selection and peer influence in an online social network. Proc. Natl. Acad. Sci. **109**(1), 68–72 (2012)
16. Newman, M.: Networks: An Introduction. Oxford University Press, Inc., Oxford (2010)
17. Newman, M.E.: Mixing patterns in networks. Phys. Rev. E **67**(2), 026126 (2003)
18. Pelechrinis, K., Wei, D.: VA-index: quantifying assortativity patterns in networks with multidimensional nodal attributes. PLoS ONE **11**(1), e0146188 (2016)
19. Pfeiffer III, J.J., Moreno, S., La Fond, T., Neville, J., Gallagher, B.: Attributed graph models: modeling network structure with correlated attributes. In: Proceedings of 23rd International Conference on World Wide Web, pp. 831–842. ACM (2014)
20. Rabbany, R., Zaïane, O.: Generalization of clustering agreements and distances for overlapping clusters and network communities. Data Min. Knowl. Disc. **29**(5), 1458–1485 (2015)
21. Silva, A., Meira, W., Zaki, M.J.: Mining attribute-structure correlated patterns in large attributed graphs. Proc. VLDB Endow. **5**(5), 466–477 (2012)
22. Traud, A.L., Mucha, P.J., Porter, M.A.: Social structure of Facebook networks. Physica A **391**(16), 4165–4180 (2012)
23. Wang, P., Xu, B., Wu, Y., Zhou, X.: Link prediction in social networks: the state-of-the-art. Sci. China Inf. Sci. **58**(1), 1–38 (2015)
24. Yin, Z., Gupta, M., Weninger, T., Han, J.: LINKREC: a unified framework for link recommendation with user attributes and graph structure. In: Proceedings of 19th International Conference on World Wide Web, pp. 1211–1212. ACM (2010)

Hierarchical Mixed Neural Network for Joint Representation Learning of Social-Attribute Network

Weizheng Chen[1(✉)], Jinpeng Wang[2], Zhuoxuan Jiang[1], Yan Zhang[1], and Xiaoming Li[1]

[1] School of Electronics Engineering and Computer Science, Peking University, Beijing, China
cwz.pku@gmail.com, jzhx1211@gmail.com, zhyzhy001@gmail.com, lxm.at.pku@gmail.com
[2] Microsoft Research, Beijing, China
jinpwa@microsoft.com

Abstract. Most existing network representation learning (NRL) methods are designed for homogeneous network, which only consider topological properties of networks. However, in real-world networks, text or categorical attributes are usually associated with nodes, providing another description for networks in a different perspective.

In this paper, we present a joint learning approach which learns the representations of nodes and attributes in the same low-dimensional vector space simultaneously. Particularly, we show that more discriminative node representations can be acquired by leveraging attribute features. The experiments conducted on three social-attribute network datasets demonstrate that our model outperforms several state-of-the-art baselines significantly for node classification task and network visualization task.

Keywords: Social-attribute network · Representation learning · Joint learning

1 Introduction

The growth of online social media produces massive amounts of user-generated content, such as tweets posted in Twitter, personal profile in LinkedIn. These data and social relations among users make up complex heterogeneous information network [16], which is an effective organization form of multi-source data. Mining such heterogeneous information network is crucial for various research tasks and commercial applications, for example, node classification [9] and product recommendation [15].

During the last few years, representation learning, also known as embedding, has become a promising and powerful tool in network analysis area. Since the density of typical social network is usually quite small in the real world,

© Springer International Publishing AG 2017
J. Kim et al. (Eds.): PAKDD 2017, Part I, LNAI 10234, pp. 238–250, 2017.
DOI: 10.1007/978-3-319-57454-7_19

traditional network representation such as adjacency matrix suffers from the data sparsity problem. Thus we can't apply most statistical machine learning algorithms to solve network analysis task directly. To overcome this problem, many NRL methods which aim to project the nodes into a low-dimensional continuous vector space have been proposed. Among those representative methods, DeepWalk [14], LINE [17], GraRep [3] and node2vec [8] learn general node representations which are not tuned for specific task for homogeneous network by only considering the structural features. However, both text and categorical attributes play an crucial role in real-life networks, e.g., papers or authors in citation networks are associated with corresponding text content, users in Twitter or Facebook have profiles with categorical attributes such as gender and job. It is necessary to uncover the potential effect of nodes' attributes in NRL process.

To utilize the rich text content information of nodes, Yang et al. [20] presents Text-associated DeepWalk (TADW) to learn network representations from both network structure and text attributes in an inductive matrix completion framework. Because of high dimensionality of text attributes (namely words), singular-value decomposition is performed on the node-attribute matrix to get robust attribute features in TADW. However, TADW has two serious weaknesses:

1. Unlike text attributes, categorical attributes space is low-dimensional, e.g., only up to dozens of demographic attributes appears in the mobile social network [6] and Twitter network [4], which makes TADW unsuitable in such a scenario.
2. The performance of TADW falls fast if texts of some nodes are missing. Note that, text information is often incomplete. For example, in online social media, some users' text features are difficult to obtain due to their privacy settings or they actually never publish any texts.

To overcome the above problems, in this paper, we propose Social-Attribute Network Representation Learning (SANRL), a scalable joint NRL framework which preserve both structural and attribute information in the unified representations. Compared with TADW, which only learn representations of nodes, our model also learn representations of attributes in the same low dimensional vector space. A hierarchical mixed neural network model is adopted to model the interactive relationship between the nodes and the attributes. We conduct experiments with three real-world network datasets, including a social network with categorical attributes, a co-author network with text attributes and a citation network with text attributes. In summary, this paper has the following three major contributions:

1. We propose a network representation learning model for the heterogeneous social-attribute networks, which can handle either categorical attributes or text attributes. Our model can make use of limited attribute information by using a coupled architecture, which makes it more flexible in real scenarios.
2. Our model can map the nodes and the attributes to the same space, which provide meaningful features for various applications.
3. The experimental results show that our model outperforms other competitive baselines significantly.

2   Related Work

Most existing unsupervised NRL models only consider the network structure information. These models are mainly based on two classical assumptions. The first is only suitable for the undirected networks, called smoothness assumption, in which two linked nodes should have close representations. Laplacian Eigenmaps [1], LINE with first-order proximity, SDNE [19] all adopt the smoothness assumption. The second is also applicable to the directed networks, in which two nodes should have close representations if they have similar neighbors. Here, the neighbors can be high-order. DeepWalk, GraRep and node2vec all adopt this assumption. To incorporate the text attribute information, TADW adopts the inductive matrix completion technique to incorporates the tfidf matrix into Deep-Walk. As far as we know, how to use the categorical attributes of nodes has not been studied yet.

Recently, learning network representations in a semi-supervised way has drawn many attentions by incorporating the label information [11,18,21]. All these models are trained in a transductive way, which means that they use a combination of an unsupervised NRL model and a classifier trained on the labeled nodes. However, all these models only learn the distributed representations of the nodes. Unlike semi-supervised NRL models which leverage the label information to get more discriminative node representation, our motivation is to enhance node representation with rich categorical or text attributes via jointly projecting the nodes and the attributes into the same vector space. In this paper, we follow the line of the unsupervised NRL models. In the future, we will explore the semi-supervised extension of our model by adding a classifier to it.

3   Problem Definition

We adopt a social-attribute network (SAN) [7] $G = (V, E, A, C)$ to represent an attributed network in this work. V is the set of social nodes, each representing a data object. E is the set of social links, each representing an edge between two social nodes. A is the set of attribute nodes, each representing an attribute. C is the set of attribute links, each representing an affiliation relationship between an attribute node and a social node. For simplification, we use $A(v) = \{a|(a, v) \in C\}$ to represent the attribute sets of node $v \in V$. All attribute links are undirected, while social links can be directed or undirected and weighted or unweighted depending upon the data type. Moreover, the weight of attribute links are defined as relative importance of certain attribute for corresponding social node.

As shown on the left side of Fig. 1, we present a sample SAN which has five social nodes and four attribute nodes. Our research target is to a learn informative continuous vector representation $u \in \mathbb{R}^d$ for each social node of a SAN, where $d \ll |V|$. In vector space $\mathbb{R}^d$, both structural information and attribute information should be preserved. Then the learned low-dimensional representations can used as input features to a variety of machine learning models such as logistic regression for classification task, k-means for clustering task. For

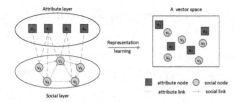

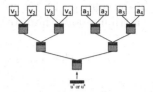

Fig. 1. Illustration of learning representations jointly for a toy social-attribute network.

Fig. 2. Mixed binary Huffman tree used in SANRL.

simplification and clarification, in the following paper, when we say "node", we refer to a social node. And when we say "attribute", we refer to an attribute node.

4 SANRL Model

4.1 Loss Function for Structural Information

We adapt DeepWalk to learn representations of social nodes based on the assumption that nodes which have similar neighborhoods will have similar representations. Here, neighborhoods refer to both direct neighbors and higher order neighbors. Next, we give a brief outline of DeepWalk.

Given a social network $G = (V, E)$ (e.g., social layer of Fig. 1), Deep-Walk generates many node sequences as training data. More specifically, staring from a node v_1, DeepWalk generates γ fixed-length sequences of nodes $S_k^{v_1} = \{v_1, v_2, \ldots, v_t\}$ through random walk for $1 \leq k \leq \gamma$. Repeat the above process for every node $v \in V$. Then we get a set S_{seq} which contains $\gamma|V|$ node sequences. By feeding shuffled S_{seq} to Skip-Gram [12], an efficient and scalable method for learning word representations, the following objective loss function will be minimized:

$$
\begin{aligned}
O_1 &= -\log P(S_{seq}) \\
&= -\sum_{v \in V} \sum_{1 \leq k \leq \gamma} \log P(S_k^v) \\
&= -\sum_{v \in V} \sum_{1 \leq k \leq \gamma} \sum_{v_i \in S_k^v} \log P(v_{i-w}, \ldots, v_{i-1}, v_{i+1}, \ldots, v_{i+w}|v_i) \quad (1) \\
&= -\sum_{v \in V} \sum_{1 \leq k \leq \gamma} \sum_{v_i \in S_k^v} \sum_{i-w \leq j \leq i+w, j \neq i} \log P(v_j|v_i),
\end{aligned}
$$

where w is the size of the sliding window and $P(v_j|v_i)$ is formulated using softmax function:

$$
P(v_j|v_i) = \frac{\exp(u_{v_j}'^T u_{v_i})}{\sum_{v \in V} \exp(u_v'^T u_{v_i})}, \quad (2)
$$

where u_v and u_v' are the "input" and "output" representations of social node v.

In the context of SAN, we replace Eq. (2) with the following Eq. (3) in O_1 to take attribute nodes into consideration:

$$P(v_j|v_i) = \frac{\exp(u'^T_{v_j} u_{v_i})}{\sum\limits_{v \in V} \exp(u'^T_v u_{v_i}) + \sum\limits_{a \in A} \exp(u'^T_a u_{v_i})}. \tag{3}$$

Here u_a and u'_a are the "input" and "output" representations of attribute node a. In addition, the original DeepWalk is only applicable to unweighted networks. Therefore, weighted random walk is adopted in the sampling process to handle weighted network in our model.

4.2   Loss Function for Attribute Information

Considering a social node v with categorical attribute sets $A(v)$, we use the co-occurrence patterns to model $P(A(v))$

$$\log P(A(v)) = \sum_{a_i \in A(v)} \{\log P(a_i|v) + \log P(v|a_i) + \sum_{a_j \in A(v), j \neq i} \log P(a_j|a_i)\}, \tag{4}$$

where $\log P(a_i|v_j)$, $\log P(v_j|a_i)$ and $\log P(a_j|a_i)$ are defined as:

$$P(a_i|v_j) = \frac{\exp(u'^T_{a_i} u_{v_j})}{\sum\limits_{v \in V} \exp(u'^T_v u_{v_j}) + \sum\limits_{a \in A} \exp(u'^T_a u_{v_j})}, \tag{5}$$

$$P(v_j|a_i) = \frac{\exp(u'^T_{v_j} u_{a_i})}{\sum\limits_{v \in V} \exp(u'^T_v u_{a_i}) + \sum\limits_{a \in A} \exp(u'^T_a u_{a_i})}, \tag{6}$$

$$P(a_j|a_i) = \frac{\exp(u'^T_{a_j} u_{a_i})}{\sum\limits_{v \in V} \exp(u'^T_v u_{a_i}) + \sum\limits_{a \in A} \exp(u'^T_a u_{a_i})}. \tag{7}$$

The above three kinds of conditional probability can capture different similarities in the SAN. Firstly, by maximizing $P(a_i|v_j)$, nodes with many similar attributes will tend to have close representations, which is the key idea of the PV-DBOW model [10]. Secondly, by maximizing $P(a_j|a_i)$, paradigmatic relations will be modeled. As a consequence, attributes with many similar contexts will tend to have close representations, which is the key idea of the Word2Vec model [12]. Finally, by maximizing $P(v_j|a_i)$, syntagmatic relations will be modeled. Then attributes often co-occur will tend to have close representations.

Unlike categorical attributes, word order and word frequency are essential properties of text attributes, which are not captured in Eq. (4). To make Eq. (4) suitable for both categorical and text attributes, we replace $A(v)$ by a attribute set sampled from $A(v)$. More specifically, given a node $v \in V$, we generate γ fixed-length sequences of attributes $T^v_k = \{a_1, \ldots, a_w\}$ through random sampling for $1 \leq k \leq \gamma$. For text attributes, T^v_k is a text window whose length is w sampled from the corresponding document of v. But for categorical attributes,

T_k^v is generated by sampling w attributes independently from $A(v)$. Using T to represent the $\gamma|V|$ attribute sequences acquired in the preceding procedure, we define the following objective loss function for attribute information:

$$
\begin{aligned}
O_2 &= -\log P(T) \\
&= -\sum_{v \in V} \sum_{1 \leq k \leq \gamma} \log P(T_k^v) \\
&= -\sum_{v \in V} \sum_{1 \leq k \leq \gamma} \sum_{a_i \in T_k^v} \{\log P(a_i|v) + \log P(v|a_i) + \sum_{a_j \in A(v), j \neq i} \log P(a_j|a_i)\}.
\end{aligned}
\tag{8}
$$

4.3   Loss Function of SANRL

To integrate both structure and attribute information into a joint representation learning framework, we use a weighted linear combination of O_1 and O_2 to formulate our objective loss function of the SANRL model:

$$
O_{joint} = O_1 + \lambda O_2, \tag{9}
$$

where λ is a trade-off parameter. If the network has rich discriminative attributes, λ should be a big number. Otherwise, a small λ is suitable. By minimizing O_{joint}, we can get two resulting vectors u_v and u_v' for each node v, and two resulting vectors u_a and u_a' for each attribute a. Finally, u_v will be used as the feature vector of v.

4.4   Learning and Complexity Analysis

The optimization scheme of SANRL model is similar to DeepWalk, in which Skip-Gram is applied to maximize the co-occurrence probability among node-node pairs, node-attribute pairs, attribute-attribute pairs and attribute-node pairs. In practice, all the representation vectors are initialized randomly at first. Then node sequences and attribute sequences are sampled alternately and iteratively as the input data streams. In our implementation, the effect of λ is reflected by controlling the sampling probability of attribute sequences. By using the back-propagation algorithm to estimate the derivatives, Eq. (9) can be optimized by adopting the asynchronous stochastic gradient descent (ASGD) algorithm. Directly computing softmax functions defined in Eqs. (3), (5), (6) and (7) is very expensive, so we use hierarchical softmax technique to speed up training.

In previous works [5,13], hierarchical softmax is widely used to train a similar neural network, in which only one type of objects such as words or nodes are mapped to the leaves of one binary Huffman tree. However, the application of hierarchical softmax in our model is very different from them. As shown in Fig. 2, every social node and attribute node are associated with one leaf in a same single mixed Huffman tree. And the input of this neural network can be either the representation vector of a node or an attribute. Thus given an arbitrary training instance whose form is an input-output pair, the computational complexity of computing $P(output|input)$ can be reduced from $O(|V|+|A|)$ to $O(\log(|V|+|A|))$. Overall, the time complexity of SANRLis $\gamma|V|d(t + \lambda w)\log(|V| + |A|)$ Because $|A|$ is much more smaller than $|V|$ in most cases, the computational complexity of SANRL is still acceptable.

5   Experiments

5.1   Dataset

Weibo Dataset. We have crawled 294,634 users' detailed information from Sina Weibo, the most popular microblogging service in China, including profiles and 3,183,187 following relationships among them. In our Weibo dataset, the user profile contains four demographic variables:

- **Gender**: male; female.
- **Age**: 1–11; 12–17; 18–29; 30–44; 45–59; 60+.
- **Education**: literature, history; natural science; engineering; economics; medicine; art.
- **Job**: Internet industry; creative design industry; cultural media industry; public service industry; manufacturing industry; scientific research industry; pharmaceutical industry; business management industry;

Here, we regard every possible value of each demographic variable as a categorical attribute. Moreover, an user may have multiple education or job attributes but values in gender and age are exclusive. We build two SANs based on Weibo dataset for node classification task, i.e., the Weibo-education network and the Weibo-job network. In the Weibo-education network, education attributes are treated as labels, which are not used in the representation learning process. The Weibo-job network is organised in the same manner.

DBLP Dataset. We use "DBLP-four-area" dataset provided in [9] to build a weighted co-author network. This data contains 20 major computer conferences from four related areas, i.e., data mining, database, information retrieval and machine learning, and 27,199 authors and all their publications in these conferences. If two authors have co-authored a paper, we add an undirected edge between them. The weight of the edge is the number of their collaborative papers. Finally, the number of edges in 66,832. The titles of all the paper published by one author is recognized as his or her text attributes. The size of the word vocabulary is 12,091. If an author publishes a paper in a certain conference, the research areas of this conference will be added to the author's label set.

5.2   Compared Algorithms

- DeepWalk [14]. DeepWalk is the first work which adopts the neural network language model to solve NRL problem.

- LINE [17]. LINE can learn two representation vectors for each node by optimizing two carefully designed objective function that preserves the first-order proximity and second-order proximity. Then the two representations are concatenated as the final representation.
- LDA [2]. Latent Dirichlet Allocation (LDA) is a classical probabilistic topic model. Each node can be represented as a topic distribution vector.
- TADW [20]. TADW is a state-of-the-art NRL algorithm based on matrix decomposition. First, Singular-value decomposition is performed on the tf-idf matrix to get robust text features of nodes. Then, the text features and a node relation matrix are fed to an inductive matrix completion framework to get node representations.
- SANRL. Our proposed method. For SANRL, the sliding window size $w = 10$, the length of each node sequence t $= 40$, number of node sequences for per node $\gamma = 80$. We set the trade-off parameter $\lambda = 2$ for two Weibo networks and $\lambda = 8$ for DBLP-author network.
- $SANRL_{doc}$. A simplified version of SANRL, in which the final objective loss function is Eq. (8). We treat the attribute sets of a node as its pseudo document. Parameter settings of $SANRL_{doc}$ is same to SANRL.

5.3   Node Classification

Table 1. Macro-F1 (%) of node classification on the Weibo-education network.

% labeled nodes	10%	20%	30%	40%	50%	60%	70%	80%	90%
DeepWalk	33.15	33.34	33.40	33.30	33.38	33.50	33.27	33.57	34.04
LINE	34.61	34.75	34.94	34.87	34.84	34.96	34.83	34.86	34.91
$SANRL_{doc}$	33.52	33.61	33.59	33.74	33.88	33.65	33.85	33.96	34.12
SANRL	**39.97**	**40.03**	**40.14**	**40.09**	**40.09**	**40.12**	**40.07**	**40.09**	**40.29**

Table 2. Macro-F1 (%) of node classification on the Weibo-job network.

% labeled nodes	10%	20%	30%	40%	50%	60%	70%	80%	90%
DeepWalk	28.69	28.85	29.09	28.72	28.89	28.75	28.69	28.71	28.52
LINE	30.08	30.26	30.37	30.57	30.36	30.57	30.76	30.55	29.56
$SANRL_{doc}$	31.09	31.07	31.01	30.98	31.05	31.04	31.04	30.96	30.99
SANRL	**38.19**	**38.33**	**38.36**	**38.37**	**38.26**	**38.39**	**38.47**	**38.42**	**38.45**

Following the settings in previous works [14,17], we also use the multi-label node classification task to evaluate the quality of the representation vectors learned by different models. The one-vs-the-rest logistic regression classifier implemented in LibLinear is used in our experiments. The Macro-F1 is chosen as the evaluation metric. We follow the suggested parameter settings in the original

Table 3. Macro-F1 (%) of node classification the DBLP-author network.

% labeled nodes	10%	20%	30%	40%	50%	60%	70%	80%	90%
DeepWalk	63.54	64.57	64.81	65.06	65.18	65.26	65.14	65.49	65.28
LINE	56.16	56.18	57.37	57.75	57.82	57.63	58.09	58.14	58.25
SANRL$_{doc}$	74.17	75.29	75.64	75.69	75.78	75.77	76.03	76.33	75.79
SANRL	**77.40**	**78.66**	**79.00**	**79.12**	**79.30**	**79.32**	**79.37**	**79.39**	**79.52**
LDA	70.35	71.39	71.70	71.73	71.90	71.92	71.95	72.00	72.14
TADW	75.27	75.70	76.00	76.16	76.15	76.31	76.50	76.87	76.49

papers for the baseline models. Note that, since there is no text attributes in the Weibo dataset, LDA and TADW are inapplicable to Weibo. Tables 1, 2, and 3 show the results of classification with different training ratios on three networks when $d = 128$ for all the models. All reported results are averaged over 10 runs.

Firstly, we observe that SANRL always significantly outperform other baselines. Compared with DeepWalk, SANRL achieves nearly 6% and 10% improvement on two Weibo networks and 13% improvement on DBLP-author network, which proves that the network represent can be enhanced with either categorical attributes or text attributes.

Secondly, on the DBLP-author network, the two content-based methods, SANRL$_{doc}$ and LDA, perform better than the structure-based methods, DeepWalk and LINE. By benefiting from incorporating attributes information and structure information, SANRL and TADW both outperform other four methods. But the relative improvement of SANRL over TADW is around 3.5% since attributes information and structure information is better balanced with a tunable trade-off parameter λ in SANRL.

5.4 Parameter Sensitivity

We also explore the sensitivity of the performance w.r.t. the dimension d and the trade-off parameter λ. Here, we take the DBLP-author network as an example. By setting the training ratio to 20%, we report the grid search results over d and λ on the DBLP-author network in Fig. 3. We observe that SANRL achieves best performance when $\lambda = 8$ and $d = 256$.

To further investigate the effect of increasing dimension d, we use Fig. 4 to show the Micro-F1 and Macro-F1 curves of different models. By varying d from 64 to 1024, we can see that SANRL consistently performs best.

In SANRL, we add a constraint to DeepWalk to let nodes with similar attributes have similar representations. Our motivation is based on the statistics shown in Fig. 5. The Jaccard coefficient is calculated between the attribute sets of two social nodes sampled from the network at random. Then we compute the conditional probability that two nodes share a same label given the corresponding discretized Jaccard coefficient. Our assumption is validated on all

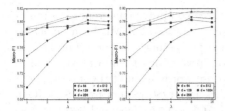

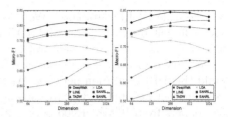

Fig. 3. Grid search over dimension d and λ.

Fig. 4. Performance over dimension d.

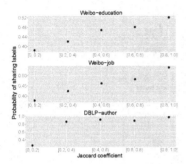

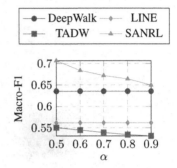

Fig. 5. Illustration of interdependencies between labels of users and attributes of users.

Fig. 6. Parameter sensitivity w.r.t α

three networks. Note that the DBLP-author network has much more discriminative attributes than those two Weibo networks. That's why a smaller λ is more suitable for the Weibo dataset.

Next, we will test the robustness of SANRL in an incomplete data scenario. We random choose some users and remove their text contents. We use α to represent the percentage of users whose texts are removed. LDA and SANRL$_{doc}$ could not work in this situation. By ranging α from 0.5 to 0.9, we report the Macro-F1 of different models when the training ratio is 0.1 in Fig. 6. Obviously, SANRL has greater robustness than TADW whose performance is very poor. The reason lies with the fact that the text content matrix calculated from the tf-idf matrix which has many rows with all zeros is uninformative. In contrast, SANRL benefits from limited text information by using a coupled design.

5.5 Network Visualization

Network visualization is an essential task in network analysis area. In this part, we use the t-SNE package which takes the node representation vectors as inputs to generate network layouts in a 2D space. Our target is to compare the properties of the layouts of DeepWalk and SANRL qualitatively. Because our previous three networks are not mono-labeled, we build a paper-citation network by using a large-scale DBLP database[1]. We select three computer conferences to represent different research fields: SIGMOD from "database", ICML from "machine learning", and ACL from "natural language processing". Papers published on these three conferences and citation relationships among them are extracted to build this DBLP-citation network. We also use words in the title of each paper as their text attributes. Finally, this network has 1280 papers from SIGMOD, 568 papers from ICML, 1,206 papers from ACL, 6,647 edges and 3,814 unique attributes.

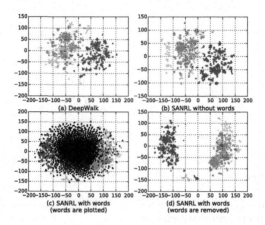

Fig. 7. t-SNE 2D representations on the DBLP-citation network. We use blue, green and red to indicate papers from SIGMOD, ICML and ACL respectively. Words are colored black (Color figure online)

Under the same parameter configuration, the 2-D layouts of the DBLP-citation network are shown in Fig. 7. Three different colors, blue, green and red are used to indicate papers from SIGMOD, ICML and ACL respectively. Attributes are colored black. In Fig. 7, (a) and (b) are generated by only feeding the node representations to the t-SNE toolkit. And (c) and (d) are generated by feeding the node representations and attributes representations together to the t-SNE toolkit. We have three observations: First, We observe that node distributions in (b) and (a) are very similar, but groups in (b) seem to be more tighter. Second, after plotting the papers and the words in a same space simultaneously in (c), the words are spread throughout the space, but most of them are densely populated in the center while different paper groups are surrounding the word groups. Finally, it is obvious that each paper groups are becoming more denser and separable in (d) which is obtained by removing words from (c). These observations show that the distinguishability of node representations is improved significantly by jointing learning the representations of their associated attributes in SANRL model.

---

[1] https://aminer.org/DBLP_Citation.

Table 4. Top-10 related words for selected papers.

Paper title	Recommended words
Dynamic multidimensional histograms	Histogram, filesystem, stholes, braid, histograms, lag, partiqle, frequencies, filtered, multidimensional
Computing weakest readings	Weakest, readings, ambiguities, scope, semantically, formalisms, polysemic, hole, dominance, distributions

5.6 Case Study

To explore the correlations among node vectors and attribute vectors learned in SANRL, we provide a case study on the DBLP-citation dataset. We recommend top-10 most related words to papers by calculating the cosine similarly between the paper vector and the word vector. As shown in Table 4, we can see that though the titles of two selected papers both have only three words, SANRL can find much richer related words by considering the citation relationship between the papers and the interactive relationship between the papers and the words.

6 Conclusion and Future Work

In this paper, we propose SANRL, an efficient model which integrates both structure and attribute information into the NRL task. By embedding nodes and attributes into the same vector space, the quality of the node representations are improved significantly. The experimental results show that SANRL outperforms competitive baselines for different data mining tasks. We strive to adapt SANRL to learn more discriminative representations by using the semi-supervised learning technique in the future.

Acknowledgments. This work is supported by 973 Program with Grant No. 2014CB340400. Yan Zhang is supported by NSFC with Grant Nos. 61532001 and 61370054, and MOE-RCOE with Grant No. 2016ZD201. We thank the anonymous reviewers for their comments.

References

1. Belkin, M., Niyogi, P.: Laplacian eigenmaps and spectral techniques for embedding and clustering. In: NIPS, vol. 14, pp. 585–591 (2001)
2. Blei, D.M., Ng, A.Y., Jordan, M.I.: Latent Dirichlet allocation. J. Mach. Learn. Res. **3**, 993–1022 (2003)
3. Cao, S., Lu, W., Xu, Q.: GraRep: learning graph representations with global structural information. In: CIKM, pp. 891–900. ACM (2015)
4. Culotta, A., Ravi, N.K., Cutler, J.: Predicting the demographics of Twitter users from website traffic data. In: ICWSM. AAAI Press, Menlo Park (2015, in press)

5. Djuric, N., Wu, H., Radosavljevic, V., Grbovic, M., Bhamidipati, N.: Hierarchical neural language models for joint representation of streaming documents and their content. In: WWW, pp. 248–255 (2015)

6. Dong, Y., Yang, Y., Tang, J., Yang, Y., Chawla, N.V.: Inferring user demographics and social strategies in mobile social networks. In: SIGKDD, pp. 15–24. ACM (2014)

7. Gong, N.Z., Talwalkar, A., Mackey, L., Huang, L., Shin, E.C.R., Stefanov, E., Shi, E.R., Song, D.: Joint link prediction and attribute inference using a social-attribute network. ACM TIST 5(2), 27 (2014)

8. Grover, A., Leskovec, J.: node2vec: scalable feature learning for networks. In: SIGKDD, pp. 855–864. ACM, New York (2016)

9. Ji, M., Sun, Y., Danilevsky, M., Han, J., Gao, J.: Graph regularized transductive classification on heterogeneous information networks. In: Balcázar, J.L., Bonchi, F., Gionis, A., Sebag, M. (eds.) ECML PKDD 2010. LNCS (LNAI), vol. 6321, pp. 570–586. Springer, Heidelberg (2010). doi:10.1007/978-3-642-15880-3_42

10. Le, Q.V., Mikolov, T.: Distributed representations of sentences and documents (2014). arXiv preprint arXiv:1405.4053

11. Li, J., Zhu, J., Zhang, B.: Discriminative deep random walk for network classification. In: ACL, pp. 1004–1013 (2016)

12. Mikolov, T., Sutskever, I., Chen, K., Corrado, G.S., Dean, J.: Distributed representations of words and phrases and their compositionality. In: NIPS, pp. 3111–3119 (2013)

13. Mnih, A., Hinton, G.E.: A scalable hierarchical distributed language model. In: Advances in Neural Information Processing Systems, pp. 1081–1088 (2009)

14. Perozzi, B., Al-Rfou, R., Skiena, S.: DeepWalk: online learning of social representations. In: SIGKDD, pp. 701–710. ACM (2014)

15. Shi, C., Zhang, Z., Luo, P., Yu, P.S., Yue, Y., Wu, B.: Semantic path based personalized recommendation on weighted heterogeneous information networks. In: CIKM, pp. 453–462. ACM (2015)

16. Sun, Y., Han, J.: Mining heterogeneous information networks: a structural analysis approach. ACM SIGKDD Explor. Newslett. 14(2), 20–28 (2013)

17. Tang, J., Qu, M., Wang, M., Zhang, M., Yan, J., Mei, Q.: Line: large-scale information network embedding. In: WWW. ACM (2015)

18. Tu, C., Zhang, W., Liu, Z., Sun, M.: Max-margin DeepWalk: discriminative learning of network representation. In: IJCAI, pp. 3889–3895 (2016)

19. Wang, D., Cui, P., Zhu, W.: Structural deep network embedding. In: SIGKDD, pp. 1225–1234, ACM, New York (2016)

20. Yang, C., Liu, Z., Zhao, D., Sun, M., Chang, E.Y.: Network representation learning with rich text information. In: IJCAI, pp. 2111–2117. AAAI Press (2015)

21. Yang, Z., Cohen, W., Salakhutdinov, R.: Revisiting semi-supervised learning with graph embeddings (2016). arXiv preprint arXiv:1603.08861

Cost-Effective Viral Marketing in the Latency Aware Independent Cascade Model

Robert Gwadera[1] and Grigorios Loukides[2(✉)]

[1] Cardiff University, Cardiff, UK
GwaderaR@cardiff.ac.uk
[2] King's College London, London, UK
grigorios.loukides@kcl.ac.uk

Abstract. A time-constrained viral marketing campaign allows a business to promote a product or event to social network users within a certain time duration. To perform a time-constrained campaign, existing works select the duration of the campaign, and then a set of k seeds that maximize the *spread* (expected number of users to which the product or event is promoted) for the selected duration. In practice, however, there are many alternative durations, which determine the monetary cost of the campaign and lead to seeds with substantially different spread. In this work, we aim to select the duration of the campaign and a set of k seeds, so that the campaign has the maximum spread-to-cost ratio (i.e., cost-effectiveness). We formulate this task as an optimization problem, under the LAIC information diffusion model. The problem is challenging to solve efficiently, particularly when there are many alternative durations. Thus, we develop an approximation algorithm that employs dynamic programming to compute the spread of seeds for several possible durations simultaneously. We also introduce a new optimization technique that is able to provide an additional performance speed-up by pruning durations that cannot lead to a solution. Experiments on real and synthetic data show the effectiveness and efficiency of our algorithm.

1 Introduction

Many businesses perform time-constrained viral marketing campaigns over social networks, such as Facebook [2,5,6,11,12]. In these campaigns, a product or event is promoted to a small set of users, who diffuse information about it, with the aim to *activate* their friends (make them aware of the product or event). The active friends of these users diffuse information, attempting to activate their own friends, and the process proceeds similarly, until the end of the campaign duration (e.g., the end of the sales period of the product, or the time the event is held). Typically, the social network is modeled as a graph whose nodes and edges correspond to users and their connections, respectively, and the initial users correspond to a subset of nodes called *seeds*.

© Springer International Publishing AG 2017
J. Kim et al. (Eds.): PAKDD 2017, Part I, LNAI 10234, pp. 251–265, 2017.
DOI: 10.1007/978-3-319-57454-7_20

Motivation. To perform a time-constrained campaign, it is necessary to determine its duration. The duration of the campaign can be modeled as a time window (interval). In practice, the start time of the window is selected by the business based on market properties, such as season, competitors' actions, and availability of products or resources to hold an event [7]. However, there are multiple choices for the end time of the window, which are determined by characteristics of the product and social network. For example, the end time of a campaign that promotes a film corresponds to a time between a few days and five weeks before the release of the film [14]. Therefore, we will assume a zero start time and model (refer to) each alternative duration as a window defined in terms of its size.

The cost of the campaign, in terms of monetary expense to a business, is a non-decreasing function of the campaign duration. For instance, a social network provider which implements a campaign on a product, as a service to the business [13], charges a fee that increases with the campaign duration. The reason is that multiple businesses compete for performing campaigns simultaneously on the same social network, and executing a campaign with large duration for a product (e.g., a comedy film) reduces the *spread* (expected number of users to which the product is promoted) [11] of other campaigns on substitute products (e.g., different comedy films). Furthermore, the spread of the campaign is also a non-decreasing function of the campaign duration [11].

Thus, a fundamental question for performing a cost-effective campaign is: "Which window (duration) offers the maximum benefit-to-cost ratio?" [16]. The need to perform cost-effective campaigns has been recognized in the marketing literature [3,7,14]. However, the problem has not been studied before. That is, existing methods [2,6,11,12] assume a fixed window that is selected by the business and aim to select a subset of k nodes, as seeds, to maximize the spread for the selected window.

Contributions. Our work makes the following contributions:

First, we formulate the *Time-constrained Spread-to-cost Maximization* (TSM) problem, as follows. Given a graph G and a set of candidate windows, each having an associated cost, select: (I) a window, and (II) a subset of k nodes of G as seeds, such that the ratio between the spread of the seeds in the window and the window cost is maximum. In the TSM problem, the spread is computed under the Latency Aware Independent Cascade (LAIC) [11] model. The model takes into account the varying delays (latencies) with which nodes may be activated in practice and generalizes other models [2,8]. Solving TSM allows implementing a cost-effective campaign. However, this is challenging because TSM is NP-hard and cannot be approximated by directly applying the greedy submodular maximization algorithm [15]. This is because the optimization function that computes the maximum spread-to-cost ratio of a seed-set over all windows is not submodular, as we show, whereas the algorithm of [15] requires its optimization function to be submodular. To illustrate TSM, we provide Example 1.

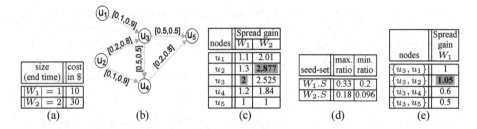

Fig. 1. (a) Size and cost of windows W_1 and W_2. (b) Graph and probability vectors of edges (the probabilities reflect how likely a node is activated by its in-neighbor with delay 0 and 1, respectively). (c) Spread gain of the empty seed-set for W_1 and W_2. The gain is caused by adding a node into the seed-set. (d) Bounds for the spread-to-cost ratio of W_1 and W_2, after k iterations. (e) Spread gain of the seed-set $\{u_3\}$ for W_1. The gain is caused by adding a node into the seed-set.

Example 1. A business plans a campaign for a new product, which starts on the day of product launch and can last one or two weeks. This is modeled with the windows W_1 and W_2, whose sizes are shown in Fig. 1a. A social network provider implements the campaign on the graph of Fig. 1b, as a service to the business. Under the LAIC model, each edge (u', u) in Fig. 1b is associated with a vector of probabilities that u is activated by u' with delay 0 and 1, respectively. The social network provider also determines the window costs as shown in Fig. 1a. The business wants to perform the campaign with the largest spread-to-cost ratio and can give away a product to two users, as an incentive to start diffusing information. Thus, the social network provider needs to solve TSM with $k = 2$.

Second, we propose a dynamic programming equation to compute the probability that a node u has been activated in $[0, W.t]$, where $W.t$ is the end time of a window W. The probability is denoted with $P_{[0,W.t]}(u)$ and computed as:

$$P_{[0,W.t]}(u) = 1 - [1 - P_{W.t}(u)] \cdot [1 - P_{[0,W.t-1]}(u)], \qquad (1)$$

where $P_{W.t}(u)$ is the probability that u becomes active at $W.t$ and $P_{[0,W.t-1]}(u)$ is the probability that u has been activated before (at any previous time point). In addition, we sum $P_{[0,W.t]}(u)$ over each node u, to compute the spread of a seed-set in W. The spread is computed by a subroutine of our algorithm for TSM. The subroutine is called *DPSC* and computes the exact value of spread, unlike existing algorithms [11,12].

Third, we propose *MASP*, an approximation algorithm for the TSM problem. The algorithm starts by associating an empty seed-set with each window. Then, it performs k iterations, where k is the input number of seeds. In each iteration j, *MASP*: (I) Computes the *spread gain* of each window's seed-set, for each *available* node (i.e., node that is not contained in the seed-set). The spread gain is the difference in spread, before and after the addition of a node into the seed-set of the window. (II) Adds into each seed-set the node that maximizes the spread gain of the seed-set. (III) Prunes windows that cannot lead to a solution.

After that, the algorithm returns the window with the largest spread-to-cost ratio, among all windows, and its associated seed-set. To improve efficiency, *MASP* creates clusters, each containing all windows with the same seed-set, and applies the *Multiple-window spread gain computation* technique to each cluster. In addition, it uses the *Pruning* technique. These techniques are summarized as follows:

Multiple-Window Spread Gain Computation. It efficiently computes the spread gain of the seed-set in each window of the cluster, for each available node. To compute the spread gain for an available node, *DPSC* is applied to the cluster and computes the spread of the seed-set in the largest window, W, of the cluster, after adding the node. Since all windows in the cluster have the same seed-set, the spread in every subwindow W' of W, with end time $W'.t$, is also obtained, by summing the probability $P_{[0,W'.t]}(u)$ of each node u, which is computed during the recursion of Eq. 1. Then, the spread gain is calculated for each window as the difference between the spread obtained by *DPSC* and the spread before adding the available node.

Example 2. *MASP* is applied in Example 1 with $k = 2$. Initially, the windows W_1 and W_2 are associated with the empty seed-set, and a single cluster $\{W_1, W_2\}$ is created. Then, the spread gain of the empty seed-set for each available node, u_1 to u_5, is computed. For instance, the spread gain for u_1 is computed as follows. First, *DPSC* is applied to the cluster $\{W_1, W_2\}$ and computes the spread of the seed-set $\{u_1\}$ in the largest window, W_2, of the cluster as $P_{[0,2]}(u_1) + \ldots + P_{[0,2]}(u_5)$. Each of these probabilities is computed recursively using Eq. 1. Thus, the spread in W_1 is also obtained as $P_{[0,1]}(u_1) + \ldots + P_{[0,1]}(u_5)$. Next, the spread gain in W_1 and in W_2 is calculated as the difference between the spread computed by *DPSC* and the spread before adding u_1.

Pruning. In iteration j, it computes, for each window, the maximum and minimum spread-to-cost ratio that the window can have after k iterations. The maximum ratio is computed for spread equal to the sum of the spread of the seed-set in the window and the spread gain for each of the top $k - j$ (i.e., remaining) available nodes, in terms of spread gain. The minimum ratio is computed for spread equal to the sum of the spread of the seed-set in the window. This corresponds to the worst case, in which the spread gain for each node is zero. Then, each window whose maximum ratio is smaller than the largest minimum ratio of all windows in the current iteration is removed.

Example 3 (continuing from Example 2). *MASP* adds u_3 into the seed-set of W_1, since u_3 maximizes the spread gain in W_1 (see Fig. 1c). The spread of the seed-set $\{u_3\}$ is 2. Thus, the maximum ratio for W_1 is $\frac{2+1.3}{10} = 0.33$. This is because the spread gain caused by u_2, the top available node in terms of spread gain, is 1.3 (see Fig. 1c), and the cost of W_1 is 10. The minimum ratio for W_1 is $\frac{2}{10} = 0.2$. The maximum and minimum ratio for W_2 is computed similarly and is equal to 0.18 and 0.096, respectively. Since the maximum ratio for W_2 is smaller than the largest minimum ratio, W_2 is pruned.

MASP produces a solution whose spread-to-cost ratio is at least $1 - \frac{1}{e} \approx 63\%$ of that of the optimal solution. This is because it applies the greedy submodular maximization algorithm [15] to the seed-set of each window, using a submodular optimization function that computes the spread in the window. As we show experimentally, our algorithm is both effective and efficient, unlike baselines that are constructed based on the existing methods for maximizing the spread of a given window in the LAIC model [11,12]. For example, it was at least one order of magnitude faster than a baseline which applies an existing approximation algorithm [11] for finding the subset of k nodes with the maximum spread to each window, and then selects the solution with the largest ratio.

2 Related Work

In [11,12], the problem of selecting k seeds that maximize the spread in a fixed window was studied under the LAIC model, and the following methods were proposed: *MC*, *ISP*, and *MISP*. These methods select a subset of k nodes as seeds, by iteratively selecting the available node that causes the maximum gain to a spread estimate. *MC* estimates the spread by performing many Monte Carlo simulations of the diffusion process. *ISP* estimates the spread assuming that a node can be activated only by a path which does not share edges with other paths and has probability at least θ to activate the node. *MISP* is a variation of *ISP* that approximates the spread gain, caused by a node, based on the spread of the node and the probability that the out-neighbors of the node are already activated. These methods are not alternatives to the *MASP* algorithm we propose, because they assume a fixed window. On the contrary, there are multiple possible windows in our TSM problem, and the challenge is to compute the spread of seeds over all windows efficiently.

In [4,10,13], the problem of seed selection when there are costs associated with nodes was studied. Specifically, in [10], each node has a given cost, while in [4,13] all nodes have the same cost. Unlike these works, we consider a time-constrained campaign where each window has an associated cost.

3 Background

Preliminaries. Let $G(V,E)$ be a directed graph, where V is a set of nodes and E is a set of edges. The set of in-neighbors of a node u is denoted with $n^-(u)$ and has size $|n^-(u)|$, which is referred to as the *in-degree* of u. The set of out-neighbors of u is denoted with $n^+(u)$ and has size $|n^+(u)|$, which is referred to as the *out-degree* of u.

A path $q = [u_1, u_2, \ldots, u_m]$ is an ordered set of nodes, which has *length* $|q| = m - 1$. A path q in which each node is unique (i.e., a path with no cycle) is a *simple* path. A path that starts and ends at the same node is a *cycle* path. We assume simple paths, unless stated otherwise.

Each window W has the following attributes: (I) seed-set $W.S$, (II) end time $W.t$, (III) spread $W.\sigma$, and (IV) spread gain $W.g()$. The spread $W.\sigma$ is defined as

$\sigma(W.S, W.t)$, where $\sigma()$ computes the expected number of nodes that are active at time $W.t$, when the seed-set is $W.S$, under the LAIC model [11]. The spread gain $W.g()$ is defined, for a given node u, as $W.g(u) = \sigma(W.S \cup \{u\}, W.t) - \sigma(W.S, W.t)$.

Let U be a universe of elements and 2^U its power set. A set function $f : 2^U \to \mathbb{R}$ is non-decreasing, if $f(X) \le f(Y)$ for all subsets $X \subseteq Y \subseteq U$, monotone, if $f(X) \le f(X \cup u)$ for each $u \notin X$, and submodular, if and only if it satisfies the diminishing returns property $f(X \cup \{u\}) - f(X) \ge f(Y \cup \{u\}) - f(Y)$, for all $X \subseteq Y \subseteq U$ and any $u \in U \setminus Y$ [9].

LAIC Model. In the LAIC model [11], each node is active or inactive. A subset $S \subseteq V$ of nodes, referred to as *seeds*, are active at the initial time 0, and all other nodes are inactive. Each edge has a probability vector $m((u', u)) = [m_0((u', u)), \dots, m_\delta((u', u))]$, where $m_i((u', u))$ is the probability that the *inactive* node u is activated by its active in-neighbor u' with delay $i \in [0, \delta]$. The probability vectors of edges are selected based on the population targeted by the campaign [11]. For example, in [11], each $m_i((u', u))$ was set to $\mathcal{P}_{u'}(i) \cdot p((u', u))$, where $\mathcal{P}_{u'}$ is a Poisson distribution with a random parameter (mean rate) λ in $[1, 20]$ that is associated with the node u' and $p((u', u)) = \frac{1}{\lceil n^-(u) \rceil}$.

The diffusion process in the LAIC model proceeds as follows. Each seed s tries to activate its out-neighbors at the initial time 0 only and, if multiple seeds have the same out-neighbor, they all try to activate it in arbitrary order. The out-neighbor u of a seed s becomes active at time $1 + i$ with probability $m_i((s, u))$, where the delay i takes each value in $[0, \delta]$. Each out-neighbor that becomes active remains active, and it tries to activate its own inactive out-neighbors. The process proceeds similarly and ends when no new node becomes active.

Let S be a seed-set and $[X_{0,u}, \dots, X_{t,u}]$ be a sequence of binary variables, such that $X_{j,u} = 1$, if the node u of the graph G becomes active at time j, and $X_{j,u} = 0$ otherwise. For brevity, we denote $P(X_{j,u} = 1)$ with $P_j(u)$ and $\sum_{j \in [0,t]} P_j(u)$ with $P_{[0,t]}(u)$. The expected number of active nodes of G at time t is given by $\sigma(S, t) = \sum_{u \in G} P_{[0,t]}(u)$ [11,12]. This equation is used in the *DPSC* subroutine.

4   Computing the Probability $P_{[0,t]}(u)$

We examine the computation of $P_{[0,t]}(u)$, the probability that a node u has been activated in $[0, t]$. $P_{[0,t]}(u)$ cannot be computed directly using Eq. 1 because $(1 - P_t(u))$, the probability that u does not become active at t, is not given. Thus, we show how to compute $(1 - P_t(u))$ by taking into account each in-neighbor of u which may activate u at t with any possible delay.

Clearly, if the node u is a seed, then $P_{[0,t]}(u) = 1$. Otherwise, $P_{[0,t]}(u)$ is given by Eq. 2:

$$P_{[0,t]}(u) = 1 - \left(\prod_{u' \in n^-(u)} \prod_{i \in [0, min(t-1, \delta)]} [1 - P_{t-1-i}(u') \cdot m_i((u', u))] \right) \cdot (1 - P_{[0,t-1]}(u))$$

$$(2)$$

Equation 2 computes $P_{[0,t]}(u)$ as the probability of the complement of the event "u does not become active at t nor before t". The probability that u does not become active at t is given in the large parentheses, and it takes into account each in-neighbor u' of u and each possible delay i. The probability that u is not active (i.e., has not been activated) before t is given by $1 - P_{[0,t-1]}(u)$. The correctness of Eq. 2 follows from Theorem 1.

Theorem 1. *Let u be a node that is not a seed. Equation 2 computes the probability u is active at time t, under the LAIC model.*

Proof (Sketch). Let $A_0, \ldots, A_{\delta'}, B$ be sets of in-neighbors of u, such that any node in A_i can activate u at t with delay $i \in [0, \delta']$, and any node in B can activate u before t. The maximum delay δ' is equal to $min(t - 1, \delta)$, since u will not become active at t if the delay is larger. Let also E_{A_i} (respectively, E_B) be the event "u became active by at least one node in A_i" (respectively, in B). Clearly, $P_{[0,t]}(u) = P(\cup_{i \in [0,\delta']} E_{A_i} \cup E_B) = 1 - P(\overline{\cup_{i \in [0,\delta']} E_{A_i} \cup E_B})$ and, by DeMorgan's laws and the multiplication rule, $P_{[0,t]}(u) = 1 - P(\overline{E_{A_{\delta'}}} \mid \cap_{i \in [0,\delta'-1]} \overline{E_{A_i}} \cap \overline{E_B}) \cdot P(\cap_{i \in [0,\delta'-1]} \overline{E_{A_i}} \cap \overline{E_B}) = 1 - P(\overline{E_{A_{\delta'}}} \mid \cap_{i \in [0,\delta'-1]} \overline{E_{A_i}} \cap \overline{E_B}) \cdot P(\overline{E_{A_{\delta'-1}}} \mid \cap_{i \in [0,\delta'-2]} \overline{E_{A_i}} \cap \overline{E_B}) \cdot \ldots \cdot P(\overline{E_{A_0}} \mid \overline{E_B}) \cdot P(\overline{E_B})$. The proof follows from: (I) $P(\overline{E_{A_i}} \mid \cap_{j \in [0,i-1]} \overline{E_{A_j}} \cap \overline{E_B}) = \prod_{u' \in n^-(u)} (1 - P_{t-1-i}(u') \cdot m_i((u', u)))$, which holds for each delay $i \in [0, \delta']$. This is because $\overline{E_{A_i}}$ occurs when each in-neighbor u' of u is contained in A_i and fails to activate u; (II) $P(\overline{E_B}) = 1 - P_{[0,t-1]}(u)$, which holds by definition. $\square$

5   The Time-Constrained Spread-to-Cost Maximization Problem

The Time-constrained Spread-to-cost Maximization problem is defined as follows.

Problem (Time-constrained Spread-to-cost Maximization (TSM)). *Given the graph $G(V, E)$, the probability vector $m(e)$ of each edge e in E, a set of windows $\mathcal{W} = \{W_1, \ldots, W_n\}$, where each W_i has a nonnegative cost $W_i.c$, and a parameter k, find a window W in $\mathcal{W}$ and a subset $S \subseteq V$ of k nodes, such that the ratio between the spread of S in W and the cost $W.c$ is maximum, over all possible windows of $\mathcal{W}$ and their corresponding subsets of k nodes.*

The set of windows $\mathcal{W}$ is determined by the business, based on characteristics of the product and social network [14], while the window costs are determined by the party performing the campaign. TSM is NP-hard, because it generalizes the NP-hard problem in [11], which requires finding a subset of k nodes with maximum spread in a fixed window. The existence of multiple windows makes our problem challenging. For example, we cannot approximate TSM using the greedy algorithm for submodular maximization [15] *with the function $f(S) = max(f_1(S), \ldots, f_n(S))$* (i.e., iteratively add into the seed-set S the node causing the largest gain $f(S \cup \{u\}) - f(S))$, where f_i outputs the spread-to-cost ratio of S in the window W_i. This is because the algorithm of [15] offers approximation guarantees only for submodular functions, whereas f is not submodular (for arbitrary window costs), as shown in Example 4.

Example 4. Consider the graph in Fig. 2a and the windows W_1 and W_2, whose sizes are 1 and 2 and costs are $W_1.c = 1$ and $W_2.c = 1.19$. The spread and the spread-to-cost ratios of different node subsets are shown in Figs. 2b and c, respectively. In Fig. 2c, f_1 (resp., f_2) computes the ratio in W_1 (resp., W_2), and the function $f = max(f_1, f_2)$ computes the maximum ratio. The function f is *not* submodular because, for $\{u_3\} \subseteq \{u_2, u_3\}$ and $u_1 \in \{u_1, \ldots, u_6\} \setminus \{u_2, u_3\}$, it holds $f(\{u_3\} \cup \{u_1\}) - f(\{u_3\}) = 3.29 - 2 = 1.29 < f(\{u_2, u_3\} \cup \{u_1\}) - f(\{u_2, u_3\}) = 4.89 - 3.29 = 1.6$.

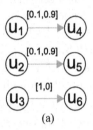

(a)

node subset	spread	
	W_1	W_2
u_1	1.1	1.91
u_2	1.1	1.91
u_3	2	2
$\{u_1, u_2\}$	2.2	3.82
$\{u_1, u_3\}$	3.1	3.91
$\{u_2, u_3\}$	3.1	3.91
$\{u_1, u_2, u_3\}$	4.2	5.82

(b)

node subset	spread-to-cost ratio		
	f_1	f_2	f
u_1	1.1	1.6	1.6
u_2	1.1	1.6	1.6
u_3	2	1.68	2
$\{u_1, u_2\}$	2.2	3.2	3.2
$\{u_1, u_3\}$	3.1	3.29	3.29
$\{u_2, u_3\}$	3.1	3.29	3.29
$\{u_1, u_2, u_3\}$	4.2	4.89	4.89

(c)

Fig. 2. (a) Graph and probability vectors of edges. (b) The spread of different subsets of nodes of a. (c) The spread-to-cost ratios of different subsets of nodes of a. The ratio in W_1 and W_2 is given by f_1 and f_2, respectively, and the maximum ratio is given by the function f.

6   The *MASP* Algorithm

In this section, we present our *MASP* algorithm and its *DPSC* and *Pruning* subroutines.

MASP initializes, for each window W_i in the given set of windows, its seed-set $W_i.S$ and spread $W_i.\sigma$ (steps 1 to 2). It also initializes a set of clusters $\mathcal{C}$ with a single cluster that contains all windows (steps 3 to 4). Then, it performs k iterations (steps 6 to 18). In each iteration, *MASP*:

I. Applies *Multiple-window spread gain computation* to each cluster, to efficiently compute the spread gain $W_i.g(v)$, for each window W_i in the cluster and each available node v (steps 7 to 11). Specifically, the largest window, W_C, in the cluster is found and each node v that is not contained in the seed-set of W_C is considered. This is without loss of generality, since all windows in the cluster contain the same seed-set. Then, *DPSC* is applied to the cluster and efficiently computes the spread of $W_i.S \cup v$ for every window W_i in the cluster (including W_C). After that, the spread gain $W_i.g(v)$ is computed as the difference between the spread that is obtained from *DPSC*, and the spread $W_i.\sigma$, which was computed before.

Algorithm: *MASP* (Multiple-window Addition Spread computation Pruning)
Input: Graph $G(V, E)$, probability vector of each edge of G, set of windows $\mathcal{W}$ and their
      costs, and parameter k
Output: A window $W \in \mathcal{W}$ and a subset $S \subseteq V$ of k nodes

```
1  foreach window Wᵢ of W do
2      Wᵢ.S ← ∅; Wᵢ.σ ← 0
3  Create cluster C comprised of all windows in W
4  Add C into the empty set of clusters C
5  j ← 1    // iteration counter
6  while j ≤ k do
7      foreach cluster C in C do
8          W_C ← the largest window in C
9          foreach node v in V \ W_C.S do
10             Apply DPSC to the cluster C and node v
11             Compute the spread gain Wᵢ.g(v), for each window Wᵢ in C
12     foreach window Wᵢ in W do
13         u ← the node u in V \ Wᵢ.S with the largest spread gain Wᵢ.g(u)
14         Wᵢ.S ← Wᵢ.S ∪ {u}
15         Wᵢ.σ ← Wᵢ.σ + Wᵢ.g(u)
16     Pruning(W)
17     C ← set of clusters, each containing all windows of W with the same seed-set
18     j ← j + 1
19 W ← the window Wᵢ in W with the maximum (Wᵢ.σ)/(Wᵢ.c)
20 S ← the seed-set of the window W
21 return {W, S}
```

II. Adds the available node with the largest spread gain into the seed-set $W_i.S$
    and updates the spread $W_i.\sigma$, for each window W_i (steps 12 to 15).

III. Applies *Pruning* to prune windows that cannot lead to a solution (step 16).

IV. Creates a new set of clusters, each containing all windows that are associated
    with the same seed-set (step 17).

Last, the algorithm finds and returns the window with the largest spread-to-cost
ratio, among all windows in $\mathcal{W}$, and its corresponding seed-set (steps 19 to 21).

Theorem 2 explains the approximation guarantee of *MASP*.

Theorem 2. MASP *finds a solution with spread-to-cost ratio at least* $1 - \frac{1}{e}$ *of
that of the optimal solution to the TSM problem, where* e *is the base of the natural
logarithm.*

Proof (Sketch). Let σ_i be the maximum spread of a subset of k nodes in a
window W_i. *MASP* constructs each seed-set $W_i.S$ using the greedy algorithm
for submodular maximization [15] *with the submodular spread function* [11] (i.e.,
iteratively adds into $W_i.S$ the node u causing the largest spread gain $W_i.g(u)$).
This guarantees that, for each W_i, $W_i.\sigma \geq (1 - \frac{1}{e}) \cdot \sigma_i$ [11]. Thus, for the window
with the maximum ratio $max_{i \in [1,n]} W_i.\sigma$, we have $max_{i \in [1,n]} W_i.\sigma \geq (1 - \frac{1}{e}) \cdot$
$max_{i \in [1,n]} \sigma_i$, which implies $max_{i \in [1,n]} \frac{W_i.\sigma}{W_i.c} \geq (1 - \frac{1}{e}) \cdot max_{i \in [1,n]} \frac{\sigma_i}{W_i.c}$. The proof
follows from observing that the spread-to-cost ratio of the solution of *MASP* is
$max_{i \in [1,n]} \frac{W_i.\sigma}{W_i.c}$ and that of the optimal solution to TSM is $max_{i \in [1,n]} \frac{\sigma_i}{W_i.c}$. □

MASP needs $O(k \cdot |\mathcal{W}| \cdot |V|^3 \cdot |W_n|)$ time, where $|W_n|$ is the size of the largest
window in $\mathcal{W}$, in the worst case in which the graph is complete, all sets contain
different seeds in each iteration, and no window is pruned.

```
Algorithm: DPSC (Dynamic Programming Spread Computation)
Input: Graph G(V, E), probability vector of each edge of G, cluster of windows C, node v
Output: Spread of the seed-set Wᵢ.S ∪ v, for each window Wᵢ in the cluster C
1  W_C ← the largest window in C
2  T ← 2D array with |V| rows and |W_C| columns, with each element equal to zero
3  S̃ ← W_C.S ∪ v      // temporary seed-set
4  foreach node s in the seed-set S̃ do
5       T[s][0] ← 1
6  t ← 1
7  R ← reachable(t)
8  while R ≠ ∅ and time t in W_C do
9       foreach node u in R do
10          T[u][t] ← the probability P_{[0,t]}(u)
11      foreach node u ∉ S̃ and u ∉ R and u may have been activated before t do
12          T[u][t] ← T[u][t − 1]
13      t ← t + 1
14      R ← reachable(t)
15 foreach window Wᵢ in C do
16      Wᵢ.σ̃ ← ∑_{u∈V} T[u][|Wᵢ|]      // spread of the seed-set Wᵢ.S ∪ v
17 return {W₁.σ̃, ..., W_C.σ̃}
```

DPSC. Given a cluster C and a node v, *DPSC* constructs a temporary seed-set by adding v into the seed-set of the largest window in C (steps 1 to 3), fills a dynamic programming array T, whose element $T[u][t]$ stores the probability $P_{[0,t]}(u)$ for a node u at time t (steps 4 to 14), and computes $W_i.\tilde{\sigma}$, the spread of the seed-set $W_i.S \cup v$, for each window W_i in C (steps 15 to 16). To improve efficiency, $P_{[0,t]}(u)$ is computed only for the set of nodes that may become active at t, which is found by a function $reachable(t)$. For all other nodes that are not seeds and may have been activated before, $P_{[0,t]}(u)$ is set to $P_{[0,t-1]}(u)$ (steps 11–12). In addition, the probability $P_{t-1-i}(u')$ in Eq. 2 is computed based on the dynamic programming array as $T[u'][t-1-i] - T[u'][t-2-i]$.

The function $reachable(t)$ finds all nodes that are reachable from the seeds through *simple* paths of length $t - i$, for each delay $i \in [0, min(t-1, \delta)]$, using a concurrent breadth-first-search (bfs). The bfs discovers only nodes that may become active at t, which is necessary to accurately compute the probability $P_{[0,t]}(u)$, for each discovered node u. Cycle paths are discarded, because the node u_1 in a cycle path $[u_1, \ldots, u_{m-1}, u_1]$ cannot be activated by the edge (u_{m-1}, u_1).

DPSC needs $O(|V|^2 \cdot |W_C|)$ time, where $|V|$ is the number of nodes of the graph and $|W_C|$ the size of the largest window in C, in the worst case when the graph is complete. In practice, social network graphs are sparse, and *DPSC* scales much better.

Pruning. This subroutine prunes windows that cannot lead to a solution of *MASP*. When applied in an iteration j, *Pruning* computes, for each window, the maximum spread-to-cost ratio that the window can have after all remaining $k - j$ iterations (steps 1 to 3). Then, it removes each window whose maximum ratio is smaller than a lower bound, which is computed as the largest spread-to-cost ratio of all windows (steps 4 to 5). The lower bound corresponds to the minimum spread-to-cost ratio of a solution. That is, we assume the worst case, in which every available node in a subsequent iteration is certainly active (i.e., each such node u has spread gain $W_i.g(u) = 0$).

Function: *Pruning*
Input: Set of windows $\mathcal{W}$
1 **foreach** *window W_i in $\mathcal{W}$* **do**

2 $\quad L \leftarrow \mathrm{argmax}_{\{u_1,\ldots,u_{k-j}\} \subseteq V \setminus W_i.S} \left(\displaystyle\sum_{u \in \{u_1,\ldots,u_{k-j}\}} W_i.g(u) \right)$

3 $\quad W_i.r \leftarrow \frac{W_i.\sigma + \sum_{u \in L} W_i.g(u)}{W_i.c}$ // max. ratio of W_i

4 $lbound \leftarrow$ largest ratio $\frac{W_i.\sigma}{W_i.c}$ of each W_i in $\mathcal{W}$

5 Remove from $\mathcal{W}$ each window W_i such that $W_i.r < lbound$

The maximum spread-to-cost ratio, $W_i.r$, of a window W_i is computed based on the following property:

– The spread, $W_i.\sigma$, of W_i cannot increase by more than $\sum_{u \in L} W_i.g(u)$ after any remaining iterations, where L is the set of $k - j$ available nodes with the largest spread gain assigned by $W_i.g()$.

The property holds because, due to the submodularity of spread [11]: (I) no node that is not contained in L can have a larger spread gain than that of a node in L, in any of the remaining $k - j$ iterations of *MASP*, and (II) after the remaining $k - j$ iterations, the spread $W_i.\sigma$ cannot increase by more than the sum of the spread gain of the nodes that are added into $W_i.S$ in the remaining iterations.

7 Experimental Evaluation

In this section, we evaluate *MASP* in terms of effectiveness and efficiency and demonstrate the benefit of its optimization techniques. Since no existing algorithms can deal with the TSM problem, we compared *MASP* against three baselines that are based on the *MC*, *ISP*, and *MISP* methods of [11,12] (see Sect. 2). The MC_B baseline applies the *MC* approximation algorithm to each window independently and then selects the solution with the largest spread-to-cost ratio. The ISP_B and $MISP_B$ baselines differ from MC_B in that they estimate the spread using *ISP* and *MISP*, respectively.

All algorithms were implemented in C++ and applied to the datasets in Table 1. All datasets are real and were used in [2,11,12], except *AB*, a synthetic dataset generated by the Albert-Barabasi model. *POL* is available at http://www-personal.umich.edu/~mejn/ and all other real datasets at http://snap.stanford.edu/data.

Table 1. Characteristics of datasets.

Dataset	Description	# of nodes ($|V|$)	# of edges ($|E|$)	Avg in-degree	Max in-degree
WI	Wikipedia adminship vote graph	7115	103689	13.7	452
PH	High energy physics citation graph	34546	421578	24.3	846
EPIN	Whom-trusts-whom graph	75879	508837	13.4	3079
POL	Graph of weblogs	1490	19090	11.9	305
AB	Synthetic dataset	10000	45040	9	9997

Following [11,12], the probability vector of each edge (u', u) was constructed by setting $p((u', u))$ to $\frac{1}{|n^-(u)|}$ and $\mathcal{P}_{u'}$ to the Poisson distribution with a random parameter (mean rate) λ in $[1, 20]$. In addition, a window set $\mathcal{W}$ of size $|\mathcal{W}|$ was comprised of the windows ending at time $1, \ldots, |\mathcal{W}|$, and δ was set to $|\mathcal{W}|-1$. The default values for k and $|\mathcal{W}|$ were 25 and 10, respectively. In addition, following [11], we set the number of Monte Carlo simulations in MC_B to 20000, and θ (minimum path probability in ISP and $MISP$) to 10^{-5}.

The window costs were assigned by the concave piece-wise linear function in Eq. 3.

$$c(W_i) = \begin{cases} |\mathcal{W}| & i = 1 \\ \frac{|\mathcal{W}|}{i} + c(W_{i-1}) & \text{otherwise} \end{cases} \tag{3}$$

Clearly, the cost of a window $c(W_i)$ increases with the end time of the window, but the increase is smaller for larger windows. Concave piece-wise linear functions model "economies of scale" (i.e., the social network provider offers discounts for longer campaigns, which makes it cheaper to extend the length of an already long campaign) [1]. All experiments ran on an Intel Xeon at 2.60 GHz with 16 GB RAM.

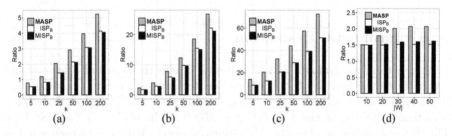

Fig. 3. Spread-to-cost ratio vs. k for (a) WI, (b) PH, and (c) $EPIN$. Spread-to-cost ratio vs. $|\mathcal{W}|$ for (d) WI.

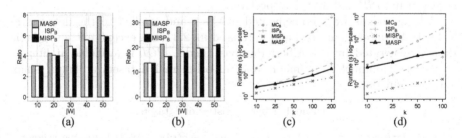

Fig. 4. Spread-to-cost ratio vs. $|\mathcal{W}|$ for (a) PH, (b) $EPIN$. Runtime vs. k for (a) POL, and (b) WI.

Effectiveness. We demonstrate that $MASP$ finds solutions with high spread-to-cost ratio, due to its exact spread computation strategy, unlike ISP_B and $MISP_B$. Figures 3a, b, and c show the result for varying k. The spread-to-cost

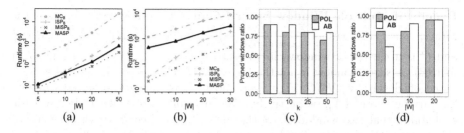

Fig. 5. Runtime vs. $|\mathcal{W}|$ for (a) *POL* and (b) *WI*. Ratio of pruned windows vs. (c) k and (d) $|\mathcal{W}|$, for *POL* and *AB*.

ratio for *MASP* was higher than that of both heuristics by 28% on average. Figures 3d, 4a, and b show the spread-to-cost ratio for varying number of windows $|\mathcal{W}|$. The spread-to-cost ratio for *MASP* was higher than that of both heuristics by 32% on average and up to 116%. MC_B found the same solutions with *MASP*, due to the large number of Monte Carlo simulations.

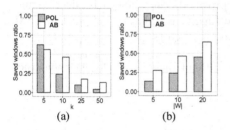

Fig. 6. Ratio of saved windows vs. (a) k and (b) $|\mathcal{W}|$, for the *POL* and *AB* datasets.

Efficiency. We demonstrate that *MASP* is significantly faster than MC_B, the only baseline that offers approximation guarantees. Figures 4c and d show the runtime for varying k. *MASP* is *at least 1, and on average 6, orders of magnitude faster* than MC_B, and it scales much better with respect to k. Figures 5a and b show the runtime for varying number of windows $|\mathcal{W}|$. *MASP* is *at least 2 orders of magnitude faster* than MC_B and scales better with respect to $|\mathcal{W}|$.

MASP is more efficient and scalable than MC_B, due to the pruning and multiple-window spread gain computation, as explained below. However, it is generally less scalable than ISP_B and $MISP_B$.

Thus, the conclusion from the effectiveness and efficiency experiments is that *MASP*: (I) finds the same solutions with MC_B, substantially outperforming ISP_B and $MISP_B$, and (II) is at least one order of magnitude faster than MC_B but less efficient than ISP_B and $MISP_B$.

Benefit of Pruning. Figure 5c shows the ratio of pruned windows, for varying k. The ratio is at least 0.7 and 0.8, for the *POL* and *AB* dataset, respectively. Figure 5d reports the ratio of pruned windows, for varying $|\mathcal{W}|$. The ratio is at least 0.77 and 0.6 for *POL* and *AB* and increases with $|\mathcal{W}|$. This is because more windows have similar ratios, due to the small increase in cost and spread, when $|\mathcal{W}|$ is large.

Benefit of Multiple-Window Spread Gain Computation. We define the
ratio of *saved* windows as $\frac{\sum_{i \in [1,k]} \sum_{C \in \mathcal{C}} (|C|-1)}{\sum_{i \in [1,k]} \sum_{C \in \mathcal{C}} |C|}$, where i is an iteration of *MASP*
and C is a cluster of windows in the set of clusters $\mathcal{C}$ (see steps 4 and 17 of
MASP). A *saved* window is not the largest in its cluster and its spread is computed efficiently by *DPSC*. Figures 6a and b show the ratio of *saved* windows
for varying k and $|\mathcal{W}|$, respectively. The ratio decreases with k, because the
probability that two windows have the same seed-set decreases with the size of
the seed-set. On the other hand, the ratio increases with $|\mathcal{W}|$, because there are
more windows that can have the same seed-set and form a cluster.

8 Conclusion

The task of performing a cost-effective, time-constrained campaign requires
selecting a window, among given alternatives, and a set of k seeds, such that
the ratio between the spread of the seeds in the window and the window cost
is maximum. In this work, we formulated this task as an optimization problem
and developed an approximation algorithm to solve it. The algorithm employs
dynamic programming and pruning to improve efficiency, and it is effective and
efficient, as shown experimentally. In the future, we plan to extend the TSM
problem when the nodes are also associated with costs.

Acknowledgments. The authors would like to thank the reviewers for their constructive comments.

References

1. Chen, D., Batson, R.G., Dang, Y.: Applied Integer Programming. Wiley, Hoboken
   (2010)
2. Chen, W., Lu, W., Zhang, N.: Time-critical influence maximization in social networks with time-delayed diffusion process. In: AAAI, pp. 592–598 (2012)
3. Cruz, D., Fill, C.: Evaluating viral marketing: isolating the key criteria. Mark.
   Intell. Plan. **26**(7), 743–758 (2008)
4. Dinh, T.N., Zhang, H., Nguyen, D.T., Thai, M.T.: Cost-effective viral marketing
   for time-critical campaigns in large-scale social networks. TON **22**(6), 2001–2011
   (2014)
5. Gomez-Rodriguez, M., Balduzzi, D., Schölkopf, B.: Uncovering the temporal
   dynamics of diffusion networks. In: ICML, pp. 561–568 (2011)
6. Gomez-Rodriguez, M., Schölkopf, B.: Influence maximization in continuous time
   diffusion networks. In: ICML, pp. 1–8 (2012)
7. Grifoni, P., D'Andrea, A., Ferri, F.: An integrated framework for on-line viral
   marketing campaign planning. Int. Bus. Res. **6**(1), 22–30 (2013)
8. Kempe, D., Kleinberg, J., Tardos, E.: Maximizing the spread of influence through
   a social network. In: KDD, pp. 137–146 (2003)
9. Krause, A., Golovin, D.: Submodular function maximization. In: Tractability
   (2013)

10. van Leeuwen, M., Ukkonen, A.: Same bang, fewer bucks: efficient discovery of the cost-influence skyline. In: SDM, pp. 19–27 (2015)
11. Liu, B., Cong, G., Xu, D., Zeng, Y.: Time constrained influence maximization in social networks. In: ICDM, pp. 439–448 (2012)
12. Liu, B., Cong, G., Zeng, Y., Xu, D., Chee, Y.M.: Influence spreading path and its application to the time constrained social influence maximization problem and beyond. TKDE **26**(8), 1904–1917 (2014)
13. Lu, W., Bonchi, F., Goyal, A., Lakshmanan, L.V.S.: The bang for the buck: fair competitive viral marketing from the host perspective. In: KDD, pp. 928–936 (2013)
14. Moore, S.: Film talk: an investigation into the use of viral videos in film marketing. J. Promot. Commun. **3**(3), 380–404 (2015)
15. Nemhauser, G.L., Wolsey, L.A., Fisher, M.L.: An analysis of approximations for maximizing submodular set functions. Math. Program. **14**(1), 265–294 (1978)
16. Saaty, T.L.: Decision Making for Leaders. RWS Publications, Pittsburgh (1990)

DSBPR: Dual Similarity Bayesian Personalized Ranking

Longfei Shi[✉], Bin Wu, Jing Zheng, Chuan Shi, and Mengxin Li

Beijing Key Lab of Intelligent Telecommunications Software and Multimedia,
Beijing University of Posts and Telecommunications, Beijing 100876, China
dalongfly@gmail.com, {wubin,shichuan,limengxin}@bupt.edu.cn,
oliviacheng@126.com

Abstract. Modern social recommendation has been steadily receiving more attention, which utilizes social relations among users to improve the efficiency of recommendation. However, most social recommendation methods only consider simple similarity information of users as social regularization and ignore the improvement of predictors of people's opinions. Meanwhile, due to the simple characteristics of data in various applications, previous works mostly leverage pointwise methods based on absolute rating assumption to solve the problem. In this paper, we propose a novel Dual Similarity Bayesian Personalized Ranking model to incorporate the similarity information of users and items into our preference predictor function. Having improved the preference predictor, we employ Bayesian Personalized Ranking model as training procedure which is a pairwise method. Empirical results on three public datasets show that our proposed model is an efficient algorithm compared with the state-of-the-art methods.

Keywords: Recommendation system · Bayesian Personalized Ranking · Heterogeneous information network

1 Introduction

With a large number of items available all the time, users have great difficulty finding the items that best match their preferences. Recommender systems appear as a natural solution to overcome the problem by learning from historical feedback. Among numerous techniques, collaborative filtering (CF) [4] has been a most popular recommender approach. Due to the efficiency and effectiveness, the low rank matrix factorization [5] becomes a primary choice for implementing CF. Despite the great success, it has been suffering inherently from cold start problem because of the sparsity of real-world datasets. In the situation, hybrid recommendation [1] has become a hot topic gradually, which can achieve better recommendation performance through combining user feedback and additional information of users and items.

Particularly, with continuous increasing popularity of social media, there is a surge of social recommendation methods [3], which utilize social relations among

J. Kim et al. (Eds.): PAKDD 2017, Part I, LNAI 10234, pp. 266–277, 2017.
DOI: 10.1007/978-3-319-57454-7_21

users or other types of information. Most social recommendation techniques leverage social regularization to restrict optimization function, but they neglect the shortcomings of the basic preference predictor. Moreover, traditional Matrix Fatorization (MF) approaches face the challenge of the vagueness of explaining non-observed feedback. Recently, pointwise [6] and pairwise [7] methods, based on MF approaches, have achieved some success. Pointwise methods suppose non-observed feedback to be intrinsically negative to certain extent while pairwise methods may be on the basis of more credible assumption. Specifically, pairwise methods assume that non-observed feedback must be less preferable than positive feedback, which directly optimize ranking item pairs instead of scoring single items (e.g., Bayesian Personalized Ranking).

In order to overcome the limitation of basic preference predictor in traditional social recommendation and pointwise methods in practical scenarios, we propose a Dual Similarity Bayesian Personalized Ranking (DSBPR). Due to the success of Heterogeneous Information Network (HIN) in many applications, we consider organizing objects and relations in a recommender system as a HIN. In the model, we integrate the heterogeneous information to generate rich similarity. Futhermore, inspired by the thought of collective intelligence, we incorporate the similarity information (i.e., the similarity between users and items) into the basic preference predictor. At last, we employ Bayesian personalized ranking as the training procedure.

Experimentally our model demonstrates significant performance improvements on three real-world datasets. The major contributions of this paper are summarized as follows:

(1) In order to inject rich heterogeneous information among users and items, we introduce a novel MF-based approach that incorporates HIN signals into the basic preference predictor.
(2) Different from traditional social recommendation, we leverage the pairwise method (i.e., Bayesian Personalized Ranking) as training procedure.
(3) We thoroughly evaluate our proposed approach on three real-world datasets and demonstrate its effectiveness in contrast to state-of-the-art recommenders.

The rest of this paper is organized as follows. We present some preliminary knowledge in Sect. 2 and the proposed DSBPR model is detailed in Sect. 3. Experiments and analysis are shown in Sect. 4. At last, we describe related work in Sect. 5 and draw the conclusion in Sect. 6.

2   Preliminary

In the section, we declare notations employed in the paper and convey some preliminary knowledge.

2.1 Dual Similarity Generated from HIN

A heterogeneous information network [13] is a special type of information network with the underneath data structure as a directed graph, which either contains multiple types of objects or multiple types of links. Fig. 1 shows the network schema of a typical heterogeneous network in a movie recommender system. The HIN contains objects from multiple types of entities: user (U), movie (M), location (L), group (G), actor (A), director (D), and type (T).

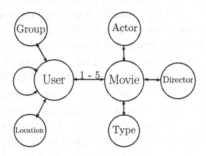

Fig. 1. Network schema of HIN example.

Two different types of objects in a HIN can be connected via different **meta path** [13], which represents a compound relation between these two types of objects. A meta path $\mathcal{P}$ is a path defined on a schema $\mathcal{S} = (\mathcal{A}, \mathcal{R})$, and is denoted in the form of $A_1 \xrightarrow{R_1} A_2 \xrightarrow{R_2} \cdots \xrightarrow{R_l} A_{l+1}$ (abbreviated as $A_1 A_2 \cdots A_{l+1}$), which defines a composite relation $R = R_1 \circ R_2 \circ \cdots \circ R_l$ between type A_1 and A_{l+1}, where $\circ$ denotes the composition operator on relations. As an example shown in Fig. 1, users can be connected via "User-User" (UU), "User-Movie-User" (UMU), "User-Group-User" (UGU) and so on. Among these, different meta paths have different semantic relations. For example, the UU path means users have social relations, while UMU path means users have watched the same movies. Therefore, we can evaluate the similarity of users (or movies) based on different meta paths. We can consider UU, UMU, UGU for users. Analogously, meaningful meta paths connecting movies include MAM, MDM, and so on.

There are several paths based similarity measures to evaluate the similarity of objects in HIN [12,13]. We define $S_U^{(p)}$ to denote the similarity matrix of users under the given meta path $\mathcal{P}_U^{(p)}$ connecting users, and $S_U^{(p)}(i,j)$ denotes the similarity of users i and j under the path $\mathcal{P}_U^{(p)}$. Similarly, $S_I^{(q)}$ denotes the similarity matrix of items under the given meta path $\mathcal{P}_I^{(q)}$ connecting items, and $S_I^{(q)}(i,j)$ denotes the similarity of items i and j under the path $\mathcal{P}_I^{(q)}$.

Since users (or items) have different similarities under different meta paths, we combine their similarities under all paths through assigning weights on these paths. For users and items, we define S_U and S_I to represent the similarity matrix of users and items on all meta paths, respectively.

$$S_U = \sum_{p=1}^{|\mathcal{P}_U|} \boldsymbol{w}_U^{(p)} S_U^{(p)}, \tag{1}$$

$$S_I = \sum_{q=1}^{|\mathcal{P}_I|} \boldsymbol{w}_I^{(q)} S_I^{(q)}, \tag{2}$$

where $\boldsymbol{w}_U^{(p)}$ denotes the weight of meta path $\mathcal{P}_U^{(p)}$ among all meta paths $\mathcal{P}_U$ connecting users, and $\boldsymbol{w}_I^{(q)}$ denotes the weight of meta path $\mathcal{P}_I^{(q)}$ among all meta paths $\mathcal{P}_I$ connecting items. In the paper, we utilize a weight learning method in [15] to learn the weights of different similarities.

2.2 Bayesian Personalized Ranking

Bayesian Personalized Ranking (BPR) [10] is a pairwise ranking optimazation model which adopts stochastic gradient descent as the training procedure. Based on the Bayesian formulation, the BPR model intends to maximize the following posterior probability:

$$p(\Theta| >_u) \propto p(>_u |\Theta)p(\Theta), \tag{3}$$

where Θ represents the parameter vector of a arbitrary model (e.g., matrix factorization). Here, there are two fundamental assumptions.

- If the user-item pair (u, i) is observed but (u, j) is not observed, it assumes that the user u prefers an item i than an item j.
- Each user is presumed to act independently from the others.

With these assumptions, the above likelihood function $p(>_u |\Theta)$ can be written as the following:

$$\prod_{u \in U} p(>_u |\Theta) = \prod_{(u,i,j) \in U \times I \times I} p(i >_u j|\Theta)^{\delta((u,i) \succ (u,j))} \\ \cdot (1 - p(i >_u j|\Theta))^{[1 - \delta((u,i) \succ (u,j))]}, \tag{4}$$

where $(u, i) \succ (u, j)$ denotes that user u prefers item i to item j.

In order to complete the Bayesian modeling approach of the personalized ranking task, it defines the individual probability that a user really prefers item i to item j as the following:

$$p(i >_u j|\Theta) = \sigma(\widehat{x}_{uij}), \tag{5}$$

where σ is the logistic sigmoid function. When using matrix factorization as the preference predictor (i.e., BPR-MF), $\widehat{x}_{uij}$ is defined as

$$\widehat{x}_{uij} = \widehat{x}_{u,i} - \widehat{x}_{u,j}. \tag{6}$$

Moreover, it introduces a general priori density $p(\Theta)$ which accords with a normal distribution with zero mean and variance-covariance matrix Σ_Θ. Then it sets $\Sigma_\Theta = \lambda_\Theta I$, so we deduced the optimization criterion (BPR-OPT):

$$\sum_{(u,i,j)\in D_s} \ln \sigma(\widehat{x}_{uij}) - \lambda_\Theta \|\Theta\|^2. \tag{7}$$

3 The DSBPR Method

In this section, we will introduce our DSBPR method which incorporates similarity information into matrix factorization based on BPR model. We firstly review the basic preference predictor based on matrix factorization. Then we introduce the improved model through incorporating dual similarity information. Finally, we develop our training procedure using BPR framework, and infer the learning algorithm of DSBPR.

3.1 Preference Predictor

Our preference predictor is based on a state-of-the-art rating predicton model, namely the low-rank matrix factorization, whose basic formulation assumes the following model to predict the preference of a user u towards an item i [4],

$$\widehat{r}_{ui} = \alpha + \beta_u + \beta_i + p_u q_i^T, \tag{8}$$

where α is the overall average rating, β_u and β_i indicate the observed deviations of user u and item i respectively, and p_u and q_i are K-dimensional vectors describing the latent factors of user i and item i. The inner product $p_u q_i^T$ captures the interaction between user u and item i (i.e., the overall interest of users in the items' characteristics).

3.2 Dual Similarity Improvement Criterion

In theory, latent factors seem to uncover any relevant dimensions. However, for the "lonely" items in the practical scenario, it's hard to estimate their latent dimensions because there are too few associated observations about them. Therefore, we consider adding more auxiliary signals (i.e., the similarity information in HIN) into the rating model to alleviate the problem mentioned above. In particular, we add the similarity of users into the left of the inner product PQ^T and then add the similarity of items into the right of the inner product as follows:

$$R = \alpha + \beta^u + \beta^i + \frac{\mu Sim^u PQ^T}{\Phi^u} + \frac{\gamma PQ^T Sim^i}{\Phi^i}, \tag{9}$$

where R is an $m \times n$ rating matrix, α and β are as in Eq. (8). P and Q are $m \times k$ and $n \times k$ latent factor matrices respectively. Sim^u is users' similarity matrix with the size of $m \times m$ and Sim^i is items' similarity matrix with the size of $n \times n$. Here, we leverage the similarity matrices that are generated from the above HIN. In order to restrict the rating function within a reasonable range, we consider Φ^u and Φ^i as the denominator in the fourth and the fifth terms of Eq. (9). Here, both Φ^u and Φ^i are $m \times n$ matrices. Specifically, we show the two terms as follows:

$$\Phi^u = \begin{pmatrix} \phi_{11}^u & \phi_{12}^u & \cdots & \phi_{1n}^u \\ \phi_{21}^u & \phi_{22}^u & \cdots & \phi_{2n}^u \\ \vdots & \vdots & \ddots & \vdots \\ \phi_{m1}^u & \phi_{m2}^u & \cdots & \phi_{mn}^u \end{pmatrix} \tag{10}$$

where $\phi_{ij}^u = \sum\limits_{l=1}^{m} Sim_{il}^u$. Similarly, we can also design the latter constrain term as follows:

$$\Phi^i = \begin{pmatrix} \phi_{11}^i & \phi_{12}^i & \cdots & \phi_{1n}^i \\ \phi_{21}^i & \phi_{22}^i & \cdots & \phi_{2n}^i \\ \vdots & \vdots & \ddots & \vdots \\ \phi_{m1}^i & \phi_{m2}^i & \cdots & \phi_{mn}^i \end{pmatrix} \tag{11}$$

where $\phi_{ij}^i = \sum\limits_{l=1}^{n} Sim_{lj}^i$. Because of using the inner product twice, we introduce the normalized parameters μ, γ, which meet:

$$\mu + \gamma = 1. \tag{12}$$

3.3 The Learning Algorithm

The learning algorithm of DSBPR is based on BPR model as mentioned above. We follow the widely used stochastic gradient descent SGD) algorithm to optimize the objective function in Eq. (7). Therefore, the learning algorithm updates parameters in the following fashion:

$$\Theta \longleftarrow \Theta + \eta(\frac{e^{-\widehat{x}_{uij}}}{1 + e^{-\widehat{x}_{uij}}} \cdot \frac{\partial}{\partial \Theta}\widehat{x}_{uij} + \lambda_\Theta \Theta). \tag{13}$$

At each iteration, one sample is uniformly drawn from D_S (i.e., the training set comprised of triples in the form of (u, i, j)) and the parameters are updated in the opposite direction of the loss function's gradient at the sampling point. Algorithm 1 shows the framework of the optimization algorithm. Moreover, $\widehat{x}_{uij}$ derivatives are:

$$\frac{\partial}{\partial \Theta}\widehat{x}_{uij} = \begin{cases} \frac{\mu Sim_u^u}{\Phi_u^u}(Q_i - Q_j) + \gamma(\frac{Sim_i^i}{\Phi_i^i}Q_i - \frac{Sim_j^i}{\Phi_j^i}Q_j) & if\ \Theta = P_u, \\ \frac{\mu Sim_u^u}{\Phi_u^u}P_u + \frac{\gamma Sim_i^i}{\Phi_i^i}P_u & if\ \Theta = Q_i, \\ -\frac{\mu Sim_u^u}{\Phi_u^u}P_u - \frac{\gamma Sim_j^i}{\Phi_j^i}P_u & if\ \Theta = Q_j, \\ 0 & else. \end{cases}$$

Algorithm 1. Alogrithm of DSBPR

Input:
D_S: training set
Θ: the parameter vector
$\mathcal{P}_U, \mathcal{P}_I$: meta path sets of users and items
η: learning rate for gradient descent

Output:
P, Q: the latent factor of users and items

1: **for** $\mathcal{P}_U^{(p)} \in \mathcal{P}_U$ and $\mathcal{P}_I^{(q)} \in \mathcal{P}_I$ **do**
2:    Calculate the similarity of users and items ($S_U^{(p)}$ and $S_I^{(q)}$)
3: **end for**
4: Initialize Θ
5: **repeat**
6:    draw (u, i, j) from D_S
7:    Update $\Theta \longleftarrow \Theta + \eta(\frac{e^{-\widehat{x}_{uij}}}{1+e^{-\widehat{x}_{uij}}} \cdot \frac{\partial}{\partial \Theta}\widehat{x}_{uij} + \lambda_\Theta\Theta)$
8: **until** convergence

Furthermore, we use three regularization constants: λ_P is used for users' features P; for the item features Q we have two regularization constants, λ_{Q+} which is used for positive updates on Q_i, and λ_{Q-} for negative updates on Q_j.

4   Experiments

In this section, we will verify the effectiveness of our model by a series of experiments compared to several state-of-the-art recommendation methods.

4.1   Datasets

Although there are many public datasets for recommendation, some of them may not contain enough objects attributes but only focus on the rating information. In order to get more available information, including rating information, attribute information of uers and items and social relations, we use three real datasets in our experiments. Movielens dataset[1] contains rating information of users on movies and attributes information of users (e.g., age, gender, occupation). Douban is a well-known social media network in China, on which users post their likes or dislikes on movies through ratings and comments, whose movie dataset (i.e., Douban Movie[2]) includes 3022 users and 6971 movies with 195493 ratings ranging 1 from 5. Stemming from the business domain, Yelp[3] is a famous user review website, which includes 14085 users and 14037 movies with 194255 ratings (scales 1–5). The detailed description can be seen in Table 1.

---

[1] https://grouplens.org/datasets/movielens/.
[2] http://www.douban.com/.
[3] http://www.yelp.com/.

Table 1. Statistics of datasets

Datasets	Relations (A-B)	Number of of A/B/A-B	Ave. degrees of A/B
MovieLens	User-Movie	6040/3952/180037	29.8/52.0
	User-Gender	6040/2/6040	1.0/3020.0
	User-Age	6040/7/6040	1.0/862.8
	User-Occupation	6040/21/6040	1.00/287.6
	Movie-Type	3952/18/3952	1.0/219.6
	Movie-Year	3952/9/3952	1.00/439.1
Douban-Movie	User-Movie	3022/6971/195493	64.69/28.04
	User-User	779/779/1366	1.75/1.75
	User-Group	2212/2269/7054	3.11/3.11
	User-Location	2491/244/2491	1.00/10.21
	Movie-Director	3014/789/3314	1.09/4.20
	Movie-Actor	5438/3004/15585	2.87/5.19
	Movie-Type	6787/36/15598	2.29/433.28
Yelp	User-Business	14085/14037/194255	4.6/20.7
	User-User	9581/9581/150532	10.0/10.0
	Business-Category	14037/575/39406	2.8/73.9
	Business-Location	14037/62/14037	1.0/236.1

4.2 Evaluation

We split these datasets into training sets and test sets by selecting a random item for each user u and the result is shown in Table 2. Moreover, we use the widely employed metric AUC (i.e., Area Under the ROC curve) to evaluate the performance of different methods. The metric AUC is defined as:

$$AUC = \frac{1}{|U|} \sum_{u \in U} \frac{1}{|E(u)|} \sum_{(i,j) \in E(u)} \delta(\hat{x}_{ui} > \hat{x}_{uj}) \tag{14}$$

where the evaluation pairs per user u are:

$$E(u) = \{(i,j)|(u,i) \in D_{test} \wedge (u,j) \notin (D_{test} \cup D_{train})\} \tag{15}$$

and $\delta(b)$ is an indicator function that returns 1 iff b is *true*. A higher value of the AUC indicates a better quality.

Table 2. Description of the datasets used in the experiments

Datasets	User-item pairs training/test
MovieLens	39054/7810
Douban-Movie	29461/5892
Yelp	28666/5733

4.3   Baseline Methods

For better evaluation of the proposed DSBPR method, we compare it with the following well-known and representative methods.

- **Random (RAND).** This baseline ranks items randomly for all users.
- **PopRank.** The baseline ranks items according to their popularity and is non-personalized.
- **PMF** [11]. Salakhutdinov and Minh proposed the basic low-rank matrix factorization method for recommendation.
- **BPR-MF** [10]. It is a pairwise method introduced by Rendle et al. and is the state-of-the-art personalized ranking for implicit feedback datasets.
- **Hete-MF** [14]. Yu et al. proposed the HIN based recommendation method through combining user ratings and items' similarity matrices.
- **DSR** [15]. Zheng et al. proposed a new dual similarity regularization to impose the constraint on users and items with high and low similarities simultaneously.

We employ HeteSim [12] to evaluate the similarity of objects. For MovieLens dataset, we use 6 meaningful meta paths for users (i.e., UAU, UGU, UOU, UMU, UMTMU, UMYMU) and 3 meaningful meta paths for movies (i.e., MTM, MYM, MUM). For Douban Movie dataset, we leverage 7 meta paths for users (i.e., UU, UGU, UMU, UMDMU, UMTMU, UMAMU) and 5 meta paths for movies (i.e., MTM, MDM, MAM, MUM, MUUM). Similarly, for Yelp dataset we utilize 4 meta paths for users (i.e., UU, UBU, UBCBU, UBLBU) and 4 meta paths for business (i.e., BUB, BCB, BLB, BUUB). These similarity data are fairly used for Hete-MF, DSR and DSBPR.

4.4   Experimental Results

In the expeiment, the tradeoff parameters are searched as the learning rate $\eta \in \{0.5, 0.05, 0.005\}$ and the regularization term $\lambda \in \{0.1, 0.01, 0.001\}$, and the iteration number is chosen from $T \in \{100, 200, 300\}$. Results in terms of the average AUC on different datasets are shown in Table 3, from which we can have the following observations:

Table 3. AUC of the sets

Dataset	Metrics AUC/(vs. c)	(a) RAND	(b) PopRank	(c) PMF	(d) Hete-MF	(e) BPR-MF	(f) DSR	(g) DSBPR
MovieLens	AUC	0.5001	0.7401	0.7800	0.8247	0.8204	0.8683	**0.8720**
	Improvement	−35.89%	−5.12%	-	5.73%	5.13%	11.32%	11.79%
Douban-Movie	AUC	0.4899	0.6069	0.9187	0.9167	0.9246	0.9365	**0.9407**
	Improvement	−46.47%	−33.93%	-	−0.21%	0.67%	1.94%	2.39%
Yelp	AUC	0.4979	0.5573	0.8296	0.8504	0.8337	**0.8687**	0.8373
	Improvement	−39.98%	−32.80%	-	2.51%	0.49%	4.71%	0.93%

1. The PopRank algorithm is not as effective as the personalized recommendation methods, because "lonely" items are inherently unpopular.
2. Building on the basis of BPR-MF, DSBPR performs better than MF and BPR-MF algorithms, which demonstates the effectiveness of incorporating the similarity of users and items via heterogeneous information network.
3. DSBPR has better performance than Hete-MF and DSR in most cases. It reveals that the improvement of the basic preference predictor instead of regularization term is also feasible. However, we also discover DSR performs better than DSBPR in Yelp, which demonstrates DSBPR is more suitable for datasets with rich relations.

Sensitivity. In addition, we conduct the experiments with different latent factors. As the number of latent factors increases (i.e., ranging from 2 to 20), BPR-MF, DSBPR perform better than the last time, which demonstrates the ability of pairwise methods in avoid overfitting as shown in Fig. 2.

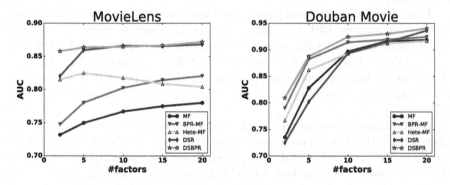

Fig. 2. AUC with varying dimensions

Training Efficiency. In Fig. 3 we show the AUC with increasing training iterations. Generally speaking, our proposed model performs better than other methods in each iteration and take no longer to converge.

5   Related Work

Through uncovering latent dimensions, MF methods relate users and items, which are the basis of many state-of-the-art recommendation approaches. When it comes to personalized ranking from implicit feedback, traditional MF approaches are challenged by the ambiguity of explaining "non-observed" feedback. Pointwise methods assume "non-observed" feedback to be inherently negative to certain degree. In contrast to pointwise methods, pairwise methods assume that positive feedback must only be more preferable than "non-observed"

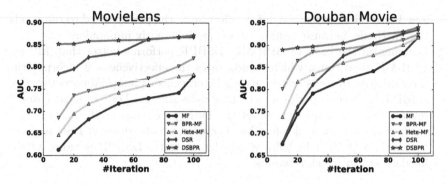

Fig. 3. AUC with training iterations (#factors = 20)

feedback, which is a little weaker but more realistic. Rendle et al. proposed a generalized BPR framework, which is the first method with such pairwise preference assumption.

Due to the great success of pairwise methods, some new algorithms have been proposed to combine BPR with some auxiliary data, such as BPR with user-side social connections [2], BPR with group preference [8] and so on. Futhermore, more and more researchers have been aware of the importance of HIN, in which objects are of different types and links. Yu et al. [14] proposed Hete-MF through combining rating information and items' similarities derived from meta paths in HIN.

However, most recommender methods only consider designing the social regularization without the improvement of the basic preference predictor. For example, Hete-MF proposed by Yu et al. merely takes the similarity information as the regularization term by the pointwise method. Qiao et al. [9] proposed the model combining heterogeneous social information with MF, which places zero-mean spherical Gaussian priors for the regularization term. Compared with the aforementioned works, DSBPR method we proposed not only incorporates the similarity of HIN information into the basic prefernce predictor but also uses the classic pairwise model (i.e., BPR) as the training procedure, which performs better to improve the recommendation results.

6   Conclusions

In this paper, we study the limitations of basic preference predictor in traditional social recommendation and pointwise methods in many applications, and design a novel algorithm called DSBPR. DSBPR introduces the similarity of users and items into the basic preference predictor based on low-rank matrix factorization framework. Moreover, our model is trained with BPR using stochastic gradient descent. Experimental results on three real-world datasets validate the effectiveness of DSBPR.

Acknowledgements. This work is supported in part by the National Key Basic Research and Department (973) Program of China (No. 2013CB329606), and the National Natural Science Foundation of China (Nos. 71231002, 61375058), and the Co-construction Project of Beijing Municipal Commission of Education.

References

1. Adomavicius, G., Tuzhilin, A.: Toward the next generation of recommender systems: a survey of the state-of-the-art and possible extensions. IEEE Trans. Knowl. Data Eng. **17**(6), 734–749 (2005)

2. Du, L., Li, X., Shen, Y.-D.: User graph regularized pairwise matrix factorization for item recommendation. In: Tang, J., King, I., Chen, L., Wang, J. (eds.) ADMA 2011. LNCS (LNAI), vol. 7121, pp. 372–385. Springer, Heidelberg (2011). doi:10. 1007/978-3-642-25856-5_28

3. Jamali, M., Ester, M.: Trustwalker: a random walk model for combining trust-based and item-based recommendation. In: Proceedings of 15th ACM SIGKDD International Conference on Knowledge Discovery and Data Mining, pp. 397–406. ACM (2009)

4. Koren, Y., Bell, R.: Advances in collaborative filtering. In: Ricci, F., Rokach, L., Shapira, B., Kantor, P.B. (eds.) Recommender Systems Handbook, pp. 145–186. Springer, Heidelberg (2011)

5. Koren, Y., Bell, R., Volinsky, C., et al.: Matrix factorization techniques for recommender systems. Computer **42**(8), 30–37 (2009)

6. Koren, Y., Sill, J.: OrdRec: an ordinal model for predicting personalized item rating distributions. In: Proceedings of 5th ACM Conference on Recommender Systems, pp. 117–124. ACM (2011)

7. Pan, W., Chen, L.: Cofiset: collaborative filtering via learning pairwise preferences over item-sets. Training **1**, 1 (2013)

8. Pan, W., Chen, L.: GBPR: group preference based Bayesian personalized ranking for one-class collaborative filtering. In: IJCAI, vol. 13, p. 2691–2697 (2013)

9. Qiao, Z., Zhang, P., Zhou, C., Cao, Y., Guo, L., Zhang, Y.: Event recommendation in event-based social networks (2014)

10. Rendle, S., Freudenthaler, C., Gantner, Z., Schmidt-Thieme, L.: BPR: Bayesian personalized ranking from implicit feedback. In: Proceedings of 25th Conference on Uncertainty in Artificial Intelligence, pp. 452–461. AUAI Press (2009)

11. Salakhutdinov, R., Mnih, A.: Probabilistic matrix factorization. In: NIPS, pp. 1257–1264 (2012)

12. Shi, C., Kong, X., Huang, Y., Philip, S.Y., Wu, B.: Hetesim: a general framework for relevance measure in heterogeneous networks. IEEE Trans. Knowl. Data Eng. **26**(10), 2479–2492 (2014)

13. Sun, Y., Han, J., Yan, X., Yu, P.S., Wu, T.: Pathsim: meta path-based top-k similarity search in heterogeneous information networks. Proc. VLDB Endow. **4**(11), 992–1003 (2011)

14. Yu, X., Ren, X., Gu, Q., Sun, Y., Han, J.: Collaborative filtering with entity similarity regularization in heterogeneous information networks. In: IJCAI-HINA Workshop (2013)

15. Zheng, J., Liu, J., Shi, C., Zhuang, F., Li, J., Wu, B.: Dual similarity regularization for recommendation. In: Bailey, J., Khan, L., Washio, T., Dobbie, G., Huang, J.Z., Wang, R. (eds.) PAKDD 2016. LNCS (LNAI), vol. 9652, pp. 542–554. Springer, Cham (2016). doi:10.1007/978-3-319-31750-2_43

Usage Based Tag Enhancement of Images

Balaji Vasan Srinivasan[1]([✉]), Noman Ahmed Sheikh[2], Roshan Kumar[3],
Saurabh Verma[4], and Niloy Ganguly[5]

[1] BigData Experience Lab, Adobe Research, Bangalore, India
balsrini@adobe.com
[2] Indian Institute of Technology, Delhi, New Delhi, India
nomanahmedsheikh11@gmail.com
[3] Indian Institute of Technology, Kanpur, Kanpur, India
roshankr1995@gmail.com
[4] Indian Institute of Technology, Rourkee, Rourkee, India
saurv4u@gmail.com
[5] Indian Institute of Technology, Kharagpur, Kharagpur, India
ganguly.niloy@gmail.com

Abstract. Appropriate tagging of images is at the heart of efficient
recommendation and retrieval and is used for indexing image content.
Existing technologies in image tagging either focus on what the image
contains based on a visual analysis or utilize the tags from the textual
content accompanying the images as the image tags. While the former
is insufficient to get a complete understanding of how the image is perceived and used in various context, the latter results in a lot of irrelevant
tags particularly when the accompanying text is large. To address this
issue, we propose an algorithm based on graph-based random walk that
extracts only image-relevant tags from the accompanying text. We perform detailed evaluation of our scheme by checking its viability using
human annotators as well as by comparing with state-of-the art algorithms. Experimental results show that the proposed algorithm outperforms base-line algorithms with respect to different metrics.

1 Introduction

A popular English idiom says "An image is worth a thousand words". Content
writers always look out for good visual supplements to enrich their content and
make it more appealing to the target audience. Fortunately, a huge repertoire of
such content (images, video, etc.) is available in the Internet - however proper
annotation with appropriate tags is necessary for their efficient retrieval. The
size of online visual data clearly calls for an automatic approach to tag them.

Existing tagging systems work towards capturing the denotational aspects of
the image, viz. what the image denotes/contains. This includes tags capturing
the various aspects present in the image. These details are either captured via
the visual features of the images or via human added tags. However, the former
tags are often generic and do not capture the entire information that is contained
in the image. Let us consider an example in Table 1 which shows an image of the

© Springer International Publishing AG 2017
J. Kim et al. (Eds.): PAKDD 2017, Part I, LNAI 10234, pp. 278–290, 2017.
DOI: 10.1007/978-3-319-57454-7_22

Table 1. Example: an image of Apple co-founder Steve Jobs along with the text from an article using a similar image in InShorts[a], an on-line news aggregator.

Apple sells its 1 billionth iPhone
Apple on Wednesday announced that it sold its one billionth iPhone last week. The news comes about two years after the company sold the 500 millionth unit of its handheld device. The iPhone was first introduced in 2007 by late Co-founder Steve Jobs and had registered its one millionth sale after 74 days of the launch.

[a]https://inshorts.com/news/apple-sells-its-1-billionth-iphone-1469693675991

Apple co-founder, Steve Jobs from the web. Figure 1(a) shows the set of tags for the image based on the visual tagging system in [18]. It can be seen that the tags thus obtained are generic in nature e.g. 'person', 'business' and do not capture any deeper information about the image e.g. Steve Jobs, Apple Inc., etc. While an author uploading these images can be expected to add some of these tags, it is not possible to cover all aspects of the image.

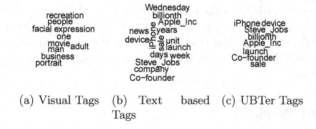

(a) Visual Tags   (b) Text based Tags   (c) UBTer Tags

Fig. 1. Tags for the image in Table 1 based on a visual tagger [18], textual parsing and our system - UBTer.

Often such images are used in different illustrations which contain valuable information about the image. To address the shortcomings of the visual tags, the accompanying content of the images can be analyzed to extract the tags. Such information can enhance both the denotational and connotational (how the image is perceived) understanding of the image. To test this hypothesis, we conducted a survey among 30 participants to rate the relevance of the text around an image in several articles on the web and its usefulness to enhance the understanding of the image. It was observed that in 91.23% of cases, the participants found the text relevant to the image. Survey respondents further opined that while the original image tags were very appropriate, the image had a different connotation when appeared along with the text, thus calling for a need to incorporate these into the image tags.

We identified an article (text included in Table 1) from InShorts[1], an on-line news aggregator using an image similar to the one in Table 1. A simple text based tagging can add a lot of noise to the tags as seen in Fig. 1(b), where the text in Table 1 was parsed to extract the textual tags e.g. "days", "week", "billionth". These noise occur primarily because of the extract textual tags that are prominent in the text but irrelevant to the image's context. The level of noise will increase with the size of the accompanying content. This calls for an automated tagging system that optimally combines the tags from accompanying text with the image tags capturing the right denotational and connotational information around the images while discarding the unrelated tags from the accompanying text resulting tags.

In this work, we propose a novel framework **UBTer** - **U**sage **B**ased image **T**agger that combines the tags derived from accompanying (usage) content with the image tags based on the visual features [18], thus integrating the information from content and usage cues. We thus achieve a balance between connotational and denotational aspects of an image. The resultant tags are shown in Fig. 1(c). We show that such a combination beats the state-of-the-art (visual and textual) tagging engines in our subjective and objective evaluations.

The paper is organized as follows. In Sect. 2, we describe the existing state of image tagging and position our framework with respect to existing systems. Section 3 introduces UBTer, - the proposed usage based tagger along with its key components. In Sect. 4 we compare the performance of UBTer against existing works via subjective and objective evaluations. We also evaluate the different parameters of UBTer to arrive at the right system configuration. Section 5 concludes the paper.

2   Related Work

Tagging and understanding textual content has been widely studied. The first step in textual tagging is extracting and detecting named entities; the popular one here is the Stanford NLP parser [11]. Once the named entities are identified, they are disambiguated and resolved into various categories [9]. Finally, the inter relationships in the content or hierarchies are identified by a semantic understanding of the text. In these works, the entities in the textual content are typically processed into a rich semantic representation (e.g. [1]) which is utilized to gain a deeper understanding of their inter-relationships.

Yang et al. [23] extract the textual tags based on a nearest-neighbor based approach and utilize the neighbors to extract the relationships between entities. Nallapatti et al. [13] use "event threading" to join different pieces of text and identify the undercurrent events in the textual topics. Shahaf and Guestrin [16] estimate the importance and "jitteriness" of the entities in the text and use it to infer the connections between different parts of the textual content.

With the advent of knowledge bases like YAGO [19], relationships from these sources are used to further enhance the understanding of the textual content.

---

[1] https://inshorts.com/news/apple-sells-its-1-billionth-iphone-1469693675991.

Kuzey et al. [6] resolve temponym based on a YAGO based entity resolution to understand textual content with temporal scopes. They develop an Integer Linear Program that jointly optimizes the mappings to knowledge base for a rounded document representation. Tandon et al. [20] mine activity knowledge from Hollywood narratives to answer questions around these activities. They capture the spatio-temporal context of the topics by constructing multiple graphs to capture relationships among activity frames which is leveraged for effective understanding. However, none of these works aim at understanding images based on a combination of visual tags and usage context which is the key challenge in our problem, where we have to combine the content and usage cues in tagging.

There also exists a large body of literature in the space of image tagging. Li et al. [8] propose methods for assignment of tags from visual aspects and use them for effective retrieval of images. Once an image is tagged, its relationships with other images have been used for further enhancing the tag set [14] or alternatively, using these tags to enhance tags of similar images [3]. The visual tags can also be enhanced and disambiguated with knowledge bases and conceptnets [22]. With the successful emergence of deep learning for image understanding, convolution neural networks have been used to find an intermediary representation *Visual Word2Vec* [5] in order to generate the image tags from this latent space. However, all these works focus on tagging the image from their visual cues/content. In our problem, we capture the usage of the images along with the visual content in the image tags to have a rounded understanding of the image.

One work that is close to the proposed solution framework is by Leong et al. [7], which relies exclusively on accompanying content for mining information relevant to the image. They construct relationships among entities based on multiple factors to arrive at the final set of tags. However, they do not use the visual tags of the images to align the accompanying content to the image and therefore have the same pitfall that we illustrated in our example in Table 1.

3 UBTer - Usage Based Tagger

We propose a novel framework, **UBTer**, which enriches the tags around an image which may not be initially contained in the set of image based tags based on the visual features. UBTer takes as input the image tags (author given and the auto tags) along with the "usage" content which uses the image for illustration. The content is processed to extract key tag candidates. Many of these tags may not be directly related with the image and hence needs to be pruned. The pruning is initiated by establishing the context of a tag. This is done by (a) scoring the importance of the tag by measuring its usage pattern in the local textual context and (b) capturing the inter-tag relationship based on certain global knowledge base. Thus we obtain a graph with weighted nodes (local importance) and weighted edges. The final tags are selected by performing a biased (based on node weight and edge weight) random walk starting from the image tags. Those nodes reached by random walk are selected in the final set. They are found not only rich and appropriate but also diverse bringing out various connotational aspects of the same image.

3.1 Tag Shortlisting

The input to UBTer is the image along with its visual tags and the accompanying text(s). The accompanying text might contain several entities that could be ambiguous e.g. Apple, Jobs in Table 1. The algorithm therefore starts with disambiguating the accompanying content for such ambiguous entities via Ambiverse [4]. Ambiverse provide a technology to automatically analyze a textual data and disambiguate named entities. It relies on the knowledge base YAGO for an accurate characterization of all the entities in the text. These entity characteristics are used along with the context of the entity in the text to disambiguate them into formal YAGO entries. We replace each occurrence of the entity with their disambiguated version. The disambiguated content is extensively parsed to identify all named entities and noun phrases using the Stanford NLP Parser [11]. Note that the image may/may not be relevant to the entirety of the entities in the accompanying text and we address this in Sect. 3.3. At the end of this step, we have a set of all candidate tags for consideration in the final tags.

3.2 Tag Importance

For each tag candidate, a score is assigned based on their importance in the local context. We calculate the total frequency of the candidate tag occurrence in the usage content accounting for the co-reference of the candidates via proper nouns by co-reference parsing. Thus, not just the direct mentions, the indirect mentions of the entities are also accounted in their local importance. We normalize the frquency counts by the counts of all entities in the text to keep the measure between 0 and 1.

For every tag candidate we also compute the average distance of the entity from the root of the corresponding dependency tree (obtained by passing the accompanying content through a dependency parser [2]). A candidate tag at the root (distance $= 1$) is the central topic of discussion in a sentence and hence is more important indicating the local relevance of the entity in the discussed subject. The inverse distance is considered as the tag importance (tags at the root gets a value of one).

The average of the two measures yields the final tag importance (n_i) whereby the tags that are in the center of discussion in the accompanying content getting higher value. We assume that a picture is added to further emphasize the central point of discussion.

3.3 Inter-tag Relationship

We build the relationships between each tag candidates leveraging two independent global knowledge base. (A) We used the Word2Vec [12] model trained on a corpus of Google News dataset with 100 billion words resulting in a final corpus of about 3 million word representations. Word2Vec yields a 300 dimensional vector for every tag candidate that represents the word in the space of the trained deep neural network. We compute the cosine-similarity between the vectors in this space which captures the semantic closeness between the tags. (B) We calculate the point-wise mutual information [21] between two entities based on their co-occurrences in the Wikipedia articles. This yields a similarity score based on how coherent the two tags are with respect to the entire Wikipedia corpus (English).

The Word2Vec based measure captures the semantic similarity between the tags because the Word2Vec space groups similarly meaning entities together. Therefore, entities closer in this space can often be interchangeably used in several context. On the other hand, the Wikipedia based measure captures the topical closeness - since entities that occur together in the several articles are closer in this space. Our final edge weight (e_{ij}) is the average of the two measures.

The edge weights along with the node importance yield a graphical representation of the candidate tags with the edge weights capturing the global relationship between the tags and the node weights indicating their local importance in the usage content.

Infusing Image Tags: To extract the usage-specific tags from the accompanying content, it is important to understand how these tag candidates relate to the visual tags. However, there may be duplicates or near duplicates to the visual tags already present within the tag set. Therefore, we first calculate the edge weight between the visual tags and every tag candidate in the graph based on the combined measure above. The tag pairs with similarity greater than a threshold (0.95 in our experiment) are merged into a single node, thus avoiding duplicity in tags. We then propagate the importance of the merged node to the adjacent nodes (at a distance of 2 edges) using an exponential decay. This ensures the propagation of the strength of the merged nodes to its neighbors and thus emphasizing the relevant pieces of the tag graphs with respect to the visual tags.

For tag pairs less than the matching threshold, an edge is added between every tag candidate whose similarity with the visual tag is significant (> 0.1 in our experiments). This ensures that the visual tags are connected to the relevant parts of the tag-graph. The series of steps is summarized in Algorithm 1.

3.4   Tag Extraction

With the graphical representation of the tags, the problem of extracting the tags that capture the context around the image boils down to identifying the top nodes in the tag graph that are closely connected to the image tags. For this we use a random walk based algorithm [15], starting the random walk from the visual tags, thus ensuring the node ranking relevant to the tag images and avoiding irrelevant tags from the accompanying text.

We define the probability of the random walk moving from a node i to another node j as, $P(tr_{i \to j}) = e_{ij} \times n_j$ where, e_{ij} is the weight of the edge (from Sect. 3.3) between tags i and j and n_j is the node importance of tag j from Sect. 3.2. The probability of the random walk staying in the same node is defined as

Algorithm 1. Tag Unifier

```
 1: procedure UNIFY(tagsFromImage, TagGraph)
 2:     tag ← normalize(tag)
 3:     for tag ∈ tagsFromImage do
 4:         for node ∈ TagGraph do
 5:             val ← similarity(tag,node) (from Sect. 3.3)
 6:             if val > σ₁ then
 7:                 MergeNodes(tag,node)
 8:                 node.weight ← MergedWeight()
 9:                 PropagateWeight(node)
10:             else if val > σ₂ then
11:                 edge ← createNewEdge(tag,node)
12:                 edge.weight ← val
13:             else
14:                 continue
15:             end if
16:         end for
17:     end for
18: end procedure
```

$P(tr_{i \to i}) = n_i$. The probabilities are normalized to conform to the requirements of a probability distribution. The final set of tags is then extracted by performing a random walk over several iteration starting from the visual/author tag nodes. This ensures that the tags selected are not just based on their importance from the accompanying text but also emphasizes on a strong relationship with the visual tags. The random walk is terminated after k (20 in our experiments) iterations and the average number of visits to a node across all runs is used as the score of the tags. The top-k tags is output as the final set of tags for the images.

4   Experimental Evaluation

We first evaluate the importance of usage tags from UBTer based on an annotator based evaluation. We then introduce 3 independent metrics that measure different aspects of the extracted tags and use them to extensively test the performance of UBTer against existing tagging baselines on the dataset from [7]. Finally, we evaluate the different parts of the UBTer to measure their significance in extracting the final tags.

4.1   Importance of Usage Tags

In order to assess the importance of usage tags over the visual tags, we conducted a survey among 45 participants to rate the overall relevance and diversity of the tags on a scale of 0–10 for the outputs from UBTer as well as those provided by the visual tagger [18] on a subset of 20 images. On a scale of 10 for tag relevance to the image, usage tags were rated at 8.08 ± 0.58 on an average against the score of 5.73 ± 1.19 for the visual tags. For diversity, usage tags received a rating of 6.79 ± 0.8, whereas the visual tags received 5.93 ± 0.73. This indicates that UBTer increases the overall relevance of the tags to the image and also performs better in terms of the diversity of the tags indicating the viability of UBTer.

4.2   Ground Truth Data Set

We utilized the dataset curated by Leong et al. [7] which contains 300 image-text pairs collected by issuing a query to Google Image API and processing one of the query results that has a significant amount of text around the images. Leong et al. [7] have also created a gold standard tag set based on manual annotations from 5 annotators via Amazon Mechanical Turk accepting annotations from annotators with approval rating > 98%. The annotators have suggested the tags about the image based on their understanding of the accompanying text. We used the Clarifai API [18] to generate the visual tags for all our experiments.

4.3   Metrics for Evaluation

Human annotations cannot be extended for a comprehensive evaluation of the tags. We therefore extend several existing metrics to measure different aspects of the tags which are described below.

The **term-significance** [10] is calculated as the significance of the tags to the textual content and is calculated by computing the Normalized Discounted Cumulative Gain (NDCG) over the term frequency of the tags from the usage content normalized based on the tag's inverse document frequency in a global corpus. The intuition here is to compute how important a tag is to the given context (usage) and normalize it with its "commonness" across a bigger corpus (as computed by the idf). We use Wikipedia as the bigger corpus similar to Leong et al. [7].

The term-significance metric purely tests the relevance of the tags to the usage content. To further capture the **tag relevance** of the tags to the gold standard tags and its overall **diversity**, we propose two additional metrics. To determine how relevant our tags are to the gold standard tags, we compute a weighted cosine similarity between the Word2Vec [12] representation of the extracted tags and the gold tags as given by,

$$sim = \frac{1}{N} \sum_i \frac{\sum_{a_j \in TopK(G_i, I_i)} cos(a_j, I_i)\gamma^j}{\sum_j \gamma^j}, \qquad (1)$$

where N is the number of tags generated for the images, I_i is the vector representation of the ith image tag and G_i is the set of all vector representations of the gold standard tags. The inner sum above computes a weighted average of the similarity between the generated tag and the most similar gold-standard tags. An average of the similarity can lead to higher relevance only when the tag is relevant to all human annotated tags. Alternatively, a max over the similarities can lead to high scores for tags even if they are similar to a single human tag. The parameter γ, $(0 \leq \gamma \leq 1)$ addresses both these scenarios via a similarity-ranked-decayed-weighted-average. The outer summation averages this measure between all the generated image tags and the gold standard tags.

Finally, for measuring the diversity in the tags, we use the **cophenet correlation coefficient** [17] (which is a measure of how faithfully a dendogram preserves the pairwise distances between the original un-modeled data points). We perform a hierarchical clustering on the tags based on their Word2Vec representation and compute the cophenet correlation coefficient as the diversity score. Cophenet correlation coefficient is then given by,

$$c = \frac{\sum_{i<j}(x(i,j) - \bar{x})(t(i,j) - \bar{t})}{\sqrt{[\sum_{i<j}(x(i,j) - \bar{x})^2][\sum_{i<j}(t(i,j) - \bar{t})^2]}} \qquad (2)$$

where, $x(i,j)$ is the distance between the ith and j^{th} tag. $t(i,j)$ is the height of the node at which the clusters corresponding to i^{th} and j^{th} clusters are first joined together. A higher value of the cophenet correlation coefficient indicates the presence of more significant clusters and hence more tag diversity.

4.4   Tagging Performance

To evaluate the proposed UBTer based tags, we compare it against the baseline algorithm in [7]. Leong et al. [7] propose 3 independent algorithms based on "Wikipedia Salience", "Flickr Picturability" and "Topic Modeling" to extract tags for an image from its accompanying textual content. In their experiments, the Wikipedia Salience based tagger was best performing in terms of the precision and recall. We used this algorithm as the baseline for our evaluations. We also compare the performance of our UBTer against the visual tagger in [18]. Figure 2 shows the Term Significance, Tag Relevance and Tag Diversity for the tags from [7,18] and UBTer.

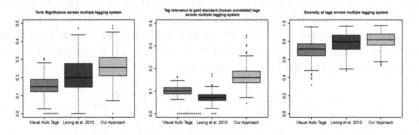

Fig. 2. Term Significance, Tag Relevance (Eq. 1) and Diversity (Eq. 2) for tags based on Clarifai [18], Wikipedia Salience [7] and UBTer

The term significance checks significance of the tags with respect to the accompanying text and hence text based taggers are expected to perform better in this measure. Along the expected lines, both the UBTer and the tagger by Leong et al. [7] perform better than the visual tagger. Between the text based taggers, the term significance is the best for UBTer indicating the superiority of the tags in capturing the local context.

The tags from UBTer are also more relevant/close to the human annotated tags based on the tag relevance (Eq. 1). A superior performance here indicate that UBTer captures the denotational aspects as well as the connotational aspects.

Capturing the connotational aspects of the images yields more diversity as indicated by the superior performances of both the text-based taggers on the scales of diversity. Here again, the tags from UBTer are marginally more diverse than the tags from Leong et al. [7].

4.5 Evaluation of Algorithmic Parameters

Finally, we independently evaluate the different parts of UBTer and their importance in extracting relevant and diverse tags capturing the image usage.

Local vs Global Context: In this experiment, we compare the local context captured by the node importance (Sect. 3.2) against the combined context captured in UBTer. We extract the top tags based on their node importance score and compare it against the UBTer tags.

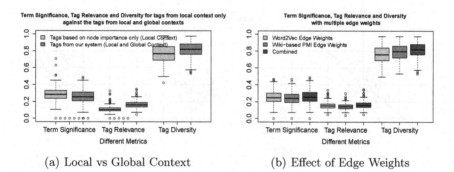

(a) Local vs Global Context   (b) Effect of Edge Weights

Fig. 3. Term Significance, Tag Relevance (Eq. 1) and Diversity (Eq. 2) for tags for different algorithmic parameters. (a) Compares the tags extracted solely based on Node Importance against the tags from UBTer (where the local and global context of the tags are jointly accounted for). (b) Compares the effects of different edge weighting mechanisms on the tagging performance

Figure 3(a) compares the Term Significance, Tag Relevance (Eq. 1) and Tag Diversity for the two cases. The term significance of the tags based on the local context with an average of 0.275 is marginally better than the term significance of UBTer (average at 0.26). Since the term significance captures the local importance of the tags in the accompanying text, hence the tags from local context is expected to be better here. However, the overall tag relevance (average of 0.1090 for local context against 0.1654 for the combined context) and tag diversity (average of 0.7554 for local context against 0.0.8155 for the combined context) is better with the combined approach since it accounts for the global relationship between the tags as well as similarity of connotational tags with visual tags. Hence better tags without compromising much on the term significance (since the difference between the two methods is not significant) is derived.

Effect of Edge Weights: In the next experiment, we compare the term significance, tag relevance and tag diversity among the edge weighting mechanisms based on Word2Vec, Wikipedia and the combined metric defined in Sect. 3.3.

From Fig. 3(b), it can be seen that while Word2Vec performs marginally better than the Wikipedia based relationship on the scales of term significance (average of 0.2516 for Word2Vec based metric against the 0.2398 average for the Wikipedia based metric) and tag relevance (average of 0.1567 for Word2Vec based metric against the 0.1451 average for the Wikipedia based metric). In terms of overall tag diversity, Wikipedia based metric is marginally better than Word2Vec (average of 0.7897 for Wikipedia based metric against the 0.7617 average for the Word2Vec based metric). This could perhaps be because Wikipedia includes more entities than the Google News Corpus on which the Word2Vec were trained, and hence aid in the extraction of diverse tags. Note that the combined approach yields the best tags across all metrics.

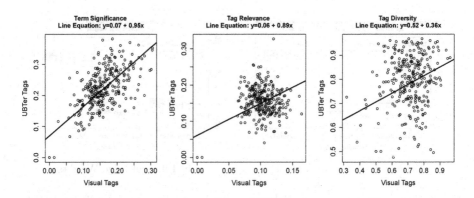

Fig. 4. Correlation between the quality of visual tags and the tags from UBTer

Effect of Visual Tag Quality: We finally compare the correlation between the quality of the visual tags and the tags from UBTer.

Figure 4 shows the correlation between the two sets of tags on the scales of Term Significance, Tag Relevance and Tag Diversity. It can be seen that there is a strong dependence of the term significance and relevance of UBTer tags with the visual tags as indicated by the slopes of 0.95 and 0.89 respectively of the corresponding line fits. This is expected since the algorithm starts the random walk from the visual tags and hence the output tag quality is directly dependent on the quality of visual tags. However the tag diversity is less dependent on the visual tags, since the diversity of the output tags is obtained more from the accompanying text than from the visual tags indicated by a lower slope of the corresponding line (0.36).

5   Conclusion

In this paper, we have proposed a novel system - UBTer to enhance the tags of an image by capturing its usage. Capturing usage through tags is not straightforward as majority of the tags describing the neighboring text of an image don't pertain to the image - our approach gleans out the relevant tags. This is done first through understanding the importance of the tag in local context (we conduct sophisticated dependency test to compute the importance) and then derive the inter-tag relationship (we use Word2Vec and Wikipedia-co-occurrence) and finally run a biased random walk to shortlist relevant tags. The tags thus obtained outperform the state-of-the art systems in the lights of several quality metrics capturing the relevance and diversity of the tags. Such a tagging system will serve well to improve the image retrieval and recommendation systems by effectively expressing the user's context.

References

1. Banarescu, L., Bonial, C., Cai, S., Georgescu, M., Griffitt, K., Hermjakob, U., Knight, K., Koehn, P., Palmer, M., Schneider, N.: Abstract meaning representation (AMR) 1.0 specification. In: Conference on Empirical Methods in Natural Language Processing. ACL (2012)
2. Chen, D., Manning, C.D.: A fast and accurate dependency parser using neural networks. In: Conference on Empirical Methods in Natural Language Processing. ACL (2014)
3. Guillaumin, M., Mensink, T., Verbeek, J., Schmid, C.: Tagprop: discriminative metric learning in nearest neighbor models for image auto-annotation. In: IEEE International Conference on Computer Vision (2009)
4. Hoffart, J., Yosef, M.A., Bordino, I., Fürstenau, H., Pinkal, M., Spaniol, M., Taneva, B., Thater, S., Weikum, G.: Robust disambiguation of named entities in text. In: Conference on Empirical Methods in Natural Language Processing. ACL (2011)
5. Kottur, S., Vedantam, R., Moura, J.M., Parikh, D.: Visual word2vec (vis-w2v): learning visually grounded word embeddings using abstract scenes. arXiv preprint arXiv:1511.07067 (2015)
6. Kuzey, E., Setty, V., Strötgen, J., Weikum, G.: As time goes by: comprehensive tagging of textual phrases with temporal scopes. In: International Conference on World Wide Web. ACM (2016)
7. Leong, C.W., Mihalcea, R., Hassan, S.: Text mining for automatic image tagging. In: International Conference on Computational Linguistics. ACL (2010)
8. Li, X., Uricchio, T., Ballan, L., Bertini, M., Snoek, C.G., Del Bimbo, A.: Image tag assignment, refinement and retrieval. In: ACM International Conference on Multimedia (2015)

9. Lieberman, M.D., Samet, H.: Adaptive context features for toponym resolution in streaming news. In: ACM SIGIR Conference on Research and Development in Information Retrieval. ACM (2012)
10. Lu, Y.T., Yu, S.I., Chang, T.C., Hsu, J.Y.J.: A content-based method to enhance tag recommendation. In: IJCAI (2009)
11. Manning, C.D., Surdeanu, M., Bauer, J., Finkel, J.R., Bethard, S., McClosky, D.: The stanford corenlp natural language processing toolkit. In: ACL (System Demonstrations) (2014)
12. Mikolov, T., Sutskever, I., Chen, K., Corrado, G.S., Dean, J.: Distributed representations of words and phrases and their compositionality. In: Advances in Neural Information Processing Systems (2013)
13. Nallapati, R., Feng, A., Peng, F., Allan, J.: Event threading within news topics. In: ACM International Conference on Information and Knowledge Management. ACM (2004)
14. Ramanathan, V., Li, C., Deng, J., Han, W., Li, Z., Gu, K., Song, Y., Bengio, S., Rossenberg, C., Fei-Fei, L.: Learning semantic relationships for better action retrieval in images. In: IEEE Conference on Computer Vision and Pattern Recognition (CVPR) (2015)
15. Sarkar, P., Moore, A.W.: Random walks in social networks and their applications: a survey. In: Aggarwal, C.C. (ed.) Social Network Data Analytics, pp. 43–77. Springer, Heidelberg (2011)
16. Shahaf, D., Guestrin, C.: Connecting the dots between news articles. In: ACM SIGKDD International Conference on Knowledge Discovery and Data Mining. ACM (2010)
17. Sokal, R.R., Rohlf, F.J.: The comparison of dendrograms by objective methods. Taxon 11, 33–40 (1962)
18. Sood, G.: clarifai: R Client for the Clarifai API (2016). R package version 0.4.0
19. Suchanek, F.M., Kasneci, G., Weikum, G.: Yago: a core of semantic knowledge. In: International Conference on World Wide Web. ACM (2007)
20. Tandon, N., de Melo, G., De, A., Weikum, G.: Knowlywood: mining activity knowledge from Hollywood narratives. In: International Conference on Information and Knowledge Management. ACM (2015)
21. Turney, P.D.: Thumbs up or thumbs down?: semantic orientation applied to unsupervised classification of reviews. In: 40th Annual Meeting on Association for Computational Linguistics (2002)
22. Xie, L., He, X.: Picture tags and world knowledge: learning tag relations from visual semantic sources. In: ACM International Conference on Multimedia (2013)
23. Yang, Y., Ault, T., Pierce, T., Lattimer, C.W.: Improving text categorization methods for event tracking. In: ACM SIGIR Conference on Research and Development in Information Retrieval. ACM (2000)

Edge Role Discovery via Higher-Order Structures

Nesreen K. Ahmed[1]([✉]), Ryan A. Rossi[2], Theodore L. Willke[1], and Rong Zhou[2]

[1] Intel Labs, Santa Clara, USA
{nesreen.k.ahmed,ted.willke}@intel.com
[2] Palo Alto Research Center (Xerox PARC), Palo Alto, USA
{rrossi,rzhou}@parc.com

Abstract. Previous work in network analysis has focused on modeling the roles of nodes in graphs. In this paper, we introduce edge role discovery and propose a framework for learning and extracting edge roles from large graphs. We also propose a general class of higher-order role models that leverage network motifs. This leads us to develop a novel edge feature learning approach for role discovery that begins with higher-order network motifs and automatically learns deeper edge features. All techniques are parallelized and shown to scale well. They are also efficient with a time complexity of $\mathcal{O}(|E|)$. The experiments demonstrate the effectiveness of our model for a variety of ML tasks such as improving classification and dynamic network analysis.

Keywords: Role discovery · Edge roles · Higher-order network analysis · Graphlets · Network motifs · Latent space models · Transfer learning

1 Introduction

In the traditional graph-based sense, roles represent node-level connectivity patterns such as star-center, star-edge nodes, near-cliques or nodes that act as bridges to different regions of the graph. Intuitively, two nodes belong to the same role if they are "similar" in the sense of graph structure. Our proposed research will broaden the framework for defining, discovering and learning network roles, by drastically increasing the degree of usefulness of the information embedded within rich graphs.

Recently, role discovery has become increasingly important for a variety of application and problem domains [5,6,9,15,19,28] including descriptive network modeling [30], classification [14], anomaly detection [30], and exploratory analysis [29]. See [28] for other applications. Despite the importance of role discovery, existing work has only focused on discovering node roles (*e.g.*, see [5,7,11,23]). We posit that discovering the roles of edges may be fundamentally more important and able to capture, represent, and summarize the key behavioral roles in the network better than existing methods that have been limited to learning only the roles of nodes in the graph. For instance, a person with malicious intent may appear normal by maintaining the vast majority of relationships and communications with

© Springer International Publishing AG 2017
J. Kim et al. (Eds.): PAKDD 2017, Part I, LNAI 10234, pp. 291–303, 2017.
DOI: 10.1007/978-3-319-57454-7_23

individuals that play normal roles in society. In this situation, techniques that reveal the role semantics of nodes would have difficulty detecting such malicious behavior since most edges are normal. However, modeling the roles (functional semantics, intent) of individual edges (relationships, communications) in the rich graph would improve our ability to identify, detect, and predict this type of malicious activity since we are modeling it directly. Nevertheless, existing work also have many other limitations, which significantly reduces the practical utility of such methods in real-world networks. One such example is that the existing work has been limited to mainly simple degree and egonet features [14,30], see [28] for other possibilities. Instead, we leverage higher-order network motifs (induced subgraphs) of size $k \in \{3,4,\ldots\}$ computed from [1,2] and other graph parameters such as the largest clique in a node (or edge) neighborhood, triangle core number, as well as the neighborhood chromatic, among other efficient and highly discriminative graph features. The main contributions are as follows:

- **Edge role discovery:** This work introduces the problem of edge role discovery and proposes a computational framework for learning and modeling edge roles in both static and dynamic networks.
- **Higher-order role discovery models:** Proposed a general class of higher-order role models that leverage network motifs and higher-order network features for learning both node and edge roles. This work is also the first to use higher-order network motifs[1] for role discovery in general.
- **Edge feature representation learning:** Proposed a novel deep graph representation learning framework that begins with higher-order network motifs and automatically learns deeper edge features.
- **Efficient and scalable:** The proposed feature and role discovery methods are efficient (linear in the number of edges) for modeling large networks. In addition, all methods are parallelized and shown to scale to massive networks.

2 Related Work

Related research is categorized into the following parts: (1) role discovery, (2) higher-order network analysis, (3) graph representation learning, (4) sparse graph features, and (5) parallel role discovery.

Role Discovery: There has been a lot of work on role discovery in general [5,6,9,14,15,19,28,30]. However, all existing approaches have focused on learning roles of *nodes* in graphs. See [28] for a recent survey on role discovery. In contrast, this work introduces the problem of *edge role discovery* and presents a computational framework for learning and extracting edge roles from large networks. Additional key differences are as follows: (1) our approach uses higher-order graphlets for discovering more intuitive and meaningful roles, and (2) the proposed role methods are parallelized and thus able to scale to extremely large real-world networks. Moreover, our approach supports graphs that are directed/undirected/bipartite, attributed, typed/heterogeneous, and signed.

---

[1] 4-vertex induced subgraphs (graphlets, motifs) and larger.

Higher-Order Network Analysis: Small induced subgraphs called graphlets (motifs) have recently been used for graph classification [36], link prediction [25], and visualization and exploratory analysis [1]. However, this work focuses on using graphlets for learning and extracting more useful and meaningful roles from large networks. Furthermore, previous feature-based role methods have been learned based on simple degree and egonet-based features. Thus, another contribution of this work is the use of higher-order network motifs (based on small k-vertex subgraph patterns called graphlets) for role discovery of nodes and edges — a key and fundamental difference between existing work.

Graph Representation Learning: While a lot of work has engineered features by hand (or manually selected them) for various ML applications, not much work has been done on learning a set of useful features automatically. Our approach is different from previous work in four fundamental ways: (1) the proposed approach learns important and useful *edge features* automatically, whereas existing approaches were designed for learning *node features*, (2) our approach is space-efficient as it learns sparse features and fast/efficient with a time complexity that is linear in the number of edges. (3) an efficient parallel implementation with strong scaling results as shown in Sect. 4 and thus well-suited for large-scale networks, and finally, (4) most graph representation learning methods were used in SRL systems for classification [12], whereas we use the proposed approach for edge role discovery.

Sparse Graph Features: We also make a significant contribution in terms of space-efficient role discovery. In particular, this work proposes the first practical space-efficient approach for feature-based role discovery by learning sparse graph features automatically. In contrast, feature-based node role methods [14,30] store hundreds/thousands of dense features in memory, which is impractical for any relatively large network, *e.g.*, they require more than 2TB of memory for a 500M node graph with 1,000 features.

Parallel Role Discovery: The existing role discovery methods are sequential, despite the practical importance of parallel role discovery algorithms that scale to massive real-world networks. This work is the first parallel role discovery approach. Furthermore, the proposed edge feature learning techniques are also parallelized and designed to be both efficient in terms of space and communication.

3   Framework

This section introduces edge role discovery along with higher-order edge role models and a computational framework for learning and extracting roles based on higher-order structures.

Extracting Higher-Order Graphlet Features: Given the graph $G = (V, E)$, we first decomposes G into its smaller subgraph components called graphlets (motifs). For this, we use parallel edge-centric graphlet decomposition methods such as [1] to compute a variety of graphlet edge features of size $k = \{3, 4, \ldots\}$ (Algorithm 1 Line 2). Moreover, our approach canleverage directed,

Algorithm 1. A framework for learning deep edge feature representations from graphs

Input:

    a directed and possibly weighted/labeled/attributed graph $G = (V, E)$

    a set of relational edge kernels/operators Φ

    a feature similarity function $\mathbb{K}\langle \cdot, \cdot \rangle$

    an upper bound on the number of feature layers to learn T

    a feature similarity threshold λ, and bin size α, $0 \leq \alpha \leq 1$

1: Set $\tau \leftarrow 1$

2: **parallel for each** $e_i \in E$ **and** subgraph $H_k \in \mathcal{H}$ **do**

3:     Compute X_{ik}, the number of instances of graphlet H_k that contain edge $e_i \in E$

4: Given G and $\mathbf{X}$, compute in/out/total/weighted *edge egonet* and *edge degree* features (feature layer $\mathcal{F}_1$ which includes the graphlet features as well). Append these to $\mathbf{X}$ and set $\mathcal{F} \leftarrow \mathcal{F}_1$

5: **repeat**   ▷ feature layers $\mathcal{F}_\tau$ for $\tau = 1, 2, ..., $ T

6:     **if** $\tau > 1$ **then**

7:         Derive candidate features using the set of relational operators Φ over each of the novel features $f_i \in \mathcal{F}_{\tau-1}$ learned in previous layers. Append the candidate features to $\mathbf{X}$ and the feature definitions to $\mathcal{F}_\tau$.

8:     For each feature $f_i \in \mathcal{F}_\tau$, sort the feature values in ascending order and then map the feature values using logarithmic binning (with a bin size of α). Given feature $f_i \in \mathcal{F}_\tau$, we set the αm edges with smallest feature values to 0, then α edges remaining are set to 1, and so on.

9:     Let $\mathcal{G}_F = (V_F, E_F)$ be the initial feature graph for feature layer $\mathcal{F}_\tau$ where V_F is the set of features from $\mathcal{F} \cup \mathcal{F}_\tau$ and $E_F = \varnothing$

10:     **parallel for each** edge feature $f_i \in \mathcal{F}_\tau$ **do**

11:         **for each** edge feature $f_j \in (\mathcal{F}_\tau \cup \mathcal{F})$ **do**

12:             **if** $\mathbb{K}(\mathbf{x}_i, \mathbf{x}_j) \geq \lambda$ **then**

13:                 Add edge (f_i, f_j) to E_F

14:     Partition the feature graph $\mathcal{G}_F$ using connected components $\mathcal{C} = \{\mathcal{C}_1, \mathcal{C}_2, ...\}$

15:     **parallel for each** $\mathcal{C}_k \in \mathcal{C}$ **do**   ▷ Prune features

16:         Find the earliest feature f_i s.t. $\forall f_j \in \mathcal{C}_k : i < j$.

17:         Remove $\mathcal{C}_k$ from $\mathcal{F}_\tau$ and set $\mathcal{F}_\tau \leftarrow \mathcal{F}_\tau \cup \{f_i\}$

18:     Discard features from $\mathbf{X}$ that were pruned (not in $\mathcal{F}_\tau$) and set $\mathcal{F} \leftarrow \mathcal{F} \cup \mathcal{F}_\tau$

19:     Set $\tau \leftarrow \tau + 1$ and initialize $\mathcal{F}_\tau$ to $\varnothing$ for next feature layer

20: **until** feature layer $\mathcal{F}_{\tau-1} = \varnothing$ (no new features emerged) **or** max layers reached ($\tau = $ T)

21: **return** $\mathbf{X}$ and the set of feature definitions $\mathcal{F}$

---

undirected, and weighted/typed graphlet counts (among other useful and discriminative graphlet edge statistics) using either exact or estimation methods. These graphlet features are then used to learn deeper higher-order edge features (see below for further details).

Edge Feature Representation Learning Framework: This section presents our deep edge feature representation learning framework (Algorithm 1). Recall that our approach leverages the previous higher-order graphlet counts as a basis for learning deeper and more discriminative higher-order edge features (Line 2–3). Next, primitive edge features are computed in Line 4, including in/out/total/weighted *edge egonet* and *edge degree* features. After computing the initial feature layer $\mathcal{F}_1$ (Line 2–4), redundant features are pruned (Line 5–20). The framework proceeds to learn a set of feature layers where each successive layer represents increasingly deeper higher-order edge features (Line 5–20), *i.e.*, $\mathcal{F}_1 < \mathcal{F}_2 < \cdots < \mathcal{F}_\tau$ such that if $i < j$ then $\mathcal{F}_j$ is said to be a deeper layer than $\mathcal{F}_i$.

The feature layers $\mathcal{F}_2, \mathcal{F}_3, \cdots, \mathcal{F}_\tau$ are learned as follows (Line 5–20): For each layer $\mathcal{F}_\tau$, we first construct and search candidate features using the set of relational edge feature operators $\boldsymbol{\Phi}$ (See Line 7), which include mean, sum, product, min, max, variance, L1, L2, and even parameterized relational kernels based on RBF, polynomial functions, among others. See Table 1 for a few examples. Now, we compute the similarity between all pairs of features and prune edges between features that are *not* significantly correlated (Line 9–13): $E_F = \{(f_i, f_j) \mid \forall (f_i, f_j) \in |\mathcal{F}| \times |\mathcal{F}| \text{ s.t. } \mathbb{K}(f_i, f_j) > \lambda\}$. This process results in a feature similarity graph where large edge weights indicate strong similarity/correlation between two features. Now, the feature similarity graph $\mathcal{G}_F$ from Line 9–13 is used to prune all redundant edge features from $\mathcal{F}_\tau$. Features are pruned by first partitioning the feature graph (Line 14) using connected components, though our approach is flexible and allows other possibilities (*e.g.*, largest clique). Intuitively, each connected component is a set of redundant edge features since edges in $\mathcal{G}_F$ represent strong dependencies between features. For each connected component $\mathcal{C}_k \in \mathcal{C}$ (Line 15–17), we identify the earliest feature in $\mathcal{C}_k = \{..., f_i, ..., f_j, ...\}$ (Line 16) and remove all others from $\mathcal{F}_\tau$ (Line 17). After pruning the feature layer $\mathcal{F}_\tau$, Line 18 ensures the pruned features are removed from $\mathbf{X}$ and updates the set of edge features learned thus far by setting $\mathcal{F} \leftarrow \mathcal{F} \cup \mathcal{F}_\tau$. Line 19 increments τ and set $\mathcal{F}_\tau \leftarrow \varnothing$. Finally, Line 20 checks for convergence, and if the stopping criterion is not satisfied, then the approach tries to learn an additional feature layer (Line 5–20).

Learning Higher-Order Edge Roles: Let $\mathbf{X} = [x_{ij}] \in \mathbb{R}^{m \times f}$ be an edge feature matrix with m rows representing edges and f columns representing higher-order graph features learned from our edge feature representation learning approach. Given $\mathbf{X} \in \mathbb{R}^{m \times f}$, the edge role discovery optimization problem is to find $\mathbf{U} \in \mathbb{R}^{m \times r}$ and

Table 1. Relational edge feat. operators

Operator		Definition
Hadamard	⊞	$\prod\limits_{e_j \in \Gamma(e_i)} f_k(e_j)$
Mean	⊡	$\frac{1}{d_i} \sum\limits_{e_j \in \Gamma(e_i)} f_k(e_j)$
Sum	⊗	$\sum\limits_{e_j \in \Gamma(e_i)} f_k(e_j)$
Wt. L_p	$\|\cdot\|_{\tilde{p}}$	$\sum\limits_{e_j \in \Gamma(e_i)} |f_k(e_i) - f_k(e_j)|^p$

$\mathbf{V} \in \mathbb{R}^{f \times r}$ where $r \ll \min(m, f)$ such that the product of two lower rank matrices $\mathbf{U}$ and $\mathbf{V}^T$ minimizes the divergence between $\mathbf{X}$ and $\mathbf{X}' = \mathbf{U}\mathbf{V}^T$. Intuitively, $\mathbf{U} \in \mathbb{R}^{m \times r}$ represents the latent *role mixed-memberships* of the edges whereas $\mathbf{V} \in \mathbb{R}^{f \times r}$ represents the contributions of the features with respect to each of the roles. Each row $\mathbf{u}_i^T \in \mathbb{R}^r$ of $\mathbf{U}$ can be interpreted as a low dimensional rank-r embedding of the ith *edge* in $\mathbf{X}$. Alternatively, each row $\mathbf{v}_j^T \in \mathbb{R}^r$ of $\mathbf{V}$ represents a r-dimensional role embedding of the jth feature in $\mathbf{X}$ using the same low rank-r dimensional space. Also, $\mathbf{u}_k \in \mathbb{R}^m$ is the kth column representing a "latent feature" of $\mathbf{U}$ and similarly $\mathbf{v}_k \in \mathbb{R}^f$ is the kth column of $\mathbf{V}$. For learning higher-order edge roles, we solve:

$$\operatorname*{arg\,min}_{(\mathbf{U},\mathbf{V})\in\mathcal{C}} \left\{ \mathbb{D}_\phi(\mathbf{X}\| \mathbf{U}\mathbf{V}^T) + \mathcal{R}(\mathbf{U},\mathbf{V}) \right\} \tag{1}$$

where $\mathbb{D}_\phi(\mathbf{X}\|\mathbf{U}\mathbf{V}^T)$ is an arbitrary Bregman divergence [10] between $\mathbf{X}$ and $\mathbf{U}\mathbf{V}^T$. Furthermore, the optimization problem in (1) imposes hard constraints $\mathcal{C}$ on $\mathbf{U}$ and $\mathbf{V}$ such as non-negativity constraints $\mathbf{U}, \mathbf{V} \geq 0$ and $\mathcal{R}(\mathbf{U},\mathbf{V})$ is a regularization penalty. In this work, we mainly focus on solving $\mathbb{D}_\phi(\mathbf{X}\|\mathbf{U}\mathbf{V}^T)$ under non-negativity constraints:

$$\operatorname*{arg\,min}_{\mathbf{U}\geq 0, \mathbf{V}\geq 0} \left\{ \mathbb{D}_\phi(\mathbf{X}\| \mathbf{U}\mathbf{V}^T) + \mathcal{R}(\mathbf{U},\mathbf{V}) \right\} \tag{2}$$

Given the edge feature matrix $\mathbf{X} \in \mathbb{R}^{m\times f}$, the edge role discovery problem is to find $\mathbf{U} \in \mathbb{R}^{m\times r}$ and $\mathbf{V} \in \mathbb{R}^{f\times r}$ such that $\mathbf{X} \approx \mathbf{X}' = \mathbf{U}\mathbf{V}^T$. To measure the quality of our edge mixed membership model, we use Bregman divergences:

$$\sum_{ij} \mathbb{D}_\phi(x_{ij}\|x'_{ij}) = \sum_{ij} \left(\phi(x_{ij}) - \phi(x'_{ij}) - \ell(x_{ij},x'_{ij}) \right) \tag{3}$$

where ϕ is a univariate smooth convex function and $\ell(x_{ij}, x'_{ij}) = \nabla\phi(x'_{ij})(x_{ij} - x'_{ij})$ where $\nabla^p\phi(x)$ is the p-order derivative operator of ϕ at x. Furthermore, let $\mathbf{X} - \mathbf{U}\mathbf{V}^T = \mathbf{X}^{(k)} - \mathbf{u}_k\mathbf{v}_k^T$ denote the residual term in the approximation $\mathbf{X} \approx \mathbf{X}' = \mathbf{U}\mathbf{V}^T$ where $\mathbf{X}^{(k)}$ is the k-residual matrix defined as:

$$\mathbf{X}^{(k)} = \mathbf{X} - \sum_{h\neq k} \mathbf{u}_h\mathbf{v}_h^T = \mathbf{X} - \mathbf{U}\mathbf{V}^T + \mathbf{u}_k\mathbf{v}_k^T, \quad \text{for } k = 1,\ldots,r \tag{4}$$

We use a fast *scalar block coordinate descent approach* that easily generalizes for heterogeneous networks [32]. The approach considers a single element in $\mathbf{U}$ and $\mathbf{V}$ as a block in the block coordinate descent framework. Replacing $\phi(y)$ with the corresponding expression from Table 2 gives rise to a fast algorithm for each Bregman divergence. Table 2 gives the updates for Frobenius norm (Fro.), KL-divergence (KL), and Itakura-Saito divergence (IS). Note that Beta divergence and many others are also easily adapted for our higher-order edge role discovery framework.

Table 2. Role divergences and update rules

	$\phi(y)$	$\nabla^2\phi(y)$	$\mathbb{D}_\phi(x\|x')$	Update ($v_{jk} =$)
Fro.	$y^2/2$	1	$(x - x')^2/2$	$\dfrac{\sum_{i=1}^{m} x_{ij}^{(k)} u_{ik}}{\sum_{i=1}^{m} u_{ik}u_{ik}}$
KL	$y\log y$	$1/y$	$x\log\frac{x}{x'} - x + x'$	$\dfrac{\sum_{i=1}^{m} x_{ij}^{(k)} u_{ik}/x'_{ij}}{\sum_{i=1}^{m} u_{ik}u_{ik}/x'_{ij}}$
IS	$-\log y$	$1/y^2$	$\frac{x}{x'} - \log\frac{x}{x'}$	$\dfrac{\sum_{i=1}^{m} x_{ij}^{(k)} u_{ik}/x'^{2}_{ij}}{\sum_{i=1}^{m} u_{ik}u_{ik}/x'^{2}_{ij}}$

Model Selection: In this section, we introduce an approach that automatically learns the appropriate role mixed-membership model. The approach is based on the Minimum Description Length (MDL) [13,26] principle; a practical formalization of Kolmogorov complexity [17]. More formally, we find the model $M_\star = (\mathbf{V}_r, \mathbf{U}_r)$ that leads to the best compression by solving:

$$M_\star = \operatorname*{arg\,min}_{M\in\mathcal{M}} \mathcal{L}(M) + \mathcal{L}(\mathbf{X}\,|\,M) \tag{5}$$

where $\mathcal{M}$ is the model space, $M_\star$ is the model given by the solving the above minimization problem, and $\mathcal{L}(M)$ as the number of bits required to encode M using code Ω, which we refer to as the description length of M with respect to Ω. Recall that MDL requires a lossless encoding. Therefore, to reconstruct $\mathbf{X}$ *exactly* from $M = (\mathbf{U}_r, \mathbf{V}_r)$ we must explicitly encode the error $\mathbf{E}$ such that $\mathbf{X} = \mathbf{U}_r \mathbf{V}_r^T + \mathbf{E}$. Hence, the total compressed size of $M = (\mathbf{U}_r, \mathbf{V}_r)$ with $M \in \mathcal{M}$ is simply $\mathcal{L}(X \mid M) = \mathcal{L}(M) + \mathcal{L}(\mathbf{E})$. Given a role mixed-membership model with r roles $M = (\mathbf{U}_r, \mathbf{V}_r) \in \mathcal{M}$, the description length is decomposed into: (1) bits required to describe the model, and (2) cost of describing the approximation errors $\mathbf{X} - \mathbf{X}_r$ where $\mathbf{X}_r = \mathbf{U}_r \mathbf{V}_r^T$ is the rank-r approximation of $\mathbf{X}$,

$$\mathbf{U}_r = \begin{bmatrix} \mathbf{u}_1 & \mathbf{u}_2 & \cdots & \mathbf{u}_r \end{bmatrix} \in \mathbb{R}^{m \times r}, \text{ and } \mathbf{V}_r = \begin{bmatrix} \mathbf{v}_1 & \mathbf{v}_2 & \cdots & \mathbf{v}_r \end{bmatrix} \in \mathbb{R}^{f \times r} \quad (6)$$

The model $M_\star$ is the model $M \in \mathcal{M}$ that minimizes the total description length: the model description cost X and the cost of correcting the errors of our model. Let $|\mathbf{U}|$ and $|\mathbf{V}|$ denote the number of nonzeros in $\mathbf{U}$ and $\mathbf{V}$, respectively. Thus, the model description cost of M is: $\kappa r(|\mathbf{U}| + |\mathbf{V}|)$ where κ is the bits per value. Similarly, if $\mathbf{U}$ and $\mathbf{V}$ are dense, then the model description cost is simply $\kappa r(m + f)$ where m and f are the number of edges and features, respectively. Assuming errors are non-uniformly distributed, one possibility is to use KL divergence (see Table 2) for the error description cost[2]. The cost of correcting a single element in the approximation is $\mathbb{D}_\phi(x \| x') = x \log \frac{x}{x'} - x + x'$ (assuming KL-divergence), and thus, the total reconstruction cost is:

$$\mathbb{D}_\phi(\mathbf{X} \| \mathbf{X}') = \sum_{ij} X_{ij} \log \frac{X_{ij}}{X'_{ij}} - X_{ij} + X'_{ij} \quad (7)$$

where $\mathbf{X}' = \mathbf{U}\mathbf{V}^T \in \mathbb{R}^{m \times f}$. Other possibilities are given in Table 2. The above assumes a particular representation scheme for encoding the models and data. Recall that the optimal code assigns $\log_2 p_i$ bits to encode a message [34]. Lloyd-Max quantization [18, 22] with Huffman codes [16, 35] are used to compress the model and data [8, 24]. Notice that we require only the length of the description using the above encoding scheme, and thus we do not need to materialize the codes themselves. This leads to the improved model description cost: $\bar{\kappa} r(|\mathbf{U}| + |\mathbf{V}|)$ where $\bar{\kappa}$ is the mean bits required to encode each value[3]. In general, the higher-order (edge) role discovery framework can easily leverage other model selection techniques such as AIC [4] and BIC [33].

4   Experiments

This section investigates the effectiveness and scalability of the proposed edge role discovery framework (Sect. 3). All network data is available at NR [27].

---

[2] The representation cost of correcting approximation errors.

[3] Note $\log_2(m)$ quantization bins are used.

Higher-Order Model Selection: We now validate our model learning approach. Figure 1 demonstrates the effectiveness of our approach for automatically selecting the "best" model from the space of models expressed in the framework (Sect. 3). In particular, our approach finds the best model with $r = 18$ roles by minimizing the description length (in bits)[4]. As expected, the model description cost is inversely proportional to the error description cost. We also demonstrate the efficiency of our approach in Fig. 2. Furthermore, Fig. 4 demonstrates the impact on the learning time, number of novel features discovered, and their sparsity, as the tolerance (ε) and bin size (α) varies.

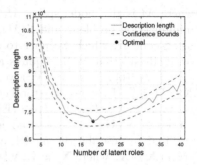

Fig. 1. The valley identifies the correct number of latent roles.

Modeling Dynamic Networks: In this section, we investigate the Enron email communication networks using the *higher-order dynamic edge role mixed-membership model*. The Enron email data consists of 151 Enron employees whom have sent 50.5k emails to other Enron employees over a 3 year period. The email communications are from 05/11/1999 to 06/21/2002. For learning we use only the first year of emails. A dynamic network $\{G_t\}_{t=1}^T$ is constructed from the remaining email communications (approximately 2 years) where each snapshot graph G_t, $t = 1, \ldots, T$ represents a month of communications. Interestingly, our higher-order dynamic *node* role mixed-membership model has 5 latent roles, whereas we learn 18 roles using the *edge* role model. Evolving edge and node mixed-

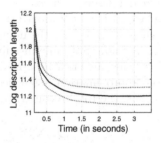

Fig. 2. Runtime of our edge role model selection. The curve is the average over 50 experiments and the dotted lines represent three standard deviations. The result reported above is from a laptop with a single core.

memberships from the Enron email communication network are shown in Fig. 3. The set of edges and nodes visualized in Fig. 3 are selected using the difference entropy rank (defined below) and correspond to the edges and nodes with largest difference entropy rank **d**. The first role in Fig. 3 represents inactivity (dark blue). The above empirical results suggest that edge roles are superior to node roles in three fundamental ways: (1) Edge roles reveal novel behavioral characteristics that are not captured by the node role models. We posit that these novel behavioral roles are intrinsic to the edge semantics (which represent communications in Fig. 3). (2) Roles learned on the edges represent behavioral characteristics at a much lower-level of granularity than those learned on nodes. (3) Edge roles are better at modeling dynamic/temporal networks and avoid

---

[4] We note that MDL is used in Fig. 1, though AIC/BIC gave similar results.

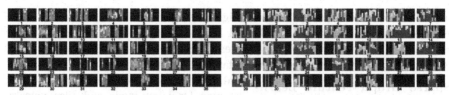

(a) Evolving *edge role* memberships   (b) Evolving *node role* memberships

Fig. 3. Temporal changes in the edge and node mixed-membership vectors. The horizontal axes of each subplot is time, whereas the vertical axes represent the components of each mixed-membership vector. Roles are represented by different colors. (Color figure online)

t/b	0.5	0.6	0.7	0.8	0.9	t/b	0.5	0.6	0.7	0.8	0.9	t/b	0.5	0.6	0.7	0.8	0.9
0.01	1.48	0.95	0.57	0.47	0.41	0.01	327	149	81	46	26	0.01	0.151	0.158	0.136	0.097	0.077
0.05	1.03	0.55	0.48	0.46	0.45	0.05	168	73	48	31	18	0.05	0.23	0.209	0.169	0.111	0.084
0.1	0.72	0.57	0.54	0.51	0.48	0.1	111	53	42	26	18	0.1	0.235	0.23	0.186	0.133	0.084
0.2	0.78	0.58	0.55	0.52	0.49	0.2	94	49	36	24	18	0.2	0.24	0.223	0.222	0.143	0.084
0.5	0.58	0.56	0.54	0.6	0.56	0.5	39	33	30	21	16	0.5	0.319	0.276	0.242	0.158	0.094

(a) Learning time   (b) Num. features learned   (c) Sparsity of features

Fig. 4. Impact on the learning time, number of features, and their sparsity, as the tolerance ε (rows) and bin size α (columns) varies.

many of the unrealistic assumptions that lie at the heart of dynamic node role mixed-membership models.

We define $\mathbf{d} = \max_t H(\mathbf{u}_t) - \min_t H(\mathbf{u}_t)$ as the difference entropy rank where $H(\mathbf{u}_t) = -\mathbf{u}_t \cdot \log(\mathbf{u}_t)$ and $\mathbf{u}_t$ is the r-dimensional mixed-membership vector for an edge (or node) at time t. Using the difference entropy rank, we are able to reveal important communications between key players involved in the Enron Scandal, such as Kenneth Lay, Jeffrey Skilling, and Louise Kitchen. In particular, anomalous relationships between these individuals appear in the top anomalies from the difference rank. Notice that when node roles are used for identifying dynamic anomalies in the graph, we are only provided with potentially malicious employees, whereas using edge roles naturally allow us to not only detect the key malicious individuals involved, but also the important relationships between them, which can be used for further analysis, among other possibilities. Many results are removed for brevity.

Fig. 5. Edge and node roles for ca-netscience. Link color represents the edge role and node color indicates the corresponding node role. (Color figure online)

Exploratory Analysis: Figure 5 visualizes the node and edge roles learned for ca-netscience. While our higher-order role edge discovery method learns a stochastic r-dimensional vector for each edge (and/or node) representing the individual role memberships, Fig. 5 assigns a single role to each link and node, *i.e.*, the role with maximum likelihood $k_\star \leftarrow \arg\max_k u_{ik}$. The higher-order edge and node roles from Fig. 5 are clearly meaningful. For instance, the red edge role represents a type of bridge relationship.

Sparse Graph Feature Learning: Recall that the proposed feature learning approach attempts to learn "sparse graph features" to improve learning and efficiency, especially in terms of space-efficiency. This section investigates the effectiveness of our sparse graph feature learning approach. Results are presented in Table 3. In all cases, our approach learns a highly compressed representation of the

Table 3. Higher-order sparse graph feature learning for latent node and edge network modeling. Recall that f is the number of features, L is the number of layers, and $\rho(\mathbf{X})$ is the sparsity of the feature matrix. Edge values are bold.

Graph	f	L	$\rho(\mathbf{X})$	$\rho(\mathbf{Z})$
socfb-MIT	**2080** (912)	**8** (9)	**0.318** (0.334)
Yahoo-msg	**1488** (405)	**7** (7)	**0.164** (0.181)
Enron	**843** (109)	**5** (4)	**0.312** (0.320)
Facebook	**1033** (136)	**7** (5)	**0.187** (0.162)
bio-DD21	**379** (723)	**6** (6)	**0.215** (0.260)

graph, requiring only a fraction of the space of current (node) approaches. Moreover, the density of edge and node feature representations learned by our approach is between $[0.164, 0.318]$ and $[0.162, 0.334]$ for nodes (See $\rho(\mathbf{X})$ and $\rho(\mathbf{Z})$ in Table 3) and up to $6x$ more space-efficient than other approaches.

Improving Classification via Link Prediction: This section demonstrates the effectiveness of edge roles for improving relational classification by predicting links between nodes in the graph. For consistency, we first construct node features from the edge role memberships using a set of relational operators (e.g., relational **mean**, **sum**, **var**, **max**, among others), as introduced in [31]. Thus, let us assume $\mathbf{x}_i$ is a k-dimensional feature vector for node $v_i \in V$. Given $\mathbf{x}_i$ and $\mathbf{x}_j$, and a positive semidefinite kernel function $K\langle\cdot,\cdot\rangle$, the relationship strength between v_i and v_j is defined as:

$$\mathbf{S} = [S_{ij}], \ \forall i, j \quad and \quad S_{ij} = \begin{cases} K\langle\mathbf{x}_i, \mathbf{x}_j\rangle & \text{if } (v_i, v_j) \notin E \wedge K\langle\mathbf{x}_i, \mathbf{x}_j\rangle > \epsilon \\ 0 & \text{otherwise} \end{cases} \quad (8)$$

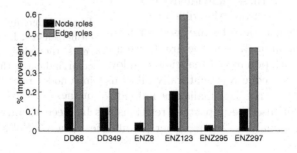

Fig. 6. Relative improvement in label consistency (homophily) — a known proxy for classification performance. In all cases, links predicted using edge roles improves the label consistency over both the initial graph as well as links predicted using node roles.

where $K\langle \mathbf{x}_i, \mathbf{x}_j \rangle$ represents the "closeness" between node v_i and v_j in the latent lower-dimensional subspace, $\mathbf{S} \in \mathbb{R}^{n \times n}$ is the (implicit) "similarity" matrix (which can be thought of as the weighted adjacency matrix for a graph G') and S_{ij} represents the relationship strength between node v_i and v_j such that $(v_i, v_j) \notin E$, and 0 otherwise. Note ϵ is a small scalar that controls sparsity. In this work, we use $K\langle \mathbf{x}_i, \mathbf{x}_j \rangle = \exp(-\|\mathbf{x}_i - \mathbf{z}_j\|^2 / 2\sigma^2)$. Given $\mathbf{S}$, let $G' = (V, E')$ denote the predicted latent graph where E' is the set of k predicted links with the largest relationship strength weights. By definition $|E| + |E'| = m + k$ and thus $E \cap E' = \varnothing$.

For quantitative evaluation of the edge roles, we use a measure of *homophily* called label consistency [21]. Let $\xi(v_i)$ be the class of v_i, then the label consistency of G is defined as: $\mathbb{L}(G) = 1/|E| \sum_{(v_i,v_j)\in E} \mathbb{L}(v_i, v_j)$ where $\mathbb{L}(v_i, v_j) = 1$ if $\xi(v_i) = \xi(v_j)$ and 0 otherwise. Hence, label consistency measures how often two connected nodes belong to the same class. It is a good proxy measure for classification performance since most existing statistical relational learning (SRL) [12] methods assume the labels of neighbors are highly correlated, *i.e.*, the network exhibits high relational autocorrelation (or homophily) [12,20]. To determine the effectiveness of edge roles for link prediction, we measure $\mathbb{L}(G)$ and $\mathbb{L}(G')$. Notice that if the higher-order edge roles (and node roles for that matter) are useful and effective, one would expect that $\mathbb{L}(G) < \mathbb{L}(G')$, that is, the predicted links resulted in higher homophily among the connected nodes since the class labels of the connected nodes in G' are more consistent than G. Results are provided in Fig. 6 for six different networks. In particular, Fig. 6 demonstrates the effectiveness of the higher-order edge roles (and node roles) for link prediction. In all cases, both the higher-order node and edge roles significantly outperform the baseline. Further, the edge role models always perform significantly better than the node roles.

Computational Complexity: Recall that m is the number of edges, f is the number of features, and r is the number of latent roles. The total computational complexity of the *higher-order latent space model* is $\mathcal{O}(f(mf + mr))$. The computational complexity is decomposed into the following main parts: Edge feature learning takes $\mathcal{O}(f(m + mf))$. Model learning takes $\mathcal{O}(mfr)$ in the worst case (which arises when $\mathbf{U}$ and $\mathbf{V}$ are completely dense). The quantization and Huffman coding terms are very small and therefore ignored. Role assignment using scalar element-wise coordinate descent has worst case complexity of $\mathcal{O}(mfr)$ per iteration which arises when $\mathbf{X}$ is completely dense. We assume the initial graphlet features are computed using fast and accurate estimation methods, seel [3].

Scalability: To evaluate the scalability of the parallel framework for modeling higher-order latent edge roles, we measure the speedup defined as $S_p = T_1/T_p$ where T_1 is the execution time of the sequential algorithm, and T_p is the execution time of the parallel algorithm with p processing units. Overall, the methods show strong scaling (See Fig. 7). Similar results were observed for other networks. The experiments used a machine with 4 Intel Xeon E5-4627 v2 3.3 GHz CPUs.

5 Conclusion

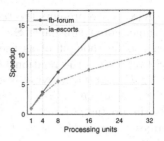

In this paper, we introduced the *edge role discovery* problem and presented a computational framework for learning and extracting edge roles from large networks. In addition, we proposed higher-order role discovery methods that leverage network motifs (including all motifs of size 3, 4, and larger) for learning more meaningful and discriminative roles. We also proposed a novel edge feature learning approach, which was used

Fig. 7. Strong parallel scaling is observed.

for our feature-based edge roles. Furthermore, all methods are space-efficient (by learning sparse features) and efficient with a runtime that is linear in the number of edges. Finally, the approach also supports graphs that are directed/undirected/bipartite, attributed, typed, and signed.

References

1. Ahmed, N.K., Neville, J., Rossi, R.A., Duffield, N.: Efficient graphlet counting for large networks. In: ICDM, p. 10 (2015)
2. Ahmed, N.K., Neville, J., Rossi, R.A., Duffield, N., Willke, T.L.: Graphlet decomposition: framework, algorithms, and applications. KAIS **50**(3), 1–32 (2016)
3. Ahmed, N.K., Willke, T.L., Rossi, R.A.: Estimation of local subgraph counts. In: IEEE BigData, pp. 1–10 (2016)
4. Akaike, H.: A new look at the statistical model identification. TOAC **19**(6), 716–723 (1974)
5. Anderson, C., Wasserman, S., Faust, K.: Building stochastic blockmodels. Soc. Netw. **14**(1), 137–161 (1992)
6. Arabie, P., Boorman, S., Levitt, P.: Constructing blockmodels: how and why. J. Math. Psychol. **17**(1), 21–63 (1978)
7. Batagelj, V., Mrvar, A., Ferligoj, A., Doreian, P.: Generalized blockmodeling with pajek. Metodoloski Zvezki **1**, 455–467 (2004)
8. Bennett, W.R.: Spectra of quantized signals. Bell Syst. Tech. **27**(3), 446–472 (1948)
9. Borgatti, S., Everett, M., Johnson, J.: Analyzing Social Networks. SAGE Publications, Thousand Oaks (2013)
10. Bregman, L.M.: The relaxation method of finding the common point of convex sets. USSR Comput. Math. Math. Phys. **7**(3), 200–217 (1967)
11. Doreian, P., Batagelj, V., Ferligoj, A.: Generalized Blockmodeling, vol. 25. Cambridge University Press, Cambridge (2005)
12. Getoor, L., Taskar, B. (eds.): Introduction to Statistical Relational Learning. MIT Press, Cambridge (2007)
13. Grünwald, P.D.: The Minimum Description Length Principle. MIT Press, Cambridge (2007)
14. Henderson, K., et al.: Rolx: structural role extraction & mining in large graphs. In: KDD, pp. 1231–1239 (2012)
15. Holland, P.W., Laskey, K.B., Leinhardt, S.: Stochastic blockmodels: first steps. Soc. Netw. **5**(2), 109–137 (1983)

16. Huffman, D.A., et al.: A method for the construction of minimum-redundancy codes. Proc. IRE **40**(9), 1098–1101 (1952)
17. Li, M., Vitányi, P.: An Introduction to Kolmogorov Complexity and Its Applications. Springer Science & Business Media, Heidelberg (2009)
18. Lloyd, S.: Least squares quantization in PCM. TOIT **28**(2), 129–137 (1982)
19. Lorrain, F., White, H.: Structural equivalence of individuals in social networks. J. Math. Sociol. **1**(1), 49–80 (1971)
20. Macskassy, S., Provost, F.: A simple relational classifier. In: KDD MRDM (2003)
21. Macskassy, S.A., Provost, F.: Classification in networked data: a toolkit and a univariate case study. JMLR **8**, 935–983 (2007)
22. Max, J.: Quantizing for minimum distortion. TOIT **6**(1), 7–12 (1960)
23. Nowicki, K., Snijders, T.: Estimation and prediction for stochastic blockstructures. J. Am. Stat. Assoc. **96**(455), 1077–1087 (2001)
24. Oliver, B., Pierce, J., Shannon, C.E.: The philosophy of PCM. IRE **36**(11), 1324–1331 (1948)
25. Rahman, M., Hasan, M.A.: Link prediction in dynamic networks using graphlet. In: Frasconi, P., Landwehr, N., Manco, G., Vreeken, J. (eds.) ECML PKDD 2016. LNCS (LNAI), vol. 9851, pp. 394–409. Springer, Cham (2016). doi:10.1007/978-3-319-46128-1_25
26. Rissanen, J.: Modeling by shortest data description. Automatica **14**(5), 465–471 (1978)
27. Rossi, R.A., Ahmed, N.K.: The network data repository with interactive graph analytics and visualization. In: AAAI (2015). http://networkrepository.com
28. Rossi, R.A., Ahmed, N.K.: Role discovery in networks. TKDE **27**(4), 1112 (2015)
29. Rossi, R.A., Gallagher, B., Neville, J., Henderson, K.: Role-dynamics: fast mining of large dynamic networks. In: WWW Companion, pp. 997–1006 (2012)
30. Rossi, R.A., Gallagher, B., Neville, J., Henderson, K.: Modeling dynamic behavior in large evolving graphs. In: WSDM, pp. 667–676 (2013)
31. Rossi, R.A., McDowell, L.K., Aha, D.W., Neville, J.: Transforming graph data for statistical relational learning. JAIR **45**(1), 363–441 (2012)
32. Rossi, R.A., Zhou, R.: Parallel collective factorization for modeling large heterogeneous networks. Soc. Netw. Anal. Mining **6**(1), 30 (2016)
33. Schwarz, G., et al.: Estimating the dimension of a model. Ann. Stat. **6**(2), 461–464 (1978)
34. Shannon, C.E.: A mathematical theory of communication. Bell Syst. Tech. **27**(1), 379–423 (1948)
35. Van Leeuwen, J.: On the construction of Huffman trees. In: ICALP, p. 382 (1976)
36. Vishwanathan, S.V.N., Schraudolph, N.N., Kondor, R., Borgwardt, K.M.: Graph kernels. JMLR **11**, 1201–1242 (2010)

Efficient Bi-level Variable Selection and Application to Estimation of Multiple Covariance Matrices

Duy Nhat Phan[1(✉)], Hoai An Le Thi[1], and Dinh Tao Pham[2]

[1] Laboratory of Theoretical and Applied Computer Science EA 3097,
University of Lorraine, Ile du Saulcy, 57045 Metz, France
{duy-nhat.phan,hoai-an.le-thi}@univ-lorraine.fr
[2] Laboratory of Mathematics, INSA–Rouen, University of Normandie,
76801 Saint-Etienne-du-Rouvray Cedex, France
pham@insa-rouen.fr

Abstract. Variable selection plays an important role in analyzing high dimensional data. When the data possesses certain group structures in which individual variables are also meaningful scientifically, we are naturally interested in selecting important groups as well as important variables. We introduce a new regularization by combining the $\ell_{p,0}$-norm and ℓ_0-norm for bi-level variable selection. Using an appropriate DC (Difference of Convex functions) approximation, the resulting problem can be solved by DC Algorithm. As an application, we implement the proposed algorithm for estimating multiple covariance matrices sharing some common structures such as the locations or weights of non-zero elements. The experimental results on both simulated and real datasets demonstrate the efficiency of our algorithm.

Keywords: Sparse group · Variable selection · Covariance matrix · DC programming · DCA

1 Introduction

Variable selection plays an important role in many applications and has drawn increased attention from many researchers in various domains such as machine learning, statistics, computational biology, signal processing and other related areas. In recent years, there are many works based on regularization methods for variable selection. When the data possesses certain group structures in which individual variables are also meaningful scientifically, we are naturally interested in selecting important groups as well as important variables within the selected groups. This is referred as bi-level variable selection. For example, in genomic data analysis, the correlations between genes sharing the biological pathway can be high. Hence these genes should be considered as a group. Moreover, we also would like to identify particularly important genes in pathways of interest. In

© Springer International Publishing AG 2017
J. Kim et al. (Eds.): PAKDD 2017, Part I, LNAI 10234, pp. 304–316, 2017.
DOI: 10.1007/978-3-319-57454-7_24

this paper, we introduce a natural approach for enforcing sparsity of groups and within each group by using the $\ell_0 + \ell_{p,0}$ regularization with $p \geq 1$.

We define the step function $s : \mathbb{R} \to \mathbb{R}$ by $s(t) = 1$ if $t \neq 0$ and $s(t) = 0$ otherwise. Assume that $x = (x_1, ..., x_d) \in \mathbb{R}^d$ is partitioned into J non-overlapping groups $x^1, ..., x^J$, then the $\ell_{p,0}$-norm and ℓ_0-norm of x are respectively defined by

$$\|x\|_{p,0} = \sum_{j=1}^{J} s(\|x^j\|_p) \text{ and } \|x\|_0 = \sum_{i=1}^{d} s(x_i). \tag{1}$$

The $\ell_0 + \ell_{p,0}$ regularization problem takes the form

$$\min \left\{ f(x) + \lambda \|x\|_0 + \gamma \|x\|_{p,0} : x \in K \subset \mathbb{R}^d \right\}, \tag{2}$$

where λ and γ are non-negative tuning parameters. The corresponding approximate problem of (2) is

$$\min \left\{ F_p(x) := f(x) + \lambda \sum_{i=1}^{d} \eta_\alpha(x_i) + \gamma \sum_{j=1}^{J} \eta_\alpha(\|x^j\|_p) : x \in K \right\}, \tag{3}$$

where $\eta_\alpha(t) = \min\{1, \alpha|t|\}$ is the Capped-ℓ_1 function [8], and α is a tuning parameter such that $\eta_\alpha(t)$ approximates the step function $s(t)$ as α tends to $+\infty$.

Many statistical modeling problems take the form of (2), for example, multiple linear/logistic/Cox regression, multiple graphical models, multiple covariance matrices estimation, and compressed sensing, etc. Several existing works have been developed for bi-level variable selection in the literature. The first work, named the group bridge method, was proposed in [5]. [2,3] proposed the composition of group-level penalties with other individual variable-level penalties for bi-level variable selection and developed a group coordinate descent algorithm for solving these problems. Using a convex approximation approach of the $\ell_0 + \ell_{2,0}$ regularization, [4,10] proposed the sparse group lasso ($\ell_1 + \ell_{2,1}$ regularization) to achieve bi-level selection.

In this paper, we investigate a DC (Difference of Convex functions) approximation approach for the general $\ell_0 + \ell_{p,0}$ regularization with $p \geq 1$. We consider the problem (2), where K is a convex set in $\mathbb{R}^d$ and f is a finite DC function on $\mathbb{R}^d$. The paper makes the following contributions.

Firstly, we develop a solution method based on DC programming and DCA (DC Algorithms), a powerful technique in nonconvex optimization [6,9], for solving the nonconvex approximate problem (3). Considering a special formulation of the approximate problem we propose a special DCA which requires to compute a proximal operator at each iteration. This proximal operator can be computed in closed form or by an inexpensive algorithm, hence the proposed algorithm is very useful in many real application problems.

Secondly, among $\ell_0 + \ell_{p,0}$ regularizations, we note that the $\ell_0 + \ell_{1,0}$ regularization is the most interesting with several useful properties from computational aspect. The DCA scheme for solving the resulting approximate problem iteratively computes a proximal operator which can be separated into independent sub-operators in many applications. This interesting feature makes our proposed approach very efficient in terms of computational complexity.

Finally, as an application, we consider the problem of simultaneous group and individual variable selection in estimation of multiple covariance matrices and perform a careful empirical experiment on both simulated and real datasets to study the performance of the proposed approach. The proposed DCA schemes move the difficulty terms to the second DC components, hence the resulting sequence of convex sub-problems is easier to solve. Especially, the convex sub-problem at each iteration of the first DCA can be decomposed into separable smaller sub-problems.

The rest of the paper is organized as follows. In Sect. 2, we present a brief introduction of DC programming and DCA for general DC programs, and illustrate how to apply DCA to solve the approximate problem. The application of the proposed algorithm to bi-level variable selection in estimation of multiple covariance matrices is described in Sect. 3. The numerical experiments are reported in Sect. 4 and Sect. 5 concludes the paper.

2   Solution Methods via DC Programming and DCA

2.1   A Brief Introduction of DC Programming and DCA

DC programming and DCA constitute the backbone of smooth/nonsmooth nonconvex programming and global optimization. They address the problem of minimizing a DC function on the whole space $\mathbb{R}^n$ or on a closed convex set $\Omega \subset \mathbb{R}^n$. Generally speaking, a standard DC program takes the form:

$$\alpha = \inf\{F(x) := G(x) - H(x) \mid x \in \mathbb{R}^n\} \quad (P_{dc}),$$

where G, H are lower semi-continuous proper convex functions on $\mathbb{R}^n$. Such a function F is called a DC function, and $G - H$ a DC decomposition of F while G and H are the DC components of F. A DC program with convex constraint $x \in \Omega$ can be equivalently expressed as an unconstrained DC program by adding the indicator function χ_Ω ($\chi_\Omega(x) = 0$ if $x \in \Omega$ and $+\infty$ otherwise) to the first DC component G.

For a convex function θ, the subdifferential of θ at $x_0 \in \mathrm{dom}\theta := \{x \in \mathbb{R}^n : \theta(x_0) < +\infty\}$, denoted by $\partial\theta(x_0)$, is defined by

$$\partial\theta(x_0) := \{y \in \mathbb{R}^n : \theta(x) \geq \theta(x_0) + \langle x - x_0, y \rangle, \forall x \in \mathbb{R}^n\}.$$

The subdifferential $\partial\theta(x_0)$ generalizes the derivative in the sense that θ is differentiable at x_0 if and only if $\partial\theta(x_0) \equiv \{\nabla_x\theta(x_0)\}$.

A point x^* is called a *critical point* of $G - H$, or a generalized Karush-Kuhn-Tucker point (KKT) of (P_{dc})) if $\partial H(x^*) \cap \partial G(x^*) \neq \emptyset$.

Starting from an initial point x^0, the DCA consists in constructing two sequences $\{x^l\}$ and $\{y^l\}$ such that, for any $l = 0, 1, 2, \ldots$

$$y^l \in \partial H(x^l) \quad \text{and} \quad x^{l+1} \in \arg\min_{x \in \mathbb{R}^n}\{G(x) - \langle y^l, x \rangle\}.$$

The sequence $\{x^l\}$ generated by DCA enjoys the following properties [6,9]:

(i) The sequence $\{F(x^l)\}$ is decreasing;
(ii) If $F(x^{l+1}) = F(x^l)$, then x^l is a critical point of (P_{dc}). In such a case, DCA terminates at l-th iteration.
(iii) Any limit point of the sequence $\{x^l\}$ is a critical point of (P_{dc}).

Note that the construction of DCA is based on G and H but not on F itself, and there are as many DCA as there are DC decompositions. This is a crucial fact in DC programming. It is important to study various equivalent DC forms of a DC problem, because each DC function F has infinitely many DC decompositions which have crucial implications for the qualities (speed of convergence, robustness, efficiency, globality of computed solutions,...) of DCA.

2.2 DCA for Solving the Approximate Problem

First of all, we assume that the DC function $f = g_1 + g_2 - h$ and there exists a nonnegative number μ such that $\frac{\mu}{2}\|x\|^2 - g_2(x)$ is convex, where g_1, g_2 and h are convex functions. In addition, the function $\eta_\alpha(t)$ can be expressed as a DC function:

$$\eta_\alpha(t) = \alpha|t| - r(t), \tag{4}$$

where $r(t) = -1 + \max\{1, \alpha|t|\}$. Hence, we have a special DC formulation of the problem (3) as follows:

$$\min_x \{F_p(x) = G_p(x) - H_p(x)\}, \tag{5}$$

where

$$G_p(x) = \chi_K(x) + g_1(x) + \frac{\mu}{2}\|x\|^2 + \lambda\alpha\|x\|_1 + \gamma\alpha\|x\|_{p,1},$$

$$H_p(x) = h(x) + \frac{\mu}{2}\|x\|^2 - g_2(x) + \lambda\sum_{i=1}^{d} r(x_i) + \gamma\sum_{j=1}^{J} r(\|x^j\|_p).$$

Following the generic DCA scheme, DCA for solving the problem (5) can be described as follows.

DCA (special DCA for solving the problem (5))

Initialization: Choose $x^0 \in K$, $l \leftarrow 0$ and a tolerance $\tau > 0$.
repeat
  1. Compute $y^l \in \partial H_p(x^l)$.
  2. Compute x^{l+1} by solving the problem:

$$\min_x \left\{ \chi_K(x) + g_1(x) + \frac{\mu}{2}\|x\|^2 + \lambda\alpha\|x\|_1 + \gamma\alpha\|x\|_{p,1} - \langle y^l, x\rangle \right\}, \quad (6)$$

  i.e., $x^{l+1} = \text{prox}_\mu^{\chi_K + g_1 + \lambda\alpha\|\cdot\|_1 + \gamma\alpha\|\cdot\|_{p,1}}(y^l/\mu)$
  3. $l \leftarrow l + 1$.
until $\|x^l - x^{l-1}\|_2 \leq \tau(\|x^{l-1}\|_2 + 1)$ or $|F_p(x^l) - F_p(x^{l-1})| \leq \tau(|F_p(x^{l-1})| + 1)$

Here, prox_μ^φ stands for the proximal operator associated to φ defined by

$$\text{prox}_\mu^\varphi(t) = \arg\min_x\{\varphi(x) + \frac{\mu}{2}\|x - t\|^2\}.$$

Remark 1. (i) We consider a special case $p = 1$, then we have $\|x\|_{1,1} \equiv \|x\|_1$ and the problem (6) can be rewritten as follows.

$$\min_x \left\{ \chi_K(x) + g_1(x) + \frac{\mu}{2}\|x\|^2 + \alpha(\lambda + \gamma)\|x\|_1 - \langle y^l, x\rangle \right\}. \quad (7)$$

This problem has the form of an ℓ_1-perturbed problem which can be found in many previous works (see [7] and referenes therein). Thanks to the ℓ_1-norm, if $\chi_K(x) + g_1(x)$ is separable in its variables, so is the problem (7). This leads to a potential massive reduction in computational complexity. Thus, we can say that the $\ell_0 + \ell_{1,0}$ is the most interesting regularization for DCA.

In addition, the $\ell_{1,0}$ regularization term can simultaneously encourage sparsity at the level of both groups and individual variables in each group. Hence, we can set $\lambda = 0$ in this case to avoid performing tuning this parameter.

(ii) Special DCA moves the difficulty terms in g_2 to the second DC component H, hence the resulting sequence of convex sub-problems are easier to solve.

3 Application to Estimation of Multiple Covariance Matrices

Estimation of sparse covariance matrices plays an important role in various areas of statistical analysis such as portfolio management and risk assessment, high dimensional classification, analysis of independence and conditional independence relationships between components in graphical models, etc. In recent years, much interest has focused on estimating a covariance matrix on the basis of an

$n \times d$ data matrix X, where n is the number of observations and d is the number of features. Suppose that the observations $x_1, ..., x_n \in \mathbb{R}^d$ are independent and identically distributed $\mathcal{N}(0, \Sigma)$, where Σ is a positive definite $d \times d$ matrix. A natural way to estimate the covariance matrix Σ is via minimizing negative log-likelihood. The resulting optimization problem is

$$\min_{\Sigma \succ 0} \{\log \det \Sigma + \operatorname{tr}(\Sigma^{-1} S)\}, \tag{8}$$

where $S = 1/n \sum_{i=1}^{n} x_i x_i^T$ is the sample covariance matrix and the notation $\Sigma \succ 0$ means that Σ is symmetric positive definite.

In this paper, we consider the case of multiple classes. Suppose that we have a dataset with Q classes. For the k-th class, let X^k be an $n_k \times d$ matrix consisting of n_k observations with the number of features d common to all classes. Furthermore, we assume that the observations within each class are independent and identically distributed according to $\mathcal{N}(0, \Sigma^k)$. Let $S^k = \frac{1}{n_k}(X^k)^T X^k$ be the sample covariance matrix for the k-th class. The Q covariance matrices are estimated via minimizing negative log-likelihood

$$\min_{\Sigma^k \succ 0} \left\{ \sum_{k=1}^{Q} n_k \left[\log \det \Sigma^k + \operatorname{tr}((\Sigma^k)^{-1} S^k) \right] \right\}. \tag{9}$$

In the problem of estimating multiple covariance matrices, the covariance matrices share some common structures such as the locations and weights of non-zero elements. Therefore, the elements (i, j) across all Q covariance matrices should be considered as groups. Moreover, each variable within each group also has different roles in its covariance matrix. Hence, we propose the bi-level variable selection problem in estimation of multiple covariance matrices which takes the form:

$$\min_{\Sigma^k \succ 0} \left\{ \sum_{k=1}^{Q} n_k \left[\log \det \Sigma^k + \operatorname{tr}((\Sigma^k)^{-1} S^k) \right] + \lambda \|\{\Sigma\}\|_0 + \gamma \|\{\Sigma\}\|_{p,0} \right\}, \tag{10}$$

where λ and γ are non-negative tuning parameters, and

$$\|\{\Sigma\}\|_0 = \sum_{k=1}^{Q} \sum_{i,j} s(\Sigma_{ij}^k), \quad \|\{\Sigma\}\|_{p,0} = \sum_{i,j} s(\|(\Sigma_{ij}^1, ..., \Sigma_{ij}^Q)\|_p).$$

If S^k is nonsingular for all k, then there exist $\delta_1, ..., \delta_Q > 0$ such that the problem (10) is equivalent to the following problem

$$\min_{\{\Sigma\} \in \Omega} \left\{ \sum_{k=1}^{Q} n_k \left[\log \det \Sigma^k + \operatorname{tr}((\Sigma^k)^{-1} S^k) \right] + \lambda \|\{\Sigma\}\|_0 + \gamma \|\{\Sigma\}\|_{p,0} \right\}, \tag{11}$$

where $\Omega = \{\{\Sigma\} := \{\Sigma^1, ..., \Sigma^Q\} : \Sigma^k \succeq \delta_k I, k = 1, ..., Q\}$. Here, I denotes the $d \times d$ identity matrix, and the notation $\Sigma^k \succeq \delta_k I$ means that $\Sigma^k - \delta_k I$ is symmetric positive semidefinite. Note that if S^k is not full rank, we can replace S^k with $S^k + \epsilon I$ for some $\epsilon > 0$.

We observe that the problem (11) takes the form of (2) where the function f is given by

$$f(\{\Sigma\}) = \sum_{k=1}^{Q} n_k \left[\log \det \Sigma^k + \text{tr}((\Sigma^k)^{-1}S^k)\right],$$

and the corresponding approximate problem is

$$\min_{\{\Sigma\} \in \Omega} \left\{ f(\{\Sigma\}) + \lambda \sum_{k=1}^{Q} \sum_{ij} \eta_\alpha(\Sigma_{ij}^k) + \gamma \sum_{ij} \eta_\alpha(\|(\Sigma_{ij}^1, ..., \Sigma_{ij}^Q)\|_p) \right\}. \quad (12)$$

We note that $\log \det \Sigma^k$ is concave while $\text{tr}((\Sigma^k)^{-1}S^k)$ is convex in Σ^k. Hence we have a natural DC decomposition of f as follows:

$$f(\{\Sigma\})) = g_1(\{\Sigma\}) + g_2(\{\Sigma\}) - h(\{\Sigma\}), \quad (13)$$

where $g_1(\{\Sigma\}) = 0$ and

$$g_2(\{\Sigma\}) = \sum_{k=1}^{Q} n_k \text{tr}((\Sigma^k)^{-1}S^k), \quad h(\{\Sigma\}) = \sum_{k=1}^{Q} -n_k \log \det \Sigma^k.$$

For estimating μ such that $\frac{\mu}{2}\|\{\Sigma\}\|^2 - g_2(\{\Sigma\})$ is convex, we have the following lemma.

Lemma 1. *If $\mu \geq \max_k n_k \|S^k\|_2 \delta_k^{-3}$, then $\frac{\mu}{2}\|\{\Sigma\}\|^2 - g_2(\{\Sigma\})$ is convex in $\{\Sigma\}$.*

Remark 2. From the Lemma 1, we can choose $\mu = \max_k n_k \|S^k\|_2 \delta_k^{-3}$.

According to DCA with $p = 1$, at each iteration l, we have to compute $\{V^l\} \in \partial H_1(\{\Sigma^l\})$, and

$$\{\Sigma^{l+1}\} = \text{prox}_{\mu}^{\chi_\Omega + \lambda\alpha\|\cdot\|_1 + \gamma\alpha\|\cdot\|_{1,1}} \left(\{V^l\}/\mu\right). \quad (14)$$

Computing $\{V^l\}$ can be explicitly given by $\{V^l\} = \{A^l\} + \{B^l\} + \{C^l\}$, where

$$(A^k)^l = \mu(\Sigma^k)^l + n_k[(\Sigma^k)^l]^{-1}S^k[(\Sigma^k)^l]^{-1} - n_k[(\Sigma^k)^l]^{-1}, \quad (15)$$

$$(B^k)_{ij}^l = \begin{cases} \lambda\alpha\,\text{sgn}(\Sigma^k)_{ij}^l & \text{if } \alpha|(\Sigma^k)_{ij}^l| \geq 1 \\ 0 & \text{otherwise} \end{cases}, \quad (16)$$

$$(C^k)_{ij}^l = \begin{cases} \gamma\alpha\,\text{sgn}(\Sigma^k)_{ij}^l & \text{if } \alpha\|(\Sigma^1)_{ij}^l, ..., (\Sigma^Q)_{ij}^l\|_1 \geq 1 \\ 0 & \text{otherwise} \end{cases}. \quad (17)$$

For computing $\{\Sigma^{l+1}\}$, we notice that $\|\{\Sigma\}\|_{1,1} = \|\{\Sigma\}\|_1$ is separable. Hence the proximal operator (14) can be separated into Q independent sub-problems of the same form

$$\min_{\Sigma^k \succeq \delta_k I} \left\{ \varphi(\Sigma^k) := \frac{\mu}{2}\|\Sigma^k\|_F^2 + \lambda\alpha\|\Sigma^k\|_1 + \gamma\alpha\|\Sigma^k\|_1 - \langle(V^k)^l, \Sigma^k\rangle \right\}. \quad (18)$$

For solving each convex sub-problem (18), we use the alternating direction method of multipliers (ADMM) [1]. The augmented Lagrangian function of this problem is

$$L_1(\Sigma^k, X, Y) = \frac{\mu}{2}\|\Sigma^k\|_F^2 - \langle(V^k)^l, \Sigma^k\rangle + (\lambda+\gamma)\alpha\|X\|_1 + \langle Y, \Sigma^k - X\rangle + \frac{\rho}{2}\|\Sigma^k - X\|_F^2.$$

More specifically, at each iteration m of ADMM, we compute

$$\Sigma^{k,l,m+1} = \arg\min_{\Sigma \succeq \delta_k I} L_1(\Sigma, X^m, Y^m) = UD_{\delta_k}U^T \quad (19)$$

$$X^{m+1} = \arg\min_{X \in \mathbb{R}^{d\times d}} L_1(\Sigma^{k,l,m+1}, X, Y^m) = \mathcal{S}\left(\Sigma^{k,l,m+1} + \frac{Y^m}{\rho}, \frac{(\lambda+\gamma)\alpha}{\rho}\right) \quad (20)$$

$$Y^{m+1} = Y^m + \rho(\Sigma^{k,l,m+1} - X^{m+1}). \quad (21)$$

where $D_{\delta_k} = \mathrm{diag}(\max(D_{ii}, \delta_k))$, $UDU^T = ((V^k)^l - Y^m + \rho X^m)/(\mu+\rho)$, and $\mathcal{S}$ is the elementwise soft-thresholding operator defined by $\mathcal{S}(A, B)_{ij} = \mathrm{sgn}(A_{ij})(|A_{ij}| - B_{ij})_+$. DCA for solving (12) with $p = 1$ is summarized in the following algorithm.

DCA1: (DCA for solving (12) with $p = 1$)

Initialization: Choose $\{\Sigma^0\} \in \Omega$, $l \leftarrow 0$ and a tolerance $\tau > 0$.
repeat
    1. Compute $(V^k)^l = (A^k)^l + (B^k)^l + (C^k)^l$ using (15)-(17).
    2. Parallel compute $(\Sigma^k)^{l+1}, k = 1, ..., Q$ by ADMM:
Set $m = 0$, choose $X^0, Y^0 \in \mathbb{R}^{d\times d}$.
    repeat
        + Compute $\Sigma^{k,l,m+1}, X^{m+1}, Y^{m+1}$ using (19)-(21),
        + $m \leftarrow m + 1$.
    until $|\varphi(\Sigma^{k,l,m}) - \varphi(\Sigma^{k,l,m-1})| \leq \tau(|\varphi(\Sigma^{k,l,m-1})| + 1)$.
    3. $l \leftarrow l + 1$.
until Stopping criterion.

According to DCA with $p = 2$, at each iteration l, we have to compute $\{V^l\} \in \partial H_2(\{\Sigma^l\})$, and

$$\{\Sigma^{l+1}\} = \mathrm{prox}_\mu^{\chi_\Omega + \lambda\alpha\|.\|_1 + \gamma\alpha\|.\|_{2,1}}(\{V^l\}/\mu). \quad (22)$$

Computing $\{V^l\}$ can be explicitly computed as $\{V^l\} = \{A^l\} + \{B^l\} + \{D^l\}$, where $\{A^l\}, \{B^l\}$ are respectively computed by using (15) and (16), and

$$
(D^k)^l_{ij} = \begin{cases} \dfrac{\gamma\alpha(\Sigma^k)^l_{ij}}{\|(\Sigma^1)^l_{ij}, \ldots, (\Sigma^Q)^l_{ij}\|_2} & \text{if } \alpha\|(\Sigma^1)^l_{ij}, \ldots, (\Sigma^Q)^l_{ij}\|_2 \geq 1 \\ 0 & \text{otherwise} \end{cases} . \qquad (23)
$$

The proximal operator (22) cannot be separated into smaller sub-operators as the previous case. Here we also apply the ADMM algorithm for computing this operator. For summary, we describe the algorithm for (12) with $p = 2$ in the algorithm below.

DCA2: (DCA for solving (12) with $p = 2$)

Initialization: Choose $\{\Sigma^0\} \in \Omega$, $l \leftarrow 0$.
repeat
    1. Compute $(V^k)^l = (A^k)^l + (B^k)^l + (D^k)^l$ using (15), (16) and (23).
    2. Compute $\{\Sigma^{l+1}\}$ by ADMM: set $m = 0$, choose $\{X^0\}, \{Y^0\} \in (\mathbb{R}^{d \times d})^Q$.
    repeat
      + Compute $\Sigma^{k,l,m+1} = U D_{\delta_k} U^T$ as in (19) for all $k = 1, \ldots, Q$.
      + Compute $(X_{ij})^{m+1} = [\|R_{ij}\|_2 - \lambda\gamma/\rho]_+ \dfrac{R_{ij}}{\|R_{ij}\|_2}$,
      where $R_{ij} = \mathcal{S}\left((\Sigma_{ij})^{l,m+1} + (Y_{ij})^m/\rho, \lambda\alpha/\rho\right)$ and $X_{ij} = (X^1_{ij}, \ldots, X^Q_{ij})$.
      + $\{Y^{m+1}\} = \{Y^m\} + \rho(\{\Sigma^{l,m+1}\} - \{X^{m+1}\})$,
      + $m \leftarrow m + 1$.
    until Stopping criterion.
    3. $l \leftarrow l + 1$.
until Stopping criterion.

Theorem 1 (Convergence properties of DCA1 and DCA2). *Let $\{\{\Sigma^l\}\}$ be the sequence generated by DCA1 (resp. DCA2)), we have*

(a) *$\{F_1(\{\Sigma^l\})\}$ (resp. $\{F_2(\{\Sigma^l\})\}$) is decreasing and $\{\{\Sigma^l\}\}$ is bounded.*
(b) *$\sum_{l=0}^{+\infty} \|\{\Sigma^l\} - \{\Sigma^{l+1}\}\|_F^2 < +\infty$, and hence $\lim_{l\to+\infty} \|\{\Sigma^l\} - \{\Sigma^{l+1}\}\|_F = 0$.*
(c) *The sequence $\{\{\Sigma^l\}\}$ has at least one limit point and every limit point of this sequence is a critical point of the problem (12).*

4 Numerical Experiments

We will compare the proposed algorithms (DCA1 and DCA2) with the approach based on the $\ell_1 + \ell_{2,1}$ regularization ($\ell_1/\ell_{2,1}$(DCA)). The $\ell_1/\ell_{2,1}$(DCA) is a convex approximation approach of the $\ell_0 + \ell_{p,0}$ regularization replaced by the $\ell_1 + \ell_{2,1}$ regularization [4,10]. The resulting problem of estimating multiple covariance matrices is

$$\min_{\{\Sigma\}\in\Omega} \left\{ \sum_{k=1}^{Q} n_k \left[\log \det \Sigma^k + \text{tr}((\Sigma^k)^{-1}S^k)\right] + \lambda\|\{\Sigma\}\|_1 + \gamma\|\{\Sigma\}\|_{2,1} \right\}. \quad (24)$$

This problem is still nonconvex and then still difficult. We propose DCA for solving it. The $\ell_1/\ell_{2,1}$(DCA) is similar to DCA2. We simply replace the computation of $\{V^l\} \in \partial H(\{\Sigma^l\})$ with $\{V^l\} = \{A^l\}$ computed by (15) in the step 1 of DCA2.

The proposed algorithms are implemented in R software and all algorithms are performed on a PC Intel i7 CPU3770, 3.40 GHz of 8 GB RAM. In experiments, we set the stop tolerance $\tau = 10^{-4}$ for DCA based algorithms and ADMM. The starting point $\{\Sigma^0\}$ of DCA is the sample covariance matrices $\{S^1, ..., S^Q\}$. The values of parameter λ, γ and ϵ are chosen through a 5-fold cross-validation procedure on training set. The approximation parameter α of the Capped-ℓ_1 is set 1. By Remark 1, we set $\lambda = 0$ in DCA1 to avoid performing tuning this parameter.

Experiment on Synthetic Datasets

We evaluate the performance of the proposed algorithms on two synthetic datasets. We consider two types of covariance graphs with three-class:

Model 1: We generate a covariance matrix for the first class as follows. $\Sigma^1 = \text{diag}(\Sigma_1, ..., \Sigma_5)$, where $\Sigma_1, ..., \Sigma_5$ are dense matrices. We create Σ^2 by resetting one of its 5 sub-network blocks to the identity, i.e., $\Sigma^2 = \text{diag}(I, \Sigma_2, ..., \Sigma_5)$. Resetting an additional sub-network block to the identity, we have $\Sigma^3 = \text{diag}(I, I, \Sigma_3, ..., \Sigma_5)$.

Model 2: $\Sigma^1 = \text{diag}(\Sigma_1, ..., \Sigma_5)$ again, however each submatrix Σ_k is zero except elements in the last row and the last column. This corresponds to a subgraph with five connected components each of which has all nodes connected to one particular node. Similarly to model 1, we create $\Sigma^2 = \text{diag}(I, \Sigma_2, ..., \Sigma_5)$ and $\Sigma^3 = \text{diag}(I, I, \Sigma_3, ..., \Sigma_5)$.

The nonzero entries of matrices $\Sigma^k, k = 1, 2, 3$ are randomly drawn in the set $\{+1, -1\}$. Finally, for each class we generate independently, identically distributed observations $X^k = [x_1^k, ..., x_{n_k}^k]$ from an $\mathcal{N}(0, \Sigma^k)$ distribution. In this experiment, for each model, we generate 10 training sets with size $n_1 = n_2 = n_3 = 200, d = 100$.

To evaluate the performance of each method, we consider three loss functions which are the average root-mean-square error (ARMSE), the average entropy loss (AEN), and the average Kullback-Leibler loss (AKL), respectively.

$$\text{ARMSE} = \frac{1}{Q} \sum_{k=1}^{Q} \frac{\|\hat{\Sigma}^k - \Sigma^k\|_F}{d},$$

$$\text{AEN} = \frac{1}{Q} \sum_{k=1}^{Q} \left[-\log \det(\hat{\Sigma}^k (\Sigma^k)^{-1}) + \text{tr}(\hat{\Sigma}^k (\Sigma^k)^{-1}) - d \right],$$

$$\text{AKL} = \frac{1}{Q} \sum_{k=1}^{Q} \left[-\log \det((\hat{\Sigma}^k)^{-1} \Sigma^k) + \text{tr}((\hat{\Sigma}^k)^{-1} \Sigma^k) - d \right],$$

where $\hat{\Sigma}^k$ is a sparse estimate of the covariance matrix Σ^k.

The experimental results on synthetic datasets are given in Table 1. In this Table, the ARMSE, AEN, AKL, number of nonzero elements on each covariance matrix (NZ1, NZ2, NZ3), and their sum (NZ), CPU time in seconds, and their standard deviations over 10 samples are reported.

We observe from Table 1 that in the both models, DCA1 gives the best results in terms of three losses. In terms of the sparsity, the number of the nonzero elements, this approach also achieves better performances than the other approaches. The second and third performing approaches with respect to the losses and the sparsity are DCA2 and $\ell_1/\ell_{2,1}$(DCA), respectively.

Regarding the training time, DCA1 is much faster than the other algorithms. This can be explained by the fact that DCA1 leads to the sequence of convex problems which can be separated into the independent sub-problems.

Experiment on Real Datasets

We illustrate the use of the sparse covariance matrix estimation problem via a real application: a classification problem based sparse quadratic discriminant analysis (SQDA). This application requires estimates of the covariance matrices. We assume that the n_k observations x_i^k ($i = 1, ..., n_k$) within the k-th class C_k are normally distributed $\mathcal{N}(\mu_k, \Sigma_k)$. We denote the prior probability of the k-th class by π_k. The quadratic discriminant function is

$$\delta_k(x) = -\frac{1}{2} \log \det \Sigma_k - \frac{1}{2}(x - \mu_k)^T \Sigma_k^{-1}(x - \mu_k) + \log \pi_k. \tag{25}$$

Then the predicted class for a new observation x is $\arg\max_k \delta_k(x)$. In practice we do not know π_k, μ_k, Σ_k, and will need to estimate them using the training data.

For the experiment, we evaluate the proposed algorithms on four datasets from UCI Machine Learning Repository[1] (Ionosphere, Waveform 2, Optical Recognition of Handwritten Digits, and Semeion Handwritten Digit). We use the cross-validation scheme to validate the performance of various approaches on these two datasets. The dataset is split into a training set containing 2/3 of the samples and a test set containing 1/3 of the samples. This process is repeated 10 times, each with a random choice of training set and test set.

The computational results are reported in Table 2. We observe that, on the Ionosphere dataset, DCA1 and DCA2 are comparable and better than $\ell_1/\ell_{2,1}$(DCA) in terms of the testing error and training error. On the Waveform 2 and Optimal datasets, DCA1 gives better testing error and training error

---

[1] https://archive.ics.uci.edu/ml/datasets.

Table 1. Comparative results of DCA1, DCA2, and $\ell_1/\ell_{2,1}$(DCA). The numbers in parentheses are standard deviations. Bold fonts indicate the best result in each row.

		DCA1	DCA2	$\ell_1/\ell_{2,1}$(DCA)
Model 1	ARMSE	**0.381** (0.004)	0.415 (0.007)	0.445 (0.002)
	AEN	**12.97** (1.07)	17.53 (1.12)	18.08 (2.62)
	AKL	**17.31** (2.24)	18.58 (3.1)	31.86 (2.75)
	NZ1	**1903.2** (298.61)	2138.6 (313.7)	2242.4 (271.5)
	NZ2	1781.6 (307.96)	2172.18 (281.6)	**1464.6** (316.4)
	NZ3	**1728.2** (288.12)	2058.37 (215.5)	1918.8 (251.2)
	NZ	**5413** (892.69)	6369.15 (810.8)	5625.8 (839.1)
	CPUs	**642.11** (3.74)	3180.56 (7.92)	3471.66 (5.18)
Model 2	ARMSE	**0.082** (0.003)	0.09 (0.007)	0.094 (0.005)
	AEN	**3.57** (0.52)	18.02 (1.31)	28.08 (2.16)
	AKL	**3.93** (0.56)	6.65 (1.66)	7.76 (1.82)
	NZ1	**352.8** (13.51)	394.18 (52.47)	448.6 (52.3)
	NZ2	**259** (13.61)	347.45 (38.52)	378.27 (36.1)
	NZ3	**255.8** (11.18)	359.72 (12.98)	264.61 (31.6)
	NZ	**867.6** (38.3)	1101.35 (103.97)	1091.48 (120)
	CPUs	**267.22** (39.33)	3843.57 (27.37)	5593.15 (24.61)

Table 2. Comparative results of real datasets. The bold font indicates the best result in each column.

		Testing error (%)	Training error (%)	Training time (s)
Ionosphere	DCA1	**5.13** (1.3)	**3.41** (0.48)	**0.094** (0.02)
	DCA2	**5.13** (0.54)	3.84 (0.72)	0.94 (0.01)
	$\ell_1/\ell_{2,1}$(DCA)	6.79 (1.78)	4.27 (0.82)	0.97 (0.04)
Waveform 2	DCA1	**13.01** (0.25)	**11.41** (1.3)	**3.28** (1.14)
	DCA2	14.64 (0.38)	12.68 (0.32)	157.64 (23.99)
	$\ell_1/\ell_{2,1}$(DCA)	15.6 (1.01)	14.57 (0.39)	259.28 (84.06)
Optical	DCA1	**2.74** (0.31)	**1.92** (0.14)	**97.9** (10.43)
	DCA2	3.64 (0.18)	2.05 (0.39)	582.92 (22.75)
	$\ell_1/\ell_{2,1}$(DCA)	3.98 (0.41)	3.18 (0.51)	574.66 (36.81)
Semeion	DCA1	7.58 (0.17)	**5.28** (0.85)	**184.13** (10.69)
	DCA2	**7.31** (0.29)	6.17 (0.73)	949.5 (29.17)
	$\ell_1/\ell_{2,1}$(DCA)	9.52 (0.84)	6.89 (0.47)	1027.73 (75.61)

than both the algorithms DCA2 and $\ell_1/\ell_{2,1}$(DCA). On the Semeion dataset, DCA2 is slightly better than DCA1 and both these algorithms are better than $\ell_1/\ell_{2,1}$(DCA) in terms of the testing error and training error. In terms of training time, DCA1 is significantly faster than DCA2 and $\ell_1/\ell_{2,1}$(DCA) on all datasets.

5   Conclusion

We have studied DC programming and DCA for bi-level variable selection problem including the $\ell_0 + \ell_{p,0}$ regularization in the objective function. Considering the special formulation of the approximate problem we have developed DCA for solving it. Concerning the bi-level variable selection in multiple covariance matrices estimation problem, numerical experiments on both simulation and real datasets have showed that DCA1 has obtained the best performance in terms of most of comparison criteria, and has taken the shortest time for training.

For the future works, we plan to study bi-level variable selection for other applications. We believe that the success of the $\ell_{1,0}$-regularization motivates and opens up a new avenue for the bi-level variable selection problems.

References

1. Boyd, S., Parikh, N., Chu, E., Peleato, B., Eckstein, P.: Distributed optimization and statistical learning via the alternating direction method of multipliers. Foundat. Trends Mach. Learn. **3**(1), 1–122 (2011)
2. Breheny, P.: The group exponential lasso for bi-level variable selection. Biometrics **71**(3), 731–740 (2015)
3. Breheny, P., Huang, J.: Penalized methods for bi-level variable selection. Stat. Interface **2**, 369–380 (2009)
4. Friedman, J., Hastie, T., Tibshirani, R.: A note on the group lasso and sparse group lasso. arXiv:1001.0736v1, pp. 1–8 (2010)
5. Huang, J., Ma, S., Xie, H., Zhang, C.H.: A group bridge approach for variable selection. Biometrika **96**(2), 339–355 (2009)
6. Le Thi, H.A., Pham Dinh, T.: The DC (difference of convex functions) programming and DCA revisited with DC models of real world nonconvex optimization problems. Ann. Oper. Res. **133**, 23–46 (2005)
7. Le Thi, H.A., Pham Dinh, T., Le Hoai, M., Vo Xuan, T.: DC approximation approaches for sparse optimization. Eur. J. Oper. Res. **244**, 26–44 (2015)
8. Peleg, D., Meir, R.: A bilinear formulation for vector sparsity optimization. Sig. Process. **88**(2), 375–389 (2008)
9. Pham Dinh, T., Le Thi, H.A.: Convex analysis approach to D.C. programming: theory, algorithms and applications. Acta Math. Vietnamica **22**(1), 289–355 (1997)
10. Wu, T.T., Lange, K.: Coordinate descent algorithms for lasso penalized regression. Ann. Appl. Stat. **2**(1), 224–244 (2008)

Entity Set Expansion with Meta Path
in Knowledge Graph

Yuyan Zheng[1], Chuan Shi[1,2(✉)], Xiaohuan Cao[1], Xiaoli Li[3], and Bin Wu[1]

[1] Beijing Key Lab of Intelligent Telecommunications Software and Multimedia,
Beijing University of Posts and Telecommunications, Beijing 100876, China
{shichuan,wubin}@bupt.edu.cn, zyy0716_source@163.com, devil_baba@126.com
[2] Beijing Advanced Innovation Center for Imaging Technology,
Capital Normal University, Beijing 100048, China
[3] Institute for Infocomm Research, A*STAR, Singapore, Singapore
xlli@i2r.a-star.edu.sg

Abstract. Entity set expansion (ESE) is the problem that expands a small set of seed entities into a more complete set, entities of which have common traits. As a popular data mining task, ESE has been widely used in many applications, such as dictionary construction and query suggestion. Contemporary ESE mainly utilizes text and Web information. That is, the intrinsic relation among entities is inferred from their occurrences in text or Web. With the surge of knowledge graph in recent years, it is possible to extend entities according to their occurrences in knowledge graph. In this paper, we consider the knowledge graph as a heterogeneous information network (HIN) that contains different types of objects and links, and propose a novel method, called MP_ESE, to extend entities in the HIN. The MP_ESE employs meta paths, a relation sequence connecting entities, in HIN to capture the implicit common traits of seed entities, and an automatic meta path generation method, called SMPG, is provided to exploit the potential relations among entities. With these generated and weighted meta paths, the MP_ESE can effectively extend entities. Experiments on real datasets validate the effectiveness of MP_ESE.

Keywords: Heterogeneous information network · Entity set expansion · Knowledge graph · Meta path

1 Introduction

Entity Set Expansion (ESE) refers to the problem of expanding a small set with a few seed entities into a more complete set, entities of which belong to a particular class. For example, given a few seeds like "China", "America" and "Russia" of country class, ESE will leverage data sources (e.g., text or Web information) to obtain other country instances, such as Japan and Korea. ESE has been used in many applications, e.g., dictionary construction [4], query refinement [6] and query suggestion [2].

© Springer International Publishing AG 2017
J. Kim et al. (Eds.): PAKDD 2017, Part I, LNAI 10234, pp. 317–329, 2017.
DOI: 10.1007/978-3-319-57454-7_25

Numerous methods have been proposed for ESE and most of them are based on the text or Web environment [5, 9, 14, 19, 20]. These methods utilize distribution information or context pattern of seeds to expand entities. For instance, Wang and Cohen [19] propose a novel approach that can be applied to semi-structured documents written in any markup language and in any human language. Recently, knowledge graph has become a popular tool to store and retrieve fact information with graph structure, such as Wikipedia and Yago. Among those text or Web based methods, some researchers also began to leverage knowledge graph as auxiliary for the performance improvement of ESE. For example, Qi et al. [12] use Wikipedia semantic knowledge to choose better seeds for ESE. However, seldom work only utilizes knowledge graph as individual data source for ESE.

In this paper, we firstly study the entity set expansion with knowledge graph. Since knowledge graph is usually constituted by $<Subject, Property, Object>$ tuples, we can consider it as a heterogeneous information network (HIN) [15], which contains different types of objects and relations. Based on this HIN, we design a novel *Meta Path* based *Entity Set Expansion* approach (called MP_ESE). Specifically, the MP_ESE employs the meta path [18], a relation sequence connecting entities, to capture the implicit common feature of seed entities, and designs an automatic meta path generation method, called SMPG, to exploit the potential relations among entities. In addition, a heuristic weight learning method is adopted to assign the importance of meta paths. With the help of weighted meta paths, MP_ESE can automatically extend entity set. Based on the Yago knowledge graph, we generate four different types of entity set expansion tasks. On almost all tasks, the proposed method outperforms other baselines.

2   Related Work

In recent years, there has been a significant amount of work on ESE and ESE has received considerable attention from both research [11, 19, 20] and industry circles (e.g., Google Sets). According to the difference of data sources utilized by ESE, these methods are based on text, Web environment and others.

For those text data source based ESE methods, they utilize the distribution information of the surrounding words of entities to expand certain class [5, 9, 14]. For those Web environment based ESE methods, proper patterns of seeds are extracted and then these patterns are used to extract new candidate entities. This kind of methods can also be used in text data source. Recently, some researchers began to take advantage of external semantic information to improve performance of set expansion for text or Web data source. Qi et al. [12] introduce the semantic knowledge by leveraging Wikipedia and reduce the seed ambiguity. Sadamitsu et al. [13] use topic information to alleviate semantic drift. Jindal and Roth [7] specify some negative examples to confine the expansion category.

More recently, HIN and knowledge graph have also been applied for related work. Yu et al. [21] propose a meta-path-based ranking model ensemble to represent semantic meaning for entity query. Different from our work, it has solved a similar but different problem (i.e., entity query), and the meta paths in their method need to be provided by domain expert users. QBEES [10] is designed for entity similarity search based on aspects of the entities, and Chen et al. [3] design a system for entity exploration and debugging. They both utilize the knowledge graph, but they do not employ the HIN method.

3 Preliminary

In this section, we describe some key concepts and present some preliminary knowledge in this paper.

Knowledge graph (KG) [16] is a large and complex graph dataset, which consists of triples of the form $<Subject, Property, Object>$, such as $<StevenSpielberg, directed, War Horse(film)>$ shown in Fig. 1. Yago [17] and DBpedia [1] are two prime examples of KG. The types of entities or relations in KG are often organized as concept hierarchy structure, which describes the sub-class relationship among entity types or relations. Figure 1(b) is a snapshot of Yago and we can see that actor is sub-class of person shown by the dashed line in Fig. 1(b). All the types share a common root called thing.

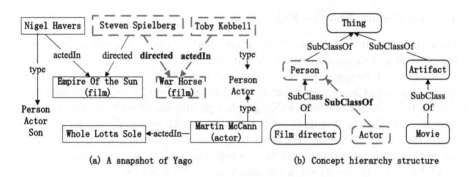

(a) A snapshot of Yago   (b) Concept hierarchy structure

Fig. 1. A snapshot of Yago with concept hierarchy structure.

Heterogeneous information network (HIN) [18] is defined as a directed graph $G = (V, E)$ with an object type mapping function $\varphi : V \rightarrow \mathcal{A}$ and a link type mapping function $\psi : E \rightarrow \mathcal{R}$, where V, E, $\mathcal{A}$ and $\mathcal{R}$ denotes object set, link set, object type set and relation type set, respectively, and the number of object types $|\mathcal{A}| > 1$ or the number of relation types $|\mathcal{R}| > 1$. In HIN, meta path [18] is widely used to capture the rich semantic meaning and is denoted in the form of $A_1 \xrightarrow{R_1} A_2 \xrightarrow{R_2} \ldots \xrightarrow{R_l} A_{l+1}$, which is a sequence of object types and link types between objects.

Since KG contain different types of objects (i.e., subject and object) and links (i.e., property), KG is a natural HIN. In Fig. 1, *actedIn* and *directed* are two kinds of links types, actor and film director are different object types.

$Person \xrightarrow{actedIn} Movie \xrightarrow{directed^{-1}} Person$ is a meta path shown by the dashed line in Fig. 1(a), $directed^{-1}$ is the opposite direction of the edge $directed$.

In addition, Toby Kebbell and Martin McCann belong to actor class. Toby Kebbell and Nigel Havers are not only the instances of actor class but also included in the actors who acted in movies Steven Spielberg directed. In order to distinguish the two kinds of sets, we call the latter as the fine grained set and the former as the coarse grained set.

4   The Proposed Method

In order to solve the problem of ESE with knowledge graph, we propose a novel approach called Meta Path based Entity Set Expansion (MP_ESE). As we have said, KG is a natural HIN, we employ the widely used meta path in HIN to exploit the potential common feature of seeds. The MP_ESE includes the following three steps. Firstly, we design a strategy of extracting candidate entities. Secondly, we develop an algorithm called Seed-based Meta Path Generation (SMPG) to automatically discover important meta paths between seeds. Finally, we get a ranking model through combining the meta paths with a heuristic strategy.

4.1   Candidate Entities Extraction

Because the number of entities in knowledge graph is extremely huge, it is unpractical and unreasonable to compute the similarity of each entity and seed. In order to reduce the number of candidate entities, we design a strategy, which leverages concept hierarchy structure introduced in Sect. 3, to get a proper set of candidate entities from knowledge graph. Specifically, it includes the following four steps as shown in Fig. 2. Step 1 obtains entity types of each seed. Step 2 generates the initial candidates types by the intersection operation. Step 3 filters the initial candidates types with the concept hierarchy structure. Step 4 extracts candidate entities of satisfying the ultimate candidates types.

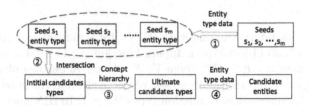

Fig. 2. The procedure of candidate entities extraction.

In order to clearly illustrate the process of candidate entities extraction, we take Fig. 1 as an example and choose Toby Kebbell and Nigel Havers as the seeds. Their entity types set is {person, actor} and {son, person, actor}, respectively. And the intersection of them is {person, actor} called the initial candidates types. These candidates types may be too general, which makes the number of candidate entities large. Therefore, we filter some candidates types using concept hierarchy structure as shown in Fig. 1(b). We choose the most specific class closest to the bottom as the ultimate candidates types. Here, we choose actor class. According to the ultimate types, we extract the candidate entities from Yago.

4.2   Seed-Based Meta Path Generation

In order to automatically discover meta paths between seeds, we design the Seed-based Meta Path Generation algorithm (SMPG). The basic idea is that SMPG begins to search the KG from all seeds and finds important meta paths that connect certain number of seed pairs, and the meta paths can reveal the implicit common character of seeds.

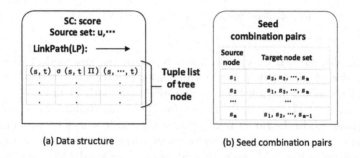

(a) Data structure   (b) Seed combination pairs

Fig. 3. Notation of data structure and seed combination pairs.

The process of meta path generation is traversing the KG in deed, and thus a novel tree structure is introduced in SMPG. SMPG works by expanding the tree structure and Fig. 3(a) shows the data structure of each tree node, which stores a tuple list of entity pairs with similarity value and the set of being visited entities. The tuple form of the list is $\langle (s, t), \sigma(s, t | \prod), (s, \cdots, t) \rangle$, where (s, t) denotes the source node and target node of the current path $\prod$. Each tree edge denotes the link type between entities. The root node of the tree contains all entity pairs composed of each seed and itself. SMPG starts to expand from the root node step by step to discover important meta paths. At each step, we check whether the score SC of the tree node is larger than the predefined threshold value ν, which guarantees that the meta path is important enough to reveal the character of seeds. If so, we pick out the corresponding meta path, otherwise make a move forward until the tree can not be further expanded. When moving forward, we choose the tree node with the maximum number of source set as well as the minimum number of tuples to expand, which indicates that the path of the tree node covers more seeds and has a better discriminability.

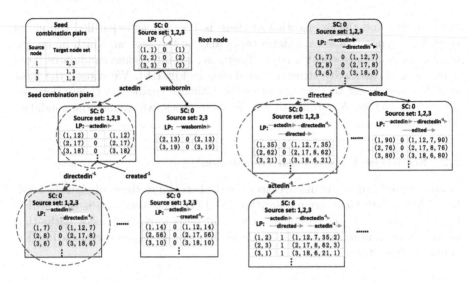

Fig. 4. Seed-based meta path generation method.

Specifically, in SMPG, we use a source set in the tree node to record the source nodes of all entity pairs in tuple list. In order to prevent the circle, we record the nodes having been visited along the path $\prod$ in $(s, \cdots, t)$ of the tuple $\langle (s, t), \sigma(s, t| \prod), (s, \cdots, t) \rangle$. Here, $\sigma(s, t| \prod)$ is the similarity that represents whether node t is in the target node set of source node s, it is 1 if so and 0 otherwise. The target node set of each source node can be found in seed combination pairs as shown in Fig. 3(b) and each seed can be combined with the other seeds. $\sigma(s, t| \prod)$ also means that whether the meta path connects the seed pair. And seed pairs that each meta path connects are also recorded. In addition, LP is the passing link path and the score SC of the tree node is the sum of all tuples similarity, which measures the importance of the tree node or path.

Let us elaborate the algorithm with an example shown in Fig. 4, where the set of seeds is {Toby Kebbell, Nigel Havers, Harrison Ford} marked as {1,2,3}. The set of seed combination pairs is {[1,(2,3)], [2,(1,3)], [3,(1,2)]} shown in Fig. 4. The root node of the tree contains all entity pairs composed of each seed and itself, and has $SC = 0$. The first expansion passes through two types of links: *actedIn* and *wasBornIn*, and gets two new tree nodes. For each new tree node, SMPG records each tuple, P and SC as well as source set. At the moment, all paths do not connect any seed pairs, so we choose the tree node with the maximum number of source set as well as the minimum number of tuples to expand. Here, we choose the tree node with link *actedIn* to expand and then get five new tree nodes. Figure 4 only demonstrates two of them. After the second expansion, there is not still path connecting seed pairs. Then we continue to choose the tree node with the maximum number of source set and the minimum number of tuples to expand, and we update the corresponding values. Except seeds 1, 2 and 3, the other marked entities such as 35, 62 denote those being visited in-between entities of the path. After several expansions, a length-4 path $Actor \xrightarrow{actedIn} Movie \xrightarrow{directed^{-1}} Person \xrightarrow{directed} Movie \xrightarrow{actedIn^{-1}} Actor$ is found shown by the dash line in Fig. 4. And we continue to repeat the process until the condition is satisfied or the tree can not be further expanded.

Algorithm 1. Seed-based Meta Path Generation Algorithm

Input: Knowledge graph G, seed set $S = \{s_1, s_2, \ldots, s_m\}$.
Output: The set of meta paths P, seed pairs SP that each meta path connects.
1 Create the root node of the tree T;
2 $sl \Leftarrow$ link types set;//the link needs connect 2 seeds or more
3 **while** T *can be expanded* **do**
4      $N \Leftarrow$ tree nodes with the maximum number of source set in T;
5      $n \Leftarrow$ tree node with the minimum number of tuples in N;
6   **for** *each tuple* $tp \in n$ **do**
7   get pair $tp.(s, t)$;
8   **for** *each neighbor* e *of* $tp.t$ *in* G **do**
9              $l \Leftarrow$ link from $tp.t$ to e;
10   **if** e *not being visited and* $l \in sl$ **then**
11   **if** l *not in* $n.child$ **then**
12   add l to $n.child$;
13   create a new child tree node with key $n.key + l$;
14   **if** $(s, e) \in$ *seed combination pairs* **then**
15                     $\sigma(s, e|\prod) \Leftarrow 1$;
16   add seed pair (s, e) to the tree node with key $n.key + l$;
17   **else**
18                     $\sigma(s, e|\prod) \Leftarrow 0$;
19   add e to the visited set $(s, \ldots, t)$;
20   insert tuple $\langle (s, e), \sigma(s, e|\prod), (s, e, \cdots, t) \rangle$ into tree node with key $n.key + l$;
21   add $tp.s$ to the source set of tree node with key $n.key + l$;
22   update the SC of tree node with key $n.key + l$;

23   **for** *each node* en *in* T **do**
24   **if** $en.SC > threshold\ \nu$ **then**
25   add meta path $\prod$ of en into P;
26   add the corresponding seed pairs that $\prod$ connects into SP;

27 return P, SP

We present the detailed steps of SMPG in Algorithm 1. Firstly, we create the root node of the tree in Step 1 and give some predefined constants in Step 2. Then we expand the tree and find the important meta paths in Steps 3–26. At each expansion, we choose the tree node with the maximum number of seeds as well as the minimum number of tuples in Steps 4, 5. Step 10 judges whether the link is in the set of the given link type, whether the neighbor node isn't visited before. If so, we make an expansion and examine whether the entity pair is in seed combination pairs in Step 14. If so, Step 15 records the connected entity pair. And we insert the new tuple into the corresponding tree node in Step 20. Meanwhile Step 21 adds the source node of the entity pair to the source set of the tree node and Step 22 updates SC. Steps 23–26 get the expected meta paths and the corresponding seed pairs.

4.3 Combination of Meta Path

SMPG discovers the important meta paths P, but the importance of each meta path is different for the further entity set expansion and it is related to the number of seed pairs that meta paths connect. Intuitively, the more seed pairs the meta path connects, the more important it is. Thus, we consider the ratio of SP_k and $m * (m - 1)$ to be the weight w'_k of meta path $p_k(p_k \in P)$, where SP_k is the number of seed pairs that meta path P_k connects, $m * (m - 1)$ denotes the total number of seed pairs and m is the number of seeds. In order to normalize w'_k, we define the final weight as follows:

$$w_k = \frac{w'_k}{\sum_{k=1}^{l} w'_k}, \tag{1}$$

where l is the number of meta paths P.

With the w_k, we can combine meta paths to get the following ranking model.

$$R(c_i, S) = \frac{1}{m} \sum_{j=1}^{m} \sum_{k=1}^{l} w_k \cdot r\{(c_i, s_j)|p_k\} \qquad s_j \in S, \ i \in \{1, 2, \cdots, n\}, \tag{2}$$

where c_i denotes the ith candidate entity, n is the number of candidates. $S = \{s_1, s_2, \cdots, s_m\}$ is the set of seeds. $r\{(c_i, s_j)|p_k\}$ denotes whether the path p_k connects c_i and s_j, it is 1 if connected and 0 otherwise.

We can compute relevance between each candidate entity and each seed using the ranking model in Eq. 2, and then rank all candidate entities.

5   Experiments

5.1   Dataset

As a typical KG, Yago is a huge semantic knowledge graph derived from Wikipedia, WordNet and GeoNames [17]. Currently, it has knowledge about more than 10 million entities and contains more than 120 million facts. We adopt "yagoFacts", "yagoSimpleTypes" and "yagoTaxonomy" parts of this dataset to conduct experiments, which contain 35 relationships, more than 1.3 million entities of 3455 instance classes. Table 1 is the description of the relevant data.

Table 1. Description of the data.

Data	Template of triples	# triples
yagoFacts	<entity relatinship entity>	4,484,914
yagoSimpleTypes	<entity rdf:type wordnet_type>	5,437,179
yagoTaxonomy	<wordnet_type rdfs:subclassof wordnet_type>	69,826

We choose four representative expansion tasks to evaluate the performance of MP_ESE. The classes used in these tasks are summarized as follows: actors of the movies Steven Spielberg directed, softwares of the companies located in Mountain View of California, movies whose director won National Film Award, and scientists of the universities located in Cambridge of Massachusetts. Four classes are written as Actor*, Software*, Movie* and Scientist*, the real number of instances in these four classes are 112, 98, 653 and 202, respectively.

5.2   Criteria

We employ two popular criteria of precision-at-k ($p@k$) and mean average precision (MAP) to evaluate the performance of our approach. $p@k$ is the percentage of top k results that belong to correct instances. Here, they are $p@30$, $p@60$ and $p@90$. MAP is the mean of the average precision (AP) of the $p@30$, $p@60$ and $p@90$. $AP = \frac{\sum_{i=1}^{k} p@i \times rel_i}{\# \text{ of correct instances}}$, where rel_i equals 1 if the result at rank i is correct instance and 0 otherwise.

5.3   Effectiveness Experiments

In this section, we will validate the effectiveness of MP_ESE on entity set expansion. Since there are no direct solutions for ESE on KG, we design three baselines. (1) Link-Based. According to the pattern-based methods in text or Web environment, we only consider 1-hop link of an entity, denoted as Link-Based. (2) Nearest-Neighbor. Inspired by QBEES [10], we consider 1-hop link and 1-hop entity at the same time, called Nearest-Neighbor. (3) PCRW. Based on the path constrained random walk [8], we only compare with length-2 path, denoted as PCRW. The reason is that the longer path needs more running time.

For each class introduced above, we randomly take three seeds from the instance set to conduct an experiment. We run algorithms 30 times and record the average results. In MP_ESE, we set the predefined threshold value ν to be $m * (m - 1)/2 + 1$, which can guarantee that the path connects half number of seeds or more, m is the number of seeds. And the max length of path is set to be 4 since meta paths with length more than 4 are almost irrelevant. The optimal parameters are set for other baselines.

The overall results of entity set expansion are given in Fig. 5. From Fig. 5, we can see that our MP_ESE approach achieves better performances than other methods on almost all conditions, especially on the Actor* and Movie* tasks. All baselines have very bad performances on Actor* and Movie*. We think the reason is that the 1-hop link or 1-hop entity can not further distinguish the character of the fine grained class but MP_ESE can distinguish well. On the Software* task, MP_ESE and PCRW have close performance. The reason is that Software* is an overlapping class and has another class label depicted by length-2 path $Software \xrightarrow{created^{-1}} Company \xrightarrow{created} Software$. Due to the fact that it has few semantic meaning, Link-Based has very bad performance. In all, MP_ESE has the best performances because it employs the important meta paths between seeds and can capture the subtle semantic meaning.

In order to intuitively observe the effectiveness of discovered meta paths, Table 2 depicts the top 3 meta paths returned by SMPG for Actor*. We observe that these meta paths reveal some common character of actor. The first meta path indicates that actors act in movies directed by the same director, which shows that SMPG can effectively mine the most important semantic meaning of Actor*. The second and the third meta paths imply that some actors act in

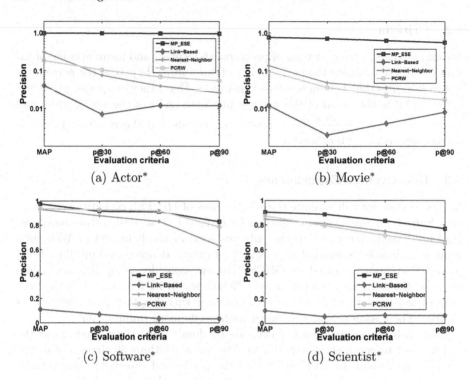

(a) Actor*   (b) Movie*

(c) Software*   (d) Scientist*

Fig. 5. The result of entity set expansion.

movies edited or composed by the same person. Through leveraging the important meta paths discovered by SMPG, we can find other entities belonging to the same class with seeds.

Table 2. Most relevant 3 meta paths for Actor*

Meta path	w
Person $\xrightarrow{actedIn}$ Movie $\xrightarrow{directed^{-1}}$ Person $\xrightarrow{directed}$ Movie $\xrightarrow{actedIn^{-1}}$ Person	0.2180
Person $\xrightarrow{actedIn}$ Movie $\xrightarrow{writeMusicFor^{-1}}$ Person $\xrightarrow{writeMusicFor}$ Movie $\xrightarrow{actedIn^{-1}}$ Person	0.1495
Person $\xrightarrow{actedIn}$ Movie $\xrightarrow{edited^{-1}}$ Person $\xrightarrow{edited}$ Movie $\xrightarrow{actedIn^{-1}}$ Person	0.1476

5.4   Impact of Seed Size

To evaluate the impact of seed size on the performance, we conduct relevant experiments in the range of seed size from 2 to 6 for Actor*. For each seed size, we randomly select the corresponding seeds from the instance set to conduct an experiment. We run our algorithm 30 times and record the maximum, the minimum and the average results, which are demonstrated in Table 3.

Table 3. The impact of seed size on Actor*.

Seed size	p@30			p@60			p@90			MAP		
	Max	Min	Mean	Max	Min	Mean	Max	Min	Mean	Max	Min	Mean
2	1.0	0.067	0.858	1.0	0.117	0.834	1.0	0.111	0.811	1.0	0.110	0.900
3	1.0	0.433	0.970	1.0	0.333	0.959	1.0	0.256	0.945	1.0	0.427	0.976
4	1.0	0.733	0.977	1.0	0.517	0.957	1.0	0.4	0.938	1.0	0.852	0.988
5	1.0	0.967	0.998	1.0	0.95	0.996	1.0	0.767	0.988	1.0	0.967	0.997
6	1.0	1.0	1.0	1.0	0.933	0.996	1.0	0.867	0.991	1.0	0.987	0.999

From Table 3, we can see that the performance has an improvement with the increasing of seed size. The performance with 2 seeds is the lowest, since 2 seeds do not contain plenty of information and may have several class labels. The performance with 3 seeds has been good enough. When the number of seeds is larger than 3, the improvement is tiny but the running time is much. Therefore, we employ 3 seeds in other experiments. In all, the proper seed size should be determined. Besides, there is a big difference between the maximum and minimum precisions because of the random seed sets.

5.5 Influence of Weight

To demonstrate the influence of the weight on the performance, We conduct experiments with different seed combinations of size 3 many times and record the average results. Table 4 reports the results with different weights on the four tasks introduced above. We can observe that the heuristic weight has better performance than average and random weights on the whole, which suggests that the importance of meta paths is different and some paths can better reflect the implicit character of seeds than others. For Software* task, weight has a tiny effect on performance, because the total number of meta paths is 5 and there exist several class labels.

Table 4. The impact of different weights.

Class	Heuristic weight				Average weight				Random weight			
	p@30	p@60	p@90	MAP	p@30	p@60	p@90	MAP	p@30	p@60	p@90	MAP
Actor*	0.970	0.959	0.945	0.976	0.948	0.934	0.917	0.958	0.941	0.915	0.889	0.949
Software*	0.915	0.911	0.830	0.926	0.911	0.908	0.824	0.925	0.918	0.903	0.812	0.930
Movie*	0.711	0.620	0.554	0.760	0.639	0.530	0.461	0.710	0.480	0.414	0.369	0.554
Scientist*	0.887	0.833	0.770	0.905	0.822	0.743	0.665	0.871	0.546	0.486	0.429	0.627

6   Conclusions

In this paper, we study the problem of entity set expansion in knowledge graph. We model knowledge graph as a heterogeneous information network and propose a *Meta Path* based *Entity Set Expansion* approach called MP_ESE, which employs the meta path to exploit the implicit common feature of seeds. In order to automatically find the important meta paths between seeds, MP_ESE designs a novel algorithm called SMPG. And then we design a heuristic strategy to assign the importance of meta paths. MP_ESE utilizes the weighted meta paths to expand entities. Experiments on Yago validate the effectiveness of MP_ESE.

Acknowledgements. This work is supported in part by the National Natural Science Foundation of China (No. 61375058), National Key Basic Research and Department (973) Program of China (No. 2013CB329606), and the Co-construction Project of Beijing Municipal Commission of Education.

References

1. Auer, S., Bizer, C., Kobilarov, G., Lehmann, J., Cyganiak, R., Ives, Z.: DBpedia: a nucleus for a web of open data. In: Aberer, K., et al. (eds.) ASWC/ISWC - 2007. LNCS, vol. 4825, pp. 722–735. Springer, Heidelberg (2007). doi:10.1007/978-3-540-76298-0_52
2. Cao, H., Jiang, D., Pei, J., He, Q., Liao, Z., Chen, E., Li, H.: Context-aware query suggestion by mining click-through and session data. In: KDD, pp. 875–883. ACM (2008)
3. Chen, J., Chen, Y., Du, X., Zhang, X., Zhou, X.: SEED: a system for entity exploration and debugging in large-scale knowledge graphs. In: ICDM, pp. 1350–1353. IEEE (2016)
4. Cohen, W.W., Sarawagi, S.: Exploiting dictionaries in named entity extraction: combining semi-Markov extraction processes and data integration methods. In: KDD, pp. 89–98. ACM (2004)
5. He, Y., Xin, D.: Seisa: set expansion by iterative similarity aggregation. In: WWW, pp. 427–436. ACM (2011)
6. Hu, J., Wang, G., Lochovsky, F., Sun, J.T., Chen, Z.: Understanding user's query intent with Wikipedia. In: WWW, pp. 471–480. ACM (2009)
7. Jindal, P., Roth, D.: Learning from negative examples in set-expansion. In: ICDM, pp. 1110–1115. IEEE (2011)
8. Lao, N., Cohen, W.W.: Relational retrieval using a combination of path-constrained random walks. Mach. Learn. **81**(1), 53–67 (2010)
9. Li, X.L., Zhang, L., Liu, B., Ng, S.K.: Distributional similarity vs. PU learning for entity set expansion. In: ACL, pp. 359–364. ACL (2010)
10. Metzger, S., Schenkel, R., Sydow, M.: Qbees: query by entity examples. In: CIKM, pp. 1829–1832. ACM (2013)
11. Pasca, M.: Weakly-supervised discovery of named entities using web search queries. In: CIKM, pp. 683–690. ACM (2007)
12. Qi, Z., Liu, K., Zhao, J.: Choosing better seeds for entity set expansion by leveraging Wikipedia semantic knowledge. In: Liu, C.-L., Zhang, C., Wang, L. (eds.) CCPR 2012. CCIS, vol. 321, pp. 655–662. Springer, Heidelberg (2012). doi:10.1007/978-3-642-33506-8_80

13. Sadamitsu, K., Saito, K., Imamura, K., Kikui, G.: Entity set expansion using topic information. In: ACL: HLT: short papers-Volume 2, pp. 726–731. ACL (2011)
14. Sarmento, L., Jijkuon, V., de Rijke, M., Oliveira, E.: More like these: growing entity classes from seeds. In: CIKM, pp. 959–962. ACM (2007)
15. Shi, C., Li, Y., Zhang, J., Sun, Y., Yu, P.S.: A survey of heterogeneous information network analysis. arXiv preprint arXiv:1511.04854 (2015)
16. Singhal, A.: Introducing the knowledge graph: things, not strings. Official Google Blog (2012)
17. Suchanek, F.M., Kasneci, G., Weikum, G.: Yago: a core of semantic knowledge. In: WWW, pp. 697–706. ACM (2007)
18. Sun, Y., Han, J., Yan, X., Yu, P.S., Wu, T.: Pathsim: meta path-based top-k similarity search in heterogeneous information networks. VLDB 4(11), 992–1003 (2011)
19. Wang, R.C., Cohen, W.W.: Language-independent set expansion of named entities using the web. In: ICDM, pp. 342–350. IEEE (2007)
20. Wang, R.C., Cohen, W.W.: Iterative set expansion of named entities using the web. In: ICDM, pp. 1091–1096. IEEE (2008)
21. Yu, X., Sun, Y., Norick, B., Mao, T., Han, J.: User guided entity similarity search using meta-path selection in heterogeneous information networks. In: CIKM, pp. 2025–2029. ACM (2012)

Using Network Flows to Identify Users Sharing Extremist Content on Social Media

Yifang Wei and Lisa Singh[✉]

Georgetown University, Washington, DC, USA
{yw255,lisa.singh}@georgetown.edu

Abstract. Social media has been leveraged by many groups to share their ideas, ideology, and other messages. Some of these posts promote extremist ideology. In this paper, we propose an approach for identifying users who engage in extremist discussions online. Our approach uses detailed feature selection to identify relevant posts and then uses a novel weighted network that models the information flow between the publishers of the relevant posts. An empirical evaluation of a post collection crawled from a web forum containing racially driven discussions and a tweet stream discussing the ISIS extremist group show that our proposed method for relevant post identification is significantly better than the state of the art and using a network flow graph for user identification leads to very accurate user identification.

Keywords: Extremism detection · Information flow network

1 Introduction

Users endorsing extremist ideology have been increasingly leveraging social media to spread their viewpoint and promote their agenda. For example, Islamic State of Iraq and Syria (ISIS) has been using social media platforms to share their ideas and recruit members/jihadists to their groups. This work presents a method for identifying users who share extremist viewpoints on social media. Our hope is that early identification will provide law enforcement options for early identification of individuals before they become dangerous.

Previous literature concerning identification of users sharing extremist content [4,6,7,14] assumes that the target user is a friend (or within a few hops of friends) of validated accounts affiliated with extremist groups [6,7], or alias accounts of these validated accounts [4,14]. While an important direction, our approach focuses on a method that does not require knowledge of the network structure in advance.

Specifically, we divide the problem into two subproblems: identifying relevant posts and then using those posts to identify individuals sharing content consistent with extremist views. Because different extremist groups use different vocabulary on social media, generic dictionaries are less effective. Therefore,

© Springer International Publishing AG 2017
J. Kim et al. (Eds.): PAKDD 2017, Part I, LNAI 10234, pp. 330–342, 2017.
DOI: 10.1007/978-3-319-57454-7_26

we propose an approach that begins by identifying features that best distinguish seed posts exhibiting extremist ideology from seed posts exhibiting anti-extremist ideology. We then use these features to identify relevant posts. The posts are then used to construct two weighted networks that model the information flow between the publishers of the identified posts. Different node centrality metrics are considered to evaluate users' contribution to spreading extremist ideology and anti-extremist ideology. Users with more contribution to sharing content and/or spreading extremist ideology than anti-extremist ideology are regarded as promoting extremist content.

The Main Contributions of Our Work Include: (1) we propose a new method for identifying relevant content that results in much higher accuracy than the state of the art; (2) we propose using information flow networks to find users sharing extremist and anti-extremist viewpoints; (3) we empirically evaluate our method on a web forum and a tweet stream and find that our method leads to accuracies above 90% in some case.

2   Related Literature

We briefly review the most recent research focusing on extremist content detection on social media, and extremist user detection on social media. We refer you to [5,15] for more general surveys of the broad area.

A primary task of extremist content detection on social media is crawling extremist contents, for which several solutions have been proposed [8,16]. Mel and Frank [16] classify a webpage into four sentiment-based classes: pro-extremist, anti-extremist, neutral, and irrelevant. They propose a web crawler capable of crawling webpages with pro-extremist sentiment, achieving 80% accuracy. Bouchard et al. [8] explore the features distinguishing terrorist websites from anti-terrorist websites, and further present a web crawler which automatically searches the Internet for extremist contents based on these features. Beyond simply crawling extremist contents, [10,12,20] analyze content exposing extremist ideology. Chatfield et al. [12] investigate the problem of how extremists leverage social media to spread their propaganda. They perform network and content analysis of tweets published by a user previously identified as an information disseminator of ISIS. Burnap et al. [10] study the propagation pattern of the information following a terrorist attack. Zhou et al. [20] analyze the hyperlink structure and the content of the extremist websites to better understand connections between extremist groups. Buntain et al. [9] study the response of social media to three terrorism attacks: the 2013 Marathon bombing in Boston, the 2014 hostage crisis in Sydney, and the 2015 Charlie Hebdo shooting in Paris. They find that the use of retweets, hashtags, and urls, along with reference to the events increases throughout the event. Rowe and Saif [17] conduct research on the behavior of Europe-based Twitter users during their transition toward pro-ISIS ideology. They show that these users exhibit significant behavioral divergence before and after their activation in terms of language usage and social interaction. We will leverage a combination of these approaches to help identify users conversing about extremist ideology.

Research on extremist user detection mainly leverages two techniques: authorship matching [4,14] and connectivity ranking [6,7]. Berger and Morgan [6] consider accounts within two hops of the verified accounts, and use different network statistics to measure these accounts' probability of being extremist-affiliated. [7] leverages an information flow network to identify accounts serving as information hubs for discussion promoting extremist ideology. Our approach is an improvement on these previous approaches in the following ways: (1) our approach pre-filters posts irrelevant to extremist ideology while [7] does not; (2) when ranking users according to their likelihood of being extremist-affiliated, our approach considers various network statistics and does not require friendship links to users exhibiting extremist behavior to identify users with this behavior. Finally, sentiment analysis is widely used to identify extremist content on social media [16,18], based on the assumption that in a discussion of extremist ideology, a post endorsing extremist ideology would contain positive sentiment. Our approach does not use sentiment, but we will compare to state of the art sentiment detectors in our evaluation.

3   Notation, Assumptions, and Definitions

This section presents definitions, assumptions, and a formal problem statement.

Notation and Assumptions. Let P be a set of posts associated with a particular forum or discussion stream on a social media site. These posts are written by a set of users U. A particular post p is written by a specific user u and is denoted p_u. For ease of exposition, we will use positive as a proxy for posts or users exhibiting extremist ideology, and negative as a proxy for posts or users exhibiting anti-extremist ideology.

We make the following assumptions about the post collection P:

1. There are features differentiating positive posts P^+ and negative posts P^-.
2. A user u might publish posts containing content that has contradictory viewpoints on extremism, e.g., a user endorsing extremist ideology might publish tweets exhibiting extremist ideology, while at the same time retweeting and replying to tweets containing content of the opposite position.
3. While user u may post a range of differing ideological messages, we assume that there is a direct relationship between the number of positive posts u propagates and the probability of promoting an extremist ideology. Similarly, we assume there is a direct relationship between the number of negative posts u propagates and the probability of promoting an anti-extremist ideology.
4. A user u that has an extremist viewpoint would play a more important role in spreading positive information than spreading negative information.

Problem Statement. Given a post collection P and the set of their publishers U, the task of identifying users sharing extremist content has two subtasks:

1. *Relevant Content Identification*: Identify the posts from P that have content consistent with extremist ideology (P^+) and anti-extremist ideology (P^-).
2. *Extremist User Identification*: Identify the users from U that have a viewpoint consistent with extremism (U^+).

4   Extremism Detection

The framework for our proposed approach is divided into two subtasks: Relevant Content Identification and Extremist User Identification. Figure 1 shows the major steps associated with each subtask: feature selection, post retrieval, network creation, user centrality calculation, and user centrality integration.

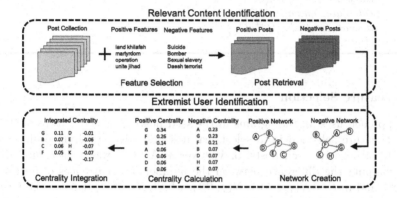

Fig. 1. The framework of our proposed approach

Algorithm 1 presents a high level view of this proposed approach. The input to our approach is a post collection P, a set of positive seed posts S^+, and a set of negative seed posts S^-. The output is a set of users U^+ identified as having a viewpoint consistent with extremism. The approach begins by identifying features F^+ best distinguishing positive seed posts from negative seed posts (Line 1), and features F^- best distinguishing negative seed posts from positive seed posts (Line 2). From P, posts containing F^+ and F^- are retrieved, respectively, denoted as P^+ and P^- (Line 3 and Line 4). An information flow network $G^+(V, E)$ is constructed, in which each node in V represents a user of a post in P^+ (Line 5). A directed edge in $G^+(V, E)$ is added to the network if a node v_i responds or reposts a message sent by node v_j. For nodes in $G^+(V, E)$, centrality metrics C^+ are calculated (Line 7). Similarly, using P^-, an information flow network $G^-(V, E)$ is constructed. For all the nodes in $G^-(V, E)$, centrality metrics C^- are calculated (Line 6 and Line 8). For all the nodes in $G^+(V, E)$ and in $G^-(V, E)$, their centrality C_u^+ and C_u^- are integrated into a single score centrality score C_u (Line 9). Users with positive integrated centrality are regarded as individuals sharing content containing extremist views.

Algorithm 1. High level algorithm for extremist user detection

---

Input:
A post collection: P
A set of positive seed posts: S^+
A set of negative seed posts: S^-

Output: *Users identified as endorsing extremism ideology*

1 $F^+ = select\_positive\_features(S^+, S^-)$
2 $F^- = select\_negative\_features(S^-, S^+)$
3 $P^+ = retrieve\_positive\_posts(F^+, P)$
4 $P^- = retrieve\_negative\_posts(F^-, P)$
5 $G^+(V, E) = create\_positive\_information\_flow\_network(P^+)$
6 $G^-(V, E) = create\_negative\_information\_flow\_network(P^-)$
7 $C^+ = calculate\_centrality(G^+)$
8 $C^- = calculate\_centrality(G^-)$
9 $C = integrate\_centrality(C^+, C^-)$
10 return $\{u|C_u > 0\}$

4.1   Feature Selection

Our approach uses all the ngrams in the seed posts S as the feature pool. The goal of feature selection is to select features best distinguishing positive seed posts from negative seed posts, or vice versa. A basic way to accomplish this is to compute the difference between the number of occurrences of an n-gram f in S^+ and in S^-: $N(f, S^+) - N(f, S^-)$, where $N(f, S^+)$ denotes the number of occurrences of f in S^+, and $N(f, S^-)$ denotes the number of occurrences of f in S^-. $N(f, S)$ only considers the intensity of a feature being used, ignoring its popularity among users. This can result in noisy features being retained. In an extreme case, a small number of users may repeatedly publish the same post. Considering only $N(f, s)$, most of words (excluding stopwords) in this post would be selected as features, including words that may be less relevant to extremism. Therefore, we consider incorporating user coverage of a feature. A higher user coverage indicates a feature being associated with posts written by different people. To evaluate user coverage of a ngram, we define a feature's Author Entropy $(\mathcal{E})$: $\mathcal{E}(f, S) = \sum_u (N(f, S_u) \log \frac{\sum_u N(f, S_u)}{N(f, S_u)})$, where $N(f, S_u)$ denotes the number of seed posts published by user u containing feature f. We use this notion in conjunction with intensity of feature usage to define a feature's importance $(\mathcal{I})$: $\mathcal{I}(f, S) = \mathcal{E}(f, S) \times N(f, S)$ We calculate each feature's importance $\mathcal{I}(f, S^+)$ for positive seed posts and $\mathcal{I}(f, S^-)$ for negative seed posts. Then we rank all the features in a descending order according to their $\mathcal{I}(f, S^+)$ score. We only consider those where $\mathcal{I}(f, S^+) > 0$ and $\mathcal{I}(f, S^-) = 0$. We select the top k features, denoted as F^+ and use these features to retrieve positive posts from P in the next step. In a similar way, we rank all the features in a descending order according to their $\mathcal{I}(f, S^-)$, and select the top k features, denoted as F^-, with the constraints that $\mathcal{I}(f, S^+) = 0$ and $\mathcal{I}(f, S^-) > 0$. These features are leveraged to retrieve negative posts from P.

4.2 Post Retrieval

This step focuses on retrieving posts containing the selected features F. To measure a post's relevance to a set of features F, we define Feature Relevance (Fr) as: $Fr(p, F) = \sum_{f \in F} Tf(f, p)Ae(f, p)$ where $Tf(f, p)$ represents the term frequency of feature f in a post p. We do not incorporate the commonly used inverse document frequency (IDF) into Feature Relevance. IDF prioritizes items most different from the rest of the corpus. Its success relies on the premise that a query term occurring less frequently in a corpus contains more information, and thus, is of more importance. However, this assumption does not hold in our case, e.g., the occurrence of white nationalist in a post would be a strong indicator of promoting white supremacy; however, it would not have a low IDF value on a forum discussing ethnic/racial issues.

For each post, we calculate its Feature Relevance to positive features $Fr(p, F^+)$ and to negative features $Fr(p, F^-)$, respectively. We retain all the posts satisfying $Fr(p, F^+) - Fr(p, F^-) > 0$, denoted as P^+, and retain all the posts satisfying $Fr(p, F^-) - Fr(p, F^+) > 0$, denoted as P^-.

4.3 Information Flow Network Creation

While there are many different representations of networks, we choose to leverage an information flow network. Posts on social media are information flows between users, e.g., on Twitter, a user can reply to, retweet, or mention other user(s) in a tweet; on a forum, a user can reply to or quote another user's post. We propose constructing information flow networks to identify the flow of extremist views.

A weighted information flow network $G = (V, E)$ is composed of a set of nodes $V(G) = v_1, ..., v_n$ and a set of edges $E(G) = e_1, ..., e_m$. Each node v_i represents a user u_i, and an edge (v_j, v_k) is added to the network G if user u_j responds to (e.g., retweets, replies to, quotes, etc.) a post of user u_k. If the post does not result in an edge between two users, a self edge is added to the graph. This is important because it allows us to capture extreme content that is being posted, but not necessarily propagating. G is also an edge weighted graph $\mathcal{W}(e_i)$, where an edge weight represents absolute information flow (described below).

While a single information flow graph can be constructed with edges containing both positive and negative post information, we choose to separately analyze positive and negative information flow by constructing two more focused graphs, G^+ and G^-. For the positive posts P^+, we construct an information flow network $G^+(V, E)$ and define the edge weight to be the the difference between the positive and negative features that are relevant to the post $Fr(p, F^+) - Fr(p, F^-)$. We choose this edge weight scheme since it reflects the *absolute* amount of positive information flowing along the edge. Similarly, for the negative posts P^-, we construct an information flow network $G^-(V, E)$ and define the edge weight to be the the difference between the negative and positive features that are relevant to the post $Fr(n, F^-) - Fr(p, F^+)$.

4.4 User Centrality Calculation

We use node centrality to measure each user's importance in sharing relevant positive and negative content. Among node centrality metrics, we consider degree (number of connections of u_i), node betweenness (fraction of shortest paths going through node v_i), pagerank (importance of node v_i based on importance of connections of v_i), and personalized pagerank (customized importance for specific types of graphs). In computing personalized pagerank [11], we need to designate a user-custom adjustment to pagerank in each iteration: $C = \alpha \mathbf{A} \times C + (1 - \alpha)C'$, where $\mathbf{A}$ denotes the transition probability matrix, C' is a user-custom vector to adjust the pagerank vector, and α is a user-custom weighing factor. We use the sum of the weight of outgoing edges incident to nodes as the user-custom vector: $C'_u = \sum_p Fr_u(p, F)$. In other words, we are increasing a user's score if he/she is sharing more content in G^+ or G^-. We calculate centrality C^+ and C^- for nodes in the positive $G^+(V, E)$ and the negative $G^-(V, E)$ networks.

4.5 User Centrality Integration

As stated in Sect. 3, we make the assumption that a user might publish posts containing content having contradicting viewpoints. We also assume that a user posting content consistent with extremist views would play a more important role in spreading positive information when compared to spreading negative information. In this step, we integrate C_u^+ and C_u^- into a single score C_u to measure a user's *absolute* importance for sharing/spreading positive information: $C_u = C_u^+ - C_u^-$. Users satisfying $C_u > 0$ are considered to be users promoting extremist views.

5 Evaluation

In this section we begin by describing the data sets, and then evaluate the different steps of our framework.

5.1 Data Sets

For our empirical analysis, we consider two distinct types of social media: microblogs and forums. The microblog data set is a Twitter stream. The forum data set is the Stromfront [2] data set.

Microblog Data Set: We work with an interdisciplinary team consisting of students, researcher, and policymakers. Some of them have years of in-field research experience in the Middle East. With help from our subject matter experts, we identified a set of hashtags that are related to ISIS. Using the Twitter API, we collected tweets containing these hashtags between September 2014 and April 2016. In total, this data set consists of 23 million tweets, published by approximately 2 million users. In previous work, we have shown that this data set contains extremist content [19]. For this evaluation, our task is to identify users sharing extremist views consistent with Islamic fundamentalism, or showing support for jihadist groups, including ISIS, Al-Qaeda, Jabhat Al Nusra, etc.

Forum Data Set: Stromfront [2] is a web forum that includes many radically driven discussions with a right-wing extremist focus. The most prevailing extremist ideologies are racism and antisemitism. The forum consists of scores of sub-forums, and each sub-forum has an explicitly-stated focus. From the sub-forums having an explicitly-stated focus on philosophy and ideology, we crawled 2.9 million posts. Our task here is to identify users promoting and/or sharing content that is racist or antisemitic.

5.2   Feature Selection

Subject matter experts on our team manually identify 1,300 tweets containing content promoting extremist ideology, 1,300 tweets containing content consistent with anti-extremist ideology, and 2,600 neutral tweets from the tweet collection. We use these 2,600 positive/negative tweets as seed posts to select features best distinguishing positive seed posts and negative seed posts. Basic pre-processings, including punctuation removal, stop word removal, and non-English word removal, are applied to these seed posts. Using the feature selection approach described in Sect. 4, the top 20 distinguishing features are identified. We set $k = 20$, since 20 is commonly regarded as an appropriate number of query terms for retrieval tasks [13]. Table 1 shows the top 5 positive features and negative features. We can see that the positive features have a clear focus on martyrdom and caliphate, while the negative features focus on terror, Daesh (a derogatory term for ISIS), and Yezidi (ISIS is holding thousands of Yezidi girls as slaves).

For the Stromfront post collection, we identify 500 seed posts containing extremist views and 500 seed posts containing anti-extremist views. Feature selection is applied to these 1,000 seed posts. The results are also shown in Table 1. We see that the positive features have a theme of white knights, while the negative features focus on evil and hate. We pause to mention that we also considered the simpler approach for feature selection that only uses the intensity of the word to identify features. Using this approach resulted in a larger number of noisy, information poor words, e.g., hey, entire, good claim, agenda.

5.3   Post Retrieval

For the tweet collection, using the top 20 positive features and top 20 negative features as query terms, 24,452 tweets and 462,436 tweets are retrieved as positive tweets and negative tweets; for the Stromfront post collection, 462,436 posts and 12,696 are retrieved as positive posts and negative posts.

Table 1. Top 5 features selected using feature importance $\mathcal{I}$

	Twitter		Stromfront	
	Positive	Negative	Positive	Negative
1	mujahideen	tcot	knights	nazis
2	martyrdom	yazidi	united white	evil nazis
3	allah accept	bombers	klux klan	hate crime
4	alhamdulilah	suicide bombers	white knights	warmongers
5	martyrdom operation	kittens daesh	jews	liberals

To better understand the accuracy of using our post retrieval method (referred to as XtremePost), we compare our method to four other methods:

1. A Naive Bayes classifier (NB Classifier) that incorporates unigrams, emoticons, urls, and POS taggers to identify extremism. It was built using the labeled ground truth data. Note, we did experiment with other classic machine learning algorithms, including Logistic Regression, Support Vector Machines, K Nearest Neighbors and Decision Trees. Naive Bayes performed better than the other models.
2. Two state of the art sentiment detection tools - Stanford CoreNLP [1] and vaderSentiment [3][1]
3. Using features generated by computing the difference in frequency intensity of positive and negative posts (referred to as Count).

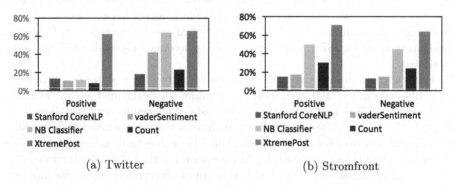

(a) Twitter   (b) Stromfront

Fig. 2. Accuracy of extremism post identification by different approaches

Due to the lack of ground truth labels of the retrieved tweets, we randomly sample 200 posts from each of the two classes. The results are shown in Fig. 2.

---

[1] vaderSentiment is also the tool employed by [18] to identify extremist users.

We can see that our approach achieves 62%/66% accuracy in identifying positive/negative posts in the tweet collection, and 71%/64% accuracy in the Stromfront post collection, significantly outperforming other methods. When analyzing the results, we find that our approach performs well because: (1) it is insensitive to abnormal grammar structures and made-up words; (2) it is insensitive to data with a skewed distribution across classes. The Naive Bayes classifier has the second best performance, but it has a very low accuracy in identifying positive posts in the tweet collection. This is not surprising since only 0.5% of the tweets in this collection are positive. The underperformance of two sentiment analyzers can be attributed to the prevalence of noise in social media posts. Social media posts tend to contain made-up words and do not to follow normal grammar rules.

5.4   Centrality Calculation

Based on the retrieved positive/negative posts, we build the information flow network. For the tweet network, an edge is added to the network if a user is retweeted, replied, or mentioned in a tweet. For a tweet that does not refer to another user, a self-edge incident to its author is added to the network. For the Stromfront post network, an edge is added to the network if a user is quoted or replied to in a post. Similarly, self-edges are added if the post does not refer to another user.

Network analysts use different measures of centrality to define importance. As mentioned in Sect. 4, we consider degree, betweenness, pagerank, and personalized pagerank. We compare all of these methods to a simple method that only considers the frequency of positive and negative posts. All the methods return a comparable number of users. For the tweet collection, the number is around 7,000; for Stromfront post collection, the number is around 20,000.

We sample 100 users from all the users identified by the different methods and evaluate what percentage of the 100 users post extremist content. The results are shown in Fig. 3. We can see that using the information flow graph and the centrality metrics of degree and personalized pagerank result in the highest accuracies. However, all of the information flow methods and the simpler count method have comparable accuracies.

In order to better understand the percentage of all the users endorsing extremist ideology, we take a 400 user random sample from each post collection, and manually check their posts. We find that about 38% users in the Stromfront post collection endorse extremist ideology in their posts, while only 0.5% users in the tweet corpus endorse extremist ideology in theirs posts.

In another experiment, we rank the identified users according to their centrality scores for each method. We are interested in the accuracy of our proposed approach at different positions along the scale. We take a 50 user sample from users identified as endorsing extremist ideology by each method at different positions along the scale, and manually evaluate what percentage of them post extremist content. For example, we take all the top 50 users, sample 50 users from the top 1,000 users, sample 50 users from the 1,001 to 2,000 ranking, etc.;

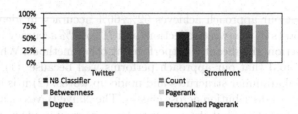

Fig. 3. Accuracy of extremist user identification by different methods

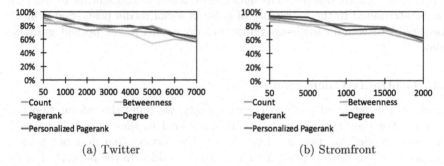

(a) Twitter   (b) Stromfront

Fig. 4. Accuracy of extremist user identification at different positions along the scale of users ranked according to different methods

then we classify them as sharing extremist content or not. The results are shown in Fig. 4. We observe a trend that the higher an identified user ranks in the scale, the more likely the user posts contain extremist views. This means that all the information flow methods can identify the most extreme users effectively.

Finally, we evaluate the efficiency of different methods. Since Betweenness, Pagerank, and Personalized Pagerank are computationally intensive, we implement them on a distributed yarn cluster, which consists of 12 nodes. Each node has 16 CPUs. In computing these metrics, we designate 10 executors, with 10 cores for each executor, which result in 100 cores. On the other hand, since both the Count and Degree method are computationally trivial and the overhead of initializing the yarn cluster would cost more than calculating these two metrics,

Table 2. Cost of centrality computation. ×100 represents 100 cores allocated

	Twitter	Stromfront
Count	0.6 s	0.9 s
Degree	0.8 s	1 s
Betweenness	744 s × 100	86 s × 100
Pagerank	911 s × 100	261 s × 100
Personalized pagerank	942 s × 100	302 s × 100

we implement them as single thread programs. Table 2 shows the time cost of the different methods. We can see that the Betweenness, Pagrank, and Personalized Pagerank methods are considerably more expensive than the Count and Degree methods. Considering both accuracy and computational cost, the Degree method ends up being the best performer.

6   Conclusions

In this paper, we propose an approach for identifying users endorsing extremist ideology on social media. Our approach first identifies posts exposing extremist ideology and posts exposing anti-extremist ideology, then constructs two weighted networks to model the information flow between the publishers of the identified posts. Different node centrality metrics are considered to evaluate users' contribution to spreading extremist ideology and anti-extremist ideology. Users with more contribution to spreading extremist ideology than anti-extremist ideology are labeled as individuals sharing extremist views. We empirically evaluate our approach on two social media post collections. We find that our approach for identifying posts that contain extremist views is significantly better than the state of the art. We also showed that using an information flow graph can achieves over 90% accuracy when identifying the top scoring users sharing extremist content.

Acknowledgments. We thank subject matter experts for labeling data and their general subject matter expertise provided throughout the process. This work was supported in part by the National Science Foundation (NSF) Grant SMA-1338507 and the Georgetown University Mass Data Institute (MDI).

References

1. Stanford corenlp sentiment analyzer. http://stanfordnlp.github.io/CoreNLP/sentiment.html
2. Stromfront. https://www.stormfront.org/forum/
3. Vader sentiment analysis. https://github.com/cjhutto/vaderSentiment
4. Abbasi, A., Chen, H.: Applying authorship analysis to extremist-group web forum messages. IEEE Intell. Syst. **20**(5), 67–75 (2005)
5. Agarwal, S., Sureka, A.: Applying social media intelligence for predicting and identifying on-line radicalization and civil unrest oriented threats. CoRR, abs/1511.06858 (2015)
6. Berger, J., Morgan, J.: The ISIS Twitter census: defining and describing the population of ISIS supporters on Twitter. Brookings Proj. US Relat. Islamic World **3**(20) (2015)
7. Berger, J., Strathearn, B.: Who matters online: measuring influence, evaluating content and countering violent extremism in online social networks (2013)
8. Bouchard, M., Joffres, K., Frank, R.: Preliminary analytical considerations in designing a terrorism and extremism online network extractor. In: Mago, V.K., Dabbaghian, V. (eds.) Computational Models of Complex Systems. ISRL, vol. 53, pp. 171–184. Springer, Cham (2014). doi:10.1007/978-3-319-01285-8_11

9. Buntain, C., Golbeck, J., Liu, B., LaFree, G.: Evaluating public response to the Boston Marathon bombing and other acts of terrorism through Twitter. In: ICWSM, pp. 555–558 (2016)

10. Burnap, P., Williams, M.L., Sloan, L., et al.: Tweeting the terror: modelling the social media reaction to the woolwich terrorist attack. Soc. Netw. Anal. Mining 4(1), 1–14 (2014)

11. Chakrabarti, S.: Dynamic personalized pagerank in entity-relation graphs. In: WWW, pp. 571–580. ACM (2007)

12. Chatfield, A.T., Reddick, C.G., Brajawidagda, U.: Tweeting propaganda, radicalization and recruitment: Islamic state supporters multi-sided Twitter networks. In: dg.o, pp. 239–249. ACM (2015)

13. Manning, C., Raghavan, P., Schtze, H.: Introduction to Information Retrieval. Cambridge University Press, Cambridge (2008)

14. Dahlin, J., Johansson, F., Kaati, L., Martenson, C., Svenson, P.: Combining entity matching techniques for detecting extremist behavior on discussion boards. In: ASONAM, pp. 850–857. IEEE (2012)

15. Hale, W.C.: Extremism on the world wide web: a research review. Crim. Justice Stud. 25(4), 343–356 (2012)

16. Mei, J., Frank, R.: Sentiment crawling: extremist content collection through a sentiment analysis guided web-crawler. In: ASONAM. ACM (2015)

17. Rowe, M., Saif, H.: Mining pro-isis radicalisation signals from social media users. In: ICWSM, pp. 329–338 (2016)

18. Scrivens, R., Davies, G., Frank, R., Mei, J.: Sentiment-based identification of radical authors. In: ICDMW, pp. 979–986. IEEE (2015)

19. Wei, Y., Singh, L., Martin, S.: Identification of extremism on Twitter. In: Social Network Analysis Surveillance Technologies (SNAT) at ASONAM (2016)

20. Zhou, Y., Reid, E., Qin, J., Chen, H., Lai, G.: Us domestic extremist groups on the web: link and content analysis. IEEE Intell. Syst. 20(5), 44–51 (2005)

MC3: A Multi-class Consensus Classification Framework

Tanmoy Chakraborty$^{(\boxtimes)}$, Des Chandhok, and V.S. Subrahmanian

University of Maryland, College Park, USA
{tanchak,vs}@umiacs.umd.edu, dchandho@terpmail.umd.edu

Abstract. In this paper, we propose **MC3**, an ensemble framework for multi-class classification. **MC3** is built on "consensus learning", a novel learning paradigm where each individual base classifier keeps on improving its classification by exploiting the outcomes obtained from other classifiers until a consensus is reached. Based on this idea, we propose two algorithms, **MC3-R** and **MC3-S** that make different trade-offs between quality and runtime. We conduct rigorous experiments comparing **MC3-R** and **MC3-S** with 12 baseline classifiers on 13 different datasets. Our algorithms perform as well or better than the best baseline classifier, achieving on average, a 5.56% performance improvement. Moreover, unlike existing baseline algorithms, our algorithms also improve the performance of individual base classifiers up to 10%. (The code is available at https://github.com/MC3-code.)

Keywords: Ensemble learning · Consensus · Multi-class classification

1 Introduction

Suppose there are multiple experts sitting together. The moderator gives an object (and its features) and asks the experts to predict its true class from a set of predefined classes. In the first round, experts use their individual heuristics to predict the true class. At the end of the round, every expert discloses her prediction, and learns the predictions made by others. If the moderator does not receive a consensus between the experts' predictions, she allows another round of predictions. In the next round, each expert uses the knowledge of others' predictions, and may modify her heuristics to come up with some other class for that object. Similarly at the end of the second round, the moderator again checks for a consensus. The iteration continues until a consensus is achieved; and finally the class obtained at the consensus is assigned to the object. This is the underlying philosophy of our proposed ensemble classification framework **MC3** (Multi-Class Consensus Classification). Experts are like base classifiers and the final prediction is achieved via consensus.

The power of ensemble classification has been widely accepted by the machine learning community [16]. Existing ensemble classifiers such as Bagging [4], Boosting [17] improve predictions of a base classifier by learning from mistakes.

© Springer International Publishing AG 2017
J. Kim et al. (Eds.): PAKDD 2017, Part I, LNAI 10234, pp. 343–355, 2017.
DOI: 10.1007/978-3-319-57454-7_27

In contrast, our ensemble algorithms support learning across multiple base classifiers in order to achieve consensus to not only achieve high prediction accuracy, but also to improve the performance of base classifiers individually. We propose two versions of **MC3**: (i) **MC3-R**, a recursive version of **MC3**, (ii) **MC3-S**, a two-stage single iterative version of **MC3**. Although **MC3-R** is computationally more expensive, it produces better predictions than **MC3-S** (which slightly trades off accuracy for lower runtime).

We conduct experiments on 13 datasets with different properties (w.r.t. size of data and feature set, class distribution etc.). We compare **MC3-R** and **MC3-S** with 12 (7 standalone and 5 ensemble) classifiers, and observe that our algorithms are either as good as the best baseline or sometimes perform even better than that, achieving an average of 5.56% higher accuracy than the best baseline. Although the best baseline varies from one dataset to another, our algorithm is a single algorithm that achieves the best performance across different datasets. Additionally, the performance of individual base classifiers is improved up to 10%. We also suggest how to select the best parameters for our classifiers.

2 Related Work

Ensemble classification has been an active research area in machine learning (see an exhaustive survey in [16]). Due to the abundance of literature in this area, we restrict our discussion to recent work. Classical ensemble classifiers such as Bagging [4], Boosting [17], Stacking [15], Random Forest [5] etc. [18] use reduced versions of training samples to train ensemble classifiers. *BPNNAdaBoost* and *BPNN-Bagging* [20] built on AdaBoost and Bagging are back-propagation neural network models for financial distress prediction. [19] used an Artificial Bee Colony algorithm for selecting the optimal base classifier and meta configuration in stacking. [8] proposed a classifier ensemble particularly for incomplete datasets. [12] used Artificial Neural Networks with Levenberg-Marquardt back propagation as base classifiers for the Rotation Forest ensemble. [9] combined bagging and rank aggregation. [22] proposed an ensemble classification approach based on supervised clustering for credit scoring. [11] designed a new ensemble pruning method which highly reduces the complexity bagging.

The philosophy behind the existing methods is that base classifiers perform well in different segments of the data and make mistakes in other segments. Ensemble methods combine predictions by balancing between quality and diversity. However, the philosophy behind our ensemble classifiers is completely different – we let each base classifier leverage the predictions made by other classifiers and train itself iteratively to come to a consensus. At the end of the iterations, we expect all the base classifiers to produce *exactly the same* prediction for an unknown instance. This in turn not only provides a strong ensemble classification in general, but also improves the performance of individual base classifiers.

Suppose we are given $\mathcal{S}_{tr}$, a set of M_{tr} training instances taken from a domain $\mathcal{D}$. The ith entry of $\mathcal{S}_{tr}$ is represented by $S_{tr}^i = (\mathbf{x}_i, y_i)$, where $\mathbf{x}_i \in \mathbb{R}^d$ is a

d-dimensional feature vector[1] and y_i is the true class of S_{tr}^i chosen from a set $\mathcal{L}$ (where $l = |\mathcal{L}| \geq 2$). We are also given $\mathcal{S}_{ts}$, a similar set of M_{ts} test instances taken from $\mathcal{D}$; however the true classes of the instances $S_{ts}^j = (\mathbf{x}_j, y_j) \in \mathcal{S}_{ts}$ are unknown, i.e., $y_j = \phi$. It is also known that each unknown instance belongs to only one class in $\mathcal{L}$. The task is to predict the true class of each instance in $\mathcal{S}_{ts}$.

Ensemble Classification: Let $\mathcal{CF} = \{CF_1, \ldots, CF_M\}$ be a set of base classifiers. Each base classifier CF_j is trained on $\mathcal{S}_{tr}$ to predict a probability distribution over possible classes for an unknown instance $S_i \in \mathcal{S}_{ts}$, i.e., $\mathbf{p}^j(S_i) = \{p^j(L_1|S_i), \ldots, p^j(L_l|S_i)\}$, based on which L_i^j is assigned to S_i such that $L_i^j = arg\,max_k p^j(L_k|S_i)$. The final class $L_i \in \mathcal{L}$ of S_i is obtained by feeding the output classes/probabilities obtained from all the base classifiers into an ensemble function $\mathcal{E}(L_i^1, \ldots, L_i^M)$ or $\mathcal{E}(\mathbf{p}^j(S_i), \ldots, \mathbf{p}^M(S_i))$. The task is to design an appropriate ensemble function to predict the final class of an unknown instance.

3   Multiclass Consensus Classification

We propose two ensemble classifiers. The first classifier, **MC3-R** is a recursive multi-class consensus classifier that achieves consensus by recursively updating each base classifier using the outcomes of other base classifiers. This classifier turns out to be most accurate, although it suffers from high computational complexity. The second classifier, **MC3-S** is a single iteration multi-class consensus classifier that approximates consensus in one iteration. **MC3-S** is much faster than **MC3-R** and is the closest competitor in terms of accuracy. In the rest of the section, we will elaborate these classifiers.

3.1   MC3-R: Recursive MC3

MC3-R (pseudo-code in Algorithm 1) takes the following inputs – training set $\mathcal{S}_{TR}$, test set $\mathcal{S}_{TS}$, a set of M base classifiers $\{\mathcal{CF}\}_{i=1}^M$, a number of iterations $Iter$, a subset selection strategy SS that selects a subset of the M base classifiers, a combination function W, and a consensus function $CONS$. It consists of two fundamental steps – achieving consensus and combining predictions of the base classifiers. **MC3-R** trains each base classifier $Iter$ times on the training set separately and selects the best parameter setting. In each iteration, **MC3-R** achieves consensus after ι levels (the value of ι varies across different iterations). Finally, **MC3-R** combines the outputs of all optimal base classifiers using a weighted function W and predicts the final classes of $\mathcal{S}_{TS}$.

Achieving Consensus: In Step 7 of Algorithm 1, **MC3-R** invokes a `getConsensus` function which starts by randomly dividing $\mathcal{S}_{TR}$ equally into $\mathcal{S}_{TR_1}$ and $\mathcal{S}_{TR_2}$ (Step 20). It then calls `getMetaFeatures` twice – in the first (*resp.* second) call each CF_i is trained on $\mathcal{S}_{TR_1}$ (*resp.* $\mathcal{S}_{TR_2}$) to predict $\mathcal{S}_{TR_2}$ (*resp.* $\mathcal{S}_{TR_1}$).

---

[1] We use boldface lower case letters for vectors (e.g., $\mathbf{x}$).

Algorithm 1. MC3-R: Recursive MC3

Data: Training set $\mathcal{S}_{TR}$, Test set $\mathcal{S}_{TS}$, No. of iterations $Iter$, Set of M base classifiers $\{CF\}_{i=1}^{M}$, Subset selection function SS, a combination function W, Consensus function $CONS$

Result: Prediction of $\mathcal{S}_{TS}$

```
 1  𝒫*(i) = φ, 1 ≤ i ≤ M          // Stores the optimal parameters of the classifiers
 2  𝒜 = φ                          // Stores the fitness error of the classifiers
 3  j = 1
 4  while j ≤ Iter do
 5  │   ι = 1
 6  │   𝒫ᵢⁱ = φ, 1 ≤ i ≤ M
 7  │   𝒜=getConsensus (CF, 𝒮_TR, ι,j)
 8  │   j + +
 9  end
    // Select best optimal parameters of base classifiers from multiple iterations
10  for (i = 1; i ≤ M; i + +) do
11  │   ĵ = argmax₁≤ⱼ≤Iter 𝒜ⱼⁱ, where 𝒜ⱼⁱ ∈ 𝒜
12  │   for (k = 1; k ≤ ι; k + +) do
13  │   │   ĈFᵢᵏ is constructed using Parameter*(i,ĵ,k), where Parameter*(i,ĵ,k) ∈ 𝒫*(i)
14  │   end
15  end
16  Use ĈFᵢᵏ (1 ≤ i ≤ M) in k different levels (where 1 ≤ k ≤ ι) to predict the classes of 𝒮_TS
17  Combine the predictions of {ĈFᵢ}ᵢ₌₁ᴹ using W to obtain final classes of 𝒮_TS
18  return Classes of 𝒮_TS    Procedure getConsensus(CF, 𝒮_TR, ι,j)
19  │   Divide 𝒮_TR equally into 𝒮_TR₁ and 𝒮_TR₂ randomly
    │   // Obtain ιᵗʰ level optimal classifiers
20  │   Parameter*(i,j,ι) = 0, 1 ≤ i ≤ M    // Parameters of CFᵢ at ιᵗʰ level of jᵗʰ iter.
21  │   𝒮ᵉ_TR₂, CF*ⁱ ← getMetaFeatures(CF, 𝒮_TR₁, 𝒮_TR₂)
22  │   𝒮ᵉ_TR₁, CF*ⁱ ← getMetaFeatures(CF, 𝒮_TR₂, 𝒮_TR₁)
23  │   Parameter*(i,j,ι) = Parameters of CFᵢ*ⁱ(j), 1 ≤ i ≤ M
24  │   𝒫*(i) = 𝒫*(i) ∪ Parameter*(i,j,ι), 1 ≤ i ≤ M
25  │   if CONS == True then
26  │   │   Errorⱼⁱ = Fitness error of CFᵢ* at the end of jᵗʰ iteration
27  │   │   𝒜ⱼⁱ = 1 − Errorⱼⁱ
28  │   │   𝒜 = 𝒜 ∪ 𝒜ⱼⁱ
29  │   │   return 𝒜
30  │   end
31  │   else
32  │   │   S = 𝒮ᵉ_TR₁ ∪ 𝒮ᵉ_TR₂
33  │   │   Use SS to select a subset of meta-features from S and augment it with the original
    │   │       feature set of 𝒮_TR to get an expanded feature set 𝒮ᵉ_TR
34  │   │   getConsensus(CF, 𝒮ᵉ_TR, ι + +)
35  │   end
36  Procedure getMetaFeatures(CF, 𝒮_tr, 𝒮_ts)
37  │   K-fold cross-validation of each CFᵢ on 𝒮_Tr to get optimal classifier CFᵢ*
38  │   Use each CFᵢ* to predict the classes of 𝒮_ts and treat them as meta-features
39  │   Augment the meta-features with the original features and create an expanded feature
    │       set for 𝒮_ts (call it 𝒮ᵉ_ts)
40  │   return 𝒮ᵉ_ts, CF*
```

The predictions of CF_is then become meta-features for $\mathcal{S}_{TR_2}$ (*resp.* $\mathcal{S}_{TR_1}$) and they are augmented with the original features to generate an expanded feature set $\mathcal{S}_{TR_2}^{e}$ (*resp.* $\mathcal{S}_{TR_1}^{e}$). In Step 26, if **MC3-R** reaches consensus based on the consensus function $CONS$ (possible definitions are given in Sect. 4), it returns the fitness error of individual classifiers which will further be used for best parameter selection (Step 11); otherwise a subset of meta-features are selected using SS (possible definitions are given in Sect. 4) and augmented with the original features of $\mathcal{S}_{TR}$ to obtain an extended set $\mathcal{S}_{TR}^{e}$ (Step 40). The entire getConsensus is repeated recursively until the consensus is achieved (Step 35).

Combining Predictions: Once consensus is achieved, the best parameters for each classifier at each level are selected based on the fitness error in different iterations (Step 13). The optimal classifiers are then used to predict the class of $\mathcal{S}_{ts}$. Finally, the predictions of the classifiers are combined using W (possible definitions are given in Sect. 4) to generate the final class of $\mathcal{S}_{ts}$ (Step 17).

3.2   MC3-S: Single Iteration MC3

MC3-S (pseudo-code in Algorithm 2) takes the same inputs as **MC3-R** (except the consensus function $CONS$ since it assumes that consensus is achieved after two levels). **MC3-S** starts by randomly dividing $\mathcal{S}_{TR}$ into two equal subsets $\mathcal{S}_{TR_1}$ and $\mathcal{S}_{TR_2}$ (Step 7). Each classifier CF_i considers $\mathcal{S}_{TR_1}$ and uses k-fold cross validation to obtain optimal parameter settings (we refer to each such optimal classifier as CF_i^*) (Step 8). Each CF_i^* is then used to predict the classes of $\mathcal{S}_{TR_2}$ (Step 9).

In the next step, the classes of $\mathcal{S}_{TR_2}$ obtained from optimal classifiers $\mathcal{CF}^*$ are used as meta-features of $\mathcal{S}_{TR_2}$. We then select a subset of meta-features using SS (Step 10) and augment them with the original features of $\mathcal{S}_{TR_2}$ to get an expanded set $\mathcal{S}_{TR_2}^{e}$ (Step 11). Note that these optimal classifiers $\mathcal{CF}^*$ will be used later for generating new features. After this, we consider each original classifier CF_i and run k-fold cross-validation on $\mathcal{S}_{TR_2}^{e}$. This step will produce another optimal set of classifiers denoted by $\{\mathcal{CF}^{**}\}_{i=1}^{n}$ (Step 12). This set of optimal classifiers will be used later for final class prediction of unknown instances.

The above steps (Steps 7–16) are repeated *Iter* times, and the optimal parameter settings for $\mathcal{CF}^*$ and $\mathcal{CF}^{**}$ are stored into *Parameter** and *Parameter***, respectively. At the same time, the accuracies of $\mathcal{CF}^{**}$ are stored in *Accuracy*.

Once *Iter* iterations are completed, we select the best parameter setting for each CF_i^* and CF_i^{**} based on the values stored in *Accuracy*. We call $\hat{\mathcal{CF}}^*$ for feature generation and $\hat{\mathcal{CF}}^{**}$ for class prediction (Step 17–20). Finally, on the test set $\mathcal{S}_{TS}$, the $\hat{\mathcal{CF}}^*$ classifiers are run to generate meta-features (Step 21), and SS is used to select a subset of meta-features (Step 22). $\hat{\mathcal{CF}}^{**}$ are then run to predict the classes (Step 23). The final class of each instance in $\mathcal{S}_{TS}$ is generated by combining the outputs of $\hat{\mathcal{CF}}^{**}$ using W (Step 24).

4   Functions Used in MC3-R and MC3-S

Here, we describe some possible definitions of the functions used in our classifiers.

• **Meta-feature Generation:** Experimental evidence from prior research [10, 15, 21] indicates that augmenting the *confidence* of base classifiers in predicting class levels as meta-features is more useful than considering the *predicted classes* directly. Ting and Witten [21] suggested using as meta-features, the probabilities

---

Algorithm 2. MC3-S: Single Iteration **MC3**

Data: Training set $\mathcal{S}_{TR}$, Test set $\mathcal{S}_{TS}$, No. of iterations $Iter$, Set of M base classifiers $\{CF\}_{i=1}^M$, Subset selection function SS, a combination function W

Result: Prediction of $\mathcal{S}_{TS}$

1 **for** $(i = 1; i \leq M; i++)$ **do**
2     $Parameter^*(i, j) = 0, 1 \leq j \leq Iter$
3     $Parameter^{**}(i, j) = 0, 1 \leq j \leq Iter$
4     $Accuracy(i, j) = 0, 1 \leq j \leq Iter$

5 $j = 1$
6 **while** $j \leq Iter$ **do**
7   Divide $\mathcal{S}_{TR}$ equally into $\mathcal{S}_{TR_1}$ and $\mathcal{S}_{TR_2}$ randomly
8   K-fold cross-validation of each CF_i on $\mathcal{S}_{TR_1}$ to get level-1 optimal classifier CF_i^*
9   Use each CF_i^* to predict the classes of $\mathcal{S}_{TR_2}$ and consider them as meta-features
10   Use SS to select a subset of features from the set of meta-features
11   Augment the selected subset of meta-features with the original features and create an expanded feature set for $\mathcal{S}_{TR_2}$ (call it $\mathcal{S}_{TR_2}^e$)
12   K-fold cross validation of each CF_i on $\mathcal{S}_{TR_2}^e$ to get level-2 optimal classifier CF_i^{**}
13     $Parameter^*(i, j)$ = Parameters of $CF_i^*, 1 \leq i \leq M$
14     $Parameter^{**}(i, j)$ = Parameters of $CF_i^{**}, 1 \leq i \leq M$
15     $Accuracy(i, j)$ = Accuracy of $CF_i^{**}, 1 \leq i \leq M$
16     $j = j + 1$

// Best CF_i^* (resp. CF_i^{**}) is used for feature generation (resp. final classification)

17 **for** $(i = 1; i \leq M; i++)$ **do**
18     $\hat{j} = \text{argmax}_{1 \leq j \leq Iter} Accuracy(i, j)$
19     $\hat{CF}_i^*$ is constructed using $Parameter^*(i, \hat{j})$
20     $\hat{CF}_i^{**}$ is constructed using $Parameter^{**}(i, \hat{j})$

// Prediction on test set

21 Use each $\hat{CF}_i^*$ to predict classes of $\mathcal{S}_{TS}$ and use them as meta-features
22 Use SS to select a subset of meta-features and augment it with the original feature set of $\mathcal{S}_{TS}$ to get an expanded feature set $\mathcal{S}_{TS}^e$
23 Predict the classes of $\mathcal{S}_{TS}^e$ using $\hat{CF}_i^{**}$
24 Combine the predictions of $\{\hat{CF}_i^{**}\}_{i=1}^M$ using W to obtain final classes of $\mathcal{S}_{TS}$
25 **return** Classes of $\mathcal{S}_{TS}$

(often used as confidence values) predicted for each possible class by each base classifier, i.e., $\mathbf{p}^j(S_i) = \{p^j(L_1|S_i), \ldots, p^j(L_l|S_i)\}$, where $j = 1, \ldots, M$ and $L_k \in \mathcal{L}$. We further extend them by augmenting two additional sets of meta-features for each instance S_i and each classifier CF_j: (i) the probability distribution multiplied by the maximum probability: $\hat{\mathbf{p}}^j(S_i) = \mathbf{p}^j(S_i) \times \max_{1 \leq k \leq l} p^j(L_k|S_i)$, (ii) the entropies of the probability distributions: $E_j(S_i) = -\sum_{k=1}^l p^j(L_k|S_i) \cdot \log_2 p^j(L_k|S_i)$. Therefore, the total number of meta-features for each instance would become $M(2l + 1)$.

• **Subset Selection:** Instead of considering *all* classifiers, we propose to use SS to select a subset of classifiers for meta-feature generation. Our selection strategies are based on two fundamental quantities – *quality* and *diversity*.

(i) **Quality (Q):** We measure the quality of each base classifier in terms of – Area under the ROC curve (AUC) (further used to measure the performance of individual classifiers in Sect. 6).

(ii) Diversity (D): We measure the diversity between the predictions of base classifiers in two ways: (i) *NMI-based measure*: $D_{nmi} = 1 - \frac{1}{M} \sum_{1 \leq i,j \leq M} NMI$ $(\mathbf{y}_i, \mathbf{y}_j)$, where Normalized Mutual Information (NMI) is a measure of similarity between two results [14], (ii) *Entropy-based measure*: $D_E = \frac{1}{|S_{tr}|} \sum_{s \in S_{tr}}$ $\frac{1}{M - \frac{M}{2}} min\{l(s), L - l(s)\}$, where $l(s) = \sum_{i=1}^{M} \delta(Y_i(s), c(s))$, and $\delta(a, b) = 1$, if $a = b$, 0 otherwise.

Greedy Strategy (G): Given the predictions of all base classifiers $\{\mathbf{y}_i\}_{i=1}^{M}$ as inputs, we select a subset by considering a trade-off between quality and diversity, which can be viewed as a multi-objective optimization problem. We choose a subset S_{CF} that maximizes the objective function:

$$J = \alpha \frac{1}{|S_{CF}|} \sum_{i=1}^{|S_{CF}|} Q(\mathbf{y}_i) + (1 - \alpha)D \tag{1}$$

The parameter α controls the trade-off between these two quantities. However, selecting a proper subset is computationally expensive. Therefore, we adopt the following greedy strategy. We start by adding the solution with highest quality and incrementally add solutions one at a time that maximizes J until local maxima is reached. We set 0.5 as the default value of α. We also consider all the features (ALL) and compare the performance of the classifiers with that of greedy strategy (see Sect. 6, Table 2(c) and (d)).

• **Output Combination:** In the final stage of our proposed classifiers (Step 17 in MC3-R and Step 24 in MC3-S), the outputs of the optimal base classifiers are aggregated through a function W. We consider two definitions for W.

(i) Majority Voting (MV): For each test instance we assign the class that the majority of base classifiers agree with. Tie breaking is resolved by assigning that class on which the base classifiers have highest confidence.

(ii) Feature-Weighted Linear Combination (FWLC): As opposed to linear stacking where each base classifier is given a weight, here we assign weights to features. Simple linear stacking defines the weighted function as $W(L_k|s) = \sum_{i=1}^{M} w_i p^i(L_k|s)$, for each $L_k \in \mathcal{L}$ and $s \in S_{tr}$. FWLC instead models the weight w_i as a linear function of features (including d original and $M(2l + 1)$ meta-features), i.e., $w_i(s) = \sum_{j=1}^{d+M(2l+1)} v_{ij} f_j(s)$ for learning weights $v_{ij} \in \mathbb{R}$. Then the weighted function yields the following objective function:

$$min_v \sum_{s \in S_{tr}} (\sum_{i=1}^{M} \sum_{j=1}^{d+M(2l+1)} (v_{ij} f_j(s) p^i(L_k|s)) - 1)^2 \tag{2}$$

The prediction is subtracted from 1 because we assume that the actual class of s is assigned the probability 1. We use linear regression to obtain the optimal weight for each feature.

• **Consensus Function:** MC3-R uses $CONS$ to reach a consensus among the base classifiers. Ideally, all the classifiers should predict the same class for an unknown instance at the end (complete consensus). However, in practice it may not be possible, and therefore we stop **MC3-R** once it reaches a certain threshold of consensus. Two possible definition of $CONS$ are as follows:

(i) Binary Consensus (BIN): For each unknown instance, we check if all pairs of classifiers agree with their predictions: $\frac{1}{|S_{tr}|}\frac{1}{M}\sum_{s\in S_{ts}}$ $\sum_{\{CF_i,CF_j\}}\delta(Y_i(s), Y_j(s))$, where $\delta(x,y) = 1$, if $x = y$, 0, otherwise.

(ii) NMI-based Consensus (NMI): We measure the average similarity between the prediction of two base classifiers using NMI: $\frac{1}{M}\sum_{\{CF_i,CF_j\}}$ $NMI(\mathbf{y}_i, \mathbf{y}_j)$.

The classifier stops once the difference between the values of the consensus function for two consecutive levels falls below a certain threshold (we take it as 0.02). Later we will see in Fig. 2(d) that **MC3-R** achieves consensus within 4–5 levels for most of the datasets.

5   Experimental Setup

Datasets: We perform our experiments on a collection of 13 datasets. These datasets are highly diverse (in terms of size, class distribution, feature size) and widely used. A summary of these datasets is shown in Table 1.

Base Classifiers: Seven (standalone) base classifiers are used in this study: (i) **DT**: CART algorithm for decision tree with Gini coefficient, (ii) **NB**: Naive

Table 1. The datasets (ordered by the size) and their properties: number of instances, number of classes, number of features, probability of the majority class (MAJ), and entropy of the class probability distribution (ENT). We further report the accuracy (AUC) of our classifiers and the best baseline for different datasets. The best baseline varies across datasets (see Sect. 6 for detailed discussion).

Properties of the dataset							Accuracy (AUC)		
	Dataset	# instances	# classes	# features	MAJ	ENT	Best baseline	MC3-R	MC3-S
Binary	Titanic [1]	2200	2	3	0.68	0.90	0.66 (SVM)	**0.67**	0.66
	Spambase [13]	4597	2	57	0.61	0.96	0.93 (RF)	**0.95**	**0.95**
	Magic [13]	19020	2	11	0.64	0.93	0.55 (RF)	**0.56**	**0.56**
	Creditcard [23]	30000	2	24	0.78	0.76	0.67 (BAG)	**0.70**	0.67
	Adults [13]	45000	2	15	0.75	0.80	0.78 (SGD)	**0.83**	0.80
	Diabetes [13]	100000	2	55	0.54	0.99	0.64 (RP)	**0.65**	**0.65**
	Susy [2]	5000000	2	18	0.52	0.99	**0.77** (BAG)	**0.77**	**0.77**
Multiclass	Iris [13]	150	3	4	0.33	1.58	0.97 (RP)	**0.98**	**0.98**
	Image [13]	2310	7	19	0.14	2.78	**0.98** (BAG)	**0.98**	**0.98**
	Waveform [13]	5000	3	21	0.34	1.58	0.89 (STA)	**0.91**	0.90
	Statlog [13]	6435	6	36	0.24	2.48	0.92 (RP)	**0.95**	0.94
	Letter recognition [13]	20000	26	16	0.04	4.69	0.49 (BOO)	**0.54**	0.50
	Sensor [13]	58509	11	49	0.09	3.45	0.98 (BOO)	**0.99**	**0.99**

Bayes algorithm with kernel density estimator, (iii) **K-NN**: K-nearest neighbor algorithm, (iv) **LR**: multinomial logistic regression, (v) **SVM**: Support Vector Machine with linear kernel, (vi) **LDA**: supervised latent Dirichlet allocation [6], (vii) **SGD**: stochastic gradient descent classifier [3]. We utilize standard grid search for hyper-parameter optimization. These algorithms are further used later as standalone baseline classifiers to compare with **MC3-R** and **MC3-S**.

Baseline Algorithms: We compare **MC3-R** and **MC3-S** with the standalone classifiers mentioned earlier. We additionally compare them with 5 state-of-the-art ensemble classifiers: (i) **Linear Stacking** (STA): stacking with multi-response linear regression [15], (ii) **Bagging** (BAG): bootstrap aggregation method [4], (iii) **AdaBoost** (BOO): Adaptive Boosting [17], (iv) **Random Forest** (RF): random forest with Gini coefficient [5], and (v) **RP:** a recently proposed random projection ensemble classifier [7]. Thus, in all, we compare our algorithms with 12 classifiers including sophisticated ensembles.

6 Experimental Results

In this section, we first present the parameter selection strategy for our classifiers. In the interest of space, we will only present the results of parameter selection for Creditcard and Waveform (as representatives of binary and multiclass datasets respectively); however exceptions will be explicitly mentioned. Following this, we will present the performance of all the algorithms for different datasets. The performance is reported after 10-fold cross validation. All experiments were performed on a cluster of 64 Xeon 2.4 GHz machines with 24 GB RAM running RedHat Linux.

Parameter Selection: Table 2 shows the performance of our classifiers for different parameter combinations. For instance, the top left entry in Table 2(a) indicates that the AUC value of **MC3-R** on the Creditcard dataset is 0.64 (*resp.* 0.63) with NMI-based greedy subset selection $G : D_{nmi}$ and binary consensus BIN (*resp.* NMI-based greedy subset selection $G : D_{nmi}$ and NMI-based consensus NMI). We observe that in general **MC3-R** and **MC3-S** perform the best with majority voting (MV) as W (exception including $FWLC$ for the Magic dataset), greedy strategy (G) with entropy-based diversity D_E as SS (exception

Table 2. Parameter selection for **MC3-R** and **MC3-S** on Creditcard and Waveform datasets (see abbreviations in Sect. 4). The accuracies are reported in terms of AUC.

(a) MC3-R (Creditcard)

| | | $SS + CONS$ | | |
		$G : D_{nmi}$+BIN(NMI)	$G : D_E$+BIN(NMI)	ALL+BIN(NMI)
W	FWLC	0.64 (0.63)	0.66 (0.65)	0.65 (0.65)
	MV	0.65 (0.66)	0.67 (**0.68**)	0.66 (0.67)

(b) MC3-R (Waveform)

| | | $SS + CONS$ | | |
		$G : D_{nmi}$+BIN(NMI)	$G : D_E$+BIN(NMI)	ALL+BIN(NMI)
W	FWLC	0.76 (0.79)	0.76 (0.78)	0.77 (0.78)
	MV	0.88 (0.88)	0.86 (**0.91**)	0.89 (0.90)

(c) MC3-S (Creditcard)

| | | SS | | |
		$G : D_{nmi}$	$G : D_E$	ALL
W	FWLC	0.66	0.66	0.67
	MV	0.67	**0.70**	0.68

(d) MC3-S (Waveform)

| | | SS | | |
		$G : D_{nmi}$	$G : D_E$	ALL
W	FWLC	0.56	0.56	0.76
	MV	0.90	**0.91**	0.89

including greedy D_{nmi}-based strategy for the Sensor dataset) and NMI-based *CONS*. Moreover, for both the classifiers, we observe in Fig. 2(c) that the performance does not change much with the number of iterations (*Iter*); therefore we take *Iter* = 1 to speedup the classifier. The rest of the experiments are conducted with these parameter settings for **MC3-R** and **MC3-S**.

Comparative Analysis: The performance of the classifiers is evaluated based on two evaluation measures – *AUC* and F-score. For better visualization, we present here the *composite performance* of all classifiers – for each evaluation measure (AUC and F-score), we separately scale the scores of the competing classifiers so that the best performing classifier has a score of 1. The composite performance of a classifier is the sum of the 2 normalized scores. If a classifier outperforms all others, then its composite performance is 2. Figure 1 shows that our classifiers outperform others, irrespective of the datasets. The composite performance of **MC3-R** and **MC3-S** is 1.99 and 1.97 respectively, followed by RF (1.92), Bagging (1.92), RP (1.87), DT (1.82), KNN (1.82), BOO (1.92), STA (1.81), LDA (1.81), SVM (1.80), LR (1.80), NB (1.73) and SGD (1.55). The absolute performance of each classifier averaged over all datasets as shown in the bottom table of Fig. 1 indicates that **MC3-R** performs 3.89% (*resp.* 5.56%) better than the best baseline in terms of AUC (*resp.* F-Score). For further comparison, the absolute accuracy of **MC3-R** and **MC3-S** along with the bast baseline is presented in Table 1. Interestingly, we observe in Table 1 that although the best baseline tends to be competitive with our classifiers, there is no particular baseline which is the best across all datasets. Therefore, *one may choose our classifiers rather than spending time deciding on which classifier to choose because our classifiers are at least as good as any existing classifier irrespective of the dataset.*

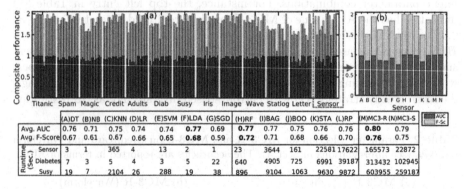

Fig. 1. (a) Composite performance of all the classifiers ((A)-(G): standalone, (H)-(L): ensemble, (M)-(N): ours) on different datasets. (b) The results on Sensor dataset are zoomed out separately. The order of classifiers on x-axis (i.e., labels on x-axis) in (a) is same as that in (b) and is omitted for better visualization. The table below presents the average accuracy of the classifiers over all the datasets, and the runtime on three largest datasets.

As expected, we observe in the bottom table of Fig. 1 that **MC3-S** is much faster than **MC3-R**. Note that the runtime is reported after running base classifiers sequentially. We further measure *overhead ratio* of an ensemble algorithm as the ratio between the total runtime of the ensemble algorithm and the total runtime taken by the base algorithms used in the ensemble algorithm. We notice that the overhead ratio of our algorithms is the minimum (around 1; while the maximum is 714 for Bagging). It essentially indicates that the runtime of our algorithms is high due to the sequential execution of the base algorithms, which can be parallelized easily.

We further study other aspects of the classifiers:

(i) *Dependency on the Feature Size*: We consider Spambase[2] having highest number of features (57) and drop 5 features at a time based on descending order of importance[3] and plot AUC in Fig. 2(a). We observe that our algorithms consistently perform well despite dropping features – **MC3-R** almost remains invariant up to 12 features. The reason might be that our classifiers produce additional meta-features to separate instances well in the feature space. This suggests that *our classifiers add high value for datasets with a small number of original features.*

(ii) *Dependency on the Size of the Training Set*: We consider Creditcard (see footnote 2) and decrease the training size from 75% to 50% (with 5% interval) of the entire dataset. Training set is selected randomly, and for each training size, the average AUC is reported in Fig. 2(b) after repeating it 20 times. We observe that our classifiers are less affected by the training size. Therefore, *one may choose our classifiers when the training size is small.*

(iii) *Dependency on the Number of Iterations*: In both **MC3-R** and **MC3-S**, we choose the best parameter setting of the base classifiers after running them *Iter* times. Figure 2(c) shows that the overall performance does not vary much with an increase in *Iter*. Therefore, we choose *Iter* = 1 to make the classifiers fast.

(iv) *Convergence of* MC3-R: **MC3-R** takes ι levels to achieve consensus. Table 2 shows that NMI-based consensus is more effective than binary consensus. Although there is no theoretical guarantee of achieving consensus since the base classifiers are treated as a black box, we empirically observe that for all the datasets **MC3-R** converges after a certain level. Figure 2(d) shows that for small datasets (e.g., Iris) consensus is achieved much faster (within 2–3 levels) than large datasets (e.g., Creditcard, Susy) for which **MC3-R** usually takes 7–8 levels of iterations (on average 4–5 levels for most of the datasets).

(v) *Dependency on the Base Classifiers*: For each dataset, we drop each base classifier in isolation and measure the change in performance of **MC3-R** and **MC3-S**. Figure 2(e) shows that LDA affects the performance the most. As mentioned in the comparative analysis, LDA seems to be the best standalone

---

[2] The patterns are exactly the same for the other datasets.

[3] We separately measure the importance of each feature by dropping it in isolation and calculate the decrease in accuracy (more decrease implies more relevance).

classifier. This may imply that incorporating strong classifiers into the base set may have a bigger impact than incorporating the weak classifiers.

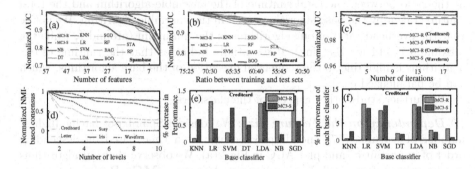

Fig. 2. Performance (normalized) of each classifier for (a) different number of features and (b) different training size. The lines corresponding to **MC3-S** and **MC3-R** are significantly different (McNemar's, $p < 0.005$) from other lines. (c) Performance of our classifiers after considering different number of iterations (*Iter*). (d) Normalized NMI-based consensus of **MC3-R** over different levels of iterations for different datasets. (e) Decrease in performance of our classifiers after dropping each base classifier in isolation. (f) Performance improvement of each base classifier due to our ensemble classifiers.

(vi) *Improvement of Individual Base Classifiers*: As opposed to traditional ensemble classifiers, our classifiers improve individual base classifiers separately once consensus is reached. Figure 2(f) shows the percentage improvement of base classifiers after incorporating meta-features generated by our classifiers. We observe that the improvement is significantly high, ranging up to 10% in some cases. Interestingly, our classifiers are able to gear up the performance of strong base classifiers (such as LDA, LR, SVM) as well.

7 Conclusion

In this paper, we have advanced the paradigm of ensemble classification by providing a new notion of "consensus learning". We have shown that there is no existing classifier which always performs the best across different datasets. Our classifiers are at the top, performing as well or better than the best existing classifier (baseline and ensembles) across all 13 datasets we considered. The rigorous study of 13 different datasets and the comparative analysis with 12 baseline classifiers allows us to assert that achieving consensus not only provides a better way of designing ensemble classifiers, but also enhances the accuracy of individual base classifiers by a significant level (up to 10%).

Acknowledgments. Parts of this work were funded by ONR grants N00014-15-R-BA010, N00014-16-R-BA01, N000141612739 and N000141612918.

References

1. Titanic dataset. https://www.kaggle.com/c/titanic. Accessed 30 Sept 2016
2. Baldi, P., Sadowski, P., Whiteson, D.: Searching for exotic particles in high-energy physics with deep learning. Nat. Commun. **5** (2014). Article No. 4308
3. Bottou, L.: Large-scale machine learning with stochastic gradient descent. In: Lechevallier, Y., Saporta, G. (eds.) Proceedings of COMPSTAT 2010, pp. 177–186. Physica-Verlag HD, Heidelberg (2010)
4. Breiman, L.: Bagging predictors. Mach. Learn. **24**(2), 123–140 (1996)
5. Breiman, L.: Random forests. Mach. Learn. **45**(1), 5–32 (2001)
6. Cai, X., Wang, H., Huang, H., Ding, C.: Simultaneous Image classification and annotation via biased random walk on tri-relational graph. In: Fitzgibbon, A., Lazebnik, S., Perona, P., Sato, Y., Schmid, C. (eds.) ECCV 2012. LNCS, vol. 7577, pp. 823–836. Springer, Heidelberg (2012). doi:10.1007/978-3-642-33783-3_59
7. Cannings, T.I., Samworth, R.J.: Random projection ensemble classification. ArXiv (2015)
8. Conroy, B., Eshelman, L., Potes, C., Xu-Wilson, M.: A dynamic ensemble approach to robust classification in the presence of missing data. Mach. Learn. **102**(3), 443–463 (2016)
9. Datta, S., Pihur, V., Datta, S.: An adaptive optimal ensemble classifier via bagging and rank aggregation with applications to high dimensional data. BMC Bioinform. **11**(1), 427 (2010)
10. Džeroski, S., Ženko, B.: Is combining classifiers with stacking better than selecting the best one? Mach. Learn. **54**(3), 255–273 (2004)
11. Guo, L., Boukir, S.: Margin-based ordered aggregation for ensemble pruning. Pattern Recognit. Lett. **34**(6), 603–609 (2013)
12. Karabulut, E.M., İbrikçi, T.: Effective diagnosis of coronary artery disease using the rotation forest ensemble method. J. Med. Syst. **36**(5), 3011–3018 (2012)
13. Lichman, M.: UCI repository (2013). http://archive.ics.uci.edu/ml
14. Manning, C.D., Raghavan, P., Schütze, H.: Introduction to Information Retrieval. Cambridge University Press, Cambridge (2008)
15. Reid, S., Grudic, G.: Regularized linear models in stacked generalization. In: Benediktsson, J.A., Kittler, J., Roli, F. (eds.) MCS 2009. LNCS, vol. 5519, pp. 112–121. Springer, Heidelberg (2009). doi:10.1007/978-3-642-02326-2_12
16. Rokach, L.: Ensemble-based classifiers. Artif. Intell. Rev. **33**(1–2), 1–39 (2010)
17. Schapire, R.E.: A brief introduction to boosting. In: IJCAI, Stockholm, Sweden, pp. 1401–1406 (1999)
18. Schclar, A., Rokach, L., Amit, A.: Ensembles of classifiers based on dimensionality reduction. CoRR abs/1305.4345 (2013)
19. Shunmugapriya, P., Kanmani, S.: Optimization of stacking ensemble configurations through Artificial Bee Colony algorithm. Swarm Evol. Comput. **12**, 24–32 (2013)
20. Sun, J., Liao, B., Li, H.: Adaboost and bagging ensemble approaches with neural network as base learner for financial distress prediction of chinese construction and real estate companies. Recent Pat. Comput. Sci. **6**(1), 47–59 (2013)
21. Ting, K.M., Witten, I.H.: Issues in stacked generalization. J. Artif. Intell. Res. (JAIR) **10**, 271–289 (1999)
22. Xiao, H., Xiao, Z., Wang, Y.: Ensemble classification based on supervised clustering for credit scoring. Appl. Soft Comput. **43**, 73–86 (2016)
23. Yeh, I.C., Lien, C.H.: The comparisons of data mining techniques for the predictive accuracy of probability of default of credit card clients. Expert Syst. Appl. **36**(2), 2473–2480 (2009)

Monte Carlo Based Incremental PageRank on Evolving Graphs

Qun Liao[1], ShuangShuang Jiang[2], Min Yu[1], Yulu Yang[1(✉)],
and Tao Li[1]

[1] College of Computer and Control Engineering,
Nankai University, Tianjin, China
liaoqun@mail.nankai.edu.cn, yumin2021@163.com,
{yangyl,litao}@nankai.edu.cn
[2] Alibaba Cloud (Aliyun), Beijing, China
highfly@mail.nankai.edu.cn

Abstract. Computing PageRank for enormous and frequently evolving real-world network consumes sizable resource and comes with large computational overhead. To address this problem, IMCPR, an incremental PageRank algorithm based on Monte Carlo method is proposed in this paper. IMCPR computes PageRank scores via updating previous results accumulatively according to the changed part of network, instead of recomputing from scratch. IMCPR effectively improves the performance and brings no additional storage overhead. Theoretical analysis shows that the time complexity of IMCPR to update PageRank scores for a network with m changed nodes and n changed edges is $O((m+n/c)/c)$, where c is reset probability. It takes $O(1)$ works to update PageRank scores as inserting/removing a node or edge. The time complexity of IMCPR is better than other existing state-of-art algorithms for most real-world graphs. We evaluate IMCPR with real-world networks from different backgrounds upon Hama, a distributed platform. Experiments demonstrate that IMCPR obtains PageRank scores with equal (or even higher) accuracy as the baseline Monte Carlo based PageRank algorithm and reduces the amount of computation significantly compared to other existing incremental algorithm.

Keywords: PageRank · Web mining · Incremental computing · Monte Carlo algorithm · Parallel and distributed processing

1 Introduction

PageRank plays an important role in Web search, social network analysis and many other application fields [1]. Nowadays, the volume of data in Internet and social networks are tremendous and evolving frequently. Computing PageRank for a large and evolving graph cost huge computational resources. Recomputing from scratch is impractical due to its considerable overhead. Many incremental PageRank algorithms are designed to improve the performance of PageRank computation for dynamic graphs. Algorithm proposed in [2] is one of the most efficient state-of-art algorithms. However, its storage overhead, due to storing all the random walk segments in previous computation, is a limitation which cannot be overlooked.

© Springer International Publishing AG 2017
J. Kim et al. (Eds.): PAKDD 2017, Part I, LNAI 10234, pp. 356–367, 2017.
DOI: 10.1007/978-3-319-57454-7_28

Aiming to compute PageRank for evolving graph efficiently, we propose IMCPR, a novel incremental PageRank algorithm based on Monte Carlo method in this paper. Inspired by previous works [2, 3], our proposed algorithm reuses pervious PageRank scores and updates them incrementally. The proposed avoids a large amount of recomputation and improves the performance significantly.

The most important characteristic of our algorithm is that it stores no previous random walk segments at all, which brings no extra storage overhead. Theoretical analysis also proves that the time complexity of our newly proposed algorithm is lower than other existing related algorithms for most real-world graphs. Moreover, it is also proved that IMCPR performs as good as the original Monte Carlo method in PageRank computation [4] in accuracy. Evaluations based on experiments of real-world graphs demonstrate that IMCPR improves the performance of PageRank significantly compared to other existing PageRank algorithms.

2 PageRank and Related Work

2.1 PageRank

Let $G = (V, E)$ be an unweighted directed graph, where V is the set of nodes and E is the set of edges. $|V|$ is the number of nodes and $|E|$ is the number of edges. For an arbitrary node j, $N(j)$ donates the set of j's outgoing neighbors. $|N(j)|$ is the number of node j's outgoing neighbors. Let A be the transition matrix, where $A(i,j) = 1/|N(i)|$ if and only if there is a direct edge $e(i, j) \in E$, and $A(i,j) = 0$ otherwise. Let π be a vector consisted of PageRank scores of all nodes in V. π is defined as Eq. (1), where α is teleport probability and θ is a vector consisted of fraction $1/|V|$.

$$\pi = \alpha A \pi + (1 - \alpha)\theta \tag{1}$$

The definition of PageRank also has an interpretation based on random walk simulation. Consider a random walk simulation on graph G defined as follows: do R random walks starting from each node of G, for a random walker, at each step it stops with probability c (here we call c as the reset probability and $c = 1 - \alpha$), or jumps to a random chosen outgoing neighbor of current node with probability α. Assume for each node v, $X(v)$ is the total number of times that all random walk segments who visits it. The approximate PageRank of node v $\tilde{\pi}(v)$ is defined as Eq. (2).

$$\tilde{\pi}(v) = cX(v)/(R|V|) \tag{2}$$

Power Iteration [1] is a fundamental algorithm for PageRank computation which computes qualified solution of Eq. (1) iteratively. It is easy to understand and implement, but it's not efficient enough in dealing with massive graph. Many improved algorithms based on this iteration method are well discussed in [5]. Another group of fundamental algorithms are Monte Carlo based algorithms [2, 4]. They simulate the random walks defined above and get accurate approximations efficiently. It is proved that even when $R = 1$, the approximations of important nodes are accurate enough for many applications [4]. These algorithms are also easy to be parallelized.

2.2 Related Work

Bahmani et al. have provided a detailed list of incremental PageRank algorithms in [2]. However, for completeness and for comparison with our own results, we concentrate on reviewing the latest PageRank algorithms for evolving graphs. There are two main categories of incremental PageRank algorithms.

- Aggregation Algorithms

These algorithms [3, 6–8] are based on the idea of graph partition and aggregation. They partition the graph into several sub-graphs and try to limit the affect of changed part of graph in the level of sub-graphs. These methods help to reduce unnecessary recomputation. However, the limitations of these algorithms are high computational load for aggregation, difficulty in partitioning real-world graphs efficiently and unstable performance depending on partitioning. A lot of evidences were discussed in detail in [2].

- Monte Carlo Based Algorithms

The most efficient incremental Monte Carlo based PageRank algorithm was proposed in [2], whose time complexity is $O(|V|\ln(n)/c^2)$ to update PageRank scores as n edges arrivals in a graph [2, 9]. Though it is efficient to update PageRank, the large storage cost for all the random walk segments in history limits the application of this algorithm for large graphs.

Our proposed algorithm doesn't suffer from the shortages of aggregation based algorithms. It handles evolving graphs with nodes and edges inserted and/or removed efficiently. Comparing to the state-of-art algorithm in [2], it requires no extra storage overhead and performs a lower time complexity to update PageRank for most of real-world graphs.

3 Incremental Monte Carlo Method for Pagerank (IMCPR)

We compute approximate PageRank scores according to Eq. (2) as initial solution. As graph evolves, IMCPR updates PageRank based on reusing previous PageRank scores and starting a proper number of random walks around the changed part. The newly started random walks help to adjust each node's times of visited by all random walk segments via adding or subtracting contribution from corresponding random walk segments. How IMCPR update PageRank scores when edges and nodes evolve are described respectively as follows.

3.1 IMCPR for Evolving Edges

Suppose an arbitrary edge $e(u, r)$ is added, we do M random walks starting from node r. For any node s, if s is passed through by any one of these random walks once, $X(s)$ is increased by one. M is a non-negative integer defined as Eq. (3). Then we do another M random walks. Each of these walks randomly picks one of u's outgoing neighbors except node r as its starting node. For any node s, if s is passed through by any one of these random walks once, $X(s)$ decreases by one.

Method for updating PageRank as edges removed is similar. Suppose an arbitrary edge $e(u, r)$ is removed, we do M random walks starting from node r. For any node s, if s is passed through by any one of these random walks once, $X(s)$ is decreased by one.

Then we do another M random walks. Each of these walks randomly picks one of u's outgoing neighbors except node r as its starting node. For any node s, if s is passed through by any one of these random walks once, $X(s)$ increases by one. Algorithm 1. presents the pseudo-code of updating PageRank for a graph with evolving edges.

$$M = \begin{cases} (1-c)X(u)/|N(u)|, & \text{edge } e(u,r) \text{ is added} \\ (1-c)X(u)/(|N(u)|-1), & \text{edge } e(u,r) \text{ is removed} \end{cases} \tag{3}$$

Algorithm 1 function ProcessingChangedEdges $(G, X, \Delta E, c, R)$

Input: A directed graph $G = (V, E)$, ΔE is set of edges newly added and/or removed, c is the reset probability, X is set of the number of random walk segments visiting each node initially, R is the number of random walks starting from each node in initial.

Output: π is the updated approximation of PageRank.

```
for each e(u,r) ∈ ΔE  do
    compute M according to Eq. (3)
    do M random walks starting from r
    for each random walk do
        if node s ∈ V is passed through by once do
            if e(u,r) is added do
                X(s)++
            else
                X(s)--
            end if
        end if
    end for
    for i = 1 to M do
        do 1 random walk starting from node v ∈ N(u) and
        v is not r
        if node s ∈ V is passed through by once do
            if e(u,r) is added do
                X(s)--
            else
                X(s)++
            end if
        end if
    end for
end for
π = cX/(|V|R)
return π
```

3.2 IMCPR for Evolving Nodes

Adding an arbitrary node r into graph G, can be regarded as two separate operations. First a node r is added, secondly edges associated with r are added. Similarly, an arbitrary node r removed means edges associated with r and the node r itself removed respectively. Thus, as an arbitrary node r added, we set $X(r)$ equal to R, times of random walks started from each node before, for initialization. Then we update PageRank scores according to the method described above. Supposes an arbitrary node r is removed, we set $X(r)$ equal to zero and update PageRank scores to process the removed edges. Adding the process for changed nodes and edges together, the pseudo-code of IMCPR algorithm is shown in Algorithm 2.

Algorithm 2 function IMCPR$(G, X, \Delta G, c, R)$

Input: A directed graph $G = (V, E)$, $\Delta G = \{\Delta E, \Delta V\}$ denotes the newly changed edges and nodes, c is the reset probability, X is set of the number of random walk segments visiting each node initially, R is the number of random walks starting from each node in initial.

Output: π is the updated approximation of PageRank.

for each $r \in \Delta V$ and r is added **do**
    $X(r) = R$
end for

for each $r \in \Delta V$ and r is removed **do**
    $X(r) = 0$
end for

return ProcessingChangedEdges$(G, X, \Delta E, c, R)$

4   Correctness and Time Complexity

4.1   Correctness Discussion

In this section, we prove that for an arbitrary node $u, \tilde{\pi}(u)$ got by IMCPR is sharply concentrated around its expectation, which is its real PageRank score $\pi(u)$.

Theorem 1. The expected PageRank score of an arbitrary node u got from IMCPR is equal to the real score. It is written as Eq. (4).

$$E[\tilde{\pi}(u)] = \pi(u) \tag{4}$$

Prove: It is proved that $E[\tilde{\pi}(u)]$ got by the original Monte Carlo based PageRank is equal to $\pi(u)$ in [4]. Thus, here we only need to prove that $E[\tilde{\pi}(u)]$ got by IMCPR is equal to the expectation got by the original algorithm.

For an arbitrary node u, IMCPR reuses $X(u)$ computed by the original Monte Carlo based PageRank in initial. It is equal to that IMCPR starting R random walks from each node of the graph G. As an edge added or removed, IMCPR updates $X(u)$ as the method defined above, which is equal to rerouting the random walks which passes through the changed edge. The fundamental idea of IMCPR is that the expectation of contribution of $X(u)$ from an arbitrary edge $e(s, u)$ is equal to $(1 - c)X(s)/|N(s)|$. Though the contribution of $X(u)$ from a particular edge varies, its expectation is steady.

The expected $X(u)$ for a node u in a Monte Carlo based algorithm can be computed as Eq. (5), where $P(v, u)$ donates the number of random walk segment which starts from v and visits u.

$$E[X(u)] = R \times E\left[\sum_{v \in V} P(v, u)\right] \tag{5}$$

The expected $P(v, u)$ for any node u and v only depend on the graph and how to choose next node in a random walk. As described in Sect. 3.1, it is straightforward that our proposed algorithm doesn't change the probability of choosing next node in random walks, so according to Eq. (5), we can tell that $E[X(u)]$ computed by IMCPR is equal to it computed by the original Monte Carlo based PageRank algorithm. Theorem 1 is proved.

Theorem 2. The PageRank score got from IMCPR is sharply concentrated around its expectation. Theorem 2 can be written as Eq. (6) for any node v, where δ is a concentrated factor and δ' is a constant depending on both δ and the reset probability c.

$$\Pr[|\tilde{\pi}(v) - \pi(v)| \geq \delta\pi(v)] \leq e^{-|V|R\pi(v)\delta'} \tag{6}$$

The proof of Theorem 2 is similar as some previous works [2, 10], but we still present the detailed derivation of the proof for the completeness of this paper.

Prove: Assuming $R = 1$(the situation that $R > 1$ can be proved like this), for an arbitrary node v, define $Z(u)$ donates c times of the visited time of node v got from the path start from node u. $Y(u)$ is the length of the random walk segment starting from u. Let $W(u) = cY(u)$, $z(u) = E[Z(u)]$. $Z(u)$ of different node u are independent. Thus,

$$\tilde{\pi}(v) = \sum_{u \in V} Z(u)/|V|, \quad \pi(v) = \sum_{u \in V} z(u)/|V| \tag{7}$$

It is obvious that $0 \leq Z(u) \leq W(u)$ and $E[W(u)] = 1$.

From the definition of expectation, Eq. (8) can be derived.

$$\left(E\left[e^{hZ(u)}\right] - 1\right)/\left(E\left[e^{hW(u)}\right] - 1\right) \leq E[Z(u)]/E[W(u)] \tag{8}$$

$$E\left[e^{hZ(u)}\right] \leq z(u)E\left[e^{hW(u)}\right] + 1 - z(u) = z(u) \times \left(E\left[e^{hW(u)}\right] - 1\right) + 1 \tag{9}$$

Because for an arbitrary y meets $1 + y \leq e^y$, thus,

$$E\left[e^{hZ(u)}\right] \leq e^{-z(u) \times \left(1 - E\left[e^{hW(u)}\right]\right)} \tag{10}$$

$$\Pr[\tilde{\pi}(v) \geq (1 + \delta)\pi(v)] \leq \frac{E\left[e^{h|V|\tilde{\pi}(v)}\right]}{e^{h|V|(1+\delta)\pi(v)}} = \frac{E\left[e^{h\sum_u Z(u)}\right]}{e^{h|V|(1+\delta)\pi(v)}} \leq \frac{\prod_u E\left[e^{-z(u) \times \left(1 - E\left[e^{hW(u)}\right]\right)}\right]}{e^{h|V|(1+\delta)\pi(v)}}$$

$$= e^{-|V|\pi(v) \times \left(1 - E\left[e^{hW(u)}\right]\right)} \Big/ e^{h|V|(1+\delta)\pi(v)} \leq e^{-|V|\delta'\pi(v)} \tag{11}$$

Similar as Eq. (11), Eq. (12) can be proved.

$$\Pr[\tilde{\pi}(v) \leq (1 - \delta)\pi(v)] \leq e^{-|V|\delta'\pi(v)} \tag{12}$$

In addition, in Eqs. (11) and (12) $\delta' = 1 + h(1 + \delta) - E\left[e^{hW}\right]$ where $W = cY$ is a random variable with Y having geometric distribution with parameter c. It means that the probability of the approximation deviated from its expectation is quite small and Theorem 2 is proved. So it is convinced that IMCPR performs as good as the original Monte Carlo based PageRank in accuracy. Above all, the correctness of IMCPR is proved.

4.2 Complexity Analysis

Supposing an arbitrary node r changed, IMCPR starts R random walks from r. Supposing an arbitrary edge $e(u, v)$ changed, there are $2 \times M$ random walks starting, including M random walks starting from node v and the same number of random walks starting from node u's other outgoing neighbors. Here we discuss the amount of operations as m nodes and n edges changed. The total number of newly started random walks in IMCPR, donated by *TotalRW*, can be calculated as Eq. (13).

$$Total\,RW = mR + 2\sum_{e(u,v) \in \Delta E} M = mR + \sum_{e(u,v) \in \Delta E} 2(1 - c)X(u)/|N(u)| \tag{13}$$

For an arbitrary node u, its outgoing neighbors must be more than zero. So Eq. (14) must be true, where $\bar{X}(u)$ refers to the average number of random walk segments visiting node u in initial.

$$Total\,RW \leq mR + 2\sum_{e(u,v) \in \Delta E} X(u) = mR + 2n\bar{X}(u) \tag{14}$$

Supposing the edge $e(u, v)$ is inserted and/or removed randomly, $\bar{X}_t(u)$ can be calculated as Eq. (15).

$$\bar{X}(u) = \sum_{u \in V} X(u)/|V| = R|V|/(c|V|) = R/c \tag{15}$$

The length of the newly started random walks is equal to the number of operation of update $X(u)$ for any arbitrary node u. So, we define the amount of computing operations of IMCPR donated by *TotalComp* as Eq. (16).

$$TotalComp = TotalRW/c \leq R/c(m + 2n/c) \qquad (16)$$

The computational cost of IMCPR is only related to the size of changed parts of the graph. Algorithm in [2] takes complexity of $O(|V| \ln(n)/c^2)$ to update PageRank scores as n edges inserted and/or removed. We compare $|E|$ and $\ln(|E|)|V|$ for each graph from Stanford Network Analysis Project [11]. It is found that $|E|$ is closed to $\ln(|E|)|V|$ in most graphs and there are only 5 graphs which have $|E|$ obviously bigger than $\ln(|E|)|V|$. Meanwhile, some graphs such as memetracker, LiveJournal, wiki-Talk, web-Google and so on, their $|E|$ are much smaller than $\ln(|E|)|V|$. So we can tell that in many real-world applications with little percentages of graph changed, IMCPR takes a lower time complexity compared to algorithm in [2].

5 Experiments and Evaluations

5.1 Experimental Setup

We perform our experiments upon a five-machine homogeneous Hama [12] cluster. Each machine in the cluster has an intel-i7 2600 CPU, 2 GB memory, 4 TB hard disk and 1 Gigabit Ethernet card. Ubuntu 14.04 and zookeeper-3.4.6 are deployed on each machine. The version of Hama is hama-0.6.4 and HDFS component is provided by hadoop-1.2.1. Eight real-world graphs from widely used datasets [11] are used in our experiments. The key parameters of these graphs are listed in Table 1.

Table 1. Main parameters of data sets

Graph	p2p-Gnutella-31	Amazon-0312	Web-NotreDame	Web-BerkStan	Higgs-twitter	Wiki-talk	Wiki-vote	Email-Enron
Nodes	63K	401K	326K	685K	457K	2.3M	7.1K	37K
Edges	148K	3.2M	1.5M	7.6M	14.9M	5.0M	104K	184K
Dangling nodes	46K	12K	187K	4.7K	0.03K	2.2M	1K	0

In the experiments, we only consider the scenarios that edges and nodes are inserted into graphs, because removal is similar. In order to generate the edges inserted, we randomly choose 10% edges in each graph as evolving edges and use the rest part of each graph as initial graph. In order to generate the inserted nodes, we randomly add a certain percentage of nodes to each dataset, and appoint a stochastic incoming and outgoing neighbor for each newly added node. We set $c = 0.15$ and $R = 20$.

5.2   Comparison with Existing Algorithms

We evaluate the performance of our approach (Algorithm 2) by comparing it to the work in [2] and the non-incremental algorithm in [4]. For simplicity, we use Bah-maniPR to refer to the algorithm proposed in [2] by Bahmani et al., and use BasicPR to refer to the original Monte Carlo based PageRank proposed in [4]. We also use PI to refer to Power Iteration [1] for short.

We evaluate the accuracy of the proposed algorithm with metric of $L1$ error which is defined as Eq. (17), where $\pi(u)$ is the "ground-truth" PageRank score of node u and $\tilde{\pi}(u)$ is the approximation. In experiments $\pi(u)$ is computed by PI algorithm (with parameter $\varepsilon = 5 \times 10^{-4}$).

$$ Err = \sum_{u \in V} |\tilde{\pi}(u) - \pi(u)| \tag{17} $$

We also evaluate the amount of computation of the proposed algorithm. We implement all the algorithms based on the BSP model [13] upon Hama. A message is sent as long as a random walker jumps to a node in these Monte Carlo based algorithms. So we get the amount of computation by counting the messages received by all machines (including messages received locally) during computation. We use Cost(Alg) to refer to the amount of computation of a particular algorithm Alg.

5.3   Accuracy

We evaluate the accuracy of the proposed algorithm. Firstly, we compare the average $L1$ error of BasicPR and IMCPR with 10% edges inserted. Table 2 describes the results. We found that the accuracy of IMCPR is roughly equal to original Monte Carlo based PageRank algorithm.

Table 2. Accuracy comparison of IMCPR and BasicPR with 10% edges inserted

Graph	Amazon0312	Web-BerkStan	Web-NotreDame	Higgs-twitter	p2p-Gnutella31	Wiki-talk
BasicPR	0.19	0.20	0.28	0.21	0.34	0.36
IMCPR	0.20	0.20	0.29	0.21	0.31	0.36

To verify that the errors do not accumulate as updating PageRank for evolving graph, we trace the accuracy of IMCPR as edges inserted continuously. There are p edges inserted respectively ($p = 1, 10, 100, 1000, 1000$). We record the errors of IMCPR and BasicPR. To make the figures intuitive, we depict Err(IMCPR)/Err (BasicPR) in the following figures. As Fig. 1(a) shows, the accuracy of IMCPR is stable and always close to the original Monte Carlo based PageRank algorithm. We also trace the accuracy as d percentages of edges inserted ($d = 1\%\%, 5\%\%, 0.1\%, 0.5\%, 1\%, 5\%, 10\%$). As Fig. 1(b) depicted, we found that the accuracy of IMCPR is always close to the baseline algorithm. In some cases (P2P with 5% and 10% edges inserted) IMCPR even gets slightly higher accuracy than the baseline algorithm does.

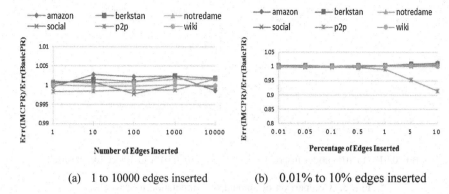

(a) 1 to 10000 edges inserted   (b) 0.01% to 10% edges inserted

Fig. 1. Comparison of accuracy

5.4   Amount of Computation

Comparison with Existing Incremental Algorithm.
To demonstrate our proposed algorithm is efficient, we compare the amount of computation of our algorithm to BahmaniPR. We insert n edges (n = 50, 100, 150) in our experiments and record the amount of computation. We found that BahmaniPR cost 8.9 to 70 times as much amount of computation compared to our proposed algorithm which is depicted in Fig. 2.

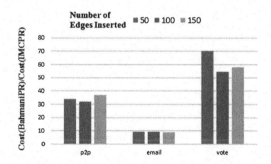

Fig. 2. Comparison of amount of computation to BahmaniPR

Comparison with Non Incremental Algorithm.
We also compare our proposed algorithm to original Monte Carlo based PageRank algorithm, we found our proposed algorithm reduces significant amount of computation compared to the original algorithm intuitively. Figure 3(a) and (b) describe the comparison of amount of computation of IMCPR and BasicPR in experiments of nodes and edges inserted respectively. As there is 1% data changed, IMCPR cuts down over 96% amount of computation at least. As 10% edges inserted IMCPR just cost 0.2 times amount of computation compared to the original algorithms at most.

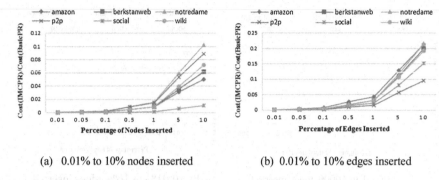

(a) 0.01% to 10% nodes inserted   (b) 0.01% to 10% edges inserted

Fig. 3. Comparison of amount of computation to BasicPR

Comparison as Different Number of Edges Changed.
Last but not least, we verifying the efficiency of IMCPR as different number of edges
evolves. We compare the amount of computation of IMCPR as 1 to 10000 edges
inserted mentioned above. We found that the amount of computation has a nearly linear
correlation with the number of the edges inserted. These results consistent with the
theoretical analysis in the previous section. Particular results are depicted in Fig. 4, to
make the figure intuitive, we take the logarithms of amount of computation as the
vertical axis.

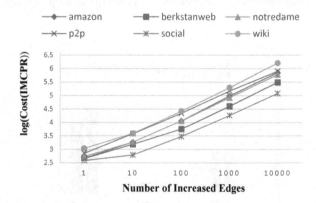

Fig. 4. Amount of computation of IMCPR as different numbers (1 to 10000) of edges inserted

6   Conclusion

In this paper, we investigate Monte Carlo based PageRank algorithms and propose an
incremental algorithm called IMCPR, which significantly reduces the amount of
computation for dynamic graphs. Both the theoretical analysis and experimental results
with several typical real-world graphs demonstrate that IMCPR performs well in
accuracy and performance. In addition, the proposed algorithm can be extended to
Monte Carlo based Personalized PageRank [14], Single-Source Shortest Paths [15] and
other random walk based algorithms.

Acknowledgement. We would like to thank Shan Shan for helpful suggestions.

References

1. Page, L., et al.: The PageRank citation ranking: bringing order to the web (1999)
2. Bahmani, B., Chowdhury, A., Goel, A.: Fast incremental and personalized pagerank. Proc. VLDB Endow. **4**(3), 173–184 (2010)
3. Desikan, P., et al.: Incremental page rank computation on evolving graphs. In: Special Interest Tracks and Posters of the 14th International Conference on World Wide Web. ACM (2005)
4. Avrachenkov, K., et al.: Monte Carlo methods in PageRank computation: when one iteration is sufficient. SIAM J. Numer. Anal. **45**(2), 890–904 (2007)
5. Langville, A.N., Meyer, C.D.: Deeper inside pagerank. Internet Math. **1**(3), 335–380 (2004)
6. Chien, S., et al.: Towards exploiting link evolution (2001)
7. Langville, A.N., Meyer. C.D.: Updating pagerank with iterative aggregation. In: Proceedings of the 13th International World Wide Web Conference on Alternate Track Papers & Posters. ACM (2004)
8. Kamvar, S., et al.: Exploiting the block structure of the web for computing pagerank. Technical report, Stanford University (2003)
9. Lofgren, P.: On the complexity of the Monte Carlo method for incremental PageRank. Inf. Process. Lett. **114**(3), 104–106 (2014)
10. Das Sarma, A., Molla, A.R., Pandurangan, G., Upfal, E.: Fast distributed PageRank computation. In: Frey, D., Raynal, M., Sarkar, S., Shyamasundar, Rudrapatna K., Sinha, P. (eds.) ICDCN 2013. LNCS, vol. 7730, pp. 11–26. Springer, Heidelberg (2013). doi:10.1007/978-3-642-35668-1_2
11. Jure, L.: Stanford Large Network Dataset Collection. http://snap.stanford.edu/data/index.html
12. Seo, S., et al.: HAMA: an efficient matrix computation with the mapreduce framework. In: 2010 IEEE Second International Conference on IEEE Cloud Computing Technology and Science (CloudCom) (2010)
13. Valiant, L.G.: A bridging model for parallel computation. Commun. ACM **33**(8), 103–111 (1990)
14. Jeh, G., Jennifer, W.: Scaling personalized web search. In: Proceedings of the 12th International Conference on World Wide Web. ACM (2003)
15. Pettie, S.: Single-source shortest paths. In: Encyclopedia of Algorithms, pp. 847–849 (2008)

Joint Weighted Nonnegative Matrix Factorization for Mining Attributed Graphs

Zhichao Huang, Yunming Ye[✉], Xutao Li, Feng Liu, and Huajie Chen

Shenzhen Key Laboratory of Internet Information Collaboration,
Shenzhen Graduate School, Harbin Institute of Technology, Shenzhen 518055, China
{huangzhichao,liufeng,chenhuajie}@stmail.hitsz.edu.cn,
{yym,lixutao}@hitsz.edu.cn

Abstract. Graph clustering has been extensively studied in the past decades, which can serve many real world applications, such as community detection, big network management and protein network analysis. However, the previous studies focus mainly on clustering with graph topology information. Recently, as the advance of social networks and Web 2.0, many graph datasets produced contain both the topology and node attribute information, which are known as attributed graphs. How to effectively utilize the two types of information for clustering thus becomes a hot research topic. In this paper, we propose a new attributed graph clustering method, *JWNMF*, which integrates topology structure and node attributes by a new collective nonnegative matrix factorization method. On the one hand, JWNMF employs a factorization for topology structure. On the other hand, it designs a weighted factorization for nodes' attributes, where the weights are automatically determined to discriminate informative and uninformative attributes for clustering. Experimental results on seven real-world datasets show that our method significantly outperforms state-of-the-art attributed graph clustering methods.

Keywords: Attributed graph · Clustering · Weight · NMF

1 Introduction

Graph clustering is a widely studied research problem and receives considerable attention in data mining and machine learning recently [1–8]. It aims to partition a given graph into several connected components based on structural similarity. Vertices from the same component are expected to be densely connected, and the ones from different components are weakly tied. Graph clustering is popularly used in community detection, protein network analysis, etc. [4–6]. The previous work focused mainly on finding clusters by exploiting the topology structures. Recently, as the advance of social networks and Web 2.0, many graph datasets appear with both the topology and node attribute information. For example, a webpage (i.e., vertex) can be associated with other webpages via hyperlinks, and it may have some inherent attributes of itself, like the text description in

© Springer International Publishing AG 2017
J. Kim et al. (Eds.): PAKDD 2017, Part I, LNAI 10234, pp. 368–380, 2017.
DOI: 10.1007/978-3-319-57454-7_29

the webpage. Such type of graphs are known as attributed graphs. Because the topology and attributes together offers us a better probability to find high-quality clusters, attributed graph clustering becomes a hot research topic.

However, finding clusters in attributed graphs is not trivial, and there are two important challenges we need to address. **Challenge 1**: how to effectively utilize the topology information and the attributes together. In conventional graph clustering methods, only the topology structure is exploited to find clusters. By contrast, conventional feature based clustering algorithms take merely the attributes into account. Different from the two types of approaches, attributed graph clustering algorithms should effectively use the two types of information together. **Challenge 2**: how to automatically determine the importance of different attributes? It is well-known that weighting features appropriately can help to find the inherent clusters, especially when there is a large portion of noisy features for clusters. We face the same challenge for attributed graph clustering. For example, in the aforementioned webpage example, each webpage may contain different textual information at different locations, e.g., title, body, advertisement, and features extracted thus may have distinct contributions to clusters. Although some methods have been put forward recently to address the first challenge [9–16], few of them notice the second challenge.

In this paper, we introduce a *Joint Weighted Nonnegative Matrix Factorization* method for clustering attributed graphs, namely JWNMF, which can address the two challenges. NMF [17,18] is a well-known technique, which could produce the promising performance in graph clustering [7,8,19]. For a given attributed graph, our method presents a mechanism by using joint-NMF to integrate the structural and attribute information. Specifically, we design two matrix factorization terms. One is modeling the topology structure and the other is for attributes. Meanwhile, we modify the NMF by introducing a weighting variable for each attribute, which can be automatically updated and determined in each iteration. Experiments are performed on seven real-world datasets, including two amazon information networks, one CMU email networks, one DBLP information network, one webpage links network and two citation information networks. Our experimental results show that the proposed JWNMF method outperforms state-of-the-art attributed graph clustering algorithms, like BAGC [11], PICS [13] and SANS [14].

The remainder of this paper is organized as follows: Sect. 2 reviews some existing work on attributed graph clustering. In Sect. 3, we introduce the proposed JWNMF method. Section 4 presents and discusses the experimental results. Finally, the conclusions are given in Sect. 5.

2   Related Work

Several clustering methods have been introduced for mining attributed graphs recently. They can mainly be categorized into two types, namely *distance-based methods* [9,10,14] and *model-based methods* [11–13,15,16]. The idea of distance-based methods is to design a unified distance which could combine and leverage structural and attribute information, and then utilize existing clustering methods, e.g., k-means or spectral clustering, to cluster attributed graphs based on the unified distance. Model-based methods leverage the interactions between edges and node attributes to construct a joint model for clustering purpose.

2.1 Distance-Based Methods

Zhou *et al.* proposed a distance-based method **SA-Cluster** [9] in 2009 and its efficient version **Inc-Cluster** [10] in 2011. The key idea of the two methods is to construct a new graph by treating node attributes as new nodes and linking the original nodes to the new attribute nodes if the corresponding attribute values are non-zeros. A unified distance for the augmented graph is designed by using a random walk process. Finally, *k-mediods* is performed to partition the new augmented graph. As the augmenting step may increase the size of graphs considerable, the two methods are hard to run on large-scale attributed graphs.

SANS was introduced in 2015 [14], which partitions attributed graph leveraging both structural and node attribute information. In the method, a weighting vector is predefined. SANS chooses the node with the largest degree (out-degree plus in-degree) as a cluster center, then other nodes connected with this node are partitioned in the cluster. As a sequel, SANS assigns the clustered nodes whose attribute similarities with those assigned nodes are larger than a threshold into the cluster. After that, the weighting vector and attribute similarities are updated. The procedure is repeated until all nodes are clustered. This method can automatically partition attributed graph without pre-defined number of clusters.

2.2 Model-Based Methods

Xu *et al.* proposed a model-based approach **BAGC** in 2012 [11]. This method introduces a Bayesian probabilistic model by assuming that the vertices in same cluster should have a common multinomial distribution for each node attribute and a Bernoulli distribution for node connections. As a result, the attributed graph clustering problem can be transformed into a standard probabilistic inference problem. The clusters can be identified by using the node-to-cluster probabilities. The drawback is that this method cannot handle weighted attributed graphs. To overcome this problem, Xu extended BAGC and proposed **GBAGC** lately [12].

PICS was proposed by Akoglu in 2012 [13]. This method is a matrix compression based model clustering approach. It treats clustering problem as a data compression problem, where the structure matrix and attribute matrix are compressed at the same time. Each cluster is regarded as a compression of a dense connected subset, and the nodes in the same cluster have similar connectivity and attribute properties. Due to less computational complexity, PICS can deal with large-scale attributed graphs.

In 2014, Perozzi proposed a user interest based attributed graph clustering method, namely **FocusCo** [15]. The method utilizes the similarities of users' interests to find an optimal clustering results for attributed graphs. **CESNA** [16] models the correlations between structures and node attributes to improve the intra-cluster similarities. The method differs from other attributed graph clustering methods in that it can detect overlapping communities in social networks.

Different from the existing studies, we propose a collective nonnegative matrix factorization method to leverage both the topology and attribute information. Moreover, we design a weighting vector to differentiate the contribution of attributes to clusters, which can be automatically determined. Our method addresses the two challenges mentioned in introduction.

3   Proposed Method

An attributed graph can be defined as $G = (V, E, A)$, where $V = \{v_1, v_2, \ldots, v_n\}$ denotes the set of nodes, $E = \{(v_i, v_j), 1 \leq i, j \leq n, i \neq j\}$ denotes the set of edges, and $A = [\mathbf{a}_1, \mathbf{a}_2, \ldots, \mathbf{a}_m]$ denotes the set of node attributes. In an attributed graph G, each node v_i in V is associated with an attribute vector $(a_1^i, a_2^i, \ldots, a_m^i)$, where each element of the vector is the attribute value of v_i on the corresponding attribute.

The key difference of attributed graph clustering to conventional graph clustering is that it needs take node attributes into account. Consequently, the ideal clustering results should follow two properties: (1) nodes in the same clusters are densely connected, and sparsely connected in different clusters; (2) and nodes in the same clusters have similar attribute values, and have diverse attribute values in different clusters.

3.1   Overview of NMF

Here, we will briefly review the *Nonnegative Matrix Factorization (NMF)* [17,18]. Let X denotes a $M \times N$ matrix whose data elements are all nonnegative. The goal of NMF is that to find two nonnegative matrix factors $V = (V_{i,j})_{M \times K}$ and $U = (U_{i,j})_{N \times K}$, where K denotes the desired reduced dimension of original matrix X. In general, $K \leq \min(M, N)$. After that, we can produce an approximation of X by $X \approx VU^T$. A commonly used objective function for NMF can be regarded as a *Frobenius norm* optimizing problem, as follows:

$$\min_{V, U \geq 0} \|X - VU^T\|_F^2$$

where $\| \cdot \|_F$ is the Frobenius norm and $V, U \geq 0$ represent the nonnegative constraints in matrix factorization.

3.2   Objective Function

Following the definition of attributed graphs above, we assume that S denotes the adjacency matrix for topology structure, and matrix A represents the attribute information where rows denote nodes and columns represent attributes. In addition, we also introduce a diagonal matrix Λ to assign a weight for each attribute. Inspired by SymNMF [7,8], which often delivers promising results for graph clustering, we apply the idea for attributed graph clustering. Specifically, we have factorizations $S \approx VV^T$ and $A\Lambda \approx VU^T$, where V is a fusing representation of topology and attribute information for nodes.

In order to integrate the two approximation into the NMF framework, we propose a weighted joint NMF optimization problem over V, U, Λ:

$$\min_{V,U,\Lambda \geq 0} \|S - VV^T\|_F^2 + \lambda\|A\Lambda - VU^T\|_F^2 \qquad (1)$$

where $S \in \mathbb{R}_+^{n \times n}$, $A \in \mathbb{R}_+^{n \times m}$, $\Lambda \in \mathbb{R}_+^{m \times m}$, $V \in \mathbb{R}_+^{n \times k}$, $U \in \mathbb{R}_+^{m \times k}$, $\mathbb{R}_+$ denotes the set of nonnegative real numbers, n denotes the number of nodes, m denotes the number of attribute categorizations, Λ is a diagonal matrix satisfying $\sum_{i=1}^m \Lambda_{i,i} = 1$ and $\lambda > 0$ is the weight to balance structural/attribute fusion and k is the number of clusters. Actually, before optimizing Eq. 1, we preprocess the adjacency matrix S and the attribute information matrix A as:

$$S = \frac{S}{\sum_{i=1}^n \sum_{j=1}^n S_{i,j}}, A = \frac{A}{\sum_{i=1}^n \sum_{j=1}^m A_{i,j}} \qquad (2)$$

Next, we will derive the updating rules of V, U and Λ.

3.3   Updating Rules

Let α, β and γ denote respectively the Lagrange multiplier matrix for the constraints $V \geq 0$, $U \geq 0$ and $\Lambda \geq 0$. By using the Lagrange formulation, we obtain the loss function without constraints:

$$L = \frac{1}{2}(\|S - VV^T\|_F^2 + \lambda\|A\Lambda - VU^T\|_F^2) + Tr(\alpha^T V) + Tr(\beta^T U) + Tr(\gamma^T \Lambda)$$

Taking partial derivatives of L with respect to V, U and Λ, we have

$$\frac{\partial L}{\partial V} = -(SV + S^T V + \lambda A\Lambda U) + (2VV^T V + \lambda VU^T U + \alpha) \qquad (3)$$

$$\frac{\partial L}{\partial U} = -\lambda \Lambda A^T V + \lambda UV^T V + \beta \qquad (4)$$

$$\frac{\partial L}{\partial \Lambda} = -\lambda A^T VU^T + \lambda A^T A\Lambda + \gamma \qquad (5)$$

In terms of Karush-Kuhn-Tucker (KKT) conditions $\alpha_{p,r} V_{p,r} = 0$, $\beta_{q,r} U_{q,r} = 0$ and $\gamma_{q,q} \Lambda_{q,q} = 0$, it follows that $\frac{\partial L}{\partial V} = 0$, $\frac{\partial L}{\partial U} = 0$ and $\frac{\partial L}{\partial \Lambda} = 0$. Base on these conditions, we can derive the following updating rules with respect to V, U and Λ:

$$V \longleftarrow V. * (SV + S^T V + \lambda A\Lambda U)./(2VV^T V + \lambda VU^T U) \qquad (6)$$

$$U \longleftarrow U. * (\Lambda A^T V)./(UV^T V) \qquad (7)$$

$$\Lambda \longleftarrow \Lambda. * (A^T V U^T)./(A^T A\Lambda) \tag{8}$$

where $.*$ and $./$ represent the elementwise multiplication and division, respectively. In order to assign the weights of Λ into a regular space, we normalize it as:

$$\Lambda = \frac{\Lambda}{\sum_{i=1}^{m} \Lambda_{i,i}} \tag{9}$$

Next, we briefly analyze the convergency and the computational complexity of above updating rules.

For proving the convergency, we just need adopt the auxiliary function described in [18]. In addition, the *KKT* conditions, which suffice the stationary point of the objective function, also imply the convergence of those updating rules.

Here, the computational complexity is discussed. Supposing the algorithm stops after t iterations, the overall cost for SymNMF [7,8] is $O(n^2 kt)$. As the objective function adds one more linear matrix factorization term, the overall cost for updating rules is $O((n^2 + m^2 + mn)kt)$.

3.4 The Joint Weighted NMF Algorithm

By combining the parts above, our attributed graph clustering algorithm JWNMF can be summarized as follows: Firstly, we preprocess the adjacency matrix S and attribute matrix A, and randomly initialize the matrices U, V and assign the values of diagonal matrix Λ with $1/m$. Then we iteratively update matrices U, V and Λ as Eqs. (6)–(9) until it converges. Finally, *LiteKmeans*[1] is performed on the factorization result V to identify k clusters.

4 Experimental Study

In this section, we evaluate the performance of our algorithm, and compare it with three state-of-the-art attributed graph clustering methods: BAGC [11], PICS [13] and SANS [14], and a benchmark clustering approach S-Cluster which is implemented by using *LiteKmeans* and focuses only on structure information. All algorithms were implemented in Matlab R2014b, and tested on a Windows 10 PC, Intel Core i5-4460 3.20 GHz CPUs with 32 GB memory.

4.1 Datasets

Seven real-world datasets are employed in our experiments, where four of them do not have ground truth and three of them have ground truth. The datasets without ground truth include two amazon information networks (Amazon Fail[2] and Disney[3]), a CMU email address network (Enron (see footnote 2)) and a

---

[1] http://www.zjucadcg.cn/dengcai/Data/Clustering.

[2] http://www.ipd.kit.edu/~muellere/consub/.

[3] http://www.perozzi.net/projects/focused-clustering/.

network of DBLP information (DBLP-4AREA (see footnote 3)). On the other hand, the datasets with ground truth (WebKB, Citeseer, Cora)[4] are all from text categorization applications. We represent all of these datasets as undirected networks. Table 1 summarizes the characteristics of the seven datasets.

Table 1. Description of seven real-world datasets

Dataset	#Nodes	#Edges	#Attributes	#Clusters
Amazon Fail	1,418	3,695	21	-
Enron	13,533	176,987	18	-
Disney	124	335	25	-
DBLP-4AREA	27,199	66,832	4	-
WebKB	877	174	1,703	5
Citeseer	3,312	117	3,703	6
Cora	2,708	151	1,433	7

4.2   Evaluation Measures

The goal of attributed graph clustering is to effectively leverage the topology and attribute information. Hence, we evaluate the attributed graph clustering based on the two aspects. Specially, to evaluate clustering results from the topology structure and the attribute points of view, we employ *modularity* and *average entropy*. Modularity [20] is a widely used evaluation measure for graph partition, and average entropy is often used in evaluating feature based clustering results.

Let $C = (C_1, C_2, \ldots, C_k)$ represents the k partitions of an attributed graph, the modularity Q and average entropy $Avg\_entropy$ are defined as:

$$Q = \sum_{i=1}^{k}(e_{i,i} - c_i^2) \tag{10}$$

$$Avg\_entropy = \sum_{t=1}^{m}\sum_{j=1}^{k}\frac{|C_j|}{nm}entropy(a_t, C_j) \tag{11}$$

where $e_{i,j}$ is the fraction of edges with the start node in cluster i and the end node in cluster j, and c_i denotes the fraction of ends of edges that are attached to nodes in cluster i, and $entropy(a_t, C_j)$ is the information entropy of attribute a_t in cluster C_j. The value with respect to modularity and average entropy falls within the range of $[-1, 1]$ and the range of $[0, +\infty)$, where higher modularity indicates dense connections between nodes within clusters but sparse connections between nodes in different clusters, and lower average entropy indicates we have similar attribute values within clusters but dissimilar attribute values in different clusters, i.e., a better clustering result.

---

[4] http://linqs.cs.umd.edu/projects//projects/lbc/index.html.

In addition to modularity and average entropy, we also utilize *Normalized Mutual Information (NMI)* to evaluate the clustering performance for the datasets with ground truth. Generally, higher NMI values indicate better clustering results.

4.3 Performance on Datasets Without Ground Truth

Effectiveness Evaluation. We show how the modularity and average entropy change with respect to different number of clusters on Amazon Fail in Fig. 1. We observe JWNMF outperforms the four baseline methods in terms of modularity when varying the number of clusters. Meanwhile, in terms of average entropy, JWNMF performs the best, except when the number of clusters is set as 8. Similar observations can be found on Enron and Disney (in Figs. 2 and 3). From Fig. 4, we can see that our method achieves the lowest average entropy on DBLP-4AREA. However, according to modularity, JWNMF is inferior to PICS. The reason is that PICS treats attributed graph clustering problem as a data compression problem, thus it prefers datasets which consist of large number of nodes but sparse topology structures. Moreover, we can see from Figs. 1, 2, 3 and 4 that average entropy has a descending trend as the number of clusters is

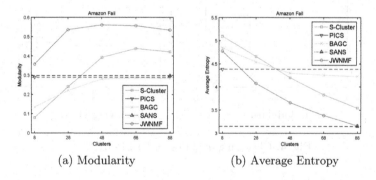

(a) Modularity   (b) Average Entropy

Fig. 1. Clustering qualities on Amazon Fail

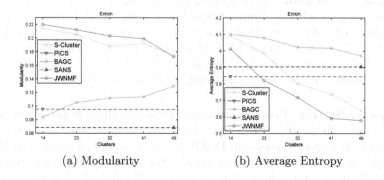

(a) Modularity   (b) Average Entropy

Fig. 2. Clustering qualities on Enron

increased. This is because increasing the number of clusters improves the chances that the nodes with similar attributes are put into the same cluster.

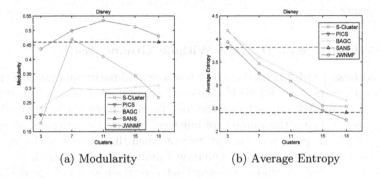

(a) Modularity   (b) Average Entropy

Fig. 3. Clustering qualities on Disney

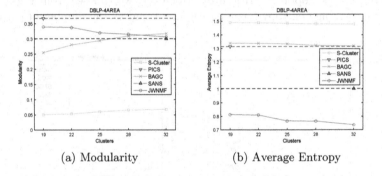

(a) Modularity   (b) Average Entropy

Fig. 4. Clustering qualities on DBLP-4AREA

Efficiency Evaluation. Table 2 shows the running time of all the methods on the four datasets without ground truth. We can see JWNMF runs much faster than three sate-of-the-art attributed graph clustering methods, PICS, BAGC and SANS. The reason is that JWNMF is a quite efficient method whose iterate computation converges very fast (usually in 100 iterations). Although S-cluster achieves the best efficiency, its clustering results can be pretty poor as in Fig. 4.

Parameter Setting. In our experiments, we search the parameter λ in the set $\{10^{-10}, 10^{-8}, 10^{-7}, 10^{-6}, 10^{-5}, 10^{-4}, 10^{-3}, 0.5\}$ to find its optimal settings on Amazon Fail, Enron, Disney and DBLP-4AREA. According to our experience, we advise to set the parameter λ in terms of the sparsity of topology structures. Specifically, it is more appropriate to use a small value of λ for datasets with dense topology structure.

Table 2. Running time (sec) on datasets without ground truth

Dataset	Clusters	S-Cluster	PICS	BAGC	SANS	JWNMF
Amazon Fail	8	0.0033	9.2490	0.6575	–	0.3254
	28	0.0075	-	1.4442	–	0.3351
	48	0.0120	-	1.7249	–	0.3755
	68	0.0133	-	1.6153	–	0.4369
	88	0.0216	-	2.0932	4.2059	0.4320
Enron	14	0.4001	385.5467	360.9523	–	101.1421
	23	0.7470	-	349.1356	–	103.0285
	32	0.8329	-	319.1372	–	73.9748
	41	1.5420	-	280.1238	–	103.1282
	49	1.2942	-	250.1443	481.8581	105.3710
Disney	3	0.0017	0.1792	0.0138	–	0.0049
	7	0.0015	-	0.0201	–	0.0062
	11	0.0017	-	0.0287	–	0.0061
	15	0.0015	-	0.0137	–	0.0081
	18	0.0014	-	0.0350	0.0414	0.0074
DBLP-4AREA	19	0.2162	762.0975	1666.9570	–	367.9434
	22	0.2454	-	1663.7425	–	306.7181
	25	0.2548	-	1601.4241	–	290.4555
	28	0.2931	-	1544.3671	–	291.9064
	32	0.3422	-	1540.0352	2182.5214	367.0382

4.4   Performance on Datasets with Ground Truth

Since PICS and SANS cannot output the ground-truth of number of clusters, we do not compare with them in this section. Table 3 reports the performance for S-Cluster, BAGC and JWNMF on the three datasets with ground truth. For JWNMF, we set $\lambda =1.5$, 0.5 and 4.5 for the three datasets, respectively. Overall, our method has better performance than the baseline methods. In particular, the improvements are significant in terms of modularity and NMI. In terms of average entropy, however, the superiority of JWNMF is slight. The reason is that the textual attribute is with huge dimensions but very sparse, which makes the computed entropies more or less equal.

In JWNMF, we introduced a weighting matrix Λ to handle noisy features. To demonstrate the merits of the weighting scheme, we inject *30%* noisy attributes of random 0/1 distribution into the three datasets. Table 4 reports the results on those noisy datasets, where JNMF represents the variant of our method by removing the weighting matrix. We find that JWNMF significantly outperforms other methods including JNMF. The results show that the weighting scheme of our model is very useful, especially in the presence of noisy attributes.

Table 3. Performance on three textual datasets (%)

Dataset	Methods	Modularity	Average entropy	NMI
WebKB	S-Cluster	0.1633	23.1949	1.4282
	BAGC	−0.0260	23.2986	0.3313
	JWNMF	**33.7672**	**23.0107**	**2.1891**
Citeseer	S-Cluster	2.2419	5.9691	0.2895
	BAGC	0	5.9791	0
	JWNMF	**23.999**	**5.9565**	**0.6178**
Cora	S-Cluster	−0.2060	8.3762	0.4014
	BAGC	0	8.3963	0
	JWNMF	**25.8493**	**8.3427**	**1.5033**

Table 4. Performance on three noisy textual datasets (%)

Dataset	Methods	Modularity	Average entropy	NMI
WebKB	S-Cluster	0.1633	35.8873	1.4282
	BAGC	0	36.1062	0
	JNMF	33.7928	35.7491	**2.3884**
	JWNMF	**37.3405**	**35.7283**	2.1879
Citeseer	S-Cluster	2.2419	21.6018	0.2895
	BAGC	0	21.6352	0
	JNMF	27.4264	21.5915	0.6725
	JWNMF	**32.1148**	**21.5890**	**0.7623**
Cora	S-Cluster	−0.2060	23.5784	0.4014
	BAGC	0	23.6321	0
	JNMF	38.7895	23.5460	1.6905
	JWNMF	**41.3449**	**23.5439**	**1.8291**

5   Conclusion

In this paper, we develop a joint weighted nonnegative factorization method, namely JWNMF, to solve the attributed graph clustering problem. By using two joint factorization terms, JWNMF nicely fuses the topology and attribute information of attributed graphs for clustering. Moreover, a weighting scheme is incorporated into JWNMF to differentiate attribute importance to clusters. An iterative algorithm is proposed to find solutions of JWNMF. Extensive experimental results show that our method outperforms state-of-the-art attribute graph clustering algorithms.

Acknowledgments. The research was supported in part by NSFC under Grant Nos. 61572158 and 61602132, and Shenzhen Science and Technology Program under Grant Nos. JCYJ20160330163900579 and JSGG20150512145714247, Research Award Foundation for Outstanding Young Scientists in Shandong Province, (Grant No. 2014BSA10016), the Scientific Research Foundation of Harbin Institute of Technology at Weihai (Grant No. HIT(WH)201412).

References

1. Van Dongen, S.M.: Graph clustering by flow simulation (2001)
2. Schaeffer, S.E.: Graph clustering. Comput. Sci. Rev. **1**(1), 27–64 (2007)
3. Ng, A.Y., Jordan, M.I., Weiss, Y., et al.: On spectral clustering: analysis and an algorithm. Adv. Neural Inf. Process. Syst. **2**, 849–856 (2002)
4. Fortunato, S.: Community detection in graphs. Phys. Rep. **486**(3), 75–174 (2010)
5. Brohee, S., Van Helden, J.: Evaluation of clustering algorithms for protein-protein interaction networks. BMC Bioinf. **7**(1), 1 (2006)
6. Wu, Z., Leahy, R.: An optimal graph theoretic approach to data clustering: theory and its application to image segmentation. IEEE Trans. Pattern Anal. Mach. Intell. **15**(11), 1101–1113 (1993)
7. Kuang, D., Ding, C., Park, H.: Symmetric nonnegative matrix factorization for graph clustering. In: SDM, vol. 12, pp. 106–117. SIAM (2012)
8. Kuang, D., Yun, S., Park, H.: SymNMF: nonnegative low-rank approximation of a similarity matrix for graph clustering. J. Glob. Optim. **62**(3), 545–574 (2015)
9. Zhou, Y., Cheng, H., Yu, J.X.: Graph clustering based on structural/attribute similarities. Proc. VLDB Endow. **2**(1), 718–729 (2009)
10. Zhou, Y., Cheng, H., Yu, J.X.: Clustering large attributed graphs: an efficient incremental approach. In: 2010 IEEE 10th International Conference on Data Mining (ICDM), pp. 689–698. IEEE (2010)
11. Xu, Z., Ke, Y., Wang, Y., Cheng, H., Cheng, J.: A model-based approach to attributed graph clustering. In: Proceedings of the 2012 ACM SIGMOD International Conference on Management of Data, pp. 505–516. ACM (2012)
12. Xu, Z., Ke, Y., Wang, Y., Cheng, H., Cheng, J.: GBAGC: a general bayesian framework for attributed graph clustering. ACM Trans. Knowl. Discov. Data (TKDD) **9**(1), 5 (2014)
13. Akoglu, L., Tong, H., Meeder, B., Faloutsos, C.: PICS: Parameter-free identification of cohesive subgroups in large attributed graphs. In: SDM, pp. 439–450. SIAM (2012)
14. Parimala, M., Lopez, D.: Graph clustering based on structural attribute neighborhood similarity (SANS). In: 2015 IEEE International Conference on Electrical, Computer and Communication Technologies (ICECCT), pp. 1–4. IEEE (2015)
15. Perozzi, B., Akoglu, L., Iglesias Sánchez, P., Müller, E.: Focused clustering and outlier detection in large attributed graphs. In: Proceedings of the 20th ACM SIGKDD International Conference on Knowledge Discovery and Data Mining, pp. 1346–1355. ACM (2014)
16. Yang, J., McAuley, J., Leskovec, J.: Community detection in networks with node attributes. In: 2013 IEEE 13th International Conference on Data mining (ICDM), pp. 1151–1156. IEEE (2013)
17. Lee, D.D., Seung, H.S.: Learning the parts of objects by non-negative matrix factorization. Nature **401**(6755), 788–791 (1999)

18. Lee, D.D.: Algorithms for non-negative matrix factorization. Adv. Neural Inf. Process. Syst. **13**(6), 556–562 (2001)
19. Jin, D., Gabrys, B., Dang, J.: Combined node and link partitions method for finding overlapping communities in complex networks. Sci. Rep. **5**, 8 p. (2015)
20. Newman, M.E.J.: Modularity and community structure in networks. Proc. Natl Acad. Sci. **103**(23), 8577–8582 (2006)

Predicting Happiness State Based on Emotion Representative Mining in Online Social Networks

Xiao Zhang[1], Wenzhong Li[1,2(✉)], Hong Huang[3], Cam-Tu Nguyen[1], Xu Chen[4], Xiaoliang Wang[1,2], and Sanglu Lu[1,2]

[1] State Key Laboratory for Novel Software Technology,
Nanjing University, Nanjing, China
`tobexiao1@dislab.nju.edu.cn, lwz@nju.edu.cn`
[2] Sino-German Institutes of Social Computing, Nanjing University, Nanjing, China
[3] Institute of Computer Science, University of Gottingen, Gottingen, Germany
[4] School of Data and Computer Science, Sun Yat-sen University, Guangzhou, China

Abstract. Online social networks (OSNs) have become a major platform for people to obtain information and to interact with their friends. People tend to post their thoughts and activities online and share their emotions with friends, which provides a good opportunity to study the role of online social networks in happiness spreading and mutual influence among the users. In this paper, we propose a framework to study the influence of happiness in OSNs. We first quantify the happiness states of users by analyzing their daily posting texts, and then conduct the statistical analysis to show that users' happiness states are influenced by their social network neighbors. Since the influence of each individual is unequal, we develop a regression model and a greedy algorithm to detect the high influence users known as emotion representatives. By using a small number of detected emotion representatives as features to train prediction models, we show that it achieves good performance in predicting the happiness states of the whole online social network users.

1 Introduction

Happiness and other emotions have been shown to be contagious: emotion states can be transferred directly from person to person via mimicry [9] and the copying of bodily actions like facial expressions [24]. Experiments have demonstrated that diverse emotions such as happiness [8], loneliness [4], and depression [17] are highly correlated between socially connected individuals in human social networks.

This work was partially supported by the National Natural Science Foundation of China (Grant Nos. 61672278, 61373128, 61321491), the EU FP7 IRSES MobileCloud Project (Grant No. 612212), the Collaborative Innovation Center of Novel Software Technology and Industrialization, and the project from State Grid Corporation of China (Research on Key Clustering Technology for Hyperscale Power Grid Control System).

© Springer International Publishing AG 2017
J. Kim et al. (Eds.): PAKDD 2017, Part I, LNAI 10234, pp. 381–394, 2017.
DOI: 10.1007/978-3-319-57454-7_30

In the recent years, online social networks (OSNs) such as Facebook and Twitter have attracted more and more users, which have become the most important source of information and the most popular platform for people to communicate with their friends. Online social network users are actively involved, frequently interacting with each other, regularly posting their status, and willingly sharing their happiness and other emotions with their friends. With the rich source of online social media and social connections, exploring the happiness states of online users and their mutual influence has caused great attention of the academic community [6,7]. Despite the evidence of emotion contagion in social networks [6–9,24], the role of online social networks in happiness influence and the key of influential users in happiness spread have not been well addressed in the past.

In this paper, we focus on the quantitative analysis of happiness influence in online social networks. We mainly focus on the following key questions: (1) How to quantify the happiness states of online social network users by exploring their daily posting texts? (2) Are the happiness states of OSN users influenced by their social network neighbors? (3) Are there high emotion influential users and how to detect them? (4) Are the happiness states of online social network users predictable by observing a small number of representatives?

To address the above issues, we propose a framework to study the influence of happiness in online social networks. As illustrated in Fig. 1, the study is based on a dataset of online social network, which is formulated as a social graph. We introduce the *happiness score* to quantify the happiness state of an individual based on analyzing his/her daily post on social media. We then analyze the happiness influence among users. Specifically, we apply a multi-linear regression model and significance test to show that users' happiness scores are correlated with their social network neighbors and some of the users have significant influence to the others' emotion states. We further detect a set of influential users called *emotion representatives* by solving an influence maximization problem. A greedy algorithm with bounded approximate ratio is proposed for representative detection. We use the detected emotion representatives as features to predict the happiness states of random users in the network, which shows better performance than the baseline algorithms.

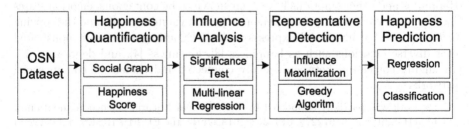

Fig. 1. The framework of happiness influence study.

The novelty and contribution of this paper are summarized as follows.

- We present the happiness influence problem in online social networks. The issues of happiness spread in large scale OSNs and the detection of emotion representatives have not been well addressed in the past.
- We propose a framework for happiness influence study in OSNs. The proposed framework provides quantitative analysis on the happiness states of OSN users and it confirms the influence of happiness between OSN users and their social network neighbors.
- We propose a greedy algorithm to detect a set of emotion representatives that maximize happiness influence. Extensive experiments show that the detected emotion representatives can be used as features to predict the happiness states of the whole online social network users with good performance.

2 Related Work

The contagion of emotion in the human society has been widely addressed in the literature. An early work showed that human emotion states can be directly transferred to the other by copying of facial expressions [24]. Data from a large real-world social network collected over 20-year period suggested that longer-lasting moods (e.g., depression, loneliness) can be transferred through networks [4,17]. Experiments in [8] showed that happiness can spread up to three hops in face-to-face human social network.

In the recent years, researchers have paid more attention to emotion analysis in the context of online social networks. The study of [7] provided quantified analysis to explain temporal variations of happiness in Twitter. Several works focused on inferring user's emotions from existing emotion labels or generated emotion scores [2,13,16]. Wang et al. proposed a constraint optimization framework to discover emotions from social media content of the users [20]. Coviello et al. proposed the instrumental variables regression method to explain the reason of emotion influence due to social contagion or homophily [6]. The "MoodCast" model [19] considered the influence from all friends as structural features to infer individuals' emotional states. Yang et al. took both user interest in text and image domains and social influence among friends into consideration for emotion prediction [22]. Despite the study of emotion contagion and prediction in online social networks, the issues such as detecting emotion influential users and predicting the emotion states of the crowd by observing a small number of representatives have not been well addressed in the massive online social networks.

Mining influential nodes in online social networks has been studied recently, whose methods can be classified into two categories: rank-based and model-based. The rank-based methods used different metrics to measure the importance of a node in the network. The most commonly used metrics are the degree centrality, betweenness centrality, and closeness centrality [23]. In addition, Page Rank [3] or PageRank-based methods (LeaderRank) [12] were also widely used in identifying influential nodes in social networks. The model-based methods developed different models for different networks. Aral and Walker used hazard models to identify not only the influential but also the susceptible members by designing randomized experiments in Facebook [1]. Sharara et al. presented a new surveying method which combines secondary data with partial knowledge from primary sources to guide the information gathering process and to identify opinion leaders in social networks [18]. However, none of the works has addressed the problem of detecting a set of key emotion representatives to maximize their influence in online social networks, as will be studied in our work.

3   Quantifying User's Happiness State

3.1   Online Social Network Dataset

Our study is based on the Twitter dataset published in the literature [21]. The dataset was collected starting from the most popular 20 users as the seed users. The snowball crawling method [11] was used to crawl the tweet data. The crawling period lasted for one month. The overall dataset contains 3,117,750 users including their profiles, social relationships and all the tweets published by the users up to Apr. 2, 2010.

The social relationships of the users can be described by a directed graph $G = (V, E)$, where V is the set of $|V| = N$ users and $E \subset V \times V$ is the set of directed edges. In Twitter, users can follow each other. If user A follows user B, A is called the *follower* of B and B is the *followee* of A, and there is a directed edge from A to B. A followee's published tweets can be viewed by all his/her followers. The dataset contains 538,726,224 published tweets which are used to analyze the happiness states of the users.

3.2   Happiness Score

To represent the happiness state of an individual, we introduce the *happiness score*, which is a quantified metric obtained by analyzing the text of user's daily tweets. Each tweet consists of a number of words reflecting the emotion states of a user. Following the work in [7], we adopt the same word list with labeled happiness score using crowdsourcing. Using word-by-word analysis, the happiness score of each tweet can be computed by averaging the happiness score of each word in the word list. And further the daily happiness score of a user can be obtained by the average score of all tweets posted in a day. To make the results more accurate, we take into account the sentence structure such as the appearing negative words (e.g., don't and isn't) which make the happiness states of the tweet reverse.

With the quantification method, the daily happiness score of an individual is represented by a real number range from 0 to 9, where 0 indicates very unhappy, and 9 indicates very happy. Applying the method to the Twitter dataset, we can find that the collective happiness states of all users approximately follow the Gaussian distribution with mean $= 5.48$ and variance $= 0.11$. For each day, we evaluate the happiness state of each user. If an individual's happiness score is larger than the mean in that day, the individual is considered as a *happy user*, otherwise, a *unhappy user*.

To study the long-term happiness states of users and their correlations, we filter out the users who rarely publish Tweets in the dataset (since there is no enough information to analyze their emotions) and only keep the *active users* defined as follows.

Active User: We denote the number of days that a user posted tweets by t. A user is active if $t \geq \alpha T$ in consecutive T days, where α is a tunable parameter with $0 \leq \alpha \leq 1$.

In our work, we set $T = 90$ days from 14/12/2009 to 13/3/2010, and choose $\alpha = 2/3$. After processing the dataset, there are 168,661 nodes and 2,923,633 directed edges remained in the social network graph.

With the quantification method, the happiness states of each user $v \in V$ can be represented by a happiness vector $\mathcal{E}(v) = [s_i(v)|i = 1, 2, \cdots, T]$, where $s_i(v)$ is the happiness score of v in the ith day.

4   Happiness Influence Analysis

With the social graph formed by the active users, we study the mutual influence of users' happiness states. We show that the happiness state of an individual is correlated with his/her followees in online social network. We apply a multi-linear regression model to infer the significance of happiness influence.

4.1   Influence of Happiness

Since Twitter is a directed social network, a user can read the tweets posted by his/her followees, but not vice versa. Intuitively, the influence of happiness is directional: the happiness states of followees can influence that of their followers. We conduct the statistical analysis to verify this claim.

We first show the correlation of happiness scores. For each active user v, we calculate v's happiness score and the average happiness score of v's followees, which results are compared in Fig. 2(a). Noted that the mean happiness score of all users is 5.48. Figure 2(a) shows high correlation between the individual's happiness score and the average score of his/her followees. It can be interpreted as: *the more happier the users you followed, the more likely you are happy.*

Given a user v, we define v's *fraction of happiness* as $\frac{\#of\ happy\ followees}{\#of\ happy\ and\ unhappy\ followees}$. Figure 2(b) shows the fraction of happiness for happy and unhappy users in five different days. It is clear that happy users have much higher fraction of happy followees (which is close to 0.6) than that of unhappy users (which is around 0.5).

To show that the higher fraction of happiness is not formed by chance, we use significance test to verify such phenomenon. We adopt a similar method

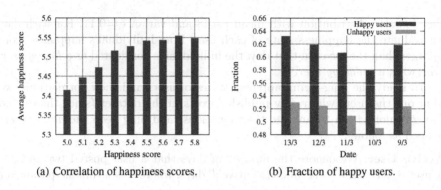

(a) Correlation of happiness scores.   (b) Fraction of happy users.

Fig. 2. Influence of happiness.

introduced in [14]. We construct N randomized networks which preserve the topology, number of happy users, and number of unhappy users. Then we randomly shuffle the happiness states of every user in every randomized network. For each randomized network i, we denote the mean fraction of happiness (over all happy users) as H_i. Then the mean of N networks is $H = \frac{1}{N}\sum_{i=1}^{N} H_i$. We conduct significance test by comparing H with the mean of the original network. The result shows that the phenomenon (happy user has more happy followees) is significant for all the days since their P-values are less than 0.01.

4.2 Multi-linear Regression Model

The above analysis shows that users' happiness states are influenced by their social network neighbors. However, the influence of each neighbor is not equal: some user may have higher influence than the others. To verify this, we introduce a multi-linear regression model to test the significance of happiness influence by different individuals.

Personalized Multi-Linear Regression Model. For a user i, we model i's happiness state as a function of the his/her previous happiness state and the happiness states of the user's followees, which is expressed by the multi-linear regression model below.

$$Y_t^i = \alpha + \beta Y_{t-1}^i + \sum_{j=1}^{K} \gamma_j Y_t^j + \epsilon, \tag{1}$$

where

- Y_t^i is the user i's happiness score at time t.
- K is the number of user i's followees in the social graph.
- γ_j is the key parameter, which means user j's influence on user i.
- α is the intercept and ϵ is the error term.

After training the multi-linear regression model, we apply the Wald test, a parametric statistical test method, to test the significance of the coefficients. Specifically, we test whether the coefficients γ_j $(j = 1, 2, \cdots, K)$ is 0 or non-zero: if γ_j is non-zero with high probability, then the happiness state of user j is considered to have significant influence on that of user i.

Using the regression model and significance test, we can divide the whole users into three sets:

Influential Users $\mathcal{I}$: The set of users who have significant influence to the other users.

Affected Users $\mathcal{A}$: The set of users who are influenced by the influential users.

Unaffected Users $\mathcal{U}$: The set of users who are neither influential users nor affected users.

Noted that a user maybe both influential and affected by the others, thus there is overlap of $\mathcal{I}$ and $\mathcal{A}$. In the Twitter dataset, we found 64,200 influential users and 84,977 affected users among the 168,661 active users in the social graph.

With the directed influence relationship, we define the *influence graph* that forms by the influential users and affected users as follows.

Influence Graph: An influence graph is denoted as $\bar{G} = (\bar{V}, \bar{E})$, where $\bar{V} = \mathcal{I} \cup \mathcal{A}$ is the set of influential and affected users, and $\bar{E} \subset \bar{V} \times \bar{V}$ is the set of directed edges from influential user to affected user.

Algorithm 1. Emotion Representatives Detection

1: $S \leftarrow \emptyset$
2: **for** i=1 **to** M **do**
3:    *Let v_i be a node with maximum degree in $\bar{B} = (\bar{L}, \bar{R}, \bar{E}')$*
4:    *Set $S \leftarrow S \cup \{v_i\}$*
5:    *Delete all nodes connected with v_i from $\bar{R}$*
6:    *Delete v_i from $\bar{L}$*
7: **end for**

Figure 3 shows the statistics of the influence graph obtained from the Twitter dataset. For a user $v \in \bar{G}$, the outdegree of v is the number of nodes affected by v. Figure 3(a) illustrates the cumulative distribution function (CDF) of outdegree in $\bar{G}$. According to the figure, about 90% users have outdrgree less than 5, and only about 1% users are highly influential who can affect more than 20 users in the social network. Figure 3(b) compares the cumulative influence scale of the top influential users, which shows that the top 20% influential users can affect about 80% other users in the influence graph.

5 Emotion Representatives Detection

According to the above analysis, some users have significant influence to the happiness states of the others. We call such influential users as *emotion*

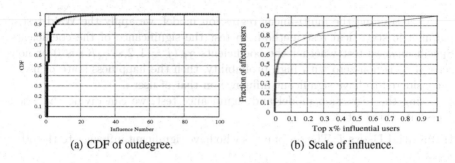

(a) CDF of outdegree.   (b) Scale of influence.

Fig. 3. Statistics of the influence graph.

representatives. In this section, we propose a method to identify the emotion representatives in online social networks. Specifically, given a number M, we want to find M users who have the greatest influence to the whole network. We formulate the problem as follows.

5.1 Problem Formulation

Given the influence graph $\bar{G} = (\bar{V}, \bar{E})$, the number of emotion representatives M, $\forall v_i \in \bar{V}$, the set of users influenced by v_i is denoted by $A(v_i)$. Let S be the set of chosen representatives. The problem can be formulated as

$$\max f(S) = |\bigcup_{v_i \in S} A(v_i)| \ s.t. \ S \subset \bar{V}, |S| \le M \tag{2}$$

where $|A(v_i)|$ indicates the cardinality of set $A(v_i)$.

5.2 Algorithm

The problem is similar to the influence maximization problem introduced by Kempe et al. in [10], but with a different influence model. According to the study of [15], finding a set S with M elements to maximize $f(S)$ is an NP-hard optimization problem. Thus we seek approximate solution to the problem. The following theorem shows some properties of the objective function f.

Theorem 1. *$f(S)$ is a monotone and submodular function.*

Proof. The monotonicity is obvious as

$$f(S + v) = |(\bigcup_{v_i \in S} A(v_i)) \cup A(v)| \ge f(S).$$

To prove the submodularity, we find that for any $S \subseteq T \subseteq V$, and $v \in V \setminus T$,

$$f(S + v) - f(S) = |A(v)/(\bigcup_{v_i \in S} A(v_i))| \ge |A(v)/(\bigcup_{v_i \in T} A(v_i))| = f(T + v) - f(T)$$

Based on the work of [15], an optimization problem with submodular property can be solved by a greedy algorithm with bounded approximate ratio. The basic idea is to choose the emotion representatives in a greedy way, and for each step a user with maximum *marginal gain* is chosen, until M users are found. To evaluate the marginal gain of a user, we transform the influence graph into a bipartite graph and then describe our algorithm.

Given the influence graph $\bar{G} = (\bar{V}, \bar{E})$ which is a directed graph, we transform it into a bipartite graph $\bar{B} = (\bar{L}, \bar{R}, \bar{E}')$, where $\bar{L} = \{v_i \mid \exists v_j \in \bar{V}, (v_i, v_j) \in \bar{E}\}$, $\bar{R} = \{v_j \mid \exists v_i \in \bar{V}, (v_i, v_j) \in \bar{E}\}$, and $\bar{E}' = \{(v_i, v_j) \mid v_i \in \bar{L}, v_j \in \bar{R}, (v_i, v_j) \in \bar{E}\}$. That is, the node set $\bar{V}$ is divided into two sets in the bipartite graph: the influential nodes $\bar{L}$ and the affected nodes $\bar{R}$. If a node is both an influencer and also affected by the others, we simply duplicate the node in the two sets. An edge in $\bar{B}$ represents that a node in $\bar{R}$ is influenced by a node in $\bar{L}$. With the derived bipartite graph $\bar{B}$, we proposed the greedy algorithm as shown in Algorithm 1.

In each iteration, we choose the user with maximum influence in $\bar{R}$ as a representative, and then delete the affected users from $\bar{R}$ to make sure that we can choose the one with maximum marginal gain in the next round. In each iteration the algorithm deletes several edges (and nodes) from $\bar{B}$, and in the worse case, all edges are deleted, which means the chosen M representatives cover all the users in $\bar{R}$. Thus the complexity of the algorithm is $O(\bar{E}')$ in the worst case.

The following theorem shows that the approximate ratio of the proposed algorithm is guaranteed.

Theorem 2. *Let S^* be the set of M nodes that maximize $f(\cdot)$. The greedy algorithm has approximate ratio $(1 - 1/e)$, i.e., the set S found by the algorithm satisfies $f(S) \geq (1 - \frac{1}{e})f(S^*)$.*

Proof. By Theorem 1, the objective function f is non-negative, monotone and submodular. According to the work [15], a greedy algorithm that always picks the element v with largest marginal gain $f(S \bigcup \{v\}) - f(S)$ is a $(1 - 1/e)$-approximation algorithm for maximizing f on M element sets S. Thus the theorem holds.

5.3 Performance Evaluation

Comparison Methods. Numerous works have been done to identify the influential users or opinion leaders in online social networks, which mainly focus on the degree centrality, betweenness centrality, closeness centrality [5], and Page Rank score [12]. In contrast with the existing methods, we detect emotion representatives by solving an influence maximization problem and apply a greedy algorithm to choose users with maximum marginal influence gains. Given the number M, we apply different methods to choose M influential users and compare their performance.

We first compare the overlapping of the set of influential users chosen by different methods. We use Jaccard coefficient to measure their similarity. Given two sets A and B, the Jaccard coefficient [23] is calculated by $coef = \frac{|A \cap B|}{|A \cup B|}$. The larger the Jaccard coefficient is, the more similar the two sets are.

We vary the value of M in 1%, 5%, 10%, 15%, 20% of the total influential user size, and calculate the Jaccard coefficient of our algorithm with the existing methods. The results are shown in Table 1(b). As shown in the table, when M is 1%, our method is most close to the Betweenness method with 60% overlap. The Degree and PageRank method have 48% and 54% overlap accordingly, which means that high degree nodes or structural important nodes are not necessary emotional influential and half of them do not have significant influence to the others' happiness. The Closeness method has very low Jaccard coefficient due to the reason that multi-hop emotion influence is not significant. With the increasing of M, most of the Jaccard coefficient decreases. The overlap is below 30% when M reaches to 20%, which means that the emotion representatives chosen by our algorithm is quite different from the existing methods.

Table 1. Comparison of Pearson correlation coefficients and Jaccard coefficient.

(a) Pearson correlation coefficient.

x%	Gre.	Deg.	Bet.	Clo.	PR
1	**0.1892**	0.0921	0.0917	0.0923	0.0912
5	**0.1866**	0.0931	0.0926	0.0933	0.0924
10	**0.1865**	0.0933	0.0929	0.0933	0.0927
15	**0.1868**	0.0934	0.093	0.0934	0.0929
20	**0.1871**	0.0934	0.0931	0.0933	0.093

(b) Jaccard coefficient.

x%	Deg.	Clo.	Bet.	PR
1	0.48	0.03	0.6	0.54
5	0.37	0.14	0.36	0.43
10	0.3	0.15	0.28	0.35
15	0.27	0.15	0.25	0.32
20	0.25	0.15	0.23	0.28

In the next, we compare the performance of our algorithm with the existing methods in terms of correlation coefficient and Euclidian distance.

Comparison of Correlation Coefficient. Assume the set of emotion representatives is $\bar{V}_R$. For a random user $v_i \in \bar{V}$, if he/she is influenced by a user $v_j \in \bar{V}_R$, we calculate their *Pearson correlation coefficient* [23] of their happiness vectors by

$$C(\mathcal{E}(v_i), \mathcal{E}(v_j)) = \frac{\sum_{t=1}^{T}(s_t(v_i) - \bar{s}(v_i))(s_t(v_j) - \bar{s}(v_j))}{\sqrt{\sum_{t=1}^{T}(s_t(v_i) - \bar{s}(v_i))^2}\sqrt{\sum_{t=1}^{T}(s_t(v_j) - \bar{s}(v_j))^2}}.$$

The larger value of Pearson correlation coefficient means the higher correlation of two data series.

We use different methods to choose $\bar{V}_R$ and calculate the average Pearson correlation coefficient of 10,000 random users in the social network. The results are compared in Table 1(a). As shown in the table, the proposed algorithm has twice correlation coefficient compared to the other methods, which means that the emotion representatives chosen by our algorithm are more correlated to the happiness states of the whole social network.

Comparison of Distance. Similarly, we can compare the distance of the happiness states of users. The Euclidian distance of two happiness scores is calculated by $|score_x - score_y|$. Again, we choose 10,000 random users in the social network and calculate their average distance to the users in $\bar{V}_R$ of 90 days. The performance of different methods is compared in Fig. 4.

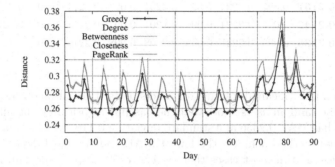

Fig. 4. Comparison of Euclidian distance.

According to the figure, the proposed algorithm has the smallest Euclidian distance in each of the 90 days, which means that the happiness score of the detected emotion representatives with our algorithm is the most close to that of the other users in the social network.

6   Happiness State Prediction

Furthermore, we use the detected emotion representatives to predict the happiness states of the whole online social network. We apply both regression and classification methods for prediction, which is discussed in the following.

Regression. Since the chosen emotion representatives have significant influence to the happiness states of other users in the social network, a natural idea is that the emotion representatives can be used to predict the happiness states of the other users. To verify this, we adopt the linear regression model to predict happiness score of random users by using the emotion representatives' happiness scores as input features. Specifically, for a random chosen user, if he/she follows some users in the set of emotion representatives, then we use the happiness scores of the corresponding emotion representatives as features for prediction; otherwise, we set the features as the happiness scores of all his/her followees (the baseline algorithm). We randomly select 10,000 users from the social graph. For each individual, we use the user's 90 day's happiness scores as the features to train the linear regression model and perform the 10-fold cross validation.

We compare the prediction performance of using the features chosen by the proposed greedy algorithm with that by the conventional methods. Table 2 shows

Table 2. MAE of 10,000 random users.

Top x%	Gre.		Deg.		Clo.		Bet.		PR	
	Ave.	SD	Ave.	SD	Ave.	SD	Ave.	SD	Ave.	SD
1	**0.2271**	0.109	0.228	0.109	0.2273	0.109	0.228	0.109	0.228	0.109
5	**0.2179**	0.099	0.2223	0.099	0.2207	0.099	0.2217	0.099	0.2222	0.099
10	**0.2157**	0.097	0.222	0.099	0.2201	0.098	0.2211	0.098	0.2219	0.098
15	**0.2142**	0.096	0.2224	0.099	0.2201	0.097	0.2211	0.097	0.2222	0.098
20	**0.2136**	0.096	0.2231	0.099	0.2208	0.098	0.2214	0.097	0.2229	0.099

the average MAE (mean absolute error) and standard deviation of the prediction results of the random users. Note that the baseline algorithm using all followees as features has MAE 0.2336 and standard deviation 0.116. As shown in the table, the proposed method achieves the lowest MAE and standard deviation among all strategies, and in most cases there is 1% to 5% improvement in predicting the happiness score. When the number of emotion representatives increases from 1% to 20%, the accuracy improves slightly, which means that choosing a small percentage of influential users is good enough to predict the happiness states of the whole social network users.

Classification. We then use classification method such as Logistic Regression (LR) and Support Vector Machine (SVM) to classify the social network users into two states: happy users (emotion scores larger than average) and unhappy users (emotion scores lower than average). Similarly, we use the emotion representatives chosen by different methods as features to predict the happiness states of random users. We select randomly 10,000 users and use their 90 day's

Table 3. MAE of 10,000 random users.

Top x%		Gre.		Deg.		Clo.		Bet.		PR	
		Ave.	SD	Ave.	SD	Ave.	SD	Ave.	SD	Ave.	SD
1	LR	**61.21**	0.131	61.16	0.13	61.12	0.131	61.16	0.13	61.17	0.13
	SVM	**58.49**	0.134	58.36	0.135	58.35	0.134	58.38	0.134	58.42	0.134
5	LR	**61.55**	0.132	61.21	0.131	61.14	0.132	61.22	0.131	61.24	0.131
	SVM	**58.51**	0.134	58.37	0.134	58.27	0.134	58.35	0.134	58.29	0.134
10	LR	**61.64**	0.132	61.18	0.131	61.13	0.132	61.19	0.131	61.23	0.131
	SVM	**58.41**	0.135	58.23	0.135	58.22	0.135	58.31	0.135	58.2	0.135
15	LR	**61.73**	0.133	61.13	0.131	61.08	0.133	61.17	0.131	61.2	0.131
	SVM	**58.45**	0.134	58.27	0.135	58.14	0.135	58.32	0.134	58.27	0.135
20	LR	**61.77**	0.133	61.09	0.132	61.06	0.132	61.14	0.132	61.15	0.131
	SVM	**58.44**	0.134	58.3	0.135	58.16	0.135	58.24	0.135	58.23	0.135

happiness scores to train the model and perform the 10-fold cross validation. Table 3 compares the average F1-score and standard deviation of different methods. Again, the proposed algorithm achieves the highest F1-score among all methods with both LR and SVM classifiers. The classification result of LR is much better than SVM, with the F1-score larger than 61 %. The F1-score is not sensitive to the number of the chosen emotion representatives. It verifies that a small number of representatives can be good indicators to infer the happiness states of the online social network users.

7 Conclusion

In this paper, we addressed the issues of happiness influence analysis in massive online social networks. We proposed quantification method to calculate the happiness scores of OSN users, based on which we showed that the happiness states of individuals in OSN are influenced by some of their social network friends (the followees in the Twitter network). To identify the high influential users, we presented a multi-linear regression model to test the significance of influence and a greedy algorithm to detect the emotion representatives. Extensive experiments showed that the detected emotion representatives can be used as features to accurately predict the happiness states of the online social network. In the future, we will further improve the performance of prediction using the combination of different features.

References

1. Aral, S., Walker, D.: Identifying influential and susceptible members of social networks. Science **337**(6092), 337–341 (2012)
2. Bao, S., Xu, S., Zhang, L., Yan, R., Su, Z., Han, D., Yu, Y.: Mining social emotions from affective text. IEEE Trans. Knowl. Data Eng. **24**(9), 1658–1670 (2012)
3. Brin, S., Page, L.: Reprint of: the anatomy of a large-scale hypertextual web search engine. Comput. Netw. **56**(18), 3825–3833 (2012)
4. Cacioppo, J.T., Fowler, J.H., Christakis, N.A.: Alone in the crowd: the structure and spread of loneliness in a large social network. J. Pers. Soc. Psychol. **97**(6), 977 (2009)
5. Chen, D., Lü, L., Shang, M.-S., Zhang, Y.-C., Zhou, T.: Identifying influential nodes in complex networks. Phys. A: Stat. Mech. Appl. **391**(4), 1777–1787 (2012)
6. Coviello, L., Sohn, Y., Kramer, A.D.I., Marlow, C., Franceschetti, M., Christakis, N.A., Fowler, J.H.: Detecting emotional contagion in massive social networks. PloS One **9**(3), 1–6 (2014)
7. Dodds, P.S., Harris, K.D., Kloumann, I.M., Bliss, C.A., Danforth, C.M.: Temporal patterns of happiness and information in a global social network: hedonometrics and twitter. PloS One **6**(12), e26752 (2011)
8. Fowler, J.H., Christakis, N.A., et al.: Dynamic spread of happiness in a large social network: longitudinal analysis over 20 years in the framingham heart study. Br. Med. J. **337**, a2338 (2008)
9. Hatfield, E., Cacioppo, J.T., Rapson, R.L.: Emotional contagion. Current Dir. Psychol. Sci. **2**(3), 96–100 (1993)

10. Kempe, D., Kleinberg, J., Tardos, É.: Maximizing the spread of influence through a social network. In: Proceedings of the 9th International Conference on Knowledge Discovery and Data Mining, pp. 137–146. ACM (2003)

11. Kwak, H., Lee, C., Park, H., Moon, S.: What is twitter, a social network or a news media? In: Proceedings of the 19th International Conference on World Wide Web, pp. 591–600. ACM (2010)

12. Lü, L., Zhang, Y.-C., Yeung, C.H., Zhou, T.: Leaders in social networks, the delicious case. PloS One 6(6), e21202 (2011)

13. Mao, Y., Lebanon, G.: Sequential models for sentiment prediction. In: ICML Workshop on Learning in Structured Output Spaces (2006)

14. Milo, R., Shen-Orr, S., Itzkovitz, S., Kashtan, N., Chklovskii, D., Alon, U.: Network motifs: simple building blocks of complex networks. Science 298(5594), 824–827 (2002)

15. Nemhauser, G.L., Wolsey, L.A., Fisher, M.L.: An analysis of approximations for maximizing submodular set functionsⅱ. Math. Program. 14(1), 265–294 (1978)

16. Nicholas, G., Rotaru, M., Litman, D.J.: Exploiting word-level features for emotion prediction. In: Spoken Language Technology Workshop, pp. 110–113. IEEE (2006)

17. Rosenquist, J.N., Fowler, J.H., Christakis, N.A.: Social network determinants of depression. Mol. Psychiatry 16(3), 273–281 (2011)

18. Sharara, H., Getoor, L., Norton, M.: Active surveying: a probabilistic approach for identifying key opinion leaders. In: Proceedings of the 22nd International Joint Conference on Artificial Intelligence, vol. 22, pp. 1485 (2011)

19. Tang, J., Zhang, Y., Sun, J., Rao, J., Yu, W., Chen, Y., Fong, A.C.M.: Quantitative study of individual emotional states in social networks. IEEE Trans. Affect. Comput. 3(2), 132–144 (2012)

20. Wang, Y., Pal, A.: Detecting emotions in social media: a constrained optimization approach. In: Proceedings of the 24th International Joint Conference on Artificial Intelligence, pp. 996–1002 (2015)

21. Xu, T., Chen, Y., Jiao, L., Zhao, B.Y., Hui, P., Fu, X.: Scaling microblogging services with divergent traffic demands. In: Proceedings of the 12th International Middleware Conference, pp. 20–39. International Federation for Information Processing (2011)

22. Yang, Y., Cui, P., Zhu, W., Yang, S.: User interest and social influence based emotion prediction for individuals. In: Proceedings of the 21st International Conference on Multimedia, pp. 785–788. ACM (2013)

23. Zafarani, R., Abbasi, M.A., Liu, H.: Social Media Mining: An Introduction. Cambridge University Press, Cambridge (2014)

24. Zajonc, R.B.: Emotion and facial efference: a theory reclaimed. Science 228(4695), 15–21 (1985)

Exploring Celebrities on Inferring User Geolocation in Twitter

Mohammad Ebrahimi[1,2(✉)], Elaheh ShafieiBavani[1,2], Raymond Wong[1,2],
and Fang Chen[1,2]

[1] Department of Computer Science and Engineering,
University of New South Wales, Sydney, NSW 2052, Australia
{mohammade,elahehs,wong,fang}@cse.unsw.edu.au
[2] Data61-CSIRO, Sydney, Australia

Abstract. Location information of social media users provides crucial
context to monitor real-time events such as natural disasters, terrorism
and epidemics. Since only a small amount of social media data are geo-
tagged, inference techniques play a substantial role to predict user spatial locations by incorporating characteristics of their behavior. Based on
utilized source of information, related works are divided into text-based
(based on text posted by users), network-based (based on the friendship
network), and some hybrid methods. In this paper, we propose a novel
approach based on the notion of celebrities to infer the location of Twitter users. We categorize highly-mentioned users (celebrities) into *local*
and *global*, and consequently utilize local celebrities as a major location
indicator for inference. A label propagation algorithm is then utilized
over a refined social network for geolocation inference. Finally, we propose a hybrid approach by merging a text-based method as a back-off
strategy into our network-based approach. Empirical experiments using
three standard Twitter benchmark datasets demonstrate the superior
performance of our approach over the state-of-the-art methods.

Keywords: Social networks · Geolocation inference · Celebrity filtering

1 Introduction

Social media provides a huge volume of data which has proven useful for predicting group behaviors and modeling populations [14]. Associating data with
a particular geolocation from which it originated creates a powerful tool for
different application such as rapid disaster response [1], opinion analysis [16],
and recommender systems [22]. However, only a small amount of social media
data are geolocation-annotated; for example, less than 1% of Twitter posts have
geolocations provided [13]. Therefore, recent work has focused on geolocation
inference (geoinference) for predicting the locations of social media posts or
users.

User geolocation inference is the task of predicting the primary (or *"home"*)
location of a user from available sources of information, such as text posted by

© Springer International Publishing AG 2017
J. Kim et al. (Eds.): PAKDD 2017, Part I, LNAI 10234, pp. 395–406, 2017.
DOI: 10.1007/978-3-319-57454-7_31

that individual, or network relationships with other individuals [10]. Geolocation inference methods usually train a model on the small set of users whose location is known (e.g., through GPS-based geotagging), and predict locations of other users using the resulting model. These models broadly fall into two categories: text-based, and network-based. Most previous researches on user geolocation inference have focused either on text-based classification approaches [8,20,26] or network-based regression approaches [13]. Recently, some methods have combined these two categories into a hybrid approach to improve the accuracy of geoinference [18,19]. They have utilized a text-based approach as a back-off strategy prior to a network-based geoinference method.

Our current work makes three key contributions: (1) we hypothesize that user geolocation inference will be improved by discriminating between different types of highly-mentioned users (celebrities). To this end, we propose a clustering-based approach to categorize users into local and global celebrities based on social network information; (2) we demonstrate that local celebrities are good location predictors for their ego networks (i.e., the part of network directly connected to them) while the global celebrities have adverse effects on geolocation inference. We utilize a label propagation algorithm to propose a network-based approach in which underlying network is based on filtering celebrities and show that it outperforms the state-of-the-art network-based approaches over three standard datasets; (3) we demonstrate that combining a text-based back-off strategy and the proposed network-based approach achieves better results than other geoinference methods.

2   Related Work

Recent increasing interest on user geolocation over social media data has caused the development of different approaches to automatic geolocation prediction based on available information sources such as the text of messages, social networks, user profile data, and temporal data. These approaches can be divided into three categories: text-based, network-based, and hybrid methods.

2.1   Text-Based Methods

The main assumption in text-based methods is that language in social media is geographically biased. It is clearly evident not only for regions speaking different languages, but also in regional dialects and use of region specific terminology [10]. In this area, early works used Gazetted expressions [15] and geographical names [17] as feature, but were shown to be sparse in coverage. In [8], a latent variable model to geolocation inference has been proposed based on the assumption that words are generated from hidden topics and geographical regions. Their model describes a sophisticated generative process from multivariate Gaussian and Latent Dirichlet Allocation (LDA) for predicting a user geolocation from tweet text via variational inference. This idea has been extended to model regions and topics jointly [11]. Similarly, another work used graphical models to jointly learn spatio-temporal topics for users [27]. Other models have used bag of word features to learn per-region classifier [20], including feature selection for location-indicative terms [3].

In [26], supervised models have been used for geolocation inference with geodesic grids having different resolutions. Information-theoretic methods have been utilized by [10] to automatically extract location-indicative words for location classification. It was reported by [25] that discriminative approaches based on hierarchical classification over adaptive grids, when optimized properly, are superior to explicit feature selection. Another research showed that sparse coding can be used to effectively learn a latent representation of tweet text to use in user geolocation inference [4]. The advantage of these generative approaches is that they are able to work with the continuous geographical space directly without any pre-discretisation, but they are algorithmically complex and do not scale well to larger datasets. A kernel-based method is used by [12] to smooth linguistic features over very small grid sizes to alleviate data sparseness. While having good results, text-based approaches are often limited to only those users who generated text that contain geographic references [13].

2.2 Network-Based Methods

Although online social networking sites allow for global interaction, users tend to befriend and interact with many of the same people online as they do offline [21]. Network-based methods exploit this property to infer the location of users from the locations of their friends [13,21]. An early work by [7] proposed an approach in which the location of a given user is inferred by simply taking the most-frequently seen location among its social network. In [13], the idea of location inference has been extended as label propagation over some form of friendship graph by interpreting location labels spatially. In his approach, locations are inferred using an iterative, multi-pass procedure. This method has been further extended by [6] to take into account edge weights in the social network and to limit the propagation of noisy locations. They weights locations as a function of how many times users interacted, thereby favoring locations of friends with whom there exists a stronger evidence of a close relationship. The main limitation of network-based models is that they completely fail to geolocate users who are not connected to geolocated components of the graph.

2.3 Hybrid Methods

As shown by [19], geolocation predictions from text can be used as a back-off for disconnected users in a network-based approach. In [18], a hybrid approach has been proposed by propagating information on a similarity graph built from user mentions in Twitter messages, together with dongle nodes corresponding to the results of a text-based geoinference method. They have reported that the performance of geolocation inference is increased by eliminating celebrities from the social network. Their approach achieved the state-of-the-art results over three standard datasets.

3   Datasets

We use three standard geotagged Twitter datasets to evaluate our approach:

GeoText is a corpus of 377,616 geotagged tweets originating within the United States by 9,475 users recorded from the Twitter API in March 2010 [8].

TwUs is a dataset of tweets compiled by [20]. In this dataset, tweets outside of a bounding box covering the contiguous United States (including parts of Canada and Mexico) were discarded, as well as users that may be spammers or robots (based on the number of followers, followees and tweets).

TwWorld is a dataset of tweets compiled by [3] in a similar fashion to TwUs; but differs in that it covers the entire Earth, and consists only of geotagged tweets. Non-English tweets and those not near a city were removed, and non-alphabetic, overly short and overly infrequent words were filtered.

We use the training, test, and development sets that come with each dataset. Table 1 summarizes descriptive statistics for the three datasets.

Table 1. Datasets details

DATASET	SCOPE	#TWEETS	#MENTIONS	#USERS	#TRAIN	#TEST	#DEV
GEOTEXT	US	378K	109K	9,475	5,685	1,895	1,895
TwUS	US	38M	3.63M	450K	430K	10K	10K
TwWORLD	World	12M	16.8M	1.4M	1.38M	10K	10K

4   The Proposed Approach

We first construct a mention network as a representative of social relationships between twitter users. Next, a novel approach is proposed to categorize celebrities to *local* and *global* types. Global celebrities are removed from the network as they do not carry useful geolocation information. Local celebrities, on the other hand, are preserved in the network since they are powerful location indicators. Considering train users and local celebrities as seeds, we run a label propagation algorithm over the mention network to infer locations of other users with unknown location. We consider this network-based approach as our baseline. Finally, we propose a hybrid approach by merging a text-based geoinference method into our baseline approach.

4.1   Mention Network

One of the prior requirements of network-based geoinference methods is a definition of what forms a relationship in Twitter to create the social network [14]. In [13], an undirected network is defined from interactions among Twitter users based on @-mentions in their tweets, a mechanism which is used for conversations between friends. Consequently these links often correspond to offline friendships, and accordingly the network will exhibit a high degree of location homophiles. In this network, nodes are all users in a dataset (train and test), as well as other external users mentioned in their tweets, and edges are created when both users mentioned one another. Since bi-directional mentions were too rare to be useful in the three datasets, we follow [19] to consider uni-directional mentions (i.e., if either user mentioned the other) as undirected edges instead. Each edge is weighted by the total number of @-mentions in tweets by either user.

4.2 Differentiating Celebrities

We consider all users that are mentioned by more than T distinct users as celebrities. Previously, [18] assumed all celebrities as global, and excluded them from the mention network (Based on the best results achieved over the development sets, T (celebrity threshold) was set to 5, 15, and 5 for GEOTEXT, TwUs, and TwWORLD respectively). They simply ignore the fact that lots of these celebrities are useful in geolocation inference. To tackle this issue, we propose a novel approach to detect location indicative celebrities (locals) and utilize them to infer location of other users. In particular, we define a celebrity as *local*, if the majority of geolocated users who mentioned it (its mentioners) are geographically close. Otherwise, it is considered as a *global* celebrity.

We utilize a density based clustering algorithm, DBSCAN [9], in order to cluster the geolocated mentioners based on their geographical coordinates. This algorithm is highly efficient and can identify arbitrary shaped clusters, where clusters are defined as dense regions separated by low dense regions. More precisely, considering a set of points to be clustered, the points are classified as *core points, (density-)reachable points* and *outliers*, as follows: (1) A point p is a core point if at least $MinPts$ points are within distance ε of it (including p). Those points are said to be directly reachable from p. Note that no points are directly reachable from a non-core point. (2) A point q is reachable from p if there is a path $p_1, \ldots, p_n$ with $p_1 = p$ and $p_n = q$, where each $p_{(i+1)}$ is directly reachable from p_i (so all the points on the path must be core points, with the possible exception of q). (3) All points not reachable from any other point are outliers. Each core point p forms a cluster together with all points (core or non-core) that are reachable from it. Each cluster contains at least one core point; non-core points can be part of a cluster, but they form its "edge", since they cannot be used to reach more points.

DBSCAN requires two parameters: ε *(eps)* and the minimum number of points required to form a dense region *(MinPts)*. The ε-*neighborhood* of a point p is defined as the set of points whose distance from p is not greater than ε. DBSCAN starts with an arbitrary starting point that has not been visited. This point's ε-neighborhood is retrieved, and if it contains sufficiently many points, a cluster is started. Otherwise, the point is labelled as noise. It should be noted that this point might later be found in a sufficiently sized ε-environment of a different point and hence be made part of a cluster. If a point is found to be a dense part of a cluster, its ε-neighborhood is also part of that cluster. Hence, all points that are found within the ε-neighborhood are added, as is their own ε-neighborhood when they are also dense. This process continues until the density-connected cluster is completely found. Then, a new unvisited point is retrieved and processed, leading to the discovery of a further cluster or noise.

Since our goal is to cluster users based on their geographical coordinate, we have modified DBSCAN algorithm to use *Haversine* formula as its distance metric. Haversine formula calculates great-circle distances between two points on a sphere from their longitudes and latitudes.

For each celebrity in the set of external users, we run DBSCAN to cluster its geolocated mentioners. If the algorithm outputs only one cluster containing more than predefined proportion (δ) of total geolocated mentioners, the celebrity is considered as a local one. Otherwise, it will be considered as a global celebrity. As a formal definition, we specify the type of celebrity C by the following equation:

$$Type(C) = \begin{cases} Local, & \text{if } ClusNo(DBSCAN(M_C, \varepsilon, MinPts)) = 1 \\ Global, & \text{otherwise} \end{cases} \quad (1)$$

where M_C denotes mentioners of C, and $ClusNo$ is the number of clusters. We set parameter ε to $70(mile)$ for GEOTEXT and TwUs datasets, and 130 for TwWORLD dataset. Since the number of points (mentioners) to be clustered are different for each celebrity, we set the $MinPts$ dynamically as η percent of total number of points (30% in our experiments). The parameter δ was set to 0.8. It should be noted that all parameters (ε, η, and δ) were tuned over the development sets. Figure 1 demonstrates sample results of the proposed clustering algorithm for mentioners of local and global celebrities in TwUs and TwWORLD datasets. Our experiments show that more than 40% of detected celebrities in all datasets are local.

In order to construct a refined mention network, we filter out the global celebrities and preserve local ones. In the next step, we utilize a label propagation algorithm over this refined mention network to infer the location of other users.

4.3 Label Propagation with Modified Adsorption

In our network-based approach, we formulate geolocation inference as label propagation over the refined mention network. Following [18], we utilize Modified Adsorption [23] as our label propagation algorithm, since it allows different levels of influence between prior/known labels and propagated label distributions.

Modified Adsorption is a graph-based semi-supervised learning algorithm which has been used for open domain class-instance acquisition. It computes a soft assignment of labels to the nodes of a graph $G = (V, E, W)$, where V is the set of nodes with $|V| = n$, E is the set of edges, and W is an edge weight matrix. Out of the $n_l + n_u = n$ nodes in G, we have prior knowledge of labels for n_l nodes (training set), while the remaining n_u nodes are unlabeled (test set). An edge $e = (a, b) \in V \times V$ indicates that the label of the two vertices $a, b \in V$ should be similar and the weight $W_{ab} \in \mathbb{R}_+$ reflects the strength of this similarity. Assume C is the set of labels, with $|C| = m$ representing the total number of labels. Y is the $n \times m$ matrix storing training label information. The l_{th} element of the

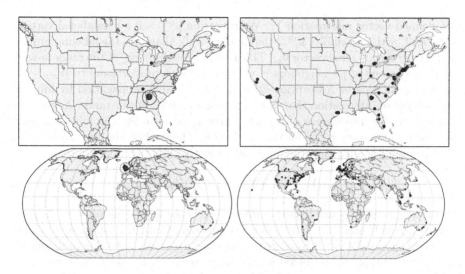

Fig. 1. Sample results for clustering mentioners of celebrities in TwUs(top) and TwWorld(buttom) datasets. When DBSCAN outputs only one cluster of mentioners (as highlighted by red points in left figures), the corresponding celebrities are identified as *local* type. On the other hand, celebrities are global if no cluster can be found by DBSCAN (right figures). (Color figure online)

vector Y_v encodes the prior knowledge for vertex v. The higher the value of Y_{vl}, the stronger we a-priori believe that the label of v should be $l \in L$ and a value of zero $Y_{vl} = 0$ indicates no prior about the label l for vertex v. Another vector, $\hat{Y} \in \mathbb{R}_+$, is the output of the algorithm, using similar semantics as Y.

Modified Adsorption is an iterative algorithm, where label estimates on node v in the $(t+1)^{th}$ iteration are updated using estimates from the tth iteration:

$$\hat{Y}^{(t+1)} = \frac{(\mu_1 \times p_v^{inj} \times Y_v + \mu_2 \times D_v^{(t)} + \mu_3 \times p^{abnd} \times r)}{M_{vv}} \qquad (2)$$

where:

$$D_v^{(t)} = \sum_u (p_v^{cont} W_{vu} + p_u^{cont} W_{uv} \hat{Y}_u) \qquad (3)$$

and:

$$M_{vv} = \mu_1 \times p_v^{inj} + \mu_2 \sum_{u \neq v} (p_v^{cont} W_{vu} + p_u^{cont} W_{uv}) + \mu_3 \qquad (4)$$

p_v^{inj}, p_v^{cont}, and p_v^{abnd} are three probabilities defined on each node $v \in V$, and r is a vector to express label uncertainty at a node. On each node v, the three probabilities sum to one, i.e., $p_v^{inj} + p_v^{cont} + p_v^{abnd} = 1$, and they are based on the random-walk interpretation of the Adsorption algorithm [24]: To label any vertex $v \in V$ (either labeled or unlabeled) a random-walk is initiated starting at v facing three options: with probability p_v^{inj} it stops and return (i.e., inject) the pre-defined vector information Y_v. We constrain $p_v^{inj} = 0$ for unlabeled vertices v. Second, with probability p_v^{abnd} the random-walk abandons the labeling process and return the all-zeros vector 0_m. Third, with probability p_v^{cont} the random-walk continues to one of v's neighbors v' with probability proportional to $W_{v'v} \geq 0$.

The main idea of Adsorption [2] is to control label propagation by limiting the amount of information that passes through a node. For instance, Adsorption can reduce the importance of a high-degree node v during the label inference process by increasing p_v^{abnd} on that node [23]. The goal of Modified Adsorption is to compute $\hat{Y}$ such that the following objective function is minimized:

$$C(\hat{Y}) = \sum_l [\mu_1 (Y_l - \hat{Y}_l)^T S(Y_l - \hat{Y}) + \mu_2 \hat{Y}_l^T L \hat{Y}_l + \mu_3 \|\hat{Y}_l - R_l\|_2^2] \tag{5}$$

where μ_1, μ_2, and μ_3 are hyperparameters; L is the Laplacian of an undirected graph derived from G, but with revised edge weights; and R is an $n \times m$ matrix of per-node label prior, with R_l representing the l_{th} column of R. The probabilities p_v^{inj}, p_v^{cont}, and p_v^{abnd} are folded inside the matrices S, L, and R, respectively.

In our experiments, we set the label confidence for training and test users to 1.0 and 0, respectively. For each local celebrity, we initialize its location to the weighted median latitude and weighted median longitude of all its geolocated mentioners. We set the label confidence for local celebrities to 0.6, so that their labels can be changed over the propagation process. Training users along with local celebrities with their corresponding labels confidences are added to the seed set. We set μ_1, μ_2, and μ_3 to 0.9, 0.15, and 0, respectively. These three parameters and label confidence have been tuned based on the best results achieved over the development sets. We run the Modified Adsorption algorithm iteratively until convergence, which usually occurs at or before 10 iterations.

4.4   Text-Based Back-Off Strategy

As reported by [19], many test users are not transitively connected to any training node (it is about 25% for GEOTEXT and TwUs, and 3% for TwWORLD). It results in label propagation fails to assign isolated users any location. This usually happens when users do not use @-mentions, or when a set of nodes constitutes a disconnected component of the graph [19]. In order to alleviate this problem, we use the tweets from each test user to estimate their location, which is then used as an initial estimation during label propagation.

Following [18], we use the text-based approach proposed by [25] as our back-off strategy. In their approach, the continuous space of geographical coordinates is discretized using a k-d tree such that each sub-region (leaf) has similar numbers of users. This results in many small regions for areas of high population density and fewer larger regions for country areas with low population density. Next, these regions are used as class labels to train a logistic regression model. We use hierarchical logistic regression with a beam search since it achieves higher results than logistic regression over a flat label set [25]. Following [18], we set the number of users in each region to 50, 2048, and 2400 for GEOTEXT, TwUs, and TwWORLD respectively. We also use a bag of unigrams (over both words and

@-mentions) and remove all features that occurred in less than 10 documents, following [25]. The features for each user are weighted using tf-idf, followed by per-user l_2 normalization [18].

In our hybrid approach, we first estimate the location for each test node using the logistic regression classifier described above, before running label propagation over the mention network. Following [18], we attach a dongle node to each test user containing its text-based estimated location. The dongle nodes with their corresponding label confidences are added to the seed set, and are treated in the same way as other labeled nodes (i.e., the training nodes and local celebrities). This iteratively adjusts the locations based on both the known training users and guessed test users, while simultaneously inferring locations for the external users. In such a way, the inferred locations of test users will better match neighboring users in their sub-graph, or in the case of disconnected nodes, will retain their initial classification estimate. For evaluation, same as [18], we use the median coordinates of all training points in the sub-region predicted by the classifier, from which we measure the error against a test user's gold standard location.

5   Experimental Results

5.1   Evaluation Metrics

In line with other work on user geolocation prediction, we use the following evaluation measures: *Acc@161*: The percentage of predicted locations which are within a 161 km (100 mile) radius of the actual location [5], as a proxy for accuracy within a metro area; *Mean*: The mean distance from the predicted location to the actual location [8]; *Median*: The median distance from the predicted location to the actual location [8]; *Post Coverage*: The percentage of tested posts (users in our experiments) for which a geoinference method can predict a location [14]. Post Coverage is a challenging metric for network-based geoinference methods, which are only able to predict locations for users in their underlying social network and therefore may be unable to infer locations for frequently-posting users that do not have social relationships (i.e., are not in the network).

So, we evaluate using the mean and median errors (in km) over all test users, and also accuracy within 161 Km of the actual location. We also evaluate our network-based approach and state-of-the-art using Coverage. Note that higher numbers are better for Acc@161 and Coverage but lower numbers are better for mean and median errors.

5.2   Results

Table 2 shows the performance of our proposed network-based (PROP-NB) and hybrid (PROP-HYB) approaches over the GEOTEXT, TwUs and TwWORLD datasets. The results are also compared with prior network-based geoinference approaches [18,19], text-based classification models [3,4,18,25], and hybrid methods [18].

Table 2. Performance of geolocation inference methods over the three Twitter datasets; CHA [4], HAN [3], WB [25], LP-RAHIMI and LR-RAHIMI [19], MADCEL-W and MADCEL-W-LR [18] ("?" signifies that no results were reported for the given metric; "-" signifies that no results were published for the given dataset)

	GeoText			TwUs			TwWorld		
	Acc@161	Mean	Median	Acc@161	Mean	Median	Acc@161	Mean	Median
CHA	?	581	425	-	-	-	-	-	-
HAN	-	-	-	45	814	260	24	1953	646
WB-UNIFORM	-	-	-	49	703	170	32	1714	490
WB-KDTREE	-	-	-	48	686	191	31	1669	509
LP-RAHIMI	45	676	255	37	747	431	56	1026	79
LR-RAHIMI	38	880	397	50	686	159	63	866	19
MADCEL-W	58	586	60	54	705	116	71	976	0
MADCEL-W-LR	59	581	57	60	529	78	72	802	0
PROP-NB	61	486	38	59	546	83	77	547	0
PROP-HYB	**64**	**476**	**32**	**66**	**438**	**56**	**79**	**491**	0

Our network-based approach outperforms the text-based models and also previous network-based models. Although network-based approach based on removing all celebrities as proposed by [18] further reduces the size of mention network; it sacrifices the accuracy by ignoring the importance of local celebrities. It also results in more isolated test users because all related edges are removed along with each celebrity node. Our network-based approach, on the other hand, improve the performance of geolocation inference in multiple ways: (1) Location information of local celebrities is propagated through the network and results in superior Acc@161, Median, and Mean; (2) As can be seen in Table 3, our network-based approach outperforms the state-of-the-art network-based approach (i.e., [18]) in terms of Coverage by 9% and 14% over the three datasets. The main reason is that we preserve and utilize local celebrities in the network, while [18] filters out all celebrities which leads more isolated users in the network.

Table 3. Post coverage of different network-based approaches

	GeoText	TwUs	TwWorld
MADCEL-W	57%	51%	72%
PROP-NB	**66%**	**65%**	**81%**

By using text-based approach as the back-off strategy, our hybrid geolocation inference model achieves the state-of-the-art results over all three datasets. The main reason is that the text-based method provides a user-specific geolocation prior for disconnected users.

6   Conclusion

We proposed a novel network-based approach for user geolocation inference in social media. We utilized a density based clustering algorithm to categorize different type of highly-mentioned users into local and global celebrities. We showed that utilizing local celebrities as powerful location indicators and eliminating global celebrities from the mention network, along with the use of Modified Adsorption for propagating location information over this network, our approach outperforms the state-of-the-art network-based approaches. We also demonstrated that by using a text-based strategy as a back-off to alleviate the problem of disconnected users, our hybrid approach achieved the best result among other user geolocation inference methods.

References

1. Ashktorab, Z., Brown, C., Nandi, M., Culotta, A.: Tweedr: mining twitter to inform disaster response. In: Proceedings of ISCRAM (2014)
2. Baluja, S., Seth, R., Sivakumar, D., Jing, Y., Yagnik, J., Kumar, S., Ravichandran, D., Aly, M.: Video suggestion and discovery for youtube: taking random walks through the view graph. In: Proceedings of WWW, pp. 895–904. ACM (2008)
3. Bo, H., Cook, P., Baldwin, T.: Geolocation prediction in social media data by finding location indicative words. In: Proceedings of COLING, pp. 1045–1062 (2012)
4. Cha, M., Gwon, Y., Kung, H.T.: Twitter geolocation and regional classification via sparse coding. In: Proceedings of ICWSM, pp. 582–585 (2015)
5. Cheng, Z., Caverlee, J., Lee, K.: You are where you tweet: a content-based approach to geo-locating twitter users. In: Proceedings of the 19th CIKM, pp. 759–768. ACM (2010)
6. Compton, R., Jurgens, D., Allen, D.: Geotagging one hundred million twitter accounts with total variation minimization. In: Proceedings of BigData, pp. 393–401. IEEE (2014)
7. Davis Jr., C.A., Pappa, G.L., de Oliveira, D.R.R., de L Arcanjo, F.: Inferring the location of twitter messages based on user relationships. Trans. GIS **15**(6), 735–751 (2011)
8. Eisenstein, J., O'Connor, B., Smith, N.A., Xing, E.P.: A latent variable model for geographic lexical variation. In: Proceedings of EMNLP, pp. 1277–1287. ACL (2010)
9. Ester, M., Kriegel, H.-P., Sander, J., Xu, X., et al.: A density-based algorithm for discovering clusters in large spatial databases with noise. In: Proceedings of KDD, vol. 96, pp. 226–231 (1996)
10. Han, B., Cook, P., Baldwin, T.: Text-based twitter user geolocation prediction. J. Artif. Intell. Res. **49**, 451–500 (2014)
11. Hong, L., Ahmed, A., Gurumurthy, S., Smola, A.J., Tsioutsiouliklis K.: Discovering geographical topics in the twitter stream. In: Proceedings of WWW, pp. 769–778. ACM (2012)
12. Hulden, M., Silfverberg, M., Francom, J.: Kernel density estimation for text-based geolocation. In: Proceedings of AAAI, pp. 145–150 (2015)
13. Jurgens, D.: That's what friends are for: inferring location in online social media platforms based on social relationships. ICWSM **13**, 273–282 (2013)

14. Jurgens, D., Finethy, T., McCorriston, J., Xu, Y.T. Ruths, D.: Geolocation prediction in twitter using social networks: a critical analysis and review of current practice. In: Proceedings of ICWSM (2015)

15. Leidner, J.L., Lieberman, M.D.: Detecting geographical references in the form of place names and associated spatial natural language. SIGSPATIAL Spec. **3**(2), 5–11 (2011)

16. Mostafa, M.M.: More than words: social networks text mining for consumer brand sentiments. Expert Syst. Appl. **40**(10), 4241–4251 (2013)

17. Quercini, G., Samet, H., Sankaranarayanan, J., Lieberman, M.D.: Determining the spatial reader scopes of news sources using local lexicons. In: Proceedings of SIGSPATIAL, pp. 43–52. ACM (2010)

18. Rahimi, A., Cohn, T., Baldwin, T.: Twitter user geolocation using a unified text and network prediction model. In: Proceedings of ACL, pp. 630–636. ACL (2015)

19. Rahimi, A., Vu, D., Cohn, T., Baldwin, T.: Exploiting text and network context for geolocation of social media users. In: Proceedings of HLT-NAACL, pp. 1362–1367. ACL (2015)

20. Roller, S., Speriosu, M., Rallapalli, S., Wing, B., Baldridge, J.: Supervised text-based geolocation using language models on an adaptive grid. In: Proceedings of EMNLP-CoNLL, pp. 1500–1510. ACL (2012)

21. Rout, D., Bontcheva, K., Preoţiuc-Pietro, D., Cohn, T.: Where's@ wally?: a classification approach to geolocating users based on their social ties. In: Proceedings of ACM-HT, pp. 11–20. ACM (2013)

22. Schedl, M., Schnitzer, D.: Location-aware music artist recommendation. In: Gurrin, C., Hopfgartner, F., Hurst, W., Johansen, H., Lee, H., O'Connor, N. (eds.) MMM 2014. LNCS, vol. 8326, pp. 205–213. Springer, Cham (2014). doi:10.1007/978-3-319-04117-9_19

23. Talukdar, P.P., Crammer, K.: New regularized algorithms for transductive learning. In: Buntine, W., Grobelnik, M., Mladenić, D., Shawe-Taylor, J. (eds.) ECML PKDD 2009. LNCS (LNAI), vol. 5782, pp. 442–457. Springer, Heidelberg (2009). doi:10.1007/978-3-642-04174-7_29

24. Talukdar, P.P., Reisinger, J., Paşca, M., Ravichandran, D., Bhagat, R., Pereira, F.: Weakly-supervised acquisition of labeled class instances using graph random walks. In: Proceedings of EMNLP, pp. 582–590. ACL (2008)

25. Wing, B., Baldridge, J.: Hierarchical discriminative classification for text-based geolocation. In: Proceedings of EMNLP, pp. 336–348 (2014)

26. Wing, B.P., Baldridge, J.: Simple supervised document geolocation with geodesic grids. In: Proceedings of ACL-HLT, pp. 955–964. ACL (2011)

27. Yuan, Q., Cong, G., Ma, Z., Sun, A., Thalmann, N.M.: Who, where, when and what: discover spatio-temporal topics for twitter users. In: Proceedings of ACM-SIGKDD, pp. 605–613. ACM (2013)

Do Rumors Diffuse Differently from Non-rumors? A Systematically Empirical Analysis in Sina Weibo for Rumor Identification

Yahui Liu[✉], Xiaolong Jin, Huawei Shen, and Xueqi Cheng

CAS Key Lab of Network Data Science and Technology,
Institute of Computing Technology, Chinese Academy of Sciences, Beijing, China
{jinxiaolong,shenhuawei,cxq}@ict.ac.cn,
liuyahui@software.ict.ac.cn

Abstract. With the prosperity of social media, online rumors become a severe social problem, which often lead to serious consequences, e.g., social panic and even chaos. Therefore, how to automatically identify rumors in social media has attracted much research attention. Most existing studies address this problem by extracting features from the contents of rumors and their reposts as well as the users involved. For these features, especially diffusion features, these works ignore systematic analysis and the exploration of difference between rumors and non-rumors, which exert targeted effect on rumor identification. In this paper, we systematically investigate this problem from a diffusion perspective using Sina Weibo data. We first extract a group of new features from the diffusion processes of messages and then make a few important observations on them. Based on these features, we develop classifiers to discriminate rumors and non-rumors. Experimental comparisons with the state-of-the-arts methods demonstrate the effectiveness of these features.

Keywords: Rumor identification · Diffusion tree · Sina Weibo

1 Introduction

Social media (e.g., Twitter and Sina Weibo), as a novel type of media, has become an important platform for people to obtain, share and spread information. According to the 38th Statistical Report on Internet Development in China released by China Internet Network Information Center (CNNIC) and Sina reports, as of June 2016, the usage of Sina Weibo was about 34% and ranked second in typical social applications. Daily active users reached 126 million. Similarly, Twitter also had 100 million daily active users. However, as a side-effect of social media, rumors, commonly defined as unconfirmed and uncertain information posted intentionally or unintentionally by some users, are also widely propagated over social media, which may lead to significant negative effects and even severe social problems, e.g., social panic and even chaos. For example, a hacker released a fake tweet about explosions in the White House through the official Twitter account of the Associated Press (AP) on April 23, 2013. This

© Springer International Publishing AG 2017
J. Kim et al. (Eds.): PAKDD 2017, Part I, LNAI 10234, pp. 407–420, 2017.
DOI: 10.1007/978-3-319-57454-7_32

rumor was widely reposted and caused an immediate social panic: S&P 500 Index instantaneously dropped 14 points, wiping out $136.5 billion in a matter of seconds, and the Dow Jones Industrial Average also dropped about 145 points within three minutes.

Therefore, how to identify rumors from non-rumors become an important research problem in recent years. The key to solve the problem of rumor identification is to consider both the generation and diffusion process of rumors. However, it is difficult to obtain a large amount of rumor data containing the whole life course of rumors and to label rumors with high credibility, which needs professional knowledge to discriminate. Existing works are mainly based on content features, user features, or their combinations to identify rumors. Some studies confirmed that temporal features [10, 13], structural features [23] may also help improve the accuracy of rumor identification. However, they suffer from two drawbacks. First, they fail to quantify the concrete differences of the proposed features between rumors and non-rumors; Second, they only take rumors and non-rumors with a great number of reposts into account, which account for only a small proportion of all messages.

We empirically observed that the diffusion patterns of rumors and non-rumors are significantly different. Particularly, the reposts of non-rumors are usually triggered directly by the source, while rumors generally have many subsequent relaying users to trigger new reposts. This key insight motivates us to study the automatic identification of rumors by systematically investigating the discriminative features of rumors and non-rumors from a diffusion perspective. Specifically, in this paper we first represent the diffusion process of messages into diffusion trees. By analyzing diffusion trees, we identify a group of new diffusion based features. For these features, we quantify their differences between rumors and non-rumors. Finally, we construct classifiers using the diffusion features to identify rumors with much fewer reposts.

This paper makes two main contributions to the field: (i) We systematically study the diffusion processes of rumors and non-rumors and discriminate a group of new diffusion features from the diffusion processes, on which rumors and non-rumors are significantly different. We further make a few important observations through empirical analysis over these features. We find that, for instance, the diffusion trees of rumors are deeper but narrower than those of non-rumors; rumors are more likely to be reposted at the first moment (i.e., within one minute). (ii) By combining the above identified features with the features widely used in existing methods, we develop classifiers to classify rumors and non-rumors, exhibiting better performance in terms of widely used metrics including accuracy and F1-measure to the state-of-the-arts baselines. This indicates that diffusion based features provide a good supplement to those existing ones for rumor identification.

2   Related Works

Existing works usually regard rumor identification as a classification problem and extract features from different perspectives for developing effective classifiers.

These works on identifying rumors can be divided into two categories according to the features used for identification.

The first category of methods is mainly based on superficial features of the original messages and the users who post them. For instance, Castillo et al. [4] extracted a wide range of features from tweets on Twitter and then used a J48 decision tree to judge the credibility of news topic. Qazvinian et al. [17] exploited three types of features, namely, content-based features, network-based features, and microblog-specific features (e.g., hashtags and URLs [18]) to identify rumors and users who endorse and spread them. Yang et al. [24] identified two new features, i.e., client program and location, to detect rumors on Sina Weibo. Sun et al. [21] particularly made use of multimedia-based features, coupled with content- and user-based features, to automatically detect event rumors on Sina Weibo. Although these features fail to achieve good classification accuracy, they can be used as important auxiliary features to identify rumors.

The other category of works identifies rumors from non-rumors mainly based on features extracted from the perspective of propagation, including the number and contents of the reposts and the users involved. First, from the contents of reposts, Mendoza et al. [15] found that on Twitter rumors often cause more questions in the propagation than non-rumors. Starbird et al. [20] and Tanaka et al. [22] observed that the emergence of corrections provides important cues in identifying or counteracting rumors. Liu et al. [11] combined these cues with verification features to debunk rumors in a real time manner. Ma et al. [12,13] used Dynamic Series-Time Structure (DSTS) and Recurrent Neural Networks (RNNs) to capture the variation of content-based and user-based features in the lifecycle of events to discriminate rumors from non-rumors, respectively. Second, from the structures of diffusion, Nadamoto et al. [16] found that the diffusion hierarchies of rumors are different between normal situations and disaster situations and in particular the hierarchies in a normal situation is higher. Through an empirical study, Friggeri et al. [7] observed that rumors get deeper cascades in social networks than reshared messages. Wu et al. [23] developed a random walk graph kernel to capture the similarity of the diffusion patterns between messages and combine message-based, user-based, and repost-based features to detect rumors. Jin et al. [9] employed the well-known SEIZ model proposed in [3] to characterize the diffusion patterns of rumors and further adopt it to detect rumors. Kwon et al. [10] used three network structure features and combined temporal and linguistic features to classify rumors and non-rumors. Although these works utilize diffusion based features, some even combined with content and user features, to improve the performance of rumor identification, there still lacks of a systematic study on the features in the diffusion process of messages, which may be very helpful to rumor identification. Therefore, in this paper we analyze the diffusion of messages from structural, temporal and user perspectives and propose a group of diffusion based features for rumor identification.

3 Data Collection

To analyze the diffusion processes of messages and examine their effect on rumor identification, a labeled dataset is needed. Sina Weibo has an official account for rumor busting. All messages announced by this account as rumors are confirmed misinformation or disinformation, which are relevant to certain events/topics and have been widely spread. Therefore, rumor busting provides high-quality rumor labels and quite diverse rumor messages. In this paper, we first crawled 453 rumor events posted by this rumor busting account, spanning from November 18, 2011 to December 31, 2014. We further adopted keyword-based search provided by "Sina Weibo advanced search" to extract all messages that are directly related to these events with human labeling as post-processing. Finally, we employed Grubbs' test to filter out outlier events which contained too much messages than other events, and retained 44,096 original messages corresponding to 178 rumor events as the rumor dataset. Sina Weibo API provides interfaces to obtain the detailed information of each message, including the message's content, its reposts and the information of all users involved in the diffusion of a message. The information of a message or a repost includes the content, the post time, the number of reposts, the number of comments, etc. The information of a user includes his/her profile, the number of followers, the number of followees, etc. Using a similar method, we extracted 180,212 original messages corresponding to 367 newsworthy non-rumor events from some news media's Weibo accounts as the non-rumor dataset of this study. Table 1 presents our datasets.

Table 1. Statistics of the datasets used in this paper

Statistics		Rumors	Non-rumors
Number of messages	All	44,096	180,212
	With reposts	14,219	51,617
	With no less than 10 reposts	3,286	16,897
	With no less than 100 reposts	805	4,790
Number of users		1,183,163	6,344,693

4 Empirical Analysis

In this section, we explore features from the perspective of diffusion to discriminate rumors from non-rumors.

4.1 Diffusion Tree

To facilitate the analysis of structural patterns of diffusion process of message, either rumor or non-rumor, we propose to use diffusion tree to represent its diffusion process. A diffusion tree of a message is represented as $T = \langle V, E \rangle$. Each node in V corresponds to a user who participates in the forwarding process of the message, and the root node is the source user who posts the original message. A link in E, from a user node u to another user node v, indicates that v forwards the message from user u. In this way, the size of the diffusion tree is the total number of reposts of the original message. Its depth is the longest path from the root node to leaf nodes, which indicates the penetrability of the message [1]. Its width is the number of nodes in the layer which has the largest number of nodes, which indicates the expansion rate of the message.

4.2   Features Extracted from the Diffusion Perspective

We originally extracted 24 features from the diffusion perspective and further applied Pearson correlation analysis for feature selection. We then employed an entropy based method [14] to identify key features. The final feature set is presented in Table 2. Those features are classified into four categories, namely, structural features extracted from diffusion trees, temporal features and repost user features obtained from all reposts of the original messages, and content features indicating the opinions and sentiment of users to the messages, reflected by questions and refutations, as well as sentiment words and emoticons in all reposts. We call these four categories of features as Diffusion Features (DFs), among which the first three categories, including 8 features, are proposed and verified in this paper. We have experimentally observed that most of the diffusion trees with less than 10 reposts have a star-shaped structure, which has very little topological information for differentiating rumors from non-rumors. Therefore, in this paper we use the dataset with no less than 10 reposts for each message to explore the diffusion characteristics of rumors and non-rumors.

A. Structural Features. Structural features are salient features of information propagation and has been leveraged for popularity prediction and cascade prediction [2]. Here, we analyze the structure of diffusion tree to acquire potential structural features for rumor identification. Figure 1(a) presents the Complementary Cumulative Distribution Function (CCDF) of the *depth* of diffusion trees corresponding to rumors and non-rumors, respectively. The CCDFs of rumors and non-rumors follow an exponential distribution with the exponents being 0.33 and 0.53, respectively. The maximal and average depth of diffusion trees are 34 and 4.96 for rumors, 25 and 3.65 for non-rumors. Moreover, the diffusion trees of rumors with the depth less than 5 account for 66.41%, while those of non-rumors occupy 84%. Figure 1(b) presents the relationship between the size and the average depth of diffusion trees. It is obvious that with the increase of the number of reposts, the average depth of diffusion trees also increases. Note that the average depth of diffusion trees of rumors is deeper than that of non-rumors in each size interval. Moreover, the average depth of diffusion trees of most of non-rumors is less than 10. All these observations suggest that the diffusion trees of the majority of rumors are deeper than those of non-rumors. It implies that the penetrability of rumors are stronger than that of non-rumors.

For diffusion trees with the same depth, we further distinguish them using the *total path length* of a diffusion tree T, defined as the sum of the length of all

Table 2. Description of features

Type	Name	Description
Structural features	Depth of diffusion tree	# of nodes in the longest path from the root to leaf nodes
	Width of diffusion tree	# of nodes in the layer containing the largest number of nodes
	Inter-layer width ratio	Sum of the ratios between the numbers of nodes of two consecutive non-root layers lower than five
	Total path length	Sum of the length of paths from the root node to all other nodes
Temporal features	Response time	Time interval between the post time of the message and the time of its first repost
	Lifecycle	Survival time of the original message
User features	Repost-participation ratio	Ratio of repost users to all users who repost/comment/thumb up for the original message
	Repost-exposure ratio	Ratio of the number of repost users to the total number of followers of all users who post or repost the message
Content features	Number of refutations	# of refutations to original message in all reposts
	Number of questions	# of questions to original message in all reposts
	Sentiment score	# of positive and negative sentiment words or emoticons in all reposts of original message

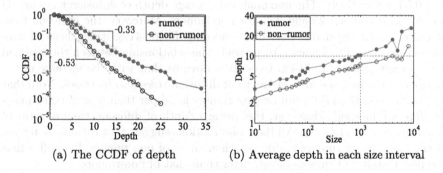

(a) The CCDF of depth   (b) Average depth in each size interval

Fig. 1. The depth of diffusion trees.

paths from the root node v_r to any other nodes v_j, i.e., $D(T) = \sum_{v_j \in V, v_j \neq v_r} d_{r,j}$, where $d_{r,j}$ is the length from v_r to v_j. In general, the metric $D(T)$ provides a measure to the structure of diffusion tree. For two diffusion trees with the same size, the higher the value of $D(T)$, the more likely that T corresponds to a rumor.

Regarding the *width* of diffusion trees, in Fig. 2(a) we present the width-to-size ratio averaged on different size intervals of diffusion trees for both rumors and non-rumors. We see that with the increase of tree size, the averaged width-to-size ratio decreases gradually. However, the averaged width-to-size ratio of non-rumors is consistently greater than that of rumors. Therefore, we can conclude that the diffusion trees of the majority of non-rumors are wider than those of rumors. Combining Figs. 2(a) and 1(b), we expect that the higher the value of width-to-size ratio, the wider and the lower the diffusion tree.

The width-indictor layer of a diffusion tree means that it is the widest layer of the tree and thus indicates its width. We found that the width-indictor layer of most diffusion trees are layer 2 to 5. Moreover, the width of the diffusion tree reflects the expansion rate of the information spreading. To measure this property, we define the *inter-layer width ratio* of a diffusion tree, T, as the sum of the ratios between the numbers of nodes of two consecutive non-root layers lower than five in the diffusion tree, i.e., $L(T) = \sum_{k=2}^{5}(l_{k+1}/l_k)$, where l_k denotes the number of nodes in the k-th layer of the diffusion tree T. The results presented in Fig. 2(b) show that $L(T)$ of most diffusion trees of rumors is greater than that of non-rumors, indicating that rumors are more probably to motivate users to further spread them, as compared to non-rumors.

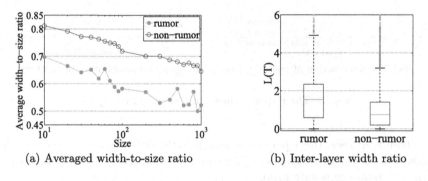

(a) Averaged width-to-size ratio   (b) Inter-layer width ratio

Fig. 2. The width and inter-layer width ratio of diffusion trees.

B. Temporal Features. Temporal pattern is the most effective feature for predicting of the popularity of messages [6, 19]. Here, we analyzed the temporal features that are potentially useful for rumor identification. The *response time* of a message indicates how fast it causes response. Figure 3(a) plots the Cumulative Distribution Function (CDF) of response time of rumors and non-rumors. It shows that the response time of 48.94% rumors and 56.58% non-rumors, respectively, is less than two minutes; up to 72.11% rumors and 83.82% non-rumors are responded within five minutes. Figure 3(a) shows a special case in the first one minute, where the curves corresponding to rumors and non-rumors intersect with each other. To explain this phenomenon, we analyzed the relationship between reposts and response time. We found that 52.3% of rumors with more than 100 reposts get the first repost within one minute, while non-rumors are 39.7%. It indicates that rumors are more likely to be responded quickly at the first moment, and get lots of reposts in subsequent time. This may be because rumors are often about issues that people greatly concern, such as child trafficking and people's livelihood. These rumors can only be obtained from a few sources, thus leading to people's anxiety and are spread widely.

The *lifecycle* of both rumors and non-rumors follows power law distributions, as shown in Fig. 3(b), with exponents as 1.79 and 2.04, respectively. There are 46.1% rumors and 50.2% non-rumors whose lifecycle is less than 3 days, while 81.9% rumors and 84.4% non-rumors have a lifecycle less than two weeks. But, the lifecycle of the majority rumors is longer than that of non-rumors. In addition, we empirically found that among rumors with a relatively long lifecycle, nearly a half are messages with a small number of reposts. The last few reposts include three possibilities: the forwarding of new fans, the forwarding of refuting rumor, or the forwarding of rumor recurrence.

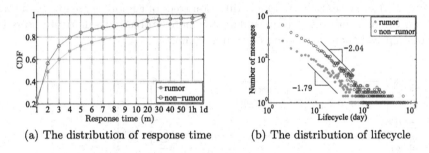

(a) The distribution of response time   (b) The distribution of lifecycle

Fig. 3. The response time and lifecycle of messages.

C. User Features. Users play a crucial role in the diffusion of messages. In what follows, we investigate two user oriented features, namely, repost-participation ratio and repost-exposure ratio.

Formally, the *repost-participation ratio* of a message is defined as $R(m_i) = r_i/(r_i + c_i + a_i)$, where r_i is the number of reposts of message m_i, c_i is the number of its comments and a_i is the number of thumb-ups for it. Empirical analysis shows that the repost-participation ratio of more than 90% of messages is greater than 0.5. Indeed, the ratio for rumors is higher than that for non-rumors, indicating that among all users participated in the diffusion of rumors and non-rumors, those who are involved in rumors are more likely to repost them.

Regarding the *repost-exposure ratio* of a message, we first assume that if a user posts or reposts a message, all his/her followers will see it. We can then formally define the repost-exposure ratio as $E(m_i) = r_i/(\sum_{j=1}^{N} f_j + f_i)$, where r_i is the number of reposts of message m_i posted by user u_i, f_i is the number of followers of u_i, and f_j is the number of followers of the repost user u_j. We find that given a rumor and a non-rumor with a similar number of reposts, the repost-exposure ratio of rumor is usually higher than that of non-rumor. It indicates that a rumor more probably motivates users to repost it as compared to a non-rumor.

5   Rumor Identification

In this section, we evaluate the effectiveness of the proposed diffusion features (DFs) on rumor identification from three aspects. Specifically, we first examine the effectiveness of each DF category and then investigate the effectiveness of all DFs combined with features developed in existing studies on both balanced and imbalanced datasets.

5.1   Effectiveness of Each DF Category

In order to better understand the effectiveness of different DF categories (see Table 2) on distinguishing rumors from non-rumors, we successively exclude each DF category and use the remained DF categories to train an SVM classifier with the RBF kernel. Table 3 presents the results of the experiment, where $(-)X$ indicates that the DF categories except X are adopted to train the SVM classifier.

Table 3. The effectiveness of different feature categories.

	Accuracy	Rumor			Non-rumor		
		Precision	Recall	F1	Precision	Recall	F1
(-)Structural features	0.679	0.676	0.716	0.696	0.683	0.641	0.661
(-)Temporal features	0.723	0.781	0.637	0.701	0.681	0.813	0.741
(-)User features	0.711	0.778	0.610	0.684	0.667	0.817	0.735
(-)Content features	0.695	0.756	0.597	0.667	0.654	0.798	0.719
All features	0.739	0.770	0.673	0.718	0.715	0.803	0.757

The results show that as compared to other DF categories, structural features improve the accuracy of rumor identification by 8.8% from 0.679 to 0.739, achieving the largest improvement among all feature categories. They are the most effective one in detecting the non-rumor messages. User features have a remarkable impact on rumor identification, which can effectively identify the rumor messages. It indicates that the participation of users directly reflects the different attractions of rumor and non-rumor messages. Temporal features have a relatively little effect on the results, as compared with structural features and user features. This is because temporal features reflect temporal characteristics of user activity. Their ability to distinguish rumors from non-rumors is weak, especially, on those messages with a small number of reposts and a short life-cycle. Content features have also strong influence on the results, which contain people's opinions on messages.

Table 4. The performance of different methods on rumor identification.

Reposts	Methods	Accuracy	Rumor			Non-rumor		
			Precision	Recall	F1	Precision	Recall	F1
≥ 100	HybridF	0.884	0.890	0.884	0.887	0.877	0.883	0.880
	HybridSVM	0.919	0.906	0.942	0.924	0.935	0.895	0.915
	HybridF+DF	0.928	0.926	0.936	0.931	0.931	0.920	0.925
	HybridSVM+DF	**0.943**	**0.928**	**0.965**	**0.946**	**0.961**	**0.920**	**0.940**
≥ 10	HybridF	0.875	0.897	0.854	0.875	0.854	0.897	0.875
	HybridSVM	0.777	0.796	0.758	0.776	0.758	0.796	0.777
	HybridF+DF	**0.906**	**0.929**	**0.883**	**0.905**	**0.883**	**0.930**	**0.906**
	HybridSVM+DF	0.810	0.815	0.814	0.814	0.805	0.806	0.806

5.2 Rumor Identification on Balanced Datasets

In order to verify the effectiveness of the proposed DFs and improve the performance of rumor identification, we combine the proposed DFs with the features adopted in the state-of-the-art study in [23]. The features and method used in [23] are employed as baselines in the following experiments, as they perform much better than those commonly used baselines presented in [4,24].

In [23], Wu et al. observed the difference of user type in diffusion trees between rumors and non-rumors, and thus developed a random walk graph kernel to calculate the similarity between diffusion trees. They further combined the graph kernel with the conventional RBF kernel to build a hybrid SVM classifier (denoted as *HybridSVM*) for rumor identification. Although their hybrid classifier achieved high classification accuracy, they only considered messages with at least 100 reposts, which accounts for only 2% to 4% of all messages. To identify rumors on a broader range, in this paper we relax the above restriction to those messages with no less than 10 reposts. We train several SVM classifiers to compare their performance on rumor identification. We denote the classifier based on the features proposed in [23] as *HybridF*, and the one combining these features with our 11 *DFs* as *HybridF + DF*. The classifier integrating our *DFs* into *HybridSVM* is denoted as *HybridSVM + DF*. We implement experiments on datasets containing messages with no less than 10 and 100 reposts, respectively.

Table 4 presents the experimental results. It can be seen that the results corresponding to the cases with no less than 100 reposts are generally better than the other cases with no less than 10 reposts. This is obvious, as the former cases have much more large diffusion trees of messages and thus the structural features are more notable. More importantly, in the former cases, both classifiers where the 11 DFs are integrated (i.e., $HybridF + DF$ and $HybridSVM + DF$) perform better than their original versions (i.e., $HybridF$ and $HybridSVM$), respectively. $HybridSVM + DF$ particularly achieve the best performance in terms of all adopted metrics. These observations justify the effectiveness and values of the proposed DFs. In the cases with no less than 10 reposts, $HybridF + DF$ shows the best performance, and $HybridF + DF$ outperforms $HybridF$ and $HybridSVM + DF$ overwhelms $HybridSVM$, respectively, which reconfirms the values of the proposed DFs. However, we can also see that in these cases, $HybridSVM$ and $HybridSVM + DF$ are no match for $HybridF$ and $HybridF + DF$, respectively, indicating that the random walk graph kernel in $HybridSVM$ does not perform well in the cases taking lots of messages with much fewer reposts into consideration. This is because the effectiveness of the random walk graph kernel highly depends on the structure of the diffusion trees of messages. In the cases with no less than 10 reposts, there are a large percentage of messages with small diffusion trees. Consequently, $HybridSVM$ cannot completely exhibit its advantages. However, DFs can well reflect the characteristics of diffusion without stringent restriction to the number of reposts.

5.3   Rumor Identification on Imbalanced Datasets

In the real world, rumors only account for a small fraction of all messages, as compared to non-rumors. Therefore, rumor identification is essentially an imbalanced classification problem. To verify whether or not DFs are robust to imbalanced data, we examine their performance by comparing $HybridF$ with $HybridF + DF$ in terms of varying proportions of rumors and non-rumors.

In the experiment, we consider five settings of data, with the proportions of rumors and non-rumors being 1:10, 1:20, 1:50, 1:80 and 1:100, respectively. There have been many methods for dealing with the problem of data imbalance [8]. In this paper, we adopt the Synthetic Minority Over-sampling Technique (SMOTE) [5] that balances different classes by oversampling the minor class until the two classes are approximately of equal size. For each proportion, we randomly extract 5 sub-datasets of rumors and non-rumors from the original data; For each sub-set, 100 runs are conducted to obtain the average values of the adopted metrics, i.e., F1, G-mean and AUC.

Figure 4 presents the results of rumor identification using $HybridF$ and $HybridF + DF$ at five data settings. Note that for the sake of space limitation, the results corresponding to AUC is not presented in Fig. 4. In both Fig. 4(a) and (b), for each proportion the result of $HybridF$ is on the left, while that of $HybridF + DF$ is on the right. The diamond symbols in the figure represents the average values of F1 and G-mean. We can observe that $HybridF + DF$ exhibits better results than $HybridF$ on each proportion of rumors to non-rumors on both F1 and G-mean. Similar to F1 and G-mean, AUC for $HybridF + DF$ ranges from 80% to 93%, and is better than that of $HybridF$ by approximately 1% to 3.4% for each setting. These observations indicate that the proposed DFs can effectively improve the performance of rumor and non-rumor classification even in the case of imbalanced datasets.

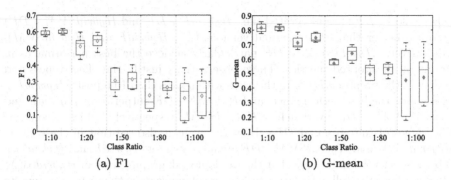

Fig. 4. Comparison between $HybridF$ and $HybridF + DF$ in terms of F1 and G-mean with different proportions of rumors and non-rumors.

6   Conclusions

Online rumors have been a severe social problem with the prosperity of social media. Therefore, in recent years, how to automatically identify rumors in social media has attracted lots of research interests in related communities. In this paper, we studied the automatic rumor identification problem from a diffusion perspective in Sina Weibo. Specifically, we first extracted 11 features of four types, including structural, temporal, user, and content features, from the diffusion processes of messages. We further observed a few interesting phenomena over these features: the diffusion trees of rumors are deeper but narrower than those of non-rumors; rumors are more likely reposted at the first moment. Coupling those new features with commonly used ones, we implemented classifiers based on SVM for classifying rumors and non-rumors. Through experiments on both balanced and imbalanced datasets and comparisons with the state-of-the-arts methods, we demonstrated the effectiveness of the new features for rumor identification. In the future, we will study the dynamic change of these features in the streaming data for real-time rumor identification.

Acknowledgments. This work was funded by the National Basic Research Program of China (973 Program) under Grant Number 2013CB329602, the National Key Research and Development Program of China under Grant Number 2016YFB1000902, and the National Natural Science Foundation of China under Grant Numbers 61572473, 61472400, 61232010. H.W. Shen is also funded by Youth Innovation Promotion Association CAS and the CCF-Tencent RAGR (No. 20160107).

References

1. Bao, P., Shen, H.W., Chen, W., Cheng, X.Q.: Cumulative effect in information diffusion: a comprehensive empirical study on microblogging network. PLOS ONE **8**(10), 1–11 (2013)
2. Bao, P., Shen, H.W., Huang, J., Cheng, X.Q.: Popularity prediction in microblogging network: a case study on sina weibo. In: Proceedings of WWW, pp. 177–178. ACM (2013)

3. Bettencourt, L.M., Cintrón-Arias, A., Kaiser, D.I., Castillo-Chávez, C.: The power of a good idea: quantitative modeling of the spread of ideas from epidemiological models. Phys. A: Stat. Mech. Appl. **364**, 513–536 (2006)
4. Castillo, C., Mendoza, M., Poblete, B.: Information credibility on twitter. In: Proceedings of WWW, pp. 675–684. ACM (2011)
5. Chawla, N.V., Bowyer, K.W., Hall, L.O., Kegelmeyer, W.P.: SMOTE: synthetic minority over-sampling technique. J. Artif. Intell. Res. **16**, 321–357 (2002)
6. Cheng, J., Adamic, L., Dow, P.A., Kleinberg, J.M., Leskovec, J.: Can cascades be predicted? In: Proceedings of WWW, pp. 925–936. ACM (2014)
7. Friggeri, A., Adamic, L.A., Eckles, D., Cheng, J.: Rumor cascades. In: The International AAAI Conference on Weblogs and Social Media (ICWSM) (2014)
8. Japkowicz, N., Stephen, S.: The class imbalance problem: a systematic study. Intell. Data Anal. **6**(5), 429–449 (2002)
9. Jin, F., Dougherty, E., Saraf, P., Cao, Y., Ramakrishnan, N., Tech, V.: Epidemiological modeling of news and rumors on twitter. In: Proceedings of the 7th Workshop on Social Network Mining and Analysis. ACM (2013)
10. Kwon, S., Cha, M., Jung, K., Chen, W., Wang, Y.: Prominent features of rumor propagation in online social media. In: Proceedings of ICDM, pp. 1103–1108. IEEE (2013)
11. Liu, X., Nourbakhsh, A., Li, Q., Fang, R., Shah, S.: Real-time rumor debunking on twitter. In: Proceedings of CIKM, pp. 1867–1870. ACM (2015)
12. Ma, J., Gao, W., Mitra, P., Kwon, S., Jansen, B.J., Wong, K.-F., Cha, M.: Detecting rumors from microblogs with recurrent neural networks. In: Proceedings of IJCAI (2016)
13. Ma, J., Wei, Z., Lu, Y.: Detect Rumors using time series of social context information on microblogging websites. In: Proceedings of CIKM, pp. 1751–1754 (2015)
14. Mei, Y., Zhong, Y., Yang, J.: Finding and analyzing principal features for measuring user influence on twitter. In: 2015 IEEE First International Conference on Big Data Computing Service and Applications, pp. 478–486. IEEE (2015)
15. Mendoza, M., Poblete, B., Castillo, C.: Twitter under crisis: can we trust what we RT? In: Proceedings of the First Workshop on Social Media Analytics, pp. 71–79. ACM (2010)
16. Nadamoto, A., Miyabe, M., Aramaki, E.: Analysis of microblog rumors and correction texts for disaster situations. In: Proceedings of IIWAS. ACM (2013)
17. Qazvinian, V., Rosengren, E., Radev, D.R., Mei, Q.: Rumor has it: identifying misinformation in microblogs. In: Proceedings of EMNLP, pp. 1589–1599. Association for Computational Linguistics (2011)
18. Ratkiewicz, J., Conover, M., Meiss, M., Goncalves, B., Patil, S., Flammini, A., Menczer, F.: Detecting and tracking the spread of astroturf memes in microblog streams. ArXiv preprint (2010)
19. Shen, H.w., Wang, D., Song, C.: Modeling and predicting popularity dynamics via reinforced poisson processes. In: Proceedings of the Twenty-Eighth AAAI Conference on Artificial Intelligence (2014)
20. Starbird, K., Maddock, J., Orand, M., Achterman, P., Mason, R.M.: Rumors, False Flags, and Digital Vigilantes: Misinformation on Twitter after the 2013 Boston Marathon Bombing. iConference (2014)
21. Sun, S., Liu, H., He, J., Du, X.: Detecting event rumors on Sina Weibo automatically. In: Ishikawa, Y., Li, J., Wang, W., Zhang, R., Zhang, W. (eds.) APWeb 2013. LNCS, vol. 7808, pp. 120–131. Springer, Heidelberg (2013). doi:10.1007/978-3-642-37401-2_14

22. Tanaka, Y., Sakamoto, Y., Matsuka, T.: Toward a social-technological system that inactivates false rumors through the critical thinking of crowds. In: HICSS, pp. 649–658. IEEE (2013)

23. Wu, K., Yang, S., Zhu, K.Q.: False rumors detection on Sina Weibo by propagation structures. In: Proceedings of ICDE, pp. 651–662. IEEE (2015)

24. Yang, F., Liu, Y., Yu, X., Yang, M.: Automatic detection of rumor on Sina Weibo. In: Proceedings of the ACM SIGKDD Workshop on Mining Data Semantics, pp. 13–19. ACM (2012)

Investigating the Dynamics of Religious Conflicts by Mining Public Opinions on Social Media

Swati Agarwal[1(✉)] and Ashish Sureka[2]

[1] Indraprastha Institute of Information Technology Delhi (IIIT-D), New Delhi, India
swatia@iiitd.ac.in
[2] ABB Corporate Research Center, Bangalore, India
ashish.sureka@in.abb.com

Abstract. The powerful emergence of religious faith and beliefs within political and social groups, now leading to discrimination and violence against other communities has become an important problem for the government and law enforcement agencies. In this paper, we address the challenges and gaps of offline surveys by mining the public opinions, sentiments and beliefs shared about various religions and communities. Due to the presence of descriptive posts, we conduct our experiments on Tumblr website- the second most popular microblogging service. Based on our survey among 3 different groups of 60 people, we define 11 dimensions of public opinion and beliefs that can identify the contrast of conflict in religious posts. We identify various linguistic features of Tumblr posts using topic modeling and linguistic inquiry and word count. We investigate the efficiency of dimensionality reduction techniques and semi-supervised classification methods for classifying the posts into various dimensions of conflicts. Our results reveal that linguistic features such as emotions, language variables, personality traits, social process, and informal language are the discriminatory features for identifying the dynamics of conflict in religious posts.

Keywords: Mining user-generated data · Public opinions · Religious conflicts · Social computing · Text classification · Tumblr · Semi-supervised learning

1 Introduction

Research shows that with the unexpected emergence of religion and faith in society, has led to the discrimination and violence against rival religious groups [1]. It is seen that the people use various different platforms (chat groups, forums, blogs, social media) to share their beliefs and opinions about their religion [6]. These people also outburst their extremist and hateful views towards other religions [3,5]. These groups of individuals take the leverage of freedom of speech and social media to post their sentiments and beliefs about a variety of sensitive topics including religion and race [3]. Despite several guidelines of social

© Springer International Publishing AG 2017
J. Kim et al. (Eds.): PAKDD 2017, Part I, LNAI 10234, pp. 421–433, 2017.
DOI: 10.1007/978-3-319-57454-7_33

media platforms[1] and constraints of freedom of speech [9], people post racist and harsh comments against other religions that can hurt religious sentiments of an individual or a community [5,11]. Figure 1 shows examples of several online posts showing the conflicts in context of Islamic religious beliefs and sentiments of authors. Figure 1 reveals that while some users posted defensive and promotional content about Islam religion; other users posted negative comments and insulting the beliefs of people believing in Islam. Further, some users only make posts to share information on real time incidents or news and not presenting any sentiment or argument for a religion. As seen in the real world, many young age people and students get influenced from social media messages and join religious wars [6]. Therefore, monitoring such content on social media and identifying religious conflicts within society, understanding the root cause of such conflicts and arguments have become an important problem for the government, social scientist and law enforcement agencies.

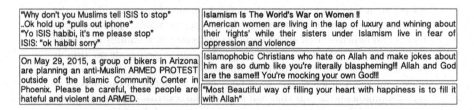

Fig. 1. A concrete example of 5 Tumblr posts showing differences and conflicts in beliefs of Tumblr bloggers on Islam religion

Background: We conduct a literature survey in the area of political and religious conflict identification on social media. We find that over the past 3 decades, social science researchers have been conducting offline surveys for identifying religious conflicts within society. Whereas, the area of identifying such conflicts by using computer science applications is not much explored. Based on our analysis, we divide our literature survey into four lines of research:

1. **Offline Data and Manual Analysis:** Swinyard et al. [13] and Wilt et al. [16] conducted surveys to examine the relationship between religious and spiritual beliefs and emotions of people such as happiness and anxiety. Yang et al. [17] present a study on the impact of low coverage of HindRAF event in media causing the religious conflicts among citizens of Malaysia.
2. **Offline Data and Automated Analysis:** Vüllers et al. [15] present a study on the religious factors of 130 developing countries. Their analysis reveal that the clashes between religious groups and attacks by religious actors are the main cause of religious conflict within state and community. Basedau et al. [7] used logistic regression approach on the same dataset to identify several discriminatory religious factors that causes conflicts, religious violence and grievances.

---

[1] https://www.tumblr.com/abuse/maliciousspeech.

3. **Online Data and Manual Analysis:** In addition to the social science researchers, various non-profit organizations like Pew Research Center[2], Berkley Center for Religion, Peace and World Affairs[3] and United States Institute of Peace[4] conduct online polls, offline statistical and text analysis on blogs and social media data to identify religious beliefs and issues within local and global regions. Some of the recent studies of Pew Research Center include the global trend and projection of population growth of various religions, gender gap in religious commitment of Muslim and Christian communities and increment and decrement rate of government restrictions on religion and social hostilities.

4. **Online Data and Automated Analysis:** Chesnevar et al. [8] propose an opinion tree using IR and argumentation technique for identifying conflicts and confronting opinions in E-Government contexts. They conduct their analysis on Twitter messages and identify the polarity (positive, negative and neutral) of contrasting arguments. In our previous study [5], we conduct a manual analysis on Tumblr posts to investigate the feasibility of content analysis for identifying religious conflicts and fill the gaps of traditional offline surveys.

Motivation: The work presented in this paper is motivated by the prior literature and a need to develop an automatic solution to identify religious conflicts among social media users. However, automatic identification of religious beliefs and faith by mining user-generated data is a technically challenging problem. In order to enhance our understanding of religious conflicts and address the challenge of local and regional data, we conduct our experiments on a wider community of Tumblr. Tumblr is the second most popular micro-blogging service that allows users to post eight different types of content including image, video, audio, chat, quote, answer, text and url [3]. Unlike Twitter, Tumblr has no character limit for tags, image captions, text body content, allowing it's users to make descriptive posts. Presence of noisy content such as misspelt words, short text, acronyms, multi-lingual text and incorrect grammar decreases the accuracy of linguistic features and Natural Language Processing tools [6]. Further, the presence of ambiguity in posts and the intent of author makes it difficult even for human annotation [3]. We, however conduct our analysis on Tumblr website because Tumblr allows users to make longer posts and express their opinions and beliefs in an open and descriptive manner which fills the gaps of offline surveys. Furthermore, Tumblr facilitates it's users to send anonymous messages and use the leverage of expressing their opinions without revealing their names [2].

Research Contributions: In contrast to the existing work our paper makes the following novel and technical contributions: (1) To the best of our knowledge, we present the first study on automated identification of religious beliefs, opinions and faith in global public communities. (2) We address the challenge of

---

[2] http://www.pewresearch.org/topics/religion-and-society/.

[3] https://berkleycenter.georgetown.edu.

[4] http://www.usip.org/about-usip.

social media content by translating multi-lingual posts into base language and extracting textual metadata of multimedia posts such as photo and video. (3) We identify various linguistic features that are discriminatory for identifying contrast in different opinions on religious posts and (4) We investigate the efficiency of multi-class semi-supervised classifier across various dimensionality reduction techniques for classifying Tumblr posts into various dimensions of conflicts.

2   Dimensions of Conflicts

We conducted a survey among 3 different groups of people- we selected 10 graduate students of our department, 30 Tumblr bloggers (followers on authors' personal Tumblr account) and 20 people from society randomly. In extension to our previous study [5], we conducted a small questionnaire consisting of questions related to their activities on social media platforms *e.g. how frequently they*

Table 1. Concrete examples of 11 dimensions and 3 polarities of religious beliefs and sentiments in Tumblr posts created about Christian religion and community

Type	Post content
IS	In a show of solidarity, Muslims are standing with Christians and giving up guilty pleasures for lent
Query	Doesn't the Bible teach us not to take a life of another? To turn the other cheek and not respond with violence? Isn't better to die and be in heaven then kill and stay on earth?
N/A	Pray for abortion access. People deserve easy access to abortion services
Defensive	I'm still over the moon about God. I'm in total awe that He not only hears me, but actually listens and does something about it. I feel so loved and acknowledged
Disappointment	If you're a Christian and voted for Trump I wanna ask you a question. What does it feel like to go against everything God wanted for us?
Annoyance	Jesus himself could crawl out of his grave, take me by the hand, and point me to salvation and heaven. I would say no. I would seriously 100% rather die as a Jew then live for even a millisecond as a Christian. So stop trying to convert me to Christianity because it is not going to happen
Insult	Burn churches not calories. Christianity is stupid!- Well I am not the only one that feels the same way
Disgust	So this dude that was running in local elections for council said women who have abortions are worse than ISIS
Ashamed	I feel like a bad Christian. I have so much hate in my heart after this election, at Drumpf, at his voters, at my country. I know I should turn the other cheek and love radically and protest without hating but I'm so angry. I feel like I can't let that hate go, not so soon. But I need to and I'm furious at myself
Disbelief	Imagine the peace we'd all have without religion. Wouldn't it be a better world?
Sarcasm	When Christ has a cold he sneesus

make religion based posts on social media or react to other religious posts. We created a set of 30 posts about different religions and asked for their opinions. Based on the dimensions proposed in our previous study and our survey, we decided 11 dimensions of opinions that can be used to define the contrast of conflict among people: Information Sharing (IS), Query, Not a religion based post (N/A), Disbelief, Defensive, Annoyance, Insult, Disappointment, Sarcasm, Ashamed and Disgust. Table 1 shows the examples of 11 Tumblr posts created about Christian religion and community reflecting the different dimensions of public opinions about the community.

3   Experimental Setup

Acquiring the Dataset: To conduct our experiments, we use the Tumblr dataset [4] extracted and made publicly available in our previous study [5]. As of November 9, 2016, this dataset is the largest dataset available of Tumblr posts and bloggers and contains all types of Tumblr posts (answer, photo, text, audio, video, url, chat, and quote) consisting of various tags frequently used in religion based posts. The published dataset contains a total of 107, 586 posts collected for 10 such tags (hinduism, islam, muslim, religion, isis, jihad, christian, islamophobia, judaism, and jews). The statistics reveal that the maximum number of posts consisting of religious tags are either posted as photo (49, 072) or text (34, 902). While, URL or link posts (10, 062) are relatively higher in comparison to chat (507), audio (390) and answer/ask box (1, 077) categories [5].

Data Pre-processing: In order to identify the religious conflicts, we conduct our analysis only on textual metadata of posts. Therefore, in this phase, we address the challenge of multi-lingual and multi-media content of the posts. In Tumblr, each type of post contains a different set of textual attributes. We acquire different metadata of all records for each type of posts available in our experimental dataset. For example, for photo and video posts, we extract only the caption and description of posts, for chat and answer posts, we extract the phrases used in the conversation. While, for URL, quote and text posts, we extract the title and body content of the posts. We discard the audio posts since these post contains only track name, artist name and album name which do not reveal any information about the content. In this paper, we conduct our experiments only on English language posts. Therefore, in order to address the challenge of multi-lingual posts, we translate all non-English posts into our base language. We use Yandex Language API[5] to detect the language of source content and translate it to English language. We further remove the posts consisting of no textual metadata. For example, photo posts with no caption or text posts consisting of only external URLs. We also remove all redundant posts from the dataset consisting of different post_id but having duplicate content. After pre-processing of the raw data, we were able to acquire a total of 89, 803 posts calling them as our experimental dataset.

---

[5] https://translate.yandex.com/developers.

Data Annotation: In order to create the ground truth for our dataset and creating a training dataset, we use 89,803 pre-processed posts for further annotation which spans only 83.4% of the acquired data. In order to remove the bias from our annotation, we hired a group of Tumblr users who had an experience of 2 to 3 years of using Tumblr website. We published a post on Tumblr and asked bloggers to volunteer for data annotation. In a span of one week, 34 bloggers replied and agreed to annotate an average of 30 posts. We declined 4 bloggers who joined Tumblr recently. Among 30 bloggers, only 23 bloggers reverted back with 690 annotated posts among which 6 posts were sampled more than once. Due to the large amount of Tumblr posts and challenge of creating ground truth [6]; we used only these 684 posts for creating our training dataset.

4   Features Identification

Topic Modeling: During our survey for identifying the dimensions of conflict, we observe that many users add religion based tags in their posts while the content of the post is irrelevant to any religion or community. Since, our experimental dataset is collected using a keyword based flagging approach, we identify the topic of each post to filter the irrelevant posts. Figure 2 shows the statistics of number of posts consisting of religion based tags and actually discussing about those religions. Figure 2 reveals that among all posts (85% of experimental dataset) consisting of seed tags related to Islam religion (islam, muslim, islamophobia, isis and jihad), only 31% of the posts are about Islam religion. Similarly, among 20,106 posts (22% of experimental dataset) consisting of judaism and jews tags, only 15% (13,695) posts belong to Judaism religion. For each post, we assign a binary value where 1 denotes the topic (religion based post) and 0 denotes the non-topic (not a religion specific post). We further extract the name of the religion being discussed in the post since a post can have content about more than one religion. For example, in the following post *"KKK burns black Churches even tho they claim Christianity as their religion and ISIS blows up mosques even tho they claim Islam as their religion."*; author mention about both Islam and Christian religion. In our experimental dataset, we find that only 40% of posts (35,799) belong to a religious topic (Islam, Hinduism, Christian and Judaism) while the remaining 60% of posts (54,004) only contains religious tags but do not contain the content related to a religious group or community.

LIWC: In order to compute the correlation between various linguistic features and sentiments, we use an open source API by LIWC- Linguistic Inquiry and Word Count [14]. We extract a total of 45 features grouped into 14 categories of linguistic dimensions. In order to identify the sentiments and emotions of the bloggers, we compute relative percentage of the emotions e.g. *sadness, anxiety, anger, happiness*. We measure the authenticity and personality traits of authors by computing summary of language variables in a post e.g. *analytical thinking and authenticity*. Further, in order to identify the personal beliefs and relation with the real world incidents, we compute the percentage of *sexual* terms, mention of *family, friends, male* and *female* references in a post. In order to identify the level of aggression and certainty of a post, we compute the percentage of use of informal language such as *swear words, slangs* and *fillers*. Apart from these features, we also compute the presence of various other linguistic dimensions such as *pronouns, negations, interrogatives words, cognitive process, perceptual process, power, time orientations* (mention of past, present or future incidents), *time* and *personal concerns* (work, religion, and death).

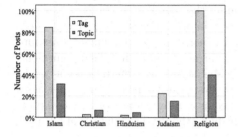

Fig. 2. Relative percentage of number of posts consisting of a religion based tag and topic

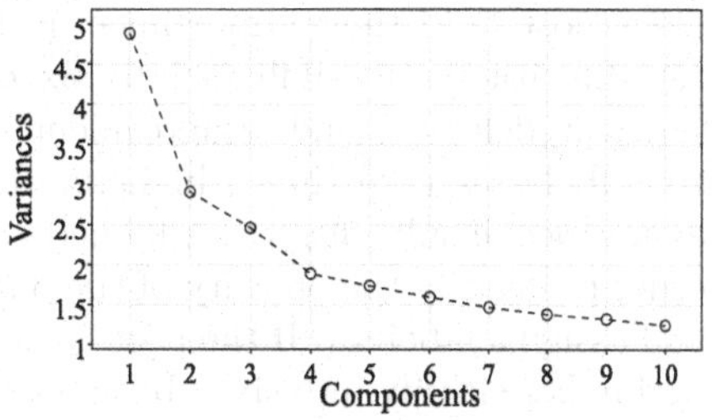

Fig. 3. Relationship between variance and components in principal component analysis

5   Features Selection

1. **Using All Features (FS1):** In first iteration, we use all 45 linguistic, sentiment and text based features extracted using LIWC. We train our model on available features and investigate the efficacy of classification of Tumblr posts into 11 dimensions of conflicts.

2. **Principal Component Analysis (FS2):** In second iteration, we use PCA-a dimensionality reduction technique to reduce the number of feature vectors. We compute the correlation among all feature vectors and identify the components chracterizing the whole data. For $n = 45$ vectors in our experimental data, we get n eigenvectors. We select the first $p = 10$ eigenvectors having maximum eigenvalues and discard the ones with less significance. We project our dataset into 10 dimensions and form our feature vector $FS2$ by taking the eigenvectors of 10 components. Figure 3 shows the distribution of variances for all 10 components selected after dimensionality reduction of the data.

3. **Attribute Selection Correlation (FS3):** In third iteration, we use Correlation Attribute Evaluation technique to identify a set of discriminatory attributes. We measure the Pearson's correlation between each attribute (feature vector) and the class. We create a correlation matrix of 45 attributes and class for each record in the dataset and compute the overall correlation by computing the weighted average of the attribute. Based on the correlation between each attribute vector and class, we create a set of 10 features having moderately higher positive and negative correlation and drop the features having correlation closer to zero. For our experimental dataset, FS3 returns

the following 10 features: mention of past and present tense, pronouns, male references, perceptual process, negative emotions, clout, presence of negation, swear words and anger.

6   Classification

Classes and Membership Groups: Based on the polarity of opinions in religious posts and their importance in defining the dynamic of conflicts, we split the dimensions of conflicts into six classes: information sharing, query, N/A, sarcasm, defensive and disagreement. We further divide disagreement class into six subclasses reflecting a higher range of negative emotions: disappointment, annoyance, insult, disgust, ashamed and disbelief. For a given data point y_m, in order to identify the polarity of the post, we first classify the post into six classes and assign a label o_m. If the post is identified as a disagreement or negative post, we further classify into six subclasses identifying the low-level details of negative emotions in a given post.

Classification Approach: Due the constraint of lack of ground truth and only a very small portion of available labeled data (2%), we use semi-supervised classification method to classify the unlabelled posts over unsupervised method. Semi-supervised classification approach uses both annotated and unlabelled data to learn the model iteratively in a snowball manner. We use 684 posts annotated by Tumblr bloggers and use them to train our model in first iteration of semi-supervised classifier. We conduct our experiments on $35,799$ posts identified as topic related (discussing about any religious group or community). Given a labeled data (X_N, C_N) where the data points are denoted by $X_N = (x_1, x_2, x_3 \ldots x_n)$ and their labels are denoted by $C_N = (c_1, c_2, c_3 \ldots c_n)$. The unlabelled data points $Y_M = (y_1, y_2, y_3 \ldots y_m)$ and their unknown labels $O_M = (o_1, o_2, o_3 \ldots o_m)$ are denoted as (Y_M, O_M).

We use the 'R' statistical language to perform classification using *"upclass"* package [12]. "Upclass"[6] is a semi-supervised classification method and an adaptive version of the model-based classification method proposed in Dean et al. [10]. Upclass uses an iterative method that initiates by using model-based classification method and uses Expectation- Maximization (EM) algorithm in further iteration until convergences. In the first iteration of classification, a set of 14 models is applied on the dataset considering different constraints (E- equal, V- variable, I- identity) upon covariance structure- volume, shape and orientation of the cluster. For example, in EEE model each cluster has equal volume, same shape and same orientation along the axis. The clustering is performed in multiple iteration by estimating group membership on unlabelled data based on the maximum likelihood of EM algorithm. In order to perform the model based discriminatory analysis on unlabelled data points, the model of data (combination of E, I, V constraints) must be known. If the model is null then Upclass fit every model to the data and identifies the best-fitted model of given data and

---

[6] http://CRAN.R-project.org/package=upclass.

attributes. To identify the best-fitted model, Upclass calculates the bayesian information criterion (BIC) value for each model. $BIC = 2\log(l) - p\log(n)$; where l is the likelihood of the data, p is the number of model parameters and n is the number of data points. The model with the highest BIC value is selected as best-fitted model for the data.

7   Empirical Analysis and Evaluation Results

In this Section, we present the classification results of Upclass semi-supervised method applied for both classes and sub-classes identification. We apply 3 iterations of Upclass supervised classification methods on all 35,799 posts for each feature vectors model (FS1, FS2 and FS3) discussed in Sect. 5. If a post is labeled as "Disagreement or Negative", we further train our model on the posts labeled under the six subclasses of disagreement and classify unknown data points into one of the six sub-groups using Upclass semi-supervised classification method. Table 2 shows the experimental results of classification performed using each feature vector model for each membership groups. Table 2 reveals that the classification model converges for each set of feature vectors. During the first step of classification both FS1 and FS3 takes similar number of iterations whereas, FS2 takes approximately 2.5 times of their iterations. Further, for FS1 and FS3, VEV is selected as the best-fitted model while for FS2 attributes, VVV showing the non-linear distribution of labels (different orientation of each cluster against the axis). Figure 4 shows the visual representation of clusters created using different models (considering the constraints on covariance structure). Table 2 also reveals that during the second set of classification (sub-groups of disagreement

Table 2. Classification results all feature selection techniques for different membership groups and observations

	FS1		FS2		FS3	
Attribute	Class	Sub-class	Class	Sub-class	Class	Sub-class
Converged	TRUE	TRUE	TRUE	TRUE	TRUE	TRUE
Iteration	272	491	604	252	207	110
Dimension	45	45	10	10	10	10
Model name	VEV	EEV	VVV	VVV	VEV	VVI

Fig. 4. Visualization of volume, shape and orientation constraints for best-fitted models for classification. V= variation, E= equal and I=identical

class), Upclass method takes a different number of iteration for each feature vector model. Further, for each feature vector, a different discriminant model is selected. While, in FS2, for both classes and sub-classes identification Upclass uses the same model i.e. VVV. While using all attributes as features, it classifies all observations into equal parts and create clusters of equal shapes varying in the orientation against axis. While, using features selected using Pearson's correlation technique, it classifies the observations in a linear manner- varying the shape and volume of the clusters while all data points aligned towards an axis. Unlike, FS1 and FS3, while using principal components as feature vectors the clusters are created in a non-linear manner- varying in size, shape and orientation of data points.

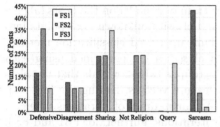

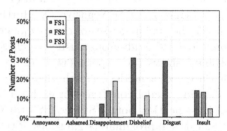

(a) Distribution of Polarity Based Classes Identified for All Observation Posts in Experimental Dataset

(b) Distribution of Sub-Classes Identified for All Posts Labelled as "Disagreement/Negative"

Fig. 5. Classification results of Upclass semi-supervised method for unknown posts categorized into polarity based classes and extreme emotions based sub-classes

Figure 5 shows the distribution of Tumblr posts classified into different groups of classes and sub-classes based on the polarity of opinions. Figure 5(a) shows the relative percentage of posts classified into each of the defined classes. Figure 5(a) reveals that while using all attributes as features, maximum number of posts (43%) are labeled as sarcasm posts which is below 10% while using dimensionality reduction techniques. While only a very small percentage of posts (5%) are classified as non-religion based posts which is significantly higher for both FS2 and FS3 (approximately 25%). The classification results shows that except FS3, using FS1 and FS2 feature vectors, the classifier does not have sufficient examples for labelling query posts. The graph in Fig. 5(a) shows that for each feature selection method, the classifier classifies 10% to 12% posts as disagreement/negative that are further classified into sub-classes. Figure 5(b) shows the relative percentage of these 10% to 12% posts further classified into sub-classes of extreme negative emotions. Figure 5(b) reveals that while taking all attributes into account, a very small percentage (~negligible) of posts are classified as "Annoyance" posts while the distribution of other classes are significantly higher. While the distribution of posts for FS2 and FS3 is varying for each category- as reflected in best-fitted model selected for classification (refer to Table 2.

The variation in distribution of all posts in different categories shows the dynamics of public opinions on religious posts. The size of each cluster (number of posts grouped in a class) for different combinations of attribute selection techniques and classification method shows the presence of religious conflicts among users on Tumblr.

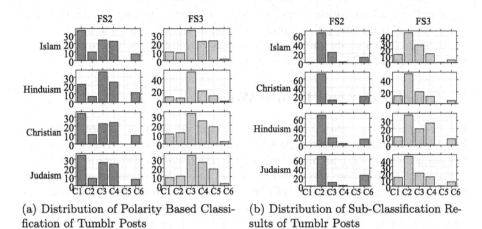

(a) Distribution of Polarity Based Classification of Tumblr Posts

(b) Distribution of Sub-Classification Results of Tumblr Posts

Fig. 6. Distribution of classification results of Tumblr posts specific to a religion.

In order to address the challenge of identifying public beliefs and opinions in Tumblr posts where authors are discussing about more than religion. We identify the name of religions being discussed in each post available in our experimental dataset. We classify each post into classes (polarity based groups) and sub-classes (extreme negative emotions based groups) and discuss the results of classification for identifying religion specific conflicts among Tumblr users. Due to the large volume size of Sarcasm cluster and no post classified as Query post, we discard the FS1 technique for identifying the conflicts among individual religious groups. Figure 6 shows the classification results and distribution of Tumblr posts classified into various dimensions of conflicts. For Fig. 6(a), $C1$, $C2$, $C3$, $C4$, $C5$ and $C6$ denote defensive, disagreement, sharing, not religion, query and sarcasm dimensions respectively. Similarly, for Fig. 6(b), $C1$, $C2$, $C3$, $C4$, $C5$ and $C6$ denote annoyance, ashamed, disappointment, disbelief, disgust and insult.

As shown in Table 2, while using principal component analysis feature vectors, the semi-supervised classification method selects VVV as best-fitted model. Figure 6(a) also reveals that for FS2 feature vectors the volume of all clusters are different making some cluster too large or too small. Further, Fig. 6(b) reveals that maximum number (more than 60%) of disagreement posts belong to "ashamed" category. Whereas, while using Pearson's correlation selection method, the posts are grouped into all classes. During the first phase of classification, there is variation in volume of observations in each cluster and while the shape of each cluster is the same. Whereas, in second phase of classification, the volume of observation and shape of each cluster varies. The classification of Tumblr posts for each religion into several dimensions of conflict shows that a lot of discussion about religious topics happen on social media where users have different opinions, beliefs and sentiments about these religions. Our results shows that various linguistic features, emotions, social presence, summary of language variables and other linguistic dimensions of user-generated data can be used to identify the conflicts within religious faith and beliefs. Furthermore, Tumblr is a rich source of collecting public opinion posted in a descriptive and open manner which is useful to study the low-level details of religious beliefs and overcome the challenges of offline data and surveys.

8   Conclusions and Future Work

Research shows that due to the rapidly growing influence of religious faith and beliefs and leading to discrimination and violence against other rivalry communities, identification of dynamics of religious conflict has become an important and challenging problem for the government and law enforcement agencies. In this paper, we address the challenge of offline surveys by mining the public opinions and beliefs from Tumblr website. We conduct our experiments on an open source dataset consisting of the largest collection of Tumblr posts. We conduct a survey among three different groups of people (graduate students, Tumblr bloggers and people from society) and define 11 dimensions of public opinion that can identify the contrast of conflicts. We investigate the feasibility and efficiency of linguistic features and different dimensionality reduction techniques and compare their results of classifying Tumblr posts into different dimensions of conflicts. Due to the small size of labelled data, we use Upclass- a semi-supervised classification method to train our model and classify unlabelled observations. Based on our results, we conclude that despite the presence of noise and ambiguity in content, linguistic features are discriminatory features for identifying the dynamics of religious conflicts. Furthermore, identifying the topic prior to the identification linguistic features can be used to disambiguate the sentiments of author while discussing about more than one religion in a single post.

Future work includes the improvement in linguistic features and making them efficient for classifying very short and short text posts. Furthermore, future work includes the identification of age and location of bloggers for identifying the collision of religious beliefs and sentiments in different age groups or different regions across the world.

References

1. Acciaioli, G.: Grounds of conflict, idioms of harmony: custom, religion, and nationalism in violence avoidance at the lindu plain, Central Sulawesi. Indonesia **72**, 81–114 (2001)
2. Agarwal, S., Sureka, A.: A topical crawler for uncovering hidden communities of extremist micro-bloggers on Tumblr. Making Sense of Microposts (#Microposts2015) (2015)

3. Agarwal, S., Sureka, A.: But i did not mean it!- intent classification of racist posts on Tumblr. In: Proceedings of European Intelligence and Security Informatics Conference (EISIC). IEEE (2016)
4. Agarwal, S., Sureka, A.: Religious beliefs on social media: large dataset of Tumblr posts and bloggers consisting of religion based tags, mendeley data, v1 (2016). http://dx.doi.org/10.17632/8hp.39rknns.1
5. Agarwal, S., Sureka, A.: A collision of beliefs: investigating linguistic features for religious conflicts identification on Tumblr. In: Krishnan, P., Radha Krishna, P., Parida, L. (eds.) ICDCIT 2017. LNCS, vol. 10109, pp. 43–57. Springer, Cham (2017). doi:10.1007/978-3-319-50472-8_4
6. Agarwal, S., Sureka, A., Goyal, V.: Open source social media analytics for intelligence and security informatics applications. In: Kumar, N., Bhatnagar, V. (eds.) BDA 2015. LNCS, vol. 9498, pp. 21–37. Springer, Cham (2015). doi:10.1007/978-3-319-27057-9_2
7. Basedau, M.: Bad religion? Religion, collective action, and the onset of armed conflict in developing countries. J. Confl. Resolut. **60**(2), 226–255 (2016)
8. Chesnevar, C.I., Maguitman, A.G.: Opinion aggregation and conflict resolution in e-government platforms: contrasting social media information. In: Interdisciplinary Perspectives on Contemporary Conflict Resolution, p. 183 (2016)
9. Cohen, J.: Freedom of expression. Philos. Public Aff. **22**(3), 207–263 (1993)
10. Dean, N., Murphy, T.B., Downey, G.: Using unlabelled data to update classification rules with applications in food authenticity studies. J. R. Stat. Soc.: Ser. C (Appl. Stat.) **55**(1), 1–14 (2006)
11. Emmons, R.A.: Assessing spirituality through personal goals: implications for research on religion and subjective well-being. Soc. Indic. Res. **45**, 391–422 (1998)
12. Russell, N., Cribbin, L., Murphy, T.B.: upclass: an R package for updating model-based classification rules (2012)
13. Swinyard, W.R., Kau, A.K., Phua, H.Y.: Happiness, materialism, and religious experience in the US and Singapore. J. Happiness Stud. **2**(1), 13–32 (2001)
14. Tausczik, Y.R.: The psychological meaning of words: LIWC and computerized text analysis methods. J. Lang. Soc. psychol. **29**(1), 24–54 (2010)
15. Vüllers, J., Pfeiffer, B.: Measuring the ambivalence of religion: introducing the religion and conflict in developing countries (RCDC) dataset. Int. Interact. **41**(5), 857–881 (2015)
16. Wilt, J.A., Grubbs, J.B., et al.: Anxiety predicts increases in struggles with religious/spiritual doubt over two weeks, one month, and one year. Int. J. Psychol. Relig. **27**, 26–34 (2016)
17. Yang, L.F., Ishak, M.S.A.: Framing interethnic conflict in Malaysia: a comparative analysis of newspapers coverage on the Hindu Rights Action Force (Hindraf). Int. J. Commun. **6**, 24 (2012)

Mining High-Utility Itemsets with Both Positive and Negative Unit Profits from Uncertain Databases

Wensheng Gan[1], Jerry Chun-Wei Lin[1(✉)], Philippe Fournier-Viger[2],
Han-Chieh Chao[1,3], and Vincent S. Tseng[4]

[1] School of Computer Science and Technology,
Harbin Institute of Technology (Shenzhen), Shenzhen, China
wsgan001@gmail.com, jerrylin@ieee.org, hcc@ndhu.edu.tw
[2] School of Natural Sciences and Humanities,
Harbin Institute of Technology (Shenzhen), Shenzhen, China
philfv@hitsz.edu.cn
[3] Department of Computer Science and Information Engineering,
National Dong Hwa University, Hualien, Taiwan
[4] Department of Computer Science,
National Chiao Tung University, Hsinchu, Taiwan
vtseng@cs.nctu.edu.tw

Abstract. Some important limitation of frequent itemset mining are that it assumes that each item cannot appear more than once in each transaction, and all items have the same importance (weight, cost, risk, unit profit or value). These assumptions often do not hold in real-world applications. For example, consider a database of customer transactions containing information about the purchase quantities of items in each transaction and the positive or negative unit profit of each item. Besides, uncertainty is commonly embedded in collected data in real-life applications. To address this issue, we propose an efficient algorithm named HUPNU (mining High-Utility itemsets with both Positive and Negative unit profits from Uncertain databases), the high qualified patterns can be discovered effectively for decision-making. Based on the designed vertical PU$^{\pm}$-list (Probability-Utility list with Positive-and-Negative profits) structure and several pruning strategies, HUPNU can directly discovers the potential high-utility itemsets without generating candidates.

Keywords: Frequent itemset · Uncertainty · Negative unit profit · PU$^{\pm}$-list

1 Introduction

Frequent itemset mining (FIM) [1,3] has become one of the core data mining tasks that is essential to a wide range of applications. However, some important limitations of FIM are that it assumes that each item cannot appear more than once in each transaction and that all items have the same importance (weight, cost, risk, unit profit or value). These assumptions often do not hold

© Springer International Publishing AG 2017
J. Kim et al. (Eds.): PAKDD 2017, Part I, LNAI 10234, pp. 434–446, 2017.
DOI: 10.1007/978-3-319-57454-7_34

in real-world applications. For example, consider a database of customer transactions containing information about the purchase quantities of items in each transaction and the unit profit of each item. All the developed FIM algorithms would discard this information and may thus discover many frequent itemsets generating a low profit. Hence, FIM fails to discover high profit patterns for many real-world applications.

To address this issue, the problem of high-utility itemset mining (HUIM) was developed [6,15]. HUIM considers the case where items can appear more than once in each transaction and where each item has a user-specified "utility" (e.g., unit profit). The goal of HUIM is to discover items/itemsets with their utility in a database is no less than the minimum utility threshold, called the high utility itemsets (HUIs), i.e., itemsets generating a high profit. HUIM plays an important role in a wide range of applications, such as website click stream analysis, cross-marketing in retail stores, and biomedical applications [4,9,14]. The problem of HUIM is more difficult than FIM, the reason is that the well-known *downward-closure property* of the support of an itemset is no longer hold in HUIM. In HUIM, however, the utility of an itemset is neither monotonic or anti-monotonic, it means that a high utility itemset may have its supersets or subsets with lower, equal or higher utility [3]. Thus, it is very difficult to prune the search space in HUIM. Many studies have been carried to develop efficient HUIM algorithms, such as Two-Phase [10], IHUP [4], UP-Growth [14], HUI-Miner [9], and FHM [8], etc.

However, these algorithms are designed under a assumption that all items having positive unit profits in a database, they cannot be applied to handle items having negative unit profits, despite that such items occur in many real-life transaction databases. For example, it is common that retail stores or supermarket sell items at a loss (e.g., printers) to stimulate the sale of other related items (e.g., proprietary printer cartridges). Although giving away a unit of some items results in a loss for supermarkets, they could provide opportunities for cross-selling and could possibly earn more money from the promotion. It was demonstrated that if classical HUIM algorithms are applied on databases containing items with negative unit profits, they can generate an incomplete set of HUIs [7]. The HUINIV-Mine [7] and FHN [12] were developed to handle the problem of HUIM with negative unit profits. In real-life applications, uncertainty is common seen when data is collected from noisy data sources such as RFID, GPS, wireless sensors, and WiFi systems [2,5]. Some algorithms of FIM have been developed to discover useful information in uncertain databases. Since utility and uncertainty are two different measures for an object (e.g., an useful pattern). The utility is a semantic measure (how "utility" of a pattern is based on the user's priori knowledge and goals), while uncertainty is an objective measure (the probability of a pattern is an objective existence). Up to now, most algorithms of HUIM have been extensively developed to handle precise data, but they are not suitable to handle the data with uncertainty. It may be useless or misleading if the discovered results of HUIs with low existential probability [11].

In light of these, in this paper, we attempt to design en efficient algorithm to discover high-utility itemsets from uncertain transaction databases by considering both positive and negative unit profits. This algorithm is named **HUPNU** (mining **H**igh-**U**tility itemsets with both **P**ositive and **N**egative unit profits from **U**ncertain databases) to mine HUIs. To the best of our knowledge, it is the first work to address this problem. The contributions of this paper are described below. (1) A vertical list structure, called PU$^{\pm}$-list (Probability-Utility list with Positive-and-Negative profits), is designed to store all the necessary information for the database. (2) A one-phase efficient algorithm named HUPNU is proposed to mine HUIs without multiple time-consuming database scan. It relies on a series of PU$^{\pm}$-lists to directly mine HUIs without generating and testing candidates. (3) Several efficient pruning strategies are further proposed to reduce the search space, a number of unpromising itemsets can be early pruned when constructing the PU$^{\pm}$-list. (4) An extensive experimental study carried on several real-life datasets shows that the complete set of HUIs can be efficiently discovered by the proposed HUPNU algorithm.

2   Preliminaries and Problem Definition

Definition 1. Let I be a set of items (symbols). An *uncertain transaction database* is a set of uncertain transactions $D = \{T_1, T_2, \ldots, T_n\}$ such that for each transaction $T_c \in I$, and T_c has a unique identifier c called its *tid*. As the attribute uncertainty model [2,5], each item i has a unique probability of existence $p(i, T_c)$. Each item $i \in I$ is associated with a positive or negative value $pr(i)$, called its external utility (e.g., unit profit). For each T_c such that $i \in T_c$, a positive number $q(i, T_c)$ is called the internal utility of i (e.g., purchase quantity). Each item i_m in D has a unique profit $pr(i_m)$, they are provided in a profit table and denoted as $ptable = \{pr(i_1), pr(i_2), \ldots, pr(i_m)\}$.

Table 1. An example uncertain quantitative database.

tid	Transaction (item: quantity, probability)	TU	RTU
T_1	(b:3, 0.85); (c:1, 1.0); (d:2, 0.70)	14	24
T_2	(a:1, 1.0); (b:1, 0.60); (c:3, 0.75); (e:1, 0.40)	19	19
T_3	(a:1, 0.55); (b:2, 0.60); (c:4, 1.0); (d:1, 0.90); (e:5, 0.40)	34	39
T_4	(b:3, 0.90); (d:1, 0.45)	16	21
T_5	(a:4, 1.0); (c:3, 0.85); (d:2, 0.70); (e:2, 0.45)	23	33

Example 1. Consider the running example w.r.t. Table 1, it contains five transactions $(T_1, T_2, \ldots, T_5)$. Transaction T_1 indicates that items $(b)1$, (c), and (d) appear in T_1 with purchase quantity as 3, 1, and 2, respectively. And assume that the unit profit of (a) to (e) are respectively defined as: $\{pr(a):6, pr(b):7, pr(c):1, pr(d):-5, pr(e):3\}$. Thus, item (d) is sold at loss.

Definition 2. The utility of an item i in a transaction T_c is denoted as $u(i, T_c)$ and defined as $pr(i) \times q(i, T_c)$. The utility of an itemset X (a group of items $X \subseteq I$) in T_c is denoted as $u(X, T_c)$ and defined as $u(X, T_c) = \sum_{i \in X} u(i, T_c)$. The utility of an itemset X in a database D is denoted as $u(X)$, it can be calculated as $u(X) = \sum_{X \subseteq T_c \wedge T_c \in D} u(X, T_c)$.

Example 2. The utility of item (e) in T_2 is $u(e, T_2) = 3 \times 1 = 3$. The utility of the itemset $\{a, e\}$ in T_2 is $u(\{a, e\}, T_2) = u(a, T_2) + u(e, T_2) = 6 \times 1 + 3 \times 1 = 9$. The utility of the itemset $\{a, e\}$ is $u(\{a, e\}) = (u(a, T_2) + u(e, T_2)) + (u(a, T_3) + u(e, T_3)) + (u(a, T_5) + u(e, T_5)) = (6+3) + (6+15) + (24+6) = 60$. The utility of the itemset $\{a, d, e\}$ is $u(\{a, d, e\}) = (u(a, T_3) + u(d, T_3)) + u(e, T_3)) + (u(a, T_5) + u(d, T_5) + u(e, T_5)) = (6 + (-5) + 15) + (24 + (-10) + 6) = 36$.

Definition 3. The probability of an itemset X (a group of items $X \subseteq I$) in T_c is denoted as $p(X, T_c)$ and defined as $p(X, T_c) = \prod_{i \in X} p(i, T_c)$. The probability of X in D is denoted as $Pro(X)$ and defined as $Pro(X) = \sum_{T_c \in D} (\prod_{i \in X} p(i, T_c))$.

Example 3. The probability of item (e) in T_2 is $p(e, T_2) = 0.40$. The probability of the itemset $\{a, e\}$ in T_2 is $p(\{a, e\}, T_2) = p(a, T_2) \times p(e, T_2) = 1.0 \times 0.40 = 0.40$. The probability of item (e) in D is $Pro(e) = 1.25$. The probability of the itemset $\{a, d, e\}$ in D is $p(\{a, d, e\}) = p(ade, T_3) + p(ade, T_5) = 0.198 + 0.315 = 0.513$.

Definition 4. An itemset X in an uncertain database D is said to be a potential high-utility itemset (PHUI) if it satisfies the following two conditions: (1) $u(X) \geq minUtil$, and (2) $Pro(X) \geq minPro \times |D|$. A PHUI is thus an itemset having both a high expected/potential probability and a high utility value.

The problem of mining high-utility itemsets with both positive and negative unit profits from uncertain databases is to discover all potential high-utility itemsets (having a high expected/existential probability and a high utility) in an uncertain database where external utility values may be positive or negative.

Example 4. If the user-specified $minPro = 0.20$ and $minUtil = 20$, ten PHUIs should be found in the running example database. They are ($\{a\}$:36, 2.55; $\{b\}$:63, 2.95; $\{e\}$:24, 1.25; $\{a, c\}$:46, 2.15; $\{a, e\}$:60, 1.07; $\{b, c\}$:52, 1.90; $\{b, d\}$:36, 1.54; $\{c, e\}$:34, 1.0825; $\{a, c, d\}$:22, 1.09; $\{b, c, d\}$:27, 1.135). $\{\{a\}$: 36, 2.55$\}$ means that the utility of $\{a\}$ is 36, and its expected probability is 2.55.

3 Proposed HUPNU Algorithm

3.1 Properties of Positive and Negative Unit Profits

According to the previous studies, the utility measure is not monotonic or anti-monotonic [9,10,14]. In other words, an itemset may have a utility lower, equal or higher than those of any of its subsets. To handle the problem for mining HUIs with both positive and negative unit profits, the HUINIV-Mine [7] and FHN [12] algorithms were developed by redefining the notion of transaction utility (TU) and the TWU measure [10] as follows.

Definition 5. The $TU(T_c) = \sum_{i \in T_c} u(i, T_c)$, but the *redefined transaction utility* (*RTU*) of T_c is defined as $RTU(T_c) = \sum_{i \in T_c \wedge pr(i)>0} u(i, T_c)$. The *redefined transaction-weighted utilization* (*RTWU*) of X is defined $RTWU(X) = \sum_{X \subseteq T_c \wedge T_c \in D} RTU(T_c)$. Thus, $RTWU(X) \geq u(X)$.

Example 5. Table 1 shows the TU and RTU of five transactions. Consider itemsets $\{a, e\}$ and $\{a, d, e\}$, the $RTWU(\{a, e\}) = 91$ and $RTWU(\{a, d, e\}) = 71$, which are over-estimations of $u(\{a, e\}) = 60$ and $u(\{a, d, e\}) = 36$.

Let $pu(X)$ and $nu(X)$ respectively denotes the sum of positive utilities and negative utilities of items in X in a transaction (or in a database). Since $u(X) = pu(X) + nu(X)$, the relationship $nu(X) \leq u(X) \leq pu(X)$ holds [12]. Thus, both $u(X)$ and $nu(X)$ cannot be used to overestimate the utility of an itemset. Although $pu(X)$ for an itemset is an upper-bound on utility, it still does not hold the downward closure of extensions with positive or negative items.

3.2   Probability-Utility List with Positive-and-Negative Profits

Definition 6. In the designed HUPNU algorithm, we define the total processing order $\succ$ such that (1) items are sorted in *RTWU*-ascending order, and (2) negative items always succeed all positive items.

Definition 7. The $PU^{\pm}$-list of an itemset X in an uncertain database D is denoted as $X.PUL$. It consisted of a set of tuples, $<tid, pro, pu, nu, rpu>$ for each transaction T_{tid} containing X. For each tuple, (1) The *tid* element is the transaction identifier; (2) The *pro* element is the existential probability of X in T_{tid}, i.e., $pro(X, T_{tid}) \geq 0$; (3) The *pu* element is the positive utility of X in T_{tid}, i.e., $u(X, T_{tid}) \geq 0$; (4) The *nu* element is the negative utility of X in T_{tid}, i.e., $u(X, T_{tid}) < 0$; (5) The *rpu* element is defined as $\sum_{i \in T_{tid} \wedge i \succ x \forall x \in X} u(i, T_{tid}) \geq 0$, such that only positive utility values of the remaining items.

Example 6. The search space of HUPNU can be represented as a $PU^{\pm}$-list based Set-enumeration tree [13], we named it as $PU^{\pm}$-tree. Since $\{RTWU(a)$: 91; $RTWU(b)$: 103; $RTWU(c)$: 115; $RTWU(d)$: 117; $RTWU(e)$: 91;$\}$, the designed processing order $\succ$ in $PU^{\pm}$-list is $\{a \succ e \succ b \succ c \succ d\}$, we have $\{a\}.PUL = \{(T_2, 1.0, 6, 0, 13), (T_3, 0.55, 6, 0, 33), (T_5, 1.0, 24, 0, 9)\}$; $\{d\}.PUL = \{(T_2, 0.70, 0, -10, 0), \quad (T_3, 0.090, 0, -5, 0), \quad (T_4, 0.45, 0, -5, 0), \quad (T_5, 0.70, 0, -10, 0)\}$; $\{a, d\}.PUL = \{T_3, 0.495, 6, -5, 0), (T_5, 0.70, 24, -10, 0)\}$.

Definition 8. Let *SUM(X.iu)*, *SUM(X.pu)*, *SUM(X.nu)*, and *SUM(X.rpu)* are respectively the sum of the utilities, the sum of *pu* values, the sum of *nu* values and the sum of *rpu* in the $PU^{\pm}$-list of X, that are: $SUM(X.pu) = \sum_{X \in T_c \wedge T_c \subseteq D} X.pu(T_c)$; $SUM(X.nu) = \sum_{X \in T_c \wedge T_c \subseteq D} X.nu(T_c)$; $SUM(X.rpu) = \sum_{X \in T_c \wedge T_c \subseteq D} X.rpu(T_c)$; $SUM(X.iu) = SUM(X.pu) + SUM(X.nu)$.

Lemma 1. Based on the $PU^{\pm}$-list, given two itemsets X and Y in a subtree in the $PU^{\pm}$-tree, if (1) $SUM(X.pu) + SUM(X.rpu) - \sum_{\forall T_c \in D, X \subseteq T_c \wedge Y \not\subseteq T_c} (X.pu + X.rpu) < minUtil$, or (2) $SUM(X.pro) - \sum_{\forall T_c \in D, X \subseteq T_c \wedge Y \not\subseteq T_c} (X.pro) < minPro \times |D|$, then neither $\{X, Y\}$ nor any of X it extensions will be a PHUI.

Strategy 1 (PU-Prune strategy). *Let X be a node of the $PU^{\pm}$-tree, and Y be the right sibling node of X. If $SUM(X.pu) + SUM(X.rpu) - \sum_{\forall T_c \in D, X \subseteq T_c \wedge Y \not\subseteq T_c} (X.pu + X.rpu) < minUtil$, or $SUM(X.pro) - \sum_{\forall T_c \in D, X \subseteq T_c \wedge Y \not\subseteq T_c} (X.pro) < minPro \times |D|$, then $\{X, Y\}$ and any of X its child nodes is not a PHUI. The construction of the $PU^{\pm}$-lists of X its children is unnecessary to be performed.*

Based on the PU-Prune strategy, a huge number of unpromising k-itemset ($k \geq 2$) can be pruned. The $PU^{\pm}$-list construct procedure with PU-Prune strategy is given in Algorithm 1. Thus, $PU^{\pm}$-list for k-itemsets ($k > 1$) can be easily constructed from $PU^{\pm}$-lists of $(k\text{-}1)$-itemsets without scanning the database.

Input: P: a pattern, Px: the extension of P with an item x, Py: the extension of P with an item y

output: The $PU^{\pm}$-list of Pxy

1 $Pxy.PUL \leftarrow \emptyset$;
2 set $Probability = SUM(X.pro)$, $Utility = SUM(X.pu) + SUM(X.rpu)$;
3 **foreach** *tuple* $ex \in Px.PUL$ **do**
4 $\quad$ **if** $\exists ey \in Py.PUL$ *and* $ex.tid = exy.tid$ **then**
5 $\quad\quad$ **if** $P.PUL \neq \emptyset$ **then**
6 $\quad\quad\quad$ Search element $e \in P.PUL$ such that $e.tid = ex.tid.$;
7 $\quad\quad\quad$ $exy \leftarrow < ex.tid, ex.pro \times ey.pro/e.pro, ex.pu + ey.pu - e.pu, ex.nu + ey.nu - e.nu, ey.rpu >$;
8 $\quad\quad$ **else**
9 $\quad\quad\quad$ $exy \leftarrow < ex.tid, ex.pro \times ey.pro, ex.pu + ey.pu, ex.nu + ey.nu, ey.rpu >$;
10 $\quad\quad$ $Pxy.PUL \leftarrow Pxy.PUL \cup \{exy\}$;
11 $\quad$ **else**
12 $\quad\quad$ $Probability = Probability - ex.pro$, $Utility = Utility - ex.pu - ex.rpu$;
13 $\quad\quad$ **if** $Probability < minPro \times |D| || Utility < minUtil$ **then**
14 $\quad\quad\quad$ **return** null;
15 **return** $Pxy.PUL$

Algorithm 1. The $PU^{\pm}$-list construct procedure with PU-Prune

---

3.3   Proposed Pruning Strategies

Based on the $PU^{\pm}$-list and the properties of probability and utility, several pruning strategies are designed in HUPNU to early prune unpromising itemsets. Assume a $(k\text{-}1)$-itemset w.r.t. a node in the Set-enumeration $PU^{\pm}$-tree be $X^{k-1}(k \geq 2)$, and any of its child nodes be denoted as X^k.

Theorem 1 (downward closure property of *RTWU* and *probability*). *In the $PU^\pm$-tree, the $Pro(X^{k-1}) \geq Pro(X^k)$ and $RTWU(X^{k-1}) \geq RTWU(X^k)$.*

Proof. Since $p(X, T_c) = \prod_{i \in X} p(i, T_c)$ for any T_c in D, it can be found that: $p(X^k, T_c) \leq p(X^{k-1}, T_c)$. X^{k-1} is subset of X^k, the *tids* of X^k is the subset of the *tids* of X^{k-1}, thus, $Pro(X^k) = \sum\limits_{X^k \subseteq T_c \land T_c \in D} p(X^k, T_c) \leq$

$\sum\limits_{X^{k-1} \subseteq T_c \land T_c \in D} p(X^{k-1}, T_c) = Pro(X^{k-1})$. It can be found that $Pro(X^{k-1})$

$\geq Pro(X^k)$. Besides, $X^{k-1} \subseteq X^k$, $RTWU(X^k) = \sum\limits_{X^k \subseteq T_c \land T_c \in D} tu(T_c) \leq$

$\sum\limits_{X^{k-1} \subseteq T_c \land T_c \in D} tu(T_c) = RTWU(X^{k-1})$.

Lemma 2 (probability upper-bound of PHUI). *The sum of all the probabilities of any node in the $PU^\pm$-tree is no less than the sum of the probabilities of any of its child nodes.*

Strategy 2. *After the first database scan, we can obtain the RTWU and probability value of each 1-item. If the RTWU of a 1-item and the sum of the probabilities of an item do not satisfy the two conditions of PHUI, this item can be directly pruned, and none of its supersets is a desired PHUI.*

Strategy 3. *When traversing the $PU^\pm$-tree based on a depth-first search strategy, if the sum of all the probabilities of a tree node X w.r.t. $Pro(X)$ in its constructed $PU^\pm$-list is less than $minPro \times |D|$, then none of the child nodes of this node is a desired PHUI.*

Lemma 3 (utility upper-bound of PHUI). *For any node X in the search space w.r.t. the $PU^\pm$-tree, the sum of $SUM(X.pu)$ and $SUM(X.rpu)$ in the $PU^\pm$-list of X is larger than or equal to utility of any one of its children.*

Thus, the sum of utilities of X^k in D w.r.t $u(X^k)$ is always less than or equals to the sum of $SUM(X^{k-1}.pu)$ and $SUM(X^{k-1}.rpu)$, it ensures that the downward closure of transitive extensions with positive or negative items. Based on these upper-bounds, we can use the following pruning conditions.

Strategy 4. *When traversing the $PU^\pm$-tree based on a depth-first search strategy, if the sum of $SUM(X^{k-1}.pu)$ and $SUM(X^{k-1}.rpu)$ of any node X is less than minUtil, any of its child node is not a PHUI, they can be regarded as irrelevant and be pruned directly.*

Strategy 5. *After constructing the $PU^\pm$-list of an itemset, if X.PUL is empty or the $Pro(X)$ value is less than $minPro \times |D|$, X is not a PHUI, and none of X its child nodes is a PHUI. The construction of the $PU^\pm$-lists for the child nodes of X is unnecessary to be performed.*

We further extend the Estimated Utility Co-occurrence Pruning (EUCP) strategy [8] in the HUPNU algorithm, a structure named Estimated Utility Co-occurrence Structure ($EUCS$) is built. $EUCS$ is a matrix that stores the $RTWU$ values of the 2-itemsets, details can be referred to [8].

Strategy 6. *Let X be an itemset (node) encountered during the depth-first search of the Set-enumeration $PU^{\pm}$-tree. If the $RTWU$ of a 2-itemset $Y \subseteq X$ according to the constructed EUCS is less than the minimum utility threshold, X would not be a PHUI; none of its child nodes is a PHUI. The construction of the $PU^{\pm}$-lists of X and its children is unnecessary to be performed.*

3.4   Main Procedure of HUPNU

As shown in Algorithm 2, the main procedure of the proposed HUPNU algorithm first scans the uncertain database to calculate the $RTWU$ (with the redefined RTU) and $Pro(i)$ of each item (Line 1). Then, it finds the set I^* of all items that not only having a existence probability no less than $minPro \times |D|$, but also having a $RTWU$ no less than $minUtil$, other items are ignored since they cannot be part of a potential HUI (Line 2). A second database scan is then performed (Line 4) after sorting the set of I^* in the designed order as $\succ$ (Line 3). During this database scan, items in transactions are reordered according to the total order $\succ$, the $PU^{\pm}$-list of each 1-item $i \in I^*$ is built and the structure named EUCS is built simultaneously. After that, the depth-first search exploration starts by calling the recursive procedure $Search$ with the empty itemset $\emptyset$, the set of single items I^*, $minPro$, $minUtil$ and the EUCS (Line 5).

Input: D: an uncertain transaction database; $minPro$, a minimum potential probability threshold; $minUtil$: a minimum utility threshold; $ptable$: a profit-table

output: The set of potential high-utility itemsets ($PHUIs$)

1 Scan D to calculate the $RTWU$ and $Pro(i)$ of single item;
2 $I^* \leftarrow$ each item i such that $Pro(i) \geq minPro \times |D| \wedge RTWU(i) \geq minUtil$;
3 Sort the set of I^* in the designed order as $\succ$;
4 Scan D again to built the $PU^{\pm}$-list for each item $i \in I^*$ and built the $EUCS$ structure;
5 call **Search** ($\emptyset$, I^*, $minPro$, $minUtil$, $EUCS$);
6 **return** $PHUIs$

Algorithm 2. The HUPNU algorithm

As shown in Algorithm 3, the search procedure operates as follows. For each extension Px of P, if the probability of Px is no less than $minPro \times |D|$, and the sum of the actual utilities values of Px in the $PU^{\pm}$-list (denoted as $SUM(X.pu) + SUM(X.nu)$) is no less than $minUtil$, then Px is a PHUI and be output (Lines 2 to 3). Then, it uses the pruning strategies 3 and 4 to determine whether the extensions of Px would be the PHUIs and should be explored (Line 4). This is performed by merging Px with all extensions Py of P such that $y \succ x$ and $RTWU(\{x,y\}) \geq minUtil$ (Line 7, pruning strategy 6), to form extensions of the form Pxy containing $|Px|+1$ items. The PU$^{\pm}$-list of Pxy is then constructed by calling the $Construct$ procedure to join the PU$^{\pm}$-lists of P, Px and Py (Lines 8 to 11). Only the promising PU$^{\pm}$-lists would be explored in next extension (Line 11, pruning strategy 5). Then, a recursive call to the $Search$ procedure with Pxy is done to calculate its utility and explore its extension(s) (Line 12).

Input: P: an itemset, $ExtensionsOfP$: a set of extensions of P, the $minPro$
threshold, the $minUtil$ threshold, the $EUCS$ structure
Output: The set of potential high-utility itemsets ($PHUIs$)

1 **foreach** *itemset* $Px \in ExtensionsOfP$ **do**
2   **if**
   $SUM(Px.pro) \geq minPro \times |D| \wedge SUM(Px.pu) + SUM(Px.nu) \geq minUtil$
   then
3   ⌊ output Px as a $PHUI$;
4   **if**
   $SUM(Px.pro) \geq minPro \times |D| \wedge SUM(Px.pu) + SUM(Px.rpu) \geq minUtil$
   then
5       $ExtensionsOfPx \leftarrow \emptyset$;
6   **foreach** *itemset* $Py \in ExtensionsOfP$ such that $y \succ x$ **do**
7   **if** $RTWU(\{x,y\}) \geq minUtil$ **then**
8             $Pxy \leftarrow Px \cup Py$;
9             $Pxy.PUL \leftarrow$ **Construct** (P, Px, Py);
10   **if** $Pxy.PUL \neq \emptyset \wedge SUM(Pxy.pro) \geq minPro \times |D|$ **then**
11   ⌊ $ExtensionsOfPx \leftarrow ExtensionsOfPx \cup Pxy$;
12   ⌊ call **Search** $(Px, ExtensionsOfPx, minPro, minUtil)$;
13 **return** $PHUIs$

Algorithm 3. The $Search$ procedure

4   Experimental Study

In this section, we evaluated the performance of the proposed HUPNU algorithm. Experiments were implemented in Java and performed on a computer with a third generation 64 bit Core i5 processor running Windows 7 operating system and 4 GB of free RAM. In the literature, note that there is none study which is related to the task of mining HUIs from uncertain database with both positive and negative profits. We compared the performance of HUPNU with the proposed several pruning strategies. Note that the HUPNU$_{P1}$ adopts the pruning strategies 2, 3 an 4, the HUPNU$_{P2}$ adopts the pruning strategies 1, 2, 3 an 4, the HUPNU$_{P123}$ adopts pruning strategies 1, 2, 3, 4 and 5, while the HUPNU$_{P1234}$ adopts all pruning strategies including EUCP strategy.

All memory measurements were done using the Java API. Experiments were carried on four real-life datasets, kosarak, accidents, psumb and mushroom which having varied characteristics. The #*transactions*, #*distinctitems*, *avg.length* and *max.length* of these four datasets are respectively as: 990002, 41270, 8.09, 2498; 340183, 468, 33.8, 51; 49046, 2113, 74, 74; 8124, 119, 23, 23. For all datasets, external utilities for items are generated between -1,000 and 1,000 by using a log-normal distribution and quantities of items are generated randomly between 1 and 5, similarly to the settings of [8,12,14]. In addition, due to the attribute uncertainty property, a unique probability value in the range of (0.0, 1.0] was assigned to each item in every transaction in these datasets.

4.1 Runtime Performance

The comparison of execution times with various *minUtil* threshold and various *minPro* are shown in Fig. 1 for all datasets. From Fig. 1, it can be observed that the runtime of all the algorithms is decreased along with the increasing of *minUtil* with a fixed *minPro*, or with the increasing of *minPro* with a fixed *minUtil*. In particular, the proposed improved algorithms are generally up to almost one or two orders of magnitude faster than the baseline one on all datasets. Among the four version algorithms, HUPNU$_{P1234}$ which adopts all pruning strategies has the best performance. It is reasonable since HUPNU$_{P1234}$ uses six pruning strategies to early prune unpromising itemsets and search space, which can avoid the costly join operations of a huge number of PU$^{\pm}$-lists for mining PHUIs. When the *minUtil* or *minPro* is set quite low, longer desired patterns are discovered, and thus more computations w.r.t. runtime are needed to process, especially in a dense dataset. Based on the PU$^{\pm}$-list, the four HUPNU algorithms directly determine the PHUIs from the Set-enumeration tree without candidate generation, it can effectively avoid the time-consuming dataset scan. Moreover, the six pruning strategies help to prune a huge of unpromising

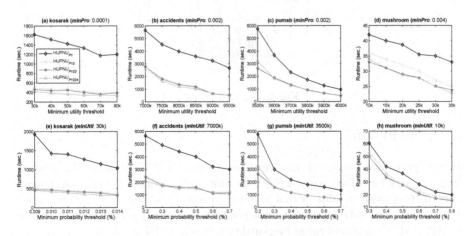

Fig. 1. Comparisons of runtime.

itemsets and to greatly reduce the computations than the baseline one. Moreover, the less memory usage is required, but we omit the detailed memory usage results due to space constraint. We can see this trend more clearly when the *minUtil* or *minPro* is set quite low. Thus, they can lead to a more compact search space and obtain the effectiveness and efficiency for mining PHUIs.

4.2   Scalability Analysis with Memory Usage and Patterns

As shown in Fig. 2, the scalability of the four algorithms is compared in the real-life dataset BMS-POS with different scales, which is set $minPro = 0.0001$, $minUtil = 10k$, and data size is set varying from 100k to 500k. It can be observed that the runtime of all compared algorithms is linear increased along with the increasing of dataset size. The runtime of $HUPNU_{P123}$ is close to that of $HUPNU_{P12}$, but significantly faster than that of $HUPNU_{P1}$. Specially, $HUPNU_{P1234}$ performs the best, and the gap of runtime among them grows wider with the increasing of dataset size. With the increasing of dataset size, the runtime of algorithms are linearly increasing as well. Figure 2(b) shows the memory usages of four algorithms which indicates the linearity in term of dataset size. In addition, $HUPNU_{P1}$ requires the most memory usage, $HUPNU_{P123}$ and $HUPNU_{P1234}$ have the similar performance on memory usage, they consume the least memory. To show the effect of the developed pruning strategies, the number of potential nodes (visited nodes in the $PU^{\pm}$-tree, denoted as N_1, N_2, N_3, and N_4) and the final derived PHUIs are further evaluated as shown in Fig. 2(c). It can be observed that $N_1 > N_2 > N_3 > N_4$, the larger dataset size is, the bigger gap among them is.

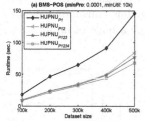 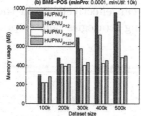 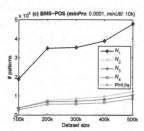

Fig. 2. Scalability test.

5   Conclusion

In this paper, we proposed an algorithm named HUPNU (mining High-utility itemsets with both Positive and Negative unit profits from Uncertain databases), it is the first work to address this problem. A novel vertical list structure, called $PU^{\pm}$-list (probability-utility list with positive-and-negative profits), is designed for HUPNU to mine potential high-utility itemsets (PHUIs) without generating candidates. Several efficient pruning strategies are further developed to reduce the search space and speed up computation. Experiments carried on several real-life datasets shows that the complete set of PHUIs can be efficiently discovered by the proposed HUPNU algorithm. HUPNU is quite efficient in terms of runtime and scalability, and the designed pruning strategies are acceptable.

Acknowledgement. This research was partially supported by the National Natural Science Foundation of China (NSFC) under grant No. 61503092 and by the Tencent Project under grant CCF-Tencent IAGR20160115.

References

1. Frequent Itemset Mining Dataset Repository. http://fimi.ua.ac.be/data/
2. Aggarwal, C.C., Yu, P.S.: A survey of uncertain data algorithms and applications. IEEE Trans. Knowl. Data Eng. **21**(5), 609–623 (2009)
3. Agrawal, R., Srikant, R.: Fast algorithms for mining association rules in large data-bases. In: Proceedings of the International Conference on Very Large Databases, pp. 487–499 (1994)
4. Ahmed, C.F., Tanbeer, S.K., Jeong, B.S., Lee, Y.K.: Efficient tree structures for high utility pattern mining in incremental databases. IEEE Trans. Knowl. Data Eng. **21**(12), 1708–1721 (2009)
5. Bernecker, T., Kriegel, H.P., Renz, M., Verhein, F., Zuefl, A.: Probabilistic frequent itemset mining in uncertain databases. In: ACM SIGKDD International Conference on Knowledge Discovery and Data Mining, pp. 119–128 (2009)
6. Chan, R., Yang, Q., Shen, Y.: Mining high utility itemsets. In: IEEE International Conference on Data Mining, pp. 19–26 (2003)
7. Chu, C.J., Tseng, V.S., Liang, T.: An efficient algorithm for mining high utility itemsets with negative item values in large databases. Appl. Math. Comput. **215**, 767–778 (2009)
8. Fournier-Viger, P., Wu, C.-W., Zida, S., Tseng, V.S.: FHM: faster high-utility itemset mining using estimated utility co-occurrence pruning. In: Andreasen, T., Christiansen, H., Cubero, J.-C., Raś, Z.W. (eds.) ISMIS 2014. LNCS (LNAI), vol. 8502, pp. 83–92. Springer, Cham (2014). doi:10.1007/978-3-319-08326-1_9
9. Liu, M., Qu, J.: Mining high utility itemsets without candidate generation. In: 21st ACM International Conference on Information and Knowledge Management, pp. 55–64 (2012)
10. Liu, Y., Liao, W., Choudhary, A.: A two-phase algorithm for fast discovery of high utility itemsets. In: Ho, T.B., Cheung, D., Liu, H. (eds.) PAKDD 2005. LNCS (LNAI), vol. 3518, pp. 689–695. Springer, Heidelberg (2005). doi:10.1007/11430919_79
11. Lin, J.C.W., Gan, W., Fournier-Viger, P., Hong, T.P., Tseng, V.S.: Mining potential high-utility itemsets over uncertain databases. In: ACM ASE BigData & Social Informatics, p. 25 (2015)
12. Fournier-Viger, P.: FHN: efficient mining of high-utility itemsets with negative unit profits. In: Luo, X., Yu, J.X., Li, Z. (eds.) ADMA 2014. LNCS (LNAI), vol. 8933, pp. 16–29. Springer, Cham (2014). doi:10.1007/978-3-319-14717-8_2
13. Rymon, R.: Search through systematic set enumeration. Technical reports (CIS), pp. 539–550 (1992)

14. Tseng, V.S., Shie, B.E., Wu, C.W., Yu, P.S.: Efficient algorithms for mining high utility itemsets from transactional databases. IEEE Trans. Knowl. Data Eng. **25**(8), 1772–1786 (2013)
15. Yao, H., Hamilton, H.J., d Butz C.J.: A foundational approach to mining itemset utilities from databases. In: SIAM International Conference on Data Mining, pp. 211–225 (2004)

Sparse Stochastic Inference with Regularization

Tung Doan[1] and Khoat Than[2(✉)]

[1] SOKENDAI The Graduate University for Advanced Studies,
Hayama, Kanagawa, Japan
tungdp@nii.ac.jp
[2] Hanoi University of Science and Technology,
No.1 Dai Co Viet Road, Hanoi, Vietnam
khoattq@soict.hust.edu.vn

Abstract. The massive amount of digital text information and delivering them in streaming manner pose challenges for traditional inference algorithms. Recently, advances in stochastic inference algorithms have made it feasible to learn topic models from very large-scale collections of documents. In this paper, we however point out that many existing approaches are prone to overfitting for extremely large/infinite datasets. The possibility of overfitting is particularly high in streaming environments. This finding suggests to use regularization for stochastic inference. We then propose a novel stochastic algorithm for learning latent Dirichlet allocation that uses regularization when updating global parameters and utilizes sparse Gibb sampling to do local inference. We study the performance of our algorithm on two massive data sets and demonstrate that it surpasses the existing algorithms in various aspects.

Keywords: Stochastic inference · Topic models · Large-scale/Stream data · Regularization

1 Introduction

Latent Dirichlet allocation (LDA) was initially presented as a graphical model for discovering topics in document collections [2]. It has then found many successful applications in wide range of fields, including bioinformatics [13,16], psychology [17], politics [7,9], to name a few.

One of the core issues in LDA is the inference of the posterior distribution of the latent variables. Unfortunately, the posterior is intractable and researchers have to approximate posterior inference. Many "batch" posterior inference algorithms have been proposed, including variational Bayes (VB) [2], collapsed variational Bayesian inference (CVB) [18], CVB0 [1], and collapsed Gibbs sampling (CGS) [8]. However, those "batch" algorithms are not practical for large scale data analysis because they often requires many sweeps through all documents in the corpus.

Recently, researchers have introduced stochastic algorithms, including stochastic variational inference (SVI) [11], moving average stochastic variational

© Springer International Publishing AG 2017
J. Kim et al. (Eds.): PAKDD 2017, Part I, LNAI 10234, pp. 447–459, 2017.
DOI: 10.1007/978-3-319-57454-7_35

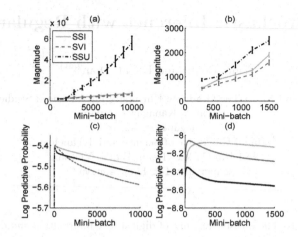

Fig. 1. Growth of global parameters ($\boldsymbol{\lambda}^t$) and overfitting possibility as mini-batch t grows (i.e., as seeing more documents). **Mis-specification** ((a) and (c)): when LDA with too many topics is learned. **Noisy condition** ((b) and (d)): when data are highly noisy. *Magnitude* shows the average magnitude of $\boldsymbol{\lambda}^t$; and *Log Predictive Probability* shows predictiveness and generalization of the learned models on unseen data. Higher predictiveness is better.

inference (MASVI) [12], sufficient statistic update (SSU) [4], sparse stochastic inference (SSI) [14], stochastic collapsed variational Bayesian inference (SCVB0) [6], and stochastic gradient Riemannian Langevin dynamics (SGRLD) [15]. These algorithms repeatedly subsample a small set of documents from the collection and then update the global parameters by stochastic gradient approximated from the subsample. Such a scheme allows them to update global parameters more frequently. In addition, they do not require multiple passes through an entire document collection and storing local variables for the full corpus. Hence, the stochastic algorithms have faster convergence rate and lower memory requirement, enabling us to deal with very large scale data. Those properties make them more preferable than traditional "batch" algorithms.

In this paper, we point out that existing stochastic methods for learning topic models are prone to overfitting, especially in streaming environments. The unconstrained nature of global parameters in existing methods is the main reason. As an example, SSU [4] updates global parameters ($\boldsymbol{\lambda}^t$) at time t from the old ($\boldsymbol{\lambda}^{t-1}$) and current statistics ($\hat{\boldsymbol{\lambda}}^t$) by $\boldsymbol{\lambda}^t \leftarrow \boldsymbol{\lambda}^{t-1} + \hat{\boldsymbol{\lambda}}^t$. Such an update scheme allows the global parameters $\boldsymbol{\lambda}^t$ grow arbitrarily large in either of the following three conditions:

- $t \rightarrow \infty$ (or the number of documents go to infinity in a text stream),
- Data is *highly noisy* such as those from Twitter [5, 10, 20], causing $\hat{\boldsymbol{\lambda}}^t$ to be uncontrolled,
- A *is-specified model* (e.g., with too many topics) is to be learned. This is common in practice of topic modeling. Mis-specification might add too much artificial information to $\hat{\boldsymbol{\lambda}}^t$.

Therefore, overfitting might easily occur. Figure 1 shows the growth of λ^t and overfitting as learning from more documents. Note that the possibility of overfitting happens not only with SSU, but also with many other methods. This finding urges us to use some regularization.

Our second contribution is a novel stochastic algorithm, namely *Regularized Sparse Stochastic Inference* (RSSI), for learning LDA from large/streaming text collections. RSSI combines online learning [3] with sparse Gibbs sampling [14] to do posterior inference for individual texts. The global parameter (topics) is regularized to belong to a simplex, which helps RSSI avoid overfitting. RSSI can deal well with real streaming environments because it does not require specifying data size in advance. In addition, taking advantages of sparse Gibbs sampling allows RSSI to perform more efficiently than most existing algorithms. Extensive experiments show the advantages of RSSI in both efficiency and predictiveness. We believe that our methodology for developing RSSI can be easily extended to a wide class of topic models.

The rest of the paper is organized as follows. In Sect. 2, we discuss possibility of overfitting in existing stochastic algorithms. We derive *regularized sparse stochastic inference* (RSSI) for LDA in Sect. 3. Empirical results are given in Sect. 4.

2   LDA and Possibility of Overfitting

Latent Dirichlet allocation (LDA) assumes that all documents in corpus $\mathcal{C}$ shares a fixed number of topics $\beta = (\beta_1, ..., \beta_K)$, each of which is a distribution over V-dimensional vocabulary. Each document is assumed to be generated from the following generative process:

For the ith word w_i in document d:

- draw topic indicator $z_i|\theta_d \sim Multinomial(\theta_d)$
- draw word $w_i|z_i, \beta \sim Multinomial(\beta_{z_i})$

where $\theta_d = (\theta_{d1}, ..., \theta_{dK})$ is topic proportion which is assumed to be drawn from $Dirichlet(\alpha)$, representing contribution of the topics to document d. One can consider β as parameters or further assumes that $\beta_k \sim Dirichlet(\eta)$.

Estimating the posterior of the latent variables given a corpus is intractable. SVI [11], MASVI [12], and SSU [4] approximate this posterior by optimizing a fully factorized variational distribution where dependence between latent variables is relaxed. SCVB0 [6] instead uses a variational distribution which still remains the dependency between z and (θ, β). SRGLD [15] is a Markov Chain Monte Carlo scheme that asymptotically produces samples from the posterior distribution. Unlike SVI, MASVI, SSU, and SCVB0 which approximate full joint posterior distribution of the latent variables, SSI [14] try to approximate posterior distribution over z and β given the corpus $P(z, \beta|\mathcal{C})$ where the variables θ are marginalized out.

2.1 Overfitting Problem

The stochastic inference algorithms mentioned above use a similar framework to learn global parameters λ (variational topic-word parameters in SVI, MASVI, SSU, SSI and SGRLD, or per-topic word token counts in SCVB0). They repeatedly subsample a mini-batch of documents from the corpus/stream, and infer local parameters which associate with each document of the mini-batch. Intermediate global parameters correspond to that mini-batch are formed, using the local parameters, in order to update the global parameters. There are two main ways for updating. After forming intermediate global parameter $(\hat{\lambda}^t)$ for the t-th mini-batch, SSU update the global parameter (λ^t) at the t-th iteration as follows:

$$\lambda^t \leftarrow \lambda^{t-1} + \hat{\lambda}^t, \tag{1}$$

while many other algorithms (e.g., SVI, SSI, MASVI) use the following update, with $\rho_t = (\tau_0 + t)^{-\kappa}, \kappa \in (0.5, 1], \tau_0 \geq 0$,

$$\lambda^t \leftarrow (1 - \rho_t)\lambda^{t-1} + \rho_t\hat{\lambda}^t. \tag{2}$$

Those update schemes allow the global parameters λ^t to grow arbitrarily large in either of the following conditions:

- *Data comes from a text stream (*$t \to \infty$*, or the number of documents go to infinity).* Since $\hat{\lambda}^t$ is always non-negative, the update formula (1) can make λ increase rapidly as $t \to \infty$. The formula (2) slows down the increase in magnitude by imposing that the new λ^t is a convex combination of λ^{t-1} in the last iteration and $\hat{\lambda}^t$. Note that $\hat{\lambda}^t$ is not always constrained to be small, as it is sometimes the variational parameter of the posterior of interest (as in SVI, SSI, MASVI,...). Therefore, when data arrive infinitely, the global parameters might grow enormously.
- *Data are highly noisy.* When working with online social networks such as Twitter and Facebook, the data are notoriously noisy [5,10,20]. Noises come from different sources including typos, ad hoc abbreviations, ungrammatical structures, etc. Those noises cause $\hat{\lambda}^t$ to be uncontrolled.
- *A mis-specified model is to be learned.* In practice, we could not know the exact number K of topics contained in a corpus. Hence mis-specification often happens. A mis-specification might cause a learning method to add too much artificial information to $\hat{\lambda}^t$.

When the global parameter grows sufficiently large, the learned LDA might not exhibit the inherent characteristics of real texts, leading to overfitting and bad generalization. Indeed, it is well-known [8] that the Dirichlet hyperparameters are often less than 1 when LDA is learned from real texts by a batch algorithm. Meanwhile, the global parameter λ_k in SSU (also SVI, SSI, MASVI) play the role as hyperparameter of the variational Dirichlet distribution that generates topic β_k. Therefore, the arbitrary growth of λ seems to be in contrary to the practice of LDA.

The arbitrary growth in magnitude of global parameters comes from the additive nature of the updates in (1) and (2), and from no limitation on those parameters. This suggests that when working with massive/streaming data, one should employ some regularization on global parameters in order to avoid overfitting.

2.2   A Simulation Study

We have seen the overfitting possibility of existing stochastic methods. This subsection investigates the behaviors of SSU, SVI, and SSI on simulated data to see further on overfitting. We designed two scenarios in this investigation:

- Scenario 1: (**Mis-specification:** *The learned model has too many topics*) The training data are generated from a model of 20 topics and 5000-dimensional vocabulary. At each iteration, the mini-batch contains 2000 documents which have average length of 200 word tokens. The number of topics of the learned model is set up to 50 topics.
- Scenario 2: (**Noisy condition:** *Data are highly noisy*) The learned model has the same configuration as the model generates data, which is remained as in scenario 1, but the training data are added 5% uniformly distributed noise.

We firstly consider magnitude of the global parameters. Figure 1a and b show how fast they increased as the three algorithms saw more data. Although all global parameters increased in both experiments, λ returned by SSU grew much faster than those fitted by SSI and SVI did. One of the main reason is that the constraint in formula (2) allows SSI and SVI significantly reduce speed of increasing magnitude of λ. This result suggests that SSU is more sensitive with overfitting than the other algorithms. Figure 1c and d shows qualities of the learned models in the two scenarios. It can clearly be seen that overfitting happened to all three algorithms in both experiments. However, SSU fell into overfitting much quicker than SSI and SVI. The constrained update formula maybe the main factor helps SSI and SVI resist overfitting longer. In addition, overfitting in the second experiment is more quickly and easily recognized. Noisy data seemed to have considerable influence on performance of those algorithms, making overfitting problem appears earlier.

3   Regularized Sparse Stochastic Inference (RSSI)

In this section we describe a new stochastic algorithm for LDA, namely *Regularized Sparse Stochastic Inference* (RSSI). Our algorithm considers topics (β) as parameters and tries to learn them from data by doing MAP estimation of topic proportions (θ). The topics (β) are learned in a stochastic manner [3]. Estimation of the intermediate topics requires computing the local variables (θ_d). We derive the approximation of θ_d from z_d, whose posterior is approximated by collapsed Gibbs sampling (CGS) which is similar as SSI [14] does. Details of RSSI are presented in Algorithm 1.

3.1   Estimation of Topics

We consider the posterior of topic proportions given the corpus $\mathcal{C}$ (whose size might be infinite): $P(\boldsymbol{\theta}|\mathcal{C}, \boldsymbol{\beta}, \alpha)$ where the topic indicators are integrated out. Since the documents are i.i.d., we have

$$P(\boldsymbol{\theta}|\mathcal{C}, \boldsymbol{\beta}, \alpha) = \prod_{d \in \mathcal{C}} P(\boldsymbol{\theta}_d|d, \boldsymbol{\beta}, \alpha) \tag{3}$$

where

$$P(\boldsymbol{\theta}_d|d, \boldsymbol{\beta}, \alpha) \propto P(\boldsymbol{\theta}_d, d|\boldsymbol{\beta}, \alpha) = P(d|\boldsymbol{\theta}_d, \boldsymbol{\beta})P(\boldsymbol{\theta}_d|\alpha) \tag{4}$$

is the posterior over topic proportion of the document $d \in \mathcal{C}$. The first term of (5) can be expressed as

$$P(d|\boldsymbol{\theta}_d, \boldsymbol{\beta}) = \prod_j P(w = j|d) = \prod_j \left(\sum_{k=1}^{K} \theta_{dk}\beta_{kj}\right)^{d_j}, \tag{5}$$

where d_j is number of times word j appears in the document d. Because $\boldsymbol{\theta}_d \sim Dirichlet(\alpha)$, the second term is $P(\boldsymbol{\theta}_d|\alpha) \propto \prod_{k=1}^{K} \theta_k^{(\alpha-1)}$. Taking the logarithm of the posterior (3) and ignoring constants, we obtain

$$\mathcal{L} = \sum_{d \in \mathcal{C}} \left(\sum_j d_j \log \sum_{k=1}^{K} \theta_{dk}\beta_{kj} + (\alpha - 1)\sum_{k=1}^{K} \log \theta_{dk}\right) \tag{6}$$

Algorithm 1. RSSI for learning LDA

Input: text data/ text stream, $K, \alpha > 0, \tau_0 \geq 0, \kappa \in (0.5, 1]$
Output: β
initialize $\beta_k^0 \in \Delta_V$ randomly.
for mini-batch $t \in 1, ..., \infty$ **do**
  $\rho_t \leftarrow (\tau_0 + t)^{-\kappa}$
  sample a subset $\mathcal{C}_t$ of documents
  for $d \in \mathcal{C}_t$ **do**
    initialize z_d^0
    discard B burn-in sweeps
    for sample $s \in 1, ..., S$ **do**
      for token $i \in 1, ..., N_d$ **do**
        $\phi_{ik}^s \propto (\alpha + N_{dk}^{-i})\beta_{kj}$
        sample z_{di}^s from $Multinomial(\phi_i^s)$
      end for
    end for
    compute $N_{dk} \approx \frac{1}{S}\sum_{s=1}^{S} N_{dk}^{(s)}$
    compute $\theta_{dk} = \frac{N_{dk}+\alpha}{N_d+K\alpha}$
  end for
  compute $\hat{\beta}_{kj}^t \propto \sum_{d \in \mathcal{C}_t} d_j \theta_{dk}$
  $\beta^t \leftarrow (1 - \rho_t)\beta^{t-1} + \rho_t \hat{\beta}^t$
end for

The task of learning the model is to estimate all topics $\boldsymbol{\beta}$, given the corpus $\mathcal{C}$ and $\alpha > 0$. $\boldsymbol{\beta}$ can be found by maximizing $\mathcal{L}$ subject to the constraint that $\boldsymbol{\beta}_k \in \Delta_V$ where $\Delta_V = \{\boldsymbol{x} \in \mathbb{R}^V : \boldsymbol{x} \geq 0, \sum_{i=1}^V x_i = 1\}$. Basing on framework used by [11] we can design a stochastic algorithm which repeats the following steps:

- *Sample a mini-batch $\mathcal{C}_t$ from the corpus $\mathcal{C}$.*
- *Estimate the local variables $\boldsymbol{\theta}_d$ for each document $d \in \mathcal{C}_t$ that maximize*

$$\mathcal{L}_{\mathcal{C}_t} = \sum_{d \in \mathcal{C}_t} \left(\sum_j d_j \log \sum_{k=1}^K \theta_{dk} \beta_{kj} + (\alpha - 1) \sum_{k=1}^K \log \theta_{dk} \right) \tag{7}$$

$$\textit{s.t.} \qquad \boldsymbol{\theta}_d \in \Delta_K \ (\forall d \in \mathcal{C}_t),$$

  given $\boldsymbol{\beta}^{t-1}$ from the previous iteration.
- *Form intermediate topics $\hat{\boldsymbol{\beta}}^t$ for $\mathcal{C}_t$ that maximize $\mathcal{L}_{\mathcal{C}_t}$ subject to $\hat{\boldsymbol{\beta}}_k^t \in \Delta_V$*
- *Update the topics*

$$\boldsymbol{\beta}^t \leftarrow (1 - \rho_t)\boldsymbol{\beta}^{t-1} + \rho_t \hat{\boldsymbol{\beta}}^t. \tag{8}$$

Note that $\rho_t = (\tau_0 + t)^{-\kappa}$ in (8) must satisfy $\sum_{t=1}^{\infty} \rho_t = \infty$ and $\sum_{t=1}^{\infty} \rho_t^2 < \infty$ to make sure that the learning algorithm will converge. In order to compute $\hat{\boldsymbol{\beta}}^t$ we use the same arguments as [19], arriving at the following formula:

$$\hat{\beta}_{kj}^t \propto \sum_{d \in \mathcal{C}_t} d_j \theta_{dk}. \tag{9}$$

3.2 Estimation of Topic Proportion

Our algorithm requires solving the following problem

$$\boldsymbol{\theta}_d^* = \arg \max_{\boldsymbol{\theta}_d \in \Delta_K} \sum_j d_j \log \sum_{k=1}^K \theta_{dk} \beta_{kj} + (\alpha - 1) \sum_{k=1}^K \log \theta_{dk}, \tag{10}$$

to infer topic proportion for each document. In the case of $\alpha \geq 1$, the problem (10) can easily be proved to be concave, therefore it can be solved in polynomial time. However, α is often set to be small in practice, and (10) is unfortunately NP-hard in the worst case when $\alpha < 1$ [21]. For that reason, instead of directly estimating the exact optimal solution of the problem (10), we try to find an approximation. In fact, if we know $\boldsymbol{z}_d$, we can recover $\boldsymbol{\theta}_d$ by the following approximation

$$\theta_{dk} = \frac{N_{dk} + \alpha}{N_d + K\alpha}, \tag{11}$$

where N_{dk} is the number of indicators z in the document having value k, and length of the document $N_d = \sum_{k=1}^K N_{dk}$. Information about topic indicator configuration can be obtained from posterior distribution over it given the document $\boldsymbol{P}(\boldsymbol{z}_d|d)$. We employ Gibbs sampling [14] to estimate $Q(\boldsymbol{z}_d)$ - the variational

Table 1. Average learning time per mini-batch in second.

	RSSI	SSI	SCVB0	SSU	SVI
New York Times	141 ± 15	92 ± 8	1663 ± 21	5891 ± 41	5974 ± 38
PubMed	21 ± 1	20 ± 1	429 ± 16	8592 ± 79	8410 ± 75

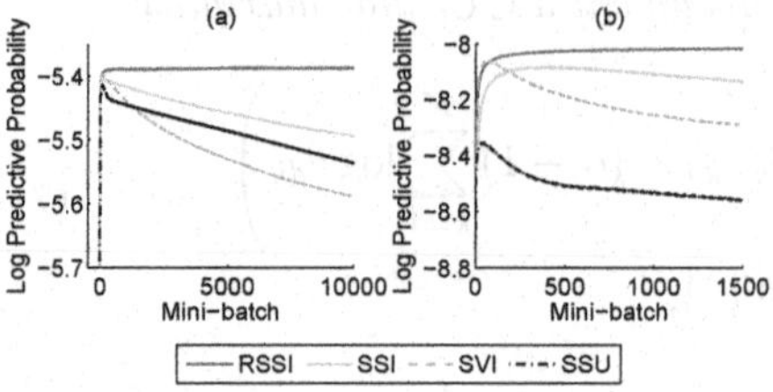

Fig. 2. Performance of RSSI compared with SSI, SVI, and SSU on simulated data (a) **Mis-specification** (b) **Noisy condition**

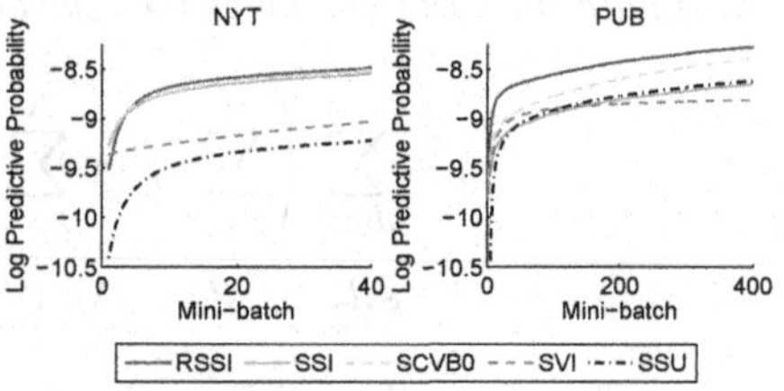

Fig. 3. Predictiveness of models learned by the five methods on *New York Time* and *PubMed* datasets as more data arrive.

distribution used to approximate the true posterior. More specifically, for each document d we iteratively resamples the topic indicator of each word token from distribution of that indicator given the current states of all the other indicators

$$Q(z_{di} = k | w_{di} = j, z_d^{-i}) \propto (\alpha + N_{dk}^{-i})\beta_{kj}, \tag{12}$$

where z_d^{-i} and N_{dk}^{-i} means that z_{di} is excluded. After B burn-in sweeps, S samples of topic indicator configurations $\{z\}^{1,\cdots,S}$ are saved. Approximation of N_{dk} in (11) can be derived as following one:

$$N_{dk} \approx \hat{N}_{dk} = \frac{1}{S} \sum_{s=1}^{S} N_{dk}^{(s)}. \tag{13}$$

This average over a finite set of samples provides a sparse approximation of N_{dk}.

3.3   Comparison with Existing Algorithms

Since both β^{t-1} and β^t belong to Δ_V, update formula (8) always regularizes new topics β^t remains in Δ_V. This regularized update scheme brings RSSI capability of avoiding overfitting which is an important advantage over existing stochastic algorithms. We turn back to the simulated experiments in Sect. 2.2 with our new algorithm. The performances of RSSI in the two experiments are depicted in Fig. 2. It shows that overfitting did not happen with RSSI in both experiments. In addition, qualities of models learned by RSSI were also improved significantly even when data are contaminated by noise.

Besides, while SVI [11], MASVI [12], SSI [14] and SGRLD [15] require specifying the number of all documents in advance, and SCVB0 [6] need to know the number of word tokens in the corpus beforehand, RSSI does not require any information about data size. It makes RSSI to be more suitable when working with data stream. In addition, while RSSI can take advances of sparse Gibbs sampling when inferring local variables, the existing algorithms (excepts SSI) never return sparse solutions. Hence, RSSI seems to be more efficient in terms of reducing runtime and memory requirement.

4   Experimental Evaluation

We compared our algorithm (RSSI) with four existing stochastic algorithms, including SVI [11], SSU [4], SSI [14], and SCVB0 [6]. To avoid possible bias in our comparison we implemented all the algorithms[1] above by Python with our best efforts. Our experiments were taken on two large data sets:

- *PubMed* (PUB)[2] contains 8.2 millions of medical articles from the pubmed central. Those documents were composed from a vocabulary of 141044 distinct words, including more than 717 million word tokens.
- *New York Time* (NYT)[2] consists of 300 thousand news with more than 65 million word tokens. The vocabulary includes more than 100 thousand distinct words.

We randomly set aside 1000 documents from each corpus for testing.

4.1   Measures for Evaluation

We evaluated our algorithm based on two quantity

- *Predictive probability:* This quantity shows the predictiveness and generalization of a model on new data. We followed the procedure in [11], randomly dividing each document in testing data set into two disjoint parts w_{obs} and w_{uobs}. We repeated the split 5 times to create 5 sets $\{w_{obs}, w_{uobs}\}^{1,\dots,5}$ with ratio of 80 : 20. We then did inference for w_{obs}^t and estimated the distribution over w_{uobs}^t given w_{obs}^t and the model. The final predictive distribution was averaged through the 5 sets.
- *Sparsity:* In practice, a document is expected to be composed from some topics rather than the whole K topics in the corpus. We define sparsity of topic proportion without accounting for the smoothing parameters α as follow

$$\mathcal{S}_\theta = \frac{\sum_{k=1}^{K} I_{\theta_k \neq 0}}{K} \tag{14}$$

where θ is computed using (12) with α removed.

---

[1] SSI was taken from http://www.cs.princeton.edu/~blei/downloads/onlineldavb.tar.
[2] The data were retrieved from http://archive.ics.uci.edu/ml/datasets/.

4.2   Parameter Settings

We set $K = 100, \alpha = \frac{1}{K}, \eta = \frac{1}{K}$ which were often used in previous studies. When inferring for a document, we terminated SVI and SSU if relative improvements of the lower bound on likelihood of that document is not better than 10^{-4} or the iterations exceed 50. We set 25 burn-in and 25 samples sweeps for SSI and RSSI. Total 50 iterations were used for SCVB0 to do inference for each document where 25 iterations were burn-in. According to [6,14], these settings are often enough to get good solutions. Based on suggestions from [11,14] we used the learning parameters: $\kappa = 0.9, \tau_0 = 1$, and mini-batch size $|\mathcal{C}_t| = 5000$.

4.3   Performance of 5 Algorithms

We firstly focus on how fast the algorithms learn models from the data sets. Average learning time for the five algorithms are reported in Table 1. SSI and RSSI worked fastest, followed by SCVB0. SVI and SSU had slowest speeds. This was due to the fact that SVI and SSU require many evaluations of Digamma, logarit and exponent functions. In addition, they have to check convergence when doing inference for each document which is very expensive. SCVB0 although contains no computationally expensive functions, it has to update all local and global variables after processing each word token, raising the total computations as the length of document increases. RSSI and SSI worked much faster than the other algorithms because of their ability of utilizing sparseness of the solutions and containing fewer evaluations of Digamma and exponent functions. RSSI was slightly slower than SSI because it has to make approximation of topic proportion after doing inference for each document.

We next investigate how good are the models returned by the five algorithms as they see more data. As we can see in Fig. 3, all the five algorithms can learn better models as they see more documents. However, after processing the same number of documents RSSI learned models with highest predictiveness levels in both data sets, followed by SCVB0. SSI only did well in NYT and SVI and SSU had inferior results in both corpora. The reason for poorer performance of SVI and SSU was that they try to optimize the same variational distributions, which relax all dependencies between the latent variables, as VB does. Hence, significant bias could be produced when approximating the true posterior. SCVB0, an incremental version of CVB0, surpassed the two algorithms as it can better approximate the posterior than VB [1,18]. Although having the similar scheme for inferring the local variables, RSSI and SSI are very different in updating global parameters. The update scheme in SSI makes global parameters grow arbitrarily, easily learning overfitted model. This might be the reason why SSI had different performances on the two data sets. Unlike the other algorithms, RSSI regularizes the global parameters (topics) to remain in a simplex Δ_V. This important feature allows RSSI to avoid overfitting, learning models that have better generalization.

We then want to see sparsity of the solutions returned by SSI and RSSI. The results are depicted in Fig. 4. We observe that both algorithms could provide

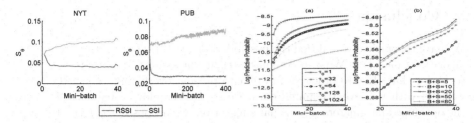

Fig. 4. Sparsity

Fig. 5. (a) τ_0 changes (b) No. samples changes

sparse solutions; however, the solutions of RSSI tended to be sparser than those of SSI. While SSI returned an average of 8 to 11 topics for each document, the figure for RSSI was 3 to 5 topics. In practice, a document written by human often relates to a few number of topics, therefore 3 to 5 topics in a document seemed to be more rational.

4.4   Sensitivity of RSSI

We now consider how the parameters affect the performance of RSSI. The parameters of RSSI include the forgetting factors and number of samples. To see the effect of a parameter, we changed its value in a finite set, but kept the other parameters fixed. We performed experiments to evaluate sensitivity of RSSI on data set NYT.

Forgetting factors: Forgetting factors κ, τ_0 appear in the learning rate $\rho_t = (\tau_0 + t)^{-\kappa}$. We found that κ did not significantly affect the performance of RSSI after investigating five settings of $\kappa \in \{0.6, 0.7, 0.8, 0.9, 1\}$, while change in τ_0 made considerable influence on the performance of RSSI. We ran RSSI with six values of $\tau_0 \in \{1, 32, 64, 128, 1024\}$. The results are depicted in Fig. 5 (a). It is worth noticing that dependence of the performance of the algorithm on τ_0 is monotonic. We recommend that τ_0 should be chosen to be small for RSSI to work well in practice.

Number of samples: RSSI employs Gibbs sampling for doing inference for each document. Gibbs sweeps performed on each document contain B burn-in sweeps which are discarded and S additional sweeps for saving the topic indicator configurations. Number of samples is the total number of sweeps $B + S$. We changed the value of $B + S$ through the set $\{5, 10, 20, 50, 80\}$. For $B + S = 5$ we considered there settings of the pairs $(B, S) : (2, 3), (3, 2)$, and $(4, 1)$. For the other values of $B + S$ we used the settings where $B = S$. The results are presented in Fig. 5(b). Performance was similar across three settings where $B + S = 5$. We observe that the algorithm learned good model when $B + S = 20$ and increasing the number of samples could not considerably improve quality of the model. These results suggest that the mixing rate of Gibbs sampling in RSSI is very fast and 15–20 sweeps is sufficient.

5   Conclusion

We have theoretically analyzed the possibility of overfitting of existing stochastic algorithms for LDA. We also demonstrated it via a simulation study with two scenarios that close to practice. In order to prevent overfitting, especially when working with massive/streaming data, we suggest employing regularization on the optimized parameters. From that suggestion, we proposed *regularized sparse stochastic inference* (RSSI) that regularizes global parameter when updating and utilizes sparse Gibbs sampling for doing local inference for LDA. That combination gives RSSI many advantages over existing algorithms. Our methodology can significantly improves quality of the fitted model and can be easily adapted to various probabilistic models.

Acknowledgments. This research is funded by Vietnam National Foundation for Science and Technology Development (NAFOSTED) under Grant Number 102.05-2014.28 and by the Air Force Office of Scientific Research (AFOSR), Asian Office of Aerospace Research & Development (AOARD), and US Army International Technology Center, Pacific (ITC-PAC) under Award Number FA2386-15-1-4011. Khoat Than is also funded by Vietnam Institute for Advanced Study in Mathematics (VIASM).

References

1. Asuncion, A., Welling, M., Smyth, P., Teh, Y.W.: On smoothing and inference for topic models. In: Proceedings of the Twenty-Fifth Conference on Uncertainty in Artificial Intelligence, pp. 27–34, (2009)
2. Blei, D.M., Ng, A.Y., Jordan, M.I.: Latent dirichlet allocation. J. Mach. Learn. Res. **3**(3), 993–1022 (2003)
3. Bottou, L.: Online Learning in Neural Networks. Online Learning and Stochastic Approximations. Cambridge University Press, Cambridge (1998)
4. Broderick, T., Boyd, N., Wibisono, A., Wilson, A. C., Jordan, M.: Streaming variational bayes. In: Advances in Neural Information Processing Systems, pp. 1727–1735 (2013)
5. Derczynski, L., Ritter, A., Clark, S., Bontcheva, K.: Twitter part-of-speech tagging for all: Overcoming sparse and noisy data. In: Proceedings of Recent Advances in Natural Language Processing, pp. 198–206 (2013)
6. Foulds, J., Boyles, L., DuBois, C., Smyth, P., Welling, M.: Stochastic collapsed variational bayesian inference for latent dirichlet allocation. In: Proceedings of the 19th ACM SIGKDD International Conference on Knowledge Discovery and Data Mining, pp. 446–454. ACM (2013)
7. Gerrish, S., Blei, D.: How they vote: Issue-adjusted models of legislative behavior. In: Advances in Neural Information Processing Systems, vol. 25, pp. 2762–2770 (2012)
8. Griffiths, T.L., Steyvers, M.: Finding scientific topics. Proc. Nat. Acad. Sci. U.S.A. **101**(Suppl. 1), 5228 (2004)
9. Grimmer, J.: A bayesian hierarchical topic model for political texts: measuring expressed agendas in senate press releases. Polit. Anal. **18**(1), 1–35 (2010)

10. Han, B., Baldwin, T.: Lexical normalisation of short text messages: Makn sens a# Twitter. In: Proceedings of the 49th Annual Meeting of the Association for Computational Linguistics, pp. 368–378. ACL (2011)
11. Hoffman, M.D., Blei, D.M., Wang, C., Paisley, J.: Stochastic variational inference. J. Mach. Learn. Res. **14**(1), 1303–1347 (2013)
12. Li, X., OuYang, J., You, L.: Topic modeling for large-scale text data. Front. IT & EE **16**(6), 457–465 (2015)
13. Liu, B., Liu, L., Tsykin, A., Goodall, G.J., Green, J.E., Zhu, M., Kim, C.H., Li, J.: Identifying functional miRNA-mRNA regulatory modules with correspondence latent dirichlet allocation. Bioinformatics **26**(24), 3105 (2010)
14. Mimno, D., Hoffman, M.D., Blei, D.M.: Sparse stochastic inference for latent dirichlet allocation. In: Proceedings of the 29th Annual International Conference on Machine Learning (2012)
15. Patterson, S., Teh, Y.W.: Stochastic gradient Riemannian Langevin dynamics on the probability simplex. In: Advances in Neural Information Processing Systems (2013)
16. Pritchard, J.K., Stephens, M., Donnelly, P.: Inference of population structure using multilocus genotype data. Genetics **155**(2), 945–959 (2000)
17. Schwartz, H.A., Eichstaedt, J.C, Dziurzynski, L., Kern, M.L., Seligman, M.E.P., Ungar, L.H., Blanco, E., Kosinski, M., Stillwell, D.: Toward personality insights from language exploration in social media. In: AAAI Spring Symposium Series (2013)
18. Teh, Y.W., Newman, D., Welling, M.: A collapsed variational bayesian inference algorithm for latent dirichlet allocation. In: Advances in Neural Information Processing Systems, vol. 19, p. 1353 (2007)
19. Than, K., Ho, T.B.: Fully sparse topic models. In: Flach, P.A., Bie, T., Cristianini, N. (eds.) ECML PKDD 2012. LNCS (LNAI), vol. 7523, pp. 490–505. Springer, Heidelberg (2012). doi:10.1007/978-3-642-33460-3_37
20. Yang, S.-H., Kolcz, A., Schlaikjer, A., Gupta, P.: Largescale high-precision topic modeling on Twitter. In: Proceedings of the 20th ACM SIGKDD International Conference on Knowledge Discovery and Data Mining, pp. 1907–1916. ACM (2014)
21. Sontag, D., Roy, D.M.: Complexity of inference in latent dirichlet allocation. In: Advances in Neural Information Processing Systems (NIPS) (2011)

Exploring Check-in Data to Infer Social Ties in Location Based Social Networks

Gunarto Sindoro Njoo[1(✉)], Min-Chia Kao[1], Kuo-Wei Hsu[2], and Wen-Chih Peng[1]

[1] National Chiao Tung University, Hsinchu, Taiwan
gunarto.cs01g@g2.nctu.edu.tw, wcpeng@cs.nctu.edu.tw
[2] National Chengchi University, Taipei, Taiwan

Abstract. Social Networking Services (SNS), such as Facebook, Twitter, and Foursquare, allow users to perform check-in and share their location data. Given the check-in data records, we can extract the features (e.g., the spatial-temporal features) to infer the social ties. The challenge of this inference task is to differentiate between real friends and strangers by solely observing their mobility patterns. In this paper, we explore the meeting events or co-occurrences from users' check-in data. We derive three key features from users' meeting events and propose a framework called SCI framework (Social Connection Inference framework) which integrates all derived features to differentiate coincidences from real friends' meetings. Extensive experiments on two location-based social network datasets show that the proposed SCI framework can outperform the state-of-the-art method.

1 Introduction

The research on mining the relationship between virtual data and physical data (e.g., users' mobilities, social links, preferences, etc.) gradually increases in the past decade. Specifically, from the check-ins in the location-based social networks (LBSNs), one can extract precious information such as the mobility patterns and the social links between users. Using check-in data is beneficial yet challenging. Various applications can be derived from using check-in data such as social strength analysis [7,11], friendship recommendation [1,3], and targeted marketing [13]. However, the challenge is that check-in data is usually sparse because some users may have more check-ins and some have fewer.

It is intuitive that friends have a greater chance to appear together at the same occasions compared to strangers because of some common events or shared interests, such as attending a wedding party of their common friends or spending time together at a coffee shop which they love. Based on this intuition, some researchers [1,3,7,11] have studied the relationship between user's mobility and social links solely based on the difference of check-ins behavior between friends and non-friends. These studies have employed *co-occurrences*, which simulates meeting events using check-in data, to solve the friendship prediction problem.

© Springer International Publishing AG 2017
J. Kim et al. (Eds.): PAKDD 2017, Part I, LNAI 10234, pp. 460–471, 2017.
DOI: 10.1007/978-3-319-57454-7_36

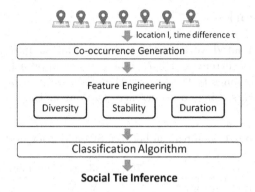

location l, time difference τ

Co-occurrence Generation

Feature Engineering

Diversity   Stability   Duration

Classification Algorithm

Social Tie Inference

Fig. 1. The overview of SCI framework

Despite the efforts made to uncover friendships using check-in data, these studies do not report the number of friendships that can be retrieved from their approaches. Further, while some studies such as [3,11] employ temporal factor in their models, they do not adopt temporal duration and stability in their methods. Whereas duration and stability can capture the general idea of how long and how often that the users have interacted with each other. Moreover, none of the studies mentioned above considers the impact of weekend data (either qualitatively or quantitatively) for depicting social links, even though it is commonly known that weekend activities usually involve social activities [1].

In this paper, we propose a unified framework called SCI framework (**S**ocial **C**onnection **I**nference framework) which consists of three stages (shown in Fig. 1). First, we extract the co-occurrences from check-in data by using the mapreduce technique. Second, we quantify three key features in the co-occurrences: diversity, stability, and duration. Finally, we aggregate co-occurrence features using machine learning algorithms to predict the social ties or friendships between users.

In summary, the contributions of this paper are as follows.

- We propose two novel features (stability and duration) in the temporal domain of the co-occurrence, which can reflect the consistency and the total duration of the meetings between users.
- We present an analysis of the benefits and the limitations of applying co-occurrence to infer the social ties and propose a map-reduce algorithm to accelerate the co-occurrence generation.
- Our experiments provide insights to the significance of weekend data for predicting social links from users' mobility data and the predictive power of each feature.

The remainders of this paper are organized as follows. Section 2 discusses related works. Section 3 formally defines the notations in our work. Section 4 describes co-occurrence generation process and the proposed features. Section 5 reports our experimental results. Finally, Sect. 6 concludes our work.

2 Related Works

We categorize the related works in the social tie analysis into two groups based on their focus: co-occurrences [1,3,7,11] and other approaches [8–10,12]. Our proposed method belongs to the former; therefore, we provide the comparisons between related works using co-occurrences and our method in Table 1.

Table 1. Comparison between SCI and prior works

Features	[7]	[11]	[3]	[1]	Our method
Location entropy	V	V	V	V	
Diversity	V				V
Time gap		V	V		
Time interval sequence				V	
Stability					V
Duration					V
Co-location graph			V		

Previous works such as [1,3,7,11] introduced a concept called *co-occurrence* to infer the social relationship between users. Both the authors in [7,11] focused on quantifying the continuous social strength between users. However, their original problem can be altered into link prediction problem by setting a threshold that differentiate friends from non-friends. Most previous works utilized location entropy concept which has been introduced by [2] in their model. Location entropy could explain the significance of a meeting event based on the location's property: meeting in a private place matters more than in a public place. However, in this paper, we use diversity concept which is also used in [7] to avoid the impact produced by frequent yet coincidental meeting events. Alternatively, the authors in [11] employed personal, global, and temporal features (PGT) to uncover social links in the co-occurrences data. They calculated the importance of meeting events by considering the user's location visit distribution, the location's popularity, and the time difference between two consecutive meeting events. Instead of using time threshold, the authors in [1] proposed a time interval sequence to identify co-occurrences; their proposed approach is on par with [7] in coarser spatial granularity but suffers from lower performance in the finer spatial granularity. Similar to [11], the authors in [3] employed the same features that are used in [11] but they built a co-occurrence graph to depict the indirect social ties between users; however, they only evaluated the top 10,000 users in their experiments and the computation cost to create the co-occurrence graph is high. It is also worth noting that although we measure the temporal feature but the proposed temporal feature is entirely different from the one suggested in [3,11]. In this paper, we measure the stability and the duration of a co-occurrence behavior by accommodating all meetings, not just the consecutive meetings. Stability is similar to the time gap that is proposed in [3,11] but is more general. On the other hand, duration represents how long the users have known each other, as in the real world we observe that friends tend to have known each other longer and meet from time to time.

Prior works have also explored social tie analysis using various methods. The authors in [12] presented the idea of using communication logs, such as call logs and message logs to depict the social link between users. However, the authors in [12] concluded that partial communication records cannot be used as a proxy to represent the social relationship. The authors in [10] took advantage of social network structure to explore the social tie strength between users, but they do not consider the correlation between spatiotemporal data and social link. The authors in [9] proposed place features and social features to predict the friendship in the location-based social network. Place features take frequent places between users into account, while social features assume that common neighbors can help on predicting the friendship. On the other hand, our work aims to accommodate spatiotemporal data without any social network information to predict the social ties. Another work [8] proposed a friendship prediction method by predicting the location that a user would visit. The authors in [8] calculated the text similarities between users from the Twitter dataset, extracted the number of common neighbors and the number of triads formed in the social graph, and derived the co-locations from each tweet; they further assumed that user will not move to other locations if he/she does not post another tweet. We argue that the assumption in [8] is too strong and may not be suitable for the real-world applications, as few users tweet on every location that they visit.

3 Notations and Problem Formulation

We denote $u \in U$ as the user in the dataset and $c \in C$ as the check-in data of the users. Each user has a sequence of check-ins $C_u = \{c_1, \ldots, c_n\}$ to represent all check-ins made by user u. Each c reflects the appearance of a user u at a specific location l at a specific time t with the form of $\{u, t, l\}$. Co-occurrence $\theta_{x,y}^z$ is a four tuple $\{u_x, u_y, t_{i,j}, l_z\}$ to reflect a situation where two users u_x and u_y meet in a particular location l_z through their check-in c_i and c_j ($c_i \in C_x$ and $c_j \in C_y$) and the time difference between c_i and c_j is lower than threshold τ, and $t_{i,j}$ is the average check-in time between c_i and c_j. Co-occurrence set $\Theta_{x,y}$ is the collection of meeting events between two users u_x and u_y among all meeting locations. $\Psi_{x,y}^z \in \Psi_{x,y}$ quantifies the meeting frequency between users u_x and u_y in the location l_z. $\Psi_{x,y} = \{\Psi_{x,y}^{z_1}, \cdots, \Psi_{x,y}^{z_m}\}$ is the meeting frequency set for all meeting locations between users u_x and u_y. Please note that every $\Psi_{x,y} = 0$ is omitted and $\sum_{z \in L} \Psi_{x,y} = |\Theta_{x,y}|$, where L is the set of all meeting locations between users u_x and u_y. Finally, social tie σ is measured between user u_x and u_y using their co-occurrence set $\Theta_{x,y}$, where $u_x, u_y \in U$ and $\sigma \in \{0, 1\}$. Finally, the problem that we aim to solve in this paper is as follow. Given $u_x, u_y \in U$ and a check-in dataset C in the form of $\{u, t, l\}$, the problem is to predict whether u_i and u_j are friends or not.

4   Methodology

In this section, we first elaborate the co-occurrence generation process. Second, we describe three features in the co-occurrences, such as how diverse the meeting location is, how regular users meet each other, and how long their encounters have occurred. Finally, we derive *diversity score* w_d, *stability score* w_s and *duration score* w_δ from each meeting event between users. Finally, we combine these scores to predict the friendship links between users.

4.1   Co-occurrence Generation

Algorithm 1 explains the process of co-occurrence generation. Co-occurrences between two users u_x and u_y are generated from the check-in sequences C_x and C_y from user u_x and u_y respectively using two parameters: distance threshold Δ and time threshold τ. As the check-ins sequences are ordered by timestamp, we can generate the co-occurrence by using two iterators ic_1 and ic_2 for both C_x and C_y. The next ic_1 and ic_2 are generated based on the timestamp order between two check-ins. Suppose c_1 is the current check-in of user u_x and c_2 is the current check-in of user u_y. Then, if the time in c_1 is larger than that in c_2, then we increment the value of ic_2. Otherwise, we increase the value of ic_1. Using this approach, we only need to select at most $|C_x| + |C_y|$ times for generating the co-occurrences between two users. Additionally, if parameter Δ is set to 0, then location loc_z is depicted from the location id. Otherwise, location loc_z is calculated using the average of two locations' coordinate. Also, co-occurrence time in $\theta^z_{x,y}$ is generated based on the average time between two users' check-ins.

The complexity of computing co-occurrences is high because we need to calculate every combination of every user pairs. However, we observe that the computation can be parallelized by splitting the user pair combinations into several subsets. Here, we apply the map-reduce technique to parallelize the co-occurrence generation.

Map. Suppose we have N users in the dataset, then we can split the users into k chunks, so each chunk has approximately N/k users. Subsequently, each chunk is processed iteratively from the first to the last user. Thus, we will have $\frac{N^2}{2\times k^2}$ computation for each chunk or $\frac{N^2}{2\times k}$ for all chunks, which theoretically can speed up the co-occurrence generation process approximately by k times.

Reduce. For each generated co-occurrence set in each chunk, we aggregate all co-occurrences. We separate the aggregation into two parts. First, we aggregate the co-occurrences which have similar user ids and location ids to obtain the frequency of the co-occurrences between two users at a particular location. Second, we can aggregate all the raw co-occurrences together in a file, without any further preprocessing. The latter serves as the generated raw co-occurrences in the dataset.

4.2   Diversity

Variation in the meeting places between users is useful for reducing the possibilities of coincidences. Supposedly we know that user u_1 often meets users u_2

Algorithm 1. Co-occurrence generation algorithm

Input: C_x, C_y, Δ, τ
Output: $\theta_{x,y}$

1 Define $\theta_{x,y} \leftarrow \emptyset$;
2 Define $ic_1, ic_2 \leftarrow 0$;
3 **while** $ic_1 < |C_x|$ *and* $ic_2 < |C_y|$ **do**
4      $c_1 = C_x[ic_1]$
5      $c_2 = C_y[ic_2]$
6   **if** *($\Delta = 0$ and $c_1.vid$!= $c_2.vid$) or (timediff $> \tau$) or (distance $> \Delta$)* **then**
7   Next ic_1 or ic_2
8   **continue**
9   **end**
10     $\theta_{x,y}^z = \{u_x, u_y, loc_z, avg(time)\}$
11     $\theta_{x,y}$.append($\theta_{x,y}^z$)
12   Next ic_1 or ic_2
13 **end**
14 **return** $\theta_{x,y}$

---

and u_3 through their check-ins. Consider the following scenarios. First, user u_1 meets user u_2 several times in the same location. Second, user u_1 meets user u_3 a few times in several locations. Thus, the meeting occasions in the former are more likely to happen by chance than those in the latter. The reason is that the possibility of meeting in more diversified locations is lower than the possibly of meeting in the same location. Here, we adapt the concept of *diversity* in co-occurrences which is introduced by recent work [7]. Shannon entropy [5] is employed to reflect the diversity as follows. Let $\Psi_{x,y}^z$ be the frequency of co-occurrence in the location z where $\Psi_{x,y}^z \in \Psi_{x,y}$, then the diversity score w_d is determined using Eq. 1.

$$w_d(x,y) = -\sum \Psi_{x,y}^z \cdot log(\Psi_{x,y}^z) \tag{1}$$

4.3 Temporal

Temporal feature indicates how long a co-occurrence between two users is and how stable it is. While strangers rarely have co-occurrences or have ones in a short time due to coincidences, friends tend to spend more time together in a more precise and less coincident fashion because they could arrange their meetings. For every meeting event between two users u_x and u_y, calculation of friendship duration using timestamps $t_{x,y}^i$ (Eq. 2) from their co-occurrence in the co-occurrence vector $\Theta_{x,y}$ is defined in Eq. 3.

$$t_{x,y}^i = time_i - time_{i-1} \tag{2}$$

where $time_i$ is the timestamp of the i-th co-occurrence in $\Theta_{x,y}$ and the co-occurrence vector $\Theta_{x,y}$ is ordered by timestamp in ascending order.

$$\delta_{x,y} = \max_i(t_{x,y}) - \min_i(t_{x,y}) \tag{3}$$

To normalize the value, we divide the duration with the maximum friendship duration of all users. For any pair of users u_x and u_y, we calculate $\max_{u \in U}(\delta_{x,y})$ for maximum friendship duration among all users. Finally, w_δ is the duration score between users from their meeting events and $\max_{u \in U}(\delta_{x,y})$ is the maximum friendship duration, and then duration score is calculated by using Eq. 4.

$$w_\delta(x,y) = \frac{\delta_{x,y}}{\max_{u \in U}(\delta_{x,y})} \tag{4}$$

Further, we need to quantify the *stability* of co-occurrence to uncover social tie between users who have numerous meeting events together. We first quantify the average time of each pair of user's meeting $\mu_{x,y}$ by dividing each w_δ with the co-occurrence frequency of a pair of users u_x and u_y in all locations, as shown in Eq. 5.

$$\mu_{x,y} = \frac{w_\delta}{\sum_z \Psi_{x,y}^z} \tag{5}$$

Subsequently, we consider standard deviation of the time of each users' meeting to understand the consistency of meeting events. Given average meeting time $\mu_{x,y}$ and time distribution of the co-occurrence $t_{x,y}$ between each pair of users u_x and u_y, we calculate the stability $\sigma_{x,y}$ using Eq. 6.

$$\sigma_{x,y} = \sum_i \left(\frac{t_{x,y}^i}{\max_i(t_{x,y}^i)} - \mu_{x,y} \right)^2 \tag{6}$$

Further, to calculate the density of meeting event among various locations, we need to normalize $\sigma_{x,y}$ by using Eq. 7.

$$\rho_{x,y} = \sqrt{\frac{\sigma_{x,y}}{|\Psi_{x,y}|}} \tag{7}$$

Finally, we deliver the calculation of stability weight w_s, as shown in Eq. 8.

$$w_s(x,y) = \exp\left(-(\mu_{x,y} + \rho_{x,y})\right) \tag{8}$$

Here, we use the average duration between meetings, by dividing $\delta_{x,y}$ with $\sum_z \Psi_{x,y}^z$. Average duration shows us the time gap between meetings, indicating cycle period of having co-occurrences. On the other hand, consistency value $\sigma_{x,y}$ shows us the *stability* of the time gap between each meeting. Lower average duration of co-occurrences and lower standard deviation value on the co-occurrences density are more desired. Moreover, the weight is delivered as an exponential function because time distributions between check-ins are sparse. Further, if the co-occurrence frequency is smaller than or equal to 1 then we set the *stability* score to 0 because we cannot quantify the temporal stability of those who only meet once.

All in all, co-occurrences score is measured by considering diversity w_d, and temporal feature. Whereas, temporal feature consists of both stability score w_s and duration score w_δ. We first normalize all the values of w_d, w_s, w_δ to the range of [0,1]. Finally, we use the tuple $\{w_d, w_s, w_\delta\}$ which is extracted from $\Theta_{x,y}$ given all combinations of user u_x and u_y as the input for the machine learning algorithms.

5   Experiments

In this section, we first present our datasets and parameter settings. Second, we analyze the impact of time threshold parameter in the co-occurrence generation. Subsequently, we show the comparison between the performance of the proposed method and the performance of the state-of-the-art method, and additionally we examine the predictive power of each feature.

5.1   Dataset and Metrics

In our experiments, we use two real datasets collected from location-based social networking services: Gowalla and Brightkite [4]. We differentiate the data into two types (all data and weekend data) to evaluate the effectiveness of the proposed method and the capability of weekend data to perceive friendship information. The statistics for all data and weekend data for each dataset are exhibited in Table 2. Here, G and B represent Gowalla dataset and Brightkite dataset respectively, while A and W represent all data and weekend data respectively. Each check-in record is a tuple of user ID, latitude, longitude, timestamp, and venue ID. Friendship information is present in both datasets, which can serve as the ground truth in our evaluation. An average number of check-ins among all users and users that have friends are presented in the Table 2 as well. Specifically, we observe that the number of users in the check-in data (*User check-ins*) is different from the number of users in the friendship data (*User friends*). Furthermore, we also notice that average number of check-ins of *User friend* [Avg. Check-ins (Friend)] is lower than that of *User check-ins* [Avg. Check-ins (All)], whereas in all data, there may be some users who have no friendships. This suggests that some of the social links cannot be captured through the mobility data alone.

Using Scikit-learn [6], we evaluate the AUC score of the friendship prediction. AUC is calculated using the area of the receiver operating characteristic (ROC) curve. Higher AUC represents that less non-relevant data is retrieved by the query. Also, we have also conducted preliminary experiments to obtain the best classifier for the experiments, and we finally select Random Forests for our classifier as it outperforms other classifiers (SVM, Decision Tree, and Logistic Regression in both datasets. Lastly, all the experiments are done by using 5-fold cross-validation.

5.2   Co-occurrence Generation

The parameters for co-occurrence generation are as follows. First, in our experiments, the time threshold τ is set to 0.5 h, 1 h, 1.5 h, or 2 h. Second, we assume that the co-occurrence happens only if the check-ins between users are associated with the same venue id. This assumption seems to be strong, but we can still retrieve friendships from the datasets[1].

---

[1] We have also evaluated the performances in the various distance thresholds Δ to $\{0\,m, 250\,m, 500\,m, 750\,m, 1000\,m\}$ in the preliminary experiments. The number of

Table 2. Statistical Information of Experimental Datasets.

	GA	GW	BA	BW
#Users (Check-ins)	107,092	92,988	51,406	35,390
#Users (Friend)	196,575	86,477	58,228	34,236
#Check-ins	6,442,890	2,013,871	4,491,143	1,334,404
#Friendships	950,327	804,878	428,156	305,671
Avg. Check-ins (All)	60.16	21.66	92.35	37.71
Avg. Check-ins (Friend)	32.77	22.54	81.53	38.57

Figure 2a shows us the percentage of the original friendships that can be retrieved through co-occurrence method alone, and Fig. 2b shows the imbalance between friends and non-friends in both datasets. We discuss several observations in this result. First, we observe that the numbers of co-occurrences generated by different time thresholds vary. With a smaller time threshold, we can only capture few friendship links because of the above assumption. Second, as explained in the Sect. 5.1, some of the friendship links cannot be captured through the mobility data alone. This issue is not discussed quantitatively in the previous works such as [3,7,11]. Here, we present that in the LBSN services, we can capture approximately 1% of the friendship links by using their mobility data alone. Third, this result promotes that weekend data is a useful indicator for predicting friendship links. Even though the percentage of check-ins in weekend data is as low as 30% of the original data, weekend data can achieve as high as 48% of the performance achieved by the original data. Fourth, it is difficult to capture 100% of the friendships because some inactive users (e.g., 45% of users in Gowalla dataset) also report their friendships, even though they have no check-ins at all. Fifth, the highest portion of friend data is only 23%, which is in Brightkite weekend data, and the rests are lower than 20%. On the light of the co-occurrence generation performance by applying map-reduce implementation, we observe that by increasing the number of chunks, we can gain higher speed-up. For example, in the Brightkite weekend dataset, by using 5 chunks, we can achieve speed-up of 2.33.

5.3   Performance Evaluation

Here, we evaluate the AUC of the proposed method and the state-of-the-art method, namely PGT [11]. PGT is a method proposed by [11] which employed the personal background mobility of the users, the global popularity of a venue, and the temporal difference between two consecutive meetings.

Figure 3 shows the AUC of the proposed method and PGT in both datasets. From Fig. 3, it is clear that our method outperforms PGT in all cases. We also

---

retrieved friendships in higher Δ is slightly higher (up to 1.5 times to $\Delta = 0$ m). However, the overall prediction performance is similar to $\Delta = 0$ m.

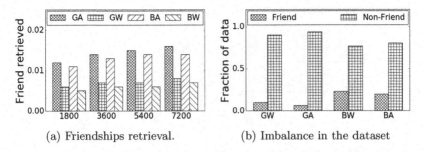

(a) Friendships retrieval.   (b) Imbalance in the dataset

Fig. 2. Percentages of friendships under various time thresholds and the degrees of imbalance in the datasets.

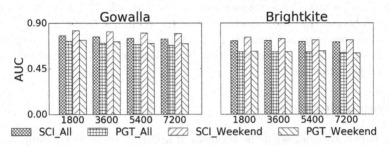

Fig. 3. AUC of SCI and PGT [11] in both datasets

observe that in general the AUC performance degrades along with the increase of the time threshold τ. This is contrast to the observation regarding the relationship between friend coverage and the time threshold. Therefore, the selection of the value of the time threshold is the trade-off between friend coverage percentage and AUC performance. In addition, the AUC values of the weekend's data are generally higher than those of the all data. This implies that the mobility data in the weekend can capture more activities relevant to social relationships.

Further, we evaluate the predictive power of each feature. In our model, we employ three features in co-occurrence: *diversity* (D), *stability* (TS), and *duration* (TD). To show the predictive power of each feature, we consider the combinations of each feature. Also, we present the predictive power of PGT's feature separately, in order to understand what combination of PGT's features performs well. Here, P0 is the sum of personal factor in PGT, P is the max of personal factor, PG is personal factor collaborate with the global factor, and finally PGT is the complete framework of PGT.

Figure 4 presents the AUC of each feature combination. The leftmost feature is the best feature in the average of the two datasets, while the rightmost feature is the worst of all. As expected, the models with more features outperform the models with fewer features, except for the diversity feature. We summarize the observations as follows. In the single feature case, diversity of meeting events performs stably in both datasets, while duration and stability cannot achieve good performance. However, we observe that the combination of stability and

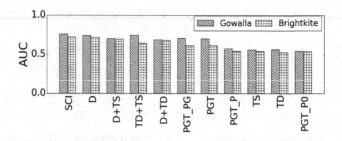

Fig. 4. Performance each feature and its combinations.

duration can achieve high performance in the Gowalla dataset. On the other hand, we observe that the improvement of the temporal factor in PGT is marginal, while our proposed temporal feature boosts the performance well. All in all, in both datasets, the combination of all features (diversity, stability, and duration) performs the best.

6   Conclusions

In this paper, we addressed the problem of inferring social ties among users in the location-based social networks. We proposed SCI framework to explore the co-occurrence concept and distinguish friends from strangers. We derive three features (diversity, temporal stability, and duration) from the co-occurrences and utilize a machine learning algorithm to predict the social ties. In addition, we identify the benefits and the limitations of co-occurrence to uncover the friendship links. On the experiments on two real datasets, we show that the SCI framework can outperform state-of-the-art method by achieving higher AUC. Further, we study the significance of the weekend data for depicting social relationship from mobility data. We show that although weekend data is only subset of the full dataset, it can retrieve the social links as much as half of the full dataset can achieve and achieve similar AUC performances.

Acknowledgements. Wen-Chih Peng was partially support by the TAIWAN MOST (104-2221-E-009-138-MY2 and 105-2634-E-009-002) and Academic Sinica Theme project No. AS-105-TP-A07.

References

1. Cheng, R., Pang, J., Zhang, Y.: Inferring friendship from check-in data of location-based social networks. In: Proceedings of the 2015 IEEE/ACM International Conference on Advances in Social Networks Analysis and Mining 2015, ASONAM 2015, pp. 1284–1291. ACM, New York (2015)
2. Cranshaw, J., Toch, E., Hong, J., Kittur, A., Sadeh, N.: Bridging the gap between physical location and online social networks. In: Proceedings of the 12th ACM International Conference on Ubiquitous Computing, UbiComp 2010, pp. 119–128. ACM, New York (2010)

3. Hsieh, H.-P., Yan, R., Li, C.-T.: Where you go reveals who you know: analyzing social ties from millions of footprints. In: Proceedings of the 24th ACM International on Conference on Information and Knowledge Management, CIKM 2015, pp. 1839–1842. ACM, New York (2015)

4. Leskovec, J., Krevl, A.: SNAP datasets: Stanford large network dataset collection. http://snap.stanford.edu/data.

5. Lin, J.: Divergence measures based on the Shannon entropy. IEEE Trans. Inf. Theor. **37**(1), 145–151 (2006)

6. Pedregosa, F., Varoquaux, G., Gramfort, A., Michel, V., Thirion, B., Grisel, O., Blondel, M., Prettenhofer, P., Weiss, R., Dubourg, V., Vanderplas, J., Passos, A., Cournapeau, D., Brucher, M., Perrot, M., Duchesnay, E.: Scikit-learn: machine learning in python. J. Mach. Learn. Res. **12**, 2825–2830 (2011)

7. Pham, H., Shahabi, C., Liu, Y.: EBM: An entropy-based model to infer social strength from spatiotemporal data. In: Proceedings of the 2013 ACM SIGMOD International Conference on Management of Data, SIGMOD 2013, pp. 265–276. ACM, New York (2013)

8. Sadilek, A., Kautz, H., Bigham, J.P.: Finding your friends and following them to where you are. In: Proceedings of the Fifth ACM International Conference on Web Search and Data Mining, WSDM 2012, pp. 723–732. ACM, New York (2012)

9. Scellato, S., Noulas, A., Mascolo, C.: Exploiting place features in link prediction on location-based social networks. In: Proceedings of the 17th ACM SIGKDD International Conference on Knowledge Discovery and Data Mining, pp. 1046–1054. ACM (2011)

10. Sintos, S., Tsaparas, P.: Using strong triadic closure to characterize ties in social networks. In: Proceedings of the 20th ACM SIGKDD International Conference on Knowledge Discovery and Data Mining, KDD 2014, pp. 1466–1475. ACM, New York (2014)

11. Wang, H., Li, Z., Lee, W.-C.: PGT: measuring mobility relationship using personal, global and temporal factors. In: 2014 IEEE International Conference on Data Mining, pp. 570–579. IEEE (2014)

12. Wiese, J., Min, J.-K., Hong, J.I., Zimmerman, J.: You never call, you never write: call and SMS logs do not always indicate tie strength. In: Proceedings of the 18th ACM Conference on Computer Supported Cooperative Work & Social Computing, pp. 765–774. ACM (2015)

13. Zhu, W.-Y., Peng, W.-C., Chen, L.-J., Zheng, K., Zhou, X.: Modeling user mobility for location promotion in location-based social networks. In: Proceedings of the 21th ACM SIGKDD International Conference on Knowledge Discovery and Data Mining, KDD 2015, pp. 1573–1582. ACM, New York (2015)

Scalable Twitter User Clustering Approach Boosted by Personalized PageRank

Anup Naik[✉], Hideyuki Maeda, Vibhor Kanojia, and Sumio Fujita

Yahoo Japan Corporation, Tokyo, Japan
anaik@yahoo-corp.jp, hidmaeda@yahoo-corp.jp, vkanojia@yahoo-corp.jp,
sufujita@yahoo-corp.jp

Abstract. Twitter has been the focus of analysis in recent years due
to various interesting and challenging problems, one of them being Clustering of its Users based on their interests. For graphs, there are many
clustering approaches which look at either the structure or at its contents.
However, when we consider real world data such as Twitter Data, structural approaches may produce many different user clusters with similar
interests. Similarly, content-based clustering approaches on Twitter Data
produce inferior results due limited length of Tweet and due to lots of
garbled data. Hence, these approaches cannot be directly used for practical applications. In this paper, we have made an effort to cluster Twitter
Users based on their interest, looking at both the structure of the graph
generated using Twitter Data, as well as its contents. By combining these
approaches, we improve our results compared to the existing techniques,
thereby generating results befitting the practical applications.

1 Introduction

There is huge fan following for idol groups or celebrities on social networks like
Twitter. These fans frequently tweet about the events of the concerned celebrity,
their latest news, videos, photos and other information; in a sense act as a *groupie*
of the celebrity/idol groups. Such users can be used as a source to obtain real-
time information about the concerned celebrity. This inspires us to cluster these
Social Influencers in the fan following social network communities.

Huge user base has made Twitter user-graph very complex, and hence, analysis of Twitter Data has become burdensome. Existing structural approaches fail
to perform effectively when we consider millions of nodes having active, inactive and spam users. On the other hand, content based approaches deteriorate
because of limited length of textual contents in a tweet and garbled data. This
observation acted as the basis of our approach to use both the structural as
well as the content aspects of Twitter, thereby nullifying the drawbacks of each.
Although follower list is available in Twitter, most of the users are inactive and
interaction among them is lacking. The users in the follower list mostly read
the tweets about the celebrity rather than posting something. Also, some of the
celebrities (especially in Japan) do not have official accounts to get the follower
list from. So for getting clusters, just taking the follower list is not effective.

© Springer International Publishing AG 2017
J. Kim et al. (Eds.): PAKDD 2017, Part I, LNAI 10234, pp. 472–485, 2017.
DOI: 10.1007/978-3-319-57454-7_37

Our standpoint is consistent with the analyses made by Cha et al. where they endorsed *the million follower fallacy* by collecting empirical evidences [8]. Our contribution in this paper includes:

- We proposed a new approach for user clustering based on both content and graph features with topical relevance and influential ranking(using Personalized PageRank score) which can be used in many areas such as online advertising, viral marketing, personalized content dissemination and so on.
- We intensively compared our approach with content based, graph based and hybrid approaches in view of topical relevance and influential measures.
- Upon empirical evaluations, we confirmed that our approach outperforms strong and the state-of-the-art baseline systems even on massive data sets. Our data consisted of one month Twitter Data (1.6 TB in its compressed form).

2 Related Work

Graph Clustering: One of the graph clustering algorithm is *SCAN* [1]. It clusters vertices based on a structural similarity measure. It uses the fact that nodes in a clusters are densely connected with other nodes in the group and sparsely connected to nodes outside. Apart from clusters, it also finds *hub* nodes, which bridge two clusters, and *outlier* nodes which are vertices marginally connected to clusters. SCAN algorithm (Sect. 3.1) gives good results when we consider the structure of the graph and hence, a modified version of it acts as the first phase of our approach. One of the drawbacks of this algorithm is that it does not look at the contents of the nodes. So some of the clusters produced may belong to the same topic but stay as different clusters in this approach which we would like to merge in ideal case. Other algorithms use number of possible "betweenness" measures to iteratively remove edges to find clusters, as used in [3]. The *min-max cut method* [2] partitions the graph into two clusters A and B, by removing the minimum number of edges needed to isolate A and B. One drawback of this approach is that one has to specify the number of clusters beforehand. The most crucial problem is that if one cuts out a single node, one may achieve the optimum solution. In practice, this approach requires some constraint, such as $|A| \approx |B|$ which are inappropriate in real social networks.

Content Clustering: Content Clustering algorithms use various features of Twitter data to cluster users, such as the approach used in paper [5]. In this approach, they have used various similarity measure as feature for k-means clustering algorithm. Even though the paper claims to successfully cluster the users based on their interest, we are unable to reproduce the results due to large size of Twitter data we use, which is much larger than the data of the Twitter public API they used, that returns only a small portion of vast Tweet data. We empirically experienced that their method is not scalable to real Tweet streams. Hayashi et al. tried to detect hijacked topics when factorizing the user-term frequency matrix [10]. Our approach solves the same problem quite differently.

Social Graph Analysis: Leskovec et al. proposed *network community profile plot* to illustrate structural properties of network communities mainly based on the *graph conductance* measure [4]. They analyzed various kinds of social and information networks and pointed out the existence of many small scale and tight communities. We have addressed the same problem from a different but practically very effective approach. Cha et al. analyzed the Twitter network introducing three influential measures of users namely indegrees, retweets and mentions [8]. Their observation endorsed the intuitions of some observers such as [9] and pointed out that each measure indicate different features of tweets and users. Their influential measure is mainly for scientific observation and analysis purposes especially for topical relevant influentials and they eliminated chronologically mature topics due to hijacked keywords by spammers. Although we are inspired by their insights on the characteristics of Twitter network, they did not propose any methods to extract topically relevant influential users from real Twitter network in view of industry usage. Weng et al. proposed an approach for finding topic-sensitive influential twitterers [11]. They evaluated only on a small dataset of less than 5000 active users (compared to 3 million in our approach). Their method works only on a toy size dataset although the LDA method on Twitter texts normally outputs junk topics due to the topic hijacking and Twitter specific text usage such as short text with many emoticons.

3   Graph Clustering

First we describe the graph clustering algorithm which partitions twitter user networks by analyzing the structural properties of the graph. We adopt SCAN algorithm [1] and propose its enhancement, namely, Weighted SCAN, which is intended to more effectively partition users according to their network activities.

3.1   SCAN Algorithm

In this section, we describe in detail, the SCAN algorithm. This algorithm acts as the first phase in our approach. It takes two parameters as input; ϵ: a threshold value to determine structural similarity between two nodes and μ: the minimal size of a cluster. Let us define some of the commonly used terms. The list of symbols and its meaning is given in Table 1.

Definition 1. *STRUCTURAL SIMILARITY: The structural similarity between node u and v, denoted by $\sigma(u, v)$, is defined as*

$$\sigma(u, v) = \frac{|\Gamma(u) \cap \Gamma(v)|}{\sqrt{|\Gamma(u)||\Gamma(v)|}} \tag{1}$$

where Γ is defined as $\Gamma(v) = \{w \in V \mid (v, w) \in E\} \cup \{v\}$.

So, $|\Gamma(u) \cap \Gamma(v)|$ becomes the number of common neighbours between u and v and $\sqrt{|\Gamma(u)||\Gamma(v)|}$ becomes the geometric mean of the two neighbourhoods' size.

Definition 2. *CORE: Node u is core iff $|N_\epsilon[u]| \geq \mu$, where N_ϵ, called ϵ-neighbourhood, is $N_\epsilon[u] = \{v \in N[u] : \sigma(u, v) \geq \epsilon\}$.*

Definition 3. *HUB AND OUTLIER: Assume node u does not belong to any cluster C. $u \in H$ iff node v and w exist in $N[u]$ such that $C[v] \neq C[u]$. Otherwise $u \in O$.*

For more information about the algorithm, please refer the paper [1].

We have modified the structural similarity formula in SCAN algorithm to incorporate the weighted edge and named the new formula as weighted structural similarity and the algorithm as Weighted SCAN (WSCAN) algorithm, the details of which is described in Sect. 3.2.

Table 1. Terms and symbols used

Symbol	Definition	Symbol	Definition
ϵ	Threshold of structural similarity, $0 \leq \epsilon \leq 1$	H	Set of hubs in G
		μ	Minimal number of nodes in a cluster
O	Set of outliers in G	$N[u]$	Set of nodes in the structure neighbourhood of node u
G	Given graph	$N_\epsilon[u]$	Set of nodes in the ϵ-neighbourhood of node u
V	Set of nodes in G	$C[u]$	Set of nodes that belong to the same cluster as node u
E	Set of edges in G	$\sigma(u, v)$	Structural similarity between node u and v

Algorithm 1. SCAN Clustering Algorithm

```
Input   : Graph G(V, E), Parameters - ϵ, μ
Output: Set of clusters, hubs, and outliers
 1: for each unclassified vertex v belongs to V do
 2:    if Core(V) then
 3:       Create new clusterID
 4:       for all structurally similar neighbors x of V do
 5:          if x is unclassified or non-member then
 6:             Assign clusterID to x
 7:          end if
 8:          if x is also a core then
 9:             Expand the graph using x also
10:          end if
11:       end for
12:    end if
13: end for
14: Further classify non-members into hubs and outliers
```

3.2 WSCAN as an Expansion of SCAN

In our approach, we construct a graph using the Reply and Re-Tweet (RT) features of tweet data. Adopting the original SCAN approach, an edge between two user nodes is created if a user has Re-Tweeted/Replied to another user, irrespective of how many times they did. However, some users RT more than once, the tweets of the same user. In the same way, some users might reply to another user many times. Clearly influential users get more than one RT.

To take these features into consideration, we have modified the structural similarity formula of SCAN algorithm to use weighted structural similarity (σ_w).

Definition 4. *WEIGHTED STRUCTURAL SIMILARITY: Let $u, v \in V$ where V is the set of nodes, where each node represents a user, $\omega_{u,v}$ is defined as a linear function of RT and Reply count between the two nodes u and v.*

$$\sigma_w(u, v) = \frac{\omega_{(u,v)}}{\omega_{(u,v)} + 1} \frac{|\Gamma(u) \cap \Gamma(v)|}{\sqrt{|\Gamma(u)|\,|\Gamma(v)|}} \tag{2}$$

where Γ is defined earlier and $\omega_{u,v} = \alpha \cdot RT + \beta \cdot R$

This weight factor allows us to retain a significant node which would, otherwise, have been marked as a hub as described in [1]. Figure 2 shows a histogram of Reply and RT count. From this graph we observed that Reply and RT counts are following the same trend, and are proportional to each other. Hence we have used $\alpha = 1$ and $\beta = 1$ in Eq. 2 for all the evaluation purposes.

We illustrate the effectiveness of weights using a toy example Graph in Fig. 1. The edge weights represent Reply and RT counts, which are considered only in WSCAN and have no significance in SCAN algorithm. Let us consider two nodes: Node 5 and 6. In case of SCAN, both these nodes are structurally similar as they have same similarity (σ) to neighbours, i.e., $\sigma(2,5) = \sigma(4,6) = 0.63$, whereas in case of WSCAN, the similarities are $\sigma_w(2,5) = 0.32$, and $\sigma_w(4,6) = 0.47$ respectively. Node 6 seems to have stronger connection with the cluster of Nodes 1, 2, 3 and 4 due to larger edge weight, thereby acting as an influential user who actually has a great influence upon a member of a strongly connected community. By appropriately selecting the value of ϵ, Node 5 can be classified as an outlier, i.e. an insignificant node, whereas, Node 6 can be included in the cluster. The steps followed for clustering the graph in WSCAN are same as SCAN, explained in Algorithm 1.

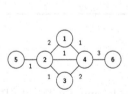

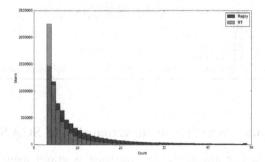

Fig. 1. Example of WSCAN (weighted structural similarity)

Fig. 2. Histogram of Reply and RT counts

4 Our Approach

Our approach has basically the following steps which are discussed in detail in subsequent sections. System diagram is shown in Fig. 3:

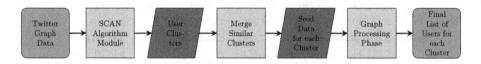

Fig. 3. System block diagram

- **Construction Phase:** Construct graph using Reply and RT feature.
- **Structural Clustering Phase:** Cluster the graph using WSCAN algorithm (refer Sect. 3.2).
- **Merging Similar Clusters:** Combine similar clusters and get the users of each cluster to be used as seed data for next step.
- **Graph Processing Phase:** Expand the list of seed users and rank them using Personalized PageRank algorithm.

4.1 Construction of Graph

Our approach starts with constructing an undirected, weighted graph from Twitter Data using Reply (R) and RT, since these are logically more meaningful than just follow feature when we consider the similarity of two users in interests. Consider two users u_1 and u_2. If u_1 has replied or re-tweeted the tweets of u_2, then (u_1, u_2) will have an edge in our graph. The sum of Reply and RT count between u_1 and u_2 becomes the weight of the edge, and is denoted by $\omega_{(u_1, u_2)}$. We have experimented with R, RT and both R+RT for constructing the graph. The results obtained after executing WSCAN on these three types of graphs with $\epsilon = 0.5, 0.45$ and 0.4 is summarized in Table 2.

Comparison of Reply (R), RT and Reply (R)+RT: Let us consider Table 2 and compare 3 types of graphs:

- **Reply Graph (R):** Users are represented by nodes and edge weight represents the reply count between two users.
- **Re-tweet Graph (RT):** Users are represented by nodes and edge weight represents the RT count between two users.
- **Reply and Re-tweet Graph (R+RT):** Users are represented by nodes and edge weight represents the sum of RT and reply count between two users.

For all the three graphs, we have considered edges whose weights are greater than 2. The observation is given below:

Table 2. Comparison of WSCAN output for the 3 types of graphs ($\mu = 2$)

ϵ	Type	#Vertices	#Hubs	#Outliers	#Clustered vertices	#W-SCAN clusters
0.5	R	2,472,492	1,250,077	402,691	819,724	340,960
	R+RT	2,596,565	1,559,828	383,581	653,156	277,340
	RT	1,121,444	649,780	292,281	179,383	78,114
0.45	R	2,472,492	1,045,536	349,296	1,077,660	410,364
	R+RT	**2,596,565**	**1,371,585**	**341,061**	**883,919**	**347,476**
	RT	1,121,444	610,051	281,454	229,939	94,720
0.4	R	2,472,492	818,927	286,833	1,366,732	457,001
	R+RT	2,596,565	1,149,285	289,819	1,157,461	406,682
	RT	1,121,444	562,266	267,904	291,274	111,709

- **Fraction of Hubs:** R+RT (0.60) > RT (0.57) > R (0.50): Consider a user U who has re-tweeted tweets of users from Cluster 1 and replied to tweets of users from Cluster 2. So in R+RT graph, U could be marked as a hub because of its involvement in both the groups. Whereas if we consider R and RT graphs, this user will be a part of Cluster 1 and Cluster 2 respectively.
- **Fraction of Outliers:** RT (0.26) > R (0.16) > R+RT (0.14): Outliers are those users which do not have any affiliation to any cluster. Nodes with very few edges to any cluster are marked as outliers. As users in R+RT graph have higher degree, the fraction of users marked as outliers are also less.
- **Fraction of WSCAN output Users:** R (0.33) > RT (0.16) > R+RT (0.10): The decrease in the fraction of clustered users in R+RT graph is because of the fact that a large fraction of them were marked as hub. This ensures that whatever users are left have strong affiliation to the cluster.

Another disadvantage of using just R or RT Graph is that the graphs produced are very sparse. When we visualized these graphs, there were many isolated clusters with just two nodes connected to each other. This depicts that there is a lot of one to one communication between pairs of nodes. So we have used R+RT Graph; R Graph to include users which are closer to each other in real-life, and RT Graph to include users which share similar content.

4.2   Find Clusters Using WSCAN

We input the graph constructed in Subsect. 4.1 to the WSCAN approach, as discussed in Sect. 3.2. This program produces clusters of users using weighted structural similarity measure as given in equation (2). We have used $\epsilon = 0.45$ and $\mu = 2$ (refer to the bold text in Table 2). This is because when we consider $\epsilon = 0.5$ and $\mu = 2$, we get very few users in clusters, as most of the users are filtered out as hubs and outliers. This leaves very few (poor quantity), but densely connected users (fine quality). On the other hand, when we consider $\epsilon = 0.4$ and $\mu = 2$, we get relatively good amount of users (fine quantity) but these users are sparsely connected (poor quality). So in order to deal with the tradeoff between quality and quantity, we have chosen $\epsilon = 0.45$ and $\mu = 2$.

4.3   Combine Similar Clusters

In this phase, we merge those clusters which are topically similar. For this we extract the HashTags of all the users in a cluster and combine them to make a document. We make document for each of the cluster produced from previous phase. Then we use Single Pass Clustering algorithm to cluster these document (which are basically clusters).

4.4   Expanding the Clusters (Graph Processing)

We use the output of each of the clusters produced from Subsect. 4.3 separately as seed data for the Graph Processing phase. For a given cluster, the Graph Processing phase basically calculates the Personalized PageRank (PPR) score [7] of all the seeds (in that cluster) and its connected nodes. It uses the Reply and RT features of Tweets to construct the graph, using which the PPR Score of the seeds and their connected nodes is calculated. We use the top 3000 nodes as per PPR Score for evaluating our result. This phase expands the nodes in a cluster by finding its topic related nodes which increases the coverage, removes most of the non-influential users and produces overlapping clusters.

5   Evaluation Experiments

In this section we describe the experiments conducted and their results. For all our experiments, we have used the Japanese Twitter data for the month of December 2015, size of which is about 1.6 TB in compressed form. From this data, we extracted the top 3,000,000 active users (using PageRank score). We have mostly used Hadoop, Pig, Java and Python in our implementations. The graphs in this paper have been generated in Python using graph-tool.[1]

5.1   Visualizing the Process and Observing the Effects

We have generated fan clusters for four celebrities/idol groups: *Arashi*[2], *AKB*[3], *Hanyu*[4] and *Yamashita*[5] using our approach.

Let us consider cluster *Arashi*. The output obtained after executing WSCAN and Single Pass Clustering is pictorially shown in Fig. 4. Here square shaped nodes represent the Seed nodes obtained after WSCAN and Single Pass Clustering. Different colours of the square nodes represent that they belong to different clusters. These different coloured square nodes are combined to a single cluster of Seed nodes (users) in the Single Pass Clustering Phase (Sect. 4.3). Figure 5 visualizes the output obtained after executing the Personalized PageRank

---

[1] https://graph-tool.skewed.de/.
[2] Arashi: A Japanese idol group.
[3] AKB (AKB48): A Japanese idol girls group.
[4] Hanyu (Yuzuru Hanyu): Japanese figure skater & 2014 Olympic champion.
[5] Yamashita (Tomohisa Yamashita): Japanese actor, singer, and TV host.

Table 3. Impact of single pass clustering

	Arashi	AKB	Hanyu	Yamashita
Clusters merged	9	10	2	4
Seed users obtained	70	79	138	109

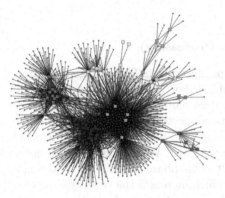 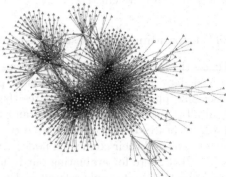

Fig. 4. Arashi cluster after executing WSCAN and Single Pass Clustering. (Color figure online)

Fig. 5. Arashi cluster after executing Personalized PageRank (Graph Processing). (Color figure online)

(Graph Processing Phase) on the seed nodes obtained after the Single Pass Clustering phase for the cluster *Arashi*. The yellow coloured nodes represent the seed nodes. The green coloured nodes represent the nodes obtained after the execution of Personalized PageRank. The red coloured nodes represent the neighbours of seed nodes which are not in the output of Graph Processing Phase. Table 3 shows the number of clusters merged by the Single Pass Clustering Phase. As seen from the table, for *Arashi*, 9 clusters (shown by different coloured square nodes in Fig. 4) were combined by the Single Pass Clustering Phase to produce 70 seed nodes. Details of other clusters can also be seen in the table.

5.2   Evaluation Design

The main problem with evaluation of the clusters is that the perfect set of users for any of these clusters is unknown. So we use crowdsourcing for evaluating our results. Crowdsourcing also eliminates biasing of test results.

Crowdsourcing: For Crowdsourcing, we frame the question so as to check whether the given user is actually interested in the group or not. We have used the topics *Arashi, AKB, Hanyu* and *Yamashita* because of the fact that these are famous in Japan and it would be easy to do crowdsourcing. 844 people participated in crowdsourcing. Each question was reviewed by three people. The average of opinion of three people was taken as the answer of a question. A sample crowdsourcing question with its options is given in Table 4.

Table 4. Sample crowdsourcing question

Question: Is this user <https://twitter.com/twitter_screen_name> interested in Arashi?			
1. Yes, this user is interested	2. No, this user is not interested	3. I don't know	4. Cannot access the account

We are using the output of the Graph Processing Phase for evaluation, considering the top 3000 users based on Personalized PageRank score. We are evaluating our approach with the following approaches:

- **Normal SCAN algorithm (NS):** This system uses SCAN [1]. We use the graph constructed using Reply and RT features (same as in our approach) of Twitter Data as input. We have used $\epsilon = 0.6$ (equivalent to $\epsilon = 0.45$ in WSCAN when we consider minimum edge weight equal to 3) and $\mu = 2$.
- **RB clustering (RB):** This system uses the HashTag information and its TF-IDF as input. We extract HashTags and tokenize them to create document for each user. So considering all the users, we have a list of documents. Then we calculate the TF-IDF of the HashTags. This acts as the feature in RB (Repeated Bisection) based clustering algorithm. We use Bayon[6] Clustering Tool for doing this and extract the users of concerned cluster for evaluation.
- **RB Clustering followed by Personalized PageRank (RB-PR):** This system is similar as the RB system. Only difference is that, using the nodes in the cluster (obtained from RB clustering) as seeds, we expand the cluster, calculating the PPR Score for the seed nodes and their connected nodes (this step finds more nodes related to the seeds). We then consider the top 3000 nodes using the PPR Score.
- **Normal SCAN followed by content clustering (NS-C):** This system is a combination of Normal SCAN (NS) and Content Clustering. Here, we use output of the NS system (described above) and perform Single Pass Clustering technique (except that the input is SCAN clusters and not WSCAN clusters).

5.3   Evaluation Metrics

User Influence Weighted Discounted Cumulative Gain: Discounted cumulative gain (DCG) is a measure of the quality of ranking of information items such as documents and used to evaluate the ranking effectiveness of, for example, the search engines [6]. We extended this in order that the measure takes user's "influenceability" into consideration because influential users are more important in our task. We defined User Influence Weighted DCG (UIWDCG) score to calculate the influence and topical relevance of users in our results. For this, we have used top 3000 users obtained after Graph Processing phase for four topics. UIWDCG score for a list of top n users is defined as:

---

[6] https://code.google.com/archive/p/bayon/.

$$UIWDCG(n) = g_1 \cdot log(IL(v_1) + 1) + \sum_{i=2}^{n} \frac{g_i \cdot log(IL(v_i) + 1)}{log(i)} \qquad (3)$$

where $g_i \in [0, 1, 2, 3]$, $IL(v)$ is the number of in-links to the vertex v representing a Twitter user. g_i is the evaluation score for vertex v at rank i, representing the number of workers out of three who marked this user as relevant to the considered topic, as per crowdsourcing results.

Table 5. Precision (correct seed nodes/total seed nodes) of seed nodes

	Arashi	AKB	Hanyu	Yamashita
Our approach	0.89(62/70)	0.82(65/79)	0.75(103/138)	0.86(94/109)
RB	0.75(758/1013)	0.45(36/80)	0.62(205/331)	0.90(101/111)

5.4   Results and Discussions

Figure 6 shows the average UIWDCG score of all the approaches. X-axis shows the Number of Users considered from top while calculating UIWDCG score. Y-axis shows the Average UIWDCG score. We have used 3 matrices to calculated the UIDWCG score, Favorite Count, Mention Count and Reply+RT count. The value of $IL(v)$ varies depending upon the matrix selected. It is clear from the graph that when we consider top 3000 users, our approach gives more influential users than other approaches. Table 6 shows the UIWDCG score of top 100, 500, 1000 and 3000 users of each cluster for various systems considered for evaluation. We observed the following points:

- Pure graph based approach, such as **NS**, is very weak, and even the least performing one. It failed to output more than 1000 users for any topic.
- Pure content based approach **RB** is also inadequate and it is not comparable with our system.
- The WSCAN step proves to be crucial one as it removes the users who either do not belong to any topic group or the ones who belong to many groups.
- As seen in Fig. 6, although **RB-PR** is better in the beginning, our approach outperforms after rank 300, finally 18.3% gain is observed at rank 3000.
- The quality of seed nodes produced is very good in our approach (ref Table 5).
- Figure 6 shows that RB has better performance than RB-PR. We conclude that good quality seeds are imperative for the good performance of Personalized Pagerank. The seeds obtained from Hashtag frequency in RB-PR were mediocre and hence, deteriorated the performance of Pagrerank. On the other hand, seeds obtained from WSCAN were superior and hence enabled Pagerank to perform effectively.

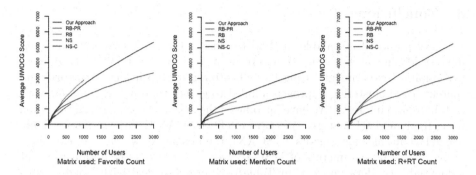

Fig. 6. Average UIWDCG (of Arashi, AKB48, Hanyu, Yamashita) vs #Users

Table 6. UIWDCG score of the four clusters that we considered.

Technique	Top 100	Top 500	Top 1000	Top 3000
Arashi				
Our approach (G+C+G)	575	**1832**	**3092**	**6865**
RB-PR (C+G)	**639**	1658	2545	6006 -
NS-C (G+C)	274	413(194)[a]	-	-
RB (C)	628	1532	2214	-
NS (G)	318	787	-	-
AKB				
Our approach (G+C+G)	473	1564	**2554**	**5981**
RB-PR (C+G)	**654**	**1619**	2439	5333
NS-C (G+C)	193(52)[a]	-	-	-
RB (C)	129(80)[a]	-	-	-
NS (G)	46(15)[a]	-	-	-
Hanyu				
Our approach (G+C+G)	605	**1807**	**2983**	**6160**
RB-PR (C+G)	**612**	1455	2451	5223
NS-C (G+C)	238(34)[a]	-	-	-
RB (C)	543	-	-	-
NS (G)	272(83)[a]	-	-	-
Yamashita				
Our approach (G+C+G)	**578**	**1556**	**2282**	**4283**
RB-PR (C+G)	575	1263	1635	3121
NS-C (G+C)	420	444(118)[a]	-	-
RB (C)	518	-	-	-
NS (G)	192(24)[a]	-	-	-

[a]Since 100/500 nodes are unavailable, number of nodes given inside () are used for calculation.

6   Conclusions

We have proposed a scalable method to cluster Twitter Users based on their interest looking at both the structure as well as the contents. According to our empirical experiments using large Twitter Data, both pure structural and content-based clustering approaches failed to gather thoroughly, the users with certain topical interests. We have introduced the notion of constructing the graph using Reply and RT features. We have modified SCAN [1] to incorporate Reply and RT count as edge weights (WSCAN), the benefits of which were explained in Sect. 3.2. The parameters of WSCAN were chosen so as to obtain few, but influential seed data, as seen in Table 5. In order to deal with isolated clusters having similar contents, we have used content-based merging using Textual Similarity. We have illustrated the effects of this step by visualizing the graph data (Figs. 4 and 5). The superiority of the proposed process to merge clusters obtained by WSCAN algorithm is observed in Table 3. The Graph Processing phase improved the coverage of our system and enabled us to obtain influential users related to the seed data. We carried out evaluations for topical relevance of clustered Twitter users by Crowdsourcing and observed significant improvement over state-of-the-art approaches on both precision-recall curves and UIWDCG measures. Our system outperforms the best performing baseline system, RB-PR, with 18.3% gain in Average UIWDCG Score for Top 3000 Users, as seen in Fig. 6. Possible future work includes looking into the contents in more detail to improve the results. Also, Topic Recognition approach needs to be improved for much better results.

References

1. Xu, X., Yuruk, N., Feng, Z., Schweiger, T.A.: Scan: a structural clustering algorithm for networks. In: Proceedings of the 13th ACM SIGKDD International Conference on Knowledge Discovery and Data Mining, pp. 824–833. ACM (2007)
2. Ding, C.H., He, X., Zha, H., Gu, M., Simon, H.D.: A min-max cut algorithm for graph partitioning and data clustering. In: 2001 Proceedings of the IEEE International Conference on Data Mining, ICDM 2001, pp. 107–114. IEEE (2001)
3. Newman, M.E., Girvan, M.: Finding and evaluating community structure in networks. Phys. Rev. E **69**(2), 026113 (2004)
4. Leskovec, J., Lang, K.J., Dasgupta, A., Mahoney, M.W.: Statistical properties of community structure in large social and information networks. In: Proceedings of the 17th International Conference on World Wide Web, pp. 695–704. ACM (2008)
5. Zhang, Y., Wu, Y., Yang, Q.: Community discovery in Twitter based on user interests. J. Comput. Inf. Syst. **8**(3), 991–1000 (2012)
6. Järvelin, K., Kekäläinen, J.: Cumulated gain-based evaluation of IR techniques. ACM Trans. Inf. Syst. (TOIS) **20**(4), 422–446 (2002)
7. Haveliwala, T.: Topic-sensitive pagerank. In: Proceedings of the 11th International Conference on World Wide Web, Honolulu, Hawaii, USA, pp. 517–526 (2002)
8. Cha, M., Haddadi, H., Benevenuto, F., Gummadi, P.K.: Measuring user influence in Twitter: the million follower fallacy. In: ICWSM, vol. 10, pp. 10–17 (2010)

9. Avnit, A.: The million followers fallacy. http://blog.pravdam.com/the-million-followers-fallacy-guest-post-by-adi-avnit/ (2009). Accessed 2 Aug 2016
10. Hayashi, K., Maehara, T., Toyoda, M., Kawarabayashi, K.-I.: Real-time top-r topic detection on Twitter with topic hijack filtering. In: Proceedings of the 21th ACM SIGKDD International Conference on Knowledge Discovery and Data Mining, KDD 2015, pp. 417–426 (2015)
11. Weng, J., Lim, E.-P., Jiang, J., He, Q.: Twitterrank: finding topic-sensitive influential twitterers. In: Proceedings of the Third ACM International Conference on Web Search and Data Mining, WSDM 2010, pp. 261–270 (2010)

An Enhanced Markov Clustering Algorithm Based on *Physarum*

Mingxin Liang[1], Chao Gao[1,2,3($\boxtimes$)], Xianghua Li[1], and Zili Zhang[1,4($\boxtimes$)]

[1] School of Computer and Information Science, Southwest University,
Chongqing 400715, China
{cgao,zhangzl}@swu.edu.cn
[2] Potsdam Institute for Climate Impact Research (PIK), 14473 Potsdam, Germany
[3] Institute of Physics, Humboldt-University zu Berlin, 12489 Berlin, Germany
[4] School of Information Technology, Deakin University,
Geelong, VIC 3220, Australia

Abstract. Community mining is a vital problem for complex network analysis. Markov chains based algorithms are known as its easy-to-implement and have provided promising solutions for community mining. Existing Markov clustering algorithms have been optimized from the aspects of parallelization and penalty strategy. However, the dynamic process for enlarging the inhomogeneity attracts little attention. As the key mechanism of Markov chains based algorithms, such process affects the qualities of divisions and computational cost directly. This paper proposes a hybrid algorithm based on *Physarum*, a kind of slime. The new algorithm enhances the dynamic process of Markov clustering algorithm by embedding the *Physarum*-inspired feedback system. Specifically, flows between vertexes can enhance the corresponding transition probability in Markov clustering algorithms, and vice versa. Some networks with known and unknown community structures are used to estimate the performance of our proposed algorithms. Extensive experiments show that the proposed algorithm has higher NMI, Q values and lower computational cost than that of the typical algorithms.

1 Introduction

Most real-world networks such as social, biological, and protein networks are organized and represented well in communities [1]. In each network with a community structure, the intra-community edges are more than the inter-community edges. Community mining seeks relevant clusters at different granular levels in a network that could affect or explain the global behavior of a whole complex system. Due to its importance for understanding potential functional and dynamical characteristics of a network, ranging from biology and engineering to social science and medicine, community mining has drawn widespread attention [2].

Although many algorithms have been proposed for mining community structures, Markov chain-based algorithms are widely used in the field of bioinformatics due to their light-weight and easy implementation in addition to the robustness of the effects of the topological noises [1,3]. And for releasing the potential

© Springer International Publishing AG 2017
J. Kim et al. (Eds.): PAKDD 2017, Part I, LNAI 10234, pp. 486–498, 2017.
DOI: 10.1007/978-3-319-57454-7_38

of Markov clustering algorithms further, researches have optimized them from many aspects, such as parallelization [4] and penalty strategy [5]. However, the key mechanism (i.e., the Markov chain-based dynamic process) plays an important role in the detected divisions and computational cost of community mining, which attracts little attention based on our survey in this field. During the Markov chain-based dynamic process, the flow simulation transfers in a network based on the expanding and contracting process alternately [3]. With transition going on, the flow within a community will flow together. In this paper, we aim to optimize the dynamic process of MCL algorithms based on a novel biological finding.

Biological studies have recently demonstrated the intelligence of a slime, *Physarum*, which shows an ability to solve maze problems and design networks without a central organ [6,7]. *Physarum* foraging behavior consists of two simultaneous self-organized processes: expansion and contraction [8]. Based on the feedback system of such processes, a *Physarum*-inspired mathematical model (denoted as PM) has been proposed [9], which shows an ability to accelerate the rate of convergence and improve the searching ability of algorithms [10].

Due to the observations we show above, we find that PM and MCL have similar flow simulations, both of which include expansion and contraction processes. However, the contraction process of MCL is based on mapping the flow to itself. In contrast, the contraction process of PM is implemented based on a feedback flow between the pseudopodia and the protoplasmic fluxes. Therefore, we are wondering can we optimize the dynamic process of MCL based on the mature feedback system in PM.

The remaining of this paper is organized as follows. Section 2 introduces the feature and basic steps of Markov clustering algorithm. Section 3 formulates our proposed *Physarum*-inspired model in which a new feedback flow is designed. Section 4 implements some experiments in order to demonstrate that our proposed algorithm could improve the qualities of detected divisions and reduce the computational cost of presentational MCL algorithms. Finally, Sect. 5 concludes this paper.

2 Related Works

According to the universal definition, a Markov chain is a sequence of a random variable X, which satisfies the Markov property. The Markov property holds that the future state of a variable depends only on its present state and is independent from its past states. Equation (1) describes a formulation of the Markov property. The time-homogeneous hypothesis, which holds that the state transition probabilities are independent from the time step ts, is a common assumption of Markov chains. A formal description of the time homogeneous hypothesis is shown in Eq. (2). Moreover, all possible values of X form a countable set of "state spaces" in the Markov chain. The probabilities of state transition are called transition probabilities.

$$P(X_{ts+1} = x | X_1, ..., X_{ts}) = P(X_{ts+1} = x | X_{ts}) \tag{1}$$

$$P(X_{ts+1} = x_j | X_{ts} = x_i) = P(X_{ts} = x_j | X_{ts-1} = x_i) \tag{2}$$

Following the Markov property and time-homogeneous hypothesis, Markov chains can be reviewed as random walks on graphs, where vertices represent the state space and edges stand for the transition probabilities. The matrices P and M represent the transition probabilities and distribution of variables, respectively. And a random walk with l time steps can be implemented based on Eq. (3). An element $p_{i,j}^l$ in the matrix P^l indicates the transition probability from v_i to v_j with l time steps, and an element $m_{i,j}^{ts}$ of m^{ts} indicates the probability of variable i falling into status j at time step ts.

$$M^{ts+l} = M^{ts} \times \prod_{i=1}^{l} P \tag{3}$$

The flow simulation of the Markov clustering algorithm is implemented by embedding a dynamic process in such a random walk [11], in which fluxes of vertices within a community will flow together with iteration going. Researches have optimized Markov clustering from many aspects, such as the parallelization and penalty strategy [4,5]. However, the Markov chain-based dynamic process, which directly affects the qualities of detected divisions and computational cost of MCL algorithms, attracts little attention of researchers. In this paper, we aim to optimize the dynamic process of the Markov clustering algorithm and maximize its potential based on the *Physarum* mathematical model.

3   *Physarum*-Inspired Markov Clustering Algorithm

3.1   Formulation of Community Mining Based on MCL

A basic assumption of MCL is a flow diffusion in a network. There are two main processes of MCL: expanding and contracting processes, which are implemented alternately based on two matrices (i.e., the flow distribution and canonical transition matrices) and three operators (i.e., expansion, inflation, and pruning operators) [3]. More specifically, the distribution of fluxes and the transition quantities of the fluxes are denoted as M and T, respectively. And $m_{i,j}$ and $t_{i,j}$ represent the flux flowing from v_j to v_i and the quantity of flux flowing from v_i to v_j in an iteration step. The initial M and T are derived from the adjacency matrix A of a network, i.e., $M_{i,j}^0 = T_{i,j}^0 = A_{i,j} / \sum_k A_{k,j}$.

Based on the matrices M and T, a typical Markov clustering algorithm (i.e., R-MCL) is used to introduced the expansion, inflation, and pruning operators of MCL. First, the expansion operator is used to spread fluxes in a network. It is implemented by multiplying the matrix by the canonical transition matrix T, as shown in Eq. (4). Then, the inflation operator is used to enlarge the inhomogeneity and prevent M from converging to the principal eigenvector of T. Such an operator raises every entry in M to the power of r and then normalizes the columns as shown in Eq. (5). As all of the entries in M are less than or equal to 1, the inhomogeneity in each column is enlarged. In other words, this operator aims to strengthen the strong flows and weaken the weak flows. Finally, the pruning operator accelerates the rate of convergence and reduces non-zero entries to save memory by removing entries under a pre-established threshold. Such thresholds are based on the average and maximum values within columns [3].

$$M^{st+1} = M^{st} \times T \tag{4}$$

$$M^{ts+1} = \frac{[M_{i,j}^{ts+1}]^r}{\sum_k [M_{k,j}^{ts+1}]^r} \tag{5}$$

In each iteration of R-MCL, the expansion, inflation, and pruning operators are executed alternately. With the iteration going on, most of the vertices will find an attractor, to which all of their fluxes flow. Each column of M has only one positive value when M converges. The row indexes of those positive values are community labels of the corresponding vertices of each column. The vertices, whose fluxes flow to the same attractor, are clustered to a community.

3.2 *Physasrum*-Inspired Feedback System

Physarum is a unicellular and multi-headed slime, which has drawn more attention due to its exhibited ability to build a self-adaptive and highly effective network for foraging [6,7]. The key mechanism of such intelligent behavior is modeled by a mathematical feedback system (named as PM) [7,9]. This PM is introduced as follows.

- The basic assumption of PM is an approximate Poiseuille flow of cytoplasmic fluxes from a source (denoted as s) to a destination (denoted as d) [7]. In the *Physarum* network, $L_{i,j}$ indicates the distance between vertices i and j. With $D_{i,j}$ standing for the conductivity of $e_{i,j}$ and p_i representing the pressure of v_i, the flux $PQ_{i,j}$ can be expressed as shown in Eq. (6) based on the Poiseuille law. Assuming that the capacity of all vertices is zero, the flux at each vertex can be expressed as Eq. (7) by considering the conservation law.

$$PQ_{i,j}^{ts} = \frac{D_{i,j}^{ts}}{L_{i,j}} |p_i^{ts} - p_j^{ts}| \tag{6}$$

$$\sum_i PQ_{i,j}^{ts} = \begin{cases} I_0, j == s \\ -I_0, j == d \\ 0, others \end{cases} \tag{7}$$

- There is a feedback between conductivities and fluxes in PM. Bio-experiments shows that the conductivities of tubes with lager fluxes are reinforced and that those with smaller fluxes degenerate. Such a feedback mechanism indicates that $D_{i,j}$ changes over time according to the flux $PQ_{i,j}$, which is expressed in Eq. (8), with a parameter u controlling the feedback [9]. In order to provide a clearer formulation and easier operation, we discretize Eqs. (8) to (9).

$$\frac{d}{dt}D_{i,j} = |PQ_{i,j}|^u - D_{i,j} \tag{8}$$

$$\frac{D_{i,j}^{ts+1} - D_{i,j}^{ts}}{\Delta ts} = |PQ_{i,j}^{ts}|^u - D_{i,j}^{ts} \tag{9}$$

As the total flux is kept constant (i.e., I_0), there is a competition among the edges. With the iteration going on, the crucial tubes survive while the others disappear. Based on the preceding steps, a highly efficient feedback system is constructed. Inspired by such feedback system, we build a novel feedback system in MCL by simulating the interaction between the fluxes and conductivities in PM.

3.3 *Physarum*-Inspired Markov Clustering Algorithm

Typical MCL is based on a flow simulation with the expansion and contraction processes, which is similar to the *Physarum* intelligent behavior. In this paper, we aim to improve the qualities of divisions and reduce the computational cost of MCL by building a feedback system based on the biological studies of *Physarum*. Specifically, the dynamic process of MCL is optimized in terms of three aspects: (1) the *Physarum*-inspired inflation operator, (2) the multi-step expansion operator, and (3) the terminal condition. The following sections introduce these modifications and then demonstrate an overview of P-MCL.

- The *Physarum*-inspired inflation operator: In this *Physarum*-inspired inflation operator, we add a feedback step in the wake of the original inflation mechanism (i.e., Eq. (5)). Regarding the flow distribution matrix M and canonical transition matrix T as fluxes and conductivities in *Physarum* networks, a new feedback interaction emerges between the distribution and transition matrices as follows. To differ the notation of canonical transition matrix T, the *Physarum*-based transition matrix is denoted as PT. And inspired by Eq. (8) in PM, the flow distribution matrix M has an effect on PT based on Eq. (10). We further discretize the Eqs. (10) to (11) for simplifying calculation, where λ stands for the time step of discretization and u is inherited from Eq. (8) in PM. And such two parameters collectively control the effect of M on PT. Because the sum of each column should still be equal to 1 for a transition matrix. After the feedback step, a normalized step is implemented based on Eq. (12). Then PT feeds back to M based on Eq. (4) in the next iteration step.

$$\frac{d}{dt}PT_{i,j} = |M_{i,j}|^u - PT_{i,j} \tag{10}$$

$$PT_{i,j}^{ts+1} = (1 - \lambda) \cdot [M_{i,j}^{ts}]^u + \lambda \cdot PT_{i,j}^{ts} \tag{11}$$

$$PT_{i,j}^{st+1} = \frac{PT_{i,j}^{st+1}}{\sum_k PT_{k,j}^{st+1}} \tag{12}$$

– The multi-step expansion operator: To solve the problem of the excessive partition of MCL [3], P-MCL also considers the effects of the flow length in each iteration step. Based on the properties of Markov chains, a l-length flow can be expressed as shown in Eq. (13). In general, a larger network requires a larger l to control the number of communities.

$$M^{st+1} = M^{st} \times \prod_{i=1}^{l} PT \tag{13}$$

– The terminal condition: As mentioned previously, the MCL algorithms will continue looping until the convergence of M. The maximal value of changes among all of the entries in M can be used to estimate the change of the whole M as well as the convergence situation. Therefore, *Energy* is defined in Eq. (14). The terminal condition is that *Energy* drops to a predetermined threshold.

$$Energy^{ts} = Max\{|M_{i,j}^{ts} - M_{i,j}^{ts-1}| \ |\forall j, i\} \tag{14}$$

– The overview of P-MCL: Fig. 1 shows an illustration of P-MCL. First, Fig. 1(a) and (b) illustrate and contrast the similarities and differences between PM and MCL based on their flow simulation and working mechanisms. Based on those similarities, Fig. 1(c) reports how to build a feedback system in MCL based on the *Physarum* model.

With an interaction between the flow distribution matrix M and *Physarum*-based transition matrix PT, a new feedback flow is constructed, which helps to improve the qualities of detected divisions. In addition to the qualities of detected divisions, computational cost is another important property of the community

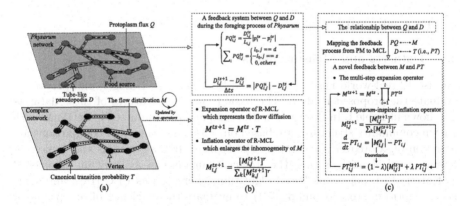

Fig. 1. (a) and (b) compare and contrast the similarities and differences between PM and MCL based on their flow simulation and working mechanisms. (c) reports how to build a feedback relationship between expansion operator and inflation operator in the MCL based on the feedback system between PQ and D during the foraging process of *Physarum*.

mining algorithm. And the cost of such feedback flow is the computational cost in expansion and inflation operators. However, it could also reduce the maximal iteration step (i.e., MI). In the next section, the computational cost of P-MCL is discussed by both theoretical and comparison analyses for verifying the superiority of P-MCL on computational cost. Comparison analyses are evaluated by (1) the convergence rate measured based on the maximal iteration step (i.e., MI) and (2) the running time measured in seconds.

The pseudo-code of P-MCL is represented in Algorithm 1. As a multilevel framework for obtaining large gains in speed is proposed in [3], we also adopt such framework for P-MCL, denoted as MLP-MCL. More detailed information about the multilevel framework can be found in [3].

Algorithm 1. P-MCL

Input: A network adjacent matrix A.
Output: A division result.
Step 1: $A=A+I$. Here, I is a unit matrix.
Step 2: Initializing M and PT.
Step 3: Implementing flows based on Eq. (13).
Step 4: Computing M^{ts+1} based on Eq. (5).
Step 5: Updating the PT^{ts+1} based on Eqs. (11) and (12).
Step 6: Deleting the elements in M^{ts}, which is smaller than
　　　　the threshold obtained based on the pruning scheme.
Step 7: **If** the terminal condition is satisfied, go to Step **8**.
　　　　Else go to Step **3**.
Step 8: Outputting the division result.

---

4 Experiments

4.1 Datasets and Measurements

We use 12 real-world networks collected by Newman[1], Batagelj and Mrvar[2] to evaluate the proposed P-MCL. The basic topological features of these datasets are shown in Table 1. In this section, the comparisons conducted in networks with known community structures are based on NMI, and the other comparisons are based on Q. Moreover, we compare the typical MCL algorithms and some representational algorithms (i.e., the stochastic-model-based algorithm (shortened as Karrer [12]), page-rank-based algorithm (shortened as PPC [13]), and some novel algorithms (e.g., Combo [15])) to evaluate the efficiency of our proposed algorithm. In the comparison, if the compared algorithm is stochastic, the results are based on 10 times repeated experiments.

---

[1] http://www-personal.umich.edu/~mejn/netdata/.
[2] http://vlado.fmf.uni-lj.si/vlado/vladonet.htm.

Table 1. The basic topological features of the real-world networks. *No* represents the used notations of those networks. k and C stand for the average degree and clustering coefficient, respectively. #C indicates the number of communities based on ground truth, in which N means the community structures are unknown.

No	Name	Vertex	Edges	k	C	#C	No	Name	Vertex	Edges	k	C	#C
G_1	Karate	34	78	4.588	0.588	2	G_7	Lesmis	77	254	6.597	0.056	N
G_2	Dolphins	62	160	5.129	0.303	2	G_8	Adjnoun	112	425	7.589	0.190	N
G_3	PolBooks	105	441	8.400	0.488	3	G_9	Celegans	297	1540	9.656	0.326	N
G_4	Football	115	613	10.660	0.403	12	G_{10}	Roget	674	613	1.819	0	N
G_5	PolBlogs	1490	19025	22.438	0.360	2	G_{11}	Netscience	1589	2742	3.451	0.878	N
G_6	YeastL	2361	7182	5.856	0.200	13	G_{12}	Power	4941	6594	2.669	0.107	N

4.2 Evaluation in Network with Known and Unknown Community Structures

NMI is used to evaluate the similarities between the divisions returned by algorithms and known community structures. The more similar to the known community structure the division is, the higher NMI value the division has. And little difference on divisions will lead to a big difference on NMI. When all the algorithms cannot find a division with $NMI = 1$, the difference of NMI just reflects part of division performances. Taking the dolphins network as an example, Fig. 2 illustrates that the division returned by P-MCL is more close to the known community structures, compared with R-MCL. In details, the communities divided by P-MCL are almost coincident with the known community structures. Meanwhile, R-MCL mixes the vertices belonging to different communities together and splits the vertices of a community into several parts. Measured by NMI, the NMI_{P-MCL} value is 0.8888, and the NMI_{R-MCL} value is 0.6470.

In addition, the football network is also a famous network for community mining. As shown in Fig. 3, P-MCL can find the basic structures of the known

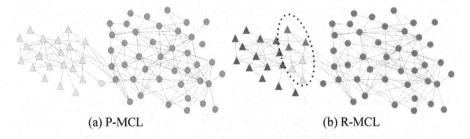

(a) P-MCL   (b) R-MCL

Fig. 2. Division results of (a) P-MCL and (b) R-MCL in the dolphins network. The shapes of the vertices represent the known community structures, and the colors indicate the division results of the algorithms. The circle emphasizes the main difference in the results returned by R-MCL and P-MCL. P-MCL is more accurate than R-MCL, where the NMI_{P-MCL} value is 0.8888, and the corresponding NMI_{R-MCL} value is 0.6470. (Color figure online)

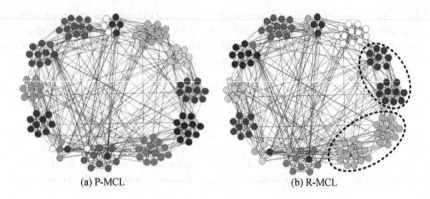

(a) P-MCL   (b) R-MCL

Fig. 3. Division results of (a) P-MCL and (b) R-MCL in the football network. The positions of the vertices represent information about the known community structures, and the colors of the vertices represent the division results generated by P-MCL and R-MCL. According to the description of NMI, the NMI_{P-MCL} value is equal to 0.9283, and the NMI_{R-MCL} value is equal to 0.8099. (Color figure online)

communities and R-MCL mixes some of the communities together, as shown in Fig. 3(b). Moreover, the NMI_{P-MCL} value shows an 12.83% improvement compared with the NMI_{R-MCL} value.

To further estimate the performance of P-MCL, a comparison among our algorithm and others is reported in Table 2, based on NMI. The results indicate that P-MCL and MLP-MCL have the highest NMI values among the algorithms. Those results indicate that divisions returned by P-MCL and MLP-MCL are more close to the known community structures. And due to the inherent structures of some datasets (e.g., G_3 and G_5), all the algorithms cannot find the division with higher NMI. However, the proposed P-MCL and MLP-MCL still have the highest NMI values on those datasets.

Table 2. Comparison of community mining in networks with known community structures in terms of NMI. The results show that MLP-MCL and P-MCL have better NMI values and exhibit a significant improvement compared with MLR-MCL and R-MCL.

NET	ALG						
	Combo	Karre	PPC	R-MCL	MLR-MCL	P-MCL	MLP-MCL
G_1	68.73	83.72	70.71	83.65	83.65	100	100
G_2	57.15	88.88	57.92	64.70	64.70	88.88	88.88
G_3	56.03	54.20	57.30	52.50	52.50	59.23	56.86
G_4	89.03	87.06	85.61	80.99	80.99	92.83	92.42
G_5	39.37	46.66	39.39	38.52	38.41	39.58	38.58
G_6	14.60	6.43	16.54	23.34	17.58	35.18	32.5

Table 3. Comparison results in networks with unknown community structures in terms of Q. The greater the modularity Q is, the better the community structure obtained. The results show that P-MCL and MLP-MCL have better modularity values and exhibit a significant improvement compared with R-MCL and MLR-MCL.

NET	ALG						
	Combo	Karre	PPC	R-MCL	MLR-MCL	P-MCL	MLP-MCL
G_7	0.560	0.457	0.454	0.465	0.465	0.615	0.654
G_8	0.302	−0.104	0.255	−0.112	−0.112	0.376	0.367
G_9	0.555	0.249	0.374	0.128	0.128	0.496	0.555
G_{10}	0.936	0.008	0.933	0.924	0.946	0.956	0.952
G_{11}	0.959	0.640	0.779	0.962	0.880	0.971	0.966
G_{12}	0.939	0.179	0.930	0.814	0.773	0.913	0.902

As not all real-world datasets have known community structures, for the datasets with unknown communities, the modularity Q is widely used to evaluate the qualities of hard division results [14]. The modularity values of the division results are listed in Table 3. As shown in the table, P-MCL and MLP-MCL have the higher Q values in all of the networks and the highest Q values in five of eight datasets. In general, these results verify that the proposed algorithm could improve the qualities of divisions returned by Markov clustering algorithms, in term of Q.

4.3   Computational Cost

The analysis of computational cost in this paper includes two parts: (1) theoretical analysis, denoted by big-O notation, and (2) comparison analysis, including running time in seconds and a coverage rate measured by the maximal iteration step (i.e., MI) defined in the over view of P-MCL.

The whole time complexity of P-MCL consists of three parts used to implement expansion, inflation, and pruning operators. The time complexity of the expansion operator, including the matrix multiplication, is $O(\sum_{i=1}^{n} d_i^2)$. Using k to indicate the average degree of vertexes, it can be expressed as $O(nk^2)$. And generally speaking, k is much smaller than n. Moreover, the time complexities of the inflation and pruning operators are both $O(m)$. As $m < n^2$, the total worst time complexity of P-MCL is presented as $O(MI \times (nk^2 + m + m)) = O(MI \times n(k^2 + n))$, where MI is the maximal iteration step. The expression of complexity of P-MCL is similar to those of other Markov clustering algorithms, such as R-MCL. The key factor in the complexities of the Markov clustering algorithms is MI. However, it is difficult to estimate MI universally. A comparison of MI between R-MCL and P-MCL is reported in the following discussions to demonstrate that the proposed feedback system can reduce MI significantly.

Besides theoretical analysis, two kinds of numerical comparisons are implemented to analyze time complexity. First, the comparison on MI is shown in

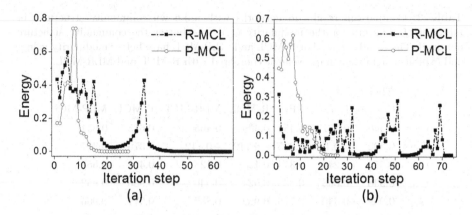

Fig. 4. Dynamic *Energies* of P-MCL and R-MCL in (a) G_5 and (b) G_{11} networks. As shown in (a) and (b), $Energy_{P-MCL}$ has a higher descent rate and a smaller fluctuation than $Energy_{R-MCL}$, which results in a lower maximal iteration step.

Fig. 4, which reports the dynamic *Energies* of R-MCL and P-MCL in increments of iteration steps. As shown in Fig. 4(a) and (b), the $Energies_{P-MCL}$ drop faster in the G_5 and G_{11} networks and satisfy the terminal condition within fewer iteration steps, compared with R-MCL. Those results demonstrate the proposed feedback flow has an ability of reducing the maximal iteration step. However, the computational cost of MCL is based on the time cost of operators and the maximal iteration step comprehensively. Thereby, we further compare the running time of R-MCL and P-MCl in Table 4, which shows the computation cost comprehensively and directly. According to this table, P-MCL has lower computational cost in seven datasets. Because our proposed algorithm increases the computational cost in each iteration step, but reduces the *MI* from a whole perspective. Therefore, the gaps between the running time of R-MCL and P-MCL are not notable in the small networks. But, with the increment of network scales, P-MCL shows a significant superiority.

Table 4. Running time of R-MCL, P-MCL in twelve real-world networks with different scales in seconds. In most datasets, P-MCL has a lower running time than that of R-MCL. The boldface indicates better results.

ALG	NET											
	G_1	G_2	G_3	G_4	G_5	G_6	G_7	G_8	G_9	G_{10}	G_{11}	G_{12}
R-MCL	**0.01**	**0.02**	0.12	**0.05**	0.12	1195.60	**0.03**	0.31	**0.24**	2.43	12.83	888.25
P-MCL	0.04	0.03	**0.12**	0.09	**0.10**	**221.42**	0.07	**0.20**	0.53	**1.62**	**3.77**	**366.53**

5   Conclusion

Markov clustering algorithm is an effective method for us to detect the community structure of a network. By mapping the flow distribution and transition matrices in a MCL to the cytoplasmic fluxes and conductivities of tubes in the *Physarum* network respectively, this paper proposes an enhanced Markov clustering algorithm, denoted as P-MCL. The new algorithm aims to optimize the dynamic process of MCL through constructing a feedback between the expansion and inflation operator in MCL. Some experiments have shown that our proposed algorithm can improve the qualities of detected divisions in terms of Q and NMI, and reduce the computational cost.

Acknowledgement. Prof. Zili Zhang and Dr. Chao Gao are the corresponding authors. This work is supported by the National Natural Science Foundation of China (Nos. 61403315, 61402379), CQ CSTC (No. cstc2015gjhz40002), Fundamental Research Funds for the Central Universities (Nos. XDJK2016A008, XDJK2016B029, XDJK2016D053) and Chongqing Graduate Student Research Innovation Project (No. CYS16067).

References

1. Fortunato, S.: Community detection in graphs. Phys. Rep. **486**(3), 75–174 (2010)
2. Centola, D.: The spread of behavior in an online social network experiment. Science **329**(5996), 1194–1197 (2010)
3. Satuluri, V., Parthasarathy, S.: Scalable graph clustering using stochastic flows: applications to community discovery. In: The 15th International Conference on Knowledge Discovery and Data Mining, pp. 737–746. ACM (2009)
4. Niu, Q., Lai, P.W., Faisal, S.M., Parthasarathy, S., Sadayappan, P.: A fast implementation of MLR-MCL algorithm on multi-core processors. In: The 21st International Conference on High Performance Computing, pp. 1–10. IEEE (2014)
5. Satuluri, V., Parthasarathy, S., Ucar, D.: Markov clustering of protein interaction networks with improved balance and scalability. In: The 1st ACM International Conference on Bioinformatics and Computational Biology, BCB 2010, pp. 247–256. ACM, New York (2010)
6. Nakagaki, T., Yamada, H., Tóth, Á.: Intelligence: Maze-solving by an amoeboid organism. Nature **407**(6803), 470 (2000)
7. Tero, A., Takagi, S., Saigusa, T., Ito, K., Bebber, D.P., Fricker, M.D., Yumiki, K., Kobayashi, R., Nakagaki, T.: Rules for biologically inspired adaptive network design. Science **327**(5964), 439–442 (2010)
8. Liu, Y., Gao, C., Zhang, Z., Wu, Y., Liang, M., Tao, L., Lu, Y.: A new multi-agent system to simulate the foraging behaviors of Physarum. Nat. Comput. **16**, 15–29 (2017)
9. Tero, A., Kobayashi, R., Nakagaki, T.: A mathematical model for adaptive transport network in path finding by true slime mold. J. Theor. Biol. **244**(4), 553–564 (2007)
10. Liu, Y.X., Gao, C., Zhang, Z.L., Lu, Y.X., Chen, S., Liang, M.X., Li, T.: Solving NP-hard problems with Physarum-based ant colony system. IEEE/ACM Trans. Comput. Biol. Bioinf. **14**(1), 108–120 (2017)

11. Van Dongen, S.M.: Graph clustering by flow simulation. Ph.D. thesis, Universiteit Utrecht (2001)
12. Karrer, B., Newman, M.E.: Stochastic blockmodels and community structure in networks. Phys. Rev. E **83**(1), 016107 (2011)
13. Tabrizi, S.A., Shakery, A., Asadpour, M., Abbasi, M., Tavallaie, M.A.: Personalized pagerank clustering: a graph clustering algorithm based on random walks. Phys. A **392**(22), 5772–5785 (2013)
14. Jin, D., Chen, Z., He, D., Zhang, W.: Modeling with node degree preservation can accurately find communities. In: The 24th International Conference on Artificial Intelligence, AAAI, pp. 160–167 (2015)
15. Sobolevsky, S., Campari, R., Belyi, A., Ratti, C.: General optimization technique for high-quality community detection in complex networks. Phys. Rev. E **90**, 012811 (2014)

Exploiting Geographical Location for Team Formation in Social Coding Sites

Yuqiang Han[1(✉)], Yao Wan[1,3], Liang Chen[2], Guandong Xu[3], and Jian Wu[1]

[1] College of Computer Science and Technology, Zhejiang University,
Hangzhou, China
{hyq2015,wanyao,wujian2000}@zju.edu.cn
[2] School of Data and Computer Science, Sun Yat-Sen University, Guangzhou, China
jasonclx@gmail.com
[3] Advanced Analytics Institute, University of Technology, Sydney, Australia
guandong.xu@uts.edu.au

Abstract. Social coding sites (SCSs) such as GitHub and BitBucket are collaborative platforms where developers from different background (e.g., culture, language, location, skills) form a team to contribute to a shared project collaboratively. One essential task of such collaborative development is how to form a optimal team where each member makes his/her greatest contribution, which may have a great effect on the efficiency of collaboration. To the best of knowledge, all existing related works model the team formation problem as minimizing the communication cost among developers or taking the workload of individuals into account, ignoring the impact of geographical location of each developer. In this paper, we aims to exploit the geographical proximity factor to improve the performance of team formation in social coding sites. Specifically, we incorporate the communication cost and geographical proximity into a unified objective function and propose a genetic algorithm to optimize it. Comprehensive experiments on a real-world dataset (e.g., GitHub) demonstrate the performance of the proposed model with the comparison of some state-of-the-art ones.

Keywords: Team formation · Geographical location · Social coding sites · Genetic algorithm

1 Introduction

With the prevalence of social networks in the world, social coding sites (SCSs) such as GitHub[1] and BitBucket[2] are changing software development toward a more collaborative manner by the way of integrating social media functionality and distributed version control tools. In SCSs, developers with different background (e.g., culture, language, location, skills) form a team and work collaboratively to contribute to a project, dramatically enhancing the efficiency of

---

[1] https://github.com.
[2] https://bitbucket.org.

© Springer International Publishing AG 2017
J. Kim et al. (Eds.): PAKDD 2017, Part I, LNAI 10234, pp. 499–510, 2017.
DOI: 10.1007/978-3-319-57454-7_39

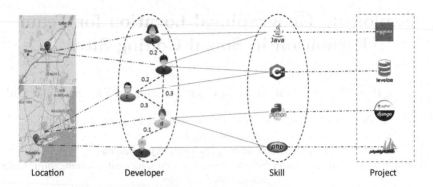

Fig. 1. Schema of developers' profiles and their corresponding skills in GitHub. The left part represents the geographical information of developers; the middle part represents a heterogeneous network among users and skills which can be constructed based on the collaborative development records of developers; the right part represents that each skill of developers can be extracted from his/her contributed projects in GitHub.

development when compared with individual development. One essential task of collaborative development is how to form a optimal team where each member makes his/her greatest contribution, which may have a great effect on the efficiency of collaboration. We called this kind of task as team formation problem.

There have been several related works [2,7,10,11,13] that try to address the team formation problem from different perspectives. In [10,13], the authors define several kinds of communication cost among teams and try to minimize the cost function. For example, the communication cost can be defined as the longest shortest path between any experts in team, the weight cost of the minimum spanning tree for subgraph, and the sum of all shortest paths between any two experts in team. This line of work optimize the team form the perspective of network structure of team. On the other hand, several works [2,7,11] take other factors such as the skill level, workload of individuals into account. The authors in these works aim to balance the workload of performing the tasks among people in the fairest possible way, on the condition that the required skills are covered. To the best of our knowledge, no existing work consider the geographical factor to boost the team formation performance especially in social coding sites. Where some works [17,18] demonstrate the importance of geographical proximity in some specific domain such as knowledge production and technological innovation. We believe that the geographical proximity may also affect the collaboration between developers in collaborative software development, and it is desirable to exploit the geographical proximity factor to improve the performance of team formation in social coding sites.

Based on this intuition, our paper proposes to integrate the conventional communication cost and geographical proximity for team formation in social coding sites such as GitHub. The challenges of our paper lies in two fold. (a) How to encode the geographical information of developers into our model. In our GitHub scenario, the developer declares his/her location attribute via a string. It is challengable to determine the impact of geographical information for team formation and then encode it mathematically (e.g., via calculating distance according to latitude and longitude or encoding it into time zone). (b) How to incorporate the geographical information and communication cost into a unified objective function and solve the optimization problem. The optimization problem in team formation issue has been proven to be NP-hard, it is challengable to devise a heuristic approach to solve the optimization problem.

To achieve this goal, in this paper, we firstly define the team formation task as finding a team of developers that cover the required skills while minimizing both the communication cost and geographical proximity, given a collaboration network and a task with a set of required skills where each skill is associated with a specific number of developers. Then, we incorporate the communication cost and geographical proximity into a unified objective function and propose a genetic algorithm to optimize it. Furthermore, we conduct comprehensive experiments on a real-world dataset to verify the effectiveness of our proposed model. Figure 1 gives an overview of developers' profile with geographical information and their corresponding skills which can be extracted from their contributed projects in GitHub. We can also note that a heterogeneous network among users and skills can be constructed based on the collaboration between users (see Sect. 3).

The main contributions of this paper are summarized as follows:

- To the best of our knowledge, this work is the first attempt to improve the performance of team formation in social coding sites by taking both communication cost and geographical proximity into consideration.
- We incorporate the communication cost and geographical location cost into a unified objective function and propose a genetic algorithm to optimize it.
- We crawl 36,701 users and 3,532,453 projects from GitHub as a real-world dataset to evaluate the performance our approach. Comprehensive experiments show the effectiveness of our model with the comparison of other baseline models

Organization. The remainder of this paper is organized as follows. In Sect. 2, we survey some works related to this paper. Section 3 shows some preliminaries. Section 4 presents details of our proposed geographical location aware model for team formation in social coding sites. Section 5 describes the real-world dataset (e.g., GitHub) we use in our experiments. Experimental results and analysis are shown in Sect. 6. Finally, we conclude this paper and propose some future directions in Sect. 7.

2 Related Work

Team Formation. The team formation problem is majorly studied in the field of collaborative social networks since it has an effect on the efficiency of collaboration. Lappas et al. [13] are the first to address the issue of social team formation by considering the communication costs for organizing a team. They also prove the team formation problem is NP-hard. Based on the perspective of communication cost, some variants have been derived. In [10], Kargar et al.

improve the communication cost function based on the sum of distances and leader distance. Ashenagar et al. [3] devise a new method to determine the distance between pairs of experts. Besides considering the communication cost between users, some other factors such as the cost of individuals [11,12] and the workload balance of team are also considered [1,2]. Majumder et al. [15] account for capacity constraints in the team formation problem so that no user is overloaded by the assignment. Farhadi et al. [7,8] suggest a skill-grading method to measure for the skill level of experts. Yang and Hu [19] propose a new cost model to solve team formation with limited time. Bhowmik et al. [4] take the Submodularity method to find a team of experts by relaxing the requirement of skill. Li and Shan [14] generalize the team formation problem by associating each required skill with a designated number of experts. Although many other factors have been considered, the geographical location of developers in social coding sites is not been considered yet.

Geographical Location. Another line of research which are related to our paper is on exploiting the geographical location. Lots of literature suggest that geographical proximity is playing an increasing important role in many domains in spite of rapid development in telecommunications technology. In [17], the authors demonstrate that the geographical proximity in the creation of economically-useful knowledge appeared to be becoming even more important. Soon and Storper [18] analyze patent citations and found that in contemporary knowledge production and innovation the role for geographical proximity was increasing. Ponds et al. [16] analyze the role of geographical proximity for collaborative scientific research and confirmed its significance. Brocco and Woerndl [5] propose two different ways to integrate location using spatial operations and utilize the location-based solution to support team composition in different computer gaming scenarios.

3 Preliminaries

We present the social coding network as an undirected graph $G = (V, E, w)$. Each vertex in V denotes an expert and the weight of each edge in E represents the communication cost between a pair of experts. We assume that (u, v) is an edge if developers u and v have participated in common projects before, and the weight of the edge is related to the fraction of projects they have worked on together, which is calculated by

$$w(u, v) = 1 - \frac{|N_u \cap N_v|}{|N_u \cup N_v|} \tag{1}$$

where N_u and N_v is the set of projects in which u and v are listed as contributors respectively. The communication cost is the sum of weights on the shortest path between two developers in G. The lower the communication cost is, the more easily they can collaborate with each other. If two experts are not connected in G(directly or indirectly), the communication cost between them is ∞. Consider the social coding network in Fig. 1, the communication costs between a and b,

a and c are 0.2, 0.4, respectively. In the social coding network, each developer possesses some skills such as programming languages. And each skill is related to some projects.

The geographical proximity is the distance between two regions, such as cities, countries and so on. It is related to the differences in culture, work habits of developers and so on. In order to quantify the geographical proximity between two developers, we extract the country of every developer, and define the geographical proximity between them as follows:

$$gp(u,v) = \begin{cases} 0, & \text{If } u \text{ and } v \text{ in he same country} \\ 1. & \text{Otherwise} \end{cases} \tag{2}$$

For example, in Fig. 1, the geographical proximity between a and b, a and e are 0, 1 respectively.

Definition 1 *(Team of Developers). Given a social coding site and a project P with some requirement of skills (e.g. programming languages), a team of developers for P is a set of developers who can meet the requirement of P.*

4   Location-Aware Model for Team Formation

In this section, we will model communication cost and geographical proximity, and then state the team formation problem followed by introducing the genetic algorithm based approach for solving the problem.

4.1   Model the Communication Cost

To evaluate the communication cost among the developers in a team T, we take the sum of communication costs among the selected developers of a team defined as follows, which is the same as [10].

Definition 2 *(Sum of Communication Costs). Given a social coding network G whose edges are weighted by the communication cost between two developers and a team T of developers from G, the sum of communication cost of T is defined as*

$$SCC(T) = \sum_{i=1}^{n} \sum_{j=i+1}^{n} cc(e_i, e_j) \tag{3}$$

where $cc(e_i, e_j)$ is the communication cost of developer e_i and e_j (as defined earlier).

4.2   Model the Geographical Location

Based on the perspective of sum of communication cost, to measure the geographical proximity of the team of experts, we define the sum of geographical proximity of a team as follows:

Definition 3 *(Sum of Geographical Proximity). Given a team T of experts, where each having a location code, the sum of geographical proximity of T is defined as*

$$SGP(T) = \sum_{i=1}^{n} \sum_{j=i+1}^{n} gp(e_i, e_j) \qquad (4)$$

where $gp(e_i, e_j)$ is the geographical proximity between expert e_i and e_j which as defined above.

4.3   Objective Function

For finding a team of developers from a social coding network that minimize the sum of communication cost as well as the sum of geographical proximity, we combine the two objective functions into a single one to convert the bi-objective optimization problem into a single objective problem and define a new combined cost function as follows which is based on the linear combination of the sum of communication cost and sum of geographical proximity.

Definition 4 *(Combined Cost Function). Given a collaboration network and a trade-off λ between the sum of communication cost and sum of geographical proximity, we define the combined cost of the team T as*

$$ComCost(T) = (1 - \lambda) \times SCC(T) + \lambda \times SGP(T) \qquad (5)$$

The parameter λ varying from 0 to 1 indicates the tradeoff between sum of communication and sum of geographical proximity.

Given the combined cost function, we now formally define the team formation problem in social coding networks as follows:

Team Formation by Minimizing the Combined Cost. Given a social coding network $G(V, E, w)$ where the developers are associated with specified skills, a project P with requirements of skills, the aim of team formation by minimizing the combined cost is to find a team $T \subseteq V$ so that each skill in P will be covered by the specified number of developers, each developer will cover and only cover one skill, and the combined cost $ComCost(T)$ defined in 4 among selected experts are as minimum as possible.

4.4   GA-based Optimization

Since the team formation by minimizing the combined cost is an NP-hard problem, we employ an genetic algorithm to find an optimal solution for the team formation problem in the context of social coding networks. The details of GA-based model are presented in the following subsections.

Encoding. We consider each candidate team as a chromosome and each developer in the team as a gene. So each candidate team is a linear vector and composes of several partitions where each one represents a skill. An example of candidate team with four required skills is represented in Fig. 2.

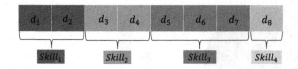

Fig. 2. An example of representation of the candidate team with four required skills

Initialization. The random method to generating the initial population ensures a good level of genetic diversity in the population and thus prevents the premature convergence of the algorithm [6,9]. So we take the random method to randomly generate the initial candidate teams fulfilling the requirement posed by the projects.

Genetic Operators. Crossover, mutation and selection are the three main types of genetic operators. They must work in conjunction with one another to ensure the success of the algorithm.

– **Crossover.** The crossover operator aims to preserve and combines the best characteristics of the parents to evolve better solutions [6,9]. We have applied a two-point crossover here with the probability P_c to generate two new offspring solutions.
– **Mutation.** This operator is applied to the encoded solutions with the probability P_m to introduce genetic diversity into the population. In this paper, we have applied two types of mutation operators - substitution mutation and swap mutation. The substitution mutation operator involves the selection of a developer in a team with skill s_m, and replacing him with a developer at random from support set of s_m. Swap mutation operator randomly selects a developer from the team and swaps him with one in the team who covers the skill in his skill set at current.
– **Selection.** The new population at generation $k + 1$ is generated by the application of genetic operators at generation of k. We combine elitism and tournament to complete the selection, which means that the best teams in generation k are automatically transferred to the population of generation $k + 1$, and the rest teams will be chosen as the parents by the tournament method to generate new teams.

Sometimes, crossover and mutation operators may produce infeasible solutions. The reparation strategy is designed to ensure the new team is feasible.

Evaluation. We apply the opposite number of combined cost as the fitness function to simultaneously optimize the sum of communication cost and sum of geographical proximity.

5   Dataset

In our paper, we conduct experiments on a real-word dataset from GitHub, which is one of the most popular social coding sites and has gained much popularity

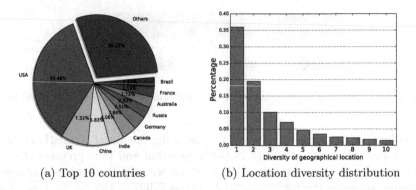

(a) Top 10 countries   (b) Location diversity distribution

Fig. 3. An overview of geographical location distribution of GitHub developers. (a) Top 10 countries with the largest number of developers. (b) Distribution of location diversity distribution, considering the composition of teams.

among a large number of software developers around the world. In GitHub, users are encouraged to contribute to a share project collaboratively, which is coincided with our scenario of team formation.

GitHub publicizes its data via APIs. Crawling GitHub website by its API, we get 28,362,019 projects, 15,647,255 users and make out the relationships between them. We then filter out users who provide the geographical location information and obtain 36,701 users and 3,532,453 their contributed projects. Constructing the network by the way described in Sect. 3, we get 1,610,072 edges. Considering the programming languages of project as required skills, we obtain 273 distinct skills.

Figure 3 presents an overview of geographical location distribution of GitHub developers. Figure 3(a) lists the top 10 countries with the largest number of developers. From this figure, we observe that more than a third of developers are from the USA, accounting for the largest part. The following parts are developers from UK and China, which are also within our expectation. Figure 3(b) shows the distribution of location diversity distribution in terms of the number of countries the developers come from in composing a team. In this figure, we observe in most teams (nearly 55%), the developers come from no more than one or two countries. And the situation that members are from many different countries is uncommon. This phenomenon just verifies our intuition.

6   Experiments

6.1   Experimental Setup

For all experiments, we set the number of skill $k = 2$ and $\lambda = 0.5$. For GA algorithm, we set the population size as 200, the number of generation as 100, the crossover probability as 0.2 and runs for 10 iterations for each experiment.

Evaluation Metrics. For evaluation, three evaluation metrics which are commonly adopted in conventional studies are used in this paper. The three evaluation metric are listed as follows:

- **Sum of Geographical Proximity.** This metric measures the geographical proximity of the team. It reveals how closely the developers of the team in terms of geographical location.
- **Sum of Communication Cost.** This metric measures the communication cost of the team. It reveals the efficiency of the communication between developers. It is also taken as an evaluation metric in some previous works.
- **Combined Cost.** This metric is the combination of sum communication cost and sum of geographical proximity.

Performance Comparison. By following [11], we compare our proposed model against the following three baselines.

- **Random Algorithm.** Random algorithm randomly creates 1,000 teams and selects the one with the minimal combined cost for the required set of skills as the optimal team.
- **Approximation Rare Algorithm.** Approximation rare algorithm selects the skill with least supporters as the initial skill. Firstly, an expert with the initial skill is selected as a seed expert followed by an expert added with the minimum communication cost to the seed expert with each of other required skills into the team. Then, the team with the minimum costs is selected among the entire candidate teams.
- **Minimum Cost Contribution Rare Algorithm.** MCC-rare algorithm chooses an expert with the skill who has rarest supporters as the initial member of candidate team, and then adds a new team member by considering its communication cost in comparison to all current team members.

All the experiments in this paper are implemented with Python 2.7, and run on a computer with an 2.2 GHz Intel Core i7 CPU and 64 GB 1600 MHz DDR3 RAM, running Debian 7.0.

6.2   Experimental Results

Figure 4 shows the performance of different models on different metrics. From this figure, we have the following observations:

- On the sum of geographical proximity evaluation metric, the proposed GA-based model achieves better performance, random algorithm gets the worst. This is because the GA-based model considers the sum of geographical proximity during the process of finding a optimal team. Other three algorithms do not consider the geographical proximity factor.
- On the sum of communication cost evaluation metric, the proposed GA-based model also achieves better performance and random algorithm worst. This is because GA-based model has a larger search space while MCC-Rare algorithm and approximation rare algorithm has a smaller one. The random algorithm do not consider the sum of communication cost factor.

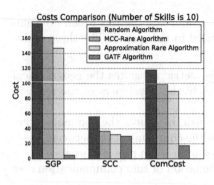

Fig. 4. Performance comparison.   **Fig. 5.** Convergence of GA.

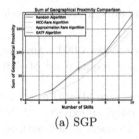   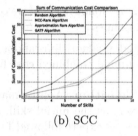   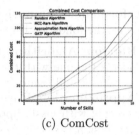

(a) SGP   (b) SCC   (c) ComCost

Fig. 6. Impact of number of skills.

- On the combined cost evaluation metric, the proposed GA-based model also achieves better performance. This is because GA-based algorithm consider both the sum of graphical proximity and sum of communication cost. MCC-Rare algorithm and approximation rare algorithm consider communication cost only. The random algorithm only covers the basic requirements of projects, including neither geographical proximity nor communication cost.

6.3   Parameter Analysis

Impact of Number of Skills. In our model, the number of skills controls the team size. To study the impact of skills number on the performance, we set skills number $k \subseteq \{2, 4, 6, 8, 10\}$. And for each k, we generate 10 random projects to take the average result. The experimental results are show in Fig. 6. Figure 6(a) shows that all algorithms will get high sum of geographical proximity with the increasing of task number. But proposed GA-based model can always achieve better performance on sum of geographical proximity. The similar are Fig. 6(b) and Fig. 6(c), where all algorithms will get high sum of communication cost and combined cost with the increasing of skills number. Our proposed GA-based model always achieve better performance.

Impact of Iterations. In our model, the iterations is directly affect the search result. To study the impact of iterations, we set the task number to 2 and generate 10 random projects to track the convergence of the algorithm. The experimental results are show in Fig. 5. As we can see from the result, the proposed GA-based model can converge after 20 iterations when the task number is 2.

7 Conclusion and Future Work

In this paper, we exploit the geographical location of developers to boost the performance of team formation in social coding sites. We incorporate the communication cost and geographical proximity into a unified objective function and propose a genetic algorithm to optimize it. Experiments on a real-world dataset (e.g., GitHub) illustrate the effectiveness of our proposed approach.

In our future work, we plan to investigate the impact of social media on the performance of team formation. For example, we can also take the social network of developers in social media (e.g., Twitter) into consideration to boost the performance of team formation. Furthermore, we will exploit the interaction patterns for the accurate interpretation of link strength between developers.

Acknowledgments. This research is partially supported by the Natural Science Foundation of China under grant of No. 61672453, the Foundation of Zhejiang Engineering Research Center of Intelligent Medicine (2016E10011) under grant of No. ZH2016007, the Fundamental Research Funds for the Central Universities, the National Science and Technology Supporting Program of China under grant of No. 2015BAH18F02, Australia Research Council (ARC) Linkage Project LP140100937.

References

1. Anagnostopoulos, A., Becchetti, L., Castillo, C., Gionis, A., Leonardi, S.: Power in unity: forming teams in large-scale community systems. In: Proceedings of the 19th ACM International Conference on Information and Knowledge Management, pp. 599–608. ACM (2010)
2. Anagnostopoulos, A., Becchetti, L., Castillo, C., Gionis, A., Leonardi, S.: Online team formation in social networks. In: Proceedings of the 21st International Conference on World Wide Web, pp. 839–848. ACM (2012)
3. Ashenagar, B., Eghlidi, N.F., Afshar, A., Hamzeh, A.: Team formation in social networks based on local distance metric. In: 2015 12th International Conference on Fuzzy Systems and Knowledge Discovery (FSKD), pp. 946–952. IEEE (2015)
4. Bhowmik, A., Borkar, V.S., Garg, D., Pallan, M.: Submodularity in team formation problem. In: SDM, pp. 893–901. SIAM (2014)
5. Brocco, M., Woerndl, W.: Location-based team recommendation in computer gaming scenarios. In: Proceedings of the 2nd ACM SIGSPATIAL International Workshop on Querying and Mining Uncertain Spatio-Temporal Data, pp. 21–28. ACM (2011)
6. Eiben, A.E., Smith, J.E.: Introduction to Evolutionary Computing, vol. 53. Springer, Heidelberg (2003)

7. Farhadi, F., Sorkhi, M., Hashemi, S., Hamzeh, A.: An effective expert team formation in social networks based on skill grading. In: 2011 IEEE 11th International Conference on Data Mining Workshops, pp. 366–372. IEEE (2011)

8. Farhadi, F., Sorkhi, M., Hashemi, S., Hamzeh, A.: An effective framework for fast expert mining in collaboration networks: a group-oriented and cost-based method. J. Comput. Sci. Technol. **27**(3), 577–590 (2012)

9. Golberg, D.E.: Genetic Algorithms in Search, Optimization, and Machine Learning, vol. 102. Addion wesley, Boston (1989)

10. Kargar, M., An, A.: Discovering top-k teams of experts with/without a leader in social networks. In: Proceedings of the 20th ACM International Conference on Information and Knowledge Management, pp. 985–994. ACM (2011)

11. Kargar, M., An, A., Zihayat, M.: Efficient bi-objective team formation in social networks. In: Flach, P.A., Bie, T., Cristianini, N. (eds.) ECML PKDD 2012. LNCS (LNAI), vol. 7524, pp. 483–498. Springer, Heidelberg (2012). doi:10.1007/978-3-642-33486-3_31

12. Kargar, M., Zihayat, M., An, A.: Finding affordable and collaborative teams from a network of experts. In: Proceedings of the SIAM International Conference on Data Mining (SDM), pp. 587–595. SIAM (2013)

13. Lappas, T., Liu, K., Terzi, E.: Finding a team of experts in social networks. In: Proceedings of the 15th ACM SIGKDD International Conference on Knowledge Discovery and Data Mining, pp. 467–476. ACM (2009)

14. Li, C.T., Shan, M.K.: Team formation for generalized tasks in expertise social networks. In: 2010 IEEE Second International Conference on Social Computing (SocialCom), pp. 9–16. IEEE (2010)

15. Majumder, A., Datta, S., Naidu, K.: Capacitated team formation problem on social networks. In: Proceedings of the 18th ACM SIGKDD International Conference on Knowledge Discovery and Data Mining, pp. 1005–1013. ACM (2012)

16. Ponds, R., Van Oort, F., Frenken, K.: The geographical and institutional proximity of research collaboration. Pap. Reg. Sci. **86**(3), 423–443 (2007)

17. Sonn, J.W., Storper, M.: The increasing importance of geographical proximity in technological innovation. What Do we Know about Innovation? in Honour of Keith Pavitt, Sussex, 13–15 (2003). November 2003

18. Sonn, J.W., Storper, M.: The increasing importance of geographical proximity in knowledge production: an analysis of us patent citations, 1975–1997. Environ. Plann. A **40**(5), 1020–1039 (2008)

19. Yang, Y., Hu, H.: Team formation with time limit in social networks. In: Proceedings 2013 International Conference on Mechatronic Sciences, Electric Engineering and Computer (MEC), pp. 1590–1594. IEEE (2013)

Weighted Simplicial Complex: A Novel Approach for Predicting Small Group Evolution

Ankit Sharma[1,3(✉)], Terrence J. Moore[2], Ananthram Swami[2], and Jaideep Srivastava[3]

[1] University of Minnesota, Minneapolis, USA
ankit@cs.umn.edu
[2] U.S. Army Research Laboratory, Adelphi, USA
{terrence.j.moore.civ,ananthram.swami.civ}@mail.mil
[3] Qatar Computing Research Institute, Doha, Qatar
jsrivastava@hbku.edu.qa

Abstract. The study of small collaborations or teams is an important endeavor both in industry and academia. The social phenomena responsible for formation or evolution of such small groups is quite different from those for dyadic relations like friendship or large size guilds (or communities). In small groups when social actors collaborate for various tasks over time, the actors common across collaborations act as bridges which connect groups into a network of groups. Evolution of groups is affected by this network structure. Building appropriate models for this network is an important problem in the study of group evolution. This work focuses on the problem of group recurrence prediction. In order to overcome the shortcomings of two traditional group network modeling approaches: hypergraph and simplicial complex, we propose a hybrid approach: *Weighted Simplicial Complex (WSC)*. We develop a *Hasse diagram* based framework to study WSCs and build several predictive models for group recurrence based on this approach. Our results demonstrate the effectiveness of our approach.

1 Introduction

With the advent of high-speed internet, collaborations are no longer restricted by physical proximity. A group of individuals, irrespective of their demographics or location, can perform a task online. This task might be writing software code (or a Wikipedia article or a Google Doc) by a group of coders (or editors), or can be a business meeting involving video chat with colleagues. Understanding the dynamics of such small (social) groups is of increasing research interest in various sub-disciplines in the social sciences [8], and is of interest to applications that require high efficiency and robustness in the performance of human groups [5].

This paper addresses the problem of *group evolution*, with specific focus on understanding the causal factors driving the evolution. The overall objective is to build a model that can predict how a group will evolve in the future, based on its history. One aspect of special interest is *group recurrence*, which can be

© Springer International Publishing AG 2017
J. Kim et al. (Eds.): PAKDD 2017, Part I, LNAI 10234, pp. 511–523, 2017.
DOI: 10.1007/978-3-319-57454-7_40

stated thus: *Which group(s) (or its subgroup(s)) among the groups observed so far, will continue to function as a group, i.e. perform some task again in the near future?*

Prior studies have demonstrated the significance of recurrence in network structure (see [15] and references therein). Most work on group evolution in social networks focuses on the evolution of arbitrary size communities or groups [7]. The sizes of these groups are usually large and the boundaries of the community depends on the definition of membership used. In this paper we study well-defined small groups which typically have size ≤20. A key difference between large groups and small groups is that membership in the former is largely based on identity, i.e. a member identifying himself with the group. In contrast, a small group is defined principally by the (regular) interaction between group members, often driven by some purpose, professional or personal. The focus of this paper is to study the evolution of small groups and, in contrast to classical social science literature, the objective is to build models that can predict future behavior, with the final goal of identifying potential causal mechanisms for small group evolution.

In contrast to prior work, we highlight the distinct nature of small groups and develop models inspired from social science theories of small groups [8]. A group can be formed depending on the requirement (fiat teams) or a set of actors can make an autonomous decision to work together (self assembly [3]). In either case, individuals find it easier to work with familiar actors [5], making *frequency* of activity by a group an important metric. Also, over time, actors build new relationships while working in different groups. A shared collaboration history is therefore created, where the same individuals are part of multiple groups, acting as bridges between groups, and resulting in a *network of groups (NOG)* (Fig. 1). This is the *network* perspective of small groups [6] where the network of groups plays a central role in the group formation process. Moreover, group formation motives and group communication processes, which are *task centered*, are very different from those involved in building friendship ties in a friendship network or joining a community, e.g., joining a news interest group, being part of a Facebook community, subscribing to a Youtube channel, or publishing within a particular research discipline. Recently, some attempts have been made to model networks as higher order relational structures such as simplicial complexes [9] and hypergraphs [10]. A hypergraph is a generalized graph where edges, now called hyperedge, instead of representing a relationship between a pair of vertices, represent a relationship between a set of vertices. If the relationship holds for every subset of the hyperedge, the hypergraph is called a *simplicial complex*. Although hypergraphs are more general, if the problem or the data has a special structure then simplicial complexes are more appropriate. For the *group recurrence* problem, we need to predict recurrence of not just observed groups but also the subgroups. Thus, simplicial complexes are more applicable to our problem.

For the *group recurrence* problem we also want our model to capture any prior knowledge associated with each group or subgroup that might indicate cohesion among group members, or the context associated with the group. We use the concept of a *weighted simplicial complex*, which is a simplicial complex where each simplex has a prior weight associated with it. We develop several schemes to generate these prior weights, modeling different prior knowledge scenarios.

We observe that a simplicial complex, from a frequent pattern mining perspective [1], is the trivial set of all the frequent patterns of frequency equal to one, mined from the transactions database of hyperedges. This motivates the use of a Hasse diagram (Fig. 1) [11] (similar to enumeration trees in pattern mining) as a graph representation for the simplicial complex. If we associate a weight with each node (representing simplicies) of the Hasse diagram it represents a weighted simplicial complex. We hypothesize that the topology of these groups plays a critical role in how past occurrences influence future occurrences of other (sub)groups.

Using the Hasse diagram, we apply a modification of the HyperPrior algorithm [12], for generating label diffusion-based machine learning models, as well as develop hierarchical label spreading algorithms for recurrence prediction. These algorithms make use of the weighted simplicial complex topology while exchanging the occurrence information between the subgroup nodes in the Hasse diagram. Our experimental analysis, conducted using the DBLP and EverQuest II datasets, shows the efficacy of the techniques developed. The main contributions of this study are:

- We present machine learning models to predict recurrence of already observed groups, which takes into account the higher order topology.
- We present a Hasse diagram-based framework to study simplicial complexes, hypergraphs, and frequent pattern mining in a unified manner.
- We show that frequent patterns can be considered as topological entities, with relationships between them guided by higher-order topological properties. To the best of our knowledge this has not been done before.

The rest of the paper is structured as follows. In Sect. 2 we describe the models of network of groups and the problem statement. Methods proposed are illustrated in Sects. 3 and 4 has experimental analysis.

2 Problem Statement and Preliminaries

2.1 Models for Network of Groups

We have a set of n *actors* $V = \{v_1, v_2, ..., v_n\}$. A subset of these *actors* can form a *group*. We have a collection of m such groups observed in the past, denoted by $G = \{g_1, g_2, ..., g_m\}$ where $g_i \subseteq V$ represents the ith *group*. The cardinality $c_i = |g_i|$ of a group is the number of actors in it. We let $R(g)$ denote the number of times group $g \in G$ has occurred. The network of groups can be modeled as a hypergraph [2] $H = (V, G)$ where the observed groups G are the hyperedges over the vertex set V of actors. We denote by $S_i = \{s_k^i, \forall k \in \{1, 2, ..., 2^{|g_i|} - 2\}\}$ the set of all proper subsets of each group $g_i \in G$. If we consider the union of all subsets

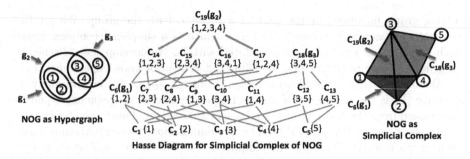

Fig. 1. Example illustrating a network of groups hypergraph (left) as a simplicial complex (right) and as a Hasse diagram (middle) corresponding to the simplicial complex, for a scenario where the actors $\{1,2,3,4,5\}$ have collaborated in the past as groups: $g_1 = \{1,2\}$, $g_2 = \{1,2,3,4\}$ and $g_3 = \{3,4,5\}$.

of the sets in G along with G itself, i.e., $C = \{G \cup (\bigcup_{i=1}^{m} S_i)\}$, then we have a (abstract) simplicial complex C and each element $c \in C$ is a simplex which represents a group or subgroup. If we also associate a weight $W(c) \in \mathbb{R}, \forall c \in C$, then we attain a *weighted simplicial complex* $\Diamond = (C, W)$. For convenience we also define the set containing the subgroups in C that were never observed in the past, i.e., $C_s = \{c|(c \in C) \wedge (c \notin G)\} = (C - G)$. Each $c \in C_s$ also has a set of groups $Q(c) \subseteq G$, of which it is a subgroup of, i.e., $Q(c) = \{x|(x \in G) \wedge (c \subset x)\}$. We define an occurrence function O which gives the occurrence count to all the groups in C as follows:

$$
O(c) = \begin{cases} R(c) + \left(\sum_{x \in Q(c)} R(x) \right) & \text{when } c \in G \\ \sum_{x \in Q(c)} R(x) & \text{when } c \in C_s \end{cases} \tag{1}
$$

In words, for an observed group we simply take the number of times it has occurred, $R(c)$, and also add the counts of the groups it has been a subset of. In the case of subgroups (those groups that haven't occurred in the past) we simply add the counts of the groups it has been a subset of. For a simplex (or (sub)group) $\alpha \in C$ we define its dimension as $dim(\alpha) = |\alpha| - 1$. If K_{max} is the maximum cardinality of any simplex in C then $(K_{max} - 1)$ is the maximum dimension of any simplex in C or simply the dimension of C.

The set of simplices of cardinality k within the simplicial complex C are defined by the set: $\pi^k = \{\sigma | \sigma \in C \wedge |\sigma| = k\}, \forall k \in \{1, ..., K_{max}\}$. For the example in Fig. 1, $C = \{C_1, ..., C_{19}\}$, $G = \{g_1, g_2, g_3\} = \{C_6, C_{19}, C_{18}\}$ and $C_s = (C - G)$.

We also define a *Hasse diagram*, T, for the simplicial complex C. The level in the diagram (Fig. 1) determines the poset relation. We use the undirected graph derived from the Hasse diagram T over the vertex set $V(T) = C$ and with a set of undirected edges $E(T) = \{(x,y) \cup (y,x)|(x,y \in V(T)) \wedge (y \subset x) \wedge (|y| = |x| - 1)\}$. In the case of a weighted simplicial complex $\Diamond = (C, W)$, we associate with each vertex the weight of the corresponding simplex it represents, i.e., $W(v), \forall v \in V(T)$. Note, we can also associate a weight with the edges but in this study we assume all edges have a unit weight. We denote $\mathbf{A}$ to be the adjacency matrix of size $(|C| \times |C|)$ associated with the graph T.

2.2   Problem Statement

We are interested in prediction of groups formed by two processes: group recurrence and subgroup recurrence. In group recurrence, a group $g_i \in G$, called a *recurring group*, observed in the past can again occur in the future. Our first problem is to predict a score for each of the groups in G. This score reflects the possibility of the given group occurring again in the future. In subgroup recurrence, a group $c_i \in C_s$ which has never been observed as a group in the past, might occur in the future. We refer to such groups as *recurring subgroups*. Our second problem is to predict a score for each of the groups in C_s, which reflects its possibility to be formed in future. We restrict ourselves to the prediction of only the recurring groups and subgroups and not groups composed of entirely new actors.

3   Methods

In this section, we first enumerate several ways of assigning prior weights. We then describe three different methods (along with several variants) to solve the problems described in the previous section. Each method models the tendency of a given group to be formed in the near future by assigning a score $\mathbf{S}(c)$ to each group in $c \in C$, returning a final vector of scores $\mathbf{S}$. The first method uses a simple group count-based approach and the next two methods consider the hierarchical structure of the higher order topology within the Hasse diagram.

3.1   Schemes for Assigning Initial Weights

Several studies on small groups have shown that social actors tend to collaborate with actors with whom they have already developed strong working relationships [5] and that repeated ties within a group positively affect its performance [3]. There are a number of ways to assign a prior weight to represent the strength of the relationships between group members. Kapoor et al. [4] defined several weights for the problem of node centrality, of which we utilize two. The first, shown in (2), corresponds to a frequency-based definition and simply counts the number of times a group has performed some task together. The second, shown in (3), enforces that the average attachment of any two individuals (or the attention span of a member towards each other member) in a group decreases in proportion to the size of the group.

$$\mathbf{W}(c) = O(c), \forall c \in C \quad (2) \qquad \mathbf{W}(c) = \frac{\log(O(c)) + 1}{|c|}, \forall c \in C \quad (3)$$

The weights in (2) and (3) initialize all groups (observed) as well as subgroups (unobserved), i.e., all the simplices. We, therefore, also design slightly different variants where we only initialize the observed groups, which emphasizes the hypergraph model of the network:

$$\mathbf{W}(c) = \begin{cases} O(c) & \text{if } c \in G \\ 0 & \text{if } c \in C_s \end{cases} \quad (4) \qquad \mathbf{W}(c) = \begin{cases} \dfrac{\log(O(c)) + 1}{|c|} & \text{if } c \in G \\ 0 & \text{if } c \in C_s \end{cases} \quad (5)$$

In the following sections we will define several algorithms which will use these four initialization schemes. We will use the suffixes: **Simp-C**, **Simp-W**, **Hyp-C**, and **Hyp-W** to refer to the initializations in (2)–(5), respectively.

3.2 Count Based Scores (CBS)

We build the first set of scores using only the occurrence information available. For this we simply take the score vector $\mathbf{S}$ as the weight defined in (2) and (3), denoted the **CBS-C** score and **CBS-W** score, respectively. The **CBS-C** score, gives each group a value which is determined by the number of times the group members have worked together in past. Whereas, **CBS-W** assigns score based upon the cohesion among the group members.

3.3 Hasse Diagram Based Models

CBS scores utilize counts of group recurrences, wherein each group was considered in isolation but do not consider the network of groups. This network encodes information about the observed groups, the unobserved groups, and the topological relations between them. Occurrences of a group affect the probability of other groups in the network to collaborate in the future. We develop two approaches applied to a Hasse diagram representation of a weighted simplicial complex to capture the local and global relational information.

Algorithm 1. GetHDSScores $(T, \mathbf{y}, K_{max}, \alpha)$

$\mathbf{f} \leftarrow \mathbf{y}, C \leftarrow V(T)$ {Get the simplicial complex corresponding to the Hasse diagram}

for $k = K_{max} - 1$ to 1 **do**

    for all $c \in \pi^k$ **do**

$$\mathbf{f}(c) \leftarrow \mathbf{f}(c) + \alpha \left(\sum_{x \in (Q(c) \cap \pi^{k+1})} y(x) \right)$$

    return f

Hasse Diagram Spread-Based Scores (HDS Scores). This class of methods is based upon the intuition that observed groups in the Hasse diagram influence the subgroups below it in the hierarchy. Influence spread can happen in a variety of ways. There are several possible counter-intuitive group phenomena. We model these in a holistic fashion by spreading scores over the Hasse diagram. We propose that if we observe a node g_i in the Hasse diagram then it spreads its score (s_i) down the hierarchy. It can send the same, more, or less of its score to its children. In general, it can send αs_i $(\alpha \geq 0)$ score to its children. These children update their scores and spread the score down the hierarchy recursively. This is shown in Algorithm 1. We initialize the algorithm using the vectors $(\mathbf{y} = \mathbf{W})$ in Eqs. (2)–(5) to get four different scores, $\mathbf{S} = GetHDSScores(T, \mathbf{y}, K_{max}, \alpha)$, which we denote as **HDSSimp-C**, **HDSSimp-W**, **HDSHyp-C** and **HDSHyp-W**, respectively.

Hasse Diagram Diffusion-Based Scores: The spread-based scores are local in the sense that the final score of a node is only determined by its initial score and the scores of its parent(s). But, in general, the nodes representing groups in the network are connected by many pathways. Therefore, it is reasonable to assume that a potential group may be affected by occurrence of non-parent groups in the network. In order to take into account this structure of the entire Hasse diagram, we apply a modification of the graph label propagation algorithm HyperPrior [12].

Each vertex (group) is initialized with a label, which encodes prior information about the recurring tendency of that node. These labels (information) then diffuse (exchange information) via random-walks through the Hasse diagram network structure. After the random-walks stabilize, the final label for each vertex is the score indicating its recurrence possibility. The final label at a given vertex represents the chances that a random walk originating from other nodes ends at this vertex. Hence, this score is a combination of both the group's initial tendency to occur plus an adjustment based on the knowledge from other groups in the network, i.e., the random walk outcomes. This adjustment models a network guided similarity between the vertex and the other nodes. Vertices that are near in the network should end up receiving similar labels/scores.

More formally, let $\mathbf{y}$ be the vector of initial labels for the vertices in the Hasse diagram T with incidence matrix $\mathbf{A}$. Vector $\mathbf{y}$ is initialized by any of the weights in (2)–(5). As in a graph-based learning task, we learn the final label (score) vector $\mathbf{f}$ by taking into account the competing aims of similar labels for vertices connected by an edge in the Hasse diagram and of similar labels between the initial and final vectors. We capture these competing aims in the following cost minimization objective:

$$\min_{\mathbf{f}} \mathbf{f}^T \mathbf{L} \mathbf{f} + \beta \|\mathbf{f} - \mathbf{y}\|^2 \tag{6}$$

where, $\mathbf{L} = \mathbf{I} - \mathbf{D_v}^{-1/2} \mathbf{A} \mathbf{D_v}^{-1/2}$ is the normalized graph Laplacian [14] and $\mathbf{D_v}$ is a diagonal matrix consisting of the vertex degrees. The first term in (6) is a smoothing term which ensures that vertices (groups) sharing an edge (having common group members) have similar scores. This term therefore, enforces the Hasse diagram structure while learning the labels. The second term measures the difference between the given initial labels and the final vertex scores. It can be shown [14] that the solution to (6) is equivalent to the solution of the following linear system:

$$\mathbf{f}^* = (1 - \mu)(\mathbf{I} - \mu\theta)^{-1}\mathbf{y}, \tag{7}$$

where $\mu = 1/(1 + \beta)$, $\theta = \mathbf{D_v}^{-1/2}\mathbf{A}\mathbf{D_v}^{-1/2}$, and $\mathbf{f}^*$ is the vector of final labels of the group nodes. Note that, $\mathbf{f}^*(c)$ is the aggregate tendency $\mathbf{S}(c)$ of a group $c \in C$ to reoccur. Therefore, we have: $\mathbf{S} = \mathbf{f}^*$.

Similar to spread-based scores, we denote the scores here by **HDDSimp-C**, **HDDSimp-W**, **HDDHyp-C**, and **HDDHyp-W** when intialized using (2)–(5). Our aim is to predict a score for the recurring groups (i.e., $g \in G$) and recurring subgroups (i.e., $c \in C_s$). For each of the methods above, we get a final vector that contains the scores for all the groups. We partition the vector $\mathbf{S}$ into two vectors $\mathbf{S_{rg}}$ and $\mathbf{S_{rs}}$ of sizes $|G|$ and $|C_s|$, respectively, such that $\mathbf{S_{rg}}(c) = \mathbf{S}(c), \forall c \in G$ and $\mathbf{S_{rs}}(c) = \mathbf{S}(c), \forall c \in C_s$. In summary, we obtain three score vectors $\mathbf{S_{rs}}$, $\mathbf{S_{rg}}$ and $\mathbf{S}$ for each of the above methods.

4   Experimental Analysis

4.1   Dataset and Statistics

Datasets: The first dataset we apply our methods to is a massive multi-player online role-playing game (MMORPG) dataset obtained from the Sony's EverQuest II (EQ II) game (www.everquest2.com). The game provides an online environment where multiple players can log in and collaborate in groups to perform various quests and missions. The server logs from this game, provided by Sony, were used to extract group interactions. Here, we treat a set of players performing a task or mission as a group in the EQ II network. The EQ II data contains logs for 21 weeks of data for training and testing. We divide them into seven training/testing splits, each of which has a two-week long training period followed by a one-week testing period.

The second dataset is the DBLP dataset (obtained from www.aminer.org) containing computer science publications from 1930–2015. The set of co-authors on a paper form a group in the DBLP network. Note that in both EQII and DBLP networks, the groups can perform multiple game tasks or co-author multiple papers. We make eleven train-test splits as follows: $(1992-95/96-98)$, $(1993-95/96-98)$, $(1993-95/96-99)$, $(1991-97/98-10)$, $(1997-00/01-03)$, $(1998-00/01-03)$, $(1998-00/01-04)$, $(2002-05/06-08)$, $(2003-05/06-08)$, $(2003-05/06-09)$ and $(2001-07/08-10)$; following the format: *(train period start year–train period end year/test period start year–test period end year)*. These splits were designed to observe the effect of varying training and testing period lengths as well as varying the entire train/test evaluation period. We have evaluated other variations of period lengths and other decades in the DBLP data, but in this paper we limit our discussion to the train/test periods we just described.

Statistics: Recall that we have two kinds of groups: (1) recurring groups that are observed in training and observed again in testing and (2) recurring subgroups

Table 1. Recurrence statistics of the various train/test periods

Dataset	Training Actors	Testing Actors	% Old Actors in Testing	% New Actors in Testing	Training Groups	Testing Groups	% Recurring Groups in Testing	% New Groups in Testing	% Groups with Old Actors	% Groups with New Actors
EQ II	3051	2215	81.67	18.33	1775	1219	67.92	32.08	88.93	11.07
Avg.		**84.06**	**15.94**				**74.01**	**25.99**	**90.51**	9.49
DBLP	677K	640K	40	60	549K	433K	12.06	87.94	84.53	15.47
Avg.		**34.73**	**65.27**				**11.65**	**88.35**	**81.17**	18.83

Table 2. Different dimension face recurrence statistics

	% Train Groups	% Test Groups	% Train Groups	% Test Groups	% Train Groups	% Test Groups	% Train Groups	% Test Groups
		EQ II Splits				DBLP Splits		
Simplex Dimension	RG+RS (Exact)		RG+RS (New Vertices)		RG+RS (Exact)		RG+RS (New Vertices)	
≥ 1	15.57	**21.25**	15.70	**21.43**	4.63	**3.09**	6.28	**4.20**
≥ 2	8.97	**12.72**	9.02	**12.80**	2.78	**1.79**	3.41	**2.19**
	RS (Exact)		RS (New Vertices)		RS (Exact)		RS (New Vertices)	
≥ 1	0.70	**0.71**	0.76	**0.77**	1.70	**0.89**	3.08	**1.61**
≥ 2	0.20	**0.23**	0.25	**0.29**	0.76	**0.38**	1.24	**0.63**
	RG (Exact)		RG (New Vertices)		RG (Exact)		RG (New Vertices)	
≥ 1	57.18	**20.55**	57.50	**20.66**	14.89	**2.21**	17.53	**2.60**
≥ 2	45.68	**12.49**	45.77	**12.51**	10.02	**1.41**	11.17	**1.57**

that are observed in testing but are only observed as a subgroup of some group that occurred in training. We shall refer to the former set as RG, the latter set as RS, and the combined set as (RG+RS). Table 1 contains several statistics for (RG+RS). However, due to space constraints, we only show statistics for the last split from each dataset, as well as the average statistics across the splits. In Table 1, an actor in the testing phase is considered "**old**" if it was observed in the training period, otherwise it is considered "**new**". Note that for any group with new actors in the testing phase, we can only test whether the subgroup with old actors is a recurring group or subgroup from the training period. These statistics are based on the distinct groups from the testing and training periods, so as to avoid any bias from the multiplicity of certain group interactions. We observe that on an average around 90% of the EQ II network groups and around 81% of the DBLP network groups formed in the test period contain at least one old actor. Only within these groups can we possibly search for recurring groups or subgroups. Note, 74% of the EQ II groups and around 12% of DBLP groups in testing period are exact recurrences and included in the set RG. This demonstrates that the recurring group process is more common in the EQ II network, whereas the recurring subgroup process is the more common feature in the DBLP network.

In Table 2, we record the statistics of the groups in training that recur in testing and of the groups in testing that are recurring groups or subgroups. We only consider groups of size ≤ 6 (i.e., faces of dimension ≤ 5) and also omit vertex recurrences since those are reported in Table 1.

For dimensions ≥ 1, the set RG+RS accounts for 20% of the testing groups in the EQ II network and 3–4% in the DBLP network. For dimensions ≥ 2, the set RG+RS accounts for approximately 12% of the testing groups in the EQ II network and only 2% in the DBLP network. These subtle observations indicate that GR and SR processes are responsible for a significant portion of future formed groups. Therefore, modeling these processes is an important step towards higher order link prediction.

4.2   Evaluation Methodology and Experimental Setup

We evaluate the performance of these methods as classifiers using the area under the curve (AUC) statistic of the receiver operating characteristics (ROC) [13]. Using the three score vectors as the model output we calculated AUC scores for two sets of prediction test scenarios. The first set includes the exact occurrences found in the testing period (referred to as "(**Exact**)") and the other set includes occurrences found with new vertices in the testing period (referred to as "(**New Vertices**)"). The following six scenarios are considered for each set:

1. **RG+RS(v):** Predicting both recurring groups and subgroups that are dyadic edges or other higher order faces. Note that for any group with new actors in the testing phase, we can only test whether the subgroup with old actors is a recurring group or subgroup from the training period.
2. **RG+RS(v+e):** Predicting both recurring groups and subgroups that are only triangles or other higher order faces. We only consider groups of size ≤ 6 and also omit vertex recurrences since those are reported in Table 1.
3. **RS(v):** Predicting only recurring subgroups that are edges or other higher order faces.
4. **RS(v+e):** Predicting only recurring subgroups that are triangles or other higher order faces.
5. **RG(v):** Predicting only recurring groups that are edges or other higher order faces.
6. **RG(v+e):** Predicting only recurring groups that are triangles or other higher order faces.

The optimal parameters were chosen for each split separately via grid search on the following parameter space: $\alpha = \{0.01, 0.1, 0.3, 0.5, 0.7, 0.9, 1, 2, 5, 10, 20\}$ and $\mu = \{10^{-7}, 10^{-6}, 10^{-5}, 10^{-4}, 10^{-3}, 0.01, 0.1, 0.3, 0.5, 0.7, 0.9, 0.99\}$. All the Hasse diagrams considered in the above methods have un-weighted edges.

Table 3. AUC scores for **EQ II** and **DBLP**

| | EQ II | | | | | | | | | | | |
| | Exact | | | | | | New Vertices | | | | | |
Method	RG+RS (v)	RG+RS (v+e)	RS (v)	RS (v+e)	RG (v)	RG (v+e)	RG+RS (v)	RG+RS (v+e)	RS (v)	RS (v+e)	RG (v)	RG (v+e)
HDDHyp-W	**0.96**	**0.98**	**0.67**	**0.87**	0.79	0.83	**0.96**	0.97	**0.68**	**0.86**	0.79	0.83
HDDHyp-C	**0.96**	**0.98**	0.63	0.78	**0.83**	0.87	**0.96**	**0.98**	0.64	0.78	0.82	**0.86**
HDDSimp-W	0.83	0.81	0.63	0.67	0.78	0.81	0.83	0.81	0.62	0.64	0.78	0.81
HDDSimp-C	0.78	0.75	0.52	0.6	0.83	0.86	0.78	0.75	0.52	0.59	**0.83**	0.85
CBS-W	0.77	0.68	0.6	0.54	0.78	0.81	0.76	0.68	0.59	0.5	0.78	0.81
CBS-C	0.76	0.72	0.5	0.49	0.82	0.85	0.76	0.72	0.5	0.47	0.82	0.85
HDSHyp-W	**0.96**	0.97	0.65	0.71	0.79	0.82	**0.96**	0.97	0.65	0.7	0.79	0.82
HDSSimp-W	0.7	0.59	0.58	0.52	0.76	0.8	0.7	0.59	0.58	0.48	0.76	0.8
HDSHyp-C	0.95	0.97	0.58	0.63	**0.83**	0.86	0.95	0.97	0.59	0.63	0.82	**0.86**
HDSSimp-C	0.68	0.61	0.49	0.48	0.82	0.85	0.67	0.6	0.48	0.44	0.82	0.85
	DBLP											
HDDHyp-W	**0.9**	**0.89**	**0.8**	**0.78**	0.69	0.68	0.82	0.85	0.73	**0.72**	0.7	0.68
HDDHyp-C	0.89	**0.89**	0.79	0.78	0.69	0.68	0.82	0.85	0.73	**0.72**	0.69	0.68
HDDSimp-W	0.77	0.79	0.73	0.73	0.7	0.69	0.77	0.77	**0.74**	0.71	0.71	0.69
HDDSimp-C	0.75	0.76	0.71	0.72	**0.71**	**0.7**	0.74	0.74	0.73	0.7	**0.72**	**0.7**
CBS-W	0.67	0.64	0.65	0.61	0.69	0.66	0.69	0.64	0.7	0.63	0.7	0.66
CBS-C	0.65	0.63	0.59	0.58	0.64	0.62	0.65	0.62	0.63	0.59	0.65	0.62
HDSHyp-W	0.89	0.88	0.75	0.73	0.69	0.66	**0.82**	0.84	0.73	0.7	0.7	0.66
HDSSimp-W	0.59	0.53	0.6	0.54	0.69	0.66	0.63	0.55	0.67	0.57	0.7	0.66
HDSHyp-C	0.88	0.87	0.72	0.71	0.64	0.62	0.8	0.83	0.68	0.67	0.65	0.62
HDSSimp-C	0.49	0.43	0.5	0.47	0.64	0.61	0.54	0.46	0.57	0.51	0.65	0.62

4.3 Results and Discussion

We compare the twelve different AUC scores, described in the prior section, for the ten methods developed in this paper. Results are reported in Table 3 for the EQ II and DBLP data. We have three different kinds of scores: CBS (Sect. 3.2), HDS (Sect. 3.3) and HDD (Sect. 3.3). Both the **CBS-W** and **CBS-C** scores are only count based and don't take into account any topological relationship between groups. On the other hand, the HDS and HDD methods take into account topology by exchanging information locally and globally, respectively. One of our main hypotheses is that topological structure affects the group recurrence behavior. We are also unaware of any methods for small group recurrence and therefore chose **CBS-W** and **CBS-C** scores as our baseline. Note, as described in Sect. 3.1, all the three genre of methods can be either count based (referred using suffix **-C**), or cohesion metric based (denoted by suffix **-W**). The count based variants do not take into account the cardinality of the (sub)groups whereas the cohesion metrics are cardinality based.

Effect of Topology: We observe from Table 3 (the best scores are highlighted in bold) that the Hasse diagram-based methods consistently outperform the count-based methods. This supports our hypothesis that Hasse diagram-based methods, which take into account topology, indeed, are more informative about the group recurrence process.

We also compare the methods against four criteria: (a) How do the prediction methods fare for recurring subgroups as compared to recurring groups?; (b) How well do the methods predict at dimensions of dyadic edges and above, i.e., the -(v) cases, compared with how well they predict at dimensions of triadic groups and above, i.e., the -(v+e) cases?; (c) How well do the methods predict the "Exact" occurrences versus the "New Vertices" occurrences?; and (d) How do the count-based "**-C**" methods compare with the cohesion-based "**-W**" methods?

We observe that in order to predict recurring subgroups, **HDDHyp-W** out-performs all other methods whether the subgroup was an "exact" occurrence or a "new vertices" occurrence in testing. This suggests that exchange of information from the groups observed in the past to the groups not observed in the past via the Hasse diagram topology and the global-based label diffusion process is more crucial for influencing the appearance of subgroups not observed in the past. In fact, the poor accuracy of the **HDDSimp** methods indicates that weights placed on (possibly unobserved) subgroups of observed groups used as prior information cause bias and hurt the predictive power of the model. Given that **HDDHyp-W** is initialized using the cohesion weights in (4), the normalization of counts only on the prior observed group occurrences in the diagram is important for recurring subgroup prediction. Moreover, the performance of predicting triangles or higher order groups ($RS(v+e)$) is higher for the EQ II data and comparable for the DBLP data to that of predicting dyadic edges or higher ($RS(v)$) across all HDD methods, implying the important role played by the Hasse diagram structure for higher order group prediction.

On the other hand for recurring group prediction the count-based methods **HDDHyp-C** and **HDDSimp-C** performed best, suggesting that the likelihood of recurrence of already-observed groups is determined more by the simple counts of past concurrences. The count-based HDS methods also give results comparable with that of the HDD methods. This implies that even the local spread of count information is sufficient for recurring group predictions. These **-Simp**-based methods using (1), which take into account the subgroup counts of the groups that occurred in training, provide good results, suggesting that the unobserved subgroups have an important influence on the potential of groups to re-occur.

Finally, we note that across both the datasets and across all the twelve experiments, the **HDD** methods generally perform better than or as good as **HDS** methods. Further results and details shall be made available in a future technical report.

5   Conclusions

We consider the problem of predicting small group evolution and focus on the sub-problem on group and subgroup recurrence. We highlight two important group recurrence processes and capture them using weighted simplicial complexes. We use a Hasse diagram corresponding to the simplicial complex as a graph whose nodes correspond to subgroups in the complex. We then build semi-supervised models on top of this graph for group recurrence prediction. We have shown that frequent patterns like small groups can be considered as topological entities, with relationships between them guided by higher order topological properties.

References

1. Aggarwal, C.C., Han, J.: Frequent Pattern Mining. Springer, Heidelberg (2014)
2. Berge, C., Minieka, E.: Graphs and Hypergraphs. North-Holland Publishing Company, Amsterdam (1973)
3. Contractor, N.: Some assembly required: leveraging web science to understand and enable team assembly. Philos. Trans. Roy. Soc. A: Math. Phys. Eng. Sci. **371**(1987), 20120385 (2013)
4. Kapoor, K., Sharma, D., Srivastava, J.: Weighted node degree centrality for hypergraphs. In: IEEE Network Science Workshop, pp. 152–155. IEEE (2013)
5. Lungeanu, A., Huang, Y., Contractor, N.S.: Understanding the assembly of interdisciplinary teams and its impact on performance. J. Informetrics **8**(1), 59–70 (2014)
6. Monge, P.R., Contractor, N.S.: Theories of Communication Networks. Oxford University Press, Oxford (2003)
7. Patil, A., Liu, J., Shen, J., Brdiczka, O., Gao, J., Hanley, J.: Modeling attrition in organizations from email communication. In: IEEE International Conference on Social Computing (SocialCom), pp. 331–338 (2013)
8. Poole, S., Poole, M.S., Hollingshead, A.B.: Theories of Small Groups: Interdisciplinary Perspectives. Sage Publications, Thousand Oaks (2004)
9. Ramanathan, R., Bar-Noy, A., Basu, P., Johnson, M., Ren, W., Swami, A., Zhao, Q.: Beyond graphs: capturing groups in networks. In: IEEE INFOCOM Workshops, pp. 870–875 (2011)
10. Sharma, A., Kuang, R., Srivastava, J., Feng, X., Singhal, K.: Predicting small group accretion in social networks: a topology based incremental approach. In: IEEE/ACM International Conference on Advances in Social Networks Analysis and Mining (ASONAM), pp. 408–415 (2015)
11. Skiena, S.: Hasse diagrams. In: Implementing Discrete Mathematics: Combinatorics and Graph Theory With Mathematica, p. 163 (1990)
12. Tian, Z., Hwang, T., Kuang, R.: A hypergraph-based learning algorithm for classifying gene expression and arrayCGH data with prior knowledge. Bioinformatics **25**(21), 2831–2838 (2009)
13. Van Trees, H.L.: Detection, Estimation and Modulation Theory. Wiley, New York (1968)
14. Zhou, D., Bousquet, O., Lal, T.N., Weston, J., Schölkopf, B.: Learning with local and global consistency. Adv. Neural Inf. Process. Syst. (NIPS) **16**(16), 321–328 (2004)
15. Zuckerman, E.W.: Do firms and markets look different? Repeat collaboration in the feature film industry, 1935–1995 (2004). Unpublished

A P-LSTM Neural Network for Sentiment Classification

Chi Lu, Heyan Huang$^{(\boxtimes)}$, Ping Jian, Dan Wang, and Yi-Di Guo

Beijing Institute of Technology, Beijing, China
{luchi,hhy63,pjian,wangdan12856,gyd409274478}@bit.edu.cn

Abstract. Neural network models have been demonstrated to be capable of achieving remarkable performance in sentiment classification. Convolutional neural network (CNN) and recurrent neural network (RNN) are two mainstream architectures for such modelling task. In this work, a novel model based on long short-term memory recurrent neural network (LSTM) called P-LSTM is proposed for sentiment classification. In P-LSTM, three-words phrase embedding is used instead of single word embedding as is often done. Besides, P-LSTM introduces the phrase factor mechanism which combines the feature vectors of the phrase embedding layer and the LSTM hidden layer to extract more exact information from the text. The experimental results show that the P-LSTM achieves excellent performance on the sentiment classification tasks.

Keywords: LSTM · Phrase-embedding · Phrase factor mechanism

1 Introduction

Text classification is an important task in many areas of nature language processing (NLP). Many different methods have been proposed for sentiment classification, such as using Support Vector Machines (SVM) with rule-based features, combining SVM with Naive Bayes (NB) [17], and building dependency trees with Conditional Random Fields (Tree-CRF) [10]. Other methods such as Maximum entropy [1] and Hidden Markov Models [13] are also widely used. Deep learning models have achieved remarkable results in NLP areas in these years. There are also many deep learning models in sentiment classification, including convolutional neural network (CNN) [5,6] and recurrent neural network (RNN) [7].

Being able to handle sequences of any length and capture long-term dependencies, RNN has great power to extract high-level information of a text, and it is widely used in NLP, especially in Neural Machine Translation (NMT). Work [7] has shown that RNN can also do a good job in sentiment classification and text classification task. RNN has many variants. In this work, we choose long short-term memory recurrent neural network (LSTM) [4] as the basic frame of the proposed P-LSTM. LSTM can solve long term dependency problem easily and can remember more information than traditional RNN.

© Springer International Publishing AG 2017
J. Kim et al. (Eds.): PAKDD 2017, Part I, LNAI 10234, pp. 524–533, 2017.
DOI: 10.1007/978-3-319-57454-7_41

It has been noted that many of neural network models focus on isolate word-embedding representation instead of the context phrase. However, texts especially short texts usually appear in sequence, therefore using information from the phrase context of the word may improve the classification accuracy. To benefit from the information of context, P-LSTM sets the context of a given word by combining the previous and the later word of the given word itself, thus a three-words phrase embedding vector is used as the basic input unit of the model, which performs well in our experiments.

In traditional RNN-based models, once the feature vectors of the high-level layer are extracted, the feature vectors of the low-level embedding layer are abandoned. P-LSTM thinks the features of low-level layer is useful too. To take advantage of both the features of low-level layer and high-level layer, P-LSTM introduces the phrase factor mechanism. The phrase factor mechanism calculates the phrase factor between low-level embedding layer and the high-level LSTM hidden layer. Then the phrase factor will be fed back into the model again to accomplish the classification task.

In this work, word vectors are initialled by the publicly available word2vec vectors that trained on 100 billion words of Google News using continuous bag-of-words architecture [9], and these vectors still need to be trained during training procedure. Previous work [6] has shown that the initialization with pre-trained word vectors can get better result than that with random word vectors. Indeed, it is obvious in our experiment that the convergence speed and the precision of model with initialized word vectors are improved.

The contributions of our work include:

- We use three-words phrase embedding instead of the isolate word embedding as the basic input unit of our model.
- To the best of our knowledge, we are the first work to introduce the phrase factor mechanism to the standard LSTM to extract richer and more exact informations from the text.

2   Related Work

Traditional methods with rule-based features models such as Naive Bayes and Support Vector Machine [17] has achieved remarkable results on sentiment classification task. However, the performance of these models depends heavily on artificial feature selection, which makes these models hard to be applied to different datasets.

Deep learning based neural network models avoid complicated feature selection and they have been widely used in many NLP tasks. Many recent works using deep learning methods have been proposed for sentiment classification. Among them, convolutional neural network (CNN) and recurrent neural network (RNN) are two popular ones. Work [5,6] proposed models based on CNN with litter hyperparameters on sentiment classification task. Owing to the CNN's capacity to capture high-level local correlations of the sentence, these works achieved good results on multiple benchmarks. Meanwhile as a sequence model,

RNN is able to deal with variable-length input sequences and discover long-term dependencies. And with the ability of explicitly modeling time-series data, RNNs are being increasingly applied to sentence modeling. Work [7,16] used a variant of RNN called Long Short Term Memory (LSTM) [4] to produce sentence representation from word presentation. Then the sentence representation will be used in its corresponding classifier to finish the sentiment classification tasks.

In this paper, we propose a model named P-LSTM which combines the slide window input mechanism in CNN model and the capacity to extract sequential text information in RNN model. What's more, motivated by the attention mechanism in NMT model [2] which calculates the attention factor between the encoder and the decoder layer, our model introduces the phrase factor mechanism to extract richer and more exact informations from the low-level embedding layer and the high-level LSTM hidden layer.

3   Model

The architecture of the P-LSTM model is shown in Fig. 1. Phrase embedding and phrase factor mechanism are introduced to a standard LSTM, which composes the main architecture of P-LSTM. P-LSTM has a range of repeated modules for each time step while processing sequential data.

Let $X_i \in \mathbb{R}^k$ be the k-dimensional word vector corresponding to the i-th word in the text. The text of length n can be represented as

$$X_{1:n} = X_1 \oplus X_2 \oplus X_3 \oplus ... \oplus X_n. \tag{1}$$

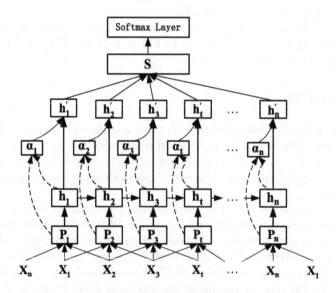

Fig. 1. The architecture of P-LSTM.

The $\oplus$ is the concatenation operation. Let $X_{i:i+j}$ be the concatenation of words $X_i X_{i+1}...X_{i+j}$. In this model, the phrase embedding vector is the concatenation of three words embedding in a row, which is the basic input unit of the P-LSTM. Given a word vector X_t, we choose the three-words phrase embedding vector $P_t = X_{t-1:t+1}$ as the t-th input vector. For a given sentence with the length of n, the previous word vector of the first word X_1 in the text is set to the vector of X_n, similarly the later word vector of the last word X_n in the text is set to the vector of X_1. At each time step, for the t-th phrase embedding vector P_t in the text, P-LSTM takes $P_t \in \mathbb{R}^{3k}$, $h_{t-1} \in \mathbb{R}^{3k}$, $c_{t-1} \in \mathbb{R}^{3k}$ as input and produces h_t, c_t based on the following formulas:

$$i_t = \sigma(W_i P_t + U_i h_{t-1} + b_i). \tag{2}$$

$$\tilde{c}_t = tanh(W_c P_t + U_c h_{t-1} + b_c). \tag{3}$$

$$f_t = \sigma(W_f P_t + U_f h_{t-1} + b_f). \tag{4}$$

$$c_t = i_t \circ \tilde{c}_t + f_t \circ c_{t-1}. \tag{5}$$

$$o_t = \sigma(W_o P_t + U_o h_{t-1} + b_o). \tag{6}$$

$$h_t = o_t \circ tanh(c_t). \tag{7}$$

Here $U_m \in \mathbb{R}^{3k \times 3k}$ and $W_m \in \mathbb{R}^{3k \times 3k}$ are all weight matrices, $b_m \in \mathbb{R}^{3k}$ are bias vectors, for $m \in \{i, f, c, o\}$. The symbols $\sigma(\cdot)$ and $tanh(\cdot)$ are LSTM activate functions which refer to the sigmoid and hyperbolic tangent functions. The symbol $\circ$ is an operation which means the element-wise multiplication. h_0 and c_0 are initialized by all zero vectors with dimension of $3k$.

After getting feature vector $h_t \in \mathbb{R}^{3k}$, the phrase mechanism calculates the phrase factor α_t by the following formula:

$$z'_t = tanh(P_t \circ h_t). \tag{8}$$

$$z_t = V_\alpha z'_t. \tag{9}$$

$$\alpha_t = sum(z_t). \tag{10}$$

Here $V_\alpha \in \mathbb{R}^{3k \times 3k}$ is the weight matrix. And $sum(\cdot)$ operation means add up all numbers of the vector. α_t is the phrase factor of the feature vector h_t, which is a real number. Once we get the phrase factor α_t, we can get the input vector $h'_t \in \mathbb{R}^{3k}$ of the pooling layer by:

$$h'_t = \alpha_t h_t. \tag{11}$$

The pooling layer takes the average of all the vector h'_t, for $t \in (1, n)$ (where n is the length of the text). It can be denoted as:

$$S = \frac{1}{n} \sum_{t=1}^{n} h'_t \tag{12}$$

Here the feature vector $S \in \mathbb{R}^{3k}$ will be passed to a fully connected softmax layer whose output y is the probability distribution over the set of l classes:

$$y = softmax(W_s S + b_s). \tag{13}$$

where $W_s \in \mathbb{R}^{l \times 3k}$ is the weight matrix, $b_s \in \mathbb{R}^l$ is bias vector.

4   Datasets and Experimental Setting

4.1   Datasets

We evaluate our model on sentiment classification task using the following datasets. Summary statistics of these datasets are in Table 1.

IMDB: A benchmark dataset for sentiment classification [8]. It is a large movie review dataset with full-length reviews; The task is to determine if the movie reviews are positive or negative. Both the training and test set have 25K reviews.
RT-2k: The standard 2000 full-length movie review dataset [11]. Classification involves detecting positive/negative reviews.
RT-s: Movie reviews with one sentence per reviews [12]. Classification involves detecting positive/negative reviews.
Subj: The subjectivity dataset consists of subjective reviews and objective plot summaries [11]. The task of subjectivity dataset is to classify the text as being subjective or objective.

Table 1. Summary statistics for the datasets. (N_+, N_-): number of positive and negative examples. l: Average number of words per example. $|V|$: Vocabulary size. Test: Test size (CV means there was no standard train/test split and 10-fold cross-validation was used)

Dataset	(N_+, N_-)	l	$|V|$	Test
IMDB	(25k,25k)	231	392k	25000
RT-2k	(1000,1000)	787	51k	CV
RT-s	(5331,5331)	21	21k	CV
Subj	(5000,5000)	24	24k	CV

4.2   Pre-trained Word Vectors

The dataset that has previously been preprocessed, and text is splitted into separate words. For each dataset, We use $20K$ words that appeared most frequently and make it as the vocabulary table of its dataset. For a given text of the a dataset, words in the vocabulary table will be initialed by word2vec vectors that are trained on 100 billion words from Google News. Each of these vectors has dimensionality of 300. Words not in the vocabulary table will be set all zero vector with the dimensionality of 300. The phrase embedding vector is the concatenation of three words, apparently it has the dimensionality of 900.

4.3   Hyper-parameters and Training

The model is trained to minimize the negative log-likelihood of predicting the correct label of the datasets, using batch gradient descent with the Adadelta update rule [19]. At each gradient descent step, weight matrices, bias vectors, and word vectors are updated. Early stopping is used on the validation set with a patience of 10 epoch.

4.4   Regulation

For regulation, we employ dropout with the rate $p = 0.5$ on the penultimate layer. Dropout prevents co-adaption of hidden units by randomly dropping out a proportion p of the hidden units during forward-backpropagation. That is, given the penultimate layer S, for output unit y in softmax layer, dropout uses:

$$y = softmax(W_s\ (S \circ r) + b_s). \tag{14}$$

where $\circ$ is the element-wise multiplication operator and $r \in \mathbb{R}^{3k}$ is a 'masking' vector of Bernoulli random variables with probability p of being 1. Gradients are backpropagated only through the unmasked units.

4.5   Model Variations

For comparison, we experiment with several variants of P-LSTM. Except for the differences of models mentioned below, tuning hyper-parameters and training procedure are kept all the same among different variants.

standard LSTM: Our baseline model where isolate word embedding is used on a standard LSTM [4] without phrase factor mechanism.
non-factor P-LSTM: The non-factor P-LSTM keeps everything exactly the same as P-LSTM except that non-factor P-LSTM has no phrase factor mechanism.
word-based P-LSTM: The word-based P-LSTM keeps everything exactly the same as P-LSTM except that word-based P-LSTM takes isolate word embedding as its input instead of phrase embedding.

5   Result and Discussion

Table 2 shows the result of our model on all four datasets against other methods. Our baseline model (standard LSTM) does not perform well on its own. P-LSTM achieves the best results comparing to other three model variations, which satisfies our expectation. The word-based P-LSTM and the non-factor P-LSTM also perform well, and we are surprised at the magnitude of the gains. Besides, we also compared our P-LSTM with several methods proposed in other papers on these four same datasets. As is shown in Table 2, our P-LSTM model achieves the best results on three of the four datasets, which is competitive against other deep learning methods (bow-CNN, seq-CNN) or the traditional feature based methods (SVM with uni-bigram, MNB with uni-bigram).

5.1   P-LSTM vs. Word-Based P-LSTM

According to the results shown in Table 2, P-LSTM outperforms the word-based P-LSTM, which is consistent with our expectations. That is to say, the phrase embedding is useful comparing to isolate word embedding.

530   C. Lu et al.

Table 2. Results of our P-LSTM model against other methods. **SVM, MNB, NBSVM**: SVM, Multinomial Naive Bayes and Naive Bayes SVM with uni-bigram [17]. **G-Dropout, F-Dropout**: Gaussian Dropout and Fast Dropout [18]. **RAE**: Recursive Autoencoders [15]. **MV-RNN**: Matrix-Vector Recursive Neural Network with parse trees [14]. **bow-CNN, seq-CNN**: bag-of-words CNN and sequence CNN [5]. **Tree-CRF**: Dependency tree with Conditional Random Fields [10]. **BoF-noDic, BoF-w/Rev**: BoF-noDic, BoF-w/Rev [10]. **BoWSVM**: SVM with bag-of-words features [11]. **WRRMB**: Word Representation Restricted Boltzmann Machine [3].

Model	IMDB	RT-2k	RT-s	Subj
Standard LSTM	90.084	83.45	77.89	91.7
Non-factor P-LSTM	90.28	85.25	78.35	91.93
Word-based P-LSTM	90.28	86.8	79.232	92.3
P-LSTM	**91.45**	89.25	**80.17**	**93.77**
SVM	89.16	87.40	77.7	91.74
MNB	86.59	85.85	79.0	93.6
NBSVM	91.22	89.45	79.4	93.2
G-Dropout	91.2	**89.7**	79.0	93.4
F-Dropout	91.1	89.5	79.1	93.6
RAE	-	-	76.8	-
MV-RNN	-	-	79.0	-
bow-CNN	91.03	-	-	-
seq-CNN	91.26	-	-	-
Tree-CRF	-	-	77.3	-
Bof-noDic	-	-	75.7	-
Bof-w/Rev	-	-	76.4	-
BoWSVM	-	87.15	-	90
WRRBM	87.42	-	-	-

Table 3 could lead us to know why the phrase embedding works. Two sentences in Table 3 have very similar format while they have opposite meaning. Both of the two sentences have the positive word "interesting". However, the word "interesting" has different phrase context in each sentence which can truly decide the label of the sentence. As is shown in Table 3, P-LSTM can judge the labels of the two sentences correctly while the word-based P-LSTM made a mistake on one of them. Apparently the word-based P-LSTM puts more emphasis on the isolate word "interesting" while ignoring the context "far from being" when judging the category of the second sentence in Table 3. However, the P-LSTM takes the three-words phrase embedding as its input, and the context of the word "interesting" is taken into consideration when judging the category of each sentence in Table 3, which is more accurate.

Table 3. Classification task on two sentences by P-LSTM and word-based P-LSTM trained on RT-s dataset. Positive sentence is labelled as 1 and negative sentence is labelled as 0.

Sentence	True label	P-LSTM	Word-based P-LSTM
"The movie is very interesting."	1	1	1
"The movie is far from being interesting."	0	0	1

5.2   P-LSTM vs. Non-factor P-LSTM

Results from all the four datasets have shown that the performance of the non-factor P-LSTM is worse than P-LSTM, which means the phrase factor mechanism works well.

According to Eqs. (10), (11), (12) and (13). We can get:

$$y = softmax(\frac{1}{n} \sum_{t=1}^{n} y_t). \tag{15}$$

for the dataset with l classes ($l = 2$ in the example below), $y_t \in \mathbb{R}^l$ in P-LSTM is:

$$y_t = \alpha_t W_s h_t + b_s. \tag{16}$$

and y_t in non-factor P-LSTM is:

$$y_t = W_s h_t + b_s. \tag{17}$$

For the first sentence in Table 3, we calculated the y_t for $t \in [1,6]$ (6 is the length of the sentence) on both the P-LSTM and the non-factor P-LSTM that are trained on the RT-s dataset, and the results are shown in Table 4. The true probability distribution y of the sentence is $[0,1]$, which means the sentence is positive.

Table 4. The results of y_t ($t \in [1,6]$) calculated on both the P-LSTM and the non-factor P-LSTM that are trained on the RT-s dataset when performing a classification task on the sentence "the movie is very interesting." **t**: sequence number of the input phrase. **t-th phrase**: the t-th input phrase of the model.

t	t-th phrase	y_t	
		P-LSTM	Non-factor P-LSTM
1	. the movie	[2.09, −2.10]	[0.30, −0.31]
2	the movie is	[−19.5, 19.5]	[0.06, −0.06]
3	movie is very	[3.98, −3.99]	[−0.74, 0.73]
4	is very interesting	[−39.9, 40.1]	[−0.23, 0.22]
5	very interesting	[−33.5, 33.5]	[1.34, −1.33]
6	interesting . the	[5.32, −5.31]	[2.21, −2.20]

According to Eq. (15) and the results in Table 4, it is clear that y_4 and y_5 in P-LSTM have the greatest impact on the softmax classification layer and the impact is positive. Similarly, y_5 and y_6 in non-factor P-LSTM have greatest impact on the softmax classification layer, but the impact is negative. Here the positive/negative impact means that the impact is positive/negative for the classifier to make the right classification decision.

On the other hand, from an intuitive point of view, for the sentence *"the movie is very interesting."*, apparently the 4-*th* phrase and the 5-*th* phrase have the most decisive effect on the true label of the sentence. It is worth to note that the 4-*th* phrase and the 5-*th* phrase are the corresponding phrases of y_4 and y_5 in P-LSTM, and y_4 and y_5 in P-LSTM have the greatest positive impact on the softmax classification layer. That is to say, the phrase factor mechanism in P-LSTM makes the classifier focus on those phrases with the most decisive effect by influencing the y_t, which improves the accuracy of the classifier greatly in comparison to the non-factor P-LSTM.

6   Conclusion and Future Work

We have described a novel model called P-LSTM. In P-LSTM, three-words phrase embedding is used instead of isolate word embedding. Besides, P-LSTM combines the feature vectors of phrase embedding layer and LSTM hidden layer by the phrase factor mechanism, which achieves satisfactory results on several sentiment classification tasks.

We would explore in the future the ways to replace the basic LSTM frame with bidirectional LSTM frame to extract the forward and the backward features of the text. We believe the bidirectional LSTM will get better result.

Acknowledgments. This work was supported by the National Basic Research Program (973) of China (No. 2013CB329303).

References

1. Ang, J., Liu, Y., Shriberg, E.: Automatic dialog act segmentation and classification in multiparty meetings. In: ICASSP (1), pp. 1061–1064 (2005)
2. Bahdanau, D., Cho, K., Bengio, Y.: Neural machine translation by jointly learning to align and translate. Computer Science (2014)
3. Dahl, G.E., Adams, R.P., Larochelle, H.: Training restricted Boltzmann machines on word observations, pp. 679–686 (2012)
4. Hochreiter, S., Schmidhuber, J.: Long short-term memory. Neural Comput. **9**(8), 1735–1780 (1997)
5. Johnson, R., Zhang, T.: Effective use of word order for text categorization with convolutional neural networks (2014, eprint arXiv)
6. Kim, Y.: Convolutional neural networks for sentence classification (2014). arXiv preprint: arXiv:1408.5882
7. Lee, J.Y., Dernoncourt, F.: Sequential short-text classification with recurrent and convolutional neural networks (2016). arXiv preprint: arXiv:1603.03827

8. Maas, A.L., Daly, R.E., Pham, P.T., Huang, D., Ng, A.Y., Potts, C.: Learning word vectors for sentiment analysis. In: Meeting of the Association for Computational Linguistics: Human Language Technologies, pp. 142–150 (2011)
9. Mikolov, T., Sutskever, I., Chen, K., Corrado, G., Dean, J.: Distributed representations of words and phrases and their compositionality. Adv. Neural Inf. Process. Syst. **26**, 3111–3119 (2013)
10. Nakagawa, T., Inui, K., Kurohashi, S.: Dependency tree-based sentiment classification using CRFs with hidden variables. In: Proceedings of the Human Language Technologies: Conference of the North American Chapter of the Association of Computational Linguistics, Los Angeles, California, USA, 2–4 June 2010, pp. 786–794 (2010)
11. Pang, B., Lee, L.: A sentimental education: sentiment analysis using subjectivity summarization based on minimum cuts. In: Meeting on Association for Computational Linguistics, pp. 271–278 (2004)
12. Pang, B., Lee, L.: Seeing stars: exploiting class relationships for sentiment categorization with respect to rating scales. In: Meeting on Association for Computational Linguistics, pp. 115–124 (2005)
13. Reithinger, N., Klesen, M.: Dialogue act classification using language models. In: EuroSpeech, Citeseer (1997)
14. Socher, R., Huval, B., Manning, C.D., Ng, A.Y.: Semantic compositionality through recursive matrix-vector spaces. In: Joint Conference on Empirical Methods in Natural Language Processing and Computational Natural Language Learning, pp. 1201–1211 (2012)
15. Socher, R., Pennington, J., Huang, E.H., Ng, A.Y., Manning, C.D.: Semi-supervised recursive autoencoders for predicting sentiment distributions. In: Conference on Empirical Methods in Natural Language Processing, EMNLP 2011, 27–31 July 2011, John Mcintyre Conference Centre, Edinburgh, UK, A Meeting of Sigdat, A Special Interest Group of the ACL, pp. 151–161 (2011)
16. Tang, D., Qin, B., Liu, T.: Document modeling with gated recurrent neural network for sentiment classification. In: Conference on Empirical Methods in Natural Language Processing, pp. 1422–1432 (2015)
17. Wang, S., Manning, C.D.: Baselines and bigrams: simple, good sentiment and topic classification. In: Proceedings of the 50th Annual Meeting of the Association for Computational Linguistics: Short Papers, vol. 2, pp. 90–94. Association for Computational Linguistics (2012)
18. Wang, S.I., Manning, C.D.: Fast dropout training. In: ICML (2), pp. 118–126 (2013)
19. Zeiler, M.D.: ADADELTA: an adaptive learning rate method. Computer Science (2012)

Learning What Matters – Sampling Interesting Patterns

Vladimir Dzyuba[1]([⊠]) and Matthijs van Leeuwen[2]

[1] Department of Computer Science, KU Leuven, Leuven, Belgium
vladimir.dzyuba@cs.kuleuven.be
[2] LIACS, Leiden University, Leiden, The Netherlands
m.van.leeuwen@liacs.leidenuniv.nl

Abstract. In the field of *exploratory data mining*, local structure in data can be described by *patterns* and discovered by mining algorithms. Although many solutions have been proposed to address the redundancy problems in pattern mining, most of them either provide succinct pattern sets or take the interests of the user into account—but not both. Consequently, the analyst has to invest substantial effort in identifying those patterns that are relevant to her specific interests and goals.

To address this problem, we propose a novel approach that combines pattern sampling with interactive data mining. In particular, we introduce the LETSIP algorithm, which builds upon recent advances in (1) weighted sampling in SAT and (2) learning to rank in interactive pattern mining. Specifically, it *exploits user feedback to directly learn the parameters of the sampling distribution that represents the user's interests.*

We compare the performance of the proposed algorithm to the state-of-the-art in interactive pattern mining by emulating the interests of a user. The resulting system allows efficient and interleaved learning and sampling, thus user-specific anytime data exploration. Finally, LETSIP demonstrates favourable trade-offs concerning both quality–diversity and exploitation–exploration when compared to existing methods.

1 Introduction

Imagine a data analyst who has access to a medical database containing information about patients, diagnoses, and treatments. Her goal is to identify novel connections between patient characteristics and treatment effects. For example, one treatment may be more effective than another for patients of a certain age and occupation, even though the latter is more effective at large. Here, age and occupation are latent factors that explain the difference in treatment effect.

In the field of *exploratory data mining*, such hypotheses are represented by *patterns* [1] and discovered by mining algorithms. Informally, a pattern is a statement in a formal language that concisely describes the structure of a subset of the data. Unfortunately, in any realistic database the *interesting* and/or *relevant* patterns tend to get lost among a humongous number of patterns.

The solutions that have been proposed to address this so-called *pattern explosion*, caused by enumerating *all* patterns satisfying given constraints, can be

© Springer International Publishing AG 2017
J. Kim et al. (Eds.): PAKDD 2017, Part I, LNAI 10234, pp. 534–546, 2017.
DOI: 10.1007/978-3-319-57454-7_42

roughly clustered into four categories: (1) *condensed representations* [9], (2) *pattern set mining* [8], (3) *pattern sampling* [5], (4) and—most recently—*interactive pattern mining* [18]. As expected, each of these categories has its own strengths and weaknesses and there is no ultimate solution as of yet.

That is, condensed representations, e.g., closed itemsets, can be lossless but usually still yield large result sets; pattern set mining and pattern sampling can provide succinct pattern sets but do not take the analyst into account; and existing interactive approaches take the user into account but do not adequately address the pattern explosion. Consequently, the analyst has to invest substantial effort in identifying those patterns that are relevant to her specific interests and goals, which often requires extensive data mining expertise.

Aims and Contributions. Our overarching aim is to enable analysts—such as the one described in the medical scenario above—to discover small sets of patterns from data that they consider interesting. This translates to the following three specific requirements. First, we require our approach to yield *concise and diverse result sets*, effectively avoiding the pattern explosion. Second, our method should *take the user's interests into account* and ensure that the results are relevant. Third, it should achieve this *with limited effort on behalf of the user*.

To satisfy these requirements, we propose an approach that combines pattern sampling with interactive data mining techniques. In particular, we introduce the LETSIP algorithm, for **Le**arn **t**o **S**ample **I**nteresting **P**atterns, which follows the *Mine, Interact, Learn, Repeat* framework [12]. It samples a small set of patterns, receives feedback from the user, exploits the feedback to learn new parameters for the sampling distribution, and repeats these steps. As a result, the user may utilize a compact diverse set of interesting patterns at any moment, blurring the boundaries between learning and discovery modes.

We satisfy the first requirement by using a sampling technique that samples high quality patterns with high probability. While sampling does not guarantee diversity *per se*, we demonstrate that it gives concise yet diverse results in practice. Moreover, sampling has the advantage that it is *anytime*, i.e., the result set can grow by user's request. LETSIP's sampling component is based on recent advances in sampling in SAT [11] and their extension to pattern sampling [14].

The second requirement is satisfied by learning what matters to the user, i.e., by interactively learning the distribution patterns are sampled from. This allows the user to steer the sampler towards subjectively interesting regions. We build upon recent work [6,12] that uses *preference learning* to learn to rank patterns.

Although user effort can partially be quantified by the total amount of input that needs to be given during the analysis, the third requirement also concerns the time that is needed to find the first interesting results. For this it is of particular interest to study the *trade-off between exploitation and exploration*. As mentioned, one of the benefits of interactive pattern sampling is that the boundaries between learning and discovery are blurred, meaning that the system keeps learning while it continuously aims to discover potentially interesting patterns.

We evaluate the performance of the proposed algorithm and compare it to the state-of-the-art in interactive pattern mining by emulating the interests of a user.

The results confirm that the proposed algorithm has the capacity to learn what matters based on little feedback from the user. More importantly, the LETSIP algorithm demonstrates favourable trade-offs concerning both quality–diversity and exploitation–exploration when compared to existing methods.

2   Interactive Pattern Mining: Problem Definition

Recall the medical analyst example. We assume that after inspecting patterns, she can judge their interestingness, e.g., by comparing two patterns. Then the primary task of interactive pattern mining consists in learning a formal model of her interests. The second task involves using this model to mine novel patterns that are subjectively interesting to the user (according to the learned model).

Formally, let $\mathcal{D}$ denote a dataset, $\mathcal{L}$ a pattern language, $\mathcal{C}$ a (possibly empty) set of constraints on patterns, and $\succ$ the unknown subjective pattern preference relation of the current user over $\mathcal{L}$, i.e., $p_1 \succ p_2$ implies that the user considers pattern p_1 subjectively more interesting than pattern p_2:

Problem 1 (Learning). Given $\mathcal{D}$, $\mathcal{L}$, and $\mathcal{C}$, dynamically collect feedback U with respect to patterns in $\mathcal{L}$ and use U to learn a (subjective) pattern interestingness function $h : \mathcal{L} \to \mathbb{R}$ such that $h(p_1) > h(p_1) \Leftrightarrow p_1 \succ p_2$.

The mining task should account for the potential diversity of user's interests. For example, the analyst may (unwittingly) be interested in several unrelated treatments with disparate latent factors. An algorithm should be able to identify and mine patterns that are representative of these diverse hypotheses.

Problem 2 (Mining). Given $\mathcal{D}$, $\mathcal{L}$, $\mathcal{C}$, and h, mine a set of patterns $\mathcal{P}_h$ that maximizes a combination of interestingness h and diversity of patterns.

The interestingness of $\mathcal{P}$ can be quantified by the average quality of its members, i.e., $\sum_{p \in \mathcal{P}} h(p) \mid /|\mathcal{P}|$. Diversity measures quantify how different patterns in a set are from each other. *Joint entropy* is a common diversity measure [21].

3   Related Work

In this paper, we focus on two classes of related work aimed at alleviating the pattern explosion, namely (1) pattern sampling and (2) interactive pattern mining.

Pattern Sampling. First pattern samplers are based on Markov Chain Monte Carlo (MCMC) random walks over the pattern lattice [4,5,16]. Their main advantage is that they support "black box" distributions, i.e., they do not require any prior knowledge about the target distribution, a property essential for interactive exploration. However, they often converge only slowly to the desired target distribution and require the selection of the "right" proposal distributions.

Samplers that are based on alternative approaches include direct two-step samplers and XOR samplers. Two-step samplers [7], while provably accurate and efficient, only support a limited number of distributions and thus cannot be easily extended to interactive settings. FLEXICS [14] is a recently proposed pattern sampler based on the latest advances in weighted constrained sampling in SAT [11]. It supports black-box target distributions, provides guarantees with respect to sampling accuracy and efficiency, and has been shown to be competitive with the state-of-the-art methods described above.

Interactive Pattern Mining. Most recent approaches to interactive pattern mining are based on learning to rank patterns. They first appeared in Xin et al. [22] and Rueping [19] and were independently extended by Boley et al. [6] and Dzyuba et al. [12]. The central idea behind these algorithms is to alternate between mining and learning. PRIIME [3] focuses on advanced feature construction for interactive mining of structured data, e.g., sequences or graphs.

To the best of our knowledge, IPM [2] is the only existing approach to interactive itemset sampling. It uses binary feedback ("likes" and "dislikes") to update weights of individual items. Itemsets are sampled proportional to the product of weights of constituent items. Thus, the model of user interests in IPM is fairly restricted; moreover, it potentially suffers from convergence issues typical for MCMC. We empirically compare LETSIP with IPM in Sect. 6.

4 Preliminaries

Pattern Mining and Sampling. We focus on *itemset mining*, i.e., pattern mining for binary data. Let $\mathcal{I} = \{1 \ldots M\}$ denote a set of items. Then, a dataset $\mathcal{D}$ is a bag of transactions over $\mathcal{I}$, where each transaction t is a subset of $\mathcal{I}$, i.e., $t \subseteq \mathcal{I}$; $\mathcal{T} = \{1 \ldots N\}$ is a set of transaction indices. The pattern language $\mathcal{L}$ also consists of sets of items, i.e., $\mathcal{L} = 2^{\mathcal{I}}$. An itemset p occurs in a transaction t, iff $p \subseteq t$. The frequency of p is the proportion of transactions in which it occurs, i.e., $freq\,(p) = |\{t \in \mathcal{D} \mid p \subseteq t\}|/N$. In labeled datasets, each transaction t has a label from $\{-, +\}$; $freq^{-,+}$ are defined accordingly.

The choice of constraints and a quality measure allows a user to express her analysis requirements. The most common constraint is *minimal frequency* $freq\,(p) \geq \theta$. In contrast to hard constraints, quality measures are used to describe soft preferences that allow to rank patterns; see Sect. 6 for examples.

While common mining algorithms return the top-k patterns w.r.t. a measure $\varphi : \mathcal{L} \to \mathbb{R}^+$, pattern sampling is a randomized procedure that 'mines' a pattern with probability proportional to its quality, i.e., $P_\varphi(p \text{ is sampled}) = \varphi(p)/Z_\varphi$, if $p \in \mathcal{L}$ satisfies $\mathcal{C}$, and 0 otherwise, where Z_φ is the (unknown) normalization constant. This is an instance of weighted constrained sampling.

Weighted Constrained Sampling. This problem has been extensively studied in the context of sampling solutions of a SAT problem. WEIGHTGEN [11] is a recent algorithm for approximate weighted sampling in SAT. The core idea consists of partitioning the solution space into a number of "cells" and sampling a solution from a random cell. Partitioning with desired properties is obtained via augmenting the SAT problem with uniformly random XOR constraints (XORs).

Algorithm 1. LETSIP

Input: Dataset $\mathcal{D}$, minimal frequency threshold θ
Parameters: Query size k, query retention l, range A, cell sampling strategy ς
    SCD: regularisation parameter λ, iterations T; FLEXICS: error tolerance κ
                                                                    ▷ *Initialization*
 1: Ranking function $h_0 = $ LOGISTIC$(\mathbf{0}, A)$ ▷ Zero weights lead to uniform sampling
 2: Feedback $U \leftarrow \emptyset$, $Q_0^* \leftarrow \emptyset$

                                                    ▷ *Mine, Interact, Learn, Repeat* loop
 3: **for** $t = 1, 2, \ldots$ **do**
 4:     $R = $ TAKEFIRST(Q_{t-1}^*, l) ▷ Retain top patterns from the previous iteration
 5:     Query $Q_t \leftarrow R \cup$ SAMPLEPATTERNS$(h_{t-1}) \times (k - |R|)$ times
 6:     $Q_t^* = $ ORDER(Q_t), $U \leftarrow U \cup Q_t^*$ ▷ Ask user to order patterns in Q_t
 7:     $h_t \leftarrow$ LOGISTIC(LEARNWEIGHTS $(U; \lambda, T)$, A)

 8: **function** SAMPLEPATTERNS(Sampling weight function $w : \mathcal{L} \to [A, 1]$)
 9:     $C = $ FLEXICSRANDOMCELL$(\mathcal{D}, freq(\cdot) \geq \theta, w; \kappa)$
10:     **if** $\varsigma = $ TOP(m) **then return** m highest-weighted patterns
11:     **else if** $\varsigma = $ RANDOM **then return** PERFECTSAMPLE(C, w)

To sample a solution, WEIGHTGEN dynamically estimates the number of XORs required to obtain a suitable cell, generates random XORs, stores the solutions of the augmented problem (i.e., a random cell), and returns a perfect weighted sample from the cell. Owing to the properties of partitioning with uniformly random XORs, WEIGHTGEN provides theoretical performance guarantees regarding quality of samples and efficiency of the sampling procedure.

For implementation purposes, WEIGHTGEN only requires an efficient oracle that enumerates solutions. Moreover, it treats the target sampling distribution as a black box: it requires neither a compact description thereof, nor the knowledge of the normalization constant. Both features are crucial in pattern sampling settings. FLEXICS [14], a recently proposed pattern sampler based on WEIGHTGEN, has been shown to be accurate and efficient. Due to the page limit, we postpone further details to the extended version of this paper [13, Appendix A].

Preference Learning. The problem of learning ranking functions is known as *object ranking*. A common solving technique involves minimizing pairwise loss, e.g., the number of discordant pairs. For example, user feedback $U = \{p_1 \succ p_3 \succ p_2, \; p_4 \succ p_2\}$ is seen as $\{(p_1 \succ p_3), (p_1 \succ p_2), (p_3 \succ p_2), (p_4 \succ p_2)\}$. Given feature representations of objects $\boldsymbol{p_i}$, object ranking is equivalent to positive-only classification of difference vectors, i.e., a ranked pair example $p_i \succ p_j$ corresponds to a classification example $(\boldsymbol{p_i} - \boldsymbol{p_j}, +)$. All pairs comprise a training dataset for a scoring classifier. Then, the predicted ranking of any set of objects can be obtained by sorting these objects by classifier score descending. For example, this formulation is adopted by SVMRANK [17].

5   Algorithm

Key questions concerning instantiations of the *Mine, interact, learn, repeat* framework include (1) the feedback format, (2) learning quality measures from feedback, (3) mining with learned measures, and crucially, (4) selecting the patterns to show to the user. As pattern sampling has been shown to be effective in mining and learning, we present LETSIP, a sampling-based instantiation of the framework which employs FLEXICS. The sequel describes the mining and learning components of LETSIP. Algorithm 1 shows its pseudocode.

Mining Patterns by Sampling. Recall that the main goal is to discover patterns that are subjectively interesting to a particular user. We use parameterised logistic functions to measure the interestingness/quality of a given pattern p:

$$\varphi_{\text{logistic}}\left(p; w, A\right) = A + \frac{1-A}{1 + e^{-\boldsymbol{w} \cdot \boldsymbol{p}}}$$

where $\boldsymbol{p}$ is the vector of pattern features for p, $\boldsymbol{w}$ are feature weights, and A is a parameter that controls the range of the interestingness measure, i.e. $\varphi_{\text{logistic}} \in (A, 1)$. Examples of pattern features include $Length\,(p) = |p|/|\mathcal{I}|$, $Frequency\,(p) = freq\,(p)/|\mathcal{D}|$, $Items\,(i, p) = [i \in p]$; and $Transactions\,(t, p) = [p \subseteq t]$, where $[\cdot]$ denotes the Iverson bracket. Weights reflect feature contributions to pattern interestingness, e.g., a user might be interested in combinations of particular items or disinterested in particular transactions. The set of features would typically be chosen by the mining system designer rather than by the user herself. We empirically evaluate several feature combinations in Sect. 6.

Specifying feature weights manually is tedious and opaque, if at all possible. Below we present an algorithm that learns the weights based on easy-to-provide feedback with respect to patterns. This motivates our choice of logistic functions: they enable efficient learning. Furthermore, their bounded range $[A, 1]$ yields distributions that allow efficient sampling directly proportional to $\varphi_{\text{logistic}}$ with FLEXICS. Parameter A essentially controls the *tilt* of the distribution [14].

User Interaction and Learning from Feedback. Following previous research [12], we use *ordered feedback*, where a user is asked to provide a total order over a (small) number of patterns according to their subjective interestingness; see Fig. 1 for an example. We assume that there exists an unknown, user-specific target ranking R^*, i.e., a total order over $\mathcal{L}$. The inductive bias is that there exists $\boldsymbol{w}^*$ such that $p \succ q \Rightarrow \varphi_{\text{logistic}}\,(p, \boldsymbol{w}^*) > \varphi_{\text{logistic}}\,(q, \boldsymbol{w}^*)$. We apply the reduction of object ranking to binary classification of difference vectors (see Sect. 4). Following Boley et al. [6], we use Stochastic Coordinate Descent (SCD) [20] for minimizing L1-regularized logistic loss. However, unlike Boley et al., we directly use the learned functions for sampling.

SCD is an anytime convex optimization algorithm, which makes it suitable for the interactive setting. Its runtime scales linearly with the number of training pairs and the dimensionality of feature vectors. It has two parameters: (1) the number of weight updates (per iteration of LETSIP) T and (2) the regularization parameter λ. However, direct learning of $\varphi_{\text{logistic}}$ is infeasible, as it results in a non-convex loss function. We therefore use SCD to optimize the standard logistic loss, which is convex, and use the learned weights w in $\varphi_{\text{logistic}}$.

Selecting Patterns to Show to the User. An interactive system seeks to ensure faster learning of accurate models by targeted selection of patterns to show to the user; this is known as *active learning* or *query selection*. Randomized methods have been successfully applied to this task [12]. Furthermore, in large pattern spaces the probability that two redundant patterns are sampled in one (small) batch is typically low. Therefore, a sampler, which produces independent samples, typically ensures diversity within batches and thus sufficient *exploration*. We directly show k patterns sampled by FLEXICS proportional to $\varphi_{\text{logistic}}$ to the user, for which she has to provide a total order as feedback.

We propose two modifications to FLEXICS, which aim at emphasising *exploitation*, i.e., biasing sampling towards higher-quality patterns. First, we employ alternative cell sampling strategies. Normally FLEXICS draws a perfect weighted random sample, once it obtains a suitable cell. We denote this strategy as $\varsigma = \text{RANDOM}$. We propose an alternative strategy $\varsigma = \text{TOP}(m)$, which picks the m highest-quality patterns from a cell (Line 10 in Algorithm 1). We hypothesize that, owing to the properties of random XOR constraints, patterns in a cell as well as in consecutive cells are expected to be sufficiently diverse and thus the modified cell sampling does not disrupt exploration.

Rigorous analysis of (unweighted) uniform sampling by Chakraborty et al. shows that re-using samples from a cell still ensures broad coverage of the solution space, i.e., diversity of samples [10]. Although as a downside, consecutive samples are not i.i.d., the effects are bounded in theory and inconsequential in practice. We use these results to take license to modify the theoretically motivated cell sampling procedure. Although we do not present a similar theoretical analysis of our modifications, we evaluate them empirically.

Second, we propose to retain the top l patterns from the previous query and only sample $k - l$ new patterns (Lines 4–5). This should help users to relate the queries to each other and possibly exploit the structure in the pattern space.

6   Experiments

The experimental evaluation focuses on (1) the accuracy of the learned user models and (2) the effectiveness of learning and sampling. Evaluating interactive algorithms is challenging, for domain experts are scarce and it is hard to gather enough experimental data to draw reliable conclusions. In order to perform extensive evaluation, we emulate users using (hidden) interest models, which the algorithm is supposed to learn from ordered feedback only.

We follow a protocol also used in previous work [12]: we assume that R^* is derived from a quality measure φ, i.e., $p \succ q \Leftrightarrow \varphi(p) > \varphi(q)$. Thus, the task is to learn to sample frequent patterns proportional to φ from (short) sample rankings.

	Iteration 1			Iteration 2			...	Iteration 30		
	$p_{1,1}$	$p_{1,2}$	$p_{1,3}$	$p_{1,3}$	$p_{2,2}$	$p_{2,3}$		$p_{29,1}$	$p_{30,2}$	$p_{30,3}$
$freq, |p|, \ldots$	52,6	49,7	48,9	48,9	53,7	54,9		73,8	60,8	54,8
Feedback U	$p_{1,3} \succ p_{1,1} \succ p_{1,2}$			$p_{1,3} \succ p_{2,2} \succ p_{2,3}$				$p_{29,1} \succ p_{30,2} \succ p_{30,3}$		
$\varphi = surp$	0.12	0.04	0.20	0.20	0.11	0.10		0.28	0.26	0.12
(pct.rank)	0.51	0.13	0.84	0.84	0.46	0.41		0.99	0.97	0.51
Regret: Max.φ	$1 - 0.84 = 0.16$			0.16				0.01		

Fig. 1. We emulate user feedback U using a hidden quality measure φ (here *surp*; the boxplot shows the distribution of φ in the given dataset). The rows above the bar show the properties of the sampled patterns that would be inspected by a user, e.g., *frequency* or *length*, and the emulated feedback. The scatter plots show the relation between φ and the learned model of user interests $\varphi_{\text{logistic}}$ after 1 and 29 iterations of feedback and learning. The performance of the learned model improves considerably as evidenced by higher values of φ of the sampled patterns (squares) and lower regret.

As φ, we use frequency $freq$, surprisingness $surp$, and discriminativity in labeled data as measured by χ^2, where $surp(p) = \max\{freq(p) - \prod_{i \in p} freq(\{i\}), 0\}$ and

$$\chi^2(p) = \sum_{c \in \{-,+\}} \frac{(freq(p)(freq^c(p) - |\mathcal{D}^c|))^2}{freq(p)|\mathcal{D}^c|} + \frac{(freq(p)(freq^c(p) - |\mathcal{D}^c|))^2}{(|\mathcal{D}| - freq(p))|\mathcal{D}^c|}$$

We investigate the performance of the algorithm on ten datasets[1]. For each dataset, we set the minimal support threshold such that there are approximately 140 000 frequent patterns. Table 1 shows dataset statistics. Each experiment involves 30 iterations (queries). We use the default values suggested by the authors of SCD and FLEXICS for the auxiliary parameters of LETSIP: $\lambda = 0.001$, $T = 1000$, and $\kappa = 0.9$.

We evaluate performance using *cumulative regret*, which is the difference between

Table 1. Dataset properties.

	$|\mathcal{I}|$	$|\mathcal{D}|$	θ	Frequent patterns
anneal	93	812	660	149 331
australian	125	653	300	141 551
german	112	1000	300	161 858
heart	95	296	115	153 214
hepatitis	68	137	48	148 289
lymph	68	148	48	146 969
primary	31	336	16	162 296
soybean	50	630	28	143 519
vote	48	435	25	142 095
zoo	36	101	10	151 806

the ideal value of a certain measure M and its observed value, summed over iterations. We use the maximal and average quality φ in a query and joint entropy H_J

---

[1] Source: https://dtai.cs.kuleuven.be/CP4IM/datasets/.

as performance measures. To allow comparison across datasets and measures, we use percentile ranks by φ as a non-parametric measure of ranking performance. We also divide joint entropy by k: thus, the ideal value of each measure is 1 (e.g., the highest possible φ over all frequent patterns has the percentile rank of 1), and the regret is defined as $\sum 1 - M\left(Q_i^*\right)$, where $M \in \{\varphi_{avg}, \varphi_{max}, H_J\}$. We repeat each experiment ten times with different random seeds and report average regret.

A Characteristic Experiment in Detail. Figure 1 illustrates the workings of LETSIP and the experimental setup. It uses the `lymph` dataset, the target quality measure $\varphi = surp$, features $= Items$, $k = 3$, $A = 0.1$, $l = 1$, $\varsigma =$ RANDOM.

LETSIP starts by sampling patterns uniformly. A human user would inspect the patterns (items not shown) and their properties, e.g., frequency or length, or visualizations thereof, and rank the patterns by their subjective interestingness; in these experiments, we order them according to their values of φ. The algorithm uses the feedback to update $\varphi_{logistic}$. At the next iteration, the patterns are sampled from an updated distribution. As $l = 1$, the top-ranked pattern from the previous iteration $(p_{1,3})$ is retained. After a number of iterations, the accuracy of the approximation increases considerably, while the regret decreases. On average, one iteration takes 0.5 s on a desktop computer.

Evaluating Components of LetSIP. We investigate the effects of the choice of features and parameter values on the performance of LETSIP, in particular query size k, query retention l, range A, and cell sampling strategy ς. We use the following feature combinations ($\|$ denotes concatenation): *Items* (I); *Items*$\|$*Length*$\|$*Frequency* (ILF); and *Items*$\|$*Length*$\|$*Frequency*$\|$*Transactions* (ILFT). Values for other parameters and aggregated results are shown in Table 2.

Increasing the query size decreases the maximal quality regret more than twofold, which indicates that the proposed learning technique is able to identify the properties of target measures from ordered lists of patterns. However, as larger queries also increase the user effort, further we use a more reasonable query size of $k = 5$. Similarly, additional features provide valuable information to the learner. Changing the range A does not affect the performance.

The choice of values for query retention l and the cell sampling strategy allows influencing the exploration-exploitation trade-off. Interestingly, retaining one highest-ranked pattern results in the lowest regret with respect to the *maximal* quality. Fully random queries ($l = 0$) do not enable sufficient exploitation, whereas higher retention ($l \geq 2$)—while ensuring higher *average* quality—prevents exploration necessary for learning accurate weights.

The cell sampling strategy is the only parameter that clearly affects joint entropy, with purely random cell sampling yielding the lowest regret. However, it also results in the highest quality regrets, which negates the gains in diversity. Taking the best pattern according to $\varphi_{logistic}$ ensures the lowest quality regrets and joint entropy equivalent to other strategies. Based on these findings, we use the following parameters in the remaining experiments: $k = 5$, features $=$ ILFT, $A = 0.5$, $l = 1$, $\varsigma =$ TOP(1).

Table 2. Effect of LETSIP's parameters on regret w.r.t. three performance measures. Results are aggregated over datasets, quality measures, and other parameters.

		Regret: avg.φ	Regret: max.φ	Regret: H_J
Query size k	5	6.35 ± 1.04	1.13 ± 0.52	$\mathbf{13.28 \pm 0.89}$
	10	$\mathbf{5.91 \pm 0.59}$	$\mathbf{0.47 \pm 0.18}$	17.44 ± 0.45
All results below are for query size of $k = 5$				
Features	I	8.17 ± 0.96	1.35 ± 0.56	13.64 ± 0.90
	ILF	6.30 ± 1.36	1.16 ± 0.59	13.15 ± 0.96
	ILFT	$\mathbf{4.60 \pm 0.78}$	$\mathbf{0.87 \pm 0.40}$	$\mathbf{13.06 \pm 0.81}$
Range A	0.5	6.43 ± 1.06	1.15 ± 0.52	$\mathbf{13.20 \pm 0.86}$
	0.1	$\mathbf{6.26 \pm 1.01}$	$\mathbf{1.11 \pm 0.51}$	13.36 ± 0.91
Query retention l	0	8.19 ± 1.21	2.53 ± 0.72	13.38 ± 0.69
	1	6.78 ± 0.99	$\mathbf{0.53 \pm 0.34}$	$\mathbf{13.06 \pm 0.72}$
	2	5.61 ± 0.94	0.61 ± 0.42	13.56 ± 1.05
	3	$\mathbf{4.80 \pm 1.00}$	0.80 ± 0.57	13.33 ± 1.22
Cell sampling ς	RANDOM	10.60 ± 0.71	1.89 ± 0.64	$\mathbf{12.15 \pm 0.59}$
	Top(1)	$\mathbf{5.14 \pm 1.13}$	$\mathbf{0.81 \pm 0.45}$	13.70 ± 1.00
	TOP(2)	5.45 ± 1.06	0.87 ± 0.47	13.60 ± 0.98
	TOP(3)	5.95 ± 1.20	0.95 ± 0.50	13.57 ± 0.96

The largest proportion of LETSIP's runtime costs is associated with sampling (costs of weight learning are low due to a relatively low number of examples). The most important factor is the number of items $|\mathcal{I}|$: the average runtime per iteration ranges from 0.8 s for lymph to 5.8 s for australian, which is suitable for online data exploration.

Comparing with Alternatives. We compare LETSIP with APLE [12], another approach based on active preference learning, and IPM [2], an MCMC-based interactive sampling framework. For the former, we use query size k and feature representation identical to LETSIP, query selector MMR ($\alpha = 0.3$, $\lambda = 0.7$), $C_{\text{RANKSVM}} = 0.005$, and 1000 frequent patterns sampled uniformly at random and sorted by $freq$ as the *source ranking*. To compute regret, we use the top-5 frequent patterns according to the learned ranking function.

To emulate binary feedback for IPM based on φ, we use a technique similar to the one used by the authors: we designate a number of items as "interesting" and "like" an itemset, if more than half of its items are "interesting". To select the items, we sort frequent patterns by φ descending and add items from the top-ranked patterns until 15% of all patterns are considered "liked".

As we were not able to obtain the code for IPM, we implemented its sampling component by materializing all frequent patterns and generating perfect samples according to the learned multiplicative distribution. Note that this approach favors IPM, as it eliminates the issues of MCMC convergence. We request 300 samples (the amount of training data roughly equivalent to that of LETSIP), partition them into 30 groups of 10 patterns each, and use the tail 5 patterns in each group for regret calculations. Following the authors' recommendations, we set the learning parameter to $b = 1.75$. For the sampling-based methods LETSIP and IPM, we also report the diversity regret as measured by joint entropy.

Table 3 shows the results. The regret of LETSIP is substantially lower than that of either of the alternatives. The advantage over IPM is due to a more powerful learning mechanism and feature representation. IPM's multiplicative weights are biased towards longer itemsets and items seen at early iterations, which may prevent sufficient exploration, as evidenced by higher joint entropy regret. Non-sampling method APLE performs the best for $\varphi = freq$, which can be represented as a linear function of the features and learned by RANKSVM with the linear kernel. It performs substantially worse in other settings and has the highest variance, which reveals the importance of informed source rankings and the cons of pool-based active learning. These results validate the design choices made in LETSIP.

Table 3. LETSIP has considerably lower regrets than alternatives w.r.t. quality and, for samplers, diversity as quantified by joint entropy. (For $\varphi = surp$ (marked by *), IPM fails for 7 out of 10 datasets due to double overflow of multiplicative weights.)

	Regret: avg.φ			Regret: joint entropy H_J		
	freq	χ^2	*surp*	*freq*	χ^2	*surp*
LETSIP	2.4 ± 0.5	2.4 ± 0.1	4.5 ± 1.4	11.7 ± 0.6	11.7 ± 0.5	15.9 ± 1.1
IPM	15.5 ± 1.8	12.8 ± 2.3	15.5 ± 1.8*	15.7 ± 1.9	15.4 ± 1.9	19.8 ± 2.1*
APLE	0.0 ± 0.0	4.5 ± 3.8	5.3 ± 3.9	–	–	–

7 Conclusion

We presented LETSIP, a sampling-based instantiation of the *Mine, interact, learn, repeat* interactive pattern mining framework. The user is asked to rank small sets of patterns according to their (subjective) interestingness. The learning component uses this feedback to build a model of user interests via active preference learning. The model directly defines the sampling distribution, which assigns higher probabilities to more interesting patterns. The sampling component uses the recently proposed FLEXICS sampler, which we modify to facilitate control over the exploration-exploitation balance in active learning.

We empirically demonstrate that LETSIP satisfies the key requirements to an interactive mining system. We apply it to itemset mining, using a well-principled method to emulate a user. The results demonstrate that LETSIP learns to sample diverse sets of interesting patterns. Furthermore, it outperforms two state-of-the-art interactive methods. This confirms that it has the capacity to tackle the pattern explosion while taking user interests into account.

Directions for future work include extending LETSIP to other pattern languages, e.g., association rules, investigating the effect of noisy user feedback on the performance, and formal analysis, e.g., with multi-armed bandits [15]. A user study is necessary to evaluate the practical aspects of the proposed approach.

Acknowledgements. Vladimir Dzyuba is supported by FWO-Vlaanderen. The authors would like to thank the anonymous reviewers for their helpful feedback.

References

1. Aggarwal, C.C., Han, J. (eds.): Frequent Pattern Mining. Springer, Heidelberg (2014)
2. Bhuiyan, M., Hasan, M.A.: Interactive knowledge discovery from hidden data through sampling of frequent patterns. Stat. Anal. Data Mining: ASA Data Sci. J. **9**(4), 205–229 (2016)
3. Bhuiyan, M., Hasan, M.A.: PRIIME: a generic framework for interactive personalized interesting pattern discovery. In: Proceedings of IEEE Big Data, pp. 606–615 (2016)
4. Boley, M., Gärtner, T., Grosskreutz, H.: Formal concept sampling for counting and threshold-free local pattern mining. In: Proceedings of SDM, pp. 177–188 (2010)
5. Boley, M., Grosskreutz, H.: Approximating the number of frequent sets in dense data. Knowl. Inf. Syst. **21**(1), 65–89 (2009)
6. Boley, M., Mampaey, M., Kang, B., Tokmakov, P., Wrobel, S.: One click mining - interactive local pattern discovery through implicit preference and performance learning. In: Workshop Proceedings of KDD, pp. 28–36 (2013)
7. Boley, M., Moens, S., Gärtner, T.: Linear space direct pattern sampling using coupling from the past. In: Proceedings of KDD, pp. 69–77 (2012)
8. Bringmann, B., Nijssen, S., Tatti, N., Vreeken, J., Zimmermann, A.: Mining sets of patterns. Tutorial at ECML/PKDD (2010)
9. Calders, T., Rigotti, C., Boulicaut, J.-F.: A survey on condensed representations for frequent sets. In: Boulicaut, J.-F., Raedt, L., Mannila, H. (eds.) Constraint-Based Mining and Inductive Databases. LNCS (LNAI), vol. 3848, pp. 64–80. Springer, Heidelberg (2006). doi:10.1007/11615576_4
10. Chakraborty, S., Fremont, D.J., Meel, K.S., Seshia, S.A., Vardi, M.Y.: On parallel scalable uniform SAT witness generation. In: Baier, C., Tinelli, C. (eds.) TACAS 2015. LNCS, vol. 9035, pp. 304–319. Springer, Heidelberg (2015). doi:10.1007/978-3-662-46681-0_25
11. Chakraborty, S., Fremont, D., Meel, K., Vardi, M.: Distribution-aware sampling and weighted model counting for SAT. In: Proceedings of AAAI, pp. 1722–1730 (2014)
12. Dzyuba, V., van Leeuwen, M., Nijssen, S., De Raedt, L.: Interactive learning of pattern rankings. Int. J. Artif. Intell. Tools **23**(06), 1460026 (2014)
13. Dzyuba, V., van Leeuwen, M.: Learning what matters - sampling interesting patterns, March 2017. http://arxiv.org/abs/1702.01975
14. Dzyuba, V., van Leeuwen, M., De Raedt, L.: Flexible constrained sampling with guarantees for pattern mining. In: Data Mining and Knowledge Discovery (in press). https://arxiv.org/abs/1610.09263
15. Filippi, S., Cappé, O., Garivier, A., Szepesvári, C.: Parametric bandits: the generalized linear case. In: Proceedings of NIPS, pp. 586–594 (2010)
16. Hasan, M.A., Zaki, M.: Output space sampling for graph patterns. In: Proceedings of VLDB, pp. 730–741 (2009)

17. Joachims, T.: Optimizing search engines using clickthrough data. In: Proceedings of KDD, pp. 133–142 (2002)
18. van Leeuwen, M.: Interactive data exploration using pattern mining. In: Holzinger, A., Jurisica, I. (eds.) Interactive Knowledge Discovery and Data Mining in Biomedical Informatics. LNCS, vol. 8401, pp. 169–182. Springer, Heidelberg (2014). doi:10.1007/978-3-662-43968-5_9
19. Rueping, S.: Ranking interesting subgroups. In: Proceedings of ICML, pp. 913–920 (2009)
20. Shalev-Shwartz, S., Tewari, A.: Stochastic methods for ℓ_1-regularized loss minimization. J. Mach. Learn. Res. **12**, 1865–1892 (2011)
21. van Leeuwen, M., Ukkonen, A.: Discovering skylines of subgroup sets. In: Blockeel, H., Kersting, K., Nijssen, S., Železný, F. (eds.) ECML PKDD 2013. LNCS (LNAI), vol. 8190, pp. 272–287. Springer, Heidelberg (2013). doi:10.1007/978-3-642-40994-3_18
22. Xin, D., Shen, X., Mei, Q., Han, J.: Discovering interesting patterns through user's interactive feedback. In: Proceedings of KDD, pp. 773–778 (2006)

PNE: Label Embedding Enhanced Network Embedding

Weizheng Chen[1(✉)], Xianling Mao[2], Xiangyu Li[3], Yan Zhang[1],
and Xiaoming Li[1]

[1] School of Electronics Engineering and Computer Science,
Peking University, Beijing, China
cwz.pku@gmail.com, zhyzhy001@gmail.com, lxm.at.pku@gmail.com
[2] Department of Computer Science,
Beijing Institute of Technology, Beijing, China
maoxl@bit.edu.cn
[3] Department of Automation, Tsinghua University, Beijing, China
lixiangyu13@mails.tsinghua.edu.cn

Abstract. Unsupervised NRL (Network Representation Learning)
methods only consider the network structure information, which makes
their learned node representations less discriminative. To utilize the label
information of the partially labeled network, several semi-supervised
NRL methods are proposed. The key idea of these methods is to merge
the representation learning step and the classifier training step together.
However, it is not flexible enough and their parameters are often hard to
tune. In this paper, we provide a new point of view for semi-supervised
NRL and present a novel model named Predictive Network Embedding
(PNE). Briefly, we embed nodes and labels into the same latent space
instead of training a classifier in the representation learning process. Thus
the discriminability of node representations is enhanced by incorporating
the label information. We conduct node classification task on four real
world datasets. The experimental results demonstrate that our model
significantly outperforms the state-of-the-art baselines.

Keywords: Network embedding · Node classification · Semi-supervised
learning

1 Introduction

Social network analysis is an important research field which has a long history.
Mining social network is crucial for many data mining applications, such as node
classification [8], information diffusion [2], and link predication [23]. When we
want to analyze a network, the first core problem to consider is how to represent
the network. Adjacency matrix is the most basic and traditional network representation, but it often suffers from the serious data sparsity problem. And we
can't feed the adjacency matrix to most statistical machine learning algorithms
directly.

© Springer International Publishing AG 2017
J. Kim et al. (Eds.): PAKDD 2017, Part I, LNAI 10234, pp. 547–560, 2017.
DOI: 10.1007/978-3-319-57454-7_43

Nowadays, embedding, also known as representation learning, has become a promising and powerful tool in the network analysis area. The primary goal of NRL task is to map the nodes of a network to meaningful low-dimensional vectors. Several representative approaches, such as DeepWalk [14], LINE [19] and GraRep [3], learn unsupervised node representations which are not tuned for specific task by only utilizing the network structure information. In reality, nodes in network are usually associated with the label information. For instance, age, gender and other demographics of social media users. These labels are usually the targets of predictions in node classification tasks. It is necessary to take the label information into consideration to enhance the quality of node representations, especially for classification.

Semi-supervised NRL models are proposed to learn discriminative node features for node classification task. Instead of training a classifier after the node representations is learned, these methods combine the representation learning step and the classifier training step together. Their objective loss functions are usually a linear combination of a first term that preserves the network structure information and second term which is a classification loss on the labeled nodes. For instance, LSHM [7] uses a regularization smoothing term to preserve the structure information and trains a liner max-margin classifier. More recently, Tu et al. [20] present MMDW (Max-Margin DeepWalk), whose loss function is a linear combination of the matrix factorization style DeepWalk and a support vector machine with Biased Gradient. However, the performance of these semi-supervised NRL models are heavily depend on their chosen classifiers. Our experiments also show that the implementation of MMDW provided by the author has a very large memory requirements.

Is there any other way to incorporate the label information except training a classifier in the representation learning process? The answer is yes. Since we can learn representations for nodes, we also can learn representations for labels. More specifically, we assign two roles to each node, one representing itself and the other is a context. Thus we can convert a partially labeled network to two bipartite networks, i.e., a node-context network and a label-context network. Then by embedding these two networks simultaneously, the label information are directly encoded into the node representations. This method is named as PNE. In summary, we have the following contributions:

1. We present a new semi-supervised NRL framework, PNE, which can map nodes and labels into the same hidden space. As far as we know, we are among the first to learn label representations for semi-supervised NRL tasks. This idea is intuitive and can be applied to many previous unsupervised NRL models.
2. We propose to use the average of context vectors of the neighbouring nodes of a node to represent this node. Node classification experiments are conducted on four real network datasets respectively. Comparing with several state-of-the-art unsupervised and semi-supervised NRL models, we discover that our model outperforms baselines remarkably.

The rest of this paper is organized as follows. Section 2 gives a discussion of the related work. Section 3 formally defines our research problem. Section 4 introduces our proposed model and its implementation in details. Section 5 presents the experimental results of node classification and parameter tuning. Finally we conclude in Sect. 6.

2 Related Work

In this section, we will give a brief review to the history of NRL research. Spectral methods, such as Locally Linear Embedding [16] and Laplacian Eigenmaps [1], are proposed to reduces the dimensionality of undirected networks. Directed Graph Embedding [4] extends Laplacian Eigenmaps to handle directed networks. However, the computational complexity of these spectral methods are too high to handle large networks.

In recent years, distributed representation learning techniques have been adopted to NRL problem. The basic assumption of distributed NRL is that the representations of nodes should be closer if they share similar contexts. We introduce several representative unsupervised NRL models here. DeepWalk [14] samples node sequences and feeds them to a Skip-Gram based Word2Vec [12] model to train node representations. Walklets [15] is an extension of DeepWalk, which takes the offsets between nodes observed in the node sequence into consideration to learn a series of representations for each node. LINE [19] consider the first-order proximity and second-order proximity between nodes to learn node representations for large scale networks. Cao et al. present GraRep [3], which integrates global structural information of the network into the learning process by optimizing k-step loss functions in a matrix factorization framework. Node2vec [6] further balances the Breadth-first Sampling and Depth-first Sampling in DeepWalk. Actually, all these models are equivalent to a special neural matrix factorization framework respectively.

Four most recent semi-supervised NRL models are LSHM, MMDW, Planetoid [24] and TriDNR [13]. The former two models consider both the structure information and the label information. Planetoid and TriDNR further use the text attributes of nodes. However, when the text information is not available, Planetoid degrades into a combination of DeepWalk and a neural network trained on the labled nodes, and TriDNR degrades into the fully unsupervised DeepWalk. In this paper, we focus on the general NRL problem based on only the structure information and the label information. And we will explore to incorporate text features into PNE in the future.

3 Problem Formulation

We first introduce some notations that will be used in our paper. Considering a partially labeled network $G = (V, E, Y)$, where V is the set of nodes, $E \subseteq (V \times V)$ is the set of edges, and Y is the label matrix of the labeled nodes. Note that we treat G as a directed network, since an undirected network can be easily transformed into a directed network by creating two directed edges for each undirected edge. We use $\{v_1, \ldots, v_L\}$ and $\{v_{L+1}, \ldots, v_{L+U}\}$ to represent the labeled nodes and unlabeled nodes respectively. Here, $|V| = L + U$. For an edge $e_{ij} \in E$, w_{ij} is its weight. If there is no edge pointed from v_i to v_j, we set $w_{ij} = 0$. We use $T = \{l_1, \ldots, l_M\}$ to represent the sets of labels, where M is the number of labels. Y is a L by M matrix whose ith row is the label vector of node v_i. If v_i is associated with label l_m, we set $y_i^m = 1$, otherwise $y_i^m = -1$.

The goal of NRL or network embedding is to map each node v_i to a continuous vector $z_i \in \mathbb{R}^d$, where $d \ll |V|$. For unsupervised NRL methods, only the network structure information should be preserved in the learned node representations. Thus, to predict labels of the unlabeled nodes, unsupervised NRL methods also need a classifier training step. A classifier such as a support vector machine or even a more complicated neural network can be trained on the labeled nodes and applied on the labeled nodes. However, the discriminative power of the node representations are tightly restricted since they are learned in a fully unsupervised setting without considering the label information.

To utilize the label information, a class of semi-supervised approaches has been proposed, such as LSHM and MMDW. The commonality between them is that they learn representations in a transductive manner, which means that the representation learning step and the classifier training step are integrated together. As far as we know, all previous semi-supervised NRL methods incorporate label information by training a classifier. Nonetheless, it is often difficult to guarantee the performance of the classifier.

Here, we explore another new direction to utilize the label information. Since we can learn node representations, why not learn label representations simultaneously? Note that, most recent unsupervised NRL methods, such as DeepWalk and LINE, assign two roles and two vectors for a node v_i. The first role is the node itself which is associated with the vector z_i. The second role is a context c_i which is used to describe other nodes. We use a set $C = \{c_1, \ldots, c_{|V|}\}$ to represent the context roles of nodes. Likewise, c_i is associated with a context vector $z_i' \in \mathbb{R}^d$. The basic motivation of node representation learning adopted in DeepWalk and LINE is that the representations of nodes should be closer if they have similar contexts. In a similar way, we build direct relationships between labels and contexts. Thus we can learn a vector $h_i \in \mathbb{R}^d$ for each label l_i. We hope that the discriminative power of node representations can be improved by modeling the interactions between labels and contexts in the node representation learning process. As a result, the learned node representations can be feed to any classifiers or used in other data mining tasks. We will introduce how to build relationships between labels and contexts later.

4   Our Model

4.1   Network Decomposition

To directly incorporate the label information into the node representation learning process, we decompose the partially labeled network G into two bipartite networks, G_{vc} and G_{lc}. As shown in Fig. 1, we provide a toy example to make the conversion process easier to understand.

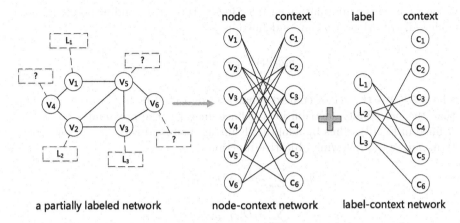

node   context   label   context

a partially labeled network   node-context network   label-context network

Fig. 1. Illustration of converting a partially labeled network to two bipartite networks. The left side is a partially labeled network which have six nodes, in which labeled nodes are colored red and unlabeled nodes are colored blue. Each node has two roles, one represents itself and the other one is a context (neighbour) of other nodes. Thus we decompose this network into two bipartite networks, i.e., a node-context network and a label-context network. The node-context network encodes the unsupervised structure information. The label-context network encodes the supervised information, capturing the label-level node co-occurrences. (Color figure online)

$G_{vc} = \{V \cup C, E_{vc}\}$ is the node-context bipartite network, which encodes the unsupervised structure information. If there is an directed edge e_{ij} in G, we just copy this edge to G_{vc} to connect v_i and c_i and the weight of e_{ij} is unchanged.

$G_{lc} = \{T \cup C, E_{lc}\}$ is the label-context network, which encodes the supervised label information. Simply put, if a node v_k is associated with a label l_i, we will link l_i and the contexts of v_k. For a label l_i and a context c_j, we use an edge $e_{ij} \in E_{lc}$ to express the relationship between them. The weight of e_{ij} is defined as:

$$f_{ij} = \sum_{k=1}^{L} \mathbf{I}(y_k^i = 1) w_{kj}, \tag{1}$$

where $I(x)$ is an indicator function. Note that the contexts are shared between G_{vc} and G_{lc}, we believe that node representations learned from these two networks can be more discriminative than only using G_{vc} as input.

4.2 Label-Context Network Embedding

We talk about how to learn label and context vectors from G_{lc} in this part. Given a label l_i, for a context c_j, the probability that c_j is observed when we see l_i is defined as the following softmax function [18]:

$$p(c_j|l_i) = \frac{\exp(z_j'^T \cdot h_i)}{\sum_{k=1}^{|V|} \exp(z_k'^T \cdot h_i)}. \tag{2}$$

Then by using the information encoded in G_{lc}, the corresponding empirical probability distribution of p can be defined as:

$$\hat{p}(c_j|l_i) = \frac{f_{ij}}{\sum_{k\in N_l(i)} f_{ik}}, \tag{3}$$

where $N_l(i)$ is the set of contexts that are connected to l_i. Naturally, we hope that the probability distributions defined in Eq. (2) approximates to the one defined in Eq. (3). Thus by adopting Kullback-Leibler divergence [9], our goal is to minimize the following objective function:

$$\begin{aligned} O_{lc} &= \sum_{i\in T} \lambda_i D_{KL}(\hat{p}(\cdot|l_i)||p(\cdot|l_i)), \\ &= \sum_{i\in T} \lambda_i D_{KL}(\hat{p}(\cdot|l_i)||p(\cdot|l_i)), \end{aligned} \tag{4}$$

where $\lambda_i = \sum_{k\in N_l(i)} f_{ik}$ is the importance of d_i in G_{lc}. After removing some constants, the above loss function can be rewritten as:

$$O_{lc} = -\sum_{(i,j)\in E_{lc}} f_{ij} \log p(c_j|l_i). \tag{5}$$

Note that directly optimizing the $p(c_j|l_i)$ term in Eq. (5) is computationally expensive since we need to iterate through all contexts. Hence, we adopt the effective negative sampling technique to reduce the computation complexity. Now we have the following loss function:

$$O_{lc} = -\sum_{(i,j)\in E_{lc}} f_{ij} \left\{ \log \sigma(z_j'^T \cdot h_i) + \sum_{k=1}^{K} E_{c_n \sim P_l(c)} \left[\log \sigma(-z_n'^T \cdot h_i) \right] \right\}, \tag{6}$$

where $\sigma(x)$ is the sigmoid function, K is the number of negative edges and $P_l(c) \propto (\sum_{i=1}^{M} f_{ic})^{0.75}$ is the noise context distribution of G_{lc}. By minimizing O_{lc}, the discriminative label information are encoded into the context vectors z_i'. For an unlabeled node, we treat its neighbouring nodes as its contexts, thus we now can use a weighted averaged context vector of these contexts to represent this node. In this way, we can create a new representation vector for every nodes no matter it is labeled or not.

Finally, for a node v_i, we use the following definition to calculate its feature vector:

$$u_i = \frac{\sum_{j\in N_v(i)} w_{ij} z_j'}{\sum_{j\in N_i(c)} w_{ij}}, \tag{7}$$

where $N_v(i)$ is the set of contexts that are connected to node v_i. We use u_i rather than z_i because we find that u_i yields better performance in node classification task. Another advantage of this strategy is that we can extend our model for online network embedding easily. For a newly arrived node, we can calculate its embedding easily if its neighbourhood information is available.

Then we can use the representation vectors of the labeled nodes to train a classifier and apply it to the unlabeled nodes. But there is a serious potential problem which may make the above process infeasible. As shown on the right side of Fig. 1, the node v_1 is not connected to any labeled nodes, it will not serve as a context in G_{lc}, which means that its context vector z_1' will never be updated. We still need to take the node-context network into consideration to ensure that every context vector is well-trained.

4.3   Node-Context Network Embedding

Now we consider the node-context network, which shares a same form with the label-context network. In order to embed this network, we can just repeat the process introduced in Sect. 4.2 and regard the nodes as labels. It is straightforward get the following objective loss function for G_{vc}:

$$O_{vc} = - \sum_{(i,j) \in E_{vc}} w_{ij} \left\{ \log \sigma(z_j'^T \cdot z_i) + \sum_{k=1}^{K} E_{c_n \sim P_v(c)} \left[\log \sigma(-z_n'^T \cdot z_i) \right] \right\},$$

(8)

where $P_v(c) \propto (\sum_{i=1}^{|V|} w_{ic})^{0.75}$ is the noise context distribution of G_{vc}. Note that $P_v(c)$ and $P_l(c)$ are two different noise context distributions determined by G_{vc} and G_{lc} respectively.

4.4   Predictive Network Embedding

To learn the embeddings of the original partially labeled network G, we use a linear combination of O_{lc} and O_{vc} to formulate our objective loss function of the PNE model:

$$O_{PNE} = O_{lc} + O_{vc}.$$

(9)

We adopt the asynchronous stochastic gradient descent (ASGD) algorithm to optimize the Eq. (9). The PNE model is trained with the unlabeled node-context network and the labeled label-context network simultaneously. The joint training strategy of PNE is summarized in Algorithm 1.

To improve the effectiveness of sampling operations, we use the alias method [21] to reduce the time complexity of sampling an edge to $O(1)$. When the number of edge samples T is large enough, the embeddings will converge.

5   Experiments

5.1   Dataset

Four real network datasets are used in our node classification experiments, including Citeseer [11], Wiki [17], Cora [10] and DBLP [8]. Citeseer is a paper

---

Algorithm 1. Joint training for PNE   •

Input: G_{vc}, G_{lc}, number of samples T, number of negative samples K
Output: node feature vectors u, node embeddings z, context embeddings z', label
    embeddings h
1: initialize all embeddings randomly from the uniform distribution [-1,1]
2: **while** $iter \leq T$ **do**
3:     sample an edge from E_{lc} and draw K negative edges, and update the label
      embeddings and the context embeddings
4:     sample an edge from E_{vc} and draw K negative edges, and update the node
      embeddings and the context embeddings
5: **end while**
6: calculate the node feature vector for each node according to the Eq. (7)

---

citation network, in which papers are categorized into 6 classes. Wiki is made up of some web pages from 17 categories and links between them. Cora is also a paper citation network, in which papers are categorized into 10 classes. DBLP is a coauthor network, in which authors are categorized into 4 classes.

Every node in Citeseer, Wiki and Cora only has one class label but authors in DBLP can have multiple labels. The weight of every edge in Citeseer, Wiki and Cora is 1 since these networks are unweighted. In contrast, the weight of an edge between two authors is the number of papers they coauthored. We follow the setting of the previous works [20,22] and treat Citeseer, Wiki, Cora as undirected networks. The statistics of the datasets are listed in Table 1.

Table 1. Statistics of our datasets.

Name	Citeseer	Wiki	Cora	DBLP
Type	Unweighted	Unweighted	Unweighted	Weighted
#Nodes	3,324	2,405	2,708	27,199
#Edges	4,732	17,981	5,429	66,832
#Labels per node	1	1	1	1.15

5.2   Compared Algorithms

– DeepWalk [14]. DeepWalk is an unsupervised NRL method. The parameters of DeepWalk are set as follows, the sliding window size is 10, the length of each node sequence is 40, the number of node sequences for per node is 80.
– LINE [19]. LINE is an unsupervised NRL method. We use LINE(1st) and LINE(2st) to represent LINE with first order proximity and second-order proximity respectively. The number of edge samples T is set to 2 million for Citeseer and Wiki. In the case of DBLP, we set $T = 20$ million.
– LSHM [7]. A semi-supervised NRL method, which trains a linear max-margin classifier in the representation learning process.

- MMDW [20]. A state-of-the-art semi-supervised NRL method based on matrix decomposition. MMDW trains a max-margin SVM as its classifier.
- PNE. We set $K = 5$ for all four datasets. For Citeseer, Wiki and Cora, we set $T = 1$ million. For DBLP, we set $T = 5$ million. We use PNE to represent the full version PNE($G_{lc} + G_{vc} + avg$) which takes G_{lc} and G_{vc} as input. We use PNE($G_{lc} + G_{vc} + ori$) to represent the variation of PNE which uses the original node embeddings z as node feature vector. We also use PNE($G_{lc} + avg$) and PNE($G_{vc} + avg$) to represent the submodels of PNE which only takes G_{lc} or G_{vc} as input.

5.3 Node Classification

We employ node classification task to quantitatively evaluate the quality of node feature vectors learned by different algorithms. Note that the state-of-the-art MMDW model trains linear SVM as its classifier, for a fair comparison, we adopt the one-vs-the-rest linear SVM implemented in Liblinear package [5] as classifier for DeepWalk, LINE and PNE. For all models, we set the length of node representations $d = 200$, which is the same setting used in [20]. For Citeseer, Wiki and Cora, we use Accuracy as the evaluation metric. In the case of DBLP, we adopt the Micro-F1 and Macro-F1 as the evaluation metrics.

Table 2. Accuracy (%) of node classification on Citeseer.

% Labeled nodes	10%	20%	30%	40%	50%	60%	70%	80%	90%
DeepWalk	52.99	55.78	57.20	59.16	58.33	60.41	58.16	58.96	58.41
LINE(1st)	45.70	51.22	54.55	56.28	57.02	58.05	58.94	59.77	59.37
LINE(2nd)	46.68	51.23	53.36	55.41	57.55	58.14	58.37	59.00	59.04
PNE($G_{vc} + avg$)	52.31	54.35	55.70	56.66	57.59	57.96	58.69	58.74	59.08
LSHM	53.67	57.73	60.10	61.61	62.69	63.43	64.09	65.51	66.02
MMDW	54.72	59.64	62.60	64.10	65.83	68.96	69.56	69.58	69.16
PNE($G_{lc} + avg$)	51.03	58.14	62.40	64.88	67.64	69.93	71.32	72.57	73.76
PNE($G_{lc} + G_{vc} + ori$)	54.10	60.05	63.40	65.75	68.53	**70.42**	71.76	72.79	**74.93**
PNE	**54.79**	**60.87**	**64.67**	**66.95**	**68.59**	70.00	**72.06**	**73.41**	74.76

In practical terms, a certain proportion of nodes is sampled from the network as labeled data to train the classifier, the remaining nodes are used for evaluation. We repeat the trial 20 times and report the averaged results. As shown in Tables 2, 3, 4 and 5, we report the performance of different models under increasing training ratios. From these tables, we mainly have the following observations and analysis:

(1) We observe that the proposed method PNE always significantly outperform other baselines. Compared with MMDW, PNE achieves nearly 2.85%, 1.99%,

Table 3. Accuracy (%) of node classification on Wiki.

% Labeled nodes	10%	20%	30%	40%	50%	60%	70%	80%	90%
DeepWalk	58.78	63.11	65.72	66.90	67.65	68.04	69.30	69.75	68.46
LINE(1st)	54.92	61.98	64.76	66.30	67.66	68.01	68.86	67.92	69.42
LINE(2nd)	57.09	59.90	62.30	62.86	63.82	64.74	64.60	65.18	65.10
PNE($G_{vc} + avg$)	57.36	60.75	62.01	63.30	64.06	64.57	65.31	65.65	65.33
LSHM	55.56	54.73	61.81	61.62	64.86	65.22	67.15	66.83	68.58
MMDW	57.25	62.01	65.04	66.67	66.89	68.23	69.22	70.18	72.61
PNE($G_{lc} + avg$)	54.10	62.16	64.66	66.37	67.89	68.83	69.71	70.28	71.14
PNE($G_{lc} + G_{vc} + ori$)	**58.98**	63.76	65.78	67.80	69.20	69.23	70.77	70.70	71.53
PNE	58.93	**64.28**	**66.79**	**68.66**	**69.44**	**70.18**	**70.79**	**70.83**	**72.78**

Table 4. Accuracy (%) of node classification on Cora.

% Labeled nodes	10%	20%	30%	40%	50%	60%	70%	80%	90%
DeepWalk	75.95	78.66	79.90	80.81	81.58	81.96	82.10	82.66	82.67
LINE(1st)	68.13	75.93	79.16	80.91	81.73	82.83	82.69	82.30	83.33
LINE(2nd)	75.02	78.55	80.17	81.11	81.75	82.76	82.88	83.12	82.61
PNE($G_{vc} + avg$)	75.34	77.98	79.21	79.94	80.11	80.53	81.60	81.22	81.71
LSHM	76.80	79.29	80.34	82.34	83.12	83.81	84.75	85.18	86.68
MMDW	73.61	79.95	82.08	82.73	83.25	84.64	**86.35**	**86.66**	87.21
PNE($G_{lc} + avg$)	74.10	78.99	81.61	83.54	84.71	85.12	85.93	86.10	86.54
PNE($G_{lc} + G_{vc} + ori$)	76.15	79.66	82.02	83.09	84.18	84.89	86.05	86.30	86.82
PNE	**77.58**	**81.22**	**82.94**	**84.54**	**84.73**	**85.55**	86.15	86.39	**87.76**

Table 5. Micro-F1 (%) of node classification on DBLP.

% Labeled nodes	10%	20%	30%	40%	50%	60%	70%	80%	90%
DeepWalk	66.85	68.03	68.57	68.79	68.85	68.95	69.10	69.31	69.50
LINE(1st)	64.06	65.73	66.39	66.77	66.67	66.89	66.81	66.96	67.20
LINE(2nd)	65.87	66.83	67.24	67.37	67.68	67.59	67.63	67.61	67.49
PNE($G_{vc} + avg$)	63.80	64.49	64.72	65.03	64.25	65.13	65.41	65.24	65.68
LSHM	70.09	73.36	76.53	78.84	81.13	84.87	85.40	87.58	87.73
MMDW	-	-	-	-	-	-	-	-	-
PNE($G_{lc} + avg$)	72.25	78.43	**83.09**	**85.28**	**87.65**	**88.79**	89.34	**91.01**	**92.10**
PNE($G_{lc} + G_{vc} + ori$)	73.47	**79.58**	82.91	**85.31**	87.05	88.16	89.04	89.48	88.67
PNE	**73.98**	79.22	81.38	85.15	87.28	88.64	**89.85**	90.01	91.40

1.81% improvement on Citeseer, Wiki and Cora respectively in the measure of Accuracy when the training ratio is 0.4. This demonstrates that incorporating the label information by label embedding may be a better strategy than training a classifier in the representation learning process.

(2) We find that MMDW encounters out-of-memory errors to handle DBLP on our Linux server which has 64G memory. But PNE only need no more than 1.2G memory on the same server to process this large network.

(3) Semi-supervised methods benefit from more labeled training samples. Compared with LINE, the relative improvement of PNE is around 8% on DBLP when the training ratio is 0.1. But it reaches up to 24% when the training ratio is 0.9.

(4) The label-context network is truly useful to improve the predictive power of the nodes. PNE always outperforms $PNE(G_{vc} + avg)$ by a large margin.

(5) For most case, PNE outperforms $PNE(G_{lc} + G_{vc} + ori)$. This demonstrates that using the average of the context vectors of the neighbouring nodes of a node to represent a node is a better and potential strategy for the node classification task.

5.4   Parameter and Convergence Sensitivity

The time and memory requirements of PNE are highly relevant to d, K and T. We test the changes in performance of PNE and LINE(2nd) on Citeseer, Wiki and DBLP when the training ratio is 0.4. When one parameter is under test, other parameters are set to their default values.

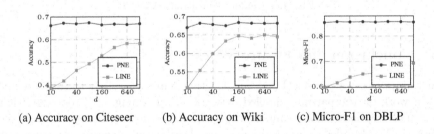

(a) Accuracy on Citeseer   (b) Accuracy on Wiki   (c) Micro-F1 on DBLP

Fig. 2. Parameter sensitivity w.r.t. d

As shown in Fig. 2, we find that when $d \in [20, 1280]$, the performance of PNE is very stable. For example, to get the best prediction performance on DBLP, PNE just need $d = 20$, but LINE need $d = 640$. This demonstrates that PNE is memory efficient by incorporating the label information into the node representations.

As shown in Fig. 3, the performance of PNE is much less sensitive to the number of negative edges K than LINE. Overall, when $K \in [3, 8]$, PNE is relatively stable. This makes it easier to select an appropriate value for K in PNE.

As shown in Fig. 4, the convergence of PNE is far faster than LINE. For example, when the number of edge samples reaches 2 million, PNE has converged already on DBLP. Meanwhile, LINE needs 15 million samples to get its best performance.

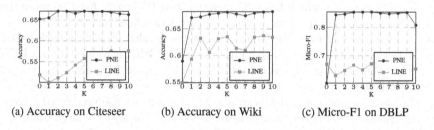

(a) Accuracy on Citeseer   (b) Accuracy on Wiki   (c) Micro-F1 on DBLP

Fig. 3. Parameter sensitivity w.r.t. K

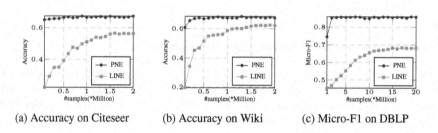

(a) Accuracy on Citeseer   (b) Accuracy on Wiki   (c) Micro-F1 on DBLP

Fig. 4. Parameter sensitivity w.r.t. T

6   Conclusion and Future Work

In this paper, we present PNE, an effective semi-supervised feature learning framework for the partially labeled networks. By learning label representations and node representations simultaneously, the discriminative node feature vectors can be obtained. The results of node classification experiments conducted on four datasets show that PNE outperforms several state-of-the-art baselines significantly.

We now consider two directions for future work. Nodes in real-world network often have abundant text information. We hope to extend PNE to further utilize rich text attributes of nodes. It is also necessary to adapt PNE model to handle heterogeneous networks, which are common but have complex structure information.

Acknowledgments. This work is supported by 973 Program with Grant No. 2014CB340400. Yan Zhang is supported by NSFC with Grant No. 61532001 and No. 61370054, and MOE-RCOE with Grant No. 2016ZD201. And we also thank the three anonymous reviewers for their valuable comments.

References

1. Belkin, M., Niyogi, P.: Laplacian eigenmaps and spectral techniques for embedding and clustering. NIPS **14**, 585–591 (2001)
2. Bourigault, S., Lagnier, C., Lamprier, S., Denoyer, L., Gallinari, P.: Learning social network embeddings for predicting information diffusion. In: WSDM, pp. 393–402. ACM (2014)
3. Cao, S., Lu, W., Xu, Q.: GraRep: learning graph representations with global structural information. In: CIKM, pp. 891–900. ACM (2015)
4. Chen, M., Yang, Q., Tang, X.: Directed graph embedding. In: IJCAI, pp. 2707–2712 (2007)
5. Fan, R.-E., Chang, K.-W., Hsieh, C.-J., Wang, X.-R., Lin, C.-J.: Liblinear: a library for large linear classification. J. Mach. Learn. Res. **9**, 1871–1874 (2008)
6. Grover, A., Leskovec, J.: Node2vec: scalable feature learning for networks. In: SIGKDD, pp. 855–864. ACM, New York (2016)
7. Jacob, Y., Denoyer, L., Gallinari, P.: Learning latent representations of nodes for classifying in heterogeneous social networks. In: WSDM, pp. 373–382. ACM (2014)
8. Ji, M., Sun, Y., Danilevsky, M., Han, J., Gao, J.: Graph regularized transductive classification on heterogeneous information networks. In: Balcázar, J.L., Bonchi, F., Gionis, A., Sebag, M. (eds.) ECML PKDD 2010. LNCS (LNAI), vol. 6321, pp. 570–586. Springer, Heidelberg (2010). doi:10.1007/978-3-642-15880-3_42
9. Kullback, S., Leibler, R.A.: On information and sufficiency. Annals Math. Stat. **22**(1), 79–86 (1951)
10. Lim, K.W., Buntine, W.L.: Bibliographic analysis with the citation network topic model. In: ACML (2014)
11. McCallum, A.K., Nigam, K., Rennie, J., Seymore, K.: Automating the construction of internet portals with machine learning. Inf. Retr. **3**(2), 127–163 (2000)
12. Mikolov, T., Sutskever, I., Chen, K., Corrado, G.S., Dean, J.: Distributed representations of words and phrases and their compositionality. In: NIPS, pp. 3111–3119 (2013)
13. Pan, S., Wu, J., Zhu, X., Zhang, C., Wang, Y.: Tri-party deep network representation. In: IJCAI-2016, pp. 1895–1901 (2016)
14. Perozzi, B., Al-Rfou, R., Skiena, S.: DeepWalk: online learning of social representations. In: SIGKDD, pp. 701–710. ACM, New York (2014)
15. Perozzi, B., Kulkarni, V., Skiena, S.: Walklets: multiscale graph embeddings for interpretable network classification. arXiv preprint arXiv:1605.02115 (2016)
16. Roweis, S.T., Saul, L.K.: Nonlinear dimensionality reduction by locally linear embedding. Science **290**(5500), 2323–2326 (2000)
17. Sen, P., Namata, G., Bilgic, M., Getoor, L., Galligher, B., Eliassi-Rad, T.: Collective classification in network data. AI Mag. **29**(3), 93 (2008)
18. Tang, J., Qu, M., Mei, Q.: PTE: predictive text embedding through large-scale heterogeneous text networks. In: SIGKDD, pp. 1165–1174. ACM, New York (2015)
19. Tang, J., Qu, M., Wang, M., Zhang, M., Yan, J., Mei, Q.: LINE: large-scale information network embedding. In: WWW, pp. 1067–1077. ACM, New York (2015)
20. Tu, C., Zhang, W., Liu, Z., Sun, M.: Max-margin DeepWalk: discriminative learning of network representation. In: IJCAI, pp. 3889–3895 (2016)
21. Walker, A.J.: New fast method for generating discrete random numbers with arbitrary frequency distributions. Electron. Lett. **8**(10), 127–128 (1974)
22. Yang, C., Liu, Z., Zhao, D., Sun, M., Chang, E.Y.: Network representation learning with rich text information. In: IJCAI, pp. 2111–2117. AAAI Press (2015)

23. Yang, Y., Chawla, N., Sun, Y., Hani, J.: Predicting links in multi-relational and heterogeneous networks. In: ICDM, pp. 755–764. IEEE (2012)
24. Yang, Z., Cohen, W., Salakhutdinov, R.: Revisiting semi-supervised learning with graph embeddings. arXiv preprint arXiv:1603.08861 (2016)

Improving Temporal Record Linkage Using Regression Classification

Yichen Hu$^{(\boxtimes)}$, Qing Wang, Dinusha Vatsalan, and Peter Christen

Research School of Computer Science, The Australian National University,
Canberra, ACT 0200, Australia
{yichen.hu,qing.wang,dinusha.vatsalan,peter.christen}@anu.edu.au

Abstract. Temporal record linkage is the process of identifying groups of records that are collected over a period of time, such as in census or voter registration databases, where records in the same group represent the same real-world entity. Such databases often contain temporal information, such as the time when a record was created or when it was modified. Unlike traditional record linkage, which considers differences between records from the same entity as errors or variations, temporal record linkage aims to capture records from entities where the attribute values are known to change over time. In this paper we propose a novel approach that extends an existing temporal approach called *decay model*, to categorically calculate probabilities of change for each attribute. Our novel method uses a regression-based machine learning model to predict decays for sets of attributes. Each such set of attributes has a principle attribute and support attributes, where values of the support attributes can affect the decay of the principle attribute. Our experimental results on a real US voter database show that our proposed approach results in better linkage quality compared to the decay model approach.

Keywords: Data matching · Temporal data · Decay · Attribute weighting · Entity resolution

1 Introduction

Record linkage (also known as data matching, entity resolution, and duplicate detection) identifies records that refer to the same real-world entity [5]. Record linkage is being used in many application domains, such as linking patient data for disease outbreak detection or clinical trails in the health industry [5], credit checking and fraud detection in the finance industry [6], and constructing population databases for social science research [11]. Challenges in record linkage are caused by the lack of unique identifiers (such as national identifier numbers), dirty data (such as misspellings and missing values), legitimate updates over time (such as changes in last name or address), and the lack of informative attributes (i.e. a dataset might not contain gender and/or date of birth).

This work was partially funded by the Australian Research Council under Discovery Project DP160101934.

© Springer International Publishing AG 2017
J. Kim et al. (Eds.): PAKDD 2017, Part I, LNAI 10234, pp. 561–573, 2017.
DOI: 10.1007/978-3-319-57454-7_44

Record linkage generally involves the following steps [5]: data preprocessing, blocking, comparison and classification, and evaluation. This paper focuses on record pair comparison and classification, especially the task of calculating a similarity value with greater effectiveness at distinguishing between matches and non-matches for temporal data compared to previous temporal and non-temporal linkage techniques. While record linkage has been studied for several decades, until recently most works in this field did not use any temporal information available in datasets [13]. However, records of the same entity can be collected over a long period of time (years or even decades), such as census data that in many countries are collected every five or ten years. During such periods, certain attribute values of an entity are likely to change, such as a person's job position, living address, and potentially their last name (if somebody gets married).

Traditional record linkage methods assume that highly similar records are most likely to belong to the same entity [5]. These techniques do not perform well on temporal data, because entities might change some of their attribute values over time. For example, when a person changes his or her last name or address, their new record is not linked to earlier records because the attribute values do not match, or their earlier records are linked by mistake to records of a different person who has the same last name and/or address [6].

Temporal record linkage aims to address the above issues by using temporal information, such as the time-stamp when a record was created or modified. These time-stamps can be used to sort records by time and calculate temporal distances between records. They therefore provide opportunities for new record linkage approaches (examples will be discussed in Sect. 2). A dataset needs to contain temporal information for each record to be used in temporal record linkage, such as the date when being entered (for medical records), the date when being published (for publication records), or the date when being collected (for datasets collected by taking snapshots of databases at different points in time).

Table 1. An example of a temporal datasets.

RecID	EntID	FName	MName	LName	Address	Sex	Age	EntryDate
r1	e2	Elsa		Clark	161 Castlereagh St, Sydney	F	24	2011-09-11
r2	e1	Ella	Rose	Taylor	456 Kent Street, Sydney	F	23	2011-10-12
r3	e1	Ella	Louise	Taylor	456 Kent Street, Sydney	F	23	2012-02-20
r4	e1	Ella	Louise	Clark	299 Elizabeth St, Sydney	F	24	2012-06-30
r5	e2	Elsa		Taylor	201 Kent Street, Sydney	F	26	2013-08-05

Example: Given five records of two entities as in Table 1, if we do not consider the temporal information *EntryDate*, records r2 and r5 will have a high similarity and will therefore be matched incorrectly, whereas records r2 and r4 will have a low similarity and this true match will be missed. Temporal record linkage aims to correctly link r1 to r5, and link r2, r3 and r4 together using the temporal information in the *EntryDate* attribute.

This paper extends an existing temporal linkage approach called *decay model* [13], which learns the probability for an attribute to change over time (*disagreement decay*) and the probability for an attribute to share the same value among different entities over time (*agreement decay*). It then uses these decays to adjust the weight given to each attribute, where the sum of adjusted attribute weights is used to calculate the similarity between a pair of records and decide if they are a match or non-match based on a similarity threshold [5]. The *decay model* assumes the probability for an attribute to change its value over a certain time period is the same for every entity, and this assumption is not always true. For example, young people are more likely to change their address than seniors, and young females are more likely to change their last name than senior males.

Contributions: We integrate a linear regression model into the *decay model* [13]. Our model uses support attributes to calculate the decay of a principle attribute whose decay is affected by the values of those support attributes. The calculated decays are therefore more specific to each entity. For example, a person's gender can affect the likelihood of changes in their last name, and when we calculate a decay for last name with gender as a support attribute we can learn a gender sensitive decay model for last name. Our intuition is that the probability for an attribute to change over time can be predicted more accurately with the help of other attributes upon which it depends. We also propose a method to adjust the impact of decay models, and evaluate our approach on four subsets of a real US voter dataset. The experimental results show that our approach improves the linkage quality compared to two baseline approaches.

2   Related Work

We discuss related work in the two areas of record linkage that are non-temporal and temporal models. The common objective of both types of models is to decide if a pair of records is a match, or if a record belongs to a cluster of records where all records refer to the same entity. Temporal models consider temporal information in addition to attribute similarities as used in non-temporal models.

Fellegi and Sunter [7] proposed a statistical non-temporal linkage model. This model weights each attribute according to two types of probabilities: (1) the probability for a pair of records that agree on an attribute to be a match; and (2) the probability for a pair of records that agree on an attribute to be a non-match. The weights of attributes are summed to calculate the matching score for each pair of records. These probabilities can be learned from training data or estimated using the Estimation-Maximization algorithm [9].

Li et al. [13] were the first to propose a temporal model which considers the probability for an attribute's value to change over time, where this probability is learned from training data. The model calculates a disagreement decay (the likelihood for an attribute to change within a certain time period) and an agreement decay (the likelihood for an entity's attribute value to be the same as another entity's within a time period). The two types of decays are used to adjust the weight of each attribute as used in the similarity calculations.

More recently, Li et al. [12] proposed a temporal model which learns the probability for each attribute value to change to some commonly occurring value over time. However, this approach requires a temporal dataset to have attributes whose values change to some commonly occurring values, such as job positions (for example, the position 'technician' can change to 'manager').

Christen and Gayler [6] modified the approach proposed by Li et al. [13] to iteratively train a temporal model using a stream of time-stamped records. Every time a certain number of records are matched, the approach uses the matching results to retrain the temporal model. The difference between this approach and the original temporal model [13] is that the latter only learns the temporal model from training data once, whereas the former continuously trains the temporal model using linkage results produced by itself.

Chiang et al. [3] proposed an algorithm which learns the probability for an attribute's value to recur within different time periods. For each value of an attribute, the algorithm constructs a transition history and uses this history to calculate the probability for a value to recur. These probabilities are used to adjust the original similarity of a pair of records.

All these existing works do not address dependencies between attributes when calculating the probability for an attribute to change over time. Although Li et al. [13] introduced a decay model, their model only calculates the probability for attribute values to change independently. We believe the decay model can be improved and made more effective to improve the linkage quality by considering dependencies between attributes.

3   Problem Statement

We now define the notation as well as the problem we aim to tackle in this paper. Let $\mathbf{R}$ be a set of records and $\mathbf{E}$ be a set of entities. Each record $r \in \mathbf{R}$ has a list of attribute values $[a_1, a_2, \ldots, a_k]$ and a time-stamp $r.t$, where each value a_i $(1 \leq i \leq k)$ is associated with an attribute A, and we use $r.A$ to denote the value a_i of A in r. Every record $r \in \mathbf{R}$ must belong to exactly one entity $e \in \mathbf{E}$. The entity to which a record r belongs to is denoted as $e(r)$.

Attribute values of an entity e can change over time, where each change (update) is represented by a new record r_i with a time-stamp $r_i.t$ and attribute value(s) that is/are different from the previous record. For example, let r_1, r_2 be two records belonging to e (in another word, $e(r_1) = e(r_2)$). If $r_1.t < r_2.t$ and $\exists A \in \mathbf{A} : r_1.A \neq r_2.A$, then we say that the value of attribute A of entity e has changed between the two time-stamps $r_1.t$ and $r_2.t$.

Given a training dataset $\mathbf{C}$ in the form of a set of clusters of records. Each cluster $C \in \mathbf{C}$ contains a set of records $\{r_1, r_2, \ldots\}$. All records in a cluster C represent the same entity, and records in different clusters represent different entities.

The temporal record linkage problem is to link all $r_a, r_b \in \mathbf{R}$, where $e(r_a) = e(r_b)$, $\exists A \in \mathbf{A} : r_a.A \neq r_b.A$. Note that it is possible to have a pair of records where $e(r_a) = e(r_b)$ and $\forall A \in \mathbf{A} : r_a.A = r_b.A$, which means no temporal update has occurred between the two records. In this case we will only keep the oldest record of the pair during data preprocessing.

The goal of our work is to address the temporal record linkage problem using a weighting strategy which adjusts the importance of attributes in order to improve the quality of linkage. Our work uses a regression model to train and predict parameters for a temporal model. Our solution is based on the assumption that adjusting the weights of each attribute A according to its probability to change over time can improve the quality of record linkage.

4   Temporal Record Linkage Framework

In this section we discuss the temporal record linkage framework used in our work. *Swoosh* is a generic record linkage method which compares records according to features (sets of attributes) selected by the user [1]. A pair of records is merged into a new record when one of their features meets the matching criteria provided by the user, and then the two original records are removed. Swoosh treats the classifier, which decides whether a pair of records is a match, as a blackbox. In this paper, we use a threshold-based classifier that classifies a pair of records as a match when its similarity is greater than a user defined similarity threshold. The objective of our proposed approach is to calculate a similarity value for a pair of records using temporal information.

The *decay model* calculates the similarity of a pair of records using the similarity of each pair of attribute values that is adjusted by weights. The weight of each attribute is calculated according to its disagreement and agreement decays [13]. Disagreement decay is the probability for an attribute to change its value within a time period, and agreement decay is the probability for multiple entities to have the same attribute value within a time period [13]. A time distance Δt refers to the difference between two time-stamps, and is measured by a time unit defined by the user, such as days, years, or hours.

A *life span* l refers to the time distance of an attribute value to be used by an entity. An attribute's life span is *full* when the value has a date when it was used first and another date when it was changed to another value. The time distance between the first and second date is a *full life span*, denoted as l_f. Similarly, if the attribute's value does not change between two time-stamps, the time distance between the time-stamps is a *partial life span*, denoted as l_p.

For example, assume that an entity has three different last names over five records, with time-stamps in the form of (year-month): 'Taylor' (2011-10) $\rightarrow$ 'Taylor' (2011-12) $\rightarrow$ 'Spire' (2012-12) $\rightarrow$ 'Spire' (2013-10) $\rightarrow$ 'Wright' (2015-10). The time distance between the first and the third records is one full life span with a length of 14 months (2011-10 to 2012-12), and the distance between the third and the fifth records is another full life span with a length of 34 months (2012-12 to 2015-10). Note that, in this example, month is being used as a time unit but this is not necessary for all datasets. From this example, the time distance between the first and the second records is a partial life span with a length of 2 months (the time distance between 2011-10 and 2011-12), and the time distance between the third and the fourth records (2012-12 and 2013-10) is another partial life span with a length of 10 months.

Let $\bar{L}_f$ denote the list of all full life spans l_f of an attribute A for all entities and let $\bar{L}_p$ denote the list of all partial life spans l_p of an attribute A for all entities. Then the disagreement decay is formally defined as below.

Definition 1. *(Disagreement decay $d^{\neq}$)* [13]: *Let Δt be a time distance, $A \in \boldsymbol{A}$ be an attribute. The disagreement decay of A over Δt is the probability $d^{\neq}(A, \Delta t)$ that an entity changes its value of A within Δt:*

$$d^{\neq}(A, \Delta t) = (|\{l \in \bar{L}_f | l \leq \Delta t\}|)/(|\bar{L}_f| + |\{l \in \bar{L}_p | l \geq \Delta t\}|) \tag{1}$$

Let $\bar{L}$ denote a list of both full and partial life spans of an attribute A for all entities. For each record, if it has the same attribute value with another record which belongs to a different entity, the time distance between the two records is added to $\bar{L}$. If no entity has the same attribute value, a life span with length ∞ is added to $\bar{L}$. Then the agreement decay is formally defined as below.

Definition 2. *(Agreement decay $d^{=}$)* [13]: *Let Δt be a time distance, $A \in \boldsymbol{A}$ be an attribute. The agreement decay of A over Δt is the probability $d^{=}(A, \Delta t)$ that two different entities share the same value of A within Δt:*

$$d^{=}(A, \Delta t) = (|\{l \in \bar{L} | l \leq \Delta t\}|)/(|\bar{L}|) \tag{2}$$

The *decay model* uses the agreement and disagreement decay to calculate w_A (weight of attribute A), as shown in (3). The comparison function s_A calculates the similarity between a pair of attribute values. s_A is defined by the user and it returns a similarity value in the range $[0, 1]$. These comparison functions can be approximate string similarity functions, such as edit-distance or Jaro-Winkler [5].

$$w_A(s_A, \Delta t) = 1 - s_A \cdot d^{=}(A, \Delta t) - (1 - s_A) \cdot d^{\neq}(A, \Delta t) \tag{3}$$

Weights are used to calculate the pair-wise similarity between two records, as shown in (4). s_r denotes the decay adjusted similarity between two records r_a and r_b. s_r is the final similarity score that is used to classify a pair of records, which decides if it is a match or non-match. s_r is in the range $[0, 1]$.

$$s_r(r_a, r_b) = \frac{\sum_{A \in \boldsymbol{A}} w_A(s_A(r_a.A, r_b.A), |r_a.t - r_b.t|) \cdot s_A(r_a.A, r_b.A)}{\sum_{A \in \boldsymbol{A}} w_A(s_A(r_a.A, r_b.A), |r_a.t - r_b.t|)} \tag{4}$$

5   Improved Decay Model

In this section we introduce an improved temporal model based on the decay model [13] described above.

5.1   Predicting Probability with a Regression Model

From the previous equations we can see that the agreement and disagreement decays are calculated using only a single attribute. For example, when the disagreement decay of attribute *last name* is calculated, the temporal model calculates the overall probability for an entity to change its last name within a given time distance Δt. However, the probability for an entity to change its last name is often associated with gender and age. The disagreement decay for last name, calculated without considering the gender and age values of an entity, would be too high for older males, and too low for younger females, because it is rare for an older man to change his last name, but more common for a young woman to change her last name when she gets married.

A set of *support attributes* is selected for *principle attributes* where the value of a support attribute may affect the probability for their attribute value to change. For example, when predicting the probability for values in attribute *address* to change, attributes *gender* and *age* can be used as support attributes to make the prediction more accurate. Support attributes are selected by the user based on their domain knowledge for each principle attribute, and each principle attribute can have zero to many support attributes. They can also be selected using a feature selection strategy that is able to explore the dependency between features [2].

To create a training dataset for each attribute, our algorithm iterates through the records of each entity. The algorithm checks if an entity has changed its principle attribute value within a time distance. The time distance ranges from 1 to the maximum time distance of the whole dataset. For each time distance, a training instance is created using the time distance and values of the support attributes as features, and using the status of value change (changed or unchanged) as class value. The training dataset is then used to train a regression model.

In this paper, we use a linear regression model, as commonly used in parameter estimation and prediction [10], to predict disagreement probability.

Disagreement Probability: We introduce a concept called *disagreement probability* $d_{prob}^{\neq}$, which has a similar definition as disagreement decay (as shown in (1)), but is modified in order to be used with a regression model. From (5), we can see that the difference between $d_{prob}^{\neq}$ and $d^{\neq}$ is that the divisor of $d_{prob}^{\neq}$ is fixed for each entity. With a fixed divisor, we can create training instances for a regression model according to if $l \leq \Delta t$. When a full life span $l \in \bar{L}_f$ is encountered, we can decide if it is lower than a certain Δt and create a training record. These training records are used to train the regression model to predict disagreement probability. For each Δt, a training record is created using: (1) the value of each support attribute of A whose model is being built; (2) the current Δt; and (3) a class value which is equal to 1 if $l \leq \Delta t$, or 0 if $l > \Delta t$ or $l \in \bar{L}_p$.

When a class value of a training record equals to 1, it means the value of A of an entity has been changed within Δt and the life span is full, whereas a class value 0 means the value of A has not been changed within Δt and the life span is partial. It needs to be noted that (5) is only relevant when we create training records. The equation provides a conceptual insight about why we create training records following the steps described above. The $d^{\neq}_{prob}$ that is being used after the training stage is predicted using the trained regression model, rather than calculated using (5).

$$d^{\neq}_{prob}(A, \Delta t) = (|\{l \in \bar{L}_f | l \le \Delta t\}|)/(|\bar{L}_f| + |\bar{L}_p|) \tag{5}$$

$d^{\neq}_{prob}$ is normalized into the range $[0, 1]$, and then it can be used as a weight to adjust attribute-wise similarities, as shown in (6). s_p denotes the similarity between a pair of records adjusted using $d^{\neq}_{prob}$.

$$s_p(r_a, r_b) = \sum_{A \in \mathbf{A}} \frac{1 - d^{\neq}_{prob}(A, |r_a.t - r_b.t|)}{\sum_{A' \in \mathbf{A}} 1 - d^{\neq}_{prob}(A', |r_a.t - r_b.t|)} \cdot s_A(r_a.A, r_b.A) \tag{6}$$

Combining Disagreement Probability with Agreement Decay: Disagreement probability can be normalized as: $d^{\neq}_{nprob}(A, \Delta t) = (d^{\neq}_{prob}(A, \Delta t))/(max(d^{\neq}_{prob}(A)))$, where $max(d^{\neq}_{prob}(A))$ is the maximum disagreement probability over all Δt. Using (3) and (4) above, with $d^{\neq}$ being replaced by $d^{\neq}_{nprob}$, a different w_A can be calculated, while (7) shows how to calculate w_A using $d^{\neq}_{nprob}$.

$$w_A(s_A, \Delta t) = 1 - s_A \cdot d^=(A, \Delta t) - (1 - s_A) \cdot d^{\neq}_{nprob}(A, \Delta t) \tag{7}$$

5.2   Adjusting the Impact of Decay Models

The intuition of using decays to adjust attribute weights is that attributes that have higher probability to change their values are less reliable than those that change less often. However, the normalized probabilities of changing values may not immediately represent the optimal weighting of attributes. For example, let *last name* have a probability to change as 10% over 3 years, and *first name* have a probability to change as 2% over the same time period. While the ratio between the two probabilities is 5:1, it does not immediately suggest that *first name* is five times more important than *last name*.

To control the impact of temporal models, a parameter $\alpha \in [0, \infty]$ is introduced during normalization, as shown in (8). When α is 0, the temporal model has the maximum impact in adjusting similarity output. When α is very large, the impact of the temporal model is close to none. The parameter α is chosen by the user based on domain knowledge. In Sect. 6, we will evaluate a range of α values.

$$s_r(r_a, r_b) = \frac{\sum_{A \in \mathbf{A}} (w_A(s_A(r_a.A, r_b.A), |r_a.t - r_b.t|) + \alpha) \cdot s_A(r_a.A, r_b.A)}{\sum_{A \in \mathbf{A}} (w_A(s_A(r_a.A, r_b.A), |r_a.t - r_b.t|) + \alpha)} \tag{8}$$

Algorithm 1. Record Linkage with Regression-based Temporal Model

Input:
- A set of temporal record clusters for training: **C**
- A set of temporal records to be linked: **R**
- For each attribute $A \in \mathbf{A}$, a set of support attributes: L_A
- A similarity threshold: t_s
- An impact adjustment value: α

Output:
- A set of merged records, each record represents an entity: $\mathbf{R}'$

```
 1: T = hashtable()   // A hashtable of training instances for each attribute
 2: for C in C do
 3:    for A in A do
 4:       Create a training instance i using L_A and the approach described in Sect. 5.1.
 5:       T[A].append(i)
 6: M_a = hashtable()   // Agreement decay models
 7: M_d = hashtable()   // Disagreement probability models
 8: for A in A do       // Train two models for each attribute
 9:    M_a[A] = decayModel(T[A])
10:    M_d[A] = linearModel(T[A])
11: Get pairs of records P from R using Swoosh described in Sect. 4.
12: for p in P do                     // For each pair of records
13:    decaysAgree = hashtable()      // Map A ∈ A to an agreement decay
14:    probsDisagree = hashtable()    // Map A ∈ A to a disagreement probability
15:    for A in A do       // Calculate agreement decays and disagreement probabilities
16:       decaysAgree[A] = M_a[A](p)
17:       probsDisagree[A] = M_d[A](p)
18:    s_r = sim(decaysAgree, probsDisagree, α)   // Calculate similarity using (8)
19:    if s_r >= t_s then
20:       R.removeRecords(p)   // Remove the two records of the pair
21:       r' = merge(p)        // Merge the pair of records into a new record
22:       R.push(r')           // Add the new record to record set
23:       Create new records pairs using r' then push the new pairs into P
24: return R as R'      // Return the merged records that cannot be merged any further
```

5.3   Algorithmic Overview of Regression-Based Temporal Linkage

Algorithm 1 describes the main steps of our approach, which integrates with our framework and produces a set of linked (merged) records from a set of temporal records **R**. From lines 1 to 5, the algorithm creates a list of training instances for each attribute. Each training instance contains a class value that indicates if the attribute value of an entity has been changed within a time period. From lines 6 to 10, the algorithm trains an agreement decay model and a disagreement probability model for each attribute, using the training instance sets created. From lines 11 to 23, the algorithm compares records in pairs, merges the record pair that is classified as a match into a new record, and compares the new record against the remaining records.

6   Experiments

In this section we first describe the datasets, baseline methods and measures used in our experiments. Then we present and discuss the experimental results.

6.1   Experimental Settings

Datasets: The real temporal datasets we used in this paper are from the North Carolina Voter Registration (NCVR) dataset collected every two months[1]. The datasets have ground truth (entity identifiers) available for all records. We selected the following attributes: *first, middle,* and *last name, name suffix, street address, city, gender,* and *age. Gender* was selected as the support attribute for *last name, age* was selected as the support attribute for *street address,* while the remaining attributes have no support attributes. The choice of support attributes was made according to domain knowledge. The NCVR dataset in total contains 8,336,205 entities, from which we randomly selected 5K, 10K, 50K, and 100K entities and their temporal records to create testing datasets. In each of the four testing datasets, 76% of the entities have one temporal record, 18.6% have two temporal records, 4% have three temporal records, and 1.4% entities have more than three temporal records. Each temporal record has one or more attribute value(s) that are different from the other records.

1K and 10K entities were randomly selected from the NCVR dataset and their temporal records are used for training. The 1K training dataset is used to train the models when using 5K and 10K testing datasets, and the 10K training dataset is used to train the models when using 50K and 100K testing datasets.

Measures: We used the standard quality measures of precision, recall, and F-measure to evaluate the record linkage quality [5] (noting recent work on how the F-measure can be misleading for record linkage when used to compare different classifiers at the same similarity threshold by weighting precision and recall differently [8]). Let $\mathbf{R}$ be a record linkage result in the form of clusters of records that are matching, and $\mathbf{S}$ be the ground truth that $\mathbf{R}$ corresponds to, which is also in the form of clusters of records. We calculate pair-wise precision $(P) = (|\mathbf{R} \cap \mathbf{S}|)/(|\mathbf{R}|)$, pair-wise recall $(R) = (|\mathbf{R} \cap \mathbf{S}|)/(|\mathbf{S}|)$, and $F_1 = 2 * P * R/(P+R)$. We used similarity thresholds $t_s = [0.6, 0.65, 0.7, 0.75, 0.8]$, and values for α in the range from 0 to 8 with an increment of 0.5. The highest F_1 score of each method was selected as the final result.

For string attributes, the similarity of a pair of attribute values was calculated using the Jaro-Winkler string comparison function [5]. The similarity of a pair of age values was calculated as: $s_{age} = 1/(|age_1 - age_2| + 1)$.

We implemented all algorithms in Python 2.7, and the experiments were conducted on a server with 64-bit Intel Xeon (2.4 GHz) CPUs, 128 GB of memory and running Ubuntu 14.04. We used the Sklearn package[2] for the linear regression classification. We implemented four algorithms for the experimental study. The first two are baselines, and the last two are the proposed approaches: (1) No model: A baseline approach with no temporal model. Weights of attributes were not adjusted by a temporal model. (2) Decay model (*Decay*): A baseline approach using the temporal model proposed by Li et al. [13] (Sect. 4). (3) Disagreement

---

[1] http://dl.ncsbe.gov/.

[2] http://scikit-learn.org.

probability regression model (*Disprob*): A temporal model which uses a regression model to predict the disagreement probability, and reduces the weights of attributes when their predicted disagreement probability is high (Sect. 5.1). (4) Disagreement probability combined with agreement decay regression model (*Mixed*): With the disagreement probability being predicted in the same way as the method above, the mixed method also calculates agreement decay from the decay model. The disagreement probability and agreement decay are combined to adjust the weight of each attribute (Sect. 5.1).

Table 2. Linkage results on the NCVR datasets with best results highlighted in bold.

Dataset	5K			10K			50K			100K		
	P	R	F_1	P	R	F_1	P	R	F_1	P	R	F_1
No model	**0.99**	0.93	0.96	**0.99**	0.90	0.94	0.94	**0.90**	0.92	0.91	**0.91**	0.90
Decay	0.96	0.95	0.96	0.95	0.93	0.92	**0.97**	0.87	0.92	**0.96**	0.88	0.91
Disprob	0.98	0.93	0.96	0.97	0.90	0.94	0.95	0.88	0.91	0.95	0.86	0.91
Mixed	0.97	**0.96**	**0.97**	0.94	**0.95**	0.94	0.96	0.90	**0.93**	0.95	0.90	**0.92**

Impact adjustment as discussed in Sect. 5.2 is implemented for *Decay* and *Disprob*, as well as the proposed algorithm *Mixed*, to allow a fair comparison.

6.2   Experimental Results

Table 2 shows the linkage results of the four algorithms on the testing datasets with impact adjustment. Results with the highest F_1 were selected. On the smaller testing datasets (5K and 10K), the *Mixed* approach achieved better recall but lower precision than the non-temporal baseline. On the larger testing datasets (50K and 100K), *Mixed* maintained similar recalls as the non-temporal baseline while performing better at precision. The result shows that our technique performs better when a dataset is large, while it does not perform worse than other techniques on smaller datasets. Because the 50K and 100K datasets used a larger training set of 10K entities the improvements would be due to this larger number of training records.

One significant difference between the smaller and the larger testing datasets is the percentage of non-matching record pairs. Even with blocking [5], the number of non-matching pairs still grows faster than linear with respect to the size of a dataset. A linkage algorithm will encounter non-match pairs more often when the testing dataset is large. As a result, we can observe that the precision of the *No model* approach decreases as the size of a dataset increases.

Figure 1 shows the effect of the impact adjustment parameter α (see Sect. 5.2). The temporal models did not perform well when the impact adjustment was not applied ($\alpha = 0$). It implies that directly applying probabilities on the weights can over-weight some attributes and decrease the linkage quality.

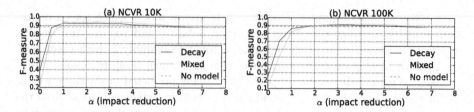

Fig. 1. The effect of different values for the impact reduction parameter (α). Without impact reduction ($\alpha = 0$), the temporal models (*Decay, Mixed*) performed poorly. At a certain point, the temporal models start to outperform the non-temporal baseline approach (*No model*). With the impact being reduced further (α increases), the temporal models eventually performed the same as the non-temporal baseline because the impact of the models has been reduced to the extent that is not significant anymore.

7   Conclusion and Future Work

In this paper we have developed a temporal model to improve the quality of temporal record linkage. Our model uses a linear regression model and multiple attribute values to predict the probability for an attribute value to change within a certain time period, and the model adjusts the weight of the attribute used in similarity calculations accordingly. The intuition of our approach is to use the dependency between attributes to predict their probability to change over time more accurately. We evaluated our approaches on four real-world datasets derived from the NCVR database. The experimental results show that our approach performed better than two baseline approaches.

In the future, we will investigate attribute dependencies for calculating the probability for two different entities to share the same attribute value (agreement probability), which can also be affected by other attributes. For example, two people who have the same phone number have a high probability to have the same address. We also aim to incorporate a frequency based weighting strategy into our framework to see if undesired high similarities can be adjusted properly. Another possible direction is to test the temporal models with different clustering techniques, such as those proposed by Li et al. [13] and Chiang et al. [4].

References

1. Benjelloun, O., Garcia-Molina, H., Menestrina, D., Su, Q., Whang, S.E., Widom, J.: Swoosh: a generic approach to entity resolution. VLDB J. **18**(1), 255–276 (2009)
2. Blum, A.L., Langley, P.: Selection of relevant features and examples in machine learning. Artif. Intell. **97**(1–2), 245–271 (1997)
3. Chiang, Y.H., Doan, A., Naughton, J.F.: Modeling entity evolution for temporal record matching. In: ACM SIGMOD, Snowbird, Utah (2014)
4. Chiang, Y.H., Doan, A., Naughton, J.F.: Tracking entities in the dynamic world: a fast algorithm for matching temporal records. PVLDB **7**(6), 469–480 (2014)
5. Christen, P.: Data Matching - Concepts and Techniques for Record Linkage, Entity Resolution, and Duplicate Detection. Springer, Heidelberg (2012)

6. Christen, P., Gayler, R.W.: Adaptive temporal entity resolution on dynamic databases. In: Pei, J., Tseng, V.S., Cao, L., Motoda, H., Xu, G. (eds.) PAKDD 2013. LNCS (LNAI), vol. 7819, pp. 558–569. Springer, Heidelberg (2013). doi:10.1007/978-3-642-37456-2_47
7. Fellegi, I.P., Sunter, A.B.: A theory for record linkage. JASA **64**(328), 1183–1210 (1969)
8. Hand, D., Christen, P.: A note on using the F-measure for evaluating data linkage algorithms. Pre-print NI16047, Isaac Newton Institute, Cambridge, UK (2016)
9. Herzog, T.N., Scheuren, F.J., Winkler, W.E.: Data Quality and Record Linkage Techniques. Springer, New York (2007)
10. Krueger, D., Montgomery, D.C., Peck, E.A., Vining, G.G.: Introduction to Linear Regression Analysis. Wiley, Hoboken (2015)
11. Kum, H.C., Krishnamurthy, A., Machanavajjhala, A., Ahalt, S.C.: Social genome: putting big data to work for population informatics. IEEE Comput. **47**(1), 56–63 (2014)
12. Li, F., Lee, M.L., Hsu, W., Tan, W.C.: Linking temporal records for profiling entities. In: ACM SIGMOD, Melbourne (2015)
13. Li, P., Dong, X.L., Maurino, A., Srivastava, D.: Linking temporal records. PVLDB **4**(11), 956–967 (2011)

Community Detection in Graph Streams by Pruning Zombie Nodes

Yue Ding[1,2], Ling Huang[1,2], Chang-Dong Wang[1,2(✉)], and Dong Huang[2,3]

[1] School of Data and Computer Science, Sun Yat-sen University, Guangzhou, China
dingy8@foxmail.com, huanglinghl@hotmail.com, changdongwang@hotmail.com
[2] Guangdong Key Laboratory of Information Security Technology,
Guangzhou, China
huangdonghere@gmail.com
[3] College of Mathematics and Informatics, South China Agricultural University,
Guangzhou, China

Abstract. Detecting communities in graph streams has attracted a large amount of attention recently. Although many algorithms have been developed from different perspectives, there is still a limitation to the existing methods, that is, most of them neglect the "zombie" nodes (or unimportant nodes) in the graph stream which may badly affect the community detection result. In this paper, we aim to deal with the zombie nodes in networks so as to enhance the robustness of the detected communities. The key here is to design a pruning strategy to remove unimportant nodes and preserve the important nodes. We propose to recognize the zombie nodes by a degree centrality calculated from the exponential time-decaying edge weights, which can be efficiently updated in the graph stream case. Based on only important and active nodes, community kernels can be constructed, from which robust community structures can be obtained. One advantage of the proposed pruning strategy is that it is able to eliminate the effect of the aforementioned "zombie" nodes, leading to robust communities. By designing an efficient way to update the degree centrality, the important and active nodes can be easily obtained at each timestamp, leading to the reduction of computational complexity. Experiments have been conducted to show the effectiveness of the proposed method.

Keywords: Community detection · Graph stream · Weighting · Pruning

1 Introduction

With the rapid development of Internet, an enormous amount of graph streams have been generated. These graphs (networks) represent the information of interactive relationships of the objects (users) over a specific period with the form of incremental graphs. Discovering communities (groups) in graph streams enables a better understanding of the natural and social structures in graph streams [1]. In the past few years, many graph stream community detection algorithms have

© Springer International Publishing AG 2017
J. Kim et al. (Eds.): PAKDD 2017, Part I, LNAI 10234, pp. 574–585, 2017.
DOI: 10.1007/978-3-319-57454-7_45

been developed aiming to address varying issues from different perspectives, such as [1–4]. For instance, for community detection in a multi-mode network, an evolutionary multi-mode clustering was proposed by utilizing the temporal information [5]. For addressing the issue that the social behavior of users varies over different graph regions at different timestamps, Wang et al. [1] designed a Local Weighted-Edge-based Pattern Summary to describe the local homogeneous region. For combining both linkage structure and content information, a random walk based method was proposed in [6].

In this paper, we identify the following unaddressed issue of community detection in graph streams. In real-world social networks, the importance of various accounts is different. There is no doubt that a huge number of accounts have little importance in networks such as Facebook and Twitter. These accounts generally have few interactions with others and this character leads to that they are also in humble status in networks. Some of them are created before long, some of them have been abandoned by their users after transient uses, and the others are just created as "Zombie fans", which occupy a huge proportion of those unimportant accounts. Different from the first two cases, "Zombie fans" are artificial accounts without authentic users, which have various functions such as being sold as fake followers to those who need a beautiful follower list, If we give equal treatment to the "zombie" and other real/healthy accounts, the detected communities will be degenerated seriously.

To address this issue, inspired by [7,8], we propose a pruning-based graph stream community detection algorithm called *PruGStream*. The basic idea is to remove unimportant nodes which are recognized by the degree centrality calculated from the exponential time-decaying edge weights in graph streams. Then an iterative strategy is designed to construct community kernels from the important and active nodes only. Finally, based on the community kernels, the complete communities which take into account all the nodes will be generated by directly assigning/updating the unimportant but active nodes to the nearest community kernels. The first advantage of the proposed method is that, by constructing community kernels from only important and active nodes, it is able to eliminate the effect of the aforementioned "zombie" nodes, leading to robust communities. By designing an efficient way to update the degree centrality, the important and active nodes can be easily obtained at each timestamp, leading to the reduce of computational complexity.

2   The Proposed Method

In this section, we will describe the proposed *PruGStream* method in detail. The algorithm is composed of three phases, namely removing unimportant nodes, generating community kernels and generating the complete communities.

2.1   Phase One: Removing Unimportant Nodes

Following [1], we adopt the incremental representation of undirected weighted graph stream as the input of our algorithm, which is defined as follows.

Definition 1 (Incremental representation of undirected weighted graph stream). *Given an undirected weighted graph stream, the incremental representation is* $G_0, G_1, G_2, \cdots, G_t, \cdots$ *where* $G_t = (\mathcal{V}_t, \mathcal{I}_t)$, *with* $\mathcal{V}_t$ *being the set of newly attached nodes at the current timestamp t, and* $\mathcal{I}_t = \left\{(v^i, v^j, Y_t^{i,j}) | v^i, v^j \in \mathcal{V}_t^{all} = \mathcal{V}_0 \bigcup \mathcal{V}_1 \bigcup \cdots \bigcup \mathcal{V}_t \right\}$ *with* $Y_t^{i,j} = Y_t^{j,i}$ *representing the number of new interactions between nodes* v^i *and* v^j *occurring between timestamps* $t - 1$ *and* t.

Following [1], we define the weight $w_t^{i,j}$ of the edge linking v^i, v^j at timestamp t as

$$w_t^{i,j} \triangleq \sum_{l=0}^{m} Y_{t_l}^{i,j} e^{-\lambda(t-t_l)} \tag{1}$$

where $\lambda \geq 0$ is the decaying constant. In our experiments, following [1], we fix λ to 0.5. In Eq. (1), it is assumed that all the interactions between nodes v^i and v^j occur only at timestamp $t_0 < t_1 < \cdots < t_m$ where t_0 is the timestamp the two nodes began to interact and $t_m = t$ is the current timestamp. By definition, it is clear that the edge weight $w_t^{i,j}$ takes into account both the number of interactions and their time-decaying properties [1].

Due to the new interactions that are generated at timestamp t, the weight that was last updated at timestamp t_e is updated as follows [1],

$$w_t^{i,j} \leftarrow w_{t_e}^{i,j} e^{-\lambda(t-t_e)} + Y_t^{i,j}. \tag{2}$$

In the real social network, there exist a large portion of unimportant users who have little interaction with others. The number of those inactive users is very large but they are inessential when detecting communities. As aforementioned, some of these unimportant nodes are healthy accounts registered by human being, while others are fake accounts. Taking into account these unimportant nodes in community detection not only wastes a lot of time, in particular in large-scale real-time graph streams, but also produces very little effect or even has the risk to damage the robustness of the detected communities. Therefore, in this paper, we divide all nodes into two opposite camps: the camp of important nodes and the camp of unimportant node, and treat them differently. We remove those unimportant nodes before community detection and do not put them back until generating the ultimate community labels. In other words, we make unimportant nodes only appear in the final result but do not participate in the process of community detection.

It is worth noticing that similar ideas have been used in data clustering. For instance, Ester et al. [7] proposed a *DBSCAN* algorithm, which for the first time uses the idea of classifying points as core points and non-core points to reduce the interference of noisy points. Only core points can form the kernels of clusters together with their reachable points, while the non-core points form the boundaries of clusters.

To divide the nodes into different sets, first of all, we need a criterion to judge the importance of nodes. Here we use degree centrality defined in [9] as follows.

Definition 2 (Degree centrality). *The degree centrality $D_c(v^i)_t$ is a criterion to judge the importance of node v^i in the graph at timestamp t,*

$$D_c(v^i)_t = \frac{deg(v^i)}{n-1} = \frac{\sum_{j=1}^{n} w_t^{i,j}}{n-1} \tag{3}$$

where $n = |\mathcal{V}_t^{all}|$ denotes the number of all nodes till timestamp t.

Degree centrality is the most direct measure to depict the centrality of nodes when analyzing networks. The higher the value of degree centrality is, the more important the node is in the network.

To distinguish the important nodes and the unimportant nodes at timestamp t, we set a parameter η and define them as follows.

Definition 3. *If the degree centrality of a node is greater than or equal to η, we regard the node as an important node and put it into the set $\bar{\mathcal{V}}$, otherwise the node belongs to the unimportant node set $\hat{\mathcal{V}}$, i.e.*

$$Important\,node\,sets: \bar{\mathcal{V}} = \{v^i \in \mathcal{V}_t^{all} | D_c(v^i)_t \geq \eta\} \tag{4}$$

$$Unimportant\,node\,sets: \hat{\mathcal{V}} = \{v^i \in \mathcal{V}_t^{all} | D_c(v^i)_t < \eta\}. \tag{5}$$

By definition, at each timestamp, we need to calculate the degree centrality for all the nodes in $\mathcal{V}_t^{all}$ because the degree centrality of all nodes may change along with the increment of $\mathcal{V}_t^{all}$ and edges, and therefore, the nodes in $\bar{\mathcal{V}}$ and $\hat{\mathcal{V}}$ need to update every timestamp. Directly recalculating the degree centrality by Eq. (3) is time-consuming and unfeasible in real-time graph stream processing. To address this issue, we expand the concept of degree centrality and design a novel strategy to update the degree centrality, as stated in the following theorem.

Theorem 1 (Degree centrality update). *The degree centrality at timestamp $t > 0$ can be updated as follows,*

$$D_c(v^i)_t = D_c(v^i)_{t-1} \times \frac{n-1-n_t}{n-1}e^{-\lambda} + \frac{\sum_{Y_t^{i,j} \in \mathcal{I}_t} Y_t^{i,j}}{n-1} \tag{6}$$

where $n > 1$ is the number of nodes in $\mathcal{V}_t^{all}$ and $n_t = |\mathcal{V}_t|$ is the number of newly attached nodes at timestamp t.

Proof. For proof purpose, we introduce another notation. Let $\ddot{\mathcal{V}}$ denote all the old nodes that exist before timestamp t but have new interactions at timestamp t, i.e., $\ddot{\mathcal{V}} = \{v^i \in \mathcal{V}_{t-1}^{all} | \exists v^j \exists Y_t^{i,j} \text{ s.t. } (v^i, v^j, Y_t^{i,j}) \in \mathcal{I}_t\}$. Therefore, $\mathcal{V}_t^{all}$ is composed of three parts: the nodes in $\mathcal{V}_t$, the nodes in $\ddot{\mathcal{V}}$ and the remaining nodes $\{v^i \in \mathcal{V}_t^{all} | v^i \notin \ddot{\mathcal{V}} \bigcup \mathcal{V}_t\}$. Besides, it is impossible for the nodes in $\mathcal{V}_t$ to appear in $\mathcal{V}_{t-1}^{all}$. Therefore, we have,

$$D_c(v^i)_t = \frac{\sum_{j=1}^n w_t^{i,j}}{n-1} \tag{7}$$

$$= \frac{\sum_{v^j \notin \check{\mathcal{V}} \cup \mathcal{V}_t} w_t^{i,j}}{n-1} + \frac{\sum_{v^j \in \check{\mathcal{V}}} w_t^{i,j}}{n-1} + \frac{\sum_{v^j \in \mathcal{V}_t} w_t^{i,j}}{n-1}$$

$$= \frac{\sum_{v^j \notin \check{\mathcal{V}} \cup \mathcal{V}_t} w_{t-1}^{i,j} e^{-\lambda}}{n-1} + \frac{\sum_{v^j \in \check{\mathcal{V}}} (w_{t-1}^{i,j} e^{-\lambda} + Y_t^{i,j})}{n-1} + \frac{\sum_{v^j \in \mathcal{V}_t} Y_t^{i,j}}{n-1}$$

$$= \frac{\sum_{v^j \notin \check{\mathcal{V}} \cup \mathcal{V}_t} w_{t-1}^{i,j} + \sum_{v^j \in \check{\mathcal{V}}} w_{t-1}^{i,j}}{n-1-n_t} \times \frac{n-1-n_t}{n-1} e^{-\lambda} + \frac{\sum_{v^j \in \check{\mathcal{V}} \cup \mathcal{V}_t} Y_t^{i,j}}{n-1}$$

$$= D_c(v^i)_{t-1} \times \frac{n-1-n_t}{n-1} e^{-\lambda} + \frac{\sum_{Y_t^{i,j} \in \mathcal{I}_t} Y_t^{i,j}}{n-1}$$

In fact, when calculating the degree centrality of a node, $n = 0$ and $n = 1$ are two trivial cases since there is no nodes in the network when $n = 0$ and there is only one node in the network leading to 0 degree centrality. What's more, for the case of $t = 0$, the degree centrality is defined according to the original definition.

2.2 Phase Two: Generating Community Kernels

By Definition 3, we can divide the nodes of $\mathcal{V}_t^{all}$ into important nodes $\bar{\mathcal{V}}$ and unimportant nodes $\check{\mathcal{V}}$ according to degree centrality. It's obvious that if a node has new interactions at timestamp t, it belongs to $\check{\mathcal{V}} \cup \mathcal{V}_t$ because $\forall v^i \in \check{\mathcal{V}} \cup \mathcal{V}_t$, $\exists v^j \exists Y_t^{i,j}$, s.t. $(v^i, v^j, Y_t^{i,j}) \in \mathcal{I}_t$. In accordance with this rule, at each timestamp, we can divide all the existing nodes $\mathcal{V}_t^{all}$ into active nodes $\check{\mathcal{V}} \cup \mathcal{V}_t$ which have new interactions at timestamp t and inactive nodes which haven't.

To ensure the robustness of the detected communities and further enhance the efficiency of the real-time graph stream processing, only the nodes which are both important and active at timestamp t are considered in discovering the kernels of communities. Denote by $\check{\mathcal{V}}$ the nodes which are both important and active at timestamp t, we have $\check{\mathcal{V}} = \bar{\mathcal{V}} \cap (\mathcal{V}_t \cup \check{\mathcal{V}})$. The remaining nodes $\mathcal{V}_t^{all} \backslash \check{\mathcal{V}}$ will be ignored temporarily in discovering the kernels of communities. The relationship of all the node sets we have defined is shown in Fig. 1.

The key of phase two is finding the community kernels, i.e. the most suitable communities for the nodes in $\check{\mathcal{V}}$. We will at first introduce the procedure for the graph when $t = 0$, which we can regard as an independent static graph, and then extend the procedure to make it applicable to all timestamps.

When $t = 0$, after each node in $\mathcal{V}_0^{all}$ has been tagged as important node or unimportant node during phase one, $\check{\mathcal{V}}$ in phase two is equal to $\bar{\mathcal{V}}$ since all the nodes in $\mathcal{V}_0^{all}$ belong to $\mathcal{V}_0$. Also we have $n_t = n$. Denote by $\mathcal{C} = \{c_1, c_2, \ldots, c_k\}$ the detected community kernels at timestamp t where k the number of communities. And at timestamp $t = 0$, $\mathcal{C}$ is initialized as $\varnothing$ and k is 0 before calculating. The final community detection result is a vector L of length n storing the detected community label for each node in network.

First of all, we calculate the similarity between the nodes in $\check{\mathcal{V}}$ and the community kernels in $\mathcal{C}$ one by one and find the most suitable community for each

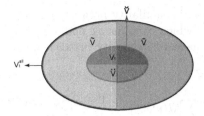

Fig. 1. The Venn graph of the node sets. $\mathcal{V}_t^{all}$ are all the nodes in the graph till timestamp t. $\mathcal{V}_t$ are the newly attached nodes at timestamp t. $\bar{\mathcal{V}}$ and $\tilde{\mathcal{V}}$ are the important nodes and unimportant nodes at timestamp t respectively. $\ddot{\mathcal{V}}$ are all the old nodes that exist before timestamp t but have new interactions at timestamp t. $\check{\mathcal{V}}$ are the nodes which are both important and active.

node. This process includes two steps for each node except the first node in $\check{\mathcal{V}}$ when $t = 0$:

1. Evaluate the similarity between the node and all community kernels in $\mathcal{C}$. The larger the value of similarity between the node and the community kernel is, the more possible the node belongs to the community.
2. Find the community kernel having the greatest similarity with the node. Then compare the value of similarity with a parameter ξ which is defined as the threshold to judge whether the community is qualified for the node. If the similarity is greater than or equal to ξ, we regard this community as the most suitable community to nodes, otherwise, we think there is no suitable community kernel in the current $\mathcal{C}$ and create a new community kernel c_{k+1} for the node, which is added to $\mathcal{C}$.

If the node is the first node in $\check{\mathcal{V}}$ when $t = 0$, since the set $\mathcal{C}$ is $\varnothing$, we make the node belong to the first generated community kernel c_1.

In the paper, we judge the similarity between one node and one community kernel using the widely used weighted Jaccard similarity. The definition of the weighted Jaccard similarity is as follows [10].

Definition 4 (Jaccard similarity). *Given two weighted vectors* $\boldsymbol{a} = \{a_1, a_2, \ldots, a_d\}$ *and* $\boldsymbol{b} = \{b_1, b_2, \ldots, b_d\}$, *with* $a_i, b_i > 0$, *the weighted Jaccard similarity is*

$$J(\boldsymbol{a}, \boldsymbol{b}) = \frac{\sum_{i=1}^{d} \min(a_i, b_i)}{\sum_{i=1}^{d} \max(a_i, b_i)} \tag{8}$$

We can define the similarity between two nodes v_x, v_y at timestamp t as the weighted Jaccard similarity between their edge weight vectors (associated with all nodes) using Eq. (8)

$$J(v_x, v_y)_t = \frac{\sum_{j=1}^{n} \min(w_t^{x,j}, w_t^{y,j})}{\sum_{j=1}^{n} \max(w_t^{x,j}, w_t^{y,j})} \tag{9}$$

Therefore, we define the similarity between a node v^i and each community kernel $c_h, \forall h = 1, \ldots, k$ which contains n_h^c nodes at timestamp t as the average value of the weighted Jaccard similarity between node v^i and all nodes in c_h where $c_h = \{v_c^1, v_c^2, \ldots, v_c^{n_h^c}\}$.

$$J(v^i, c_h)_t = \frac{1}{n_h^c} \times \sum_{l=1}^{n_h^c} J(v^i, v_c^l) = \frac{1}{n_h^c} \times \sum_{l=1}^{n_h^c} \frac{\sum_{j=1}^n \min(w_t^{i,j}, w_t^{v_c^l,j})}{\sum_{j=1}^n \max(w_t^{i,j}, w_t^{v_c^l,j})} \quad (10)$$

After the first traversal of all nodes in $\check{\mathcal{V}}$, we can get some preliminary community kernels. However, for each node in $\check{\mathcal{V}}$, the community it belongs to is assigned by considering only the similarity before assigning the node and as the update of the community kernels, the similarity between the nodes and their associated community kernels would change, which implies that the weighted Jaccard similarity may be smaller than ξ again. In this case, the assigned community kernels should also be changed accordingly. To solve this problem, we design an iterative "Check and Repair" strategy as follows.

Firstly, we clear the set $\check{\mathcal{V}}$, and for each node in $\check{\mathcal{V}}$ we check whether the similarity between the node and the community kernel it stays is still no less than ξ. If the similarity is still no less than ξ, the community kernel is still qualified to the node, otherwise it is unqualified. We pick up all the nodes which are in unqualified community kernels and put them into $\check{\mathcal{V}}$, then repeat the procedure of generating community kernels until all nodes in $\check{\mathcal{V}}$ find their qualified community kernels and $\check{\mathcal{V}} = \varnothing$.

The community kernel generation procedure is similar at other timestamps. When $t \neq 0$, the community kernel set $\mathcal{C} \neq \varnothing$ anymore and $\check{\mathcal{V}} = \bar{\mathcal{V}} \bigcap (\mathcal{V}_t \bigcup \check{\mathcal{V}})$ rather than $\check{\mathcal{V}} = \bar{\mathcal{V}}$. Thus, in the step of finding the preliminary community kernels at timestamp t, we just update the community kernels by considering only $\check{\mathcal{V}}$ rather than reconstruct the community kernels from scratch, which is able to make a balance between the accuracy and efficiency.

2.3   Phase Three: Generating the Complete Communities

In phase two, we just construct the community kernels using the important nodes. That means we ignore the nodes which belong to unimportant nodes $\tilde{\mathcal{V}}$. Although these unimportant nodes have little influence on generating communities, we still need to find community labels for them. Therefore, in this phase, we want to directly label the nodes in $\tilde{\mathcal{V}}$ but ignore their influence to the communities. In this way, we can concentrate more on important nodes and eliminate interference of marginal and unimportant nodes. Among all unimportant nodes, at timestamp t, the nodes belonging to $\tilde{\mathcal{V}} \bigcap \mathcal{V}_t$ i.e. the newly attached unimportant nodes with no community labels should be considered in this phase. Besides, due to the newly occurred interactions between timestamps $t - 1$ and t, nodes belonging to $\tilde{\mathcal{V}} \bigcap \check{\mathcal{V}}$ have more possibility to change their community labels than other inactive and unimportant nodes. Therefore, we only update the nodes in $\tilde{\mathcal{V}} \bigcap (\check{\mathcal{V}} \bigcup \mathcal{V}_t)$ in this phase.

Accordingly, at timestamp t, for each node in $\tilde{\mathcal{V}} \bigcap (\ddot{\mathcal{V}} \bigcup \mathcal{V}_t)$, we assign it to the community with the maximal Jaccard similarity. However, the assignment of the nodes in $\tilde{\mathcal{V}} \bigcap (\ddot{\mathcal{V}} \bigcup \mathcal{V}_t)$ should not change the community kernels obtained in phase two, because the community kernels would be used at the next timestamp. So in this phase, we use a vector L of length n to store the community labels of all the nodes $\mathcal{V}_t^{all}$. What's more, if some of the previous important nodes become unimportant at the current timestamp due to the time-decaying strategy, they will be removed from the community kernels $\mathcal{C}$ so as to not only reflect the evolving of communities but also reduce the computational complexity.

3 Experiments

In this section, extensive experiments are conducted to confirm the effectiveness of our method. Firstly, parameter analysis is conducted to show how the parameters affect the community detection performance of our method on two static networks. Then, comparison experiments are conducted to compare our algorithm with 4 state-of-the-art community detection algorithms on 7 different datasets including both static networks and graph streams.

3.1 Datasets and Evaluation Metrics

Datasets. In our experiments, both static networks and graph streams are used.

1. **Static networks:**
   (a) **Zachary karate club network:** It contains the friendships between 34 members of a karate club at a US university, as described by Wayne Zachary in 1977 [11]. The members in the network could be divided into two communities: administrators and instructors.
   (b) **American college football network:** It is a labeled network of American football games between Division IA colleges, which is collected by M. Girvan and M. Newman [12]. There are 115 vertices indicating teams and 613 edges representing regular-season games between the two vertices in the graph. The labels of nodes indicate one of the 12 conferences to which those football teams belong.
   (c) **Books about US politics:** The dataset, also called polbooks, collected by V. Krebs, contains 105 nodes representing books sold on Amazon and 441 edges representing the frequency of co-purchasing of books by the same buyers. The labels indicate the political tendency of the books.
   (d) **LFR:** LFR benchmark network [13] is one of the most widely used benchmark networks in the study of community detection. Compared with the real networks, the LFR networks are more standardized and flexible. Here we use two LFR networks with different sizes, which are LFR-1000 with 1000 nodes/15484 edges and LFR-3000 with 3000 nodes/153380 edges respectively.

2. **Dynamic graph streams:**
   (a) **LFR graph stream:** The LFR graph stream is extended from the LFR static graph. To gain the dynamic network, we separate the nodes into 10 sections evenly, and at each timestamp we append the network with some part of nodes as well as the associated edges to simulate the expansion of dynamic network. Here, two graph streams, namely LFR-1000 graph stream and LFR-3000 graph stream, are used in our experiments.

Evaluation Metrics. Two popular evaluation metrics are used in our experiments, which are Normalized Mutual Information (NMI) [14] and Rand Index (RI) [15]. All the metrics are used to evaluate the similarity between the detected community labels and the ground truth community labels, where higher value indicates better community detection results.

3.2 Parameter Analysis

In this subsection, parameter analysis is conducted to show the effect of the two parameters ξ and η on the performance of the proposed method on two evaluation metrics, namely NMI and RI. Both ξ and η vary from 0 to 1.

Analysis on ξ. The parameter ξ determines whether a node will be put into the existing community kernels. If the weighted Jaccard similarity between the node and the closest community kernel is not less than ξ, then the node will belong to the community kernel temporarily.

Figure 2(a) and (b) show the performance of our algorithm in terms of NMI and RI versus different ξ. The parameter η is set as 0.1 on Karate and 0.09 on Football. From the results, it is clear that, the proposed algorithm exhibits a stable performance in the testing range. In particular, on the Football dataset, very stable but high community detection results in terms of both NMI and RI can be obtained.

Analysis on η. η is used to judge whether a vertice belongs to $\bar{\mathcal{V}}$ or not.

Figure 2(c) and (d) show the performance of our algorithm in terms of NMI and RI as a function of η, on the karate club network when $\xi = 0.1$ and on American football network when $\xi = 0.22$. The optimum value of η is related to the degree distribution of the network. Similar to ξ, it is not suitable for η to be too large, but it can be zero when all the active nodes are in $\bar{\mathcal{V}}$. However, compared with choosing η as 0, using a suitable η can enhance the performance of the algorithm, in particular, on the Karate dataset. On the Football dataset, it is clear that relative stable community results in terms of both NMI and RI can be obtained in the range of [0 0.09].

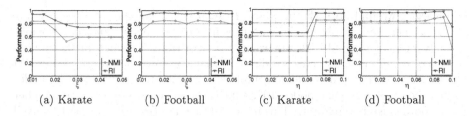

(a) Karate   (b) Football   (c) Karate   (d) Football

Fig. 2. Analysis of the effect of ξ and η on the community detection performance on different real static networks. On each dataset, the values of NMI or ARI are plotted as a function of ξ respectively.

3.3   Comparison Experiments

In this subsection, we conduct experiments to compare the performance of the proposed approach with the existing algorithms on both static networks and graph streams. The compared algorithms are summarized as follows.

1. Comparison algorithms in static networks
   (a) **Modularity** [16] is the most well-known community detection algorithm based on modularity.
   (b) **Ncut** [17], as one kind of spectral clustering, detects communities by optimizing the normalized cut criterion.
   (c) **Alink-Jaccard** [18] is a kind of hierarchical clustering method, which uses the jaccard similarity as the distance between nodes.
2. Comparison algorithms in dynamic networks
   (a) **Facetnet** [2] is a classical framework for analyzing communities and evolutions in dynamic networks.

In what follows, we will report the comparison results on five different static networks and two dynamic networks in Table 1 and Fig. 3 respectively.

Table 1. The performance of different community detection algorithms on the five static networks. The best results are marked in bold.

Algorithms	Karate		Football		Poolbooks		LFR-1000		LFR-3000	
	NMI	RI	NMI	RI	NMI	RI	NMI	RI	NMI	RI
PruGStream	**0.836**	**0.941**	**0.837**	**0.975**	**0.601**	0.834	**0.945**	**0.989**	**0.930**	**0.990**
alinkjaccard	0.477	0.631	0.508	0.877	0.450	0.741	0.840	0.938	0.748	0.892
Ncut	**0.836**	**0.941**	0.503	0.896	0.574	**0.843**	0.921	0.963	0.861	0.924
Modularity	0.454	0.700	0.511	0.893	0.559	0.801	0.880	0.943	0.833	0.958

1. Performance on static networks
   The comparison results are reported in Table 1. Overall, the results show that our method achieves the best performance compared with the existing methods on almost all the testing datasets. As for the performance on

Karate, we find both the proposed *PruGStream* algorithm and the existing *Ncut* method have obtained the best results on both metrics and show obvious advantage compared with the other two algorithms. Through further analysis of experimental results, we discover that the clustering results generated by *PruGStream* and *Ncut* are completely consistent. They classify all nodes correct except a noisy node "10", which has one connection to one member of both clusters separately. This indicates that *PruGStream* can perform well in lightweight datasets and even execute error-free judgments. From the performance on other networks, it can be seen that the proposed *PruGStream* approach still generates better clustering results than the compared methods in terms of NMI on all networks. Besides, the secondary winner is *Ncut*, of which the performance is slightly behind *PruGStream*. As for *PruGStream*, this superiority can be due to the proposed pruning strategy, which is able to construct robust community kernels. To sum up with comparison results, we can conclude that our *PruGStream* performs well and stably in community detection on static networks.

2. Performance on graph streams

   Figure 3 show the values of NMI and RI obtained by *PurGStream* and *Facetnet* on the two graph streams, namely LFR-1000 graph stream and LRF-3000 graph stream. From the figure, it is clear that the proposed *PurGStream* method generates better results than *Facetnet* in all experiments. Although the two algorithms have similar trends on the performance, our method are more stable and performs better. The comparison results have confirmed the effectiveness of our method on community detection on graph streams.

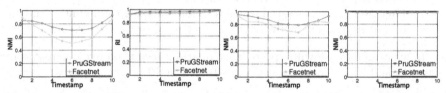

(a) LFR-1000 stream (b) LFR-1000 stream (c) LFR-3000 stream (d) LFR-3000 stream

Fig. 3. Comparison results on the graph streams in terms of both NMI and RI as a function of timestamps.

4   Conclusions

In this paper, we have purposed a novel community detection algorithm in graph streams termed *PruGStream* to address the issue of "zombie" nodes. The basic idea is to develop an efficient pruning strategy to construct community kernels from only important and active nodes at each timestamp. Based on the community kernels, robust community structure can be obtained. The proposed method can not only discover robust community structure that is insensitive to the "zombie" nodes but also dramatically reduce computational complexity. Experiments have been conducted to confirm the effectiveness of the proposed method.

Acknowledgment. This work was supported by NSFC (No. 61502543 and No. 61602189) and the PhD Start-up Fund of Natural Science Foundation of Guangdong Province, China (2016A030310457 and 2014A030310180).

References

1. Wang, C.D., Lai, J.H., Yu, P.: Dynamic community detection in weighted graph streams. In: Proceedings of SDM, SIAM, pp. 151–161 (2013)
2. Lin, Y.R., Chi, Y., Zhu, S., Sundaram, H., Tseng, B.L.: Facetnet: a framework for analyzing communities and their evolutions in dynamic networks. In: Proceedings of the 17th International Conference on World Wide Web, pp. 685–694. ACM (2008)
3. Kim, M.S., Han, J.: A particle-and-density based evolutionary clustering method for dynamic networks. Proc. VLDB Endowment **2**(1), 622–633 (2009)
4. Gudkov, V., Montealegre, V., Nussinov, S., Nussinov, Z.: Community detection in complex networks by dynamical simplex evolution. Phys. Rev. E **78**(1), 016113 (2008)
5. Tang, L., Liu, H., Zhang, J., Nazeri, Z.: Community evolution in dynamic multimode networks. In: KDD, pp. 677–685 (2008)
6. Wang, C.D., Lai, J.H., Yu, P.S.: NEIWalk: community discovery in dynamic content-based networks. IEEE Trans. Knowl. Data Eng. **26**(7), 1734–1748 (2013)
7. Ester, M., Kriegel, H.P., Sander, J., Xu, X., et al.: A density-based algorithm for discovering clusters in large spatial databases with noise. KDD **96**, 226–231 (1996)
8. Rodriguez, A., Laio, A.: Clustering by fast search and find of density peaks. Science **344**(6191), 1492–1496 (2014)
9. Freeman, L.C.: Centrality in social networks: I. conceptual clarification. Soc. Netw. **1**(3), 215–239 (1979)
10. Jaccard, P.: Etude de la distribution florale dans une portion des alpes et du jura. Bull. De La Soc. Vaudoise Des Sci. Nat. **37**(142), 547–579 (1901)
11. Zachary, W.W.: An information flow model for conflict and fission in small groups. J. Anthropol. Res. **33**, 452–473 (1977)
12. Girvan, M., Newman, M.E.: Community structure in social and biological networks. Proc. Natl. Acad. Sci. **99**(12), 7821–7826 (2002)
13. Lancichinetti, A., Fortunato, S., Radicchi, F.: Benchmark graphs for testing community detection algorithms. Phys. Rev. E Stat. Nonlinear Soft Matter Phys. **78**(2), 561–570 (2008)
14. Strehl, A., Ghosh, J.: Cluster ensembles-a knowledge reuse framework for combining multiple partitions. J. Mach. Learn. Res. **3**(Dec), 583–617 (2002)
15. Rand, W.M.: Objective criteria for the evaluation of clustering methods. J. Am. Stat. Assoc. **66**(336), 846–850 (1971)
16. Newman, M.E.: Modularity and community structure in networks. Proc. Natl. Acad. Sci. Unit. States Am. **103**(23), 8577–8582 (2006)
17. Shi, J., Malik, J.: Normalized cuts and image segmentation. IEEE Trans. Pattern Anal. Mach. Intell. **22**(8), 888–905 (2000)
18. Iamon, N., Boongoen, T., Garrett, S., Price, C.: A link-based cluster ensemble approach for categorical data clustering. IEEE Trans. Knowl. Data Eng. **24**(3), 413–425 (2010)

Bilingual Lexicon Extraction from Comparable Corpora Based on Closed Concepts Mining

Mohamed Chebel[1]([✉]), Chiraz Latiri[1], and Eric Gaussier[2]

[1] Research Laboratory LIPAH, Faculty of Sciences of Tunis,
University Tunis El Manar, Tunis, Tunisia
mohammedchebel@gmail.com, chiraz.latiri@gnet.tn
[2] Research Laboratory LIG, AMA Group,
University Joseph Fourier, Grenoble I, France
eric.gaussier@imag.fr

Abstract. In this paper, we propose to complement the context vectors used in bilingual lexicon extraction from comparable corpora with concept vectors, that aim at capturing all the words related to the concepts associated with a given word. This allows one to rely on a representation that is less sparse, especially in specialized domains where the use of a general bilingual lexicon leaves many words untranslated. The concept vectors we are considering are based on closed concepts mining developed in Formal Concept Analysis (FCA). The obtained results on two different comparable corpora show that enriching context vectors with concept vectors leads to lexicons of higher quality, especially in specialized domains.

1 Introduction

Bilingual lexicon extraction using parallel[1] and comparable[2] corpora has been the subject of many studies. However, the scarcity of multilingual parallel corpora, particularly for specialized areas, has led researches in bilingual lexicon extraction to use comparable corpora [15]. The exploitation of comparable corpora marked a turning point in the task of bilingual lexicon extraction, and raises a constant interest thanks to the abundance, the continuous growth and the availability of such corpora [4,7,12,13].

Most state-of-the-art approaches using comparable corpora to extract bilingual lexicons are based on the assumption that a word and its translations tend to appear in similar contexts across languages [4]. The context of a given word, which we will refer to as the *base word*, is usually represented by words surrounding it, *i.e.*, words co-occurring within a contextual window [4,7], or words related through syntactic dependency relations [13]. Furthermore, each word in

---

[1] A parallel corpus is a collection of texts that are translation of one another.
[2] A comparable corpus is a collection of multilingual documents dealing with the same topics and generally produced at the same time. They are not necessarily translation of each other.

© Springer International Publishing AG 2017
J. Kim et al. (Eds.): PAKDD 2017, Part I, LNAI 10234, pp. 586–598, 2017.
DOI: 10.1007/978-3-319-57454-7_46

the context of a base word can be weighted according to the strength of its association with the base word. Recently, context vectors are extracted using *word embedding* techniques [16] using the *word2vec* toolkit[3]. This tool induces vector based word representations by two ways [11]: by using the current word to predict the surrounding window of context words or by trying to predict a target word from the words surrounding it within a neural network architecture. In the rest of the paper and in our work, *context vectors* denote *word embedding based context vectors*.

One thus obtains, for each source and target word in the comparable corpus, a context vector that provides an implicit semantic representation of the word. From these context vectors, the *standard approach* of bilingual lexicon extraction from comparable corpora proceeds by translating all the elements of the context vector of each target word into the source language[4], using an existing bilingual dictionary, usually a general one as specialized ones are more difficult to obtain. Candidate translations of a given source word are then those target words for which their translated context vectors are the closest to the one of the source word.

If this approach has proven successful in many different studies, it nevertheless suffers from the fact that the context vectors, once translated, are sparse, especially in specialized domains where the use of a generic bilingual dictionary is not appropriate. We thus conjecture here that one can gain by relying on an enriched representation of the word. In particular, *if a word appears in the context vector of a base word, then it is likely that all the words that belong to the same concept should appear as well*. However, as only one word representing a concept is usually present in a given context (bearing in mind that the contexts are generally limited to few words around the base word), context vectors fail to integrate all the words related to a given concept.

We thus propose here to rely on Formal Concept Analysis (FCA) [5] to extract concept vectors that capture all the words related to closed formal concepts associated with the base word. Such concept vectors can then be used to enrich the word embedding based context vectors and improve the standard approach to bilingual lexicon extraction from comparable corpora.

The remainder of the paper is organized as follows: Sect. 2 presents the related work on bilingual lexicon extraction from comparable corpora. Section 3 details our FCA based approach to enhance bilingual lexicon extraction through the construction of concept vectors. An experimental study illustrating the gain given by concept vectors is then presented in Sect. 4. Lastly, Sect. 5 concludes the paper and provides some perspectives to this work.

---

[3] https://code.google.com/p/word2vec/.

[4] One can also translate each element of the source context vectors into the target language.

2   Related Work

Most studies addressing the task of bilingual lexicon extraction from comparable corpora are based on the standard approach, that was developed at the end of the 1990s [4]. They can be classified in three families of approaches, as follows:

1. *Statical approaches:* They are based on contextual windows centered around the base word. The relation between a base word and its context is thus mainly a co-occurrence relation, within a relatively short window. In [15], authors discuss how the window size can be set, whereas several approaches have tried to improve over the standard approach, either by incorporating a third language as pivot language [9] or considering different types of corpora (specialized or general corpora) and different corpora sizes (balanced and unbalanced corpora) [12]. Recently, context vectors are extracted using *word embedding* techniques. These latter represent each word as a d-dimensional vector of real numbers, and vectors that are close to each other are shown to be semantically related. In particular, in [11], authors proposed an efficient embedding algorithm that provides state-of-the-art results on various linguistic tasks. It was popularized via *word2vec*, a toolkit for creating word embeddings vectors, as used in [6] to extract context vectors.
2. *Syntactic approaches:* In these approaches, the context is formed by the words with which the base word is syntactically related [13]. The main motivation is to rely on less noisy context. However, the restriction to syntactic dependencies worsens the sparsity problem mentioned above and such approaches can only be deployed on very large corpora.
3. *Hybrid approaches:* Such approaches aim at a trade-off between the preceding approaches. In [1], authors proposed a combinatorial approach between a co-occurrence and a syntactic contextual representation. It combines four statistical models and compares the lexical dependencies to identify candidates translations. We notice that such approaches again suffer from the same sparsity problem.

Our approach can be seen as a kind of hybrid approach, since it aims at complementing word embedding based context vectors with additional knowledge represented as closed formal concepts of terms. Unlike to syntactic approaches, that rely on a syntactic analysis to extract this additional information, our approach is based on formal concept analysis that neither necessitates syntactic parsers (not readily available for many languages) nor very large corpora.

In the next section, we detail the extraction and use of concept vectors.

3   An FCA-Based Approach for Bilingual Lexicon Extraction

In the remainder, $\mathcal{C}^{comp} = \mathcal{C}^s \cup \mathcal{C}^t$ will denote a comparable corpus, usually unbalanced in the amount of source ($\mathcal{C}^s$) and target ($\mathcal{C}^t$) texts. More generally, the notation $\mathcal{C}$ will be used to refer to any monolingual corpus, of source or target language. As aforementioned, we propose here to rely on concept vectors in addition to context vectors, for bilingual lexicon extraction from comparable corpora. Our proposed method is based on the following steps:

1. **Context vector extraction:** We use the *word2vec* toolkit in order to extract context vectors, by learning 300-dimensional representations of words with *The Skip-Gram* model introduced by *Mikolov et al.* in [11]. This latter induces vector based word representations by using the current word to predict the surrounding window of context words with a neural network. We have relied in this study on standard parameter settings, without trying to optimize parameter values: the size of the contextual window is equal to 5 [7] and the subsampling option to 1e-05; additionally, the negative sampling method is used to estimate the probability of a target word. Note that a word not occurring in the context of a base word receives a weight of 0. One finally obtains, for each word t, a vector as follows: $\vec{V}_{t_{(t \in C)}} = (w_1, \ldots, w_{|C|})^T$, where T denotes the transpose, $|C|$ the number of words in C and w_i the weight of the association of the ith word with t, measured here by the cosine between the learned word representations. Examples of French, English and Italian context vectors computed from the comparable corpus SDA95_french, GlasgowHerald95 and SDA95_italian (CLEF'2003) are given in Table 1.

2. **Concept vector extraction:** This step, which is be described in detail in Sect. 3.2, consists in mining formal closed concepts from comparable corpora, from which concept vectors are built.

3. **Similarity calculation:** Once the source context and concept vectors are translated, dimension by dimension, the weights being preserved, to the target language, we make use of a combined cosine similarity between context and concept vectors in order to compute translation candidates for each source word.

Table 1. Examples of *English, French and Italian context vectors*, extracted from the comparable corpora SDA95_french, GlasgowHerald95 and SDA95_italian (CLEF'2003)

French context vectors
économie = {mondiale, nationale, marché, croissance, crise, ... }
politique = {classe, internationale, relations, vie, européenne, ... }

English context vectors
economy = {problems, global, crisis, difficulty, stability, ... }
politic = {international, european, social, finance, strategy, ... }

Italian context vectors
economia = {crescita, stabilita, indicatore, difficoltà, competitivita, ... }
politica = {internazionale, sociale, orientamento, problametica, europeo, ... }

We now address the problem of extracting concept vectors from texts. To do so, we first adapt the fundamental elements of Formal Concept Analysis (FCA) presented in [5] to our setting.

3.1   Key FCA Elements

First, we formalize an extraction context made up of documents and index terms, called textual context.

Definition 1. A *textual context* is a triplet $\mathfrak{M} := (\mathcal{C}, \mathcal{T}, \mathcal{I})$ **where:**

- $\mathcal{C} := \{d_1, d_2, \ldots, d_n\}$ *is a finite set of n documents.*
- $\mathcal{T} := \{t_1, t_2, \ldots, t_m\}$ *is a finite set of m distinct words in the corpus. The set $\mathcal{T}$ comprises the words of the different documents in $\mathcal{C}$.*
- $\mathcal{I} \subseteq \mathcal{C} \times \mathcal{T}$ *is a binary (incidence) relation. Each couple $(d, t) \in \mathcal{I}$ indicates that document $d \in \mathcal{C}$ contains term $t \in \mathcal{T}$.*

In the following, we recall basic definitions of the Galois lattice-based paradigm in FCA [5] and its applications to closed concepts mining.

Definition 2. *A **concept** $C = (T, D)$ is defined by two sets, a set of terms T and a set of documents D, respectively called "intension" and "extension" of the concept, and such that all terms in T co-occur in all documents of D. The support of C in $\mathfrak{M}$ is equal to the number of documents in $\mathcal{C}$ containing all the term of T. The support is formally defined as follows[5]:*

$$Supp(C) = |\{d | d \in \mathcal{C} \ \wedge \ \forall \, t \in T : (d, \ t) \in \mathcal{I}\}| \qquad (1)$$

A concept is said *frequent* (*aka large* or *covering*) if its terms co-occur in the corpus a number of times greater than or equal to a user-defined support threshold, denoted *minsupp*. Otherwise, it is said *unfrequent* (*aka rare*).

Definition 3. *Galois Closure Operator Let $C = (D, T)$ be a concept. Two functions are defined in order to map sets of documents to sets of terms and vice versa:*

$$\Psi : \mathcal{P}(\mathcal{T}) \rightarrow \mathcal{P}(\mathcal{C}) \ \ and \ \ \Psi(T) := \{d | d \in \mathcal{C} \ \wedge \ \forall \, t \in T : (d, \ t) \in \mathcal{I}\} \qquad (2)$$

$$\Phi : \mathcal{P}(\mathcal{C}) \rightarrow \mathcal{P}(\mathcal{T}) \ \ and \ \ \Phi(D) := \{t | t \in \mathcal{T} \ \wedge \ \forall \, d \in D : (d, \ t) \in \mathcal{I}\} \qquad (3)$$

where $\mathcal{P}(X)$ denotes the power set of X. Both functions Ψ and Φ constitute Galois operators. $\Psi(T)$ is equal to the set of documents containing all the word of T. Its cardinality is then equal to $Supp(T)$. On the other hand, $\Phi(D)$ is equal to the set of words appearing in all the documents of D. Consequently, the compound operator $\Omega := \Phi \circ \Psi$ is a Galois closure operator which associates to a set of words T the set of words which appear in all the documents in which the words of T co-occur.

---

[5] In this paper, we denote by $|X|$ the cardinality of the set X.

Thus, according to the Galois closure operator, a closed concept is defined as follows:

Definition 4. *A concept $C = (D, T)$ is said to be closed if $\Omega(C) = C$. A closed concept is then the maximal set of common words to a given set of documents. A closed concept is said to be frequent w.r.t. the minsupp threshold if $Supp(C) = |\Psi(T)| \geq minsupp$ [14]. Hereafter, we denote by C_c a closed concept.*

It is worth noting that in our work, each closed concept represents a class of documents grouped by a set of representative terms. So, a closed concept represents a maximal terms group sharing the same documents and includes the most specific expression describing the associated documents.

3.2 Concept Vector Extraction from Comparable Corpora

In order to extract the most representative terms, a linguistic preprocessing is performed on the document collections by using a morpho-syntactic tagger such as TREETAGGER[6]. In this work, we focus only on the common nouns, the proper nouns and adjectives. The rationale for this focus is that nouns and adjectives are the most informative grammatical categories and are most likely to represent the content of documents [2].

The set T is thus the set of all the nouns and adjectives selected from each corpus. In order to extract closed concepts, we adapted the CHARM-L algorithm [18] to our textual context $\mathfrak{M}$, which generates all the frequent closed concepts greater than or equal to a minimal threshold of the support. This latter which allows eliminating marginal terms occurring in few documents, is a user-tuned parameter that can be set according to the word distributions in the considered corpus (*e.g.* by selection from the Zipfian word distribution). At the end of this step, all source and target closed concepts whose support is greater than or equal to *minsupp* are extracted. Examples of such closed concepts obtained from the comparable corpus SDA95_french, GlasgowHerald95 and SDA95_italian (CLEF'2003) are given in Table 2.

Let $\#(Cc)$ denote the number of closed concepts in $\mathcal{C}$. For any term t and any closed concept Cc in $\mathcal{C}$, we propose to associate a weight $\mu(t, Cc)$ reflecting the importance of t in Cc. We rely here on the weight proposed in [3] and that is based on the weighting schema $tf \times idf$ [17].

The concept vector for t is then defined by: $\overrightarrow{Vc_t} = (\mu(t, Cc_1), .., \mu(t, Cc_{\#(Cc)}))^T$. As μ is null when the term does not co-occur with all the terms in the intension of a closed concept, only the closed concepts *containing* t are taken into account. The extracted concept vectors are then combined with context vectors. We propose in this paper two possible combinations, which will enable us to weigh the impact of each vector type on the final result. We rely here on [12]:

---

[6] http://www.ims.uni-stuttgart.de/projekte/corplex/TreeTagger/.

Table 2. Examples of English, French and Italian closed concepts extracted from the comparable corpus SDA95French, GlasgowHerald95 and SDA95Italian (CLEF'2003)

French Closed Concepts
{ économie, chômage, dépenses, initiative, croissance ,...}, {125, ..., 39002}
{ politique, famille, politicien, résponsabilité, financement, ... }, {20, ..., 45910}

English Closed Concepts
{$economy, european, development, sanctions, unemployment, ...$}, {44,...,55411}
{$politic, corruption, political, funding, class, ...$}, {50,...,54409}

Italian Closed Concepts
{$economia, europeo, disoccupazione, finanziario, migratorio, ...$}, {10,...,52411}
{$politica, corruzione, politico, finanziamento, internazionale, ...$}, {28,...,56409}

1. *Direct combination*: For each base word, a single vector of dimension $C + \#(Cc)$ is build from its concept and context vectors, denoted in the following as the *combined* vector. It contains both local co-occurrences information provided by the context vector, and global information provided by the concept vector. Once the source combined vectors have been translated, they are compared to target combined vectors via the standard cosine similarity measure;

2. *Weighted combination*: Context and concept vectors are treated as distinct vectors, which are translated separately and which are then compared via a weighted linear combination. Thus, similarity between two words t (from the source corpus) and t' (from the target corpus) is assessed as follows: $SIM(t,t') = \lambda cos(\overrightarrow{V}_t^{trans}, \overrightarrow{V}_{t'}) + (1-\lambda)cos(\overrightarrow{Vc}_t^{trans}, \overrightarrow{Vc}_{t'})$, where $\lambda \in [0;1]$ is a parameter weighing the relative importance of context and concept vectors, which can be learned by *e.g.* cross-validation (see Sect. 4); "trans" denotes here that the vector has been translated, meaning that the weight of a source word is transferred to its translation(s), as provided by a bilingual dictionary.

4   Experiments and Results

We evaluate our bilingual lexicon extraction approach on two different corpora:

1. *CLEF'2003*, which is a subset of the multilingual collection used in the Cross-Language Evaluation Forum CLEF'2003[7]. We rely here on the news articles (from newspapers of news agencies) SDA95 in French (42,615 documents), GlasgowHerald95 in English (56,472 documents), and SDA95 in Italian (48 980 documents). This corpus is an *unspecialized* comparable corpus;
2. *The Breast Cancer corpus*, which is a specialized, unbalanced corpus composed of documents collected from the Elsevier website[8]. The documents

[7] http://www.clef-campaign.org/.
[8] www.elsevier.com.

were retrieved from the medical domain, within the sub-domain of "breast cancer". We use the same corpus as in [12]. The corpus involves 130 French documents (about 530,000 words) and 1,640 English documents (about 7.4 million words).

Each corpus has been preprocessed with TREETAGGER, only nouns and adjectives are used for building the concept vectors. Table 3 summarizes the main statistics of each corpus. As bilingual dictionary, we used the general French-English bilingual dictionary of [8] that contains 74921 entries. The number of dictionary entries that are present in CLEF'2003 is 20432, whereas it is of 6861 for Breast Cancer. We used also an Italian-English bilingual dictionary that contains 28744 entries, the number of dictionary entries that are present in CLEF'2003 is 7011. Lastly, by selection from the Zipfian word distribution; we set *minsupp* for concept vector extraction to 30 for CLEF'2003, and 20 for Breast Cancer to focus on informative closed concepts (as low values of *minsupp* tend to yield non-informative concepts).

Table 3. Statistical features of the comparable corpora CLEF'2003 and BREAST CANCER after preprocessing

Corpora	Features	English	French	Italian
CLEF' 2003	Number of documents	56472	42615	48980
	Vocabulary size	227301	105010	123259
Breast cancer	Number of documents	1640	130	-
	Vocabulary size	93653	12411	-

4.1 Evaluation

It is difficult to compare the results of different state of the art works in bilingual lexicon extraction from comparable corpora, due to differences between the used corpora or the linguistic resources such as bilingual dictionaries [15]. Nowadays, no dataset that can serve as a reference has been set up. For this reason, we use the results of the standard approach based context vectors, as a reference, and we evaluate the performance of our approach using precision (P), recall (R), $F1$ score and Mean Average Precision (MAP) as defined in [12] at the top N terms of the ranked candidates translations list. We will compare our achieved results on the Breast Cancer corpus to those of [12].

The precision assesses the proportion of lists containing the correct translation, whereas the recall gives the proportion of translations that are recovered in the candidate list. The $F1$ score is the harmonic mean between precision and recall. In case of multiple translations, a list is deemed to contain the correct translation as soon as one of the possible translations is present [8]. The MAP defined in [10], is used to show the ability of the algorithm to precisely rank the selected candidate translations. Assuming the total number of English words in the reference list is m, let r_i be the rank of the first correct translation in the candidate translation list for the ith term in the evaluation set. The MAP is then defined by: $\mathrm{MAP}(m) = \frac{1}{m} \sum_{i=1}^{m} \frac{1}{r_i}$, with the convention that if the correct translation does not appear in the top N candidates, $\frac{1}{r_i}$ is set to 0. The MAP is our primary measure to compare proposed methods.

To evaluate the quality of the lexicons extracted, we used 10-fold cross-validation for CLEF'2003 and the reference list of 169 French/English single words used in [12] for Breast Cancer. We divided the bilingual dictionaries into 3 parts, namely: 10% of the source words together with their translations were randomly chosen and used as the test set, 10% of the source words together with their translations were randomly chosen and used as the validation set for learning the parameter λ for weighting the combined model and the rest is devoted to the training corpus on which the context and concept vectors are extracted.

We notice that source words not present in source context/concept vectors or with no translation in target context/concept vectors are excluded from the evaluation and validation set. The value of λ obtained through cross-validation on CLEF'2003 is 0.7 for $SDA95\_French$ and GlasgowHerald95 (FR-EN) and 0.6 for $SDA95\_Italian$ and GlasgowHerald95 (IT-EN). For Breast Cancer the value of λ is 0.6. We make use of this values in all our experiments. Table 4 shows the evolution of different evaluation measures (precision, recall, $F1$ score and MAP) for N equal to 200 according to different values of λ for the two corpora (the evaluation is computed on the test set for CLEF'2003). We notice that, the values obtained through cross-validation fall within the range ($[0, 6; 0, 8]$) of the best possible values on each corpus as shown by the values in bold in Table 4.

Table 4. Evolution of performance acc. to λ on CLEF'2003 and Breast Cancer

	0.0	0.1	0.2	0.3	0.4	0.5	0.6	0.7	0.8	0.9	1.0
CLEF'2003 corpus (FR-EN)											
Precision	0.190	0.254	0.312	0.380	0.387	0.393	0.396	**0.397**	0.395	0.394	0.391
Recall	0.112	0.146	0.195	0.237	0.239	0.244	0.245	0.249	**0.251**	0.252	0.226
$F1$ score	0.140	0.185	0.240	0.273	0.295	0.301	0.302	**0.306**	0.306	0.304	0.286
MAP	0.106	0.125	0.146	0.162	0.163	0.170	0.170	**0.172**	0.168	0.166	0.155
Breast cancer corpus (FR-EN)											
Precision	0.389	0.462	0.501	0.526	0.530	0.532	**0.537**	0.534	0.533	0.531	0.531
Recall	0.311	0.366	0.403	0.422	0.424	0.425	**0.426**	0.426	0.434	0.431	0.430
$F1$ score	0.357	0.408	0.446	0.470	0.469	0.472	**0.475**	0.473	0.478	0.475	0.475
MAP	0.227	0.263	0.300	0.317	0.320	0.326	**0.328**	0.327	0.324	0.322	0.321
CLEF'2003 corpus (IT-EN)											
Precision	0.239	0.331	0.384	0.448	0.464	0.471	**0.476**	0.476	0.472	0.470	0.466
Recall	0.121	0.169	0.221	0.232	0.237	0.241	0.243	**0.244**	0.242	0.238	0.230
$F1$ score	0.160	0.223	0.280	0.305	0.313	0.319	**0.322**	0.321	0.320	0.316	0.308
MAP	0.148	0.171	0.189	0.201	0.204	0.208	**0.210**	0.209	0.207	0.201	0.194

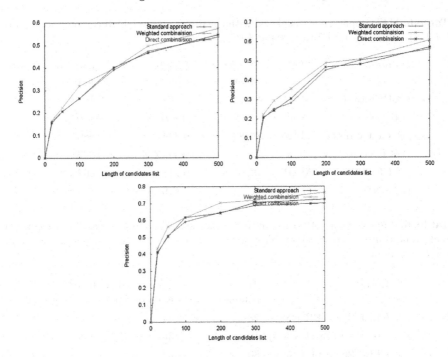

Fig. 1. Results (precision) of the standard approach, direct and weighted combinations using the best value of λ for CLEF'2003 (FR-EN) (left), Breast Cancer (FR-EN) (right) and CLEF'2003 (IT-EN)(down)

4.2 Results and Discussion

Figure 1 highlights the different precision values obtained with respectively the standard approach, the weighted combination and the direct combination, using the best value of λ for the two comparable corpora. We consider different lengths N of the candidates list, varying from 20 to 500. We can see that the weighted combination approach significantly outperforms the direct combination approach. Indeed, the overall precision of the weighted combination approach increases, especially for medium and large lengths of candidates list as $N = 100$, 300 and 500. The method seems to become effective in the case where the length of the candidate list grows, this can be explained by the fact that the probability of obtaining correct translations increases with the candidate list growth. Moreover, we notice that the direct combination is more sensitive to noise that can be caused by further information compared to the weighted combination. This is due to the fact that the direct combination directly modifies the context vectors, whereas the weighted combination allows one to better control the impact of concept vectors.

Table 5 displays the results obtained in terms of MAP. From these results, we notice that the overall MAP is improved by the concept vectors on the two comparable corpora (+8.6% for CLEF'2003 (FR-EN), +8.9% for CLEF'2003

Table 5. MAP improvement achieved with the direct and weighted combinations using the best value of λ for CLEF'2003 and Breast Cancer

Corpora	Measure	Standard approach	Weighted combination	Direct combination
CLEF'2003	MAP	0.161	0.175	0.164
(FR-EN)	Gain	-	**+8.6 %**	**+1.9 %**
Breast cancer	MAP	0.383	0.424	0.389
(FR-EN)	Gain	-	**+10.7 %**	**+1.5%**
CLEF'2003	MAP	0.201	0.218	0.202
(IT-EN)	Gain	-	**+8.9 %**	**+0.4 %**

Table 6. Improvement of vector average size achieved after translation of context vector and context vector combined with concept vector for CLEF'2003 and the Breast Cancer corpora

Corpora	Translated context vector	Translated context and concept vector
CLEF'2003 (FR-EN)	28	41
Breast cancer (FR-EN)	19	36
CLEF'2003 (IT-EN)	30	44

(IT-EN) and +10.7% for Breast Cancer (FR-EN)). This demonstrates that the information in concept vectors is relevant to represent words in a bilingual lexicon extraction setting: the quality of the extracted bilingual lexicons improves with the integration of concept vectors. Moreover, experiments show that the Breast Cancer corpus yields better results than CLEF'2003 corpus, which can be explained by the fact that the vocabulary used in the breast cancer field is more specific and less ambiguous than the one used in journalistic corpora. For Breast Cancer, the obtained results (0.424% for the weighted combination) are comparable to the ones reported in [12] (0.423% is the best MAP assessed on the unbalanced version of the corpus). It is worth noting here that we do not use the same bilingual dictionary and that the two approaches are different and could efficiently complement each other.

As we conjectured earlier, our approach allows to obtain a representation of vectors that is less sparse than the one obtained with context vectors. Indeed, as shown in Table 6, the average size of vectors for CLEF'2003 has increased from 28 to 41 words (FR-EN) and from 30 to 44 (IT-EN) when considering concept vectors, whereas it has increased from 19 to 36 words for Breast Cancer. This increase is important and shows that the similarity between a base word and its candidate translations relies on more information. This information is valuable as illustrated in the results discussed above.

5 Conclusion

In this paper, we proposed a new approach for bilingual lexicon extraction based on formal closed concepts. The extracted concepts are then used to build concept vectors that complement the embedding based context vectors used in bilingual lexicon extraction from comparable corpora. The experimental study, conducted on two comparable corpora (a specialized one from the medical domain and a general one made of news articles), showed that the concept vectors retained provide a partial solution to the sparsity problem encountered with context vectors. Furthermore, the quality of the lexicons extracted with both concept and context vectors is higher than the quality of the lexicons extracted with only context vectors. In the future, we plan to investigate the combination of concept vectors and context vectors also obtained from a syntactic analysis.

References

1. Andrade, D., Matsuzaki, T., Tsujii, J: Effective use of dependency structure for bilingual Lexicon creation. In: Gelbukh, A. (ed.) CICLing 2011. LNCS, vol. 6609, pp. 80–92. Springer, Heidelberg (2011). doi:10.1007/978-3-642-19437-5_7
2. Barker, K., Cornacchia, N.: Using noun phrase heads to extract document keyphrases. In: Hamilton, H.J. (ed.) AI 2000. LNCS (LNAI), vol. 1822, pp. 40–52. Springer, Heidelberg (2000). doi:10.1007/3-540-45486-1_4
3. Chebel, M., Latiri, C., Gaussier, E.: Extraction of interlingual documents clusters based on closed concepts mining. In: 19th International Conference KES 2015, Singapore, pp. 537–546 (2015)
4. Fung, P.: A statistical view on bilingual Lexicon extraction: from parallel corpora to non-parallel corpora. In: Farwell, D., Gerber, L., Hovy, E. (eds.) AMTA 1998. LNCS (LNAI), vol. 1529, pp. 1–17. Springer, Heidelberg (1998). doi:10.1007/3-540-49478-2_1
5. Ganter, B., Wille, R.: Formal Concept Analysis. Springer, Heidelberg (1999)
6. Baroni, M., Georgiana, D., Kruszewski, G.: Don't count, predict! a systematic comparison of context-counting vs. context-predicting semantic vectors. In: 52nd Annual Meeting ACL 2014, Baltimore, Maryland (2014)
7. Laroche, A., Langlais, P.: Revisiting context-based projection methods for term-translation spotting in comparable corpora. In: 23rd International Conference COLING 2010, Beijing, China, pp. 617–625 (2010)
8. Li, B., Gaussier, E.: An information-based cross-language information retrieval model. In: Baeza-Yates, R., Vries, A.P., Zaragoza, H., Cambazoglu, B.B., Murdock, V., Lempel, R., Silvestri, F. (eds.) ECIR 2012. LNCS, vol. 7224, pp. 281–292. Springer, Heidelberg (2012). doi:10.1007/978-3-642-28997-2_24
9. Linard, A., Daille, B., Emmanuel, M.: Attempting to bypass alignment from comparable corpora via pivot language. In: 8th Workshop on BUCC, Beijing, pp. 32–37 (2015)
10. Manning, C.D., Raghavan, P., Schütze, H.: Introduction to Information Retrieval. Cambridge University Press, Cambridge (2008)
11. Mikolov, T., Sutskever, I., Chen, K., Corrado, G.S., Dean, J.: Distributed representations of words and phrases and their compositionality. In: NIPS, vol. 2013, pp. 3111–3119 (2013)

12. Morin, E., Hazem, A.: Looking at unbalanced specialized comparable corpora for bilingual Lexicon extraction. In: ACL 2014, Baltimore, USA, pp. 284–293 (2014)

13. Gamallo Otero, P.: Comparing window and syntax based strategies for semantic extraction. In: Teixeira, A., Lima, V.L.S., Oliveira, L.C., Quaresma, P. (eds.) PROPOR 2008. LNCS (LNAI), vol. 5190, pp. 41–50. Springer, Heidelberg (2008). doi:10.1007/978-3-540-85980-2_5

14. Pasquier, N., Taouil, R., Bastide, Y., Stumme, G., Lakhal, L.: Generating a condensed representation for association rules. J. Intell. Inf. Syst. **2005**, 29–60 (2005)

15. Prochasson, E., Morin, E.l., Kageura, K.: Anchor points for bilingual Lexicon extraction from small comparable corpora. In: Machine Translation Summit, France (2009)

16. Ronan, C., Jason, W.: A unified architecture for natural language processing: deep neural networks with multitask learning. In: ICML2008, pp. 160–167 (2008)

17. Salton, G., Buckley, C.: Term-weighting Approaches in Automatic Text Retrieval. Information Processing Management. Pergamon Press Inc, Tarrytown (1988)

18. Zaki, M.J., Hsiao, C.J.: Efficient algorithms for mining closed itemsets and their lattice structure. IEEE Trans. Knowl. Data Eng. **17**, 462–478 (2005)

Collective Geographical Embedding
for Geolocating Social Network Users

Fengjiao Wang[1(✉)], Chun-Ta Lu[1], Yongzhi Qu[2], and Philip S. Yu[1]

[1] Univeristy of Illinois at Chicago, Chicago, USA
{fwang27,clu29,psyu}@uic.edu
[2] Wuhan University of Technology, Wuhan, China
quwong@whut.edu.cn

Abstract. Inferring the physical locations of social network users is one of the core tasks in many online services, such as targeted advertisement, recommending local events, and urban computing. In this paper, we introduce the Collective Geographical Embedding (CGE) algorithm to embed multiple information sources into a low dimensional space, such that the distance in the embedding space reflects the physical distance in the real world. To achieve this, we introduced an embedding method with a location affinity matrix as a constraint for heterogeneous user network. The experiments demonstrate that the proposed algorithm not only outperforms traditional user geolocation prediction algorithms by collectively extracting relations hidden in the heterogeneous user network, but also outperforms state-of-the-art embedding algorithms by appropriately casting geographical information of check-in.

Keywords: Geolocation · Geometrical embedding · Geometric regularization

1 Introduction

Urban computing has attracted many research attentions [22]. Cross-domain data can be fused together to aid this task [19,21]. One of the core tasks towards these services is to infer the physical location of participants, as it not only advances the recognition of individual behavioural patterns but also facilitates the analysis of the crowd mobility and communication.

Intuitively, friendships between users provide a valuable hint since people tend to live close to their friends. As a partial observation of users' social relations, online social networks (OSNs) shed a light on the problem of geolocating individuals [9,14]. Another useful information is the online footprints shared in OSNs, which can be observed through the geotagged contents generated by users. Unfortunately, most of existing approaches only focus on one single data source, either the social network of the online friendships [7,8] or the content of the online footprints [1,4]. There are several crucial challenges that hinder the performance of the existing methods: (1) **Data Sparsity:** Due to privacy

© Springer International Publishing AG 2017
J. Kim et al. (Eds.): PAKDD 2017, Part I, LNAI 10234, pp. 599–611, 2017.
DOI: 10.1007/978-3-319-57454-7_47

concern, not many users choose to reveal their location information. Research in Twitter suggest that only 16% of users registered city level locations in their profiles [12], and the percentage of tweets with geographical coordinates was merely 1% [18]. (2) **Noisy Signals:** The signals retrieved from OSNs may not conform the assumption that the friends and footprints of a user will be close to the user's physical location. Reasons lead to noisy signals include global online friendships, frequent relocation, and posting geotagged contents during travel, etc. Such sparse and noisy data constitute a major challenge for label propagation based methods [7,8] and probability estimation based methods [2]. (3) **Scalability:** Since OSNs often contain millions nodes and links, how to handle such a large scale data poses another challenge. In particular, most methods that involve sophisticated NLP techniques [1] require a huge amount of computational resources and may not be applicable to large-scale datasets.

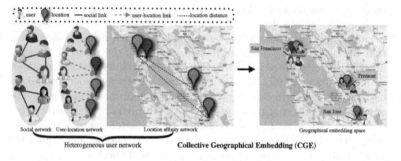

Fig. 1. Example of learning the geographical embedding space from heterogeneous networks.

Recently, network embedding techniques [5,16,17] are introduced to embed network data into a low dimensional space while preserving the neighborhood closeness of the network data. Through embedding all objects into a common low dimensional space, it is possible to calculate the similarity between each pair of objects to mitigate the sparsity problem in network data. Although several studies [5,17] have been proposed to model multiple networks concurrently, these methods do not differentiate each type of the objects involved. Furthermore, the embeddings learned by the existing methods do not have any physical meanings.

Since each tagged location is associated with a geographic coordinate (e.g., latitude and longitude), the distance between the embeddings of any pair of locations should be able to reflect the geographical distance. In this paper, we propose a Collective Geometrical Embedding (CGE) algorithm that can effectively infer the geolocation of social network users, by jointly learning the embeddings of users and check-ins with respect to the real-world geometrical space. In other words, the real geometrical distance between any pair of objects (i.e., users or locations) is resembled by euclidean distance of two vectors in the low dimensional space. Figure (1) illustrates the main concept of the geometrical embedding learning, where the left figure shows an example of a heterogeneous user network, the right figure depicts a snapshot of the geographical embedding space learned through the proposed algorithm. The heterogeneous user network shown includes a user network, a user-location network, and a location affinity network. By collectively embedding the heterogeneous network into a common subspace while preserving the geometrical distances between users and locations, the goal of inferring users' geolocations can be achieved without difficulty.

The main contributions of this paper can be summarized as follows:

1. We directly leverage multiple information sources by embedding a heterogeneous network, which alleviates the problem of sparse and noisy data.
2. We propose a collective geometrical embedding (CGE) method that integrates the geometrical regularization into the process of network embedding, which makes the learned embeddings preserving not only the neighborhood closeness of network data but also the geometrical closeness of locations. To the best of our knowledge, this work is the first to learn an embedding space that can reflect the real-world geolocation characteristics.
3. Through the extensive empirical studies on real-world datasets, we demonstrate that the proposed CGE method significantly outperforms other state-of-the-art algorithms in addressing the problem of geolocating individuals.

2 Preliminaries

In this section, we first introduce the definition of each source for the heterogeneous network and present the problem statement of this study.

Definition 1. Social Network *A social network can be represented by $G_{uu} = (\mathcal{U}, \mathcal{E}_{uu})$, where $\mathcal{U} = \{u_1, u_2, ...u_N\}$ denotes the set of users, and $\mathcal{E}_{uu}$ denotes the set of edges. Each $e_{ij} \in \mathcal{E}_{uu}$ is a social link between user i and user j.*

Next, we present the definition of user-location network, in which the frequency of visit was used to set the weight of edges between users and locations.

Definition 2. User-Location Network *A user-location network is represented by $G_{up} = (\mathcal{U} \cup \mathcal{P}, \mathcal{E}_{up})$, where $\mathcal{U} = \{u_1, u_2, \ldots, u_N\}$ denotes the set of users, $\mathcal{P} = \{p_1, p_2, \ldots, p_M\}$ denotes the set of locations, and the weight w_{ik} on the edge $e_{ik} \in \mathcal{E}_{up}$ is the number of times that the user u_i visited the location p_k.*

Definition 3. Location Affinity Network *A location affinity network can be represented by $G_{pp} = (\mathcal{P}, \mathcal{E}_{pp})$, where $\mathcal{P} = \{p_1, p_2, \ldots, p_M\}$ denotes the set of locations, and the weight w_{ij} on the edge $e_{ij} \in \mathcal{E}_{pp}$ indicates the location closeness between the locations p_i and p_j.*

Definition 4. Heterogeneous User Network *A heterogeneous user network can be represented by $G_u = G_{uu} \cup G_{up} \cup G_{pp}$, which consists of the social network G_{uu}, the user-location network G_{up} and the location affinity network G_{pp}. The same sets of users and locations are shared in G_u.*

Definition 5. Geolocating Social Network Users *Given a heterogeneous user network G_u, estimate a location $\hat{p}_{u_i}$ for each user u_i in $\mathcal{U}$ such that the estimated location $\hat{p}_{u_i}$ close to u_i's physical location p_{u_i}.*

3   Methodology

In this section, we introduce the proposed method that learns the geographical embeddings of users and locations through the heterogeneous user network w.r.t. the real-world geometrical space. Since the heterogeneous user network consists of multiple bipartite networks, we first present how to learn the network embedding from a single bipartite network.

3.1   Bipartite Network Embedding

Given a bipartite network $G = (\mathcal{V}_A \cup \mathcal{V}_B, \mathcal{E})$, the goal of network embedding is to embed each vertex $v_i \in \mathcal{V}_A \cup \mathcal{V}_B$ into a low dimensional vector $\boldsymbol{v}_i \in \mathbb{R}^d$, where d is the dimension of the embedding vector. Inspired by [17], we consider to learn the embeddings by preserving the second-order proximity, which means two nodes are similar to each other if they have similar neighbors. In the following, we take the user-location network $G_{up} = (\mathcal{U} \cup \mathcal{P}, \mathcal{E}_{up})$ as an example to illustrate the learning process of embeddings. To begin with, we use a softmax function to define the conditional probability of a user $u_i \in \mathcal{U}$ visits a location $p_j \in \mathcal{P}$:

$$P(p_j|u_i) = \frac{e^{\boldsymbol{p}_j^T \boldsymbol{u}_i}}{\sum_{k=1}^{M} e^{\boldsymbol{p}_k^T \boldsymbol{u}_i}} \tag{1}$$

To preserve the weight w_{ui} on edge e_{ui}, we make the conditional distribution $P(\cdot|u_i)$ close to its empirical distribution $\hat{P}(\cdot|u_i)$, which can be defined as $\hat{P}(p_j|u_i) = \frac{w_{ij}}{o_i}$, where $o_i = \sum_{p_k \in N(u_i)} w_{ik}$ is the out-degree of u_i, and $N(u_i)$ is the set of the u_i's neighbors, i.e., the locations that u_i have visited.

By minimizing the Kullback-Keibler (KL) divergence between two distributions $P(\cdot|u_i)$ and $\hat{P}(\cdot|u_i)$ and omitting some constants, we can obtain the objective function for embedding the bipartite graph G_{up} as follows:

$$\mathcal{J}_{up} = - \sum_{e_{ij} \in \mathcal{E}_{up}} w_{ij} \log P(p_j|u_i) \tag{2}$$

Since a homogeneous network can be easily converted to a bipartite network, we can derive similar objective for embedding social network G_{uu} as follows:

$$\mathcal{J}_{uu} = - \sum_{e_{ij} \in \mathcal{E}_{uu}} w_{ij} \log P(u_j|u_i) \tag{3}$$

By jointly learning $\{\boldsymbol{u}_i\}_{i=1,\dots,N}$ and $\{\boldsymbol{p}_j\}_{j=1,\dots,M}$ that minimize the objectives Eqs. (2) and (3), we are able to represent social network users and locations in low dimensional vectors. By far, the embeddings are learned only from the network structure. Next, we introduce the collective geometrical embedding algorithm to preserve the geometric structure w.r.t. the physical closeness in between different objects.

3.2   Collective Geometrical Embedding

According to the local invariance assumption [3], if two samples p_i, p_j are close in the intrinsic geometric with regard to the data distribution, then their embeddings $\boldsymbol{p}_i$ and $\boldsymbol{p}_j$ should also be close. In this work, we consider to preserve the geometric structure of locations by incorporating the following geometric regularization in the learning process:

$$\mathcal{R}(\mathbf{P}) = \sum_{i,j=1}^{M} w_{ij}(\boldsymbol{p}_i - \boldsymbol{p}_j)^2 \tag{4}$$

where the w_{ij} represents the geometric closeness between locations p_i and p_j, which can be obtained with the RBF kernel.

To ease the subsequent derivation, we rewrite Eq. (4) in trace form. Let matrix $\mathbf{U}$ and matrix $\mathbf{P}$ denote the user embedding matrix and the location embedding matrix, respectively, where each row within $\mathbf{U}$ and $\mathbf{P}$ is the embedding vector of a user and a location. Using the weight matrix $\mathbf{W}$ whose element w_{ij} is the weight between two locations and the diagonal matrix $\mathbf{D}$ whose elements $d_{ii} = \sum_{j=1}^{M} w_{ij}$, the Laplacian matrix $\mathbf{L}$ is defined as $\mathbf{L} = \mathbf{D} - \mathbf{W}$. Then $\mathcal{R}(\mathbf{P})$ can be reduced into the trace form:

$$\mathcal{R}(\mathbf{P}) = \frac{1}{2} \sum_{i,j=1}^{M} w_{ij}(\boldsymbol{p}_i - \boldsymbol{p}_j)^2 = \frac{1}{2} Tr(\mathbf{P}^T(\mathbf{D} - \mathbf{S})\mathbf{P}) = \frac{1}{2} Tr(\mathbf{P}^T \mathbf{L} \mathbf{P}) \tag{5}$$

To learn the geometrical embeddings from the heterogeneous user network, we minimize overall objective function as follows:

$$\min_{\mathbf{U},\mathbf{P}} \mathcal{J} = \mathcal{J}_{uu} + \mathcal{J}_{up} + \lambda \mathcal{R}(\mathbf{P}) \tag{6}$$

where λ is the regularization parameter that controls the importance of the geometric regularization.

Since the edges in different networks have different meanings and the weights are not comparable to each other, we alternatively minimize the objective of each network independently to optimize Eq. (6). The same strategy has also been applied in literature [17], while the geometrical regularization is not considered in previous works. For the objective term of each network, taking $\mathcal{J}_{up}$ as an example, it is time-consuming to directly evaluate as it requires to sum over the entire set of edges when calculating the conditional probability $P(\cdot|u_i)$. We adopt the techniques of negative sampling [13] to approximate the evaluation, where multiple negative edges are sampled from some noisy distribution. More specifically, it specifies the following objective function for each edge e_{ij}:

$$\log \sigma(\boldsymbol{p}_j^T \cdot \boldsymbol{u}_i) + \sum_{u=1}^{k} E_{p_n \sim P_n(p)} [\log \sigma(-\boldsymbol{p}_n^T \cdot \boldsymbol{u}_i)] \tag{7}$$

Algorithm 1. Collective Geographical Embedding Algorithm

Input: Heterogeneous user network $G_u = G_{uu} \cup G_{up} \cup G_{pp}$, parameter λ, the
      embedding dimension d, the maximum number of iterations $iter$;
Output: Geographical embedding matrix $\mathbf{U}$ and $\mathbf{P}$.
Initialization: user embedding $\boldsymbol{u}$, location embedding $\boldsymbol{p}$;
while $j \leq iter$ **do**
   | Sample an edge from $\mathcal{E}_{uu}$, draw k negative edges and update user
   | embeddings;
   | Sample an edge from $\mathcal{E}_{up}$, draw k negative edges and update user
   | embeddings and location embeddings;
   | Sample a location p_i from $\mathcal{P}$, update the location embedding $\boldsymbol{p}_i$ using the
   | partial derivative in Eq. 8.
end

---

where $\sigma(x) = \frac{1}{1+exp(-x)}$ is the sigmoid function, and k is the number of negative edges. The first term shows that if there is a link between vertices u_i and p_j, then force two vectors close to each other. The second term shows after sampling negative links from whole sets of vertices, force two vectors u_i and $\boldsymbol{p}_n$ far away from each other if there is no link between u_i and p_n. We set the sampling distribution $P_n(p) \propto o_i^{3/4}$ as proposed in [13], where o_i is the out-degree of vertex u_i. For the detailed optimization process, readers can refer to [16]. We can minimize the objective term of the social network, $\mathcal{J}_{uu}$, in a similar way.

As for minimizing the geometrical regularization, $\mathcal{R}(\mathbf{P})$, it is to enforce the embedding of each location to be as similar to the locations close to it as possible. Thus, we can sample a location $p_i \in \mathcal{P}$ at each iteration and update its embedding $\boldsymbol{p}_i$ by gradient descent. The gradient of $\mathcal{R}(\mathbf{P})$ w.r.t. $\boldsymbol{p}_i$ can be derived as follows:

$$\frac{\partial \mathcal{R}(\mathbf{P})}{\partial \mathbf{p}_i} = \sum_j w_{ij}(\mathbf{p}_i - \mathbf{p}_j) = (\sum_j w_{ij} - w_{ii})\mathbf{p}_i - \sum_{j \neq i} w_{ij}\mathbf{p}_j = [(\mathbf{D} - \mathbf{W})\mathbf{P}]_{i*} = [\mathbf{L}\mathbf{P}]_{i*},$$
$$(8)$$

where $[\cdot]_{i*}$ means the i-th row of the given matrix.

The detailed process of the proposed algorithm is summarized in Algorithm 1. After obtaining the geometric embeddings of users and locations, we can train any classifier (e.g., SVM or logistic regression) by feeding the embeddings as feature vectors and the associated geographic regions at the desired scale (such as city-scale or state-scale) as the labels.

4 Experiments

4.1 Experiment Setup

To evaluate the performance of the proposed CGE algorithm, we conduct extensive experiments on the following two datasets. The statistics of each dataset is summarized in Table 1. For both datasets, the social network is

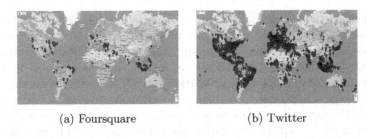

(a) Foursquare   (b) Twitter

Fig. 2. Distribution of users' locations in Foursquare and Twitter networks.

constructed from bi-directional friendships between social network users, user-location network is constructed by the users' check-in logs, and users' physical locations reported in their profiles are used as ground truth. We aim to predict users' home location to the city level, since many users only report city-level addresses. City-level location information in text format is converted into city-level coordinates according to geolocators[1]. Note that such coordinates are being canonicalized with each city district corresponding to exactly the same coordinate. Distribution of users' home locations in two datasets is shown in Fig. 2. Instead of only focusing on users lived in the US, we are tackling users globally, which creates more challenge for the learning task.

Table 1. Datasets

Dataset	Users	Locations	Social links	User-location links
Foursquare	15,799	141,444	38,197	212,588
Twitter	25,355	403,770	156,060	564,298

We compared the proposed approach with three state-of-the-art user geolocation prediction algorithms and two network embedding algorithms.

1. **FIND** [2] selects the location that maximizes the probability of friendships given the distance between the location candidates and the friends' home locations.
2. **LP** [7] selects the most popular location among the given user's friends' home locations by a simple majority voting algorithm, while the user's friends network were rebuilt via the depth-first search algorithm.
3. **SLP** [8] refers to Spatial Label Propagation. It spatially propagates location labels through the social network, using a small number of initial locations, which is an extension of the idea of label propagation.
4. **LINE** [16] embeds a homogeneous network into a low dimensional space.
5. **PTE** [17] learns the embeddings of a heterogeneous network by joint learning the embeddings of each sub-network.
6. **CGE** is the proposed method in this paper.

---

[1] https://github.com/networkdynamics/geoinference.

To evaluate the performance of the different approaches, we randomly sample 50% of user instances as the training set and use the other 50% of user instances as the testing set. This random sampling experiment is repeated 10 times. For the FIND algorithm, three coefficients are set the same as in paper [2]. For the LP algorithm, the minimal number of friends is set to 1, the maximum number of friends is set to 10000, and the minimal location votes is set to 2. For the SLP algorithm, the number of iterations is set to 5 and the other parameters' settings follow paper [8]. For all the embedding algorithms (LINE, PTE, and CGE), the embedding dimensionality is set to 100. We tried dimensionalities in the range [50, 200] and found that 100 generally gives the best results. To simplify the comparison, we simply set the regularization parameter λ in CGE to 1. For the other parameters in the network embedding algorithms, we follow the setting in the paper [17]. The learned embeddings are used as feature vectors to train an SVM classifier with the RBF kernel.

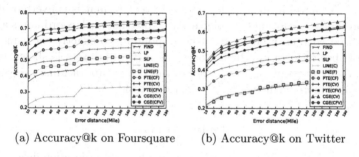

(a) Accuracy@k on Foursquare   (b) Accuracy@k on Twitter

Fig. 3. Performance comparison on Foursquare and Twitter datasets

To study the contribution of different sources, different combinations of sub-networks in the heterogeneous user network are fed into the algorithms as denoted in the following manner. For CGE taking three networks as inputs, we denote this setting as CGE(CFV), where **C** (check-in) stands for user-location network, **F** (friend) denotes friendship network, and **V** (venue) represents location affinity network. If only one or two networks were taken as inputs, we denote them as (C) or (CV), etc.

Three metrics are used to evaluate the performance of the compared methods. The first metric is **Accuracy**@k, which measures the percentage of predictions that are within k miles of the true location. We report multiple values of k to compare different approaches in a comprehensive manner. The second metric is **Average Error Distance (AED)**, where a smaller value of which indicates better performance. The third metric is **Area Under Curve (AUC)** under a cumulative distribution function $F(x) = P(distance \leq x)$, where $F(x)$ shows the percentage of inferences having an error distance less than x miles away from the true location [9]. Higher AUC scores indicate better performance.

4.2 Quantitative Results

Figure 3 shows the performance of user geolocation algorithms on two datasets. From the comparison results with regard to Accuracy@k, we make three observations as follows. Firstly, embedding-based algorithms consistently outperform non-embedding based benchmarks. For instance, if we consider Accuracy@30, in Fig. 3a, CGE(CFV) correctly predicts 66.5% of users, while the best performance of non-embedding based algorithms SLP only predicts 49.1% of users. Because embedding-based algorithms can fully explore the network structure of the given information, which alleviates the issues of sparse and noisy signals, embedding-based methods (LINE, PTE and CGE) outperform non-embedding based methods. Secondly, among embedding-based algorithms, algorithms such as PTE and CGE which are capable of handling heterogeneous networks perform better than LINE which is only applicable to homogeneous networks. Thirdly, we can observe that CGE consistently achieves the best performance in both datasets, as shown in Fig. 3a and 3b. With exactly the same amount of information, the proposed CGE always outperforms PTE for a variety of error distance k. For example, in Fig. 3a, with user-location network and location affinity network, CGE(CV) correctly predicts 61% of users' home locations within 10 miles, while PTE(CV) correctly predicts 56% of users' home locations within the same distance. These results indicate the robustness of the proposed CGE algorithm.

Table 2. The classification performance "mean ± standard deviation" on user geolocation prediction task. "↑" indicates the larger the value the better the performance. "↓" indicates the smaller the value the better the performance.

	Foursquare		Twitter	
	AED ↓	AUC ↑	AED ↓	AUC ↑
LP	2526.21 ± 34.05	45.52% ± 0.37%	4924.64 ± 18.24	19.30% ± 0.12%
SLP	1673.31 ±0.73	61.21% ± 0.03%	2172.99 ±2.40	53.21% ± 0.04%
FIND	1805.88 ± 28.25	57.41% ± 0.39%	2647.07 ± 16.84	42.53% ± 0.20%
LINE(C)	2018.94 ± 30.15	58.60% ± 0.28%	2759.46 ± 20.62	41.92% ± 0.03%
LINE(F)	1308.49 ± 19.04	63.83% ± 0.47%	2474.04 ± 15.23	44.36% ± 0.19%
PTE(CF)	1006.31 ± 21.41	68.80% ± 0.32%	1634.34 ± 16.48	54.30% ± 0.29%
PTE(CV)	1065.06 ± 24.30	71.56% ± 0.20%	1192.38 ± 133.4	63.80% ± 1.22%
PTE(CFV)	935.17± 11.50	72.35% ± 0.19%	1247.78 ±4.79	61.26% ± 0.11%
CGE(CV)	779.94± 29.15	75.93% ± 0.35%	**991.22 ± 17.77**	**65.27% ± 0.26%**
CGE(CFV)	**773.31 ± 20.55**	**77.13% ± 0.17%**	1000.47 ±8.97	64.24% ± 0.07%

Table 2 shows the AED and AUC scores of various algorithms on two datasets. Similar observations can be made as above. CGE(CFV) algorithm achieves the smallest error distance and the highest AUC scores for the Foursquare dataset, while CGE(CV) achieves the best performance for the Twitter dataset. This is primarily due to the fact that Twitter relationships mixes friendship relationships with other kinds of unbalanced, asymmetrical relationships [7]. More importantly, when using the same data sources, CGE always performs better than PTE. This shows that the proposed graph regularization is more suitable for modeling geographical information in user geolocation problem.

To evaluate the contribution of different sub-networks, we compare the results using partial information with the results using complete information. The comparisons are performed using CGE algorithm on both datasets. As can be seen in Fig. 4a, without user-location network (green line), the performance deteriorates the most (around 19%). Without location affinity network (purple line), performance drops around 13%. Without friend network information, the algorithm drops the least compared with other cases (around 3%). Note that, without friend network information, CGE achieves slightly higher accuracy on Twitter dataset, as shown in Fig. 4b, because Twitter relationships contain heavy noise. It can be concluded that: (1) Compared with friend information and location affinity network, user-location network plays the most important role in user geolocation prediction. (2) Considering the geometrical information in location affinity network can significantly improve the prediction performance. (3) Friend network can also be a valuable complementary source.

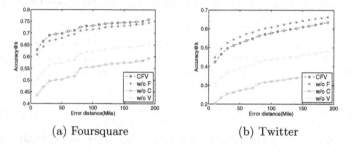

(a) Foursquare   (b) Twitter

Fig. 4. Performance contribution of sub-networks. "w/o" means without certain sub-network.

The robustness of the proposed algorithm is also tested by varying the size of the training users. Note that, when decreasing the size of the training users, we use locations' embedding vectors as additional training data to balance training samples across different settings. As can be seen in Fig. 5, when the size of the training users decreases from 50% to 20%, accuracy@k only drops around 5%. The evaluation results on the size of training set indicate that CGE(CFV) is capable of producing high-quality embedding vectors of users and locations.

Visualization of users' embedding vectors learned by different algorithms are shown in Fig. 6. Due to limited space, only the results of Foursquare dataset are shown. We pick users who reside in three different countries as three different classes. Users' embedding vectors (in 100-dimensional space) are further mapped to two-dimensional space with Isomap. Compared with other algorithms, CGE(CFV) generates the most meaningful layout, as shown in Fig. 6e,

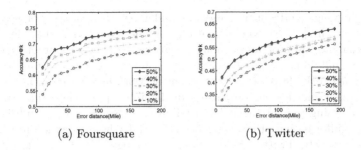

(a) Foursquare   (b) Twitter

Fig. 5. Performance comparison with varied training size.

in the sense that it naturally forms three clusters and pulls the centers of the different clusters far away from each other. This indicates that the proposed CGE algorithm leveraged different source information effectively. Running time of various algorithms are shown in Fig. 6f. The run time of CGE algorithms are modestly longer compared with other embedding methods, but provides the best prediction performance.

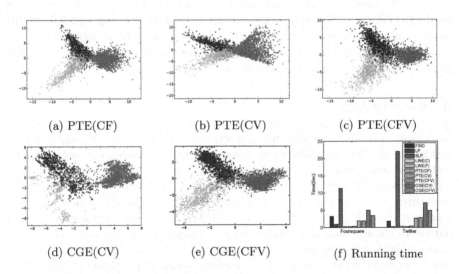

(a) PTE(CF)   (b) PTE(CV)   (c) PTE(CFV)

(d) CGE(CV)   (e) CGE(CFV)   (f) Running time

Fig. 6. Visualization of users reside in three different countries (Blue: US, Green: Brazil, Red: Malaysia) in Foursquare. Running time comparison (f). (Color figure online)

5   Related Work

Location Prediction: Works on identifying users' home locations [20] can be roughly divided into two categories based on the information used. One category of related works focus on extracting text information [1,4] from tweets. The general idea is to extract location-related text information (words, phrase, topic) through language model or probabilistic model. Another category of works focus on social graphs [2,7,8], where they rely on the assumption that tie strength is a strong indicator of users' home locations. [2] aims to predict the location of an individual by leveraging geographic and social relationships in the Facebook network. [9] reviews most recent network-based approaches, and proposes two new metrics on comparison of different approaches. [14] studies the problem of using publicly available attributes (mayorship, tips, and likes) and geographic information of locatable friends to infer home location in three networks respectively, Twitter, Foursquare, and Google+. Other works [10,11,15] consider text and network information simultaneously. [11] propose an algorithm derived from a generative model. [10,15] provide two ways of combining the results from network-based approaches and text-based algorithms. However, most of the above-mentioned algorithms were either inefficient or based on simple combination of different source information.

Network Embedding: Recently, network embedding technique ([5,6,16,17]) drew lots of attention due to the merit of distributed representation learning. Embedding objects into a mutually related common space can mitigate the sparsity problem to a large extent. Moreover, by jointly modeling multiple networks, it is able to capture complex interaction among heterogeneous objects in the connected networks. Different from existing network embedding algorithms, this paper treats the guidance information (locations' geographical information) discriminately as a geometric regularization term to smoothly encode the local geometrical structure into the embedding space.

6   Conclusion

This paper proposed a collective geometrical embedding (CGE) algorithm to tackle the problem of geolocating users. Multiple heterogeneous networks are embedded into a low dimensional space through two strategies: the first is to embed the social network and the user-location network by preserving local structures; while the other is to incorporate the geographical information as the guidance through graph regularization. Evaluation on two different real-world datasets demonstrated the effectiveness of the proposed approach. For future work, multiple types of social links and multiple types of user-location relations can be included in the proposed framework. Besides, the proposed embedding method can be further extend for location recommendation.

Acknowledgments. This work is supported in part by NSF through grants IIS-1526499, and CNS-1626432, and NSFC 61672313. Yongzhi Qu would like to acknowledge national natural science foundation of China (NSFC 51505353).

References

1. Ahmed, A., Hong, L., Smola, A.J.: Hierarchical geographical modeling of user locations from social media posts. In: WWW 2013 (2013)

2. Backstrom, L., Sun, E., Marlow, C.: Find me if you can: improving geographical prediction with social and spatial proximity. In: WWW 2010 (2010)
3. Belkin, M., Niyogi, P.: Laplacian eigenmaps and spectral techniques for embedding and clustering. In: NIPS 2001 (2001)
4. Cha, M., Gwon, Y., Kung, H.T.: Twitter geolocation and regional classification via sparse coding. In: ICWSM 2015 (2015)
5. Chang, S., Han, W., Tang, J., Qi, G.J., Aggarwal, C.C., Huang, T.S.: Heterogeneous network embedding via deep architectures. In: KDD 2015 (2015)
6. Feng, S., Li, X., Zeng, Y., Cong, G., Chee, Y.M., Yuan, Q.: Personalized ranking metric embedding for next new POI recommendation. In: IJCAI 2015 (2015)
7. Davis Jr., C.A., Pappa, G.L., de Oliveira, D.R.R., de Lima Arcanjo, F.: Inferring the location of Twitter messages based on user relationships. T. GIS **15**, 735–751 (2011)
8. Jurgens, D.: That's what friends are for: inferring location in online social media platforms based on social relationships. ICWSM **13**, 273–282 (2013)
9. Jurgens, D., Finethy, T., McCorriston, J., Xu, Y.T., Ruths, D.: Geolocation prediction in Twitter using social networks: a critical analysis and review of current practice. In: ICWSM 2015 (2015)
10. Kotzias, D., Lappas, T., Gunopulos, D.: Addressing the sparsity of location information on Twitter. In: EDBT/ICDT 2014 Workshops (2014)
11. Li, R., Wang, S., Chang, K.C.: Multiple location profiling for users and relationships from social network and content. PVLDB **5**, 1603–1614 (2012)
12. Li, R., Wang, S., Deng, H., Wang, R., Chang, K.C.C.: Towards social user profiling: unified and discriminative influence model for inferring home locations. In: KDD 2012 (2012)
13. Mikolov, T., Sutskever, I., Chen, K., Corrado, G.S., Dean, J.: Distributed representations of words and phrases and their compositionality. In: NIPS 2013 (2013)
14. Pontes, T., Magno, G., Vasconcelos, M., Gupta, A., Almeida, J., Kumaraguru, P., Almeida, V.: Beware of what you share: inferring home location in social networks. In: ICDMW 2012 (2012)
15. Rahimi, A., Vu, D., Cohn, T., Baldwin, T.: Exploiting text and network context for geolocation of social media users. In: HLT 2015 (2015)
16. Tang, J., Qu, M., Wang, M., Zhang, M., Yan, J., Mei, Q.: Line: large-scale information network embedding. In: WWW 2015 (2015)
17. Tang, J., Qu, M., Wang, M., Zhang, M., Yan, J., Mei, Q.: PTE: predictive text embedding through large-scale heterogeneous text networks. In: KDD 2015 (2015)
18. Valkanas, G., Gunopulos, D.: Location extraction from social networks with commodity software and online data. In: ICDMW 2012 (2012)
19. Wang, F., Lin, S., Yu, P.S.: Collaborative co-clustering across multiple social media. In: MDM 2016 (2016)
20. Zheng, Y.: Location-based social networks: users. In: Computing with Spatial Trajectories (2011)
21. Zheng, Y.: Methodologies for cross-domain data fusion: an overview. IEEE Trans. Big Data **1**, 16–34 (2015)
22. Zheng, Y., Capra, L., Wolfson, O., Yang, H.: Urban computing: concepts, methodologies, and applications. ACM Trans. Intell. Syst. Technol. **5**, 38 (2014)

Privacy-Preserving Mining and Security/Risk Applications

Partitioning-Based Mechanisms Under Personalized Differential Privacy

Haoran Li[1(✉)], Li Xiong[1], Zhanglong Ji[2], and Xiaoqian Jiang[2]

[1] Emory University, Atlanta, USA
{hli57,lxiong}@emory.edu
[2] University of California at San Diego, San Diego, USA
{z1ji,x1jiang}@ucsd.edu

Abstract. Differential privacy has recently emerged in private statistical aggregate analysis as one of the strongest privacy guarantees. A limitation of the model is that it provides the same privacy protection for all individuals in the database. However, it is common that data owners may have different privacy preferences for their data. Consequently, a global differential privacy parameter may provide excessive privacy protection for some users, while insufficient for others. In this paper, we propose two partitioning-based mechanisms, privacy-aware and utility-based partitioning, to handle personalized differential privacy parameters for each individual in a dataset while maximizing utility of the differentially private computation. The privacy-aware partitioning is to minimize the privacy budget waste, while utility-based partitioning is to maximize the utility for a given aggregate analysis. We also develop a t-round partitioning to take full advantage of remaining privacy budgets. Extensive experiments using real datasets show the effectiveness of our partitioning mechanisms.

1 Introduction

Differential privacy [6] is one of the strongest privacy guarantees for aggregate data analysis. A statistical aggregation or computation satisfies differential privacy (DP) if the outcome is formally indistinguishable when run with and without any particular record in the dataset. One common mechanism for achieving differential privacy is to inject random noise, that is calibrated by the sensitivity of the computation (i.e. the maximum influence of any record on the outcome) and a global privacy parameter or budget ϵ. A lower privacy parameter requires larger noise to be added and provides a higher level of privacy.

One important limitation of DP is that it provides the same level of privacy protection for all data subjects in a database. This approach ignores the reality that different individuals may have very different privacy requirements for their personal data, as shown in Fig. 1. In the medical domain, some patients may openly consent their data for studies or have a low privacy restriction while others may have a high privacy restriction of their medical records. The privacy setting where users in a dataset could set their own privacy preferences is considered as

© Springer International Publishing AG 2017
J. Kim et al. (Eds.): PAKDD 2017, Part I, LNAI 10234, pp. 615–627, 2017.
DOI: 10.1007/978-3-319-57454-7_48

"personalized differential privacy" (PDP) [10]. One possible approach to achieve PDP is to use the minimal privacy budget among all records, called *minimum mechanism* [10]. But this may introduce an unacceptable amount of noise into the outputs because of under-utilized (wasted) privacy budget for most users, resulting in poor utility. Another possible approach, called *threshold mechanism* [10], is to set a privacy threshold and select records with privacy budgets no less than the threshold as a subset, which is then used for a target DP aggregate computation. However the threshold is difficult to choose due to the tradeoff between the perturbation error and the sampling error. A higher privacy budget threshold will result in less perturbation error but at the cost of fewer number of records and a potentially higher sampling error, and vice versa.

Name	Age	Zip	Salary	Budget α_i
Alice	22	02152	70000	0.01
Emily	32	02112	180000	0.02
John	31	02130	105000	0.05
Olga	27	02114	110000	0.07
Frank	36	02232	90000	0.09
Bob	35	01245	140000	0.11
Mark	33	04323	110000	0.14
Cecilia	39	02121	100000	0.15

Fig. 1. Dataset with personalized privacy parameters

Our Contributions. This paper investigates two novel partitioning mechanisms for achieving PDP while fully utilizing the privacy budgets of different individuals and maximizing the utility of the target DP computation: privacy-aware and utility-based partitioning. Given any DP aggregate computation M, our partitioning mechanisms group records with various privacy budgets into k partitions, apply M on each partition using its minimum privacy budget, then bag perturbed results from k partitions to compute the final output. To maximally utilize all leftover privacy budgets, we also develop a t-round partitioning and prove its convergence theoretically. The privacy-aware mechanism considers all privacy budgets as a histogram and groups histogram bins with similar values to minimize privacy waste. The utility-based mechanism partitions all privacy parameters with the goal of maximizing the utility of target computation M. In particular, we find that the utility-based mechanism has superior performance for many important DP aggregate analysis, such as count queries, logistic regression and support vector machine. This is because it considers both privacy budget waste and the number of records in each partition, which significantly impact the utility of target DP aggregate mechanisms. Extensive experiments demonstrate the general applicability and superior performance of our methods.

2   Related Work

Differential privacy has attracted increasing attention in recent years as one of the strongest privacy guarantees for statistical data analysis [6]. Alaggan et al.

[1] proposed *heterogeneous differential privacy*, which to our knowledge is the first work to consider various privacy preferences of data subjects. They proposed a "stretching" mechanism, based on the Laplace mechanism by rescaling the input values due to corresponding privacy parameters. But it cannot be applied to many commonly used functions (e.g. *median, min/max*), and counting queries which count the number of non-zero values in a dataset. Jorgensen et al. [10] proposed two PDP mechanisms. The first one, sampling mechanism, samples a subset of original dataset by assigning each record a weight determined by its own privacy budget and a predefined threshold, then uses the sampled subset for DP aggregate mechanisms. The second one, PDP-exponential mechanism, is based on the exponential mechanism [14], and develops a special utility function for a given aggregate analysis to satisfy PDP particularly. While the PDP-exponential mechanism provides better utility for simple count queries, it is not easily applicable to remove for complex aggregate computations (e.g. logistic regression). In our experiments, we compare our methods with the sampling mechanism [10].

3   Preliminaries

Personalized Differential Privacy. A mechanism is differentially private if its outcome is not significantly affected by the removal or addition of a single user. An adversary thus learns approximately the same information about any individual, irrespective of his/her presence or absence in the original dataset. We give formal definition of differential privacy as below:

Definition 1 (ϵ-differential privacy [5]). *A randomized mechanism $\mathcal{A}$ gives ϵ-differential privacy if for any dataset D and D' differing in at most one record, and for an arbitrary set of possible outputs of $\mathcal{A}$, we have $Pr[\mathcal{A}(D) \in \mathcal{O}] \leq e^{\epsilon} Pr[\mathcal{A}(D') \in \mathcal{O}]$.*

The privacy parameter ϵ, also called the privacy budget, specifies the privacy protection level. A common mechanism to achieve differential privacy is the Laplace mechanism [5] that injects a small amount of independent noise to the output of a numeric function f to fulfill ϵ-differential privacy. The noise is drawn from $Lap(b)$ with pdf $Pr[\eta = x] = \frac{1}{2b}e^{-\frac{|x|}{b}}$, and $b = \Delta_f/\epsilon$, where Δ_f is the sensitivity defined as the maximal L_1-norm distance between the outputs of f over D and D'. A lower value of ϵ requires a larger perturbation noise with less accuracy, and vice versa.

Personalized differential privacy allows each individual in a database to set their own privacy parameter ϵ of their data. We assume in this paper that the personalized privacy parameters are public and not correlated with any sensitive information. For example, in Fig. 1, a sensitive attribute Salary is not correlated with the privacy budget. We give formal definition of PDP as below:

Definition 2 (Personalized Differential Privacy [10]). *For a privacy preference $\phi = (\epsilon_1, \ldots, \epsilon_n)$ of a set of users U, a randomized mechanism $\mathcal{A}$ gives*

ϕ-PDP if for any dataset D and D' differing in at most one arbitrary user u, and for an arbitrary set of possible outputs of $\mathcal{A}$, we have $Pr[\mathcal{A}(D) \in \mathcal{O}] \leq e^{\phi^u} Pr[\mathcal{A}(D') \in \mathcal{O}]$, where ϕ^u is the privacy preference corresponding to user $u \in U$.

Sampling Mechanism. The sampling mechanism [10] for PDP first samples a subset D' due to privacy preference vector, then applies DP aggregate computations on D'. Consider a function $f : D \to R$, a dataset D with n records of n individual data owners, and a privacy preference vector $\phi = (\epsilon_1, \ldots, \epsilon_n)$. Given ϵ_T ($\epsilon_{min} \leq \epsilon_T \leq \epsilon_{max}$), the sampling mechanism selects each record $x_j \in D$ ($1 \leq j \leq n$) with probability $p_j = 1$ if $\epsilon_j \geq \epsilon_T$, and samples other records i.i.d. with probability $p_j = \frac{e^{\epsilon_j} - 1}{e^{\epsilon_T} - 1}$ if $\epsilon_j < \epsilon_T$.

4   Partitioning Mechanisms

In this section, we propose two partitioning mechanisms to fully utilize the privacy budget of individuals and maximizing the utility of target DP computations. The general partitioning mechanism includes: (1) partition records of D horizontally into k groups $(D_1, \ldots, D_k)$ due to various privacy budgets; (2) compute noisy output q_i of target aggregate mechanism M for each D_i with ϵ_i-differential privacy, and (3) ensemble $(q_1, \ldots, q_k)$ to compute q. We define the general partitioning mechanism as below:

Definition 3 (The General Partitioning Mechanism). *For an aggregate function $f : D \to R$, a dataset D with n records of n individual users, and a privacy preference $\phi = (\epsilon_1, \ldots, \epsilon_n)$ ($\epsilon_1 \leq \ldots \leq \epsilon_n$). Let $Partition(D, \phi, k)$ be a procedure that partitions the original dataset D into k partitions $(D_1, \ldots, D_k)$. The partitioning mechanism is defined as $PM = B(DP_{\epsilon_1}^f(D_1), \ldots, DP_{\epsilon_k}^f(D_k))$ where $DP_{\epsilon_i}^f$ is any target ϵ_i-differentially private aggregate mechanism for f, B is an ensemble algorithm.*

The partitioning mechanisms have no privacy risk because it is computed directly from public information, privacy budget of each record. The target aggregate mechanism guarantees ϵ_i-DP for each partition, with ϵ_i as the minimum privacy parameter value of the records in that partition.

4.1   Privacy-Aware Partitioning Mechanism

We develop privacy-aware partitioning mechanism with the goal of grouping records with similar privacy budgets, such that the amount of wasted budget is minimized. Formally, we formulate the privacy budget waste of a partition D_i as $W_i = W(\epsilon_{i,1}, \ldots, \epsilon_{i,n_i}) = \sum_{j=1}^{n_i} (\epsilon_{i,j} - min(\epsilon_{i,j}))^2$, where n_i is number of records in D_i, $\epsilon_{i,j}$ is the privacy budget of jth-record of D_i, and $min(\epsilon_{i,j})$ ensures ϵ_i-DP for D_i. We define privacy-aware partitioning algorithm as follows:

Definition 4 (Privacy-aware partitioning). *In a sorted privacy budget vector* $\phi = (\epsilon_1, \ldots, \epsilon_n)$, *where* $\epsilon_1 \leq \epsilon_2 \leq \ldots \leq \epsilon_n$, *we want to split* ϕ *into* k *partitions such that* $W(\phi) = \sum_{i=1}^{k} W_i$ *is minimized, where* $W_i = \sum_{j=1}^{n_i} (\epsilon_{i,j} - min(\epsilon_{i,j}))^2$.

With a predefined k, we find the optimal k-partitioning using dynamic programming and present the privacy-aware partitioning algorithm in Algorithm 1.

Algorithm 1. Privacy-aware partitioning mechanism $W*$ of finding the optimal k-partition of $(\epsilon_1, \ldots, \epsilon_n)$ for a given definition of the function W

Require: Sorted $\phi = (\epsilon_1, \ldots, \epsilon_n)$ and k
Ensure: k partitions of original dataset
1. if $k = 0$ then return 0
2. $minW = $ inf
3. foreach $j \in \{k - 1, \ldots, n\}$ do
   $currentW = W^*((\epsilon_1, \ldots, \epsilon_j), k - 1) + W(\epsilon_{j+1}, \ldots, \epsilon_n)$
   if $currentW < minW$ **then**
      $minW = currentW$
      $partitions[k - 1] = (\epsilon_{j+1}, \ldots, \epsilon_n)$
4. return $minW$ and indexes of k partitions

Before running Algorithm 1, we first sort all privacy budgets in ascending order. Sorting records in the descending order of privacy budgets generates the same partition. When we sort privacy budgets, the sequence of corresponding data records follows the order of privacy budgets. Therefore, we know which records are included in which partition. To simplify the algorithm, we do not include representation of data records. In step 3, we use dynamic programming to find the optimal partition for a given definition of the function W. The goal is to minimize the waste of privacy budgets in each partition by computing the distance between individual budget and the minimum budget of the current partition. Note that we represent Algorithm 1 as W^*, and $currentW = W^*((\epsilon_1, \ldots, \epsilon_j), k-1) + W(\epsilon_{j+1}, \ldots, \epsilon_n)$ means that we recursively use Algorithm 1 to compute $k - 1$ partitions.

Optimal Number of Partitions. Algorithm 1 finds an optimal k-partitioning given a predefined k. To choose an optimal k, let us consider two extreme cases: (i) we can have n partitions where each record is its own partition and no privacy budget is wasted, or (ii) all data records can be grouped as one partition to maximize the number of records in the partition. The amount of generated noise could be significant in the previous case, while large amount of privacy budget waste may be incurred in the latter case. We need to consider the trade-off between n and ϵ to find the optimal k by building the following objective function:

$$\min_{k} \sum_{i=1}^{k} [\frac{1}{n_i} \sum_{j=1}^{n_i} (\epsilon_{i,j} - min(\epsilon_{i,j}))^2] \qquad (1)$$

Equation (1) implies a tradeoff between the partition size and privacy budget waste. Due to Eq. (1), neither extreme case (i) nor (ii) can lead to optimal value of Eq. (1). If we set a minimum threshold T of partition size n_i for the target differentially private mechanism, we can search different number of partitions from 1 to $\frac{n}{T}$, and find the optimal partition number. The minimum number of records n_i required in one partition is reasonable because many aggregate mechanisms (e.g. logistic regression, support vector machine) require a minimum training data size to ensure acceptable performance, due to machine learning theory. For example, Shalev-Shwartz and Srebro [16] show that for a given classifier with expected loss defined on a differentiable loss function, the excess loss of the classifier will be upper bounded if training data size is larger than a threshold.

Complexity. Sorting all privacy budgets takes $O(n \log n)$. Computing optimal k takes $O(n)$, since we need to scan privacy vector at most $m = \frac{n}{T}$ times ($\frac{n}{T}$ is constant here since we control T to make $\frac{n}{T}$ constant for complexity reduction). The privacy-aware partitioning takes $O(mn \log n)$ complexity using dynamic programming with intermediate results saved and optimization tricks. The overall complexity is $O(mn \log n)$.

4.2   Utility-Based Partitioning Mechanism

The privacy-aware partitioning mechanism aims to fully utilize the privacy budget of individual users which will indirectly optimize the utility of the target DP computation. In this section, we present a utility-based partitioning mechanism explicitly optimized for target DP computations. The utility-based partitioning is inspired by an observation that many DP machine learning algorithms (e.g. [5,7,9,19]) have their performance related with n, ϵ for a dataset of n records with ϵ-DP. We give definition of utility-based partitioning below.

Definition 5 (Utility-based partitioning). *In a sorted privacy budget vector* $\phi = (\epsilon_1, \ldots, \epsilon_n)$, *where* $\epsilon_1 \leq \epsilon_2 \leq \ldots \leq \epsilon_n$, *and let* n_i *denote the number of records in* D_i, *we want to split* ϕ *into* k *partitions to maximize* $\sum_{j=1}^{n_i} U(n_i, min(\epsilon_{i,j}))$, *where* $U(n_i, min(\epsilon_{i,j}))$ *is a utility function of target DP computation, which is related with* n_i *and* $min(\epsilon_{i,j})$.

Algorithm 2 presents the utility-based partitioning. We observe that $U(n, \epsilon)$ can be considered as a general utility form in a series of existing state-of-the-art DP algorithms (e.g. [3,4,8,11–13,15,17,18,20]). (i) *Count query* In the Laplace mechanism, the noisy result of a function f can be represented as $f(D) + \nu$, where ν follows $Lap(\frac{\Delta_f}{\epsilon})$, and Δ_f is the sensitivity related to number of records n. If we normalize $f(D)$ by n, Δ_f would become $\frac{\Delta_f}{n}$. Thus, the variance of Laplace distribution can be considered as the utility function $U(n, \epsilon) = 2(\frac{\Delta_f}{n\epsilon})^2$. Maximizing $n\epsilon$ will lead to best utility with a high probability. (ii) *Empirical risk minimization.* We take for example the DP empirical risk minimization mechanism (DPERM) proposed by Chaudhuri and Sarwate [11]. The reason is

Algorithm 2. Utility-based partitioning mechanism U^* of finding the optimal k-partition of $(\epsilon_1, \ldots, \epsilon_n)$ for a given definition of utility function U

Require: $(\epsilon_1, \ldots, \epsilon_n)$ and k
Ensure: k partitions of original data records
1. **if** $k = 0$ **then return** $U(n, \epsilon_{min})$
2. $maxUtility = 0$
3. **foreach** $j \in \{k - 1, \ldots, n\}$ **do**
   $currentUtility = U^*((\epsilon_1, \ldots, \epsilon_j), k - 1) + U(min(\epsilon_{j+1}, \ldots, \epsilon_n), n - j)$
   if $currentUtlity > maxUtility$ **then**
       $maxUtility = currentUtility, partitions[k - 1] = (\epsilon_{j+1}, \ldots, \epsilon_n)$
4. **return** $maxUtility$

---

that DPERM can be easily generalized to important machine learning tasks, such as logistic regression and support vector machine, which have a convex loss function as the optimization objective. Our utility function form can be extended to a class of DP machine learning mechanisms.

Assume that n records in a dataset D are drawn i.i.d. from a fixed distribution $F(X, y)$. Given F, the performance of privacy preserving empirical risk minimization algorithms in [11] can be measured by the expected loss $L(f)$ for a classifier f, defined as $L(f) = E_{(X,y) \sim F}[l(f^T x, y)]$, where the loss function l is differentiable and continuous, the derivative l' is c-Lipschitz. By [11], the expected loss of the private classifier f_p can be bounded as below

$$L(f_p) \leq L(f_0) + \frac{16\|f_0\|^4 d^2 \log^2(d/\sigma)(c+e_g/\|f_0\|^2)}{n^2 e_g^2 \epsilon^2} + O(\|f_0\|^2 \frac{log(1/\sigma)}{ne_g}) + \frac{e_g}{2} \quad (2)$$

where $L(f_0)$ is the expected loss of the true classifier f_0, ϵ is the privacy budget, e_g is the generalization error, and d is the number of dimensions of input data. If we consider the second part of Eq. (2), we can build a utility function as $U(n, \epsilon) = \frac{16\|f_0\|^4 d^2 \log^2(d/\sigma)(c+e_g/\|f_0\|^2)}{n^2 e_g^2 \epsilon^2} + \|f_0\|^2 \frac{log(1/\sigma)}{ne_g} + \frac{e_g}{2}$, where only n and ϵ are variables.

Optimal Number of Partitions. Akin to privacy-aware partitioning mechanism, we need to select an optimal value for k, in order to maximize the sum of utility function value over all partitions.

$$\max_k \sum_{i=1}^{k} U(n_i, \min_{1 \leq j \leq n_i} (\epsilon_{i,j})) \quad (3)$$

Here, a minimum threshold T of each partition size is also required for a differentially private task. Theoretically, we can search different number of partitions from 1 to $\frac{n}{T}$ to find the optimal number of partitions with the maximum value of objective function (3).

Complexity. Sorting all privacy budgets is $O(n \log n)$. Finding the optimal partitioning takes $O(n)$, due to complexity of Algorithm 1. The utility-based partitioning takes $O(n)$. The overall complexity of Algorithm 2 is $O(n \log n)$.

4.3   T-Round Partitioning

After the first round of partitioning, we may still have records with remaining budgets. Extra rounds of partitioning can be applied iteratively on the remaining records with leftover privacy budgets. In this part, we prove by iteratively apply our algorithm to the leftover budget from previous iterations, the leftover budget will decrease exponentially, which means all input budgets will be used up soon.

Here we define a T-round partitioning as iteratively grouping n records into k partitions according to the objective function in Definition 3, then consume the smallest budget in each group and update the leftover budget. The leftover budget for the l-th record in the t-th round is denoted as ϵ_l^t.

Theorem: $\sum_{l=1}^{n}(\epsilon_l^T)^2 \leq \left(\frac{n}{n-1+k^2}\right)^T \sum_{l=1}^{n}(\epsilon_l)^2$, which means the leftover privacy budget converges to 0 exponentially.

Proof. Without loss of generality, we assume ϵ_n is the largest among all input privacy budgets, and select the partition that partitions the interval $[0, \epsilon_n]$ into k intervals with equal length ϵ_n/k. In this case, for the leftover budget ϵ_l^{1*} we have $\epsilon_l^{1*} \leq \epsilon_n/k, \epsilon_n^{1*} = \epsilon_n/k$ for all $1 \leq l \leq n$. Thus $\sum_{l=1}^{n}(\epsilon_l^{1*})^2 \leq \sum_{l=1}^{n}(\epsilon_n/k)^2 = n(\epsilon_n/k)^2$. Furthermore, since we have $\epsilon_l \geq \epsilon_l^{1*}$, there is $\sum_{l=1}^{n}(\epsilon_l)^2 - \sum_{l=1}^{n}(\epsilon_l^{1*})^2 \geq \sum_{l=1}^{n}[(\epsilon_l)^2 - (\epsilon_l^{1*})^2] \geq (\epsilon_n)^2 - (\epsilon_n^{1*})^2 = (\epsilon_n)^2\left(1 - \frac{1}{k^2}\right)$. Combining them together, we conclude $\frac{\sum_{l=1}^{n}(\epsilon_l)^2}{\sum_{l=1}^{n}(\epsilon_l^{1*})^2} = \frac{\sum_{l=1}^{n}(\epsilon_l)^2 - \sum_{l=1}^{n}(\epsilon_l^{1*})^2}{\sum_{l=1}^{n}(\epsilon_l^{1*})^2} + 1 \geq \frac{(\epsilon_n)^2\left(1-\frac{1}{k^2}\right)}{n(\epsilon_n/k)^2} + 1 = \frac{k^2-1}{n} + 1 \frac{\sum_{l=1}^{n}(\epsilon_l^{1*})^2}{\sum_{l=1}^{n}(\epsilon_l)^2} \leq \frac{n}{k^2-1+n}$. Since the optimal partition must have smaller $\sum_{l=1}^{n}(\epsilon_l^1)^2$ than this very naive partition, there must be $\frac{\sum_{l=1}^{n}(\epsilon_l^1)^2}{\sum_{l=1}^{n}(\epsilon_l)^2} \leq \frac{n}{k^2-1+n}$. Similarly, if we take ϵ_l^1 as input to the next round, we can get $\frac{\sum_{l=1}^{n}(\epsilon_l^2)^2}{\sum_{l=1}^{n}(\epsilon_l^1)^2} \leq \frac{n}{k^2-1+n}$, etc. When we multiply these inequalities together, we conclude $\sum_{l=1}^{n}(\epsilon_l^T)^2 \leq \left(\frac{n}{n-1+k^2}\right)^T \sum_{l=1}^{n}(\epsilon_l)^2$.

4.4   Ensemble

Once we have partitions, we run DP mechanism on each partition, and then use ensemble methods to aggregate the result from each partition. Due to conclusions of [2], our ensemble rule is that the private output of partition with equal number of records but smaller privacy budgets than other partitions would be dropped out. We also consider types of learning problems. For numerical situation, like bagging multiple linear regression or count queries, we aggregate all private predicted values from all partitions. The weights will depend on $O(n_i, \epsilon_i)$. Assume the numerical task is P, the aggregated result would be $\tilde{Y} = \sum_{i=1}^{k} w_i P(D_i)$. For classification tasks, we use majority voting.

5   Experiment

In this section, we experimentally evaluate partitioning-based mechanisms and compare it with the sampling mechanism in [10]. Partitioning-based mechanisms are implemented in MATLAB R2010b and Java, and all experiments were performed on a PC with 2.8 GHz CPU and 8 G RAM.

5.1   Experiment Setup

Datasets. We use two datasets from the Integrated Public Use Microdata Series[1], US and Brazil, with $370K$ and $190K$ census records collected in the US and Brazil, respectively. There are 13 attributes in each dataset, namely, Age, Gender, Martial Status, Education, Disability, Nativity, Working Hours per Week, Number of Years Residing in the Current Location, Ownership of Dwelling, Family Size, Number of Children, Number of Automobiles, and Annual Income. Among these attributes, Marital status is the only categorical attribute with 3 values. We categorize Marital Status into two binary attributes. With this transformation, both of our datasets become 14 dimensions.

Privacy Specification. For personalized differential privacy, we generate the privacy budgets for all records randomly from uniform distribution and normal distribution. We set the range of privacy budget value ϵ from 0.01 to 1.0, with $\epsilon = 0.01$ being users with high privacy concern, and sample i.i.d. privacy budgets from $Uniform(0.01, 0.1)$ and $Normal(0.1, 1)$.

Comparison. We evaluate the utility of our mechanisms using random range-count queries, support vector machine, and logistic regression, and compare it with the sampling mechanism [10] and baseline *Minimum*.

Metrics. For count query evaluation, we generated random range-count queries with random query predicates covering all attributes defined as "Select COUNT(*) from D Where $A_1 \in I_1$ and $A_2 \in I_2$ and ... and $A_m \in I_m$". For each attribute A_i, I_i is a random interval generated from the domain of A_i.

We measure the count query accuracy by the relative frequency error $RFE(q) = (A(q) - A'(q))/n$, where for a query q, $A(q)$ is the true answer. $A'(q)$ is the noisy answer, n is number of records in the original dataset. Here we use relative frequency error to scale query errors based on n, because sampling mechanism generates a partial number of records from original datasets.

For the support vector machine, we use the area under the curve (AUC), and higher AUC value means better discrimination. For logistic regression, annual income is converted into a binary attribute: values higher than mean are mapped to 1, and 0 otherwise. To be consistent with [10], we measure the accuracy

---

[1] Minnesota Population Center. Integrated public use microdata series-international: Version 5.0. 2009. https://international.ipums.org.

of logistic regression with misclassification rate, the fraction of tuples that are incorrectly classified. For space limitation, we only show experiment results of support vector machine, and the performance of logistic regression has the same trend with count query.

5.2   Experimental Results

Partitioning-Based Mechanisms for Count Query. Figures 2 and 3 investigate the relative frequency error between partitioning mechanisms and the sampling mechanism under normal and uniform distribution of privacy preferences. We vary the privacy budget thresholds of the sampling mechanism. The errors of the partitioning mechanisms remain at a horizontal line since it does not need to set privacy budget threshold. The accuracy of sampling mechanism reaches optimal when the budget threshold attains the mean of all privacy budget values, which is consistent with the experimental conclusion in [10]. We can observe that the accuracy of sampling mechanism deteriorates sharply when threshold value is smaller than the mean privacy budget. This is because when the number of records is sufficiently large, the privacy budget dominates the performance. Our partitioning mechanisms remain stable and perform almost the same with the optimal performance of sampling mechanism. Utility-based partitioning has slightly better performance in the experiments, since it considers both privacy and utility of the target DP computation. The baseline *Minimum* performs similarly with the privacy budget threshold being the smallest. This is because when the threshold becomes the smallest value, sampling mechanism is equal to *Minimum*. This conclusion remains the same for the following experiments.

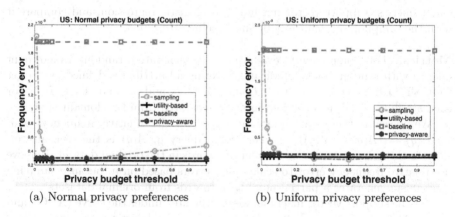

(a) Normal privacy preferences   (b) Uniform privacy preferences

Fig. 2. Relative frequency error for the count task (US)

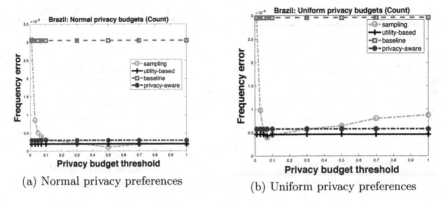

(a) Normal privacy preferences

(b) Uniform privacy preferences

Fig. 3. Relative frequency error for the count task (Brazil)

Partitioning-Based Mechanisms for Support Vector Machine (SVM).
Figures 4 and 5 illustrate the performances of different mechanisms for SVM
classification. There is no obvious pattern for sampling mechanism on which
privacy budget threshold has the optimal utility, and it is difficult to choose
the threshold for an optimal utility. However, our partitioning mechanisms have
superior performance than sampling mechanism. The performance of sampling
mechanism under uniform privacy budgets fluctuates, because the number of
records in the experiment is small for SVM, and as a result, it is difficult to
select an optimal threshold before running private SVM. The performance of
sampling mechanism under normal privacy budgets arrives the best when the
threshold value is around 0.5, which approximates the average of all privacy
budgets.

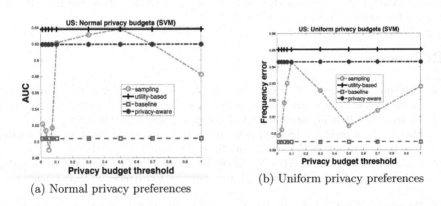

(a) Normal privacy preferences

(b) Uniform privacy preferences

Fig. 4. AUC for support vector machine (US)

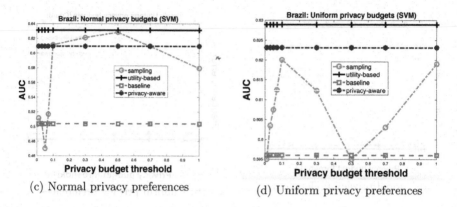

(c) Normal privacy preferences

(d) Uniform privacy preferences

Fig. 5. AUC for support vector machine (Brazil)

6   Conclusions

In this paper, we developed two partitioning-based mechanisms for PDP that aims to fully utilize the privacy budgets of different individuals and maximize the utility of target DP computations. Privacy-aware partitioning minimizes privacy budget waste, and utility-based partitioning maximizes a utility function of target mechanism. For future work, it will be useful to evaluate the utility of partitioning mechanisms for different aggregations or analytical tasks. It will also be of interest to extend notions of personalized differential privacy to social networks, where the individuals are nodes, and edges represent connections between pairs.

Acknowledgement. This research was supported by the Patient-Centered Outcomes Research Institute (PCORI) under contract ME-1310-07058, the National Institute of Health (NIH) under award number R01GM114612, R01GM118609, and the National Science Foundation under award CNS-1618932.

References

1. Alaggan, M., Gambs, S., Kermarrec, A.: Heterogeneous differential privacy. In: Workshop on Theory and Practice of Differential Privacy Alongside ETAPS (2015)
2. Breiman, L.: Bagging predictors. Mach. Learn. **24**(2), 123–140 (1996)
3. Cao, Y., Masatoshi, Y.: Differentially private real-time data publishing over infinite trajectory streams. IEICE Trans. Inf. Syst. **99**(1), 163–175 (2016)
4. Cao, Y., Yoshikawa, M., Xiao, Y., Xiong, L.: Quantifying differential privacy under temporal correlations. In: 33rd IEEE International Conference on Data Engineering (2017)
5. Dwork, C., McSherry, F., Nissim, K., Smith, A.D.: Calibrating noise to sensitivity in private data analysis. In: Halevi, S., Rabin, T. (eds.) TCC 2006. LNCS, vol. 3876, pp. 265–284. Springer, Heidelberg (2006). doi:10.1007/11681878_14

6. Dwork, C., Roth, A.: The algorithmic foundations of differential privacy. Found. Trends Theor. Comput. Sci. **9**(3–4), 211–407 (2014)
7. Fletcher, S., Islam, M.Z.: A differentially private random decision forest using reliable signal-to-noise ratios. In: Pfahringer, B., Renz, J. (eds.) AI 2015. LNCS (LNAI), vol. 9457, pp. 192–203. Springer, Cham (2015). doi:10.1007/978-3-319-26350-2_17
8. Friedman, A., Schuster, A.: Data mining with differential privacy. In: The 16th ACM International Conference on Knowledge Discovery and Data Mining (2010)
9. Jagannathan, G., Monteleoni, C., Pillaipakkamnatt, K.: A semi-supervised learning approach to differential privacy. In: 13th IEEE International Conference on Data Mining Workshops, ICDM Workshops, pp. 841–848 (2013)
10. Jorgensen, Z., Yu, T., Cormode, G.: Conservative or liberal? Personalized differential privacy. In: 31st IEEE International Conference on Data Engineering (ICDE), pp. 1023–1034 (2015)
11. Chaudhuri, C.M.K., Sarwate, A.D.: Differentially private empirical risk minimization. J. Mach. Learn. Res. **12**, 1069–1109 (2011)
12. Li, H., Xiong, L., Jiang, X.: Differentially private synthesization of multi-dimensional data using copula functions. In: The 17th International Conference on Extending Database Technology, pp. 475–486 (2014)
13. Li, H., Xiong, L., Jiang, X., Liu, J.: Differentially private histogram publication for dynamic datasets: an adaptive sampling approach. In: The 24th ACM International Conference on Information and Knowledge Management (2015)
14. McSherry, F., Talwar, K.: Mechanism design via differential privacy. In: IEEE Symposium on Foundations of Computer Science (2007)
15. Fletcher, S., Islam, M.Z.: A differentially private decision forest. In: Proceedings of the 13th Australasian Data Mining Conference (2015)
16. Shalev-Shwartz, S., Srebro, N.: SVM optimization: inverse dependence on training set size. In: The 25th International Conference on Machine Learning (2008)
17. Xiao, Y., Xiong, L., Fan, L., Goryczka, S., Li, H.: DPCube: differentially private histogram release through multidimensional partitioning. Trans. Data Priv. **7**(3), 195–222 (2014)
18. Xu, S., Cheng, X., Su, S., Xiao, K., Xiong, L.: Differentially private frequent sequence mining. IEEE Trans. Knowl. Data Eng. **28**(11), 2910–2926 (2016)
19. Yang, C.: Rigorous and flexible privacy models for utilizing personal spatiotemporal data. In: The 42nd International Conference on Very Large Databases (2016)
20. Yang, C., Yoshikawa, M.: Differentially private real-time data release over infinite trajectory streams. In: 16th IEEE International Conference on Mobile Data Management (2015)

Efficient Cryptanalysis of Bloom Filters for Privacy-Preserving Record Linkage

Peter Christen[1(✉)], Rainer Schnell[2], Dinusha Vatsalan[1],
and Thilina Ranbaduge[1]

[1] Research School of Computer Science, The Australian National University,
Canberra, Australia
{peter.christen,dinusha.vatsalan,thilina.ranbaduge}@anu.edu.au
[2] Methodology Research Group, University Duisburg-Essen, Duisburg, Germany
rainer.schnell@uni-due.de

Abstract. Privacy-preserving record linkage (PPRL) is the process of identifying records that represent the same entity across databases held by different organizations without revealing any sensitive information about these entities. A popular technique used in PPRL is Bloom filter encoding, which has shown to be an efficient and effective way to encode sensitive information into bit vectors while still enabling approximate matching of attribute values. However, the encoded values in Bloom filters are vulnerable to cryptanalysis attacks. Under specific conditions, these attacks are successful in that some frequent sensitive attribute values can be re-identified. In this paper we propose and evaluate on real databases a novel efficient attack on Bloom filters. Our approach is based on the construction principle of Bloom filters of hashing elements of sets into bit positions. The attack is independent of the encoding function and its parameters used, it can correctly re-identify sensitive attribute values even when various recently proposed hardening techniques have been applied, and it runs in a few seconds instead of hours.

Keywords: Privacy · Re-identification · Frequency analysis · Data linkage · Entity resolution · Data matching

1 Introduction

Integrating data from different sources with the aim to remove duplicates, enrich data, and correct errors and inconsistencies is a crucial data pre-processing task for many data mining and analytics applications [4]. Example applications include healthcare, business analytics, national censuses, population informatics, fraud detection, government services, and national security.

The authors would like to thank the Isaac Newton Institute for Mathematical Sciences, Cambridge, for support and hospitality during the programme *Data Linkage and Anonymisation* where this work was conducted (EPSRC grant EP/K032208/1). Peter Christen was also supported by a grant from the Simons Foundation. The work was also partially funded by the Australian Research Council (DP130101801).

J. Kim et al. (Eds.): PAKDD 2017, Part I, LNAI 10234, pp. 628–640, 2017.
DOI: 10.1007/978-3-319-57454-7_49

However, growing concerns about privacy and confidentiality increasingly preclude the exchange or sharing of personal identifying attributes, such as names, dates of birth, and addresses, which are generally required for linking databases due to the non-existence of common unique entity identifiers [4,18]. Work in privacy-preserving record linkage (PPRL) aims to develop techniques for identifying records that correspond to the same entity across several databases while not compromising the privacy and confidentiality of the entities [18].

PPRL is achieved by conducting linkage on the encoded (masked) values of the identifying attributes of records across two or more databases. Several data encoding techniques for PPRL have been developed. These can be categorized into cryptographic secure multi-party computation (SMC) and perturbation-based techniques [18]. The former are accurate and provably secure, but they incur expensive computation and communication costs. Most PPRL techniques are therefore based on perturbation-based techniques that provide adequate privacy protection while achieving acceptable linkage quality [11,18].

Bloom filter (BF) encoding is one such perturbation-based technique that has successfully been used in several recent practical PPRL applications [2,12]. A BF is a binary vector with bits initially set to 0. A value can be encoded into a BF using a set of hash functions by setting corresponding bits to 1 [13], and the approximate similarity between two BFs can be calculated by counting the number of positions where both BFs have 1-bits in common.

As we discuss in detail in the next section, BFs can be susceptible to cryptanalysis attacks that aim to re-identify the encoded sensitive attribute values [7–10]. Using frequency counts and patterns in a set of BFs, these attacks iteratively map bit patterns to known attribute values. These existing attacks are however not practical as they require knowledge of certain parameters used during the BF encoding phase, and they have high computational costs.

Our contribution in this paper is an efficient frequency-based approach for attacking BFs that exploits the fundamental property of how the elements of sets are hashed into BFs, as we describe in Sect. 3. In contrast to existing attack methods, our novel approach does not require any assumption on the BF parameters used when sensitive attribute values were encoded. It is also significantly faster, making it a viable attack on large sets of BFs to evaluate whether they provide adequate privacy protection. We experimentally evaluate our attack method on two real-world data sets, showing its efficiency and effectiveness.

Given BF encoding is now being employed in real-world PPRL applications [2,12], it is crucial to study possible attacks on BFs to ensure their security and to make users of such systems aware of the weaknesses of BF encoding. Our novel attack method allows data custodians to identify such weaknesses of BF encoding that otherwise could be exploited by an attacker.

2   Prior Attacks on Bloom Filter Based PPRL

Schnell et al. [13] were the first to introduce an approximate matching approach for PPRL using Bloom filters (BFs), as we describe in detail in Sect. 3. A recent study by Randall et al. [12] has shown that PPRL based on BF encoding can achieve similar linkage quality as can be achieved with traditional linkage methods on the unencoded attribute values. As a result, BF encoding has been the PPRL technique of choice in several recent practical PPRL applications [2,12]. However, BFs are prone to different attacks [7–10]. Therefore, a comprehensive analysis of the weaknesses of BFs in the PPRL context is required. We now describe the few studies that have been conducted on attacks on BFs.

Kuzu et al. [8] formulated a cryptanalysis attack on BFs as a constraint satisfaction problem (CSP). CSP is characterized by a set of variables and a set of constraints on these variables. The aim of this attack is to identify a set of values from a given domain that can be assigned to each variable such that the constraints are satisfied. This is achieved by a frequency analysis of the sensitive attribute values and the BF encodings of the records in a database. However, this attack requires that the attacker has access to a global database where the encoded records are drawn from. This is unlikely in practical applications.

In 2013, Kuzu et al. [9] investigated the accuracy of their CSP attack with two real-world databases. The authors tried to re-identify the personal details of patients in an encoded medical database by using a frequency analysis of BF encodings of a public voter registration database that has a different frequency distribution to the medical database. Their results indicated that although the CSP attack might be feasible in such situations, it is less likely to be accurate in identifying original attribute values and it requires more computational resources. The attack re-identified four out of 20 frequent names correctly.

Niedermeyer et al. [10] more recently proposed an attack on BFs built from German surnames. The attack was based on the frequencies of sub-strings of length 2 extracted from frequent surnames. Of 7, 580 surnames, the authors re-identified the 934 most frequent ones (about 12%) before stopping the attack. In contrast to the approach in [9], this attack only depends on the availability of a list of the most common surnames. This work was extended by Kroll and Steinmetzer [7] into a cryptanalysis on BF encodings of several attributes, which was able to re-identify 44% of all attribute values correctly. However, both attacks are based on the specific double hashing scheme used by Schnell et al. [13].

Existing cryptanalysis attacks are feasible only for certain settings and assumptions used in the BF encoding phase. They also require excessive computational resources making them not practical in real settings. Our novel attack method, described next, improves on both these drawbacks of existing methods.

3   Overview and Preliminaries

We now provide an overview of our attack on BFs, as illustrated in Fig. 1. As with other attacks on BFs [7–10], our approach exploits the frequency distribution of a set of BFs that were generated from a large database. As for notation, we use bold letters for sets (with upper-case bold letters for lists or sets of sets) and italics type letters for integer or string values. We denote sets with curly and lists with square brackets, where lists have an order while sets do not.

Public database			BF database		
First name	Freq		Bloom filter	Freq	
karen	231	BF$_1$	[1,0,1,1,0,1]	242	
mary	171	BF$_2$	[1,1,0,0,1,0]	184	
kate	109	BF$_3$	[0,0,1,0,1,1]	115	
mareo	42	BF$_4$	[0,1,0,1,1,1]	48	
...	...		$p_1\ p_2\ p_3\ \cdots\ p_6$	...	

(1a) Position candidate sets

$\hat{c}[p_1]$ = {ka,ar,re,en,ma,ry}
$\bar{c}[p_1]$ = {ka,at,te,ma,ar,re,eo}
$\hat{c}[p_2]$ = {ma,ar,ry,re,eo}
$\bar{c}[p_2]$ = {ka,ar,re,en,at,te}
$\hat{c}[p_3]$ = {ka,ar,re,en,at,te}
$\bar{c}[p_3]$ = {ma,ar,ry,re,eo}
...

(1b) Position q-gram sets

$c[p_1] = \hat{c}[p_1] \setminus \bar{c}[p_1]$ = {en,ry}
$c[p_2] = \hat{c}[p_2] \setminus \bar{c}[p_2]$ = {ma,ry,eo}
$c[p_3] = \hat{c}[p_3] \setminus \bar{c}[p_3]$ = {ka,en,at,te}
...

(2) Re-identify attribute values (BF$_1$)

G = {karen, mary, kate, mareo}
$g_{p_1} = G \odot c[p_1]$ = {karen, mary}
$g_{p_3} = g_{p_1} \odot c[p_3]$ = { karen}

Fig. 1. Outline of the proposed cryptanalysis attack, which is based on a set of BFs and a set attribute values, both sorted according to their frequencies. In steps (1a) and (1b), the attack exploits the bit patterns in BFs to identify the sets of q-grams that are possible, $\mathbf{c}^+[p]$, and *not* possible, $\mathbf{c}^-[p]$, respectively, for each bit position p. In step (2), the re-identification of a set of (sensitive) attribute values, $\mathbf{G}$, is conducted by intersecting the q-gram sets of values in $\mathbf{G}$ with the sets $\mathbf{c}$ (illustrated using $\odot$).

We assume the attacker has access to a set of encoded BFs, $\mathbf{B}$, and their frequencies, but he does not know anything about the parameters used in the encoding process (such as the number of hash functions used, or the actual hashing mechanism). We assume these BFs represent a set of records that encode sensitive values from one or a few attributes. Based on the frequency distribution of the Hamming weights (number of 1-bits) in $\mathbf{B}$, the attacker can guess which attribute(s) have been encoded, because different attributes (such as first name, surname, city name, or postcode) have distinctive distributions of Hamming weights. The distribution of Hamming weights is independent of the (unknown) secret key. Therefore, an attacker can sample attribute values from a publicly available population database (such as a telephone directory) and select a set of frequent values, $\mathbf{V}$, from an attribute that has a frequency distribution that is similar to the distribution of the BF set to be attacked.

In step (1), we first align BFs and attribute values according to their frequencies, and consider the set of most frequent values in both. For each bit position p in the BFs, for all corresponding attribute values that have this bit set to 1 we add their q-grams (sub-strings of length q generated from attribute values) to the set $\mathbf{c}^+[p]$ of possible q-grams for that position. The reasoning is that a 1-bit means at least one q-gram of an attribute value was hashed to this position. For all attribute values with a value of 0 at bit position p we add their q-grams to the set $\mathbf{c}^-[p]$ of *not* possible q-grams for that position, because a 0-bit means no q-gram of an attribute value could have been mapped to this position.

At the end of step (1), for each position p we obtain the set $\mathbf{c}[p] = \mathbf{c}^+[p] \setminus \mathbf{c}^-[p]$ of q-grams that potentially could have been hashed to position p. Based on the list $\mathbf{C} = [\mathbf{c}[1], \ldots, \mathbf{c}[l]]$, where l is the length of the BFs, and a set $\mathbf{G}$ of attribute values we aim to re-identify (i.e. learn which BF possibly encodes which value in $\mathbf{G}$), in step (2) we analyze each BF in $\mathbf{B}$ and remove those attribute values from $\mathbf{G}$ that are not possible matches according to $\mathbf{C}$ because they do not contain any q-grams that would have been hashed to a certain 1-bit.

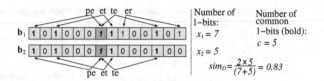

Fig. 2. An example Dice coefficient similarity calculation of the two first names 'peter' and 'pete' encoded in BFs, as described in Sect. 3. The dark bit shows a hash collision.

For example, in Fig. 1, for the most frequent BF_1, 'kate' is not a possible value because in order to obtain a 1-bit in position p_1, it would have to contain either the q-gram 'en' or 'ry'; while 'mary' is also not possible because it would need to contain one of the q-grams 'ka', 'en', 'at', or 'te' in position p_3.

Before we formalize and present our approach in detail in Sect. 4, we first describe BF encoding, as well as some recent approaches to harden them.

Bloom Filter Encoding: Proposed by Bloom [1] in 1970 for the space and time efficient representation of sets, a BF **b** is a bit vector of length l where all bits are initially set to 0. k independent hash functions, $h_1, \ldots, h_k$, each with range $1, \ldots, l$, are used to map the elements s in a set **s** into the BF by setting the bit positions $\mathbf{b}[h_j(s)] = 1$, with $1 \leq j \leq k$.

For PPRL, the set **s** of q-grams generated from string values [13], or neighboring values for numerical values [17], can be hash-mapped into a BF. These BFs are then either sent to a linkage unit (LU, an external party that conducts the linkage) to calculate the similarity between BFs in order to classify them as matches or non-matches [13], or they are partially exchanged among the database owners to distributively calculate the similarities between BFs [16].

The Dice coefficient has been used for comparing BFs since it is insensitive to many matching zeros in long BFs [13]. For two BFs, $\mathbf{b}_1$ and $\mathbf{b}_2$, the Dice coefficient similarity is: $sim_D(\mathbf{b}_1, \mathbf{b}_2) = 2c/(x_1 + x_2)$, where c is the number of bit positions that are set to 1 in both BFs (common 1-bits), and x_1 and x_2 are the number of bit positions set to 1 in $\mathbf{b}_1$ and $\mathbf{b}_2$, respectively. Figure 2 shows the encoding of bigrams ($q = 2$) of two string values into $l = 14$ bits long BFs using $k = 2$ hash functions, and their Dice coefficient similarity calculation.

Different encoding methods have been proposed for BFs. Hashing several attributes of a record into one BF is a method known as cryptographic long term key (CLK). It is used to improve privacy [14]. Another record-level BF encoding (RBF) was proposed to improve linkage quality [5]. In RBF, attribute values are first hashed into different BFs and then bits are selected from each attribute-level BF into a RBF according to attribute weights.

The initial proposal of BFs for PPRL used a double hashing scheme [13], where the k individual bit positions for an element s to be hashed are determined by the sum of the integer representation of two independent hash functions that are mapped into the range $1 - l$. Random hashing has recently been proposed as an improvement over double hashing to prevent against cryptanalysis attacks, where k random numbers are drawn for every element s to be hashed [10,15].

Bloom Filter Hardening: Several BF hardening methods have been studied in recent times to reduce the vulnerability of BFs against cryptanalysis attacks [15]. Compared to attribute-level BFs, record-level BF encodings such as CLK and RBF reduce the risk of re-identification by such attacks [8,10].

Exploiting the fact that data sets with (near) uniform Hamming weight distribution of BFs are more difficult to attack (with existing attack methods) than data sets with non-uniform distributions, balancing BFs with constant Hamming weight has been proposed [15]. Balanced BFs can be constructed by concatenating a BF of length l with its negated copy (all bits flipped) and then permuting the $2l$ bits. Another proposed approach to harden BFs is XOR-folding, where a BF of length l is split into two halves of length $l/2$ each, and then bit-wise exclusive OR is applied to combine the two shorter BFs [15].

While balancing and XOR-folding are easy and data independent hardening techniques, salting with record-specific values has been suggested as an alternative hardening method where an additional (record-specific) value is concatenated with attribute values before being hashed into the BF [10]. A cryptanalysis attack is unlikely to be successful without knowing the salting key, however attributes suitable for salting might not be available in a database. Other, more experimental hardening techniques, include random bits and fake record injection [5,18], as well as BLIP (BLoom-and-flIP) which flips bits (noise addition) in a BF according to a differential privacy model [15]. Many of these hardening techniques improve security against attacks at the cost of a reduction in linkage quality [15]. In the experiments in Sect. 5 we will investigate if balancing and XOR-folding make BFs more resistant to our proposed attack.

4   Frequency-Based Bloom Filter Cryptanalysis

We now describe our frequency-based attack on BFs in detail. As shown in Fig. 1, the attack consists of two main steps. First, for each BF position we find its set of possible and *not* possible q-grams (steps (1a) and (1b) in Fig. 1). Next, we re-identify for each BF the set of attribute values that possibly were hashed into this BF (step (2) in Fig. 1). Algorithms 1 and 2 show the details of steps (1a) and (1b), and (2), respectively, which we describe in the next two sub-sections. We then provide an analysis of our attack and discuss its limitations.

4.1   Candidate Q-Gram Set Generation

For our attack we require as input a set of BFs, **B**, that we assume to come from a sensitive database (the one from which we aim to re-identify its sensitive values), and a set of attribute values, **V**, assumed to come from a public database. Each BF $\mathbf{b}_i \in \mathbf{B}$ and each attribute value $v_i \in \mathbf{V}$ has a frequency attached to it, denoted with $\mathbf{b}_i.f$ and $v_i.f$, respectively. Unlike previous attacks on BFs [8–10], we do not require any other information about how the BFs were encoded.

Algorithm 1 starts by initializing two empty sets for each BF position p: $\mathbf{c}^+[p]$ of possible q-grams at that position, and $\mathbf{c}^-[p]$ of *not* possible q-grams

---

Algorithm 1: *Candidate q-gram set generation* – Steps (1a) and (1b)

Input:
- $\mathbf{V}$: Set of attribute values and their frequencies from a public database
- $\mathbf{B}$: Set of BFs and their frequencies from the sensitive database
- l: Length of Bloom filters
- q: Length of sub-strings to extract from attribute values
- m: Minimum frequency for BFs and attribute values

Output:
- $\mathbf{C}$: List of possible q-grams for each BF position

```
 1:  c⁺[p] = {}, c⁻[p] = {}, 1 ≤ p ≤ l       // Initialize list of candidate q-gram sets
 2:  V_F = {v ∈ V : v.f ≥ m}                 // Attribute values with frequency of at least m
 3:  B_F = {b ∈ B : b.f ≥ m}                 // BFs with frequency of at least m
 4:  revSort(B_F), revSort(V_F)              // Sort according to frequencies, highest first
 5:  A = [(b_i, v_i) : b_i ∈ B_F, v_i ∈ V_F : b_i.f > b_j.f ∧ v_i.f > v_j.f : 1 ≤ i < j ≤ min(|V_F|, |B_F|)]
                                             // Align V_F and B_F as long as their frequencies are unique
 6:  for (b_i, v_i) ∈ A do:                  // Step (1a): Get candidate sets of q-grams
 7:      q_i = genQGramSet(v_i, q)           // Convert attribute value into its q-gram set
 8:      for 1 ≤ p ≤ l do:                   // Loop over all BF positions
 9:          if b_i[p] == 1 then:            // Bit at position p is 1
10:              c⁺[p] = c⁺[p] ∪ q_i         // Add to set of possible q-grams
11:          else:                           // Bit at position p is 0
12:              c⁻[p] = c⁻[p] ∪ q_i         // Add to set of not possible q-grams
13:  C = []                                  // Step (1b): Initialize empty list of q-gram sets
14:  for 1 ≤ p ≤ l do:                       // Loop over all BF positions to combine q-gram sets
15:      C.append(c⁺[p] \ c⁻[p])             // Remove not possible from possible q-grams
16:  return C
```

---

at that position. Next, in lines 2 and 3, we find all BFs and attribute values that occur at least m times. In line 4 we sort both the BFs and attribute values according to their frequencies in reverse order (most frequent first), and then we align a BF $\mathbf{b}_i$ and an attribute value v_i into the sorted list $\mathbf{A}$ of pairs $(\mathbf{b}_i, v_i)$. We do this as long as both the ith BF $\mathbf{b}_i$ and ith attribute value v_i have a unique frequency compared to the next, less frequent, BF or attribute value, respectively. This stopping criterion ensures that we do not have a BF that could correspond to two or more attribute values, and vice versa, as this would lead to more uncertainty in the mapping of q-grams into BF positions.

Step (1a) of our approach starts in line 6 and loops over pairs of aligned BFs and attribute values, $(\mathbf{b}_i, v_i) \in \mathbf{A}$. First, we convert v_i into its set of q-grams, $\mathbf{q}_i$, in line 7. Then we loop over all BF positions, $1 \leq p \leq l$, in line 8, and if the bit at position p in BF is 1 (i.e., $\mathbf{b}_i[p] = 1$), we add the q-gram set $\mathbf{q}_i$ to the set $\mathbf{c}^+[p]$ of possible q-grams at that position (line 10) because a 1-bit means any q-gram from $\mathbf{q}_i$ could have been hashed to that position. Conversely, if the BF bit at position p is 0, then no q-gram from $\mathbf{q}_i$ could have been hashed to that position and so we add the q-grams in $\mathbf{q}_i$ to $\mathbf{c}^-[p]$ (line 12).

In step (1b), line 13 onwards in Algorithm 1, we get the final set of q-grams for each BF position p as the set of possible q-grams ($\mathbf{c}^+[p]$) minus the set of not possible q-grams ($\mathbf{c}^-[p]$), and add these sets to the list $\mathbf{C}$ in line 15.

4.2 Attribute Value Re-identification

The re-identification step, shown in Algorithm 2, has as input the same set of BFs, $\mathbf{B}$, as used in the first step, as well as the list of candidate q-grams sets, $\mathbf{C}$,

Algorithm 2: *Attribute value re-identification* – Step (2)

Input:
- **B**: Set of BFs from the sensitive database (same as in Algorithm 1)
- **C**: List of possible q-grams for each BF position (from Algorithm 1)
- **G**: Set of attribute values we aim to re-identify in the set of BFs **B**

Output:
- **R**: List of possible attribute values re-identified for each BF in **B**

```
 1:  R = []                           // Initialize empty list of re-identified attribute values
 2:  for bᵢ ∈ B do:                   // Loop over all BFs
 3:      gᵢ = G                        // Initialize set of candidate attribute values as all possible values
 4:      for 1 ≤ p ≤ l do:            // Loop over all BF positions
 5:          if bᵢ[p] == 1 then:       // Bit at position p is 1
 6:              c = C[p]              // Set of possible q-grams at this bit position
 7:              for vⱼ ∈ gᵢ do:       // Check all candidate attribute values
 8:                  if (∀q ∈ c : q ∉ vⱼ) then:   // Check if no q-gram from c in attribute value
 9:                      gᵢ = gᵢ \ vⱼ   // Value could not have been hashed into this BF
10:      R.append(gᵢ)                 // Append possible attribute values for this BF
11:  return R
```

generated in Algorithm 1. A set of frequent attribute values, $\mathbf{G}$, also needs to be provided. These are the values we aim to re-identify (guess) in $\mathbf{B}$ using $\mathbf{C}$ (as shown in Fig. 1). The output of the algorithm is a set of one or more re-identified attribute value(s), $\mathbf{g}_i \subset \mathbf{G}$, for each BF $\mathbf{b}_i \in \mathbf{B}$, collected in the result list $\mathbf{R}$.

The algorithm loops over all BFs $\mathbf{b}_i$ in $\mathbf{B}$ (line 2 onwards), and for each $\mathbf{b}_i$ it initializes the set of possible candidate attribute values $\mathbf{g}_i$ (that potentially have been hashed into this BF) as all values in $\mathbf{G}$ (line 3). Next we loop over all BF positions p (line 4), and for any position that has a 1-bit ($\mathbf{b}_i[p] = 1$) we retrieve the set of possible q-grams $\mathbf{c} = \mathbf{C}[p]$ at that position (line 6).

Using $\mathbf{c}$, we now check for each attribute value v_j in $\mathbf{g}_i$ if *at least one of the q-grams* in $\mathbf{c}$ occurs in that value (lines 7 and 8). If v_j does not contain at least one q-gram from $\mathbf{c}$, it could not have generated the 1-bit at position p. If this is the case, in line 9 we remove v_j from the set of candidates $\mathbf{g}_i$. At the end of this loop (line 10), the set $\mathbf{g}_i$ will contain all attribute values from $\mathbf{G}$ that possibly have generated the BF $\mathbf{b}_i$, and we append this set to the results list $\mathbf{R}$. Note that positions with 0-bits do not help us to remove values $v_j \in \mathbf{g}_i$ because any q-gram in a value v_j is potentially hashed into other positions.

4.3   Analysis and Limitations

We now discuss the complexity and limitations of our cryptanalysis attack.

Complexity: We assume $N = |\mathbf{B}|$ is the number of BFs coming from the sensitive database, Q is the average number of q-grams per value in the set of attribute values $\mathbf{V}$ from the public database, k is the number of hash functions used to hash q-grams into BFs, and l is the length of a BF.

In step (1) of our attack, we first find all frequent BFs and attribute values that occur at least m times. This requires a linear scan through $\mathbf{B}$ and $\mathbf{V}$, respectively, which has a complexity of $O(N)$ and $O(|\mathbf{V}|)$. In line 4 of Algorithm 1, the sorting function used on values in $\mathbf{B}_F$ and $\mathbf{V}_F$ has a complexity of $O(M \cdot \log M)$, where $M = min(|\mathbf{B}_F|, |\mathbf{V}_F|)$. In step (1a), each attribute value in $\mathbf{A}$ is converted into its set of q-grams which are then added to the sets $\mathbf{c}^+[p]$ and $\mathbf{c}^-[p]$ for all BF bit positions $1 \leq b \leq l$. This step has a computational complexity of $O(M \cdot Q \cdot l)$. In step (1b), the sets of possible q-grams are added to list $\mathbf{C}$ by performing a set difference operation between $\mathbf{c}^+[p]$ and $\mathbf{c}^-[p]$. Assuming $\mathbf{Q}$ is the set of all unique q-grams generated from all values $v_i \in \mathbf{V}$, then on average $t = |\mathbf{Q}| \cdot k/l$ q-grams are hashed into each bit position. Therefore each set difference is of $O(t)$, and so step (1b) has an overall complexity of $O(t \cdot l)$ for l bit positions.

In step (2) of our attack, Algorithm 2 iterates through each BF $\mathbf{b}_i \in \mathbf{B}$ to re-identify the possible attribute values $\mathbf{g}_i \in \mathbf{G}$ that map to $\mathbf{b}_i$. For each bit position p that is set to 1, we check if any q-gram in $\mathbf{C}[p]$ occurs in a value $v_j \in \mathbf{g}_i$. This leads to a total complexity of $O(N \cdot |\mathbf{G}| \cdot l \cdot t)$, assuming the average size of a $\mathbf{C}[p]$ is t. Finally, the worst case space complexity of the list $\mathbf{R}$ returned by Algorithm 2 is $O(N \cdot |\mathbf{G}|)$ if every value in $\mathbf{G}$ is mapped to each $\mathbf{b}_i \in \mathbf{B}$.

Limitations: Our frequency-based cryptanalysis attack on BFs depends on several assumptions. First, we assume the attacker has access to a publicly available population database from where the set $\mathbf{V}$ of attribute values and their frequencies can be extracted. We also assume that $\mathbf{B}$ does contain a sub-set of BFs that occur several times, and that the frequency distribution of BFs in $\mathbf{B}$ is similar to the frequency distribution of a single or a sub-set of attribute values in $\mathbf{V}$. Without such frequency information, that can be aligned, our attack (like previous cryptanalysis attacks on BFs [7–10]) would not be possible.

5   Experiments and Results

We conducted our experimental study using real data sets from two domains. The first are a pair of North Carolina Voter Registration (NCVR) data sets (ftp://alt.ncsbe.gov/data/) collected in June 2014 (the database to be attacked) and October 2016 (the public database). These data sets contain over five million records of voters including their first names, surnames, and addresses. We present results of our attack individually on the first name attribute, as well as on the concatenation of the first name and surname attributes.

The second are a pair of census data sets (named UKCD) collected from the years 1851 and 1901 (used as the sensitive and public databases, respectively) for the town of Rawtenstall in England [6]. These data sets contain around 50,000 records with personal details of individuals. We again use the first name attribute, and the concatenation of the first name and surname attributes.

The parameter settings we use in our experiments are $q = [2, 3, 4]$ (length of sub-strings used in BF encoding), BF length $l = [250, 500, 1000, 2000]$, and either the double or random hashing method [15] as described in Sect. 3. We calculate the number of hash functions k based on l and q such that the false positive rate is minimized [16]. We also apply the BF hardening techniques balancing and XOR-folding [15]. We use different numbers for the most frequent attribute values (ranked by their frequencies) to be re-identified: $|\mathbf{G}| = [10, 20, 50, 100]$.

We evaluate the accuracy of our attack by calculating (1) the percentage of correct guesses with 1-to-1 matching, (2) the percentage of correct guesses

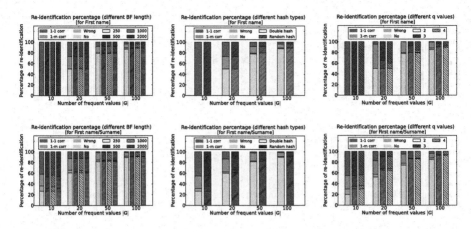

Fig. 3. Results for the NCVR data sets with first name (top) and combination of first and surname (bottom) for different BF lengths (left), different hashing methods (middle), and different values of q (right). No BF hardening was applied.

with 1-to-m (many) matching, (3) the percentage of wrong guesses, and (4) the percentage of no guesses, where these four percentages sum to 100. These four categories are labeled as '1-1 corr', '1-m corr', 'Wrong', and 'No' in the plots, respectively. We evaluate the efficiency of our attack using run time.

We implemented our attack using Python 2.7 and ran all experiments on a server with 64-bit Intel Xeon 2.4 GHz CPUs, 128 GB of memory and running Ubuntu 14.04. The programs and data sets are available from the authors.

Discussion: In Fig. 3 we show the results for the NCVR data sets. As can be seen, when the BF length l increases (left column) the percentage of correct re-identifications mostly increases, as with larger l the number of q-grams mapped to a certain bit position decreases. All values can be correctly re-identified for individual attributes when the number of frequent values is 10. Around half of all values can still be re-identified even when values from two attributes are combined. The middle column shows that random hashing (which supposedly improves privacy on BFs compared to double hashing [7,10]) does not provide improved protection against our attack, as a similar percentage of values can be correctly re-identified for both hashing approaches. As for using different values of q (right column) the accuracy of an attack somewhat improves when q is increased because larger values of q result in more unique q-grams.

The accuracy results for the UKCD data sets are shown in Fig. 4 for the same attributes as for the NCVR data sets. Similar re-identification patterns as with the NCVR data sets can be seen for different BF lengths (left) and hashing methods (middle). We also study how different BF hardening methods affect the re-identification accuracy, as shown in the right column plots in Fig. 4. For the single attribute case (top right), neither of the hardening techniques is capable of reducing the re-identification accuracy, however for the combined attribute case both hardening techniques improve the privacy of BF encoding.

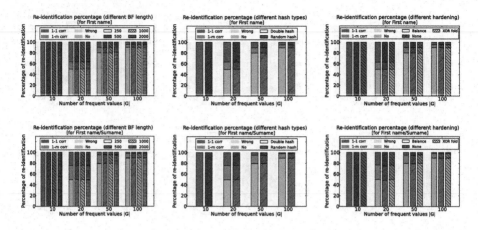

Fig. 4. Results for the UKCD data sets with first name (top) and combination of first and surname (bottom) for different BF lengths (left), different hashing methods (middle), and different BF hardening methods (right).

Table 1. Comparison of re-identification results with existing BF attack methods.

Publication	Data set	Num BFs	1-1 corr	Run time
Kuzu et al. [8]	NCVR first names	3,500	400	1,000 s
Kuzu et al. [9]	Patient names	20	4	Few seconds
”	”	42	0	> week
Niedermeyer et al. [10]	German surnames	7,580	934	> days
Kroll and Steinmetzer [7]	German names and locations	100K	44K	> days
Our approach	NCVR first names	10–100	7–10	0.73–0.75 s
”	NCVR first and surnames	10–100	3–6	1.5–1.9 s

In Table 1 we compare our attack method with existing approaches in terms of accuracy and efficiency (with results taken from the corresponding papers). As can be seen, our method is both more efficient and effective in re-identification, having both higher accuracy and reduced computational requirements.

These results show the vulnerability of basic BF encoding to our novel attack. They highlight the need for improved hardening techniques to overcome such attacks. Our attack provides data custodians with an efficient method to evaluate the privacy of their BF encoded databases before using them for PPRL.

Recommendations: As a set of guidelines for the practical application of BF based PPRL systems, to limit the vulnerability of such systems to known attack methods we recommend to use record-level BF encoding (CLK or RBF), apply advanced BF hardening methods [15], and reduce the frequency of bit patterns (for example, by salting) to prevent any frequency analysis.

6   Conclusions and Future Work

We have presented a novel efficient frequency-based attack on Bloom filters (BFs) that contain encoded sensitive attribute values intended for privacy-preserving record linkage (PPRL). Unlike earlier attacks on BFs for PPRL, our approach only requires an attacker to have access to a public database of attribute values, but no information about the BF encoding used. Our approach is faster than earlier attacks, making it feasible for database owners to efficiently validate the security of their encoded sensitive databases before they are being sent to other parties for conducting PPRL. We believe our attack is an important component to making PPRL more secure for practical applications.

As future work we will study how our approach can be modified for attacking composite and record-level BFs. We also plan to investigate the risk of re-identification when advanced hardening techniques, such as BLIP or BF salting, have been applied. Finally, our attack can be accelerated by further analyzing the re-identified attribute values, while correlations between 1-bits and q-gram sets across BFs could be identified using association rule mining techniques [3].

References

1. Bloom, B.: Space/time trade-offs in hash coding with allowable errors. Commun. ACM **13**(7), 422–426 (1970)
2. Boyd, J.H., Randall, S.M., Ferrante, A.M.: Application of privacy-preserving techniques in operational record linkage centres. In: Gkoulalas-Divanis, A., Loukides, G. (eds.) Medical Data Privacy Handbook, pp. 267–287. Springer, Cham (2015). doi:10.1007/978-3-319-23633-9_11
3. Ceglar, A., Roddick, J.F.: Association mining. ACM CSUR **3**(2), 5 (2006)
4. Christen, P.: Data Matching – Concepts and Techniques for Record Linkage, Entity Resolution, and Duplicate Detection. Springer, Heidelberg (2012)
5. Durham, E.A., Kantarcioglu, M., Xue, Y., Toth, C., Kuzu, M., Malin, B.: Composite Bloom filters for secure record linkage. IEEE TKDE **26**(12), 2956–2968 (2014)
6. Fu, Z., Christen, P., Zhou, J.: A graph matching method for historical census household linkage. In: Tseng, V.S., Ho, T.B., Zhou, Z.-H., Chen, A.L.P., Kao, H.-Y. (eds.) PAKDD 2014. LNCS (LNAI), vol. 8443, pp. 485–496. Springer, Cham (2014). doi:10.1007/978-3-319-06608-0_40
7. Kroll, M., Steinmetzer, S.: Who is 1011011111...1110110010? Automated cryptanalysis of Bloom filter encryptions of databases with several personal identifiers. In: Fred, A., Gamboa, H., Elias, D. (eds.) BIOSTEC 2015. CCIS, vol. 574, pp. 341–356. Springer, Cham (2015). doi:10.1007/978-3-319-27707-3_21
8. Kuzu, M., Kantarcioglu, M., Durham, E., Malin, B.: A constraint satisfaction cryptanalysis of Bloom filters in private record linkage. In: Fischer-Hübner, S., Hopper, N. (eds.) PETS 2011. LNCS, vol. 6794, pp. 226–245. Springer, Heidelberg (2011). doi:10.1007/978-3-642-22263-4_13
9. Kuzu, M., Kantarcioglu, M., Durham, E., Toth, C., Malin, B.: A practical approach to achieve private medical record linkage in light of public resources. JAMIA **20**(2), 285–292 (2013)
10. Niedermeyer, F., Steinmetzer, S., Kroll, M., Schnell, R.: Cryptanalysis of basic Bloom filters used for privacy preserving record linkage. JPC **6**(2), 59–79 (2014)

640   P. Christen et al.

11. Ranbaduge, T., Vatsalan, D., Christen, P., Verykios, V.: Hashing-based distributed multi-party blocking for privacy-preserving record linkage. In: Bailey, J., Khan, L., Washio, T., Dobbie, G., Huang, J.Z., Wang, R. (eds.) PAKDD 2016. LNCS (LNAI), vol. 9652, pp. 415–427. Springer, Cham (2016). doi:10.1007/978-3-319-31750-2_33
12. Randall, S., Ferrante, A., Boyd, J., Bauer, J., Semmens, J.: Privacy-preserving record linkage on large real world datasets. JBI **50**, 205–212 (2014)
13. Schnell, R., Bachteler, T., Reiher, J.: Privacy-preserving record linkage using Bloom filters. BMC Med. Inform. Decis. Mak. **9**(1), 41 (2009)
14. Schnell, R., Bachteler, T., Reiher, J.: A novel error-tolerant anonymous linking code. In: Working Paper, German Record Linkage Center (2011)
15. Schnell, R., Borgs, C.: Randomized response and balanced Bloom filters for privacy preserving record linkage. In: ICDMW DINA, pp. 218–224 (2016). doi:10.1109/ICDMW.2016.0038
16. Vatsalan, D., Christen, P.: Scalable privacy-preserving record linkage for multiple databases. In: ACM CIKM, Shanghai (2014)
17. Vatsalan, D., Christen, P.: Privacy-preserving matching of similar patients. JBI **59**, 285–298 (2016)
18. Vatsalan, D., Christen, P., Verykios, V.: A taxonomy of privacy-preserving record linkage techniques. IS **38**(6), 946–969 (2013)

Energy-Based Localized Anomaly Detection in Video Surveillance

Hung Vu[1], Tu Dinh Nguyen[1], Anthony Travers[2(✉)], Svetha Venkatesh[1], and Dinh Phung[1]

[1] Center for Pattern Recognition and Data Analytics,
Deakin University, Geelong, Australia
{hungv,tu.nguyen,svetha.venkatesh,dinh.phung}@deakin.edu.au
[2] Defence Science and Technology Organization (DSTO), Melbourne, Australia
Anthony.Travers@dsto.defence.gov.au

Abstract. Automated detection of abnormal events in video surveillance is an important task in research and practical applications. This is, however, a challenging problem due to the growing collection of data without the knowledge of what to be defined as "abnormal", and the expensive feature engineering procedure. In this paper we introduce a unified framework for anomaly detection in video based on the restricted Boltzmann machine (RBM), a recent powerful method for unsupervised learning and representation learning. Our proposed system works directly on the image pixels rather than hand-crafted features, it learns new representations for data in a completely unsupervised manner without the need for labels, and then reconstructs the data to recognize the locations of abnormal events based on the reconstruction errors. More importantly, our approach can be deployed in both offline and streaming settings, in which trained parameters of the model are fixed in offline setting whilst are updated incrementally with video data arriving in a stream. Experiments on three publicly benchmark video datasets show that our proposed method can detect and localize the abnormalities at pixel level with better accuracy than those of baselines, and achieve competitive performance compared with state-of-the-art approaches. Moreover, as RBM belongs to a wider class of deep generative models, our framework lays the groundwork towards a more powerful deep unsupervised abnormality detection framework.

1 Introduction

Developing intelligent video surveillance systems has been attracting research and application interest in computer vision community [11,15]. One of the most important surveillance problems is to automatically detect and analyze the abnormal events in video streams. The anomalous events are commonly assumed to be rare, irregular or significantly different from the others [15]. Examples include accesses

This work was partially supported by the Australian Research Council under the Discovery Project DP150100031 and the DST Group.

J. Kim et al. (Eds.): PAKDD 2017, Part I, LNAI 10234, pp. 641–653, 2017.
DOI: 10.1007/978-3-319-57454-7_50

to restricted area, leaving strange packages, movements in wrong direction, which can be captured by the camera monitoring systems in airports, car parks, stations and public spaces in general. Identifying the anomaly behaviors allows early intervention and in-time support to reduce the consequent cost.

The existing literature of anomaly detection on video data offers two approaches: supervised learning and unsupervised learning. Typical supervised methods include support vector data description [17], mixture of dynamic texture models [7] and supervised sparse coding [8], that use data labeled as *normal* to learn the model parameters and then judge the testing data as *abnormal* based on their probabilities or distances to the model. The methods in this approach, however, require the training data annotated with labels which are labor-intensive for large-scale data, rendering them inapplicable to the video streaming from surveillance systems where the amount of data grows super-abundantly. Moreover, it is also infeasible to model the diversity of normal event types in practice.

The unsupervised learning approach overcomes this issue by modeling the data without the need for labels. Typical methods include principle component analysis (PCA) [13], one-class support vector machines (OC-SVM) [1,16], Gaussian mixture models (GMM) [2,9], dynamic sparse coding [18], Bayesian non-parametric factor analysis (BNF) [10] and scan statistics [6]. The PCA learns a linear transformation to a lower dimensional linear space called "residual subspace", and then detect the anomalies using the residual signals of the projection of this data onto the residual subspace. The OC-SVM learns a hyperplane that achieves maximum separation between the normal data points and the origin, and then use the distance from a data point to this hyperplane to determine the abnormality. Alternatively, the GMM is a probabilistic method that models the data distribution, and use the posterior as the signal for anomaly detection. Other methods, such as sparse coding [18], compute the anomaly signal as the error of reconstructing data from a learned dictionary. Meanwhile, the BNF detects anomaly events using rareness scores that are based on the contributions of latent factors to reconstruct the scene. Scan statistics [6] measures the difference between statistical information inside and outside a region to discover anomalous objects. These methods, however, critically depend on the hand-crafted, low-level features extracted for video and image, such as histograms of oriented gradients (HOG) [18], optical flow features [6,9,13,16,17] and histograms of optical flow (HOF) [18]. The hand-crafted features rely on the design of preprocessing pipeline and data transformation, which is labor-intensive and normally requires exhaustive prior knowledge.

Recently there have been several studies that use deep learning techniques to automatically learn high-level representations for data to avoid the requirement of domain experts in designing features. When applying to anomaly detection for video data, the common approach is to extract features at the first stage (cf. autoencoders in [12,16]), and then use a separate classifier (e.g., OC-SVM) for detection at the second stage. An alternative method is to use the convolutional autoencoder (ConvAE) [4] to optimize the error when reconstructing the training data, and then use the reconstruction errors to recognize the abnormalities in testing data. Training these methods, however, is non-trivial due to their complicated architectures with multiple models or multiple layers.

In this paper, we propose a unified framework for anomaly detection in video based on the restricted Boltzmann machine (RBM) [3,5], a recent powerful energy-based method for unsupervised learning and representation learning. Our proposed system employs RBMs as core modules to model the complex distribution of data, capture the data regularity and variations, as a result effectively reconstruct the normal events that occur frequently in the data. The idea is to use the errors of reconstructed data to recognize the abnormal objects or behaviors that deviate significantly from the common. This is similar to the idea of using ConvAE. However, the key difference between our method and that approach is the ConvAE is a deterministic method that faces difficulty in modeling data which follow a set of probabilistic distributions, whilst ours is based on RBM, is probabilistic energy-based method that directly models the data distribution and captures data regularity.

Our framework is trained in a completely unsupervised manner that does not involve any explicit labels or implicit knowledge of what to be defined as abnormal. In addition, it can work directly on raw pixels without the need for expensive feature engineering procedure. Another advantage of our method is the capability of detecting the exact boundary of local abnormality in the video frame. To handle the video data coming in a stream, we further extend our method to incrementally update parameters without retraining the models from scratch. Our solution can be easily deployed in arbitrary surveillance streaming setting without the expensive calibration requirement.

We qualitatively and quantitatively evaluate the performance of our anomaly detection framework through comprehensive experiments on three real-world datasets. Our primary target is to investigate the capabilities of capturing data regularity, reconstructing the data and detecting local abnormalities of our system. The experimental results show that our proposed method can effectively reconstruct the data regularity, and thus detect and localize the abnormalities at pixel level with better accuracies than those of baselines, and competitive performance compared with state-of-the-art approaches.

In short, our contributions are: (i) a novel unified RBM-based framework that can act as a completely unsupervised model on raw pixels; thus there is no need to extract hand-crafted features; (ii) an incremental version of our system that can efficiently work in a streaming setting; and (iii) a comprehensive evaluation of the effectiveness of our method on real-world video surveillance application.

2 Framework

We now describe our energy-based framework to detect abnormal events in video surveillance data. First we briefly review restricted Boltzmann machines that are the key components in our proposed system. We then present our framework and the extension for streaming video data.

2.1 Restricted Boltzmann Machine

A restricted Boltzmann machine (RBM) [3,14] is a bipartite undirected graphical model wherein the bottom layer contains observed variables called visible units and the top layer consists of latent *representational variables*, known as hidden units. Two layers are fully connected but there is no connection within layers.

Model Representation. More formally, assume a binary RBM with M visible units and K hidden units, let $\boldsymbol{x}$ denote the set of visible variables: $\boldsymbol{x} = [x_1, x_2, \ldots, x_M]^\top \in \{0,1\}^M$ and $\boldsymbol{h}$ indicate the set of hidden ones: $\boldsymbol{h} = [h_1, h_2, \ldots, h_K]^\top \in \{0,1\}^K$. The RBM assigns an energy function for a joint configuration over the state $(\boldsymbol{x}, \boldsymbol{h})$ as:

$$E\left(\boldsymbol{x}, \boldsymbol{h}; \psi\right) = -\left(\boldsymbol{a}^\top \boldsymbol{x} + \boldsymbol{b}^\top \boldsymbol{h} + \boldsymbol{x}^\top \boldsymbol{W} \boldsymbol{h}\right) \tag{1}$$

where $\psi = \{\boldsymbol{a}, \boldsymbol{b}, \boldsymbol{W}\}$ is the set of parameters. $\boldsymbol{a} = [a_m]_M \in \mathbb{R}^M, \boldsymbol{b} = [b_k]_K \in \mathbb{R}^K$ are the biases of hidden and visible units respectively, and $\boldsymbol{W} = [w_{mk}]_{M \times K} \in \mathbb{R}^{M \times K}$ represents the weights connecting the hidden and visible units. The model admits a Boltzmann distribution (also known as Gibbs distribution) as follows:

$$p\left(\boldsymbol{x}, \boldsymbol{h}; \psi\right) = \frac{1}{\mathcal{Z}\left(\psi\right)} \exp\left\{-E\left(\boldsymbol{x}, \boldsymbol{h}; \psi\right)\right\} \tag{2}$$

where $\mathcal{Z}\left(\psi\right) = \sum_{\boldsymbol{x},\boldsymbol{h}} \exp\left\{-E\left(\boldsymbol{x}, \boldsymbol{h}; \psi\right)\right\}$ is the normalization constant, also called partition function. This guarantees that the $p\left(\boldsymbol{x}, \boldsymbol{h}; \psi\right)$ is a proper density function.

Since the network has no intra-layer connections, units in one layer become conditionally independent given the other layer. Thus the conditional distributions over visible and hidden units are factorized as:

$$p\left(\boldsymbol{h} \mid \boldsymbol{x}; \psi\right) = \prod_{k=1}^{K} p\left(h_k \mid \boldsymbol{x}; \psi\right) \tag{3} \quad p\left(\boldsymbol{x} \mid \boldsymbol{h}; \psi\right) = \prod_{m=1}^{M} p\left(x_m \mid \boldsymbol{h}; \psi\right) \tag{4}$$

Parameter Estimation. As an energy-based model, the learning goal of RBM is to minimize the energy in Eq. (1) of the observed data. As the visible probability is inversely proportional to the energy as shown in Eq. (2), it is equivalent to maximize the following log-likelihood of data: $\log p\left(\boldsymbol{x}; \psi\right) = \log \sum_{\boldsymbol{h}} p\left(\boldsymbol{x}, \boldsymbol{h}; \psi\right)$. The parameters are updated in a gradient ascent fashion as follows:

$$\psi \leftarrow \psi + \eta\left(\mathbb{E}_{p(\boldsymbol{x},\boldsymbol{h};\psi)}\left[\nabla_\psi E\left(\boldsymbol{x}, \boldsymbol{h}; \psi\right)\right] - \mathbb{E}_{p(\boldsymbol{h}\mid\boldsymbol{x};\psi)}\left[\nabla_\psi E\left(\boldsymbol{x}, \boldsymbol{h}; \psi\right)\right]\right)$$

for a learning rate $\eta > 0$. Here $\mathbb{E}_{p(\boldsymbol{x},\boldsymbol{h};\psi)}$ denotes the expectation with respect to the full model distribution and $\mathbb{E}_{p(\boldsymbol{h}\mid\boldsymbol{x};\psi)}$ the data expectation with respect to the conditional distribution given the observed $\boldsymbol{x}$. Whilst $\mathbb{E}_{p(\boldsymbol{h}\mid\boldsymbol{x};\psi)}$ can be computed efficiently, $\mathbb{E}_{p(\boldsymbol{x},\boldsymbol{h};\psi)}$ is generally intractable. Thus we must resort to approximate methods, and in this paper, we choose contrastive divergence (CD) [5] as it proves to be fast and accurate.

Data Reconstruction. Once the model parameters ψ has been learned, the RBM can project an input data $\boldsymbol{x}$ onto the hidden space to obtain the new representation $\tilde{\boldsymbol{h}} = [\tilde{h}_1, \tilde{h}_2, \ldots, \tilde{h}_K]^\top$ where $\tilde{h}_k$ is shorthand for the posterior

$\tilde{h}_k = p(h_k = 1 \mid \boldsymbol{x}) = \sigma(b_k + \sum_m w_{mk}x_m)$, in which $\sigma(x)$ is the sigmoid function $\sigma(x) = (1 + e^{-x})^{-1}$. This hidden posterior vector is then mapped back into the input space to form the reconstructed data $\tilde{\boldsymbol{x}} = [\tilde{x}_1, \tilde{x}_2, \ldots, \tilde{x}_M]^{\top}$ where $\tilde{x}_m = p\left(x_m = 1 \mid \tilde{\boldsymbol{h}}; \psi\right) = \sigma\left(a_m + \sum_k w_{mk}\tilde{h}_k\right)$, similarly to the hidden posterior. These projection and mapping are very efficient due to the nice factorizations in Eqs. (3, 4).

2.2 Anomaly Detection Using RBM

We now describe our proposed framework that is based on the RBM to detect anomaly events for each frame in video data. In general, our system is a two-phase pipeline: training phase and detecting phase. Particularly in the training phase, our model: (i) takes a series of video frames in the training data as a collection of images, (ii) divides each image into patches, (iii) gathers similar patches into clusters, and (iv) learns separate RBM for each cluster using the image patches. The detecting phase consists of three steps: (i) collecting image patches in the testing video for each cluster, and then using the learned RBM to reconstruct the data for the corresponding cluster of patches, (ii) proposing the regions that are *potential* to be abnormal by applying a predefined threshold to reconstruction errors, and then finding connected components of these candidates and filtering out those too small, and (iii) updating the model incrementally for the data stream. The overview of our framework is illustrated in Fig. 1. In what follows, we describe training and detecting phases in more details.

Training Phase. Assume that the training data consists of N video frames with the size of $H \times W$ pixels, let denote $\mathcal{D} = \{\boldsymbol{x}_t \in \mathbb{R}^{H \times W}\}_{t=1}^{N}$. In real-life video surveillance data, $H \times W$ is usually very large (e.g., hundreds of thousand pixels), hence it is often infeasible for a single RBM to handle such high-dimensional image. This is because the high-dimensional input requires a more complex model with an extremely large number of parameters (i.e., millions). This makes the parameter learning more difficult and less robust since it is hard to control the bounding of hidden activation values. Thus the hidden posteriors are easily collapsed into either zeros or ones, and no more learning occurs.

To tackle this issue, one can reduce the data dimension using dimensionality reduction techniques or by subsampling the image to smaller size. This solution, however, is computational demanding and may lose much information of the original data. In this work we choose to apply RBMs directly to raw imaginary pixels whilst try to preserve information. To that end, we train our model on $h \times w$ patches where we divide each image $\boldsymbol{x}_t$ into a grid of $N_h \times N_w$ patches: $\boldsymbol{x}_t = \{\boldsymbol{x}_t^{i,j} \mid 1 \le i \le N_h, 1 \le j \le N_w\}$. This approach greatly reduces the data dimensionality and hence requires smaller models. One way is to learn independent RBMs on patches at each location (i, j). However, this would result in an excessive number of models, for example, 400 RBMs to work on the 240×360 image resolution and 12×18 patch size, hence leading to very high computational complexity and memory demand.

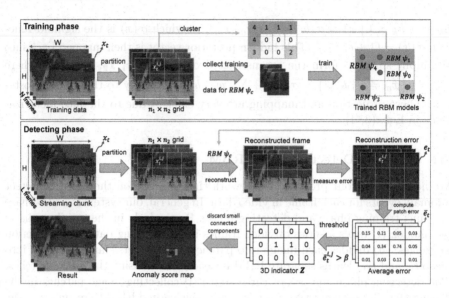

Fig. 1. The overview of our proposed framework.

Our solution is to reduce the number of models by grouping all similar patches from different locations for learning a single model. We observe that it is redundant to train a separate model for each location of patches since most adjacent patches such as pathways, walls and nature strips in surveillance scenes have similar appearance and texture. Thus we first train a RBM with a small number of hidden units ($K = 4$) on all patches $\{x_t^{i,j}\}$ of all video frames. We then compute the hidden posterior $\tilde{h}$ for each image patch $x_t^{i,j}$ and binarize it to obtain the binary vector: $\tilde{h} = \left[\mathbb{I}\left(\tilde{h}_1 > 0.5\right), \ldots, \mathbb{I}\left(\tilde{h}_K > 0.5\right)\right]$ where $\mathbb{I}(\bullet)$ is the indicator function. Next this binary vector is converted to an integer value in decimal system, e.g., 0101 converted to 5, which we use as the *pseudo-label* $\lambda_t^{i,j}$ of the cluster of the image patch $x_t^{i,j}$. The cluster label $c^{i,j}$ for all patches at location (i,j) is chosen by voting the pseudo-labels over all N frames: $\lambda_1^{i,j}, \lambda_2^{i,j}, \ldots, \lambda_N^{i,j}$. Let C denote the number of unique cluster labels in the set $\{c^{i,j} \mid 1 \leq i \leq N_h, 1 \leq j \leq N_w\}$, we finally train C independent RBMs with a larger number of hidden units ($K = 100$), each with parameter set ψ_c for all patches with the same cluster label c.

Detecting Phase. Once all RBMs have been learned using the training data, they are used to reveal the irregular events in the testing data. The pseudocode of this phase is given in Algorithm 1. Overall, there are three main steps: reconstructing the data, detecting local abnormal objects and updating models incrementally. In particular, the stream of video data is first split into chunks of L non-overlapping frames, each denoted by $\{x_t\}_{t=1}^{L}$. Each patch $x_t^{i,j}$ is then reconstructed to obtain the reconstruction $\tilde{x}_t^{i,j}$ using the learned RBM with parameters $\psi_{c^{i,j}}$, and all together form the reconstructed data $\tilde{x}_t$ of the frame x_t. The reconstruction error $e_t = [e_t^{i,j}] \in \mathbb{R}^{H \times W}$ is then computed as: $e_t^{i,j} = |x_t^{i,j} - \tilde{x}_t^{i,j}|$.

To detect anomal pixels, one can compare the reconstruction error e_t with a given threshold. This approach, however, may produce many false alarms when normal pixels are reconstructed with high errors, and may fail to cover the entire anomaly objects in such a case that they are fragmented into isolated high error parts. Our solution is to work on the average error $\bar{e}_t^{i,j} = ||e_t^{i,j}||_2 / (h \times w)$ over patches rather than individual pixels. These errors are then compared with a predefined threshold β. All pixels in $x_t^{i,j}$ are considered abnormal if $\bar{e}_t^{i,j} \geq \beta$.

Applying the above procedure and then concatenating L frames, we obtain a binary 3D rectangle $Z \in \{0,1\}^{L \times H \times W}$ wherein $z_{i,j,k} = 1$ indicates the abnormal voxel whilst $z_{i,j,k} = 0$ the normal one. Throughout the experiments, we observe that most of abnormal voxels in Z are detected correctly, but there still exist several small groups of voxels are incorrect. We further filter out these false positive voxels by connecting all their *related* neighbors. More specifically, we first build a sparse graph whose nodes are abnormal voxels $z_{i,j,k} = 1$ and edges are the connections of these voxels with their abnormal neighbors $z_{i+u,j+v,k+t} = 1$ where $u, v, t \in \{-1, 0, 1\}$ and $|u| + |v| + |t| > 0$. We then find all connected components in this graph, and discard small components spanning less than γ contiguous frames. The average error $\bar{e}_t^{i,j}$ after this component filtering step can be used as final anomaly score.

Algorithm 1. RBM anomaly detection

Input: Video chunk $\{x_t\}_{t=1}^L$, models $\{\psi_c\}_{c=1}^C$, thresholds β and γ
Output: Detection Z, score $\{\bar{e}_t^{i,j}\}$
1: **for** $t \leftarrow 1, \ldots, L$ **do**
2:     **for** $x_t^{i,j} \in x_t$ **do**
3:         $\tilde{x}_t^{i,j} \leftarrow$ reconstruct$(x_t^{i,j}, \psi_{c^{i,j}})$
4:         $e_t^{i,j} \leftarrow |x_t^{i,j} - \tilde{x}_t^{i,j}|$
5:         $\bar{e}_t^{i,j} \leftarrow \frac{1}{h \times w} ||e_t^{i,j}||_2$
6:         **if** $\bar{e}_t^{i,j} \geq \beta$ **then**
7:             **for** $p \in x_t^{i,j}$ **do**
8:                 $Z(p) \leftarrow 1$
9:             **end for**
10:        **else**
11:            **for** $p \in x_t^{i,j}$ **do**
12:                $Z(p) \leftarrow 0$
13:            **end for**
14:        **end if**
15:     **end for**
16:     **for** $c \leftarrow 1, \ldots, C$ **do**
17:         $X_t^c \leftarrow \{x_t^{i,j} \mid c^{i,j} = c\}$
18:         $\psi_c \leftarrow$ updateRBM(X_t^c, ψ_c)
19:     **end for**
20: **end for**
21: $Z \leftarrow$ remove_small_components(Z, γ)

In the scenario of streaming videos, the scene frequently changes over time and it could be significantly different from those are used to train RBMs. To tackle this issue, we extend our proposed framework to enable the RBMs to adapt themselves to the new video frames. For every incoming frame t, we extract the image patches and update the parameters $\psi_{1:C}$ of C RBMs in our framework following the procedure in the training phase. Recall that the RBM parameters are updated iteratively using gradient ascent, thus here we use several epochs to ensure the information of new data are sufficiently captured by the models.

One problem is the anomalous objects can be presented in different sizes in the video. To deal with this issue, we apply our framework to the video data at different scales whilst keeping the same patch size $h \times w$. This would help the patch partially or entirely cover objects at certain scales. To that end, we rescale the original video into different resolutions, then employ the same procedure above to compute the average reconstruction error map $\bar{e}_t$ and 3D rectangular indicators $\boldsymbol{Z}$. The average error maps are then aggregated into one matrix using max operation. Likewise, indicator tensors are merged into one before finding the connected components. We also use overlapping patches to localize anomalous objects more accurately. Pixels in the overlapping regions are averaged when combining patches into the whole map.

3   Experiment

In this section, we empirically evaluate the performance of our anomaly detection framework both qualitatively and quantitatively. Our aim is to investigate the capabilities of capturing data regularity, reconstructing the data and detecting local abnormalities of our system. For quantitative analysis, we compare our proposed method with several up-to-date baselines.

We use 3 public datasets: UCSD Ped 1, Ped 2 [7] and Avenue [8]. Under the unsupervised setting, we disregard labels in the training videos and train all methods on these videos. The learned models are then evaluated on the testing videos by computing 2 measures: area under ROC curve (AUC) and equal error rate (EER) at frame-level (no anomaly object localization evaluation) and pixel-level (40% of ground-truth anomaly pixels are covered by detection), following the evaluation protocol used in [7] and at dual-pixel level (pixel-level constraint above and at least α percent of detection is true anomaly pixels) in [12]. Note that pixel-level is a special case of dual-pixel where $\alpha = 0$. Since the videos are provided at different resolution, we first resize all into the same size of 240×360.

For our framework, we duplicate and rescale video frames to multiscale copies with the ratios of 1.0, 0.5 and 0.25, and then use 12×18 image patches with 50% overlapping between two adjacent patches. Each RBM now consists of 216 visible units and 4 hidden units for clustering step whilst 100 hidden units for training and detecting phases. All RBMs are trained using CD_1 with learning rate $\eta = 0.1$. To simulate the streaming setting, we split testing videos in non-overlapping chunks of $L = 20$ contiguous frames and use 20 epochs to incrementally update parameters of RBMs. The thresholds β and γ to determine anomaly are set to 0.003 and 10 respectively. Those hyperparameters have been tuned to reduce false alarms and to achieve the best balanced AUC and EER scores.

3.1   Region Clustering

In the first experiment, we examine the clustering performance of RBM. Figure 2 shows the cluster maps discovered by RBM on three datasets. Using 4 hidden units, the RBM can produce a maximum of 16 clusters, but in fact, the model returns less and varied number of clusters for different datasets at different scales. For example, (6, 7, 10) similar regions at scales (1.0, 0.5, 0.25) are found for Ped 1 dataset, whilst these numbers for Ped 2 and Avenue dataset are (9, 9, 8) and (6, 9, 9) respectively. This suggests the capability of automatically selecting the appropriate number of clusters of RBM.

Fig. 2. Clustering result on some surveillance scenes at the first scale: (first column) example frames; (second) cluster maps produced by RBM; (third) filters learned by RBM; and (fourth) cluster maps produced by k-means.

For comparison, we run k-means algorithm with $k = 8$ clusters, the average number of clusters of RBM. It can be seen from Fig. 2 that the k-means fails to connect large regions which are fragmented by the surrounding and dynamic objects, for example, the shadow of tree on the footpath (Case 1), pedestrians walking at the upper side of the footpath (Case 2). It also assigns several wrong labels to small patches inside a larger area as shown in Case 3. By contrast, the RBM is more robust to the influence of environmental factors and dynamic foreground objects, and thus produces more accurate clustering results. Taking a closer look at the filters learned by RBM at the third column in the figure, we can agree that the RBM learns the basic features such as homogeneous regions, vertical, horizontal, diagonal edges and corners, which then can be combined to construct the entire scene.

3.2   Data Reconstruction

We next demonstrate the capability of our framework on the data reconstruction. Figure 3 shows an example of reconstructing the video frame in Avenue dataset. Here the abnormal object is a girl walking toward the camera. It can be seen that our model can correctly locate this outlier behavior based on the reconstruction errors shown in Figure 3(c) and (d). This is because the RBM can capture the data regularity, thus produces low reconstruction errors for regular objects and high errors for irregular or anomalous ones as shown in Figure 3(b) and (c).

To examine the change of reconstruction errors in a stream of video frames, we visualize the maximum average reconstruction error in a frame as a function of frame index as shown in Fig. 4. The test video #1 in UCSD Ped 1 dataset contains some normal frames of walking on a footpath, followed by the appearance of a cyclist moving towards the camera. Our system could not detect the emergence of the cyclist since the object is too small and cluttered by many surrounding

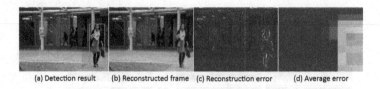

(a) Detection result   (b) Reconstructed frame   (c) Reconstruction error   (d) Average error

Fig. 3. Data reconstruction of our method on Avenue dataset: (a) the original frame with detected o2utlier female (yellow region) and ground-truth (red rectangle), (b) reconstructed frame, (c) reconstruction error image, (d) average reconstruction errors of patches. (Color figure online)

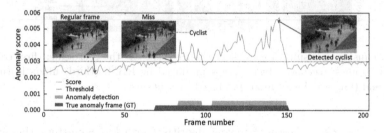

Fig. 4. Average reconstruction error per frame in test video #1 of UCSD Ped 1 dataset. The shaded green region illustrates anomalous frames in the ground truth, while the yellow anomalous frames detected by our method. The blue line shows the threshold. (Color figure online)

pedestrians. However, after several frames, the cyclist is properly spotted by our system with the reconstruction errors far higher than the threshold.

3.3 Anomaly Detection Performance

In the last investigation, we compare our offline RBM framework and its streaming version (called S-RBM) with the unsupervised methods for anomaly detection in the literature. We use 4 baselines for comparison: principal component analysis (PCA), one-class support vector machine (OC-SVM) [1], gaussian mixture models (GMM), and convolutional autoencoder (ConvAE) [4]. We use the variant of PCA with optical flow features from [13], and adopt the results of ConvAE from the original work [4]. The results of ConvAE are already compared with recent state-of-the-art baselines including supervised methods.

We follow similar procedures to what of our proposed framework for OC-SVM and GMM, but apply these baselines on image patches clustered by k-means. The kernel width and lower bound of the fraction of support vectors of OC-SVM are set to 0.1 and 10^{-4} respectively. In GMM model, the number of Gaussian components is set to 20 and the anomaly threshold is -50. These hyperparameters are also tuned to obtain the best cross-validation results. It is noteworthy that it is not straightforward to implement the incremental versions of the baselines, thus we do not include them here.

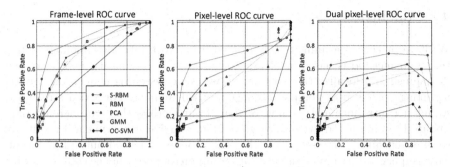

Fig. 5. Comparison ROC curves on UCSD Ped 2. Three figures share the same legend. Higher curves indicate better performance. It is notable that, unlike frame and pixel-level evaluations, dual-pixel level curves may end at any points lower than (1,1).

The ROC curves are shown in Fig. 5 whilst AUC and EER scores are reported in Table 1. Both RBM and S-RBM outperform the PCA, OC-SVM, GMM with higher AUC and lower EER scores. Specially, our methods can produce higher AUC scores at dual pixel-level which shows better quality in localizing anomaly regions. Additionally, S-RBM achieves fairly comparable results with the ConvAE. It is noteworthy that the ConvAE is a 12-layer deep architecture consisting of sophisticated connections between its convolutional and pooling layers. On the other hand, our RBM anomaly detector has only two layers, but obtains a respectable performance. We believe that our proposed framework is a promising system to detect abnormalities in video surveillance applications.

Table 1. Anomaly detection results (AUC and EER) at frame-level, pixel-level and dual pixel-level ($\alpha = 5\%$) on 3 datasets. Higher AUC and lower EER indicate better performance. Meanwhile, high dual-pixel values point out more accurate localization. We do not report EER for dual-pixel level because this number do not always exist. Best scores are in bold. Note that the frame-level results of ConvAE are taken from [4], but the pixel-level and dual-pixel level results are not available.

	Ped1					Ped2					Avenue				
	Frame		Pixel		Dual	Frame		Pixel		Dual	Frame		Pixel		Dual
	AUC	EER	AUC	EER	AUC	AUC	EER	AUC	EER	AUC	AUC	EER	AUC	EER	AUC
PCA	60.28	43.18	25.39	39.56	8.76	73.98	29.20	55.83	24.88	44.24	74.64	30.04	52.90	37.73	43.74
OC-SVM	59.06	42.97	21.78	37.47	11.72	61.01	44.43	26.27	26.47	19.23	71.66	33.87	33.16	47.55	33.15
GMM	60.33	38.88	36.64	35.07	13.60	75.20	30.95	51.93	18.46	40.33	67.27	35.84	43.06	43.13	41.64
ConvAE	**81.00**	**27.90**	-	-	-	**90.00**	21.70	-	-	-	70.20	**25.10**	-	-	-
RBM	64.83	37.94	41.87	36.54	16.06	76.70	28.56	59.95	19.75	46.13	74.88	32.49	43.72	43.83	41.57
S-RBM	70.25	35.40	**48.87**	**33.31**	**22.07**	86.43	**16.47**	**72.05**	**15.32**	**66.14**	**78.76**	27.21	**56.08**	**34.40**	**53.40**

4   Conclusion

We have presented a unified energy-based framework for video anomaly detection. Our method is based on RBMs to capture data regularity, and hence can distinguish and localize the irregular events. Our system is trained directly on the image pixels in a completely unsupervised manner. For video streaming, we further introduce a streaming version of our method that can incrementally update the parameters when new video frames arrive. Experimental results on several benchmark datasets show that the proposed method outperforms typical unsupervised baselines and achieves competitive performance compared with state-of-the-art method for anomaly detection.

Finally we note that our proposed approach is designed so that multiple RBMs are trained to capture different image statistics localized at different regions. Thus it is immediately amendable to a distributed and parallel implementation for a scalable system. Furthermore, as RBM belongs to a wider class of deep generative models, our framework is readily generalized to a more powerful deep unsupervised abnormality detection framework.

References

1. Amer, M., Goldstein, M., Abdennadher, S.: Enhancing one-class support vector machines for unsupervised anomaly detection. In: SIGKDD, pp. 8–15 (2013)
2. Basharat, A., Gritai, A., Shah, M.: Learning object motion patterns for anomaly detection and improved object detection. In: CVPR (2008)
3. Freund, Y., Haussler, D.: Unsupervised learning of distributions on binary vectors using two layer networks. Technical report, Santa Cruz, CA, USA (1994)
4. Hasan, M., Choi, J., Neumann, J., Roy-Chowdhury, A.K., Davis, L.S.: Learning temporal regularity in video sequences. In: CVPR 2016 (2016)
5. Hinton, G.: Training products of experts by minimizing contrastive divergence. Neural Comput. **14**(8), 1771–1800 (2002)
6. Hu, Y., Zhang, Y., Davis, L.S.: Unsupervised abnormal crowd activity detection using semiparametric scan statistic. In: CVPRW, pp. 767–774 (2013)
7. Li, W.X., Mahadevan, V., Vasconcelos, N.: Anomaly detection and localization in crowded scenes. PAMI **36**(1), 18–32 (2014)
8. Lu, C., Shi, J., Jia, J.: Abnormal event detection at 150 fps in matlab. In: ICCV (2013)
9. Lu, T., Wu, L., Ma, X., Shivakumara, P., Tan, C.L.: Anomaly detection through spatio-temporal context modeling in crowded scenes. In: ICPR (2014)
10. Nguyen, V., Phung, D., Pham, D.S., Venkatesh, S.: Bayesian nonparametric approaches to abnormality detection in video surveillance. Ann. Data Sci. (AoDS) **2**(1), 21–41 (2015)
11. Oluwatoyin, P.P., Wang, K.: Video-based abnormal human behavior recognition - a review. IEEE Trans. Syst. Man Cybern. 865–878 (2012)
12. Sabokrou, M., Fathy, M., Hoseini, M.: Real-time anomalous behavior detection and localization in crowded scenes. In: CVPRW (2015)
13. Saha, B., Pham, D.S., Lazarescu, M., Venkatesh, S.: Effective anomaly detection in sensor networks data streams. In: ICDM, pp. 722–727 (2009)

14. Smolensky, P.: Information processing in dynamical systems: foundations of harmony theory. In: Parallel Distributed Processing: Explorations in the Microstructure of Cognition, vol. 1, pp. 194–281. MIT Press, Cambridge (1986)

15. Sodemann, A.A., Ross, M.P., Borghetti, B.J.: A review of anomaly detection in automated surveillance. IEEE Trans. Syst. Man Cybern. Part C (Appl. Rev.) **42**(6), 1257–1272 (2012)

16. Xu, D., Ricci, E., Yan, Y., Song, J., Sebe, N.: Learning deep representations of appearance and motion for anomalous event detection. In: BMVC (2015)

17. Zhang, Y., Lu, H., Zhang, L., Ruan, X.: Combining motion and appearance cues for anomaly detection. Pattern Recogn. **51**, 443–452 (2016)

18. Zhao, B., Fei-Fei, L., Xing, E.P.: Online detection of unusual events in videos via dynamic sparse coding. In: CVPR, Washington, DC, USA, pp. 3313–3320 (2011)

A Fast Fourier Transform-Coupled Machine Learning-Based Ensemble Model for Disease Risk Prediction Using a Real-Life Dataset

Raid Lafta[1,2]($\boxtimes$), Ji Zhang[1], Xiaohui Tao[1], Yan Li[1], Wessam Abbas[1],
Yonglong Luo[3], Fulong Chen[3], and Vincent S. Tseng[4]

[1] Faculty of Health, Engineering and Sciences, University of Southern Queensland,
Toowoomba, Australia
{RaidLuaibi.Lafta,ji.zhang,xtao,yan.li}@usq.edu.au
[2] Computer Center, University of Thi-Qar, Thi-Qar, Iraq
[3] School of Mathematics and Computer Science,
Anhui Normal University, Wuhu, China
ylluo@ustc.edu.cn, long005@mail.ahnu.edu.cn
[4] Department of Computer Science, National Chiao Tung University,
Hsinchu, Taiwan
vtseng@cs.nctu.edu.tw

Abstract. The use of intelligent technologies in clinical decision making have started playing a vital role in improving the quality of patients' life and helping in reduce cost and workload involved in their daily healthcare. In this paper, a novel fast Fourier transform-coupled machine learning based ensemble model is adopted for advising patients concerning whether they need to take the body test today or not based on the analysis of their medical data during the past a few days. The weighted-vote based ensemble attempts to predict the patients condition one day in advance by analyzing medical measurements of patient for the past k days. A combination of three algorithms namely neural networks, support vector machine and Naive Bayes are utilized to make an ensemble framework. A time series telehealth data recorded from patients is used for experimentations, evaluation and validation. The Tunstall dataset were collected from May to October 2012, from industry collaborator Tunstall. The experimental evaluation shows that the proposed model yields satisfactory recommendation accuracy, offers a promising way for reducing the risk of incorrect recommendations and also saving the workload for patients to conduct body tests every day. The proposed method is, therefore, a promising tool for analysis of time series data and providing appropriate recommendations to patients suffering chronic diseases with improved prediction accuracy.

Keywords: Fast Fourier transformation · Ensemble model · Recommender system · Heart failure · Time series prediction · Telehealth

© Springer International Publishing AG 2017
J. Kim et al. (Eds.): PAKDD 2017, Part I, LNAI 10234, pp. 654–670, 2017.
DOI: 10.1007/978-3-319-57454-7_51

1 Introduction

The chronical diseases such as heart disease have become the main public health issue worldwide which accounting for 50% of global mortality burden [1]. Due to lack of chronical diseases prediction tools, most of the populations around the world may be suffering from chronical diseases [2]. Recently, the survival rates have been noticeably increased due to technological improvements in diseases prediction models.

One of the important problems in medical science is accurate prediction of disease based on analysing historical data of patients. The data mining techniques and statistical analysis have been extensively used to provide major assistance to experts in disease prediction [27].

Recommendation systems can be defined as computer applications that assist and support medical practitioners in improved decision-making recommendation [3,28,29]. Those systems can help in minimizing medical errors and providing more detailed data analysis in shorter time [4].

In the recent years, the ensemble methods have been very robust for the blend of various predictive models. The major purpose of ensemble model is to improve the overall accuracy of prediction model. An ensemble is a set of base learners that use to enhance the prediction performance of low-quality data [5]. Bagging is an ensemble algorithm that was proposed by Breiman in mid-1990's [6]. Empirical results showed that both regression and classification problem ensemble are often more accurate than individual classifier that make them up [5]. Therefore, much research efforts have been invested using machine learning ensemble for chronical diseases prediction.

Least square-support vector machine (LS-SVMs) are a relatively new kind of machine learning techniques that was proposed by [17]. They have been recently used in the field of disease prediction. There are several studies in disease prediction filed where LS-SVMs are used. LS-SVM has been used by [18] for heart disease prediction. Muscle fatigue prediction in electromyogram (sEMG) signal is implemented using LS-SVM that proposed by [19]. Finally, LS-SVM has also been successfully applied by [20] to predict breast cancer.

Due to the ensemble outperforms individual classifiers, several such ensemble approaches have been proposed recently. A combination of different data mining techniques have been applied on different datasets. An ensemble framework based on different classifiers has been used by Das et al. and Helmy et al. [7,8] to generate high prediction accuracy for heart disease patients. The results show that heterogeneous ensemble has better results as compared to individual classifiers. A novel ensemble has been proposed by Bashir et al. [9] to improve the classification and heart disease prediction. The proposed ensemble used a bagging algorithm with a multi-objective optimized weighted voting that contacted on heart disease datasets. Verma et al. [10] developed a novel hybrid model using data mining methods. In their model, the proposed ensemble was used to predict coronary artery disease cases using non-invasive clinical data of patients.

Fast Fourier transform, an efficient technique to compute the discrete and the inverse, is an emerging tool for prediction. It has recently been applied to: analyze and predict electricity consumption in buildings [11,12], to fore-cast water demand [13,14], to detect epileptic seizure in electroencephalography (EEG) [15,16].

Since the importance of the prediction in medical domain as well as the urgency of demanding more powerful analytic tools in this regard, further efforts are definitely needed to enhance evidence-based decisions quality. In this work, we propose a novel fast Fourier transform-coupled a machine learning ensemble to predict and assess the short-term risk of disease and provide patients with appropriate recommendations for necessity of taking a medical test in the incoming day.

The remainder of this paper is organised as follows. Section 2 explains the details of fast Fourier transformation and machine learning classifiers that constructing the proposed ensemble model. Section 3 briefly defines the proposed methodology including predictive model development and describes the used data set. Section 4 discussed in details the experimental evaluation results. Finally we conclude the paper and highlight the future work in Sect. 5.

2 Theoretical Background

2.1 Bootstrap Aggregation (Bagging)

An ensemble method is one of the combination approaches used to overcome the limited generalization performance of individual models and to generate more accurate predictions than single models. Bagging is a machine learning ensemble used to solve problems by combining the decisions of multiple classifiers [21]. During a bootstrap method, in a bagging method, classifiers are trained independently and then aggregated by an appropriate combination strategy. The proposed ensemble model can be divided into two phases. At the first phase, bagging uses bootstrap sampling to generate a number of training sets. At the second phase, training the base classifiers is performed using bootstrap training sets generated during the first step. The generic flowchart of bagging algorithm is shown in Fig. 1. In this study, the training set was divided into multiple datasets using bootstrap aggregation approach, and then individually classifiers are applied on these datasets to generate the final prediction. We argue that each individual classifier in the weighted-bagging approach should has a different individual performance level. We proposed, therefore, to assign a weight for each classifier's vote based on how well the classifier performed. The classifier's wight is calculated based on it's error rate so that the classifier that has the lower error rate is more accurate, and therefore, it should be assigned the higher weight for that classifier. The weight of classifier C_i's vote is calculated as follows [22]:

$$\log \frac{1 - error(C_i)}{error(C_i)} \tag{1}$$

The proposed weighted-bagging ensemble can be easily understood by the following example:

1. Suppose that the classifier training is performed for training data and error rate is calculated.
2. Neural Networks (NN), Support Vector Machine (SVM) and Naive Bayes (NB) are used as individual classifiers. Following error rate results are generated for each classifier: NN = 0.25, SVM = 0.14, NB = 0.30.
3. Now, according to the formula given in *Eq.* (6), the resultant weights are as follows: NN = 0.47, SVM = 0.78, NB = 0.36.
4. Suppose, the algorithms have predicted the following classes for a test day: NN = 0, SVM = 1, NB = 0. (0: no test required; 1: test needed).
5. Based ensemble classifier, the weighted vote will be generated the following prediction results:
   Class 0: $NN + NB \longrightarrow 0.47 + 0.36 \longrightarrow 0.83$,
   Class 1: $SVM \longrightarrow 0.78$.
6. Finally, according to weighted vote, the class 0 has higher value as compared to class 1. Therefore, the ensemble classifier for this test day will be classified as Class 0.

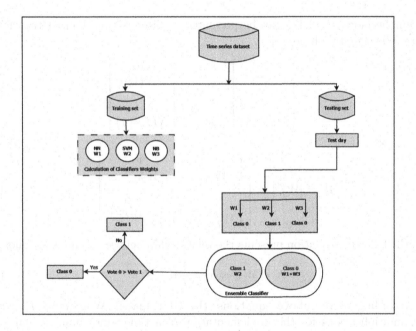

Fig. 1. The generic flowchart of bagging algorithm

2.2 Fast Fourier Transform-Coupled Machine Learning Based Ensemble (FFT-MLE) Model

The major purpose of this study was to demonstrate the effectiveness of fast Fourier transform-coupled machine learning based ensemble model for advising patients concerning whether they need to take the body test today or not based on analysis of their medical data during the past a few days. After developing a ensemble model that consists from three predictive models, Fast Fourier transformation algorithm, which is a data-preprocessing tool for non-stationary singles, is proposed. A fast Fourier transform (FFT) is an efficient algorithm to calculate the discrete Fourier transform (DFT) and the inverse. DFT decomposes the input sequence values to extract the frequency information in order to predict the next day. The discrete-time Fourier transform of a time series x(t) can be defined as:

$$X(c^{jw}) = \sum_{t=-\infty}^{\infty} x(t)c^{-jwt} \tag{2}$$

where t is the time index of discrete, and w refers to the frequency. There are T input time series x(t), so the transform pair of DFT can be defined as:

$$X(P) = \sum_{t=0}^{T-1} x(t)W_T^{tp} \Leftrightarrow x(t) = \frac{1}{T}\sum_{p=0}^{T-1} X(P)W_T^{-tp}, where \quad W = c^{-j2\Pi/T} \tag{3}$$

Furthermore, the DFT can be presented as discrete-time Fourier transform of a cyclic signal with period T.

$$x = \begin{bmatrix} x(0) \\ x(1) \\ \vdots \\ x(T-1) \end{bmatrix}, \quad X = \begin{bmatrix} X(0) \\ X(1) \\ \vdots \\ X(T-1) \end{bmatrix} \tag{4}$$

$$W = [W_T^{pt}] = \begin{bmatrix} 1 & 1 & \cdots & 1 \\ 1 & W_T & \cdots & W_T^{T-1} \\ \vdots & \vdots & \vdots & \vdots \\ 1 & W_T^{T-1} & \cdots & W_T^{(T-1)(T-1)} \end{bmatrix} \tag{5}$$

and the following equation presents the relationship between x and X as follows:

$$X = Wx \Leftrightarrow x = \frac{1}{T}W^H X \tag{6}$$

According to the above equations, the DFT matrix W requires T^2 complex multiplications for the implementation of a time series input signal x(t) with length T. Therefore, a required implementation cost for factorizing the fast Fourier transform W into a matrix is lower than the direct DFT. For each stage of fast Fourier transform requires $T/2$ multiplications and T additions [23].

In practical sense, the input time series data segments into set of slide windows with a length of k (the size of the sliding window used in time series data analysis). The input time series are decomposed using fast Fourier transform to extract the frequency information included in the input data in order to predict the following medical test day.

3 Methodology

3.1 Predictive Model Development

The predictive models are developed in MATALB environment on a desktop computer with the configurations of a 3.40 GHz Intel core i7 CPU processor with 8.00 GB RAM. The major purpose of this study was to investigate the performance of the fast Fourier transform-coupled machine learning ensemble to predict the short-term risk of disease and provide patients with appropriate recommendations for necessity of taking a medical test in the incoming day. The training data is used to train the classifiers that construct the ensemble and testing data to evaluate the performance of predictive model. In this study, the time series medical data were partitioned into about 75% as training data and 25% as testing data.

Figure 2 illustrates the different stages of the fast Fourier transform coupled-ensemble model. Basically, the input time series data was segmented into a set of sub-segments with overlapping of m based on a predefined value of k to identify the window size of sub-segment. Let $X = \{y_1, y_2, y_3, \ldots, y_n\}$ is a time series of n test measurements. The main idea is to separate X into a number of overlapping sub-segments. The overlapping value m is set to be a test measurement of one day. Then, each sub-segment is passed through fast Fourier transform in order to obtain the desired information. A resulting set of fast Fourier transform coefficients of 28 levels is tested to figure out the desired FFT level. Different combinations of statistical features from each level are tested and analysed the performance of the proposed model with different FFT characteristics. The purpose of using FFT in this paper is to study the properties of time series in frequency domain which could be difficult to obtain in time domain. The basic idea of frequency analysis is to re-express the original time series as a new sequence which determine the important of each frequency components. The fast Fourier transform was used to decompose each time series slide window to acquire five $(\alpha, \beta, \gamma, \delta, \text{and } \theta)$ frequency bands Fig. 2. Based on the literature, the high frequency band could be able to capture the desired information, therefore, the high frequency band was divided into 8 sub-frequency bands. In addition, the original time series slide window was also added, as a reference, to the feature extraction test. As a result, the total number of frequency bands sets is $(5 + 8 + 1 = 14)$ for each slide window. Furthermore, the power of the FFT coefficients was calculated for all the 14 sets of frequency bands. It allows to compute the square of the absolute value of the Fourier coefficients. These frequencies were grouped together and used as input to the proposed ensemble. As a result, the features were extracted from 28 set of frequency bands $(14 + 14 = 28)$. All the bands features were tested and analysed to figure out best combination of features. Figure 3 shows an example of decomposing a slide window of time series into 28 bands.

Two sets of statistical features were extracted from each band to find out the best combination of features to present the data. Two and four statistical features were extracted from each band. The mentioned 28 bands were tested and

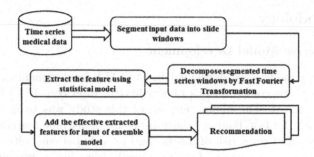

Fig. 2. The stages of developing a Fast Fourier Transform-Coupled Machine Learning based Ensemble (FFT-MLE) model

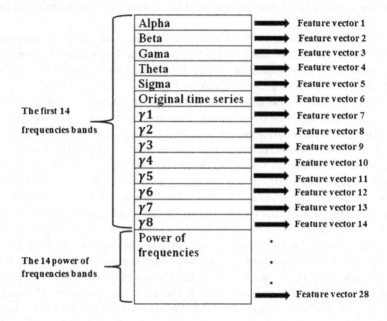

Fig. 3. The generic decomposition of slide window time series into 28 bands

analysed with those features sets at each stage, and the results were recorded. However, it was observed that using four statistical features yield better results, in term of accuracy, compared with using two features. The extracted features from all the 28 bands were used as key features to training the ensemble's classifiers in order to predict the following day.

The two features sets were included (max and min), while the four features sets were included (max, min, standard deviation, and median). The features were denoted $(X_{min}, X_{max}, X_{std}, X_{med})$. The short explanations of statistical features are provided in Table 1. To test the relationship between the features and the risk prediction, extensive experiments were conducted. As a result, two vectors of 2X28 and 4X28 were extracted and used to identify whether the patient requires to take a test today or not.

The features were tested separately by testing each band and by using all features of bands as a one vector. Our findings showed that the 4 features set gives a better results than those of two features set. The extracted features from each band, four feature set, were grouped in one vector and used as the features set to predict the next day. All the detailed are discussed in simulation and experiment results section.

In the experiments, we use a statistical approach to extract the statistical features from each band, and then put all the features from one segment in a vector to present the window. It was found that some of time series data are symmetric distribution and other skewed distribution. The min and max are considered appropriate measures for a time series with symmetric distribution, whilst, for a skewed distribution, mean and standard division are used to measure the center and spread of dataset [24, 25].

Table 1. Short explanations of statistical features

Feature name	Formula	Description
Maximum value	$X_{max} = Max[x_n]$	Where $x_n = 1, 2, 3, \ldots, n$ is a time series, N is the size of slide window, AM is the mean of slide window
Minimum value	$X_{min} = Min[x_n]$	
Mean	$X_{mean} = \dfrac{1}{n} \displaystyle\sum_{1}^{n} x_i$	
Standard division	$X_{SD} = \sqrt{\displaystyle\sum_{n=1}^{N}(x_n - AM)\dfrac{2}{n-1}}$	

3.2   Evaluation Design

In this section, we offer details concerning the strategy of our experimental evaluation including datasets, performance metrics and the experimental platform.

As the predictive performance of the FFT-MLE model is quite important, assessment of potential predictions is critically dependent on the quality of the used dataset. For this reason, telehealth data from Tunstall dataset will be conducted in this work. We use a real-life dataset obtained from our industry collaborator Tunstall to test the practical applicability of the FFF-MLE model. A Tunstall dataset obtained from a pilot study has been conducted on a group of heart failure patients and the resulting data were collected for their day-to-day medical readings of different measurements in a tele-health care environment. The Tunstall database employed in the development of the algorithm consists of data from six patients with a total of 7,147 different time series records. Data were acquired between May and January 2012, using a remote telehealth collaborator. The dataset is by nature in a time series and contains a set of measurements taken from the patients on different days. Each record in the dataset consists of a few different meta-data attributes about the patients such as patient-id, visit-id, measurement type, measurement unit, measurement value, measurement question, date and date-received. The characteristics of the features of the dataset are shown in Table 2.

Table 2. Characteristic features of the dataset

Feature name	Feature type
id	Numeric
id-patient	Numeric
hcn	Numeric
visit-id	Numeric
measurement type	Nominal
measurement unit	Nominal
measurement value	Numeric
measurement question	Nominal
date	Numeric
date-received	Numeric

In addition, each record contains a few medical attributes including Ankles, Chest Pain, and Heart Rate, Diastolic Blood Pressure (DBP), Mean Arterial Pressure (MAP), Systolic Blood Pressure (SBP), Oxygen Saturation (SO2), Blood Glucose, and Weight. Ethical clearance was obtained from the University of Southern Queensland (USQ) Human Research Ethics Committee (HREC) prior to the onset of the study. This dataset is used as the ground truth result to test the performance of our recommendation system. The recommendations produced by our system will be compared with the actual readings of the measurement in question recorded in the dataset to see how accurate our recommendations are.

Due to the patient's historical medical data has often class-imbalanced problem (i.e. the number of normal data is much more than that of abnormal data), we are carefully dealt with the class-imbalanced problem for classifier building. The over-sampling and under-sampling have been proposed as a good means to address this problem. The predictive accuracy is usually used to evaluate the performance of machine learning algorithms. However, this measure is not appropriate when the used data is imbalanced [26].

The performance of individual classifiers as well as the proposed ensemble is evaluated by calculating the *accuracy, workload saving*, and *risk*. Accuracy refers to the percentage of correctly recommended days against the total number of days that recommendations are provided; workload saving refers to the percentage of the total number of days when recommendations are provided against the total number of days in the dataset, while risk refers to the percentage of incorrectly recommended days that recommendations are no test needed. Mathematically, Accuracy, workload saving and risk are defined as follows:

$$Accuracy = \frac{NN}{NN + NA} \times 100\% \tag{7}$$

$$Saving = \frac{NN + NA}{|\mathcal{D}|} \times 100\% \tag{8}$$

$$Risk = \frac{NR}{|\mathcal{D}|} \times 100\% \tag{9}$$

Where NN denotes the number of days with correct recommendations, NA denotes the number of days with incorrect recommendations, NR denotes the number of days with incorrect days that recommendations are no test needed, and $|\mathcal{D}|$ refers to the total number of days in the dataset. Here, a correct recommendation means that the model produces the recommendation of "no test required" for the following day and the actual reading for that day in the dataset is normal. If this is a case, the recommendation is considered accurate.

4   Result Analysis

The using of the fast Fourier transform-coupled machine learning based ensemble (FFT-MLE) model aims at short-term risk assessment in patients based on analytic of a patient's historical medical data using fast Fourier transform. As mentioned above, the time series slide windows were decomposed by using the FFT. Then, the suitable features were selected as input for the ensemble model. The new time series selected were employed as input of the ensemble's classifiers instead of the original time series data. Different sets of statistical features, as mentioned above, were used to determine the best number of features for each slide window. The detailed results are discussed in the following sub sections:

4.1   Prediction Accuracy with Different Number of Features

To evaluate the relationship between the number of the extracted features and the prediction accuracy, several experiments were conducted using different sets of features. Based on the experiment results, When the number of the statistical features is increased, the predictive accuracy of the proposed model is more significant. The three classifiers of the FFT-MLE model, neural networks (NN), least square-support vector machine (LS-SVM) and naive Bayes (NB), were trained with different sets of features.

Typically, an ensemble model is a supervised learning technique for combining multiple weak learners or models to produce a strong model [9]. Based on our findings, a group of classifiers is likely to make better decisions compared to individuals. The experimental results showed that the proposed method using an ensemble classifier gives a satisfactory recommendation accuracy compared to a single classifier.

Table 3. Prediction accuracy of Ensemble model based on the first 5 frequencies bands with two features.

Tunstall dataset (%)				
Measurement	Ensembles	Accuracy (%)	Saving (%)	Risk (%)
Heart rate	Neural network	70.64	54.33	09.92
	LS-SVM	75.37	62.34	07.21
	Naive Bayes	71.87	53.54	09.78
	Ensemble model	**85.38**	**60.60**	**5.90**
DBP	Neural network	71.30	58.18	09.76
	LS-SVM	78.45	64.75	07.44
	Naive Bayes	72.11	57.66	09.78
	Ensemble model	**86.27**	**65.57**	**5.88**
MAP	Neural network	69.70	50.33	09.92
	LS-SVM	77.98	64.34	07.21
	Naive Bayes	70.17	56.54	12.78
	Ensemble model	**83.33**	**62.60**	**06.45**
SO2	Neural network	70.30	54.33	09.98
	LS-SVM	74.39	65.34	07.95
	Naive Bayes	68.78	50.54	12.78
	Ensemble model	**84.38**	**66.47**	**5.90**

Two-Features Set. In this experiment, first, the first main frequencies bands $(\alpha, \beta, \gamma, \delta,$ and $\theta)$ were selected from each slide window and then the two features of $(X_{min}$ and $X_{max})$ for each band were utilized to evaluate the performance of the proposed model. Through this process, each slide window was converted to a vector of 10 extracted features. The extracted features were randomly divided into training and testing sets. Each classifier is trained on whole training set. Then, we individually apply the basic *bagging* algorithm on each day in test set by assigning a weighted vote for each classifier based on it's performance in the training stage. The classifier that has the lower error rate is more accurate, and therefore, it should be assigned the higher weight for that classifier. The final prediction of each day in test set is calculated based on the sum of weights for each class. As a result, the class that has the highest weight will be selected as class label for that day. Each day in test set is classified into "need test" or "no test needed" labels. Table 3 presents the comparison of accuracy, workload saving and risk results of ensemble model with individual classifier techniques for the four measurements. It is compared with different classifiers such as Neural network (NN), Least square- Support Vector Machine (LS-SVM) and Naive Bayes (NB).

From the obtained results in this table, although the proposed FFT-MLE model is achieved noticeably significant results using five frequencies bands.

Table 4. Prediction accuracy of Ensemble model based on 28 frequencies bands with two features

Tunstall dataset (%)				
Measurement	Ensembles	Accuracy (%)	Saving (%)	Risk (%)
Heart rate	Neural network	74.13	50.38	07.80
	LS-SVM	80.37	60.14	05.21
	Naive Bayes	75.20	50.55	06.10
	Ensemble model	**90.20**	**60.30**	**4.20**
DBP	Neural network	75.74	55.11	07.75
	LS-SVM	82.88	60.24	05.65
	Naive Bayes	77.74	55.74	06.95
	Ensemble model	**91.40**	**65.57**	**4.10**
MAP	Neural network	75.45	56.20	07.95
	LS-SVM	81.93	67.22	05.77
	Naive Bayes	76.28	53.13	07.30
	Ensemble model	**91.85**	**65.20**	**04.00**
SO2	Neural network	73.25	51.50	07.95
	LS-SVM	80.55	60.38	05.10
	Naive Bayes	74.11	54.36	07.82
	Ensemble model	**90.50**	**60.25**	**4.15**

However, the five bands were not enough to represent the slide windows because they did not appear the appropriate characteristics of slide windows. Therefore, all of 28 bands instead of the first five bands were considered to represent each slide window. In this case, $28 \times 2 = 56$ statistical features for each slide window were extracted and then used in the training of ensemble model.

From a observation the results in Table 4, it can be noticeably seen that the performance of the proposed FFT-MLE model, for all measurements, is improved compared with the previous results. This is because that the accuracies have significantly increased using all frequencies bands instead of the top five bands. According to the results in Table 4, the prediction accuracy of the proposed FFT-MLE model is increased by more than 5% compared to the obtained results in Table 3.

Four-Features Set. To improve the predictive performance of the FFT-MLE model, a four-features set of $(X_{min}, X_{max}, X_{std}, X_{med})$ was selected, tested to analyse the time series medical data. For each slide window, 5×4 statistical features were extracted and then used to evaluate the performance of the proposed FFT-MLE model, where 5 refers to the main bands of FFT that including: $\alpha, \beta, \gamma, \delta$, and θ, and the 4 indicates the number of selected features. The obtained results were showed that using four features set can be improved the prediction

Table 5. Prediction accuracy of Ensemble model based on the first 5 frequencies bands with a four-features set.

Tunstall dataset (%)				
Measurement	Ensembles	Accuracy (%)	Saving (%)	Risk (%)
Heart rate	Neural network	74.30	49.33	07.60
	LS-SVM	78.39	60.34	07.40
	Naive Bayes	74.73	52.42	07.50
	Ensemble model	**88.75**	**60.20**	**4.80**
DBP	Neural network	75.30	50.13	07.76
	LS-SVM	81.62	61.25	06.30
	Naive Bayes	74.45	57.30	07.82
	Ensemble model	**89.41**	**62.54**	**4.50**
MAP	Neural network	73.62	53.33	07.95
	LS-SVM	82.98	60.40	06.21
	Naive Bayes	75.33	51.40	07.50
	Ensemble model	**90.20**	**60.10**	**04.10**
SO2	Neural network	76.20	56.55	07.70
	LS-SVM	82.50	63.64	06.30
	Naive Bayes	74.60	55.50	07.5
	Ensemble model	**90.33**	**62.48**	**4.00**

accuracy of the proposed FFT-MLE model. As a result, an accuracy of 94%, for all measurements, was attained. Table 5 shows the prediction accuracies results and risk assessment using a four-features set with the five selected bands for all measurements. According to the obtained results in Table 5, the prediction accuracies for all measurements were noticeably improved using a four-features set. The obtained results proved that the four selected features have significantly improved the predictive performance of the proposed FFT-MLE model. The prediction accuracy of all measurements was increased by more than 6% compared with the two-features set results.

However, in order to further increase the prediction accuracy of the proposed model, all of 28 bands instead of the five bands were also selected, used to represent each slide window. The $28 \times 4 = 112$ statistical features for each slide window were extracted and then used in the training of ensemble model. From the obtained results in Table 6, it can be noticed that the performance of the proposed FFT-MLE model significantly improved for all measurements after having increased the number of selected bands with more features. Figure 4 shows the averaged accuracies using both of 5 and 28 bands with 2 and 4 features sets for all measurements.

Table 6. Prediction accuracy of Ensemble model based on 28 frequencies bands with a four-features set

Tunstall dataset (%)				
Measurement	Ensembles	Accuracy (%)	Saving (%)	Risk (%)
Heart rate	Neural network	79.57	53.88	06.90
	LS-SVM	86.50	64.75	04.95
	Naive Bayes	78.40	54.54	07.05
	Ensemble model	**94.15**	**63.25**	**3.30**
DBP	Neural network	78.35	56.14	07.10
	LS-SVM	90.50	64.50	04.10
	Naive Bayes	77.44	59.30	07.75
	Ensemble model	**95.30**	**64.13**	**2.85**
MAP	Neural network	80.30	55.50	06.65
	LS-SVM	89.50	63.40	05.50
	Naive Bayes	78.60	52.30	07.20
	Ensemble model	**94.50**	**64.30**	**03.00**
SO2	Neural network	79.25	54.60	06.95
	LS-SVM	89.40	64.50	05.25
	Naive Bayes	78.55	56.25	07.30
	Ensemble model	**95.20**	**63.13**	**02.95**

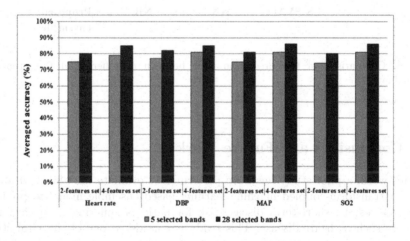

Fig. 4. The obtained results of two and four features sets after applying 5 and 28 FFT decomposition for all measurements

4.2 Prediction Time

In this experiment, the prediction time including training time and execution time of classifiers was proposed. Figure 5 shows the prediction time for each clas-

sifier and the proposed ensemble as well. From the results in Fig. 5, we observed
that although the proposed model took more time compared with individual classifiers, it provided more accurate recommendation to patients suffering chronic
diseases. On other hand, the least square-support vector machine (LS-SVM) was
recorded the lowest prediction time compared with other individual classifiers.

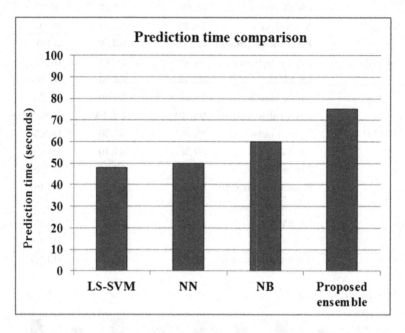

Fig. 5. Comparison of the prediction time between classifiers and the proposed
ensemble

5   Conclusions and Future Work

In this work, a pilot study has been performed to evaluate the ability of a fast
Fourier transform-coupled machine learning based ensemble model to predicts
and assesses the short-term disease risk for patients suffering from chronical
diseases such as heart disease. This study is considered one of the vital studies to use medical measurements of patient in the assessment and prediction
of the short-term disease risk. This research presents a machine learning based
ensemble model which incorporates fast Fourier transformation algorithm for
pre-processing of input time series data. This ensemble is based on three heterogeneous learners named neural networks, least square support vector machine
and naive Bayes in order to generate appropriate recommendations. The prediction model is developed aiming at improving the quality of clinical evidence-
based decisions and helping reduce financial and timing cost taken by patients.

The experimental results showed that the four-features set yields a better predictive performance for all measurements compared to the two-features set. In addition, it is also mentioned that using the all of 28 bands give reasonable prediction accuracies under all measurements.

Future research directions include application of the proposed model on different datasets for more validations. We also plan to incorporate wavelet transformation with fast Fourier translation in order to pre-prossing the time series data.

Acknowledgement. The authors would like to thank the support from National Science Foundation of China through the research projects (Nos. 61572036, 61370050, and 61672039) and Guangxi Key Laboratory of Trusted Software (No. kx201615).

References

1. Kuh, D., Shlomo, Y.B.: A Life Course Approach to Chronic Disease Epidemiology. Inem Oxford University Press, London (2004)
2. Atlas, I.D.: International Diabetes Federation Diabetes Atlas, 6th edn. International Diabetes Federation, Basel (2013)
3. Thong, N.T.: HIFCF: an effective hybrid model between picture fuzzy clustering and intuitionistic fuzzy recommender systems for medical diagnosis. Expert Syst. Appl. **42**(7), 3682–3701 (2015)
4. Chen, D., Jin, D., Goh, T.-T., Li, N., Wei, L.: Context-awareness based personalized recommendation of anti-hypertension drugs. J. Med. Syst. **40**(9), 202 (2016)
5. Valentini, G., Masulli, F.: Ensembles of learning machines. In: Marinaro, M., Tagliaferri, R. (eds.) WIRN 2002. LNCS, vol. 2486, pp. 3–20. Springer, Heidelberg (2002). doi:10.1007/3-540-45808-5_1
6. Breiman, L.: Bagging predictors. Mach. Learn. **24**(2), 123–140 (1996)
7. Das, R., Turkoglu, I., Sengur, A.: Effective diagnosis of heart disease through neural networks ensembles. Expert Syst. Appl. **36**(4), 7675–7680 (2009)
8. Helmy, T., Rahman, S., Hossain, M.I., Abdelraheem, A.: Non-linear heterogeneous ensemble model for permeability prediction of oil reservoirs. Arab. J. Sci. Eng. **38**(6), 1379–1395 (2013)
9. Bashir, S., Qamar, U., Khan, F.H.: BagMOOV: a novel ensemble for heart disease prediction bootstrap aggregation with multi-objective optimized voting. Australas. Phys. Eng. Sci. Med. **38**(2), 305–323 (2015)
10. Verma, L., Srivastava, S., Negi, P.: A hybrid data mining model to predict coronary artery disease cases using non-invasive clinical data. J. Med. Syst. **40**(7), 1–7 (2016)
11. Tsai, C.-L., Chen, W.T., Chang, C.-S.: Polynomial-Fourier series model for analyzing and predicting electricity consumption in buildings. Energy Build. **127**, 301–312 (2016)
12. Ji, Y., Xu, P., Ye, Y.: HVAC terminal hourly end-use disaggregation in commercial buildings with Fourier series model. Energy Build. **97**, 33–46 (2015)
13. Brentan, B.M., Luvizotto Jr., E., Herrera, M., Izquierdo, J., Prez-Garca, R.: Hybrid regression model for near real-time urban water demand forecasting. J. Comput. Appl. Math. **309**, 532–541 (2016)
14. Odan, F.K., Reis, L.F.R.: Hybrid water demand forecasting model associating artificial neural network with Fourier series. J. Water Resour. Plan. Manag. **138**(3), 245–256 (2012)

15. Samiee, K., Kovcs, P., Gabbouj, M.: Epileptic seizure classification of EEG time-series using rational discrete short-time Fourier transform. IEEE Trans. Biomed. Eng. **62**(2), 541–552 (2015)
16. Kovacs, P., Samiee, K., Gabbouj, M.: On application of rational discrete short time Fourier transform in epileptic seizure classification. IEEE Trans. Biomed. Eng. 5839–5843 (2014)
17. Suykens, J.A., Vandewalle, J.: Least squares support vector machine classifiers. Neural Process. Lett. **9**(3), 293–300 (1999)
18. Bai, Y., Han, X., Chen, T., Yu, H.: Quadratic kernel-free least squares support vector machine for target diseases classification. J. Comb. Optim. **30**(4), 850–870 (2015)
19. Sharawardi, N.A., Choo, Y.-H., Chong, S.-H., Muda, A.K., Goh, O.S.: Single channel sEMG muscle fatigue prediction: an implementation using least square support vector machine. In: Information and Communication Technologies (WICT), pp. 320–325 (2014)
20. Li, S., Tang, B., He, H.: An imbalanced learning based MDR-TB early warning system. J. Med. Syst. **40**(7), 1–9 (2016)
21. Gao, H., Jian, S., Peng, Y., Liu, X.: A subspace ensemble framework for classification with high dimensional missing data. Multidimens. Syst. Sig. Process. 1–16 (2016)
22. Han, J., Kamber, M., Pei, J.: Data Mining: Concepts and Techniques. Elsevier, Amsterdam (2011)
23. Alfred, M.: Signal Analysis Wavelets, Filter Banks, Time-Frequency Transforms and Applications. Wiley, New York (1999)
24. Şen, B., Peker, M., Çavuşoğlu, A., Çelebi, F.V.: A comparative study on classification of sleep stage based on EEG signals using feature selection and classification algorithms. J. Med. Syst. **38**(3), 1–21 (2014)
25. Diykh, M., Li, Y.: Complex networks approach for EEG signal sleep stages classification. Expert Syst. Appl. **63**, 241–248 (2016)
26. Bach, M., Werner, A., Żywiec, J., Pluskiewicz, W.: The study of under-and over-sampling methods' utility in analysis of highly imbalanced data on osteoporosis. Inf. Sci. **384**, 174–190 (2016)
27. Weng, C.-H., Huang, T.C.-K., Han, R.-P.: Disease prediction with different types of neural network classifiers. Telemat. Inform. **33**(2), 277–292 (2016)
28. Zhang, J., Li, H., Gao, Q., Wang, H., Luo, Y.: Detecting anomalies from big network traffic data using an adaptive detection approach. Inf. Sci. **318**, 91–110 (2015). Elsevier Publisher
29. Zhang, J., Gao, Q., Wang, H.: SPOT: a system for detecting projected outliers from high-dimensional data streams. In: 24th IEEE International Conference on Data Engineering (ICDE 2008), pp. 1628–1631. IEEE Computer Society, Cancun, April 2008

Assessing Death Risk of Patients with Cardiovascular Disease from Long-Term Electrocardiogram Streams Summarization

Shenda Hong[1,2], Meng Wu[1,2], Jinbo Zhang[1,2], and Hongyan Li[1,2(✉)]

[1] Key Laboratory of Machine Perception (Peking University),
Ministry of Education, Beijing, China
lihy@cis.pku.edu.cn
[2] School of Electronics Engineering and Computer Science, Peking University,
Beijing, China

Abstract. Cardiovascular disease (CVD) is the leading cause of death around the world. Researches on assessing patients death risk from Electrocardiographic (ECG) data has attracted increasing attention recently. In this paper, we summarize long-term overwhelming ECG data using morphological concern of overall evolution. And then assessing patients death risk from high value density ECG summarization instead of raw data. Our method is totally unsupervised without the help of expert knowledge. Moreover, it can assist in clinical practice without any additional burden like buy new devices or add more caregivers. Comprehensive results show effectiveness of our method.

1 Introduction

Cardiovascular disease (CVD) is the leading cause of death around the world. In 2012, an estimated 17.5 million people died from CVD, a number that is expected to grow to more than 23.6 million by 2030. CVD deaths represented about 3 of every 10 deaths of all global death. In 2011, the age-standardized death rate attributable to all CVD was 229.6 per 100 000 [1]. New research looking at the costs of cardiovascular disease in six EU member states (France, Germany, Spain, Italy, Sweden and the United Kingdom) concludes that the financial burden will rise to 122.6 billion by 2020, up from 102.1 billion in 2014 [2].

Physicians use a variety of measurements to assess patients risk and to take proper treatment. Blood tests including Cholesterol test, C-reactive protein, Lipoprotein (a), Natriuretic peptides aim to determine biochemical states [3]. Magnetic Resonance Imaging (MRI) to measure cardiac volumes and derive ejection fraction (EF) [4]. Analytics based on these clinical data is very limited. First, clinical data is inadequate. It has to be recorded by caregivers. Second, it can only reflect a snapshot of patient without continuous pathology evolution recording. Last, these measurements are invasive that leading to vessels insertion, radiation and allergy.

© Springer International Publishing AG 2017
J. Kim et al. (Eds.): PAKDD 2017, Part I, LNAI 10234, pp. 671–682, 2017.
DOI: 10.1007/978-3-319-57454-7_52

Electrocardiographic (ECG) data is an interpretation of the electrical activity of the heart, it can reflect the overall evolution of the patients pathophysiology over time. Moreover, ECG data is easy to acquire and it is a non-invasive recording procedure. Bedside monitors and portable ECG machine can help with data recording both in hospital and out of hospital. Decision making with the help of ECG data has attracted increasing attention recently.

However, discovering pathology under ECG data, we faced up with following challenges:

- **Overwhelming data:** Common 1-lead 125 Hz ECG monitors generate over 10 million data points per day of one patient. Not to speak of 12-lead or 500 Hz high resolution ECG monitors. Such overwhelming data poses problem both on limited device storage and procession capacity.
- **Low value density:** On the one hand, we have to discard current data after procession because of storage shortage. On the other hand, useful information like critical shifts or morphology variation appear rarely on long term ECG data. Therefore, precisely extracting and effectively keeping useful information from fragment of low value density ECG data are of great concern.
- **Expert knowledge:** A large majority of ECG analysis methods require identify P, Q, R, S, T, U segments at first. Identification methods usually have high error rate. In practice, only expert physician can mark these segments correctly and efficiently. But it is still a huge burden of man power which is unrealistic on overwhelming ECG data.
- **Morphological concern:** Recent work on ECG analysis has shown the association between morphology and patient condition stages [5]. It reveals that pathology is determined by the values in a certain period represented as a wave. Methods focus on Duration (Q to S, Q to T, etc.), Amplitude (QRS, QT, etc.) and Slope (RS, ST, etc.) cant capture overall evolution. Moreover, identify P, Q, R, S, T, U waves is also a challenge which is posed above.

In this paper, we summarize long-term overwhelming ECG data using morphological concern of overall evolution. And then assessing patients risk from high value density ECG summarization instead of raw data. Our method is totally unsupervised without the help of expert knowledge. Moreover, it can assist in clinical practice without any additional burden like buy new devices or add more caregivers.

We first using pattern growth graph (PGG) to summarize trend, morphology variation and critical shifts of original data. PGG is capable of reducing the storage size of patterns to about 0.3% with less than 5% relative error. Next, we propose ConverGence Index (CGI) to measure convergence of graph. Then, we extract biomarkers including CGI from PGG summarization to assess death risk of patients with CVD. Finally, we conduct thorough experiments on ICU clinical data, results show effectiveness of our method.

2   Background

2.1   ECG Diagram

Electrocardiogram (ECG) is a non-invasive representation of the heart's cardiac cellular electrical activity recorded from electrodes on the body surface. It provides wealth of information related to the electrical patterns proper, the geometry of heart tissue and the metabolic state of the heart. The diagram of healthy sinus ECG waves is shown as Fig. 1.

The ECG is a pseudo periodical time series with x-axis representing standard time and y-axis representing voltage measures. Following wave segments are basic features of the ECG. The P wave segment (P) is associated with sequential depolarization of the right and left atria. The QRS complex (QRS) is associated with right and left ventricular depolarization. The ST wave segment (ST) is associated with ventricular repolarization. The U wave segment (U) is an electrical-mechanical event at beginning of diastole. Some typical features of ECG are also illustrated in the figure.

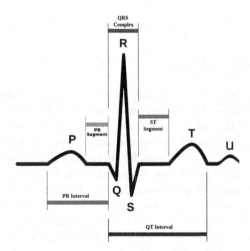

Fig. 1. Normal sinus ECG diagram

2.2   CVD Evolutionary

Healthy people with normal ECG is composed by repeated normal sinus rhythm. Cardiovascular diseases, especially chronic heart diseases, are often related to long-term variation of duration, interval and morphology [6]. Large volume ECG data is required for analysis by variation accumulation. In clinical practice, subtle variations of snapshot ECG data can only be interpreted by expert physicians [7]—not for patient's self diagnosis. It is a big loss for early CVD detection.

3 Summarizing ECG Streams

In this section, we first split original ECG streams into short segments with the help of QRS complex. Then we use patterns to represent the outline of segments and reduce their storage cost. Finally, we organize patterns together by generating pattern growth graph.

3.1 Wave Splitting

Given ECG time series $X = \{X(t) : t = 1, 2, ..., N\}$ composed by N data points for a certain patient. We first split point series into consecutive waves. A special digital bandpass filter is required to reduces false detections caused by the various types of interference present in ECG signals while detect QRS complex correctly. An open source QRS detection [8] is used to achieve this task. We then split in the end of peak point.

Finally, we get M wave of short time series linked together to a long time series. Formally representing as

$$X = \{S_1, S_2, ..., S_M\} \tag{1}$$

Where $S_i = \{S_i(t) : t = 1, 2, ...\}$.

3.2 Pattern Representation

After the waves in the ECG streams are obtained, concise representation of these waves are required since waves contains large amount of data points. It is not realistic to store all the details of a wave, but summary data points to smaller patterns with acceptable error bound can save space and reduce computational cost. There are some existing methods for represent ECG data, like PAA [9], SAX [10], Zigzag [11]. Piecewise Linear Representation (PLR) is the most effective method without losing important morphological variation [12].

Intuitively, given a series of data points $S_i(t)$, PLR produce the best linear representation (defined as pattern) such that the maximum error for any wave does not exceed the user specified threshold. There existing two ways to generate the pattern of a wave: (1) Sliding Window: It processes along the timeline, a new line segment created when the sum of residual error exceeds the predefined threshold. (2) Bottom Up: It creates all finest possible approximation of a wave. Then merge pairs of lowest error cost iteratively when the sum of residual error below the predefined threshold. In [13], the author proposed a new mixed method Sliding Window And Bottom-up (SWAB), which take advantages of both sliding window (high efficiency) and bottom up (high quality). An example of PLR result on a single ECG wave is shown in Fig. 2.

3.3 Constructing Pattern Growth Graph

Our goal is to summarize ECG stream using pattern growth graph (PGG). Next, we will construct PGG from individual pattern obtained by PLR.

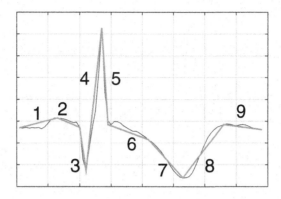

Fig. 2. PLR represent of a ECG wave

Definition 1. *PGG*: *PGG is a directed graph $PGG = (V, E)$. Vertices V represent line segments in PLR representation, linked with directed edges E in time order.*

Definition 2. *Segment Match*: *Given error threshold ε, ECG subsequence $L = \{(x_1, t_1), (x_2, t_2), ..., (x_m, t_m)\}$, line segment $Seg = \{(X(t), t), X(t) = kt + b\}$. Defining that subsequence L matches line segment Seg if : $(length(L) - length(Seg))/length(Seg) < \varepsilon$ and $\sum_{i=1}^{m} |(x_i - kt_i - b)/x_i| < \varepsilon$, where $length()$ the time duration of a sequence or line segment.*

Growth means add vertices or directed edges to existing PGG when new pattern P coming. There are three possible cases:

1. Un-matched: No segments in P matches segments in PGG. It will grow PGG with P.
2. Partially matched: Not all segments in P matches segments in PGG. The matched segments will be reused and new segments grow from matched segments.
3. Totally matched: All segments in P matches segments in PGG. Thus, no segments are generated, means no changes on PGG.

Finally we get summarization of ECG stream using PGG. A simple example is presented in Fig. 3.

3.4   Characteristics of PGG

Some important characteristics of PGG include:

- Convergence: Our experiments show that the number of un-matched patterns and partially matched patterns are decreased along with time.
- Graph scale: PGG shows excellent compression effects. In our experiment, it consumes about 0.3 % storage space with less than 5% relative error.
- Retrospection: If we also store the patterns occurrence timestamps. PGG can retrospect full time ECG evolution by reconstructing an approximate stream view.

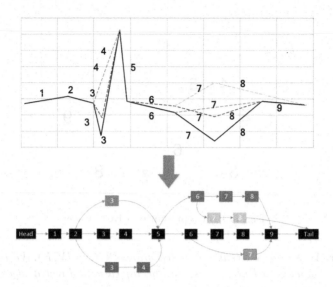

Fig. 3. PLR represented ECG waves (above) and constructed PGG (below). Each node in PGG is one segment in ECG waves, nodes linked by one edge in PGG represents two consecutive segments in ECG waves.

4   PGG Convergence

In this section, we will define a new measurement of a graph—Convergence Index (CGI). We will use CGI to assess death risk of CVD patients. Result are illustrated in section experiments.

4.1   Convergence Index

The CGI is computed on a directed graph. Here are some useful definitions.

Definition 3. *Independent Set of Directed Graph: A subset S of the vertices V of a directed graph $G = (V, E)$ is independent if no edge in the graph has both startpoint and endpoint in S.*

Definition 4. *Entropy of Vertex Set: Let X be a subset of vertex set V of a directed graph $G = (V, E)$. Defining*

$$H(X) = \sum_{i=1}^{n} p_i log \frac{1}{p_i} \tag{2}$$

as the entropy of vertex set X in G, where $\sum_{i=1}^{n} p_i = 1$. Each p_i represent the possibility of one possible vertex choice. So that n is the number of all possible vertex choice in a given X.

Normally, for all i, p_i are equal. So $H(X) = log(n)$.

Definition 5. *Mutual Information of Two Vertex Set:* Let $I(X \wedge Y)$ be *mutual information between two vertex set X and Y of a directed graph $G = (V, E)$. Where $I(X \wedge Y) = H(X) - H(X|Y)$.*

Finally, given a graph PGG, define the CGI of PGG

$$CGI = min_{X,Y} I(X \wedge Y) \qquad (3)$$

where the minimum is taken over all pairs of random variables X, Y such that: X is a uniformly random vertex in PGG. Y is an independent set containing X.

Intuitively, for node X and subset Y in PGG, X is chosen with maximal stochasticity (uniform distribution), and Y is the set of distinguishable nodes. If they have an edge, they are not distinguishable. That's why Y is an independent set. What CGI do is trying to quantify the stochasticity of graph for such an arbitrary Y. Low stochasticity of Y indicates low CGI index while high stochasticity of Y indicates high CGI index.

Why use CGI to evaluate the death risk of CVD patients? The reason is that CVD patients usually have high abnormal probability of their heart's cardiac cellular electrical activity, which can be directly reflected by ECG records. PGG summarizes long term ECG evolution and CGI measures the convergence index of PGG. Briefly, CGI quantifies the pathology of CVD patients.

4.2 Properties of CGI

Here are two useful properties and we illustrate them without proof:

– Subadditivity: $G_1 = (V, E_1)$ and $G_2 = (V, E_2)$, construct $G_1 = (V, E_1 \cup E_2)$. Then, $CGI(G) \leq CGI(G_1) + CGI(G_2)$
– Monotonicity: $G_1 = (V, E)$ and $G_2 = (V, E')$, $E \subset E'$. Then, $CGI(G_1) \leq CGI(G_2)$

Considering two special graph. The one is no edge graph: $G = (V, E)$, $|V| = n$ and $|E| = 0$. Then if X is a uniformly random vertex, and Y is fixed to be the vertex set V so that it can contain all possible X. We get $CGI \leq I(X \wedge Y) = 0$. But $CGI \geq 0$, so that $CGI = 0$. The other one is complete graph: $G = (V, E)$, $|V| = n$ and $|E| = n(n-1)/2$. Then the only independent set Y containing a given vertex X has to be $Y = X$. So that $H(X|Y) = 0$. Thus $CGI = min_{X,Y} I(X \wedge Y) = log(n) - 0 = log(n)$.

Since CGI is monotonic, we can also conclude that the range of CGI is $[0, log(n)]$ where n is the number of vertices in PGG.

5   Experiments

In this section, we depict the experimental settings and ICU data to test the proposed method. In addition, we report our experimental results using different performance measures.

5.1   Experimental Setup

MIMIC (Medical Information Mart for Intensive Care) is an open-access database includes Clinical Databases (contain electronic medical records) and Waveform Databases (contain biomedical streams) for over 58,000 ICU patients at the Beth Israel Deaconess Medical Center (BIDMC) from June 2001 to October 2012 [14]. We use subset of databases among 1022 CVD patients that contain all Waveform Database records that have been associated with Clinical Database records.

We view each patient as an instance, and extract useful features from both Clinical Databases and Waveform Databases. Three categories of biomarkers are taken into account in our experiments: General demographics—Gender, Age, BMI, Systolic Blood Pressure (SBP); expertise of clinical staff—Glasgow Coma Scale (GCS); computational measurements from ECG—Heart Rate Variation (HRV), Respiration Rate Variation (RRV); and our CGI. Meanwhile, five outcomes are also taken into consideration: hospital expire (HE), within 30 days to death (DTD30), 30 to 90 days to death (DTD90), 90 to 365 days to death (DTD365), days to death (DTD). The details of the these biomarkers can be referred to the related work section.

5.2   Correlations Between Biomarkers and Outcomes

We first study the correlations among all related biomarkers and outcomes. We conduct Pearson Correlation Coefficient as correlation measurement. Since not all of the patients have record of death date, we only use partial of the entire data. The result is shown in Fig. 4. The correlation heatmap is generated by a symmetric matrix, means that the column name order is identical with row name order. We see that correlations are strong within biomarkers. Particular in BMI, SBP and GCS, the reason is that high BMI people with hypertension co-morbidity are considered high danger patients by caregivers. In the last row of heatmap, we also find that CGI has the most correlation coefficients with DTD30, DTD90 and DTD365 than other biomarkers. Since these outcomes can reflect patients death risk directly, the result shows that CGI best fit the CVD patients death risk.

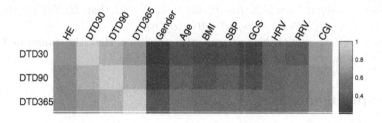

Fig. 4. All parameter correlations

5.3   CGI Performance

Now we are going to further investigate CGI performance on assessing patients death risk. We divide patients into four groups by days to death. Apart from formal introduced DTD30, DTD90, DTD365, we also regard days out of 1000 as Alive. Note that some records have missing field on days to death. These incomplete records should be excluded because we are not sure whether the patients are dead or alive. Apparently, patients in group DTD30 have the highest death risk, and patients in group Alive have the lowest death risk. Result of CGI distribution on different groups are illustrated with boxplot in Fig. 5. The black bold lines in middle of the rectangle are average CGIs of groups. We can see that average CGIs are decreased from DTD30 to Alive. Besides, the 75% quantile of CGI in DTD30 group is higher than 25% quantile of CGI in Alive group, which shows a good separation. These results are accord with actual death risk of patients. There are some other interesting discovers. For example, DTD30 group have the highest CGI value than other groups, but it also have the lowest CGI variance. A possible explaination is that CGI is computed using log function, leading to smaller variation when value increased.

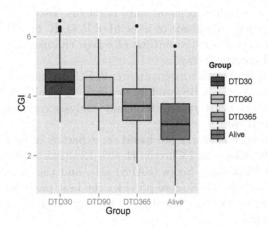

Fig. 5. Distribution of CGI in different patient groups

Furthermore, we employ Kaplan-Meier survival analysis to compare the risk degree evaluation. This is done by calculating the hazard ratio (HR) and p-values for each biomarkers between DTD30 and Alive. Since Kaplan-Meier survival analysis can handle with patients whose death dates are not recorded, we used all avaliable patient data. We compare CGI to three categories of risk assessing methods comprehensively. Details are introduced in related work section. It can be seen from Table 1 that SBP, GCS, HRV and CGI are correlated with the death risk, with a high statistical significance ($p \leq 0.05$). Yet, CGI clearly outperforms the others, it has the lowest p-value and the highest hazard ratio. The result shows that CGI best quantifies the pathology of CVD patients.

Table 1. Association between different measurements and death risk

Category	Biomarkers	Hazard ratio	P value
Demographic	Age > 65	1.82	0.062
	Female	1.68	0.092
	BMI	1.21	0.130
	SBP	2.66	0.044
Expert	GCS	2.47	0.035
Computational	HRV	2.53	0.024
	RRV	1.48	0.082
Proposed	CGI	3.41	0.012

6   Related Work

Generally, our work can be seen as extracting biomarkers (feature engineering on ECG streams) from long term ECG streams. Related techniques can be loosely divided into three categories.

A number of researches focus on wave segments feature extraction [15–17]. They extract duration, amplitude or interval of P, Q, R, S, T, U wave segments. These features varied when the outline of waves changes, in order to classify normal wave and abnormal wave.

Another category of work concern on overall ECG streams. [18] compare Consecutive Beats Similarity (CBS) and accumulate them all through. [19] obtain R?R intervals to get heart rate (HRV) and analysis correlation between HR variation and death risk.

The third category of works are based on expertise of clinical staff, including the Acute Physiology and Chronic Health Evaluation (APACHE) [20], the Simplified Acute Physiology Score (SAPS) [21], and the Glasgow Coma Scale (GCS) [22]. Particularly, GCS score provides the best performance [23,24]. But all of them require an expert clinical panel to select variables and denote levels of severity for each.

7   Conclusion

In this paper, we summarize long-term overwhelming ECG data and assessing death risk of patients with cardiovascular disease. Our major innovative work include:

- We summarize ECG stream using morphological concern of overall evolution. And assessing patients risk from high value density ECG summarization instead of raw data.
- Our method is totally unsupervised without the help of expert knowledge. It can assist in clinical practice without any additional burden like buy new devices or add more caregivers.

– We conduct comprehensive experiments on real ICU datasets. Experiments demonstrate the effectiveness of the proposed method.

In the future, we will further consider more effective biomarkers on PGG, and extend our work on Electroencephalography (EEG) to assessing Parkinsons Disease.

Acknowledgement. This work was supported by Natural Science Foundation of China (No. 61170003).

References

1. Mozaffarian, D., Benjamin, E.J., Go, A.S., Arnett, D.K., Blaha, M.J., Cushman, M., de Ferranti, S., Despres, J.P., Fullerton, H.J., Howard, V.J., et al.: Heart disease and stroke statistics-2015 update: a report from the american heart association. Circulation **131**(4), e29 (2015)
2. Nichols, M., Townsend, N., Scarborough, P., Rayner, M.: Cardiovascular disease in Europe 2014: epidemiological update. Eur. Heart J. **35**(42), 2950–2959 (2014)
3. Ballantyne, C.M., Hoogeveen, R.C., Bang, H., Coresh, J., Folsom, A.R., Heiss, G., Sharrett, A.R.: Lipoprotein-associated phospholipase A2, high-sensitivity C-reactive protein, and risk for incident coronary heart disease in middle-aged men and women in the Atherosclerosis Risk in Communities (aric) study. Circulation **109**(7), 837–842 (2004)
4. Vasan, R.S., Larson, M.G., Benjamin, E.J., Evans, J.C., Reiss, C.K., Levy, D.: Congestive heart failure in subjects with normal versus reduced left ventricular ejection fraction: prevalence and mortality in a population-based cohort. J. Am. Coll. Cardiol. **33**(7), 1948–1955 (1999)
5. Chia, C.C., Blum, J., Karam, Z., Singh, S., Syed, Z.: Predicting postoperative atrial fibrillation from independent ECG components. In: Twenty-Eighth AAAI Conference on Artificial Intelligence (2014)
6. Bender, J., Russell, K., Rosenfeld, L., Chaudry, S.: Oxford American Handbook of Cardiology. Oxford University Press, New York (2010)
7. Moody, G.B., Mark, R.G.: The impact of the MIT-BIH arrhythmia database. IEEE Eng. Med. Biol. Mag. **20**(3), 45–50 (2001)
8. Zong, W., Moody, G., Jiang, D.: A robust open-source algorithm to detect onset and duration of QRS complexes. In: Computers in Cardiology, pp. 737–740. IEEE (2003)
9. Yi, B.K., Faloutsos, C.: Fast time sequence indexing for arbitrary Lp norms. In: VLDB (2000)
10. Keogh, E., Lin, J., Fu, A.: HOT SAX: efficiently finding the most unusual time series subsequence. In: Fifth IEEE International Conference on Data Mining, 8 p. IEEE (2005)
11. Wu, H., Salzberg, B., Zhang, D.: Online event-driven subsequence matching over financial data streams. In: Proceedings of the 2004 ACM SIGMOD International Conference on Management of Data, pp. 23–34. ACM (2004)
12. Tang, L.a., Cui, B., Li, H., Miao, G., Yang, D., Zhou, X.: Effective variation management for pseudo periodical streams. In: Proceedings of the 2007 ACM SIGMOD International Conference on Management of Data, pp. 257–268. ACM (2007)

682   S. Hong et al.

13. Keogh, E., Chu, S., Hart, D., Pazzani, M.: An online algorithm for segmenting time series. In: Proceedings IEEE International Conference on Data Mining, ICDM 2001, pp. 289–296. IEEE (2001)
14. Goldberger, A.L., Amaral, L.A., Glass, L., Hausdorff, J.M., Ivanov, P.C., Mark, R.G., Mietus, J.E., et al.: PhysioBank, PhysioToolkit, and PhysioNet – components of a new research resource for complex physiologic signals. Circulation 101(23), e215–e220 (2000)
15. Chiang, H.S., Shih, D.H., Lin, B., Shih, M.H.: An APN model for arrhythmic beat classification. Bioinformatics 30(12), 1739–1746 (2014)
16. Tafreshi, R., Jaleel, A., Lim, J., Tafreshi, L.: Automated analysis of ECG waveforms with atypical QRS complex morphologies. Biomed. Sig. Process. Control 10, 41–49 (2014)
17. Mar, T., Zaunseder, S., Martínez, J.P., Llamedo, M., Poll, R.: Optimization of ECG classification by means of feature selection. IEEE Trans. Biomed. Eng. 58(8), 2168–2177 (2011)
18. Chia, C.C., Syed, Z.: Scalable noise mining in long-term electrocardiographic timeseries to predict death following heart attacks. In: Proceedings of the 20th ACM SIGKDD International Conference on Knowledge Discovery and Data Mining, pp. 125–134. ACM (2014)
19. Huikuri, H.V., Stein, P.K.: Heart rate variability in risk stratification of cardiac patients. Prog. Cardiovasc. Dis. 56(2), 153–159 (2013)
20. Knaus, W.A., Zimmerman, J.E., Wagner, D.P., Draper, E.A., Lawrence, D.E.: Apache-acute physiology and chronic health evaluation: a physiologically based classification system. Crit. Care Med. 9(8), 591–597 (1981)
21. Le Gall, J.R., Loirat, P., Alperovitch, A., Glaser, P., Granthil, C., Mathieu, D., Mercier, P., Thomas, R., Villers, D.: A simplified acute physiology score for ICU patients. Crit. Care Med. 12(11), 975–977 (1984)
22. Teasdale, G., Jennett, B.: Assessment of coma and impaired consciousness: a practical scale. Lancet 304(7872), 81–84 (1974)
23. Grmec, S., Gasparovic, V.: Comparison of APACHE II, MEES and Glasgow Coma Scale in patients with nontraumatic coma for prediction of mortality. CRITICAL CARE-LONDON- 5(1), 19–23 (2001)
24. Sakr, Y., Krauss, C., Amaral, A., Réa-Neto, A., Specht, M., Reinhart, K., Marx, G.: Comparison of the performance of SAPS II, SAPS 3, APACHE II, and their customized prognostic models in a surgical intensive care unit. Br. J. Anaesth. 101(6), 798–803 (2008)

Spatio-Temporal and Sequential Data Mining

On the Robustness of Decision Tree Learning Under Label Noise

Aritra Ghosh[1], Naresh Manwani[2], and P.S. Sastry[3(✉)]

[1] Microsoft, Bangalore, India
arghosh@microsoft.com
[2] International Institute of Information Technology, Hyderabad, India
naresh.manwani@iiit.ac.in
[3] Indian Institute of Science, Bangalore, India
sastry@ee.iisc.ernet.in

Abstract. In most practical problems of classifier learning, the training data suffers from label noise. Most theoretical results on robustness to label noise involve either estimation of noise rates or non-convex optimization. Further, none of these results are applicable to standard decision tree learning algorithms. This paper presents some theoretical analysis to show that, under some assumptions, many popular decision tree learning algorithms are inherently robust to label noise. We also present some sample complexity results which provide some bounds on the sample size for the robustness to hold with a high probability. Through extensive simulations we illustrate this robustness.

Keywords: Robust learning · Decision trees · Label noise

1 Introduction

For supervised learning of a classifier, we make use of labeled training data. When the class labels in the training data may be incorrect, it is referred to as label noise. Subjectivity and other errors in human labeling, measurement errors, insufficient feature space are some of the main reasons behind label noise. In many large data problems, labeled samples are often obtained through crowd sourcing and the unreliability of such labels is another reason for label noise. Learning from positive and unlabeled samples can also be cast as a problem of learning under label noise [5]. Thus, learning classifiers in the presence of label noise is an important problem [6].

Decision tree is among the most widely used machine learning approaches [19]. However, not many results are known about the robustness of decision tree learning in presence of label noise. It is observed that label noise in the training data increases size of the learnt tree; detecting and removing noisy examples improves the learnt tree [3]. Through an extensive empirical study it is observed that decision tree learning is fairly robust to label noise [13]. In this paper, we present a theoretical study of robustness of decision tree learning.

© Springer International Publishing AG 2017
J. Kim et al. (Eds.): PAKDD 2017, Part I, LNAI 10234, pp. 685–697, 2017.
DOI: 10.1007/978-3-319-57454-7_53

Most theoretical analyses of learning classifiers under label noise are in the context of risk minimization. The robustness of risk minimization depends on the loss function used. It is proved that any convex potential loss is not robust to uniform or symmetric label noise [9]. While most standard convex loss functions are not robust to symmetric label noise, the 0–1 loss is [11]. A general sufficient condition on the loss function for risk minimization to be robust is derived in [7]. The 0–1 loss, sigmoid loss and ramp loss are shown to satisfy this condition while convex losses such as hinge loss and the logistic loss do not satisfy this condition. Interestingly, we can have a convex loss (which is not a convex potential) that satisfies this sufficient condition and the corresponding risk minimization essentially amounts to a highly regularized SVM [18]. Robust risk minimization strategies under the so called class-conditional (or asymmetric) label noise are also proposed [12,17]. None of these results are applicable for the popular decision tree learning algorithms because they cannot be cast as risk minimization.

In this paper, we analyze learning of decision trees under label noise. We consider some of the popular impurity function based methods for learning of decision trees. We show, in the large sample limit, that under symmetric or uniform label noise the split rule that optimizes the objective function under noisy data is the same as that under noise-free data. We explain how this results in the learning algorithm being robust to label noise (under the large sample limit). We also derive some sample complexity bounds to indicate how large a sample we need at a node. We explain how these results indicate robustness of random forest also. We present empirical results to illustrate this robustness of decision trees and random forests. For comparison we also present results obtained with SVM algorithm.

2   Label Noise and Decision Tree Robustness

In this paper, we only consider binary decision trees for binary classification. We use the same notion of noise tolerance as in [11,18].

2.1   Label Noise

Let $\mathcal{X} \subset \mathcal{R}^d$ be the feature space and let $\mathcal{Y} = \{1, -1\}$ be the class labels. Let $S = \{(\mathbf{x}_1, y_{\mathbf{x}_1}), \ldots, (\mathbf{x}_N, y_{\mathbf{x}_N})\} \in (\mathcal{X} \times \mathcal{Y})^N$ be the *ideal* noise-free data drawn *iid* from a fixed but unknown distribution $\mathcal{D}$ over $\mathcal{X} \times \mathcal{Y}$. The learning algorithm does not have access to this data. The noisy training data given to the algorithm is $S^\eta = \{(\mathbf{x}_i, \tilde{y}_{\mathbf{x}_i}), i = 1, \cdots, N\}$, where $\tilde{y}_{\mathbf{x}_i} = y_{\mathbf{x}_i}$ with probability $(1 - \eta_{\mathbf{x}_i})$ and $\tilde{y}_{\mathbf{x}_i} = -y_{\mathbf{x}_i}$ with probability $\eta_{\mathbf{x}_i}$. As a notation, for any $\mathbf{x}$, $y_{\mathbf{x}}$ denotes its 'true' label while $\tilde{y}_{\mathbf{x}}$ denotes the noisy label. Thus, $\eta_{\mathbf{x}} = \Pr[y_{\mathbf{x}} \neq \tilde{y}_{\mathbf{x}} \mid \mathbf{x}]$. We use $\mathcal{D}^\eta$ to denote the joint probability distribution of $\mathbf{x}$ and $\tilde{y}_{\mathbf{x}}$.

We say that the noise is *uniform* or *symmetric* if $\eta_{\mathbf{x}} = \eta$, $\forall \mathbf{x}$. Note that, under symmetric noise, a sample having wrong label is independent of the feature vector and the 'true' class of the sample. Noise is said to be *class conditional* or *asymmetric* if $\eta_{\mathbf{x}} = \eta_+$, for all patterns of class $+1$ and $\eta_{\mathbf{x}} = \eta_-$, for all patterns of class -1. When noise rate $\eta_{\mathbf{x}}$ is a general function of $\mathbf{x}$, it is termed as *non-uniform* noise. Note that the value of η is unknown to the learning algorithm.

2.2   Criteria for Learning Split Rule at a Node of Decision Trees

Most decision tree learning algorithms grow the tree in top down fashion starting with all training data at the root node. At any node, the algorithm selects a split rule to optimize a criterion and uses that split rule to split the data into the left and right children of this node; then the same process is recursively applied to the children nodes till the node satisfies the criterion to become a leaf. Let $\mathcal{F}$ denote a set of split rules. Suppose, a split rule $f \in \mathcal{F}$ at a node v, sends a fraction a of the samples at v to the left child v_l and the remaining fraction $(1 - a)$ to the right child v_r. Then many algorithms select a $f \in \mathcal{F}$ to maximize

$$C(f) = G(v) - (aG(v_l) + (1 - a)G(v_r)) \tag{1}$$

where $G(\cdot)$ is a so called impurity measure. There are many such impurity measures. Of the samples at any node v, suppose a fraction p are of positive class and a fraction $q = (1 - p)$ are of negative class. Then the Gini impurity is defined by $G_{\text{Gini}} = 2pq$ [1]; entropy based impurity is defined as $G_{\text{Entropy}} = -p \log p - q \log q$ [16]; and misclassification impurity is defined as $G_{\text{MC}} = \min\{p, q\}$. Often the criterion C is called the *gain*. Hence, we also use $\text{gain}_{\text{Gini}}(f)$ to refer to $C(f)$ when G is G_{Gini} and similarly for others.

A split criterion different from impurity is twoing rule [1]. Let p, q, a be as above and let p_l (p_r), q_l (q_r) be the corresponding fractions at the left (right) child v_l (v_r) under split rule f. Then twoing rule selects $f \in \mathcal{F}$ which maximizes $G_{\text{Twoing}}(f) = a(1 - a)[|p_l - p_r| + |q_l - q_r|]^2/4$.

2.3   Noise Tolerance of Decision Tree

We want the decision tree learnt with noisy labels to have the same error on noise-free test set as that of the tree learnt using noise-free data. Since label noise is random, on any specific noisey training data, the tree learnt would also be random. Hence, we say the learning method is robust if, in the limit as training set size goes to infinity, the above holds. We now formalize this notion.

Definition 1. *A split criterion C is said to be* noise-tolerant *if*

$$\arg\min_{f \in \mathcal{F}} C(f) = \arg\min_{f \in \mathcal{F}} C^{\eta}(f)$$

where $C(f)$ is the value of the split criterion C for a split rule $f \in \mathcal{F}$ on noise free data and $C^{\eta}(f)$ is the value of the criterion function for f on noisy data, in the limit as the data size goes to infinity.

Let the decision tree learnt from training sample S be represented as $LearnTree(S)$ and let the classification of any $\mathbf{x}$ by this tree be represented as $LearnTree(S)(\mathbf{x})$.

Definition 2. *A decision tree learning algorithm LearnTree is said to be noise-tolerant if*

$$P_D(LearnTree(S)(\mathbf{x}) \neq y_{\mathbf{x}}) = P_D(LearnTree(S^\eta)(\mathbf{x}) \neq y_{\mathbf{x}})$$

Note that for the above to hold it is sufficient if *LearnTree(S)* is same as *LearnTree(S^η)*. That is, if the tree learnt with noisy samples is same as that learnt with noise-free samples.[1]

3   Theoretical Results

Robustness of decision tree learning requires the robustness of the split criterion at each non-leaf node and robustness of the labeling rule at each leaf node.

3.1   Robustness of Split Rules

We are interested in comparing, for any specific split rule f, the value of $C(f)$ with its value (in the large sample limit) when there is symmetric label noise.

Let the noise-free samples at a node v be $\{(\mathbf{x}_i, y_i), i = 1, \cdots, n\}$. Under label noise, the samples at this node would become $\{(\mathbf{x}_i, \tilde{y}_i), i = 1, \cdots, n\}$. Suppose in the noise-free case a split rule f sends n_l of these n samples to the left child, v_l, and $n_r = n - n_l$ to right child, v_r. Since the split rule depends only on the feature vector $\mathbf{x}$ and not the labels, the points that go to v_l and v_r would be the same for the noisy samples also. However, what changes with label noise are the class labels and hence the number of examples of different classes at a node.

Let n^+ and $n^- = n - n^+$ be the number of samples of the two classes at node v in the noise-free case. Let these numbers for v_l and v_r be n_l^+, n_l^- and n_r^+, n_r^-. Let these quantities in the noisy case be denoted by $\tilde{n}^+, \tilde{n}^-, \tilde{n}_l^+, \tilde{n}_l^-$ etc. Define binary random variables, Z_i, $i = 1, \cdots, n$, by $Z_i = 1$ iff $\tilde{y}_i \neq y_i$. By definition of symmetric label noise, Z_i are *iid* Bernoulli random variables with expectation η.

Let $p = n^+/n, q = n^-/n = (1 - p)$. Let p_l, q_l and p_r, q_r be these fractions for v_l and v_r. Let the corresponding quantities for the noisy case be $\tilde{p}, \tilde{q}, \tilde{p}_l, \tilde{q}_l$ etc. Let p^η, q^η, p_l^η etc. be the values of $\tilde{p}, \tilde{q}, \tilde{p}_l$ in the large sample limit. We have

$$\tilde{p} = \frac{\tilde{n}^+}{n} = \frac{1}{n}\left(\sum_{i:\tilde{y}_i=+1} 1\right) = \frac{1}{n}\left(\sum_{i:y_i=+1}(1 - Z_i) + \sum_{i:y_i=-1} Z_i\right) \qquad (2)$$

$$\tilde{p}_l = \frac{\tilde{n}_l^+}{n_l} = \frac{1}{n_l}\left(\sum_{i:\mathbf{x}_i \in v_l, \tilde{y}_i=+1} 1\right) = \frac{1}{n_l}\left(\sum_{i:\mathbf{x}_i \in v_l, y_i=+1}(1 - Z_i) + \sum_{i:\mathbf{x}_i \in v_l, y_i=-1} Z_i\right)$$

---

[1] For simplicity, we do not consider pruning of the tree.

All the above expressions involve sums of independent random variables. Hence, by law of large numbers, in the large sample limit we get

$$p^\eta = p(1 - \eta) + q\eta = p(1 - 2\eta) + \eta; \quad p_l^\eta = p_l(1 - \eta) + q_l\eta = p_l(1 - 2\eta) + \eta \quad (3)$$

Note that, under symmetric label noise, $\Pr[Z_i = 1] = \Pr[Z_i = 1|y_i] = \Pr[Z_i = 1|\mathbf{x}_i \in B, y_i] = \eta$, for any subset B of the feature space and this fact is used in deriving Eq. (3).

To find the large sample limit of criterion $C(f)$ under label noise, we need values of the impurity function which in turn needs p^η, q^η, p_l^η etc. which are as given above. For example, the Gini impurity is given by $G(v) = 2pq$ for the noise free case. For the noisy sample, its value can be written as $\tilde{G}(v) = 2\tilde{p}\tilde{q}$. Its value in the large sample limit would be $G^\eta(v) = 2p^\eta q^\eta$. Using the above we can now prove the following theorem about robustness of split criteria.

Theorem 1. *Splitting criterion based on Gini impurity, mis-classification rate and twoing rule are noise-tolerant to symmetric label noise given $\eta \neq 0.5$.*

Proof. We prove robustness of Gini impurity here. Robustness under other criteria can similarly be proved.

Let p and q be the fractions of the two classes at a node v and let a be the fraction of points (under a split rule) at the left child, v_l. Recall that the fraction a is same for noisy and noise-free data. The Gini impurity is $G_{\text{Gini}}(v) = 2pq$. Under symmetric label noise, Gini impurity (under large sample limit) becomes (using Eq. (3)),

$$G_{\text{Gini}}^\eta(v) = 2p^\eta q^\eta = 2[((1 - 2\eta)p + \eta)((1 - 2\eta)q + \eta)]$$
$$= 2pq(1 - 2\eta)^2 + (\eta - \eta^2) = G_{\text{Gini}}(v)(1 - 2\eta)^2 + (\eta - \eta^2)$$

Similar expressions hold for $G_{\text{Gini}}^\eta(v_l)$ and $G_{\text{Gini}}^\eta(v_r)$. The (large sample) value of criterion or impurity gain of f under label noise can be written as

$$\text{gain}_{\text{Gini}}^\eta(f) = G_{\text{Gini}}^\eta(v) - [a\, G_{\text{Gini}}^\eta(v_l) + (1 - a)G_{\text{Gini}}^\eta(v_r)]$$
$$= (1 - 2\eta)^2[G_{\text{Gini}}(v) - a\, G_{\text{Gini}}(v_l) - (1 - a)\text{Gini}(v_r)] = (1 - 2\eta)^2\text{gain}_{\text{Gini}}(f)$$

Thus for any $\eta \neq 0.5$, if $\text{gain}_{\text{Gini}}(f^1) > \text{gain}_{\text{Gini}}(f^2)$, then $\text{gain}_{\text{Gini}}^\eta(f^1) > \text{gain}_{\text{Gini}}^\eta(f^2)$. Which means that a maximizer of impurity gain based on Gini index under noise-free samples will be also a maximizer of gain under symmetric label noise, under large sample limit.

Remark: Another popular criterion is impurity gain based on entropy which is not considered in the above theorem. The impurity gain based on entropy is not noise-tolerant as can be shown by a counterexample. Consider a case where a node has n samples (n is large). Suppose, under split rule $f_1(f_2)$ we get $n_l = 0.5n(0.3n)$, $n_l^+ = 0.05n(0.003n)$ and $n_r^+ = 0.25n(0.297n)$. Then it can be easily shown that the best split under no-noise (f_2) does not remain best under 40% noise. However, such counter examples may not be generic and entropy based method may also be robust to label noise in practice.

3.2   Robustness of Labeling Rule at Leaf Nodes

We next consider the robustness of criterion to assign a class label to a leaf node. A popular approach is to take majority vote at the leaf node. We prove that, majority voting is robust to symmetric label noise. We also show that it can be robust to non-uniform noise also under a restrictive condition.

Theorem 2. *Let* $\eta_{\mathbf{x}} < 0.5, \forall \mathbf{x}$. *(a). Then, majority voting at a leaf node is robust to symmetric label noise. (b). It is also robust to nonuniform label noise if all the points at the leaf node belong to one class in the noise free data.*

Proof. Let p and $q = 1 - p$ be the fraction of positive and negative samples at leaf node v.

(a) Under symmetric label noise, the relevant fractions are $p^{\eta} = (1-\eta)p + \eta q$ and $q^{\eta} = (1 - \eta)q + \eta p$. Thus, $p^{\eta} - q^{\eta} = (1 - 2\eta)(p - q)$. Since $\eta < 0.5$, $(p^{\eta} - q^{\eta})$ will have the same sign as $(p - q)$, proving robustness of the majority voting.

(b) Let v contain all the points from the positive class. Thus, $p = 1, q = 0$. Let $\mathbf{x}_1, \cdots, \mathbf{x}_n$ be the samples at v. Under non-uniform noise (with $\eta_{\mathbf{x}} < 0.5, \forall \mathbf{x}$),

$$p^{\eta} = \frac{1}{n}\sum_{i=1}^{n}(1 - \eta_{\mathbf{x}_i}) > \frac{0.5}{n}\sum_{i=1}^{n}1 = 0.5 \qquad (4)$$

Thus, the majority vote will assign positive label to the leaf node v. This proves the second part of the theorem.

3.3   Robustness of Decision Tree Learning Under Symmetric Label Noise: Large Sample Analysis

We have shown that the split rule that maximizes the criterion function under symmetric label noise is same as the one which maximizes it under noise-free case (under large sample limit). This means, under large sample assumption, the same split rule would be learnt at any node irrespective of whether the labels come from noise-free data or noisy data. (Here we assume for simplicity that there is a unique maximizer of the criterion at each node. Otherwise we need some prefixed rule to break ties. We are assuming that the $\mathbf{x}_i$ at a node are same in the noisy and noise-free cases. These are same at the root. If we learn the same split at the root, then at both its children the samples would be same in the two cases and so on).

Our result for leaf node labeling implies that, under large sample assumption, with majority rule, a leaf node would get the same label under noisy or noise-free data. To conclude that we learn the same tree, we need to examine the rule for deciding when a node becomes a leaf. If this is determined by the depth of the node or number of samples at the node then it is easy to see that the same tree would be learnt with noisy and noise-free data. In many algorithms one makes a node as leaf if no split rule gives positive value to the gain. This will also lead to learning of the same tree with noisy samples as with noise-free samples, because we showed that the gain under noisy case is a linear function of the gain under noise-free case.

Robustness Against General Noise: In our analysis, we have only considered symmetric label noise. In the case of class-conditional noise, noise rate is same for all feature vectors of a class though it may be different for different classes. In the risk minimization framework, class conditional noise can be handled when the noise rates are known (or can be estimated) [7,12,14,17]. We can extend the analysis presented in Sect. 3.1 to relate expected fraction of examples of a class in the noisy and noise-free cases using the two noise rates. Thus, if the noise rates are assumed known (or can be reliably estimated) it should be possible to extend the analysis here to the case of class-conditional noise. In the general case when noise rates are not known (and cannot be reliably estimated), it appears difficult to establish robustness of impurity based split criteria.

3.4 Sample Complexity Under Noise

We established robustness of decision tree learning algorithms under large sample limit. Hence an interesting question is that of how large the sample size should be for our assertions about robustness to hold with a large probability. We provide some sample complexity bounds in this subsection. (Due to space constraint, we provide proof sketch in Appendix).

Lemma 1. *Let leaf node v have n samples. Under symmetric label noise with $\eta < 0.5$, majority voting will not fail with probability at least $1 - \delta$ when $n \geq \frac{2}{\rho^2(1-2\eta)^2} \ln(\frac{1}{\delta})$, where ρ is the difference between fraction of positive and negative samples in the noise-free case.*

The sample size needed increases with increasing noise (η) and decreasing ρ (which can be viewed as 'margin of majority'), which is intuitively clear.

Lemma 2. *Let there be n samples at a non-leaf node v. Given two splits f_1 and f_2, suppose gain (Gini, misclassification, twoing rule) for f_1 is higher than that for f_2. Under symmetric label noise with $\eta \neq 0.5$, gain from f_1 will be higher with probability $1-\delta$ when $n \geq \mathcal{O}(\frac{1}{\rho^2(1-2\eta)^2} \ln(\frac{1}{\delta}))$, where ρ denotes the difference between gain of the two splits in the noise-free case.*

While these results shed some lights on sample complexity, we emphasize that these bounds are loose and are obtained using concentration inequalities. In experimental section, we provide results on how many training samples are needed for robust learning of decision trees on a synthetic dataset.

3.5 Noise Robustness in Random Forest

A random forest [2] is a collection of randomized tree classifiers. We represent the set of trees as $g_n = \{g_n(\mathbf{x}, \pi_1), \cdots, g_n(\mathbf{x}, \pi_m)\}$. Here $\pi_1, \cdots, \pi_m$ are *iid* random variables, conditioned on data, which are used for partitioning the nodes. Finally, majority vote is taken among the random tree classifiers for prediction. We denote this classifier as $\bar{g}_n$.

In a ***purely random forest classifier***, partitioning does not depend on the class labels. At each step, a node is chosen randomly and a feature is selected randomly for the split. A split threshold is chosen uniformly randomly from the interval of the selected feature. This procedure is done k times. In ***a greedily grown random forest classifier*** each tree is grown greedily by improving impurity with some randomization. At each node, a random subset of features are chosen. Tree is grown by computing the best split among those random features only. Breiman's random forest classifier uses Gini impurity gain [2].

A purely random forest classifier/greedily grown random forest, $\bar{g}_n$, is robust to symmetric label noise with $\eta < 0.5$ under large sample assumption. In purely random forest, randomization is on the partitions and the partitions do not depend on class labels (which may be noisy). We proved robustness of majority vote at leaf nodes under symmetric label noise. Thus, for a purely random forest, the classifier learnt with noisy labels would be same as that learnt with noise-free samples. Similarly for a greedily grown trees with Gini impurity measure, we showed that each tree is robust because of both split rule robustness and majority voting robustness. Thus when large sample assumption holds, greedily grown random forest will also be robust to symmetric label noise. The sample complexity for random forests should be less than that for single decision tree because the ensemble classifier results in some variance reduction. Empirically we observe that, often random forest has better robustness than a single decision tree in finite sample cases.

4   Empirical Illustration

In this section, we illustrate our robustness results for learning of decision trees and random forest. We also present results with SVM whose sensitivity towards noise widely varies [9,11,13,18].

4.1   Dataset Description

We used four 2D synthetic datasets. Details are given below. (Here n denotes total number of samples, p_+, p_- represent the class conditional densities, and $\mathcal{U}(\mathcal{A})$ denotes uniform distribution over set $\mathcal{A}$).

- Dataset 1: Checker board 2×2 Pattern: Data uniform over $[0,2] \times [0,2]$ and one class region being $([0,1] \times [0,1]) \cup ([1,2] \times [1,2])$ and $n = 30000$.
- Dataset 2: Checker board 4×4: Extension of the above to a 4×4 grid.
- Dataset 3: Imbalance Linear Data. $p_+ = \mathcal{U}([0,0.5] \times [0,1])$ and $p_- = \mathcal{U}([0.5,1] \times [0,1])$. Prior probabilities of classes are 0.9 & 0.1, and $n = 40000$.
- Dataset 4: Imbalance and Asymmetric Linear Data. $p_+ = \mathcal{U}([0,0.5] \times [0,1])$ and $p_- = \mathcal{U}([0.5,0.7] \times [0.4,0.6])$. Prior probabilities are 0.8 & 0.2, and $n = 40000$.

We also present results for 6 UCI datasets [8].

4.2 Experimental Setup

We used decision tree/random forest (RF) implementation in scikit learn library
[15]. We present results only with Gini impurity based decision tree classifier.
Number of trees in random forest was set to 100. For SVM we used libsvm
package [4]. For the results presented in Sect. 4.4, the following setup is used.
Minimum leaf size is the only user-chosen parameter in random forest and decision trees. For synthetic datasets, minimum samples in leaf node was restricted
to 250. For UCI datasets, it was restricted to 50. For SVM, we used linear kernel
(l) for Synthetic Datasets 3, 4 and quadratic kernel (p) for Checker board 2 × 2
data. In all other datasets we used gaussian kernel (g). For SVM, we selected
hyper-parameters using validation data. (Validation range for C is 0.01–500 and
for γ in the Gaussian kernel it is 0.001–10). We used 20% data for testing and
20% for validation. Noise rate was varied from 0%–40%. As synthetic datasets
are separable, we also experimented with class conditional noise with the two
noise rates for the two classes being 40% and 20%. In all experiments, noise was
introduced only on training and validation data. Test set was noise free.

4.3 Effect of Sample Size on Robustness of Learning

Here we present experimental results on the test accuracy for different sample
sizes using the 2 × 2 checker board data. We choose a leaf sample size and learn
decision tree and random forest with different noise levels. (The training set size
is fixed at 20000). We do this for a number of choices for leaf sample size. The
test accuracies in all these cases are shown in Fig. 1(a). As can be seen from
the figure, even when training data size is huge, we do not get robustness if leaf
sample size is small. This is in accordance with our analysis (as in Lemma 1)
because minimum sample size is needed for the majority rule to be correct with
a large probability. A leaf sample size of 50 seems sufficient to take care of even
30% noise.

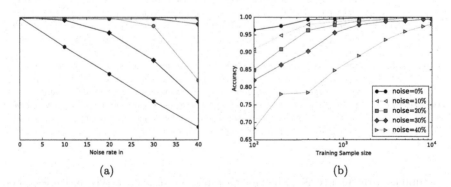

(a)   (b)

Fig. 1. For 2 × 2 Checker board data variation of accuracy with (a) Minimum leaf size,
(b) Training data size, for different noise levels for DT

Next we experiment with varying the (noisy) training data size. The results are shown in Fig. 1(b). It can be seen that with 400/4000 sample size decision tree learnt has good test accuracy (95%) at 20%/40% noise (the sample ratio is close to $\frac{(1-2\times0.4)^2}{(1-2\times0.2)^2} = 1/9$ as provided in Lemma 1). We need larger sample size for higher level of noise. This is also as expected from our analysis.

4.4 Comparison of Accuracies of Learnt Classifiers

The average test accuracy and standard deviation (over 10 runs) on different data sets under different levels of noise are shown in Table 1 for synthetic datasets and in Table 2 for UCI datasets. In Table 2 we also indicate the dimension of feature vector (d), the number of positive and negative samples in the data (n^+, n^-).

For synthetic datasets, the sample sizes are large and hence we expect good robustness. As can be seen from Table 1, for noise-free data, all classifiers (decision tree, random forest and SVM) perform equally. However, with 30% or 40% noise, the accuracies of SVM are much poorer than those of decision tree and random forest. For example, for synthetic datasets 3 and 4, the average accuracies of decision tree and random forest classifiers continue to be 99% even at 40% noise while those of SVM drop to about 90% and 80% respectively. Note that even with very large sample sizes, we do not get robustness in SVM. It can be seen that decision tree and random forest classifiers are robust to class conditional noise also, even without knowledge about noise rate (as indicated by last column in the table). Our current analysis does not prove this robustness; this is one possible extension of the theoretical analysis presented here.

Table 1. Comparison of accuracies on synthetic datasets

Data	Method	$\eta = 0\%$	$\eta = 10\%$	$\eta = 20\%$	$\eta = 30\%$	$\eta = 40\%$	$\eta_+ = 40\%$ $\eta_- = 20\%$
2 × 2 CB	Gini	99.95 ± 0.05	99.9 ± 0.06	99.91 ± 0.1	99.82 ± 0.16	98.97 ± 0.83	99.45 ± 0.83
	RF	99.99 ± 0.02	99.96 ± 0.02	99.91 ± 0.05	99.87 ± 0.06	99.16 ± 0.18	99.11 ± 0.45
	SVM (p)	99.83 ± 0.12	97.38 ±1.21	91.88 ± 2.65	87.96 ± 5.52	76.42 ± 4.43	68.78 ± 0.97
4 × 4 CB	Gini	99.76 ± 0.18	99.72 ± 0.16	99.46 ± 0.18	98.71 ± 0.32	95.21 ± 1.08	97.36 ± 1.23
	RF	99.94 ± 0.02	99.9 ± 0.02	99.78 ± 0.04	99.35 ± 0.15	96.23 ± 0.91	95.41 ± 0.53
	SVM (g)	99.6 ± 0.05	98.58 ± 0.23	97.81 ± 0.24	96.83 ± 0.46	92.22 ± 2.5	91.24 ± 0.85
Dataset 3	Gini	100.0 ± 0.01	100.0 ± 0.01	99.99 ± 0.01	99.99 ± 0.02	99.92 ± 0.07	99.92 ± 0.18
	RF	100.0 ± 0.01	100.0 ± 0.01	99.99 ± 0.01	99.98 ± 0.02	99.86 ± 0.12	99.9 ± 0.13
	SVM (l)	99.89 ± 0.04	96.65 ± 0.26	90.02 ± 0.3	90.02 ± 0.3	90.02 ± 0.3	90.1 ± 0.31
Dataset 4	Gini	100.0 ± 0.0	99.99 ± 0.01	99.99 ± 0.01	99.98 ± 0.03	99.73 ± 0.54	99.88 ± 0.26
	RF	100.0 ± 0.0	99.99 ± 0.01	99.99 ± 0.01	99.93 ± 0.09	99.91 ± 0.11	99.7 ± 0.31
	SVM (l)	99.86 ± 0.03	99.21 ± 0.24	96.55 ± 4.05	79.96 ± 0.34	79.96 ± 0.34	79.96 ± 0.34

Similar performance is seen on UCI datasets also as shown in Table 2. For breast cancer dataset, there is a small drop in the average accuracy of decision tree with increasing noise rate while for random forest the drop is significantly less. This is also expected because the total sample size here is less. Although

SVM has significantly higher average accuracy than decision tree in 0% noise, at 40% noise its average accuracy drops more than that of decision tree. In all other data sets also, decision tree and random forest are more robust than SVM as can be seen from the table.

Table 2. Comparison of accuracies on UCI datasets

Data (d, n^+, n^-)	Method	$\eta = 0\%$	$\eta = 10\%$	$\eta = 20\%$	$\eta = 30\%$	$\eta = 40\%$
Breast cancer	Gini	92.04 ± 3.0	90.36 ± 3.02	90.0 ± 2.24	90.22 ± 2.38	87.23 ± 7.72
$(10, 239, 444)$	RF	96.64 ± 0.93	96.79 ± 1.23	96.64 ± 1.82	95.91 ± 1.47	96.13 ± 1.39
	SVM	96.79 ± 1.67	96.06 ± 1.91	95.91 ± 2.27	93.72 ± 4.55	92.48 ± 3.62
German	Gini	71.2 ± 3.47	71.7 ± 2.5	71.25 ± 3.16	70.25 ± 2.75	64.65 ± 6.29
$(24, 300, 700)$	RF	70.75 ± 2.71	70.8 ± 2.94	70.9 ± 2.84	71.05 ± 2.44	69.35 ± 3.41
	SVM	75.25 ± 5.45	74.45 ± 3.68	72.1 ± 2.37	69.45 ± 3.06	64.55 ± 7.18
Splice	Gini	91.26 ± 1.65	91.23 ± 1.61	90.22 ± 1.53	86.22 ± 4.11	74.38 ± 5.54
$(60, 1648, 1527)$	RF	94.76 ± 0.68	93.94 ± 0.76	93.87 ± 1.39	91.97 ± 1.82	82.69 ± 3.05
	SVM	91.1 ± 0.77	88.83 ± 1.08	87.67 ± 1.09	83.04 ± 1.36	70.47 ± 6.58
Spam	Gini	89.74 ± 1.15	89.01 ± 1.86	87.61 ± 2.05	84.57 ± 1.83	80.8 ± 3.0
$(57, 1813, 2788)$	RF	92.07 ± 1.1	92.2 ± 0.91	92.06 ± 1.15	91.04 ± 1.95	88.81 ± 1.5
	SVM	89.2 ± 1.02	86.41 ± 0.88	82.55 ± 1.72	76.64 ± 2.28	68.02 ± 3.95
Wine (white)	Gini	75.36 ± 0.76	74.72 ± 1.69	73.56 ± 1.34	73.08 ± 1.94	69.4 ± 5.72
$(11, 3258, 1640)$	RF	76.4 ± 1.38	76.74 ± 1.22	76.45 ± 1.18	74.74 ± 3.27	72.89 ± 1.89
	SVM	75.34 ± 0.76	72.43 ± 1.73	71.08 ± 2.0	68.07 ± 2.18	65.24 ± 2.71
Magic	Gini	83.75 ± 0.42	83.58 ± 0.49	82.33 ± 0.56	81.36 ± 1.08	78.0 ± 1.74
$(10, 12332, 6688)$	RF	85.24 ± 0.58	85.37 ± 0.61	85.3 ± 0.58	84.83 ± 0.71	82.37 ± 1.34
	SVM	82.7 ± 0.43	82.24 ± 0.45	81.0 ± 0.34	79.16 ± 0.43	69.5 ± 3.33

5   Conclusion

In this paper, we investigated the robustness of decision tree learning under label noise. We proved that decision tree algorithms based on Gini or misclassification impurity and the twoing rule algorithm are all robust to symmetric label noise. We also provided some sample complexity results for the robustness. Through empirical investigations we illustrated the robust learning of decision tree and random forest. Decision tree approach is very popular in many practical applications. Hence, the robustness results presented in this paper are interesting. Though we considered only impurity based methods, there are other algorithms for learning decision trees (e.g., [10]). Extending such robustness results to other decision tree learning algorithms is an interesting problem. All the results we proved are for symmetric noise. Extending these results to class conditional and non-uniform noise is another important direction for future research.

A   Proof Sketch of Lemmas 1, 2

Let $n^+(\tilde{n}^+)$ and $n^-(\tilde{n}^-)$ denote the positive and negative samples at the node under noise-free case (noisy case). Taking positive class as majority, we note

$\rho = (n^+ - n^-)/n$. Using Hoeffding bound it is easy to show $\Pr[\tilde{n}^+ - \tilde{n}^- < 0] \leq \exp\left(-\frac{\rho^2 n(1-2\eta)^2}{2}\right)$. This gives bound for samples needed as $n > \frac{2}{\rho^2(1-2\eta)^2}\ln(\frac{1}{\delta})$, completing proof of Lemma 1.

Let n, n_l, n_r be the number of samples at v, v_l, v_r and recall $n_l = an$ and $n_r = (1-a)n$. Recall that $\tilde{p}, \tilde{p}_l, \tilde{p}_r$ are fraction of positive samples at v, v_l, v_r and $p^\eta, p_l^\eta, p_r^\eta$ are their large sample values. Then, using Hoeffding bounds we get (with $\epsilon_1 = \epsilon$, $\epsilon_2 = \epsilon/\sqrt{a}$ and $\epsilon_3 = \epsilon/\sqrt{1-a}$),

$$\Pr\left[\left(|\tilde{p} - p^\eta| \geq \epsilon_1\right) \cup \left(|\tilde{p}_l - p_l^\eta| \geq \epsilon_2\right) \cup \left(|\tilde{p}_r - p_r^\eta| \geq \epsilon_3\right)\right] \leq 6e^{-2n\epsilon^2} \quad (5)$$

When this event happens, with some algebraic manipulation, one can show for Gini impurity, $|\hat{\text{gain}}^\eta_{\text{Gini}}(f) - \text{gain}^\eta_{\text{Gini}}(f)| \leq 6(1-2\eta)\epsilon$ where $\hat{\text{gain}}^\eta_{\text{Gini}}$ is the random Gini-gain under noise with sample size n and $\text{gain}^\eta_{\text{Gini}}$ is its large sample limit. This gives us the bound as needed in Lemma 2. We can prove the lemma for other criteria also similarly.

References

1. Breiman, L., Friedman, J., Olshen, R., Stone, C.: Classification and Regression Trees. Wadsworth and Brooks, Monterey (1984)
2. Breiman, L.: Random forests. Mach. Learn. **45**(1), 5–32 (2001)
3. Brodley, C.E., Friedl, M.A.: Identifying mislabeled training data. J. Artif. Intell. Res. **11**, 131–167 (1999)
4. Chang, C.-C., Lin, C.-J.: LIBSVM: a library for support vector machines. ACM Trans. Intell. Syst. Technol. **2**, 27 (2011)
5. du Plessis, M.C., Niu, G, Sugiyama, M.: Analysis of learning from positive and unlabeled data. In: Advances in Neural Information Processing Systems (2014)
6. Frénay, B., Verleysen, M.: Classification in the presence of label noise: a survey. IEEE Trans. Neural Netw. Learn. Syst. **25**, 845–869 (2014)
7. Ghosh, A., Manwani, N., Sastry, P.S.: Making risk minimization tolerant to label noise. Neurocomputing **160**, 93–107 (2015)
8. Lichman, M.: UCI machine learning repository (2013)
9. Long, P.M., Servedio, R.A.: Random classification noise defeats all convex potential boosters. Mach. Learn. **78**(3), 287–304 (2010)
10. Manwani, N., Sastry, P.S.: Geometric decision tree. IEEE Trans. Syst. Man Cybern. **42**(1), 181–192 (2012)
11. Manwani, N., Sastry, P.S.: Noise tolerance under risk minimization. IEEE Trans. Cybern. **43**(3), 1146–1151 (2013)
12. Natarajan, N., Dhillon, I.S., Ravikumar, P.K., Tewari, A.: Learning with noisy labels. In: Advances in Neural Information Processing Systems (2013)
13. Nettleton, D.F., Orriols-Puig, A., Fornells, A.: A study of the effect of different types of noise on the precision of supervised learning techniques. Artif. Intell. Rev. **33**(4), 275–306 (2010)
14. Patrini, G., Nielsen, F., Nock, R., Carioni, M.: Loss factorization, weakly supervised learning and label noise robustness. In: Proceedings of The 33rd International Conference on Machine Learning, pp. 708–717 (2016)
15. Pedregosa, F., et al.: Scikit-learn: machine learning in Python. J. Mach. Learn. Res. **12**, 2825–2830 (2011)

16. Quinlan, J.R.: Induction of decision trees. Mach. Learn. **1**(1), 81–106 (1986)
17. Scott, C., Blanchard, G., Handy, G.: Classification with asymmetric label noise: consistency and maximal denoising. In: The 26th Annual Conference on Learning Theory, 12–14 June 2013, pp. 489–511 (2013)
18. van Rooyen, B., Menon, A., Williamson, R.C.: Learning with symmetric label noise: the importance of being unhinged. In: Advances in Neural Information Processing Systems, pp. 10–18 (2015)
19. Wu, X., et al.: Top 10 algorithms in data mining. Knowl. Inf. Syst. **14**(1), 1–37 (2007)

Relevance-Based Evaluation Metrics for Multi-class Imbalanced Domains

Paula Branco[1,2(✉)], Luís Torgo[1,2], and Rita P. Ribeiro[1,2]

[1] LIAAD - INESC TEC, Porto, Portugal
{paula.branco,ltorgo,rpribeiro}@dcc.fc.up.pt
[2] DCC - Faculdade de Ciências - Universidade do Porto, Porto, Portugal

Abstract. The class imbalance problem is a key issue that has received much attention. This attention has been mostly focused on two-classes problems. Fewer solutions exist for the multi-classes imbalance problem. From an evaluation point of view, the class imbalance problem is challenging because a non-uniform importance is assigned to the classes. In this paper, we propose a relevance-based evaluation framework that incorporates user preferences by allowing the assignment of differentiated importance values to each class. The presented solution is able to overcome difficulties detected in existing measures and increases discrimination capability. The proposed framework requires the assignment of a relevance score to the problem classes. To deal with cases where the user is not able to specify each class relevance, we describe three mechanisms to incorporate the existing domain knowledge into the relevance framework. These mechanisms differ in the amount of information available and assumptions made regarding the domain. They also allow the use of our framework in common settings of multi-class imbalanced problems with different levels of information available.

1 Introduction

The class imbalance problem is a relevant problem with extensive research literature. It occurs in many application domains like medical, financial, meteorological, and others. Assessing performance in these contexts has been studied and several metrics were proposed. However, most proposals for this type of problems are only applicable to binary classification problems [7]. Recently, the multi-class imbalance problem has received increased attention.

In this paper, we address the key issue of performance assessment for multi-class imbalanced domains. These domains require special purpose evaluation metrics that are able to adequately reflect the preference biases of the users concerning prediction errors. In imbalanced domains, the user is typically more interested in the minority class(es) while the majority class(es) are usually less relevant. Therefore, traditionally used measures, such as Accuracy, are not suitable for this type of problems due to their inability of taking into account the user preferences. For multi-class imbalanced domains the few solutions that exist are essentially extensions of metrics used for the binary case.

© Springer International Publishing AG 2017
J. Kim et al. (Eds.): PAKDD 2017, Part I, LNAI 10234, pp. 698–710, 2017.
DOI: 10.1007/978-3-319-57454-7_54

There is a direct connection between imbalanced domains and cost-sensitive learning. However, when we face a cost-sensitive problem we have a cost matrix defined for the task at hand that is used to assess the models performance. The model with minimum cost (or maximum benefit) is the best. The tasks we are addressing in this paper are different because there is a class imbalance but no cost matrix is available. This is the usual setting when dealing with imbalanced classes. Typically, the only information available regarding the user preferences is informal and can be expressed as: *"the minority class(es) is(are) the most important one(s)"*. This is an important class of applications as it is well known that cost/benefit information is frequently hard to obtain or simply not available.

When the user preference bias is not uniform across the domain of the target variable it is important to transfer this information to the evaluation metrics. We propose a new evaluation framework that incorporates this information. The proposed measures are based on the existence of different relevance/importance scores for the problem classes and try to mirror the user preference bias in the evaluation of the predictions of a model. This means that the same errors made in two different classes with different importance scores can have different weights in the final evaluation score. We also propose three mechanisms for estimating the expected domain preferences in a typical imbalanced multi-class setting. These mechanisms can be used when the user is not able to precisely specify each class relevance. The proposed mechanisms differ in the assumptions regarding the domain and amount of information that the user is able to provide.

The main contributions of this work are: (i) highlight that existing metrics for handling multi-class imbalanced domains are not always adequate; (ii) propose a new evaluation framework that accounts for user preferences in multi-class imbalanced domains; (iii) propose three mechanisms for estimating the preference bias in typical multi-class imbalance settings; and (iv) compare the discrimination capability of existing and new proposed metrics for this problem. This paper is organized as follows. Section 2 describes existing metrics for handling multi-class imbalance domains. Section 3 explains why these metrics are unsuitable for this problem providing three examples where those metrics show unreliable results. Section 4 presents our framework for performance assessment on multi-class imbalance problems, and mechanisms to deal with different information levels. Section 5 evaluates our framework regarding performance and discrimination capability under different scenarios. Section 6 concludes the paper.

2   Evaluation Metrics for Multi-class Imbalanced Learning

Several metrics have been proposed to evaluate the performance within the problem of class imbalance for two classes. However, only a few have been successfully adapted to address the more difficult problem of multi-class imbalanced domains.

Let C represent the total number of classes of a problem. Consider a $C \times C$ confusion matrix, mat, for which $mat_{k,l}$ represents the examples of the true class k that were predicted as class l. For a class i, tp_i represents the true positives for class i; tn_i are the true negatives for class i, i.e., all the examples that were correctly predicted and are not from class i; fp_i is the number of false positive for class i, i.e. all the examples incorrectly predicted as class i, and fn_i are the false negatives for class i. We use t_i and p_i for the total number of true and predicted examples for class i respectively, i.e., $t_i = tp_i + fn_i$ and $p_i = tp_i + fp_i$. The indexes M and μ represent respectively a Macro and Micro averaging strategy for a metric, where the first strategy averages the metric results over all classes while the second uses the pooled results. With this notation, we define the following metrics for a class i:

$$recall_i = \frac{tp_i}{t_i} \tag{1}$$

$$precision_i = \frac{tp_i}{p_i} \tag{2}$$

$$F_{\beta i} = \frac{(1+\beta^2)precision_i \cdot recall_i}{\beta^2 \cdot precision_i + recall_i} \tag{3}$$

where β sets the relative importance of $recall_i$ in comparison with $precision_i$.

Table 1 presents a description of the existing metrics for multi-class imbalance tasks. For a more comprehensive overview, we also include some multi-class measures which were not specifically developed for imbalanced domains. The Area Under the ROC Curve (AUC) is not considered in this paper. Although some attempts have been made to also adapt AUC to a multi-class context [9] we opted not to include it here for two reasons. The first reason is related to the demonstrated incoherence of AUC metric [8]. The second reason concerns the nonexistence of a well-developed ROC analysis for multi-class problems [14].

The metrics described in Table 1 can be clustered into recall-based ($MAvG$, Rec_M, Rec_μ), precision-based ($Prec_M$, $Prec_\mu$) or general metrics ($AvAcc$, $F_{\beta M}$ $F_{\beta\mu}$, AvF_β, CBA, MCC, RCI and CEN) depending on the information used. Thus, each type of metric presents a different evaluation perspective. While recall-based metrics are focused on the true class labels, precision-based metrics consider the predicted class labels and the general metrics aggregate both perspectives into a single value providing a global performance overview. An alternative solution to Table 1 metrics consist of not aggregating the $precision_i$, $recall_i$ and $F_{\beta i}$ measures. However, this has the disadvantage of generating a large number of results increasing the complexity of the analysis of the results.

The metrics in Table 1 present differences in both the range of values they may take and the representation of the best performing classifier. For a straightforward comparison we present the metric **value** and a **normalized value**. This **normalized value** corresponds to the metric value in a percentage, where 0% matches the worst possible performance and 100% the best.

3   Unsuitability of the Existing Evaluation Metrics

The so-called "imbalanced problems" are based on the assumption that the user has a differentiated interest in the problem classes. In two-class problems the user preference bias is, usually, towards the minority class. This also happens in the multi-class context.

Table 1. Performance assessment metrics for imbalanced domains with C classes.

Metric	Description	Definition
$AvAcc$	Classes average accuracy.	$\frac{1}{C} \sum_{i=1}^{C} \frac{tp_i + tn_i}{tp_i + tn_i + fp_i + fn_i}$
$MAvG$	Geometric average of recall in each class [15].	$\sqrt[C]{\prod_{i=1}^{C} recall_i}$
Rec_M	Arithmetic macro-average of recall in each class.	$\frac{1}{C} \sum_{i=1}^{C} recall_i$
$Prec_M$	Arithmetic macro-average of precision in each class.	$\frac{1}{C} \sum_{i=1}^{C} precision_i$
Rec_μ	Arithmetic micro-average of recall in each class.	$\sum_{i=1}^{C} tp_i / \sum_{i=1}^{C} t_i$
$Prec_\mu$	Arithmetic micro-average of precision in each class.	$\sum_{i=1}^{C} tp_i / \sum_{i=1}^{C} p_i$
$F_{\beta M}$	Mean F_β measure evaluated with macro-averaged precision and recall [14].	$\frac{(1 + \beta^2) \cdot Prec_M \cdot Rec_M}{\beta^2 \cdot Prec_M + Rec_M}$
$F_{\beta \mu}$	Mean F_β measure evaluated with micro-averaged precision and recall [14].	$\frac{(1 + \beta^2) \cdot Prec_\mu \cdot Rec_\mu}{\beta^2 \cdot Prec_\mu + Rec_\mu}$
AvF_β	Extension for any value of β of the definition for F_1 measure to multi-class [4].	$\frac{1}{C} \sum_{i=1}^{C} \frac{(1 + \beta^2) \cdot precision_i \cdot recall_i}{\beta^2 \cdot precision_i + recall_i}$
CBA	Class balance accuracy [12].	$\frac{\sum_{i=1}^{C} \frac{mat_{i,i}}{max\left(\sum_{j=1}^{C} mat_{i,j}, \sum_{j=1}^{C} mat_{j,i}\right)}}{C}$
MCC	Matthews correlation coefficient introduced for two-class problems and extended to multi-class [6,11].	$\frac{X}{YZ}$, where $X = \sum_{k,l,m=1}^{C} (mat_{k,k} mat_{m,l} - mat_{l,k} mat_{k,m})$ $Y = \sqrt{\sum_{k=1}^{C} \left(\sum_{l=1}^{C} mat_{l,k} \right) \left(\sum_{\substack{f,g=1 \\ f \neq k}}^{C} mat_{g,f} \right)}$ $Z = \sqrt{\sum_{k=1}^{C} \left(\sum_{l=1}^{C} mat_{k,l} \right) \left(\sum_{\substack{f,g=1 \\ f \neq k}}^{C} mat_{f,g} \right)}$
RCI	Relative classifier information [13]	$\frac{H_d - H_o}{H_d}$, where $H_d = - \sum_{i=1}^{C} \left(\frac{\sum_{l=1}^{C} mat_{i,l}}{C} log \frac{\sum_{l=1}^{C} mat_{i,l}}{C} \right)$ $H_o = \sum_{j=1}^{C} \left(\frac{\sum_{k=1}^{C} mat_{k,j}}{C} H_{oj} \right)$ and $H_{oj} = - \sum_{i=1}^{C} \left(\frac{mat_{i,j}}{\sum_{k=1}^{C} mat_{k,j}} log \frac{mat_{i,j}}{\sum_{k=1}^{C} mat_{k,j}} \right)$
CEN	Confusion entropy [16].	$\sum_{j=1}^{C} (P_j CEN_j)$, where $P_j = \frac{\sum_{k=1}^{C} mat_{j,k} + mat_{k,j}}{2 * \sum_{k,l=1}^{C} mat_{k,l}}$, $CEN_j = - \sum_{\substack{k=1 \\ k \neq j}}^{C} (P_{j,k}^j \log_{2(C-1)}(P_{j,k}^j) + P_{k,j}^j \log_{2(C-1)}(P_{k,j}^j))$ $P_{i,i}^i = 0, \quad P_{i,j}^i = mat_{i,j} / \left(\sum_{k=1}^{C} (mat_{i,k} + mat_{k,i}) \right), i \neq j$

Several metrics have been proposed (cf. Table 1) to assess the performance in multi-class imbalanced domains. We claim that these solutions are not adequate for these domains because they fail to reflect the user preferences in several situations and therefore can be misleading. To demonstrate this, we use the three cases described below. The user can also follow the strategy of observing each class precision, recall and F_β. To show this perspective, we also include the evaluation provided by these measures for each class in the next examples.

Multi-class imbalance problems can be grouped into: multi-minority, multi-majority and complete. In a multi-minority scenario one class has significantly more examples than the mean number of examples of all classes, i.e., $t_{maj} >> \bar{t}$, where $\bar{t} = \sum_{i=1}^{C} t_i/C$ is the mean number of examples of all classes. On a multi-majority case a single class is significantly less frequent than the others, i.e., $t_{min} << \bar{t}$. In the complete case, several classes can have a significantly larger size than other classes which have a significantly smaller size relatively to $\bar{t}$.

The cases described below exemplify the depicted scenarios. They illustrate the unsuitability of the existing metrics and show the need of a more adequate framework for this context. We assume that the most relevant classes are the less populated. Tables 2 and 3 describe these cases.

Table 2. Cases 1 to 3 confusion matrix (top) and $prec_i$, rec_i and F_{1i} (bottom).

		Case 1 preds				Case 2 preds				Case 3 preds			
		c_1	c_2	c_3		c_1	c_2	c_3		c_1	c_2	c_3	c_4
trues	c_1	5	0	0	c_1	1	0	3	c_1	1	3	0	0
	c_2	0	10	0	c_2	0	100	0	c_2	9	1	0	0
	c_3	0	300	0	c_3	0	0	200	c_3	0	0	100	0
									c_4	0	0	0	200

Class rec_i $prec_i$ F_{1i}				Class rec_i $prec_i$ F_{1i}				Class rec_i $prec_i$ F_{1i}			
c_1	1	1	1	c_1	0.25	1	0.4	c_1	0.25	0.1	0.14
c_2	1	0.032	0.063	c_2	1	1	1	c_2	0.1	0.25	0.14
c_3	0	n. def.	n. def.	c_3	1	0.985	0.993	c_3	1	1	1
								c_4	1	1	1

Case 1: Multi-minority Example - In this case, the two minority classes are correctly predicted and the majority class is completely mispredicted.

Case 2: Multi-majority Example - In this case both majority classes are correctly predicted and the minority class is nearly always mispredicted.

Case 3: Complete Example - In this case two majority classes are correctly predicted while the two minority classes are almost always mispredicted.

Table 3 includes a summary of the misleading metrics for the cases presented. Generally, we observe that the metrics fail to correctly represent the user preferences. Either by providing an over- or under-estimated value, the metrics are not able to correctly incorporate the domain knowledge, and therefore, the results obtained are not reliable. In more detail, for case 1, $MAvG$, provides a result of zero which is clearly not adequate given that both minority classes have a

Table 3. Performance assessment metrics in Case 1, 2 and 3. N.Val: normalized value; Ac.: Accordance with user preferences (misleading: ×, suitable: ✓).

Metric	Case 1			Case 2			Case 3		
	N.Val.(%)	Value	Ac.	N.Val.(%)	Value	Ac.	N.Val.(%)	Value	Ac.
$AvAcc$	36.5	0.365	×	99.3	0.993	×	98.1	0.981	×
$MAvG$	0.0	0.000	×	63.0	0.630	✓	39.8	0.398	✓
Rec_M	66.7	0.667	✓	75.0	0.750	×	58.8	0.588	×
$Prec_M$	Not defined		×	99.5	0.995	✓	58.8	0.588	×
Rec_μ	4.8	0.048	×	99.0	0.990	×	96.2	0.962	×
$Prec_\mu$	4.8	0.048	×	99.0	0.990	✓	96.2	0.962	×
F_{1M}	Not defined		×	85.5	0.855	×	58.8	0.588	×
$F_{1\mu}$	4.8	0.048	×	99.0	0.990	×	96.2	0.962	×
AvF_1	Not defined		×	79.8	0.798	×	57.1	0.571	×
CBA	34.4	0.344	×	74.5	0.745	✓	55.0	0.550	×
MCC	65.1	0.301	✓	98.9	0.978	×	96.2	0.923	×
RCI	36.8	0.368	×	92.6	0.926	×	97.9	0.979	×
CEN	97.8	0.022	✓	98.1	0.019	×	98.5	0.015	×

perfect score regarding the recall metric, a problem also observed by [12]. The remaining metrics marked in Table 3 for case 1 are misleading because they present a normalized value approximately below 45%. In case 2, the minority and most important class was almost always incorrectly predicted. However, all metrics, with exception of $MAvG$, CBA, $Prec_M$ and $Prec_\mu$, over-estimate the value of the confusion matrix which can be misleading. In case 3 the metrics are unable to show that both minority and important classes were almost always incorrectly predicted. Although big mistakes occur on all minority classes, most metrics normalized value is high or moderate which is misleading.

The cases described show that no metric provides reliable results in all situations. When the classes have a distinct relevance to the user it is unavoidable to consider this relevance in the evaluation. Thus, a new framework is required for embedding the relevance into the existing metrics. This framework should also be usable when the user has a more informal information. So, mechanisms for embedding different levels of information provided by the user are necessary.

4   A Framework for Relevance-Based Evaluation

4.1   Relevance-Based Metrics for Multi-class Imbalance Learning

Our proposal is based on the assumption that classes have different relevance for the user. A certain number of classes may be extremely important while the performance on other classes may be negligible. The key idea is to use the relevance values as weights for the classes when evaluating the models performance.

The use of weights is a well-known strategy. However, only two metrics were proposed using this notion. A weighted macro-averaging recall [2] was proposed for multi-class although it was only used in binary classification. Moreover, no guidelines for defining/choosing the weights were provided. A weighted AUC for multi-class was presented [10], with weights determined by the classes prevalence.

Our relevance-based metrics proposal assumes that the user assigns an importance score to each problem class. Let us suppose that this domain information is converted into a function $\phi()$ that maps each class into a relevance score in the interval $[0, 1]$. The value 0 is assigned to a class with zero relevance, and the value 1 is assigned to a class with maximum relevance to the user. For instance, a relevance function for a four-class problem can be define as: $\phi(c_1) = 0.2$, $\phi(c_2) = 0$, $\phi(c_3) = 0.9$ and $\phi(c_4) = 1$. From this illustrative $\phi()$ function, class c_1 has a very low relevance, class c_2 is irrelevant, and classes c_3 and c_4 are very relevant.

Our proposal incorporates the user preference bias, expressed by the definition of a relevance function, in the metrics definition in the form of weights. This means that, if a class is very important to the user, then the performance on that class will also have a large weight in the evaluation. On the other hand, misclassification errors of less relevant classes have a reduced impact on the final evaluation. Eqs. 4 to 8 present an adaptation of recall, precision, $F_\beta - measure$ and CBA to incorporate relevance.

$$Rec^\phi = \frac{1}{\sum\limits_{i=1}^{C} \phi(i)} \sum_{i=1}^{C} \phi(i) \cdot recall_i \tag{4}$$

$$Prec^\phi = \frac{1}{\sum\limits_{i=1}^{C} \phi(i)} \sum_{i=1}^{C} \phi(i) \cdot precision_i \tag{5}$$

$$F_\beta^\phi = \frac{(1+\beta^2) \cdot Prec^\phi \cdot Rec^\phi}{(\beta^2 \cdot Prec^\phi) + Rec^\phi} \tag{6}$$

$$AvF_\beta^\phi = \frac{1}{\sum\limits_{i=1}^{C} \phi(i)} \sum_{i=1}^{C} \frac{\phi(i) \cdot (1+\beta^2) \cdot precision_i \cdot recall_i}{(\beta^2 \cdot precision_i) + recall_i} = \frac{1}{\sum\limits_{i=1}^{C} \phi(i)} \sum_{i=1}^{C} \frac{\phi(i) \cdot (1+\beta^2) \cdot tp_i}{\beta^2 \cdot t_i + p_i} \tag{7}$$

$$CBA^\phi = \sum_{i=1}^{C} \phi(i) \cdot \frac{mat_{i,i}}{max\left(\sum\limits_{j=1}^{C} mat_{i,j}, \sum\limits_{j=1}^{C} mat_{j,i}\right)} \tag{8}$$

where $\phi(i)$ is the relevance of class i; t_i and p_i are the total number of true and predicted examples for class i; and tp_i is the number of true positives for class i.

With this framework we obtain the three evaluation perspectives: recall-based, precision-based and general measures. These metrics were selected because they cover all perspectives under a simple formulation.

4.2 Mechanisms for Relevance Estimation

The above evaluation framework depends on the availability of domain information regarding the classes relevance. However, this information may exist with different levels of detail. We will consider 4 types of information:

- **Informal**: characterized by completely informal domain knowledge. This is typical in imbalanced domains where no quantification regarding the importance of each class exists. Frequently, it is only stated that "the minority classes are the most important". This creates serious problems to the performance evaluation because the user does not specify the classes non-uniform importance.
- **Intermediate informal**: more information available although very limited. We assume the user provides a partial order of the classes by their importance.
- **Intermediate formal**: more complete information available. We consider that the user is able to provide a total order of the classes.
- **Formal**: the user provides a full specification of the relevance function. Although being the ideal setting, this is not so common in real world domains.

We will present mechanisms to estimate the relevance function from these different levels of available information. If the user fully specifies the relevance function (formal level) no mechanism is needed. To denote this situation we will add ϕ to the metrics name. The proposed mechanisms are pertinent because for most imbalance domains the full relevance function is unknown. Our goal is to incorporate the available domain knowledge in the evaluation framework.

Informal Level - Using Classes Prevalence (PREV)

When no preferences regarding the domain are provided, it is possible to use the observed frequency of the classes to obtain valid relevance scores. Our proposal sets the relevance of a class to be inversely proportional to its observed frequency in the available data:

$$\hat{\phi}(i) = \frac{1/t_i}{\sum_{i=1}^{C} 1/t_i} \tag{9}$$

where t_i is class i total number of examples. Using the estimated relevance we may obtain any of the proposed relevance-based metrics. We stress that the use of this method is not mandatory for applying our framework, provided that the user gives more domain information or specifies a relevance function.

Intermediate Informal Level - Using Classes Partial Order (PO)

A partial order specifies a binary relation which may hold between some pairs of classes. This relation is denoted as $c_1 < c_2$ and is read as "c_1 precedes c_2". In the context of relevance-based metrics, the relation $c_1 < c_2$ represents that c_1 has a lower relevance value than c_2, i.e. c_1 is less important than c_2. The relation is named partial because it does not provide a full relation between all the classes, i.e., there are pairs of classes named incomparable because the relation between both was not specified. More details regarding partially ordered sets can be obtained in [3]. Figure 1 shows on the left side an example of a partial order on a problem with 7 classes. Several studies have been conducted to estimate rankings from a partial order (e.g. [1]). However, as far as we known, no attempt has been made to use the partial order of classes to estimate their relevance. The main advantage of this method is that it is less demanding for the user when compared to a full specification of the relevance function. Moreover, to use a partial order of classes is preferable to not having any information at all.

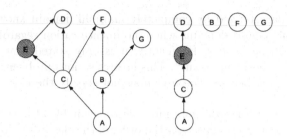

Fig. 1. An example of a partially ordered set (left hand side) and the construction of a LPOM for class E (right hand side).

To estimate the classes relevance using a partial order we will apply the US-model [1]. This method builds a Local Partial Order Model (LPOM) for each class. A LPOM for a class node X represents all the successor (S), predecessor (P) and incomparable (U) nodes in relation to X. Then, the estimated average rank of node X is defined as $Rank(X) = \frac{(|S|+1)+(|S|+1+|U|)}{2} = |S| + 1 + \frac{|U|}{2}$. Figure 1 (on the right) shows the LPOM for node E. In this example node E has 2 successors (nodes A and C), 1 predecessor (node D) and 3 incomparable nodes (B, F and G). Node E ranking, according to the proposed US-model, is $Rank(E) = 4.5$. Our proposal, uses the classes ranks derived from the partial order provided by the user and estimates the relevance of each class i as follows:

$$\hat{\phi}(i) = \frac{Rank(i)}{\max_{\forall i \in C} Rank(i)} \tag{10}$$

Intermediate Formal Level - Using Classes Total Order (TO)
This mechanism is similar to the previous one, but now the user is required to provide a total order of the problem classes. This is a more demanding task for the user because no pair of classes can remain incomparable. Still, it is less demanding than fully specifying the relevance function. Given a total order, only the magnitude of the classes relevance remains unspecified. We use the US-model [1] previously used in PO mechanism. For a node X, $Rank(X) = |S| + 1$ because X has no incomparable nodes. The relevance is estimated with Eq. 10. The $\phi()$ function values are equidistant and range from $\frac{1}{C}$ to 1, where C is the number of classes. The metrics obtained by each described mechanism, have respectively $PREV$, PO and TO appended to their name.

4.3   Implementation Issues

To maximize the number of valid results supplied by the metrics, we exclude from the calculations of precision and recall-based metrics, all classes i for which $recall_i$ or $precision_i$ are not defined and use the AvF_1^ϕ extension presented in Eq. 7. This way, we can always obtain Rec^ϕ, $Prec^\phi$ and F_1^ϕ and maximize the number of obtained results for AvF_1^ϕ. With the extension proposed in [5] AvF_1^ϕ is only undefined when class i has neither true values nor predictions. To allow a fairer comparison we also applied these strategies to existing metrics.

5   Experimental Evaluation

5.1   Agreement with User Preferences

We will now present the performance of the proposed metrics in the cases described in Sect. 3. Table 4 provides the user-defined relevance and also the relevance inferred from a simulation of incomplete user information using the mechanisms defined in Sect. 4.2. The performance results are shown in Table 5. Generally, we observe that the metrics considering the proposed evaluation framework are able to overcome the difficulties detected on the other existing metrics. The new metrics are capable of reflecting the user preferences independently of the level of information considered. It is noteworthy that for the most informal levels of information (PREV and PO) the results obtained for all the cases are preferable to those of the other existing metrics. Moreover, the results become more adjusted to the user preferences with the increase of the information level. In summary, all the proposed mechanisms show results that are more in accordance with the user preferences than the previous existing metrics.

Table 4. Case 1, 2 and 3 information for each mechanism.

	Case 1				Case 2				Case 3				
	$\phi(c_1)$	$\phi(c_2)$	$\phi(c_3)$	Order	$\phi(c_1)$	$\phi(c_2)$	$\phi(c_3)$	Order	$\phi(c_1)$	$\phi(c_2)$	$\phi(c_3)$	$\phi(c_4)$	Order
PREV	0.66	0.33	0.01		0.94	0.04	0.02		0.64	0.32	0.03	0.02	
PO	1	1	0.4	$c_3 < c_1$	1	0.5	0.5	$c_3 < c_1$	1	0.86	0.57	0.42	$c_3 < c_1$
													$c_4 < c_1$
				$c_3 < c_2$				$c_2 < c_1$					$c_4 < c_2$
TO	1	0.67	0.33	$c_3 < c_2 < c_1$	1	0.67	0.33	$c_3 < c_2 < c_1$	1	0.75	0.5	0.25	$c_4 < c_3 <$
													$c_2 < c_1$
ϕ	1	0.9	0.1		1	0.2	0.1		1	0.9	0.2	0.1	

We also tested the proposed metrics on 16 real world data sets[1]. Although we observe differences in the metrics results, it is not possible to assess the agreement with the user preferences because, in this case, we lack a ground truth.

5.2   Discrimination Capability

In this section we assess how well the metrics recognize different situations expressed in the confusion matrix. We consider problems with 3 or 4 classes and determine the percentage of different scores obtained by each metric in all possible confusion matrices for a problem.

We tested the multi-minority, multi-majority and complete scenarios, with problems with 3, 3 and 4 classes respectively. A problem with 3 classes with i, j, and k examples is denoted by $i - j - k$. For instance, problem 2-4-15 has 2, 4

---

[1] The experimental framework, code and results of this evaluation is available in https://github.com/paobranco/Relevance-basedMulticlassImbalanceMetrics.

Table 5. Performance assessment metrics normalized value for Cases 1, 2 and 3 (in bold: values in accordance with user preferences).

Metric	Case			Metric	Case			Metric	Case		
	1	2	3		1	2	3		1	2	3
$AvAcc$	36.5	99.3	98.1	Rec^{PREV}	98.9	29.2	24	Rec^{TO}	**83.3**	62.5	43
$MAvG$	0	**63**	**39.8**	$Prec^{PREV}$	67.7	**100**	17.8	$Prec^{TO}$	61.3	**99.8**	41.5
Rec_M	**66.7**	75	58.8	F_1^{PREV}	80.4	45.3	20.4	F_1^{TO}	**70.6**	**76.9**	42.2
$Prec_M$	34.4 [a]	**99.5**	58.8	AvF_1^{PREV}	68	43.4	17.8	AvF_1^{TO}	52.1	69.9	40
Rec_μ	4.8	99	96.2	CBA^{PREV}	67	29.2	13.7	CBA^{TO}	51.1	62.3	37
$Prec_\mu$	4.8	99	96.2	Rec^{PO}	83.3	**62.5**	**46.7**	Rec^{ϕ}	95	42.3	29.1
F_{1M}	45.4 [a]	85.5	58.8	$Prec^{PO}$	51.6	**99.6**	46	$Prec^{\phi}$	54.2	**99.9**	28.4
$F_{1\mu}$	4.8	99	96.2	F_1^{PO}	63.7	**76.8**	**46.4**	F_1^{ϕ}	69	59.4	28.7
AvF_1	35.4 [a]	79.8	57.1	AvF_1^{PO}	44.3	69.8	44.3	AvF_1^{ϕ}	52.8	53.8	26
CBA	34.4	**74.5**	55	CBA^{PO}	43	62.1	41.5	CBA^{ϕ}	51.5	42.2	22.3
MCC	**65.1**	98.9	96.2								
RCI	36.8	92.6	97.9								
CEN	**97.8**	98.1	98.5								

[a] Evaluated using the strategies described in Section 4.3

and 15 examples of classes c_1, c_2 and c_3. We tested multi-minority $(i - j - k)$ and multi-majority $(i - k - l)$ problems with $i \in \{2, 3\}$, $j \in \{4, 5\}$, $k \in \{15, 16\}$ and $l \in \{17, 18\}$. We only analysed problems *2-3-9-10* and *2-3-9-11* on the complete scenario due to the exponential number of confusion matrices generated.

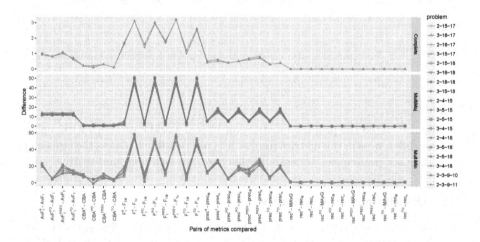

Fig. 2. Differences in the percentage of discrimination achieved between existing and corresponding proposed metrics in each scenario.

Figure 2 shows the difference between the discrimination percentage of pairs of metrics (a relevance-based metric and an existing metric). Relevance-based metrics achieve a higher discrimination capability when compared to their corresponding initial proposals. Only metrics based on recall and CBA present some difficulty in improving the discrimination capability where we obtain differences of zero or negative in 5% and 2% of the results respectively. The results show that the proposed evaluation framework is able to better discriminate different setups in multi-class imbalanced problems. In summary, our experiments show that our proposal provides an enhanced discrimination capability and results more in accordance with the user preferences.

6 Conclusions

Class imbalance is a problem appearing in many relevant application domains. Performance assessment under this situation is a key issue that has been addressed, mainly, for the two-classes case. For the multi-class imbalance problem, only a few solutions exist. We have shown that existing metrics for multi-class imbalance domains are not adequate in certain cases. We propose a new relevance-based evaluation framework that integrates the notion of a non-uniform importance across the target variable domain through a relevance function.

The evaluation of imbalanced domains is still an open issue in two-classes and multi-class problems. Relevance-based metrics are suitable for evaluating predictive tasks on imbalanced domains because they are able to reflect the user preferences. Such metrics easily adapt to different types of domain knowledge. We provide three mechanisms to facilitate the users task of embedding domain knowledge into the proposed relevance framework for performance assessment. This integration boosts the capability of correctly reflecting the performance of cases that other measures are not able to capture. We also show that these metrics present an enhanced discrimination capability. For reproducibility purposes, all the code used in this paper is available in https://github.com/paobranco/Relevance-basedMulticlassImbalanceMetrics.

Acknowledgements. This work is financed by the ERDF – European Regional Development Fund through the Operational Programme for Competitiveness and Internationalisation - COMPETE 2020 Programme within project POCI-01-0145-FEDER-006961, and by National Funds through the FCT – Fundação para a Ciência e a Tecnologia (Portuguese Foundation for Science and Technology) as part of project UID/EEA/50014/2013. The work of P. Branco is supported by a Ph.D. scholarship of FCT (PD/BD/105788/2014).

References

1. Brüggemann, R., Sørensen, P.B., Lerche, D., Carlsen, L.: Estimation of averaged ranks by a local partial order model#. J. Chem. Inf. Comput. Sci. **44**(2), 618–625 (2004)
2. Cohen, G., Hilario, M., Sax, H., Hugonnet, S., Geissbuhler, A.: Learning from imbalanced data in surveillance of nosocomial infection. Artif. Intell. Med. **37**(1), 7–18 (2006)

3. Dushnik, B., Miller, E.W.: Partially ordered sets. Am. J. Math. **63**(3), 600–610 (1941)
4. Ferri, C., Hernández-Orallo, J., Modroiu, R.: An experimental comparison of performance measures for classification. Pattern Recognit. Lett. **30**(1), 27–38 (2009)
5. Forman, G., Scholz, M.: Apples-to-apples in cross-validation studies: pitfalls in classifier performance measurement. SIGKDD Explor. Newsl. **12**(1), 49–57 (2010)
6. Gorodkin, J.: Comparing two K-category assignments by a K-category correlation coefficient. Comput. Biol. Chem. **28**(5), 367–374 (2004)
7. Gu, Q., Zhu, L., Cai, Z.: Evaluation measures of the classification performance of imbalanced data sets. In: Cai, Z., Li, Z., Kang, Z., Liu, Y. (eds.) ISICA 2009. CCIS, vol. 51, pp. 461–471. Springer, Heidelberg (2009). doi:10.1007/978-3-642-04962-0_53
8. Hand, D.J.: Measuring classifier performance: a coherent alternative to the area under the ROC curve. Mach. Learn. **77**(1), 103–123 (2009)
9. Hand, D.J., Till, R.J.: A simple generalisation of the area under the ROC curve for multiple class classification problems. Mach **45**(2), 171–186 (2001)
10. Hempstalk, K., Frank, E.: Discriminating against new classes: one-class versus multi-class classification. In: Wobcke, W., Zhang, M. (eds.) AI 2008. LNCS (LNAI), vol. 5360, pp. 325–336. Springer, Heidelberg (2008). doi:10.1007/978-3-540-89378-3_32
11. Matthews, B.W.: Comparison of the predicted and observed secondary structure of T4 phage lysozyme. BBA-Protein Struct. **405**(2), 442–451 (1975)
12. Mosley, L.: A balanced approach to the multi-class imbalance problem. Graduate Theses and Dissertations, Paper 13537 (2013)
13. Sindhwani, V., Bhattacharya, P., Rakshit, S.: Information theoretic feature crediting in multiclass support vector machines. In: SDM, pp. 1–18. SIAM (2001)
14. Sokolova, M., Lapalme, G.: A systematic analysis of performance measures for classification tasks. Inf. Process. Manag. **45**(4), 427–437 (2009)
15. Sun, Y., Kamel, M.S., Wang, Y.: Boosting for learning multiple classes with imbalanced class distribution. In: ICDM, pp. 592–602. IEEE (2006)
16. Wei, J.M., Yuan, X.J., Hu, Q.H., Wang, S.Q.: A novel measure for evaluating classifiers. Expert Syst. Appl. **37**(5), 3799–3809 (2010)

Location Prediction Through Activity Purpose: Integrating Temporal and Sequential Models

Dongliang Liao[1(✉)], Yuan Zhong[2], and Jing Li[1]

[1] University of Science and Technology of China, Hefei, China
liaodl@mail.ustc.edu.cn, lj@ustc.edu.cn
[2] Northeastern University, Boston, MA, USA
yzhong@ccs.neu.edu

Abstract. Based on the growing popularity of smart mobile devices, location-aware services become indispensable in human daily life. Location prediction makes these services more intelligent and attractive. However, due to the limited energy of mobile devices and privacy issues, the captured mobility data is typically sparse. This inherent challenge deteriorates significant principles in mobility modeling, i.e. temporal regularity and sequential dependency. To tackle these challenges, by utilizing temporal regularity and sequential dependency, we present a location prediction model with a two-stage fashion. Firstly, it extracts predictive features to effectively target the better performer from sequential and temporal models. Secondly, according to the inferred activity, it adopts non-parametric Kernel Density Estimation for posterior location prediction. Extensive experiments on two public check-in datasets demonstrate that the proposed model outperforms state-of-the-art baselines by 10.1% for activity prediction and 12.9% for location prediction.

Keywords: Location prediction · Activity prediction · Mobility modeling · Context-Aware Hybrid approach · Kernel Density Estimation

1 Introduction

With the ubiquity of smart mobile devices and the development of positioning technology, an overwhelming number of location-aware services have gained increasing popularity in recent years. These services have offered an unprecedented opportunity for both academia and industry to study human mobility behavior with access to various kinds of data, such as GPS trajectories, WiFi records, cellular phone logs, smart card transactions and social network check-ins, etc. They also shed light on a myriad of potential applications like user profiling, location understanding, urban planning and mobility modeling [11,19].

Among them, location prediction plays a key role. Generally, scholars handle this task with two-broad-category approaches, sequential modeling and temporal regularity modeling. Viewing that user activity serves as mobility motivation, activity prediction [5,8,9,16,17] is introduced as auxiliary to reduce vast

© Springer International Publishing AG 2017
J. Kim et al. (Eds.): PAKDD 2017, Part I, LNAI 10234, pp. 711–723, 2017.
DOI: 10.1007/978-3-319-57454-7_55

location candidate space (million magnitude from population level and thousand magnitude from individual level). Unfortunately, many unresolved difficulties remain tough in location prediction: (1) Sensitive to sequential dependency, sequential models [1,3,7,16] deteriorate when the timespan of consecutive mobility records being far like days or even months [3,15]. This is often the case due to device energy limitation and user privacy concern. (2) Temporal model performs poorly at night or during weekends [13] owing to decayed regularity; (3) Even for the same activity purpose, people still conceive different preferences under different contexts. E.g., at midnight, Alice would buy snacks from 7-Eleven near home, instead of Stop&Shop where she usually visits in the daytime. One thing worth noting is that these three problems are non-trivial. Simply take the last example for illustration. Because of data sparsity, directly estimating the user location preference for the specific time is obsessed with under-fitting. In addition, "contexts" are highly diversified and even only for the time context, modeling "location open hour", "user rest period", etc. simultaneously can be overwhelming.

In this paper, we tackle above challenges by decomposing location prediction into two subtasks [5,8,16], user activity inference and location inference based on activity. For activity inference, sequential and temporal models can fit respectively. However, as previously indicated, both are ineffective in certain circumstances. Here we design a Context-Aware Hybrid (CAH) module to integrate temporal regularity and sequential dependency models dynamically. More specifically, a set of elaborate evaluation features (e.g. density of recent records, regularity strength of user historical activities) are extracted as context features and based on that, a supervised classifier is applied to select the better performer between sequential and temporal models. For location inference, we adopt a time-aware approach for posterior location distribution calculation. Technically, instead of employing parameterized models which usually fall into a training dilemma, Kernel Density Estimation is applied to capture the visit time distribution at specific locations. Last but not the least, we summarize these two phases to leverage final location prediction.

Our main contributions are summarized as follows:

1. With a set of features assessing the performance of sequential and temporal models, we develop a Context-Aware Hybrid approach to combine them for user activity prediction.
2. We introduce Kernel Density Estimation to model the time variation of location preference for a given user, and construct a two-stage model to predict future locations based on the inferred activity.
3. The experimental results on two public datasets validate that our model significantly outperforms state-of-the-art baselines in terms of both activity prediction accuracy and locations prediction accuracy.

The rest of paper is structured as follows: Sect. 2 reviews related mobility prediction works. Section 3 formulates the prediction problem and introduces the notations. Our proposed model is presented in Sect. 4. Experimental results based on two real world public datasets are presented in Sect. 5. Finally, the conclusion, limitation and future work outlook are offered in Sect. 6.

2 Related Work

2.1 Mobility Pattern Model

We categorize relevant mobility prediction models into sequential model, temporal model and hybrid model.

Sequential Model. Song et al. [12] found that Order-2 Markov with fallback had the best performance on the location prediction. Applied in mobility prediction by Cheng et al. [1], Factorizing Personalized Markov Chain extended the Markov Chain via factorization of transition matrix. Zhang et al. [18] extracted users' mobility sequential pattern from historical check-ins as a Location-Location Transition Graph. The problem of sequential models lies in that when adjacent mobility records gap for a long time like several days or even months, the performance becomes undesirable [1,3].

Temporal Model. Cho et al. [2] proposed a time-aware Gaussian Mixture model combining periodic short-range movements and sporadic long-distance travels. Wang et al. [13] provided a Regularity Conformity Heterogeneous (RCH) model to predict user location at specific time, considering both the regularity and conformity. Yang et al. [15] employed a Tensor Factorization model to capture the user temporal activity preference. However, these methods depend heavily on temporal regularity and data with decayed mobility regularity (e.g. at night or during weekends) leads to low accuracy [13].

Hybrid Model. Lian et al. [6] incorporated Markov model and temporal regularity model into the hidden Markov framework to predict user regular locations. This method suffered the same drawback as sequential model. Feng et al. [3] developed Personalized Ranking Metric Embedding (PRME) method to balance sequential dependency and user preference, by a threshold of transition timespan. PRME ignored the temporal regularity and a fix threshold cannot satisfy all the scenarios. In contrast to these methods, the proposed CAH approach combine temporal and sequential models flexibly depending on mobility context.

2.2 Location Prediction with Activity Information

Some researchers exploited activity information to improve the location predictability [5,8,9,16,17]. Noulas et al. [9] captured factors driving user movements, including the activity preference and activity transition. Yuan et al. [17] came up with a unified model W^4 (who, when, where, what) to discover individual mobility behaviors from spatial, temporal and activity aspects. Ye et al. [16], Li et al. [5] and Liu et al. [8] modeled activity sequential pattern, and predicted locations based on above activity. However, none of them absorb temporal and sequential model simultaneously to infer user activity preference.

In addition, given the activity distribution, Yuan et al. [17] and Li et al. [5] assumed the user location preference followed multinomial distribution. Ye et al. [16] ranked locations based on check-in frequency. Liu et al. [8] applied Matrix Factorization to predict user preference of specific locations. However, these methods fail to capture the time variation of user location preference. Instead, we adopt a generative approach to model the time variation pattern.

3   Problem Formulation

Let $\mathcal{V} = \{v_1, v_2, \ldots, v_{|\mathcal{V}|}\}$ and $\mathcal{C} = \{c_1, c_2, \ldots, c_{|\mathcal{C}|}\}$ represent locations and categories. Each location belongs to a certain category indicating the activity purpose of users. Given a set of users $\mathcal{U}$, each mobility record can be defined as a quadruple $r = (u, v, c, t)$, representing that user u visits location v at time t for activity c. Here, for the ease of calculation, t is discretized from continuity to discrete by 24 h. Our goal is to predict user u's next location $\hat{v}$, given the next visit time $\hat{t}$ and the recent visit sequence before $\hat{t}$, $\tau_{\hat{t}}^u$.

4   Methodology

4.1   Overview

We construct a two-stage model to predict activities and locations. The overall framework of the proposed model is presented in Fig. 1(a). It consists of two stages for activity and location prediction respectively and each stage incorporates offline model training and online prediction. In the first stage, Context-Aware Hybrid (CAH) approach is adopted to dynamically select the better performer from sequential and temporal models for activity prediction, i.e. inferring

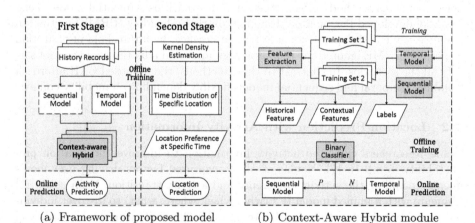

(a) Framework of proposed model   (b) Context-Aware Hybrid module

Fig. 1. Overall framework and CAH module

$P_u(c|\tau_{\hat{t}}^u, \hat{t})$. In second stage, based on inferred user activity, Kernel Density Estimation is exploited to approximate $P_u(v|c, \tau_{\hat{t}}^u, \hat{t})$. Finally, location prediction is achieved by $\hat{v} = \arg \max_v P_u(v|\tau_{\hat{t}}^u, \hat{t})$, where

$$P_u(v|\tau_{\hat{t}}^u, \hat{t}) = \sum_{c_j} P_u(v|c_j, \tau_{\hat{t}}^u, \hat{t}) P_u(c_j|\tau_{\hat{t}}^u, \hat{t}) = P_u(v|c_v, \hat{t}) P_u(c_v|\tau_{\hat{t}}^u, \hat{t}) \quad (1)$$

Note that during the second phrase, we ignore the sequential pattern of location by simplifying $P_u(v|c_v, \tau_{\hat{t}}^u, \hat{t})$ to $P_u(v|c_v, \hat{t})$. The reason is that the sequential dependency of the user mobility has been captured in the first stage. Although the geo-distance may influence user location preference, introducing distance does not significantly improve the prediction performance [6], due to the highly uncertain timespans between adjacent records and the convenient transportation in the modern world.

Figure 1(b) shows specific details of Context-Aware Hybrid module. We partition the training data into training set 1 and training set 2. The former is utilized for learning sequential and temporal models, and the latter is employed to evaluate the performances of them. In this work, we assign Tensor Factorization as temporal model and smoothed Order-1 Markov Chain as sequential model. With features of user contextual and historical factors and labels of the better performer between sequential and temporal models, we build a binary classifier for online prediction.

4.2   User Activity Prediction

Sequential Model. Markov model has been proved effective in mobility prediction [12]. Due to the data sparsity, we filter the transitions with timespans larger than threshold ε, and merely consider Order-1 Markov Chain. The transition probability is estimated by Kneser-Ney smoothing technique [6]. In particular, let $n_\varepsilon^u(c_i, c_j)$ indicate the times of user u transferring from activity c_i to c_j within ε. The transition probability is derived as:

$$P_u(c_j|c_i) = \frac{\max\{n_\varepsilon^u(c_i, c_j) - \delta, 0\}}{\sum_k n_\varepsilon^u(c_i, c_k)} + \frac{\delta \sum_k \mathbf{I}\{n_\varepsilon^u(c_i, c_k) > 0\} \cdot \sum_k \mathbf{I}\{n_\varepsilon^u(c_k, c_j) > 0\}}{\sum_k n_\varepsilon^u(c_i, c_k) \cdot \sum_k \sum_l \mathbf{I}\{n_\varepsilon^u(c_l, c_k) > 0\}}$$

where $\mathbf{I}\{\cdot\}$ is an indicator function and δ is the discount parameter. The basic intuition of this equation is to discount the observed times of transition from c_i to c_j, and turn them over to low frequency transitions.

Temporal Model. We adopt the non-negative Tensor Factorization (TF) method for inferring the activity preference at specific time [15]. A user-time-activity tensor $\mathbf{T} \in \mathbb{R}^{|\mathcal{U}| \times 24 \times |\mathcal{C}|}$ is built, in which the element $\mathbf{T}_{u,h,c}$ equals to the frequency of activity c at hour of day (HOD) h by user u. Using Canonical decomposition model [4], $\mathbf{T}$ is decomposed into three matrices, user feature matrix $\hat{U} \in \mathbb{R}^{|\mathcal{U}| \times L}$, time feature matrix $\hat{T} \in \mathbb{R}^{24 \times L}$, and activity feature matrix

$\hat{A} \in \mathbb{R}^{|\mathcal{C}| \times L}$ (L is the latent space dimension). User u's preference of activity c at h could be described as: $Pref_{u,h,c} = \sum_{i=1}^{L} \sum_{j=1}^{L} \sum_{k=1}^{L} \hat{U}_{u,i} \cdot \hat{T}_{h,j} \cdot \hat{A}_{c,k}$. For user u, the probability of activity c given time t is formulated as follows, where t_h is HOD of t.

$$P_u(c|t) = \frac{Pref_{u,t_h,c}}{\sum_{c' \in \mathcal{C}} Pref_{u,t_h,c'}} \tag{2}$$

Context-Aware Features Extraction. Given the Markov and TF models, we extract several kinds of features determining the accuracies of these two models, including temporal contextual features, sequential contextual features and historical features.

Temporal Contextual Features: This group of features refer to factors severely affecting temporal model at next visit HOD h, i.e. temporal regularity strength and data density at h. (1) Temporal regularity strength determines the limit of predictability, measured by *entropy* [11], defined as: $H(Z) = -\sum_i P(z_i) \log P(z_i)$ over random variable Z. We introduce random variable A_h^u, activity at h of user u, whose entropy $H(A_h^u)$ can be calculated based on u's history records Γ_u. Moreover, the number of distinct activities at h in Γ_u, correlating with $H(A_h^u)$, is also considered here, signified by $N_a^u(h)$. (2) Data density is represented by $N_r^u(h)$, the number of history records at h of user u. In summary, $H(A_h^u)$, $N_a^u(h)$ and $N_r^u(h)$ constitute temporal contextual features.

Sequential Contextual Features: The accuracy of sequential model depends on contexts of user recent records, i.e. the timespan and recent record frequency. (1) Timespan feature: As the sequence dependency decays over time, $D_1^u(\hat{t})$, the interval between $\hat{t}$ and nearest record time of user u, is introduced to model it. (2) Recent record frequency features: Data sparsity means missing latest activities and reduced performance of sequential model. Thus we propose two features: the length of $S_{\hat{t}}^u$, the longest mobility sequence ending by $\hat{t}$, satisfying that timespans between any adjacent records is less than ε; $D_2^u(\hat{t})$, the timespan between $\hat{t}$ and the earliest record time in $S_{\hat{t}}^u$. In summary, we use $D_1^u(\hat{t})$, $D_2^u(\hat{t})$, $|S_{\hat{t}}^u|$ as sequential contextual features.

Historical Features: From the whole mobility historical sequences, we consider user specific features (independent of context) of temporal regularity, sequential dependency and activity regularity strengths. (1) User specific temporal regularity strength is defined by $E_h(N_a^u(h))$ and $E_h(H(A_h^u))$, where $N_a^u(h)$ and $H(A_h^u)$ are defined above, and $E_h(Y) = \sum_{i=1}^{24} P_u(h_i)Y(h_i)$. (2) User specific sequential dependency strength is captured by $E_c(M_a^u(c))$ and $E_c(H(A_c^u))$, where $M_a^u(c)$ is the number of distinct activities of u after activity c, A_c^u is a random variable of the activity for user u after activity c, and $E_c(Y) = \sum_i P_u(c_i)Y(c_i)$. (3) User specific activity regularity strength is measured by the number of distinct activities N^u and the activity entropy $H(A^u)$ in history records Γ_u, where A^u is a random variable of the activity for user u. In summary, the historical features include $E_h(N_a^u(h))$, $E_h(H(A_h^u))$, $E_c(H(A_c^u))$, $E_c(M_a^u(c))$, N^u and $H(A^u)$.

Context-Aware Hybrid. Given the user u, the visit sequence $\tau_{\hat{t}}^u$ and the next activity time $\hat{t}$, feature vector X is calculated as mentioned before. We build a binary classifier to target at the better performer between TF and Markov models, taking feature vector X as input. Let positive class represent that Markov model is more effective, then $P_u(c_i|\tau_{\hat{t}}^u, \hat{t})$ is estimated as follows, where c_n is the latest activity in $\tau_{\hat{t}}^u$ and y is the output of classifier:

$$P_u(c_i|\tau_{\hat{t}}^u, \hat{t}) = \begin{cases} P_u(c_i|c_n), & \text{if } y = 1 \\ P_u(c_i|\hat{t}), & \text{if } y = -1 \end{cases} \tag{3}$$

We split user u's history records Γ_u into two parts, $\Gamma_u^{(1)}$ for training Markov and TF models, and $\Gamma_u^{(2)}$ for training the classifier. For the record $r : (u, t_r, c_r, v_r)$ in $\Gamma_u^{(2)}$, let $Rank_m(c_r)$ represent the probability rank of actual activity c_r generated by Markov model, and $Rank_t(c_r)$ is the probability rank generated by TF model. Then the record can be labeled as positive or negative depending on the sign of $Rank_t(c_r) - Rank_m(c_r)$. However, apart from contextual and historical features, the capacity of these two models may also be slightly affected by some random factors, such as the stochastic error. When the Markov and TF models perform similarly on the activity prediction, these random errors lead to wrong labeling. Therefore, we only take the records satisfying $|Rank_t(c_r) - Rank_m(c_r)| > \xi$ as training examples of the classifier. ξ is called filtering parameter. At last, considering that the numbers of positive and negative examples may be unbalanced, we set the negative-rate of training examples as the weight of positive class and the positive-rate as the weight of negative class.

4.3 User Location Prediction

As we have discussed in Sect. 1, the user preference of specific location changes over time. Without sufficient training data, directly estimating the probability $P_u(v|t, c_v)$ leads to the under fitting problem. The generative approach is more effective to address this missing data situation than the discriminative approach. If the time variation pattern of the user location preference could be modeled as probability distribution $P_u(t_h|v)$, we can approximate the probability of location v to be visited at time t as follows, where t_h is the HOD of t:

$$P_u(v|t, c_v) = \frac{P_u(t_h|v)P_u(v)}{\sum_{v' \in c_v} P_u(t_h|v')P_u(v')} \tag{4}$$

However, the time variation pattern varies from location to location. For example, some restaurants have three peak periods in a day including breakfast time, lunch time and dinner time, while some other restaurants only focus on dinner time. Due to this case, we perform non-parametric Kernel Density Estimation to reckon $P_u(t_h|v)$, which is widely used to estimate the shape of unknown probability density. The density of location v at t_h is formulated by

$$P_u(t_h|v) = \frac{1}{n_h d} \sum_{i=1}^{n} K\left(\frac{\Delta(t_h, h_i)}{d}\right) \tag{5}$$

where $\Delta(t_h, h_i) = min(|t_h - h_i|, 24 - |t_h - h_i|)$ is the interval between HOD t_h and h_i, $K(\cdot)$ is the kernel function, d is the bandwidth, and n_h is the number of distinct record hours on this location.

5   Experiments

5.1   Datasets

We evaluate our model on public check-in datasets in two big cities (New York and Tokyo), collected by Yang et al. [15]. In these two datasets, check-in records last from Apr 2012 to Feb 2013, and locations are classified into 251 categories. The statistics description is shown in Table 1. We do not study other public datasets due to the lack of activity information, such as the Gowalla dataset [2].

Table 1. Datasets statistic

	#User	#Location	#Check-in	#Location per user	#Category per user
NYC	1,083	38,333	227,420	84.04	40.22
TKY	2,293	61,858	573,703	92.43	32.40

5.2   Experiment Setting

Evaluation Plan. In the following experiments, we set the proportion of training set $\Gamma_u^{(1)}$, $\Gamma_u^{(2)}$ and test dataset as 7:2:1. For more convincing results, we repeat each experiment 10 times and take the average of metrics into comparison.

Parameter Setting. We set the timespan threshold ε as 6 h following the empirical rule [1,3], and the discount parameter as empirical formula $\delta = \frac{n_1}{n_1 + 2n_2}$ (n_1 and n_2 are the number of one-time transitions and two-times transitions)[6]. The latent space dimension L of TF model is recommended as 64 on these datasets by Yang et al. [15]. We select the standard normal kernel function and rule-of-thumb bandwidth $d = (4\hat{\sigma}/3n)^{\frac{1}{5}} \approx 1.06\hat{\sigma}^{-\frac{1}{5}}$ for KDE [10]. We study the effect of filtering parameter ξ in Sect. 5.3 and set it as 60.

5.3   Activity Prediction Evaluation

Effect of Features and Parameters. Firstly, we study the performance of binary classifier with different features. After attempting several methods such as logistic regression, decision tree and SVM, we apply the one with high performance and low training cost: Classification and Regression Tree (CART). The classification performance is measured by accuracy Acc and weighted average F-score F, following [14].

The classification performance evaluation based on different feature groups is shown in Table 2, where Seq, Tem and His are the abbreviation of sequential

Table 2. Features evaluation

	Seq	Tem	His	His+Seq	Tem+His	Tem+Seq	All
Acc of NYC	0.7154	0.7440	0.7511	0.7587	0.7534	0.7570	**0.7593**
F of NYC	0.7362	0.7677	0.7794	0.7701	0.7747	0.7666	**0.7803**
Acc of TKY	0.6822	0.7040	0.6998	0.7044	0.7054	0.7062	**0.7158**
F of TKY	0.6719	0.7022	0.6719	0.7074	0.7019	0.7029	**0.7150**

contextual features, temporal contextual features and historical features. We can observe that every paired feature groups combination outperforms the individual one, and combining all the features gets the best performance, implying that all three feature groups are effective and necessary.

Figure 2(a) describes the importance of features. Sequential contextual features and the sequential entropy take a larger proportion in TKY dataset. One possible reason is that the sequence regularities of users are stronger in TKY dataset, which makes sequential model more important in CAH approach.

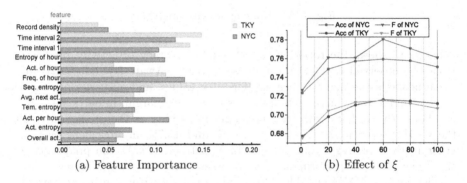

(a) Feature Importance   (b) Effect of ξ

Fig. 2. Feature importance and parameter effect

Besides, Fig. 2(b) reports the effect of filtering parameter ξ. As ξ increases, the labels of training examples become more credible. Thus the performance gets better when ξ varies from 0 to 60. However, there is a negative correlation between ξ and the number of classifier training examples. Owing to the insufficiency of training examples, the classification accuracy will fall back when ξ is bigger than 60.

Activity Prediction. After training the classifier, we apply the most frequently used metric of mobility prediction performance, Acc@top$k$, to contrasting the performance of proposed CAH approach with following 5 baselines:

1. Most Frequent: This method assigns the most frequent activity of user u at time t as the result of prediction.
2. Fallback Markov: Order-2 Markov with fallback has been utilized widely in mobility prediction on GPS trajectories and WiFi network [12].

3. Smooth Markov and Tensor Factorization: The sequential and temporal models we used, which have been introduced in Sect. 4.2.
4. HMM of CEPR: This model integrates temporal regularity and Markov models into a hidden Markov framework [6].

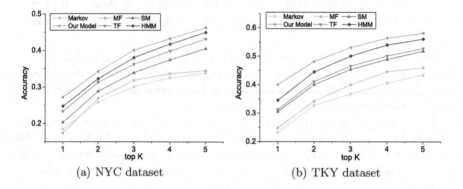

(a) NYC dataset   (b) TKY dataset

Fig. 3. Acc@top$k$ of activity prediction

Figure 3 shows the top-k ($k = 1, 2, 3, 4, 5$) accuracy of activity prediction. It can be observed that (1) the proposed Context-Aware Hybrid (CAH) approach achieves the highest accuracy for all k values, and outperforms Smooth Markov and Tensor Factorization models by a large margin. In particular, when we choose the activity with the maximum probability as the prediction result, the CAH approach shows at least 10.1% and 15.7% improvement over any other method on NYC dataset and TKY dataset; (2) As the prediction list size k increases, the performance gaps between CAH and some baselines become smaller, such as HMM and TF. This result is not surprising since a user usually prefer about 30–40 activities according to Table 1. In addition, it is clear that a large prediction list size k is meaningless for practical applications, thus getting higher accuracy with a small k is much more valuable.

5.4   Location Prediction Evaluation

For location prediction evaluation, we use the same metric as activity prediction (i.e. Acc@top$k$) and study following methods for comparison:

1. Most Frequent: Returning the most frequent locations of user as result.
2. KDE: Predicting locations only with generative method of the second stage, without the first stage.
3. PRME: This method [3] constructs a metric embedding model to balance sequential information and individual preference.
4. HMM of CEPR: We provide two versions of this approach. HMM represents the original approach of [6], predicting locations without activity information. HMM&KDE uses the hidden Markov framework of [6] to predict activities and the proposed generative approach to predict locations.

5. CAH&Rank/CAH&MFT: We apply other two methods to predict user future location based on CAH's results. CAH&Rank ranks locations by overall frequency [16]. CAH&MFT(Most Frequent of Time) directly estimates $P_u(v|t, c_v)$ based on frequency of location v at time t by user u.
6. CAH&KDE: The integrity version of the proposed model in this article.

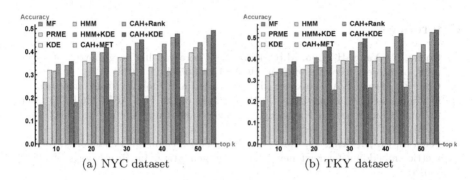

(a) NYC dataset   (b) TKY dataset

Fig. 4. Acc@top$k$ of Location Prediction

Figure 4 depicts the Acc@top$k$ ($k = 10, 20, 30, 40, 50$) of above methods. We can learn from that: (1) the integrity version of the proposed model (CAH&KDE) gets the best results for all the k values. Specifically, it shows 12.9% and 14.4% improvement over HMM of CEPR and 28.7% and 20.1% improvement over PRME, when $k = 10$; (2) the proposed generative approach (CAH&KDE) outperforms any other location prediction method based on CAH's results(i.e. CAH&Rank/CAH&MFT). Note that CAH&MFT gets the worst result, which is in line with the discussion in Sect. 4.2; (3) the performances of CAH&KDE and HMM&KDE are obviously better than KDE and HMM, implying that exploiting activity information facilitates location prediction. In addition, it also proves, to some extent, our two-stage framework is suitable for other activity prediction approaches; (4) the comparison of CAH&KDE and HMM&KDE indicates that improving activity prediction accuracy is beneficial to location prediction.

6   Conclusion

In this article, we propose a two-stage method to predict locations. In the first stage, we study the contextual and historical features that impact the prediction accuracy of sequential and temporal models, then we adopt a binary classifier to switch between these two models depending on predicting context. In the second stage, Kernel Density Estimation is performed to capture the time variation of the user location preference. Based on the evaluation results, our model significantly outperforms existing approaches.

Several interesting future directions exist for further exploration. For example, the sequential dependency and temporal regularity of user activities may affect each other, which makes it possible to improve the predictability.

Aknowledgements. The research is funded by National Key Research and Development Program, investigation on global optimization of resource scheduling method based on the specific application under the contract number No. 2016YFB0201402.

References

1. Cheng, C., Yang, H., Lyu, M.R., King, I.: Where you like to go next: successive point-of-interest recommendation. IJCAI **13**, 2605–2611 (2013)
2. Cho, E., Myers, S.A., Leskovec, J.: Friendship and mobility: user movement in location-based social networks. In: Proceedings of the 17th ACM SIGKDD International Conference on Knowledge Discovery and Data Mining, pp. 1082–1090. ACM (2011)
3. Feng, S., Li, X., Zeng, Y., Cong, G., Chee, Y.M., Yuan, Q.: Personalized ranking metric embedding for next new POI recommendation. In: Proceedings of IJCAI (2015)
4. Kim, S., Park, H.: Fast nonnegative tensor factorization with an active-set-like method. High-Performance Scientific Computing. Springer, Heidalberg (2012). doi:10.1007/978-1-4471-2437-5_16
5. Li, X., Lian, D., Xie, X., Sun, G.: Lifting the predictability of human mobility on activity trajectories. In: 2015 IEEE International Conference on Data Mining Workshop (ICDMW), pp. 1063–1069. IEEE (2015)
6. Lian, D., Xie, X., Zheng, V.W., Yuan, N.J., Zhang, F., Chen, E.: CEPR: a collaborative exploration and periodically returning model for location prediction. ACM Trans. Intell. Syst. Technol. (TIST) **6**(1), 8 (2015)
7. Liu, Q., Wu, S., Wang, L., Tan, T.: Predicting the next location: a recurrent model with spatial and temporal contexts. In: Thirtieth AAAI Conference on Artificial Intelligence (2016)
8. Liu, X., Liu, Y., Aberer, K., Miao, C.: Personalized point-of-interest recommendation by mining users' preference transition. In: Proceedings of the 22nd ACM International Conference on Information & Knowledge Management, pp. 733–738. ACM (2013)
9. Noulas, A., Scellato, S., Lathia, N., Mascolo, C.: Mining user mobility features for next place prediction in location-based services. In: 2012 IEEE 12th International Conference on Data Mining, pp. 1038–1043. IEEE (2012)
10. Silverman, B.W.: Density Estimation for Statistics and Data Analysis, vol. 26. CRC Press, Boca Raton (1986)
11. Song, C., Qu, Z., Blumm, N., Barabási, A.L.: Limits of predictability in human mobility. Science **327**(5968), 1018–1021 (2010)
12. Song, L., Kotz, D., Jain, R., He, X.: Evaluating location predictors with extensive wi-fi mobility data. In: INFOCOM 2004, Twenty-Third Annual Joint Conference of the IEEE Computer and Communications Societies, vol. 2, pp. 1414–1424. IEEE (2004)
13. Wang, Y., Yuan, N.J., Lian, D., Xu, L., Xie, X., Chen, E., Rui, Y.: Regularity and conformity: location prediction using heterogeneous mobility data. In: Proceedings of the 21th ACM SIGKDD International Conference on Knowledge Discovery and Data Mining, pp. 1275–1284. ACM (2015)

14. Witten, I.H., Frank, E.: Data Mining: Practical Machine Learning Tools and Techniques. Morgan Kaufmann, Burlington (2005)

15. Yang, D., Zhang, D., Zheng, V.W., Yu, Z.: Modeling user activity preference by leveraging user spatial temporal characteristics in LBSNs. IEEE Trans. Syst. Man Cybern.: Syst. **45**(1), 129–142 (2015)

16. Ye, J., Zhu, Z., Cheng, H.: Whats your next move: user activity prediction in location-based social networks. In: Proceedings of the SIAM International Conference on Data Mining. SIAM (2013)

17. Yuan, Q., Cong, G., Ma, Z., Sun, A., Thalmann, N.M.: Who, where, when and what: discover spatio-temporal topics for twitter users. In: Proceedings of the 19th ACM SIGKDD International Conference on Knowledge Discovery and Data Mining, pp. 605–613. ACM (2013)

18. Zhang, J.D., Chow, C.Y.: Spatiotemporal sequential influence modeling for location recommendations: a gravity-based approach. ACM Trans. Intell. Syst. Technol. (TIST) **7**(1), 11 (2015)

19. Zhong, Y., Yuan, N.J., Zhong, W., Zhang, F., Xie, X.: You are where you go: inferring demographic attributes from location check-ins. In: Proceedings of the Eighth ACM International Conference on Web Search and Data Mining, pp. 295–304. ACM (2015)

Modeling Temporal Behavior of Awards Effect on Viewership of Movies

Basmah Altaf[1], Faisal Kamiran[2], and Xiangliang Zhang[1(✉)]

[1] King Abdullah University of Science and Technology, Thuwal, Saudi Arabia
{basmah.altaf,xiangliang.zhang}@kaust.edu.sa
[2] Information Technology, University of the Punjab, Lahore, Pakistan
faisal.kamiran@itu.edu.pk

Abstract. The "rich get richer" effect is well-known in recommendation system. Popular items are recommended more, then purchased more, resulting in becoming even more popular over time. For example, we observe in Netflix data that awarded movies are more popular than non-awarded movies. Unlike other work focusing on making fair/neutralized recommendation, in this paper, we target on modeling the effect of awards on the viewership of movies. The main challenge of building such a model is that the effect on popularity changes over time with different intensity from movie to movie. Our proposed approach explicitly models the award effects for each movie and enables the recommendation system to provide a better ranked list of recommended movies. The results of an extensive empirical validation on Netflix and MovieLens data demonstrate the effectiveness of our model.

Keywords: Awards effect estimation · Popularity bias · Recommender systems

1 Introduction

Recommendation systems have been widely used in e-commerce to assist users in finding their potentially interested products [10], e.g., in Amazon, Ebay, NetFlix, LinkedIn, YouTube, and IMDB. The ranked list of products suggested by recommendation systems can be based on the overall popularity (number of views/ratings/downloads) for the product, or based on the overall average rating, or based on user purchase history. Often, the ranked list of items in recommendation system is dominated by the most popular items [11] due to the "rich get richer" effect. Popular items are recommended more, thence purchased more, resulting in becoming even more popular over time. Though "rich get richer" is a well-known phenomenon, there has been little attention paid in studying the temporal behavior of such effects in popularity.

In this work, we model the varying effect of awards on the popularity of items over time in the context of movies domain[1]. The most popular awards for

---

[1] The same framework can be applied to study award effect in other domains like songs (Grammy awards), scientific papers (best paper award), books (best seller book) and others.

© Springer International Publishing AG 2017
J. Kim et al. (Eds.): PAKDD 2017, Part I, LNAI 10234, pp. 724–736, 2017.
DOI: 10.1007/978-3-319-57454-7_56

movies include Oscar Award, Golden Globe Award, Bafta Award and Satellite Award. In general awards are considered a way to acknowledge the contribution of extraordinary movies to the industry. However many critics question the integrity of the award's mechanism and consider them as boosting mechanisms for movies to find lucrative marketing re-birth or launch actors and actresses to super-stardom. For instance in case of Oscar award, millions of dollars are spent on efforts to promote nominees to members of the Academy[2]. Considering such skepticism about different movie awards, it is very important to estimate the boosting effects of awards over the popularity of different movies.

In the past few years, there is a growing recognition of popularity bias in recommendation systems [1,3,5,6]. Different techniques have been proposed to penalize the discrimination of popularity on user ratings in recommendation systems [4,8,11]. They target on making neutralized recommendation. However, none of these work explicitly models the popularity and the influence of the "rich get richer" effect, e.g., contributed by movies awards.

Estimating the awards effect on popularity of items over time is important for several reasons. It can help in predicting revenues, future earnings, ranking on recommender systems of awarded items, e.g., airlines, hotels, movies, songs, scientific papers, etc. Our proposed framework models the popularity of items by two aspects, the base merit and the awards' effects, both of which vary with time due to the variation of subscribed users to movie system and the decrease of attractiveness of movies to users over time. The award effect model $\mathcal{D}$ explicitly estimates the extra viewership of awarded movies, while the base merit model $\mathcal{M}$ is learned for predicting movie popularity without award effect. The two models are shown to be effective for predicting the popularity of movies during March to Sept 2005 in Netflix (Jan-2006 to Dec-2008 in MovieLens), after trained on Netflix data from earlier stage: Dec-1999 to Feb-2005 (MovieLens data from Jan-1996 to Dec-2005). In addition, the model $\mathcal{D}$ is used to analyze how award effect varies on movies with different genres. Moreover, model $\mathcal{M}$ is applied to produce a ranked list of recommended movies without award effect. Through checking the ratings of recommended movies in IMDB and NetFlix, we show that our recommended movies without award effect have higher average rating than the recommended movies with award effect.

The remaining part of this paper is organized as follows. Section 2 discusses related work. Section 3 introduces the proposed model. Section 4 presents experimental evaluation. Section 5 concludes the paper.

2   Related Work

Work related to this study falls into two categories. From the application perspective, this study aligns with the popularity bias in recommendation systems [1,3–6,11] and from the problem setting perspective, it is connected to uplift modeling [7,9].

---

[2] http://www.politicalcampaigningtips.com/oscar-campaigning-the-politics-of-the-academy-awards/.

Kamishima et al. [4] propose to design an information neutral recommendation system, making the recommendations neutral from a specific view point, i.e., a specific feature of user (e.g., gender) or item (e.g., brand or popularity). They propose a penalty term to address the filter bubble problem of personalized recommendation systems. Their proposed penalty term ensures the statistical independence between the considered neutrality view point and preference scores. In this paper, we do not restrict our study to personalized recommendations and quantify the effects of different awards over the popularity of movies.

While it is clear that collaborative filtering algorithms outperform popularity-based recommendations in terms of accuracy and sales diversity, it has been pointed out that they still suffer from bias. Zhao et al. study a similar subject of popularity bias in recommendation system [11]. They mainly follow an opinion-based weighting function. If popular items are similarly rated by two neighboring users, these popular items are given the weight as the inverse log of the popularity. Otherwise (e.g., two users have different opinions), popular items are given the weights as log of the popularity. For less popular items, the assigned weight is 1. Their purpose is to improve diversity and accuracy in recommendation. However, our goal is to estimate the effect of awards on the popularity of items over time. Our work differentiates the user response and popularity of movies with and without awards.

In uplift modeling [7,9], two data sets (with and without action) are formed. The task is to model the effect of a particular action for a given instance and to identify the instances for which it is worth to do the action. The primary difference in our settings is that our uplift event is not revocable (i.e., awards are perpetual) and in practice it also becomes hard to have two exact movies with and without certain awards. In addition, we study the temporal effect of awards on the popularity of items over time, rather than only considering popularity with or without award.

3   Modeling the Award Effect on Popularity of Items

3.1   Problem Setting

Suppose we have a set of items (e.g., movies) where each item is viewed or rated at a given time t. Each item i can be described by a vector $X(i)$ with non-temporal attributes (e.g. movie genre), and additionally a temporal attribute vector $Z^t(i)$ that records time related attributes (e.g., average rating of a movie per week) at a certain time t. The popularity of an item (i.e., views or sales count) varies with time, and attains an artificial increase after the occurrence of an award event s, which is also a time varying attribute,

$s^t(i) = 0$ when there was no boosting event for item i at t;
$s^t(i) = 1$ when there was a boosting event for item i at t.

The popularity of each item i can be defined by grouping the rating counts into n identically sized time bins, where each time bin represents an instance of the calendar, e.g., day, week, and month. Denoting the popularity of an item i

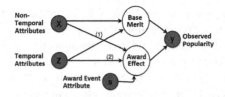

Fig. 1. A model of award affected popularity.

in time bin t by y_{it}, we define the **item instance** e_{it} to be an instance of an item i in time bin t, described as:

$$e_{it} = \{X(i), Z^t(i), s^t(i), y_{it}\} \tag{1}$$

The goal is to model the popularity score y_{it} of an item i at time t (e.g., how many viewers a movie will attract in a given week) given X_i, Z_i^t, and $s^t(i)$.

3.2   Award Effect Modeling

Our proposed model is shown in Fig. 1. The observed popularity is a result of the base merit (depending on X and Z) and the awards effect (introduced by s and affected by X and Z). The observed popularities are modeled as:

$$y = \mathcal{M} + s\mathcal{D}, \qquad s \in \{0, 1\}, \tag{2}$$

where $\mathcal{M}$ is the model learned on non-awarded data that predicts base merit of a movie, $\mathcal{D}$ is the model for estimating the award nomination effect or award winning effect. The intuition behind this model is as follows. When the items are not yet awarded/never awarded ($s = 0$), the observed popularity originates from the base merit model ($\mathcal{M}$) only. When the award event occurs ($s = 1$), the observed popularity comes from two terms: the base merit model ($\mathcal{M}$) and the award effect term $\mathcal{D}$. Both models $\mathcal{M}$ and $\mathcal{D}$ can be learned by a regression algorithm, e.g., Support Vector Regression. We focus on designing the learning procedure of using different training data with their corresponding targets. Note that we will consider award nomination and award winning differently. Model $\mathcal{D}_{nom}$ and $\mathcal{D}_{win}$ are learned and applied to nominated items and awarded items, respectively, but following the same learning procedure. Thus we only use subscriptions $_{nom}$ and $_{win}$ when necessary.

For learning $\mathcal{M}$, $\mathcal{D}_{nom}$ and $\mathcal{D}_{win}$, we divide training data into three subsets:

– set **A** contains instances of the movies that were neither nominated nor awarded ($s = 0$), or movie instances not yet nominated at time $t < t_{nom}$.
– set **B** contains all the award nominated movie instances after nominations were announced $t \geq t_{nom}$, and instances of award winning movies before winning award but after nomination at time $t_{nom} \leq t < t_{win}$.
– set **C** contains the instances of award winning movies after winning award(s) at time $t \geq t_{win}$.

The learning strategy is generally as follows: Model $\mathcal{M}$ is learned on data **A** and its corresponding observed popularity $y^{(A)}$. The award nomination effect model $\mathcal{D}_{nom}$ is learned on data **B** and its corresponding effect $d^{(B)} = y^{(B)} - \hat{y}^{(B)}$, where $y^{(B)}$ is the observed popularity and $\hat{y}^{(B)}$ is the predicted popularity when applying the above learned $\mathcal{M}$ on **B**. Similarly, the award winning effect model $\mathcal{D}_{win}$ is learned on data **C** and its corresponding effect $d^{(C)} = y^{(C)} - \hat{y}^{(C)}$, where $y^{(C)}$ is the observed popularity and $\hat{y}^{(C)}$ is the predicted popularity when applying the above learned $\mathcal{M}$ on **C**. As Model $\mathcal{M}$ is free from award nomination/winning effect, $\hat{y}^{(B)}$ and $\hat{y}^{(C)}$ are the non-awarded predictions of popularity for instances in **B** and **C**, respectively. The observed values $y^{(B)}$ are affected due to the award nomination and $y^{(C)}$ are influenced due to award winning. Taking $d^{(B)}$ and $d^{(C)}$ as the regression targets, the learned $\mathcal{D}_{nom}$ and $\mathcal{D}_{win}$ thus model the award nomination and award winning effect in **B** and **C**, respectively.

With different forms of award effect, we study three different models given in the following text:

(1) **Modeling Constant Effect of Awards** $(\mathcal{D}^{(0)})$ assumes that the effect introduced by the awards is constant for each item instance for all the time, i.e., an award introduces a fixed amount of publicity to viewers who want to see the winning movie independently of its content all the time. The constant effect can be defined as the mean of d, $\mathcal{D}^{(0)} = \frac{\sum_{i \in B} d_i^{(B)}}{|B|}$ for nomination instances, and $\mathcal{D}^{(0)} = \frac{\sum_{i \in C} d_i^{(C)}}{|C|}$ for awarded instances.

(2) **Modeling Awards Effect Based on** X $(\mathcal{D}^{(X)})$ assumes that the awards effect depends on the movie non-temporal attributes X, but is independent from temporal attributes, e.g., additional viewers watch the award winning movie because of the genre of this movie. The effect of the awards can be modeled as a function of item non-temporal attributes, $\mathcal{D}^{(X)} = f(X|s = 1)$.

(3) **Modeling Awards Effect Based on** X **and** Z $(\mathcal{D}^{(XZ)})$ assumes that the awards effect depends on the movie non-temporal attributes X as well as temporal attributes Z^t. We consider it as the most realistic assumption about the real data. For award nominated or award winning movies, additional viewership will depend on the time since the movie was released and the content of the movie. For instance, if a movie was released eleven months before the award ceremony, maybe most of the people have already seen it anyway or consider it to be an old movie and do not want to see it regardless of the award. The item and time related award effect can be modeled as a function of item and time attributes, $\mathcal{D}^{(XZ)} = f(X, Z^t|s = 1)$.

3.3   Temporal Behavior Award Effect Modeling

We propose Temporal behavior Award Effect Modeling (TAEM) algorithm to solve the learning problem in Eq. (2). Algorithm 1 gives the pseudo code for the learning of $\mathcal{M}$, $\mathcal{D}_{nom}$ and $\mathcal{D}_{win}$ as discussed in Sect. 3.2, and the prediction after learning. Note that Algorithm 1 applies to the learning of $\mathcal{D}^{(0)}$, $\mathcal{D}^{(X)}$ and $\mathcal{D}^{(XZ)}$, which only differ on the usage of selected feature sets.

Algorithm 1. Temporal Behavior Award Effect Modeling (TAEM)

---

1 **begin** Model learning

    Input: Data sets $Train\,\mathbf{A}$, $Train\,\mathbf{B}$, $Train\,\mathbf{C}$

    Output: Base merit model $\mathcal{M}$,

                Award effect models $\mathcal{D}_{nom}$ and $\mathcal{D}_{win}$

2 Learn $\mathcal{M}$ on training set $\mathbf{A}$:

    $\mathcal{M} : e_{it} \mapsto y_{it},$ $\{e_{it}, y_{it}\} \in Train\,\mathbf{A};$

3 Make predictions on training set $\mathbf{B}$ using model $\mathcal{M}$:

    $\hat{y_{it}}^{(B)} = \mathcal{M}(e_{it}),$ $e_{it} \in Train\,\mathbf{B};$

4 Make predictions on training set $\mathbf{C}$ using model $\mathcal{M}$:

    $\hat{y_{it}}^{(C)} = \mathcal{M}(e_{it}),$ $e_{it} \in Train\,\mathbf{C};$

5 Calculate the residuals $d_{it}^{(B)} = y_{it}^{(B)} - \hat{y_{it}}^{(B)};$

6 Calculate the residuals $d_{it}^{(C)} = y_{it}^{(C)} - \hat{y_{it}}^{(C)};$

7 Construct a new set $\mathbf{B}'$ with $d_{it}^{(B)}$ as new targets;

8 Construct a new set $\mathbf{C}'$ with $d_{it}^{(C)}$ as new targets;

9 Learn the award nomination effect estimation model $\mathcal{D}_{nom}$ on set $\mathbf{B}'$:

    $\mathcal{D}_{nom} : e_{it}^{(B)} \mapsto d_{it}^{(B)};$

10 Learn the award winning effect estimation model $\mathcal{D}_{win}$ on set $\mathbf{C}'$:

    $\mathcal{D}_{win} : e_{it}^{(C)} \mapsto d_{it}^{(C)};$

11 **begin** Evaluation and Recommendation

    Input: Data sets $Test\,\mathbf{A}$, $Test\,\mathbf{B}$ and $Test\,\mathbf{C}$,

             Award event attribute s_{nom}, s_{win}

    Output: Estimated y, ranking of test data sets

12 $y_{it} = \mathcal{M} + s_{it}\mathcal{D};$

13 Rank testing data by y_{it} in each time bin t

According to the application need, our proposed algorithmic solution can be applied in fashion of offline, online and adaptive settings.

Offline Learning: In this setting, we learn models $\mathcal{M}$, $\mathcal{D}_{nom}$ and $\mathcal{D}_{win}$ from a given training data that include instances collected until time t (line 1–10 in Algorithm 1). Predictions are made for the items from week $t + 1$ on-wards (line 11–13 in Algorithm 1). The learned models $\mathcal{M}$, $\mathcal{D}_{nom}$ and $\mathcal{D}_{win}$ are fixed without updating. This setting is a realistic scenario when there are enough data to learn accurate models and the relation between models and targets is fixed and will not change over time.

Online Learning: In this setting, models $\mathcal{M}$, $\mathcal{D}_{nom}$ and $\mathcal{D}_{win}$ are updated over time as more data arrive. The initial model is learned on all data up to time t inclusive. Predictions are made for time $t + 1$. At time $t + 1$, the models are updated or relearned using all the available data up until time $t + 1$ inclusive (run again line 1–10 in Algorithm 1). The new models are used to make predictions for time $t + 2$. This process is repeated for all coming weeks. This scenario is useful when there are not enough training data to learn an accurate initial model. More accurate models are learned as more data arrive.

Adaptive Learning: This scenario is useful when the underlying concept is changing over time, thus the relation between the inputs and the target needs to be updated. Before making prediction for $t + 1$, we re-learn the models from data in the last h (a fixed window size) weeks. These most up-to-date data enable the model to capture the dynamic, usually hidden, and important factors affecting the popularity.

4   Experimental Evaluation

4.1   Data Sets and Experimental Setting

We use two real world data sets Netflix [2] and MovieLens 10 million ratings data set[3]. The movies selected for our experiments are released from Dec-1999 to Sep-2005 in Netflix and from Jan-1996 to Dec-2008 in MovieLens when ratings are available for analysis. From Netflix (MovieLens), 134 (99) award winning movies, 53 (98) award nominated movies and 178 (102) movies that were never nominated nor awarded are selected. As described in Sect. 3.2, instances of these movies then define three sets **A**, **B** and **C**.

Each movie instance is described by 47 non-temporal attributes (forming a vector X) and 14 time related attributes (forming a vector Z). The attributes of X extracted from IMDB ⟨www.imdb.com⟩, Rotten Tomatoes ⟨http://www.rottentomatoes.com⟩ and Box Office Mojo ⟨http://www.boxofficemojo.com⟩ contain 9 numerical attributes: *budget, languages released count, run time, overall average rating, opening gross, total gross, opening theaters count, total theaters count, box office ranking in release year,* and 38 binary attributes, which include an attribute of *is Adult (i.e., is this movie only for adults), and is Sequel Or Adaptation,* nine binary attributes showing the genre of a movie, ten binary attributes for showing the studio in which movie is released, five binary attributes for showing the release year of movie, twelve binary attributes to show release month of movie. The Z includes 14 numerical attributes: *movie age, DVD release age, weekly average rating, popularity of previous week, popularity of second last week, popularity of third last week, popularity of fourth last week, average popularity of last two weeks, average popularity of last three weeks, number of ratings since release, total number of views (of all movies in the previous week), users joined the system in given week, total users in the system till this week,* and *Netflix subscribers growth.* We have eventually 58,059 instances from Netflix binned by week, and 33,138 instances from MovieLens binned by month.

In Netflix data sets **A**, **B** and **C**, we use instances from the first 275 weeks (Dec-1999 to Feb-2005) for training and instances of the last 32 weeks (March to Sept 2005) for testing. The selection is for the purpose of evaluating the predictive power of our models. In MovieLens data set, instances of first 120 months (Jan-1996 to Dec-2005) are used for training and the last 36 months (Jan-2006 to Dec-2008) are used for testing. We report only the results obtained on

---

[3] at https://grouplens.org/datasets/movielens/10m/.

NetFlix data set here due to the space limitation. Similar results and observations are found in MovieLens data set.

In all the experiments, we use Mean Absolute Error (MAE) as an evaluation measure. It measures how close the predicted values are from the observed ones.

4.2 Award Effect Quantification

In this section, we show the temporal award effect estimation capacity of our proposed method at individual instance level in Figs. 2, 3, and 4, where x-axis is the index of test instances (ordered by their actual popularity) and y-axis is the popularity. The model $\mathcal{D}$ for award effect quantification here is the one $\mathcal{D}^{(XZ)}$ (learned from both feature X and Z), as $\mathcal{D}^{(XZ)}$ performs better than $\mathcal{D}^{(X)}$ and $\mathcal{D}^{(0)}$ (see the comparison in next section).

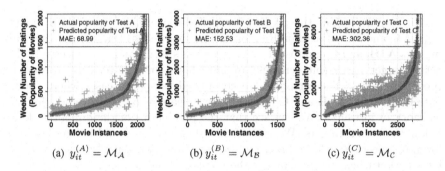

(a) $y_{it}^{(A)} = \mathcal{M}_{\mathcal{A}}$ (b) $y_{it}^{(B)} = \mathcal{M}_{\mathcal{B}}$ (c) $y_{it}^{(C)} = \mathcal{M}_{\mathcal{C}}$

Fig. 2. Prediction in increasing order of popularity of three models learned from training sets of **A**, **B**, and **C** separately. (Color figure online)

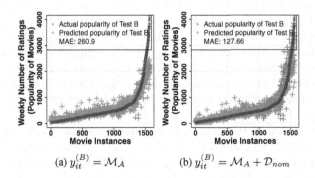

(a) $y_{it}^{(B)} = \mathcal{M}_{\mathcal{A}}$ (b) $y_{it}^{(B)} = \mathcal{M}_{\mathcal{A}} + \mathcal{D}_{nom}$

Fig. 3. Prediction of TAEM for award nominated instances in set **B**, estimation without award nomination effect (a) and with individual award nomination effect (b).

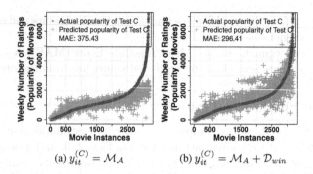

(a) $y_{it}^{(C)} = \mathcal{M}_{\mathcal{A}}$ (b) $y_{it}^{(C)} = \mathcal{M}_{\mathcal{A}} + \mathcal{D}_{win}$

Fig. 4. Prediction of TAEM for awarded winning instances in set **C**, estimation without award winning effect (a) and with individual award winning effect (b).

Figure 2(a–c) show the predictions (red stars) of three specialized models, $\mathcal{M}_{\mathcal{A}}, \mathcal{M}_{\mathcal{B}}$ and $\mathcal{M}_{\mathcal{C}}$, (based on Random Forest Regression) learned over the training sets **A**, **B**, and **C** and tested over the test sets **A**, **B**, and **C**, respectively. The actual target values (blue crosses) are also shown for a better comparison with predicted values. These models are considered as special cases where enough data in data sets **A**, **B** and **C** are available to learn a specialized model for each category of instances. We can see that the predictions of these models closely follow the actual targets. We expect that such models would be the best performing ones over their respective data sets. Note that our test set is from the later period when subscribers and the number of online movies have increased. Such a growth thus brings difficulties in prediction. Hence, high MAE scores are intuitive and acceptable.

Figure 3(a) shows the performance of our proposed TAEM over test set **B** but ignoring the nomination award effect, i.e., predicting by $\mathcal{M}_{\mathcal{A}}$ only. As expected, such predictions are lower than the actually observed popularity that includes different amount of award effect for different instances. Figure 3(b) shows the prediction of $\mathcal{M}_A + \mathcal{D}_{nom}$, which includes the estimated award effect. We see that our proposed method quantifies accurately the award effect in the popularity of inflated instances due to award nomination's boosting effect. When we add this calculated award effect to the non-awarded prediction in Fig. 3(a), it becomes very close in Fig. 3(b) to the predictions of specialized but award effect estimation model learned over the award nomination affected popularity data set **B** (shown in Fig. 2(b)).

Figure 4(a–b) shows the award effect quantification of award winning movies (set **C**) by TAEM. Similar to Fig. 3, it can be seen that the predictions with ignorance of awards are lower than the award affected popularity observation. Moreover, the award effect due to winning is greater than that due to nomination. When adding the estimated award effect to the non-awarded prediction in Fig. 4(a), as shown in Fig. 4(b), the scores are close to the predictions made by specialized model in Fig. 2(c). We can conclude from Figs. 3 and 4 that TAEM well quantifies the award effect in the popularity of each individual instance.

4.3   Comparison of Different Award Effect Estimation Models

We now compare different award effect models presented in Sect. 3.2 and three TAEM learning settings in Sect. 3.3. We evaluate the performance of these models by their predictive power measured in MAE.

1. **Single model:** prediction is made by a single model learned from the combination of all training sets **A**, **B**, and **C**.
2. TAEM with constant award effect (**TAEM with $\mathcal{D}^{(0)}$**).
3. TAEM with award effect learned from feature X only (**TAEM with $\mathcal{D}^{(X)}$**).
4. TAEM with award effect learned from feature X and Z (**TAEM with $\mathcal{D}^{(XZ)}$**).
5. **Three models**, which is built in an ideal scenario where enough data are available to build three separate models: $\mathcal{M}_{\mathcal{A}}$, $\mathcal{M}_{\mathcal{B}}$, and $\mathcal{M}_{\mathcal{C}}$ from the training sets **A**, **B**, and **C**, respectively. These three models are expected to achieve the highest accuracy when each of them is applied only to instances in its own set, as shown in Fig. 2. However, they failed to quantify and correct the award effect.

Figure 5 compares the MAE (y-axis) of different award effect models (x-axis) tested over the aggregation of test **A**, **B** and **C** in offline, online and adaptive settings. Generally, TAEM with different award effect estimation settings performs better than the single model. The advanced TAEM with award effect learned from feature X and Z (TAEM with $\mathcal{D}^{(XZ)}$) performs as good as the ideal three-models. Importantly, our method has the capability to estimate the award effect for each privileged item (as already shown in Sect. 4.2). In addition, we see that the online and adaptive settings ($h = 100$) have better performance than the offline setting.

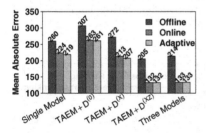

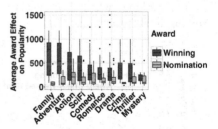

Fig. 5. Comparison of different award effect estimation models.

Fig. 6. Award nomination and winning effect by genre of movies

4.4 Award Effect on Different Movie Genres

Since we are able to estimate the award effect by $\mathcal{D}$, it is interesting to study how the award effect varies on movies with different genres. Figure 6 shows the average effect of award nomination and award winning over all the test weeks, for movies with different genres. Generally, we can see that award winning has higher influence than award nomination on movie popularity, especially for *Family, Adventure, Action and Science Fiction* movies. This conforms with the behavior of movie viewers. Movies in these genres can be accepted by a large population of viewers. Award winnings usually cause people's interest and curiosity to watch the awarded movies.

4.5 Recommendation of Top-k Based on Model $\mathcal{M}$

Given the task of generating top-k movies for recommendation, in this section, we validate the usefulness of our model by comparing the quality of top-k movies recommended according to the observed popularity (with award effect) and the predicted popularity based on model $\mathcal{M}$ (base merit of movies, without award effects). In other words, we apply model $\mathcal{M}$ on all movies for evaluating their base popularity by excluding the extra viewership introduced by awards. It is interesting to see whether our selection based on popularity estimated by $\mathcal{M}$ without award effect is better than the selection based on observed popularity with award effect. To evaluate if the ranked top-k movies are *good*, we check their actual ratings given by users not only in NetFlix (1–5 scale), but also in IMDB (1–10 scale) for checking the ratings given by users in a different system, IMDB. High ratings indicate *good* movies.

Figure 7 shows the average rating of top 20 movies in the ranking list produced by our model estimation (Award Effect Treated Ranking) and observed popularity (Observed Ranking). Two independent movie rating systems (IMBD and NetFlix) consistently verify the *better* quality of movies in award effect treated ranking, which has significantly higher average rating than observed ranking in all test weeks.

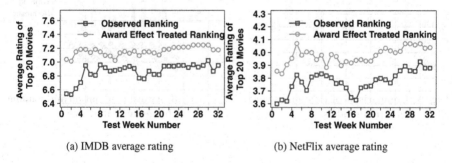

(a) IMDB average rating   (b) NetFlix average rating

Fig. 7. Average rating of the top-20 movies ranked by observed popularity (with award effect) and by the award effect treated popularity in each test week.

To study if *better* movies are always ranked higher in different settings of k, we take average rating of the top-k movies in all test weeks for each setting of k, and show the results in Fig. 8. The x-axis represents the varying k, while the y-axis is the average ratings of the top-k movies over all 32 test weeks. We see consistently in IMBD and NetFlix that award effect treated ranking generates movies list with *better* quality than observed ranking. Thus, our proposed model serves well the purpose of recommendation based true merit of items.

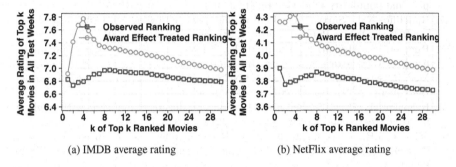

(a) IMDB average rating

(b) NetFlix average rating

Fig. 8. Average rating of top-k movies when varying the value of k.

5   Conclusion and Future Work

In this paper, we study the temporal effects of award nomination and award winning as popularity boosting events, and propose a framework for modeling the movie popularity in terms of base merit and the award effect. To model the popularity at individual movie level, we collected 47 non-temporal attributes and 14 time-related attributes for each movie. The experimental evaluation confirms the effectiveness of our proposed model on learning the temporal effect of award nomination and award winning. Also, the model can be used for producing a list of top-k recommended movies in better quality.

In this paper, we have assumed a simplified scenario that our task is to determine the award context-aware list of the best items at certain time for a group of users. In future, we plan to extend our methodology to personalized award context-aware recommendations and study the influence of multiple boosting events occurring at the same time.

References

1. Adomavicius, G., Kwon, Y.: Improving aggregate recommendation diversity using ranking-based techniques. IEEE TKDE **24**(5), 896–911 (2012)
2. Bennet, J., Lanning, S.: The netflix prize. KDD Cup and Workshop (2007)

3. Breese, J.S., Heckerman, D., Kadie, C.: Empirical analysis of predictive algorithms for collaborative filtering. In: Proceedings of the Fourteenth Conference on UAI, pp. 43–52 (1998)
4. Kamishima, T., Akaho, S., Asoh, H., Sakuma, J.: Enhancement of the neutrality in recommendation. In: Proceedings of the RecSys Workshop on Human Decision Making, pp. 8–14 (2012)
5. Lai, S., Liu, Y., Gu, H., Xu, L., et al.: Hybrid recommendation models for binary user preference prediction problem. In: KDD Cup, pp. 137–151 (2012)
6. Oh, J., Park, S., Yu, H., Song, M., Park, S.T.: Novel recommendation based on personal popularity tendency. In: IEEE ICDM, pp. 507–516 (2011)
7. Radcliffe, N.J., Surry, P.D.: Real-world uplift modelling with significance-based uplift trees. White Paper TR-2011-1, Stochastic Solutions (2011)
8. Ruggieri, S., Hajian, S., Kamiran, F., Zhang, X.: Anti-discrimination analysis using privacy attack strategies. In: Calders, T., Esposito, F., Hüllermeier, E., Meo, R. (eds.) ECML PKDD 2014. LNCS (LNAI), vol. 8725, pp. 694–710. Springer, Heidelberg (2014). doi:10.1007/978-3-662-44851-9_44
9. Rzepakowski, P., Jaroszewicz, S.: Decision trees for uplift modeling with single and multiple treatments. Knowl. Inf. Syst. **32**(2), 303–327 (2012)
10. Schafer, J.B., Konstan, J., Riedl, J.: Recommender systems in e-commerce. In: Proceedings of the 1st ACM Conference on Electronic Commerce, pp. 158–166 (1999)
11. Zhao, X., Niu, Z., Chen, W.: Opinion-based collaborative filtering to solve popularity bias in recommender systems. In: Proceedings of International Conference on Database and Expert Systems Applications, pp. 426–433 (2013)

A *Physarum*-Inspired Ant Colony Optimization for Community Mining

Mingxin Liang[1], Chao Gao[1], Xianghua Li[1,2,3(✉)], and Zili Zhang[1,4(✉)]

[1] School of Computer and Information Science, Southwest University,
Chongqing 400715, China
li_xianghua@163.com, zhangzl@swu.edu.cn
[2] Potsdam Institute for Climate Impact Research (PIK),
14473 Potsdam, Germany
[3] Institute of Physics, Humboldt-University zu Berlin,
12489 Berlin, Germany
[4] School of Information Technology, Deakin University,
Geelong, VIC 3220, Australia

Abstract. Community mining is a powerful tool for discovering the knowledge of networks and has a wide application. The modularity is one of very popular measurements for evaluating the efficiency of community divisions. However, the modularity maximization is a NP-complete problem. As an effective optimization algorithm for solving NP-complete problems, ant colony based community detection algorithm has been proposed to deal with such task. However the low accuracy and premature still limit its performance. Aiming to overcome those shortcomings, this paper proposes a novel nature-inspired optimization for the community mining based on the *Physarum*, a kind of slime molds cells. In the proposed strategy, the *Physarum*-inspired model optimizes the heuristic factor of ant colony algorithm by endowing edges with weights. With the information of weights provided by the *Physarum*-inspired model, the optimized heuristic factor can improve the searching abilities of ant colony algorithms. Four real-world networks and two typical kinds of ant colony optimization algorithms are used for estimating the efficiency of proposed strategy. Experiments show that the optimized ant colony optimization algorithms can achieve a better performance in terms of robustness and accuracy with a lower computational cost.

Keywords: Community mining · Ant colony algorithm · *Physarum*

1 Introduction

Community mining is associated with the graph clustering that is a powerful tool for knowledge discovering in many real-world complex systems [1]. Identifying the structural characteristics of a network has a wide application in knowledge discovery, such as the function prediction in the protein-protein networks [2], the real-time recommendation systems construction [3], and the information diffusion analysis [4].

© Springer International Publishing AG 2017
J. Kim et al. (Eds.): PAKDD 2017, Part I, LNAI 10234, pp. 737–749, 2017.
DOI: 10.1007/978-3-319-57454-7_57

Many algorithms have been proposed for mining such structures in networks, such as optimization-based algorithms [5] and stochastic model based algorithms [6]. Currently, a modularity measure Q has been proposed and widely used for estimating the qualities of community divisions [7]. Specifically, the modularity maximization is a NP-complete problem, which is intractable for traditional optimization algorithms, such as mathematical programming [8]. In the field of heuristics algorithms, ant colony optimization (ACO) algorithm is popular in dealing with the NP-complete problems [9]. But low accuracy and robustness limit its performance and application.

Recently, a kind of slime molds cell, *Physarum*, has shown an intelligence of network designing and path finding in biological experiments [10,11]. Moreover, for uncovering the key mechanism of the intelligent behavior of *Physarum*, a mathematical model has been proposed by Tero et al. [12]. This *Physarum*-inspired model has been used for optimizing the heuristic algorithms [13]. Based on the characters of *Physarum*-inspired model, we wonder can the *Physarum* model optimize the ant colony optimization algorithm for community mining?

Based on the above motivation, the main contributions of this paper are as follows. Taking advantages of *Physarum*-inspired model, which could recognize the inter-community edges coarsely, a novel nature-inspired optimization algorithm has been proposed based on ant colony optimization. In the new algorithm, the heuristics factor of traditional ACO is optimized based on the recognition of *Physarum*-inspired model, which could instruct the ants to find better solutions and improve the efficiency of algorithms. Meanwhile, four real-world networks and two representative kinds of ant colony algorithms are used to demonstrate the efficiency of the proposed nature-inspired optimization algorithm, in terms of accuracy and robustness.

The remaining of this paper is organized as follows. Section 2 formulates the community mining and introduces the ant colony optimization for community mining. And then, the nature-inspired optimization is proposed based on *Physarum*-inspired model in Sect. 3. Section 4 reports the experiments on four real-world networks and two typical kinds of ant colony clustering algorithms. Finally, Sect. 5 concludes this paper.

2   Related Work

2.1   Formulation of Community Detection

Community mining is to divide the vertexes in a network into communities, where vertexes across communities are sparsely connected, and vertexes within a community are relatively densely connected. Based on inherent structural features, a modularity measure, denoted as Q, is proposed to evaluate the qualities of community divisions [1]. Therefore, the community mining problem can be formulated to an optimization problem, which is to maximize the modularity value. The formulation of community mining is shown as follows.

Considering a network $G(V, E)$, where V and E stand for the sets of vertexes and edges respectively. And a community is a subset of V, where vertexes have a common certain feature. With NC indicating the number of communities, a community division is a set of communities, $C = \{C_1, C_2, \ldots, C_{NC}\}$, in which $C_i \neq \emptyset$, $C_i \neq C_j$, and $C_i \cap C_j = \emptyset$, for all i and j. After that, the mission of community mining can be represented as Eq. (1).

$$C^* = \arg \max_C Q(G, C) \tag{1}$$

The modularity Q is computed based on the topological structure of a network and its divisions, which is defined as Eq. (2). Specifically, $\delta(i, j)$ indicates the community relationship between vertexes i and j, while A and d_i stand for the adjacent matrix of a network and degree of vertex i, respectively. $A_{i,j}$ is equal to 1, if there is an edge connecting vertexes i and j. Otherwise, $A_{i,j}$ is equal to 0. Moreover, the degree of vertex i can be expressed as $d_i = \sum_j A_{i,j}$. In details, $\delta(i, j)$ is equal to 1, if and only if vertexes i and j belong to the same community. Otherwise, $\delta(i, j)$ is equal to 0.

$$Q = \frac{1}{2|E|}(A_{ij} - \frac{d_i d_j}{2|E|})\delta(i, j) \tag{2}$$

2.2   Ant Colony Algorithms for Community Mining

Ant colony optimization is under a general category of nature-inspired algorithm, which is inspired by the collective behaviors of ants. In the ant colony optimization algorithm, each ant finds a community division based on a probability directed by the pheromone matrix and the heuristic factor. The most important parts of a ant colony algorithm are searching, mutating and updating pheromone matrix. Here, we take a typical ant colony optimization for clustering, denoted as ACOC, as an example to introduce the basic parts of an ant colony algorithm for community mining [14].

Searching Strategy: In each iteration, every ant finds a community division based on a probability matrix, which is as shown in Eq. (3). P_{i,c_j} indicates the probability of vertex i belonging to community C_j. And c_j stands for the label of community C_j.

$$P_{i,c_j} = \frac{(\eta_{i,c_j})^\beta (Tau_{i,c_j})^\alpha}{\sum\limits_{k=1}^{NC} (\eta_{i,c_k})^\beta (Tau_{i,c_k})^\alpha} \tag{3}$$

In Eq. (3), η_{i,c_j} is the heuristic factor, which helps improve the search ability of ants based on the adjacent matrix of networks. For example, the heuristic factor in ACOC indicates the number of edges connecting the vertexes in community C_j from vertex i. Based on the character of community structure, the more edges connecting vertexes in community C_j vertex i joints, the larger probability of vertex i belonging to community C_j is. And the expression of η_{i,c_j} is shown in Eq. (4), in which n_{i,c_j} indicates the number of edges connecting vertex i and vertexes in community C_j. Here, C_j is based on the best community division found by algorithm. And Tau stands for the pheromone matrix, which is updated by the ants in each iteration based on the qualities of solutions.

$$\eta_{i,c_j} = \frac{n_{i,c_j}}{\sum\limits_{k=1}^{NC} n_{i,c_k}} \tag{4}$$

Based on the P matrix, the details in ACOC of assigning a community label to a vertex is introduced as follows. Taking the vertex i as an example, the community label c_j with maximal P value (i.e., $arg \max_{c_j} P_{i,c_j}$) is assigned to vertex i with a probability p_0. Meanwhile, the community label of vertex i is assigned based on the roulette way, with a probability $1 - p_0$. As shown in Fig. 1, assigning community labels for all the vertexes in such way, a community division emerges.

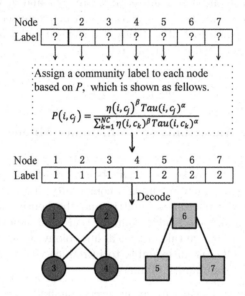

Fig. 1. The formulation of community division based on ACOC. Each community division is coded as a string of integers, which represents the community label of corresponding vertex.

Mutation Strategy: Mutation operator is a kind of random searching process, which aims to improve the diversity of solutions and protect ant colony clustering algorithms from premature. In the adopted mutation strategy of ACOC, each vertex is reassigned by a random community label with a probability P_m. And the mutation of a solution is accepted, if and only if the reassigning improves the modularity value of corresponding community division. Figure 2 shows an simple example of such mutation strategy.

Pheromone Matrix Updating Strategy: There are two phases for updating the pheromone matrix Tau. The first phase is implemented when an ant finishes

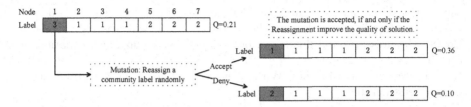

Fig. 2. An example of the mutation strategy in ACOC. Each mutated vertex is reassigned by a random community label. And the mutation is accepted, if and only if the reassigning improves the modularity value of corresponding community division.

its searching in an iteration. In this phase, every ant updates the pheromone matrix based on Eq. (5). Specifically, ρ is the volatile coefficient of pheromone, and Q indicates the quality of community division that the ant finds. The second phase executes after all the ants finish their local searching. Based on the community divisions with the top Q values in the current iteration, Eq. (6) is implemented for enhancing the effects of better community divisions. In Eq. (6), $\phi(i, c_j)$ equals to 1, if and only if vertex i within community C_j based on the corresponding community division. Otherwise, $\phi(i, c_j)$ equals to 0.

$$Tau_{i,c_j} = (1 - \rho) \cdot Tau_{i,c_j} + 2\rho \cdot Q \cdot \phi(i, c_j) \tag{5}$$

$$Tau_{i,c_j} = (1 - \rho) \cdot Tau_{i,c_j} + Q_{top} \cdot \phi(i, c_j) \tag{6}$$

With the local searching and updating for Tau, ants will aggregate to certain community divisions with higher Q values. And the solution with the highest Q value will be outputted as the optimal community division.

3   A Novel Nature-Inspired Optimization Algorithm for Community Mining

3.1   Edges Endowed with Weights Based on the *Physarum* Model

Physarum is a kind of slime with the abilities of designing networks and solving maze [10,11]. Moreover, inspired by the bio-experiments of *Physarum*, a mathematical model is proposed and used for optimizing the heuristic algorithms [13]. In this paper, the *Physarum* model (PM) is modified to endow edges with weights, which could be used to recognize the intra-community edges in a network.

$$PQ^t_{i,j} = \frac{D^{t-1}_{i,j}}{L_{i,j}} |p^t_i - p^t_j| \tag{7}$$

The basic hypothesis of PM is a Poiseuille flow in a network. And the core mechanism of PM is the feedback system between the cytoplasmic fluxes and conductivities of tubes in the Poiseuille's flow. This feedback system has two main processes. First, $PQ^t_{i,j}$, $D^t_{i,j}$, $L_{i,j}$ and p^t_i denote the flux, the conductivity, the length of $e_{i,j}$, and the pressure of v_i at time step t, respectively. Then, the relationship among the flux, conductivity, length, and pressure can be represented as Eq. (7). According to the Kirchhoff's law, which is represented in Eq. (8), the pressures and fluxes can be obtained, by solving such equations at each iteration step. And then, $PQ^t_{i,j}$ feeds back to $D^t_{i,j}$ based on Eq. (9). After that, an iteration step finishes. With such feedback going on, a highly efficient network is generated.

$$\sum_i PQ^{t-1}_{i,j} = \begin{cases} I_0, & if\ v_j\ is\ an\ inlet \\ -I_0, & if\ v_j\ is\ an\ oulet \\ 0, & others \end{cases} \tag{8}$$

$$D^t_{i,j} = \frac{PQ^t_{i,j} + D^{t-1}_{i,j}}{k} \tag{9}$$

The major modification of PM is the scheme of choosing inlets/outlets in each iteration. In such model, when a vertex is chosen as an inlet, the others are chosen as outlets. More specifically, Eq. (8) is modified as Eq. (10), in which D and L are known. With a certain inlet and outlet, we can construct a set of equations based on Eq. (10). By solving such equations, p_i can be obtained. And, in each iteration step of PM, every vertex is chosen as the inlet once. When v_l is chosen as the inlet, a local conductivity matrix, denoted as $D^t(v_l)$, is calculated based on the feedback system (i.e., Eqs. (7), (8) and (9)). Finally, after all the local conductivity matrixes are obtained, the global conductivity matrix is updated by the average of local conductivity matrixes based on Eq. (11). A detailed description of PM is represented in Algorithm 1.

$$\sum_i \frac{D^{t-1}(v_l)_{i,j}}{L_{i,j}}|p^t_i - p^t_j| = \begin{cases} I_0, & if\ v_j\ is\ an\ inlet \\ \frac{-I_0}{|V|-1}, & others \end{cases} \tag{10}$$

$$D^t = \frac{1}{|V|} \sum_{l=1}^{|V|} D^t(v_l) \tag{11}$$

With such modifications, the conductivities computed by *Physarum* model contain the information about inter-community edges recognition. *Physarum* model tends to endow inter-community edges with larger conductivities, vice versa. Figure 3 shows the edges with the top 20 percent conductivities in two networks based on such *Physarum* model. As it reported, the most of the edges with 20 percent conductivities connect vertexes in different communities. And there is almost no edge within the communities.

3.2   Nature-Inspired Optimization for Community Mining

Utilizing the character of conductivities computed by *Physarum* model, a novel ant colony optimization algorithm is proposed in this section, which aims to overcome the shortcomings of the low accuracy and premature. Through endowing edges with weights based on the *Physarum* model, the heuristic factor of proposed ant colony algorithm is optimized for improving the computational efficiency.

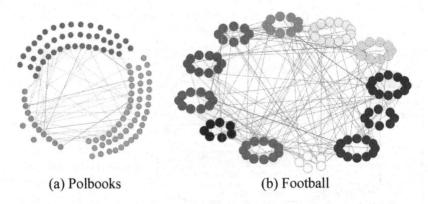

(a) Polbooks   (b) Football

Fig. 3. Edges with top 20 percent conductivities in two networks based on the *Physarum* model. The different colors are used to label communities.

Algorithm 1. *Physarum* network mathematical model for community mining

---

Input: An adjacent matrix A
Output: A conductivity matrix D
1. Initializing D^0 and the maximal iteration step T
2. For t from 1 to T
3.   For all vertexes in V
4.   Choosing v_l as the inlet
5.   Calculating $p^t_i, \forall i$, based on Eq. (10)
6.   Calculating $PQ^t_{i,j}, \forall i, j$, based on Eq. (7)
7.   Updating $D^t(v_l)$ based on Eq. (9)
8.   End for
9.   Updating D^t based on Eq. (11)
10. End for
11. Outputting D^T

$$\eta^*_{i,c_j} = \frac{n^*_{i,c_j}}{\sum\limits_{k=1}^{NC} n^*_{i,c_k}} = \frac{\sum\limits_{h\in C_j} 1/w_{i,h}}{\sum\limits_{k=1}^{NC} \sum\limits_{h\in C_k} 1/w_{i,h}} \qquad (12)$$

Taking the advantages of *Physarum* model, the intra-community edges tend to have larger weights. In contrast, the inter-community edges tend to have a small ones. Based on such character, we can optimize the heuristic factor of ant colony clustering algorithms in order to improve the search ability of such algorithms during the process of community detection. Takeing ACOC as an example, we can optimize heuristic factor based on Eq. (12), in which $w_{i,k}$ indicates the conductivity of $e_{i,k}$ and $n^*_{i,c_j} = \sum_{h\in C_j} \frac{1}{w_{i,h}}$. With such expression, η^*_{i,c_j} has a larger value when there are the same intra-community edges connecting vertex i and vertexes in community C_j, compared with the original η_{i,c_j}, vice versa. Such nature-inspired optimization exaggerates the inhomogeneity of

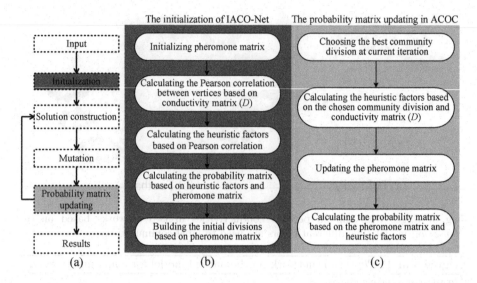

Fig. 4. Optimizing the heuristic factor in ACOC and IACO-Net based on the conductivity matrix D returned by *Physarum* model. (a) A basic framework of ant colony algorithm. (b) The optimized initialization of IACO-Net. (c) The optimized probability matrix updating process of ACOC.

original heuristic factor, and offers a more obvious information to ants, which leads to a higher accuracy and better robustness.

Although the heuristic factor is common and important in ant colony algorithms for community mining, the heuristic factor is used in different way in various ant colony algorithms. However, the *Physarum* based optimization strategy adapts various of ant colony algorithms easily. Here we employ two representative ant colony algorithms (i.e., ACOC [14] and IACO-Net [15]) to show the flexibility of our proposed method. Figure 4 illustrates the flowchart of optimizing the heuristic factors in ACOC and IACO.

4   Experiments

4.1   Datasets

Four real-world networks collected by Newman[1] and two ant colony clustering algorithms (i.e., ACOC [14] and IACO-Net [15]) are used to estimate the proposed algorithm. The basic topological features of those networks are shown in Table 1. For a clear expression, a prefix (i.e., $P-$) adds to the name of algorithm with the proposed strategy. And all the experiments are implemented in the same environment, which means that comparing algorithms have a same parameter setting and running environment. Moreover, the results are based on 20 repeated experiments to eliminate the fluctuation and evaluate the robustness.

---

[1] http://www-personal.umich.edu/~mejn/netdata/.

Table 1. The basic topological features of the real-world networks. N and E denote the number of vertexes and edges in a network. k and C stand for the average degree and clustering coefficient of networks, respectively. NC indicates the number of communities in networks based on the background.

Name	N	E	k	C	NC
Krateclub	34	78	4.588	0.588	2
Dolphins	62	160	5.129	0.303	2
Football	115	613	10.660	0.403	12
Polbooks	105	441	8.400	0.488	3

4.2 Experiment Results

Table 2 shows the box charts of modularity values returned by ACOC, IACO-Net, and their optimized algorithms on four real-world networks, which reports the distributions of results based on 20 repeated experiments. Due to the randomness of maximum and minimum, the comparison of those algorithms focuses on the first and third quartiles, and average. As is shown in such figure, the P-ACOC and P-IACO-Net have a higher average on all the four networks. And the first and third quartiles of optimized algorithms are also higher than that of original algorithms on all of four networks. Meanwhile, the distribution ranges of optimized algorithms are smaller, compared with that of original ones. It means the proposed strategy can enhance the robustness of ant colony algorithms.

Table 2. Results returned by ACOC, IACO-Net and their optimized algorithms on four networks in term of Q based on 20 repeated experiments. $Q1$ and $Q3$ indicate the first and third quartiles, respectively. And AVE stands for the average of those results on four networks.

Metrics	Algorithm	Karate	Dolphins	Football	Polbooks
$Q1$	ACOC	0.2907	0.3031	0.1613	0.4060
	P-ACOC	0.3286	0.3350	0.1943	0.4275
	ACO-Net	0.4198	0.5078	0.5769	0.5086
	P-IACO-Net	0.4198	0.5154	0.5866	0.5094
$Q3$	ACOC	0.3386	0.3751	0.2294	0.4363
	P-ACOC	0.3718	0.3778	0.2508	0.4475
	ACO-Net	0.4198	0.5170	0.5893	0.5169
	P-IACO-Net	0.4198	0.5195	0.5945	0.5168
AVE	ACOC	0.3149	0.3363	0.1947	0.4218
	P-ACOC	0.3512	0.3564	0.2261	0.4401
	ACO-Net	0.4196	0.5122	0.5824	0.5112
	P-IACO-Net	0.4197	0.5176	0.5903	0.5133

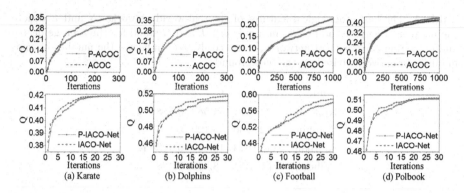

Fig. 5. The dynamic averages of Q with the increment of iteration in four networks. The results show that the proposed optimized strategy can obviously improve the search abilities of ACOs.

For further estimating the efficiency of proposed strategy, Fig. 5 reports the dynamic changes of average modularity values with the increments of iterations. As is shown in such figure, at the initial phase, the Q values of the optimized algorithms are close to those of original algorithms. However, the algorithms with the proposed strategy have a higher growth rate, compared with the original ant colony algorithms. With the growing of iterations, the difference between original and optimized algorithms emerges. There is a distinct gap between the lines of original and optimized algorithms at the end of iterations.

Other optimization and heuristic algorithms are also used to evaluate the efficiency of proposed algorithms for community mining. The compared algorithms include the evolution algorithm (i.e., GA-Net [16]), swarm intelligence algorithm (i.e., RWACO [17]), hierarchical clustering algorithm (i.e., FN [18]), and label propagation based algorithms (i.e., LPA [19]). Table 3 reports the modularity values returned by those algorithms. As shown in such table, P-IACO-Net has the highest Q values on three of four networks. Meanwhile, the Q values of P-ACOC have significant improvements, compared with that of ACOC.

The cost of such *Physarum*-inspired optimized strategy is the computational cost. And the time complexity of *Physarum* model is analyzed as follows. For *Physarum* model, at each iterative step, every vertex should be chosen as the inlet once. When a vertex is chosen, a corresponding system of equations needs to be solved. In other words, there are N equations to solve in each iteration step. The worst computation complexity of solving a system of equations is $O(N^3)$. With an empirical setting (i.e., $T = 1$), the total computation complexity of *Physarum*-inspired optimized strategy is $O(N^4)$. For a NP-complete problem, this computation complexity is acceptable. Moreover, Table 4 shows the running time of ACOC, IACOC-Net and their optimized algorithms in seconds, which also verifies that the proposed *Physarum*-inspired optimized strategy does not increase the computational complexity noticeably.

Table 3. Comparison the proposed algorithms with other optimization and heuristic algorithms. The community divisions are evaluated by modularity value Q.

Alg.	Net.			
	KarateClub	Dolphins	Football	Polbooks
ACOC	0.314	0.336	0.194	0.421
P-ACOC	0.350	0.365	0.226	0.440
IACO-Net	0.419	0.512	0.582	0.511
P-IACO-Net	0.419	0.517	0.590	0.512
FN	0.252	0.371	0.454	0.502
LPA	0.370	0.480	0.588	0.504
GA-Net	0.406	0.467	0.598	0.490
RWACO	0.371	0.377	0.601	0.456

Table 4. The running time of optimization-based algorithms in seconds. From this table, we can conclude that our proposed computational framework does not bring more computational burden for original algorithms.

Alg.	Net.			
	Karate	Dolphins	Polbooks	Football
P-ACOC	2.2814	4.3298	8.5971	9.9121
ACOC	2.1862	4.2678	8.4494	9.8882
P-IACO-Net	1.1906	2.4156	7.5031	15.4844
IACO-Net	1.0875	2.2906	7.2562	14.8844

5   Conclusion

Inspired by the *Physarum*-inspired model, a novel nature-inspired optimization algorithm for community mining is proposed in this paper based on the optimized ant colony optimization. In the proposed novel algorithm, the heuristic factor is optimized by a *Physarum*-inspired strategy. The proposed strategy integrates the knowledge of *Physarum*-inspired model into the heuristic factor for exaggerating the inhomogeneity of original ones and offering more extra knowledge for ants. Experiments on four real-world networks and two typical kinds of ant colony algorithms show the improvements of optimized algorithm in terms of accuracy and robustness. Moreover, the time complexity analysis shows that the proposed strategy does not increase the computational complexity of ant colony clustering algorithm noticeably.

Acknowledgement. Prof. Zili Zhang and Dr. Xianghua Li are the corresponding authors. This work is supported by the National Natural Science Foundation of China (Nos. 61403315,61402379), CQ CSTC (No. cstc2015gjhz40002), Fundamental Research Funds for the Central Universities (Nos. XDJK2016A008, XDJK2016B029, XDJK2016D053) and Chongqing Graduate Student Research Innovation Project (No. CYS16067).

References

1. Fortunato, S.: Community detection in graphs. Phys. Rep. **486**(3), 75–174 (2010)
2. Lee, J., Lee, J.: Hidden information revealed by optimal community structure from a protein-complex bipartite network improves protein function prediction. PLoS ONE **8**(4), 1–11 (2013)
3. Li, D., Lv, Q., Xie, X., Shang, L., Xia, H., Lu, T., Gu, N.: Interest-based real-time content recommendation in online social communities. Knowl.-Based Syst. **28**, 1–12 (2012)
4. Weng, L., Menczer, F., Ahn, Y.Y.: Virality prediction and community structure in social networks. Sci. Rep. **3**, 2522 (2013)
5. Gong, M., Cai, Q., Chen, X., Ma, L.: Complex network clustering by multiobjective discrete particle swarm optimization based on decomposition. IEEE Trans. Evol. Comput. **18**(1), 82–97 (2014)
6. Karrer, B., Newman, M.E.: Stochastic blockmodels and community structure in networks. Phys. Rev. E **83**(1), 016107 (2011)
7. Newman, M.E.: Modularity and community structure in networks. Proc. Nat. Acad. Sci. **103**(23), 8577–8582 (2006)
8. Mandala, S.R., Kumara, S.R., Rao, C.R., Albert, R.: Clustering social networks using ant colony optimization. Oper. Res. **13**(1), 47–65 (2013)
9. Mohan, B.C., Baskaran, R.: A survey: ant colony optimization based recent research and implementation on several engineering domain. Expert Syst. Appl. **39**(4), 4618–4627 (2012)
10. Nakagaki, T., Yamada, H., Tóth, Á.: Intelligence: maze-solving by an amoeboid organism. Nature **407**(6803), 470–470 (2000)
11. Tero, A., Takagi, S., Saigusa, T., Ito, K., Bebber, D.P., Fricker, M.D., Yumiki, K., Kobayashi, R., Nakagaki, T.: Rules for biologically inspired adaptive network design. Science **327**(5964), 439–442 (2010)
12. Tero, A., Kobayashi, R., Nakagaki, T.: A mathematical model for adaptive transport network in path finding by true slime mold. J. Theor. Biol. **244**(4), 553–564 (2007)
13. Liu, Y.X., Gao, C., Zhang, Z.L., Lu, Y.X., Chen, S., Liang, M.X., Li, T.: Solving NP-hard problems with Physarum-based ant colony system. IEEE/ACM Trans. Comput. Biol. Bioinf. **14**(1), 108–120 (2017)
14. Shelokar, P., Jayaraman, V., Kulkarni, B.: An ant colony approach for clustering. Anal. Chim. Acta **509**(2), 187–195 (2004)
15. Mu, C., Zhang, J., Jiao, L.: An intelligent ant colony optimization for community detection in complex networks. In: 2014 IEEE Congress on Evolutionary Computation, pp. 700–706 (2014)
16. Pizzuti, C.: GA-Net: a genetic algorithm for community detection in social networks. In: Rudolph, G., Jansen, T., Beume, N., Lucas, S., Poloni, C. (eds.) PPSN 2008. LNCS, vol. 5199, pp. 1081–1090. Springer, Heidelberg (2008). doi:10.1007/978-3-540-87700-4_107

17. Jin, D., Yang, B., Liu, J., Liu, D.Y., He, D.X.: Ant colony optimization based on random walk for community detection in complex networks. Ruanjian Xuebao/J. Softw. **23**(3), 451–464 (2012)
18. Newman, M.E.J.: Fast algorithm for detecting community structure in networks. Phys. Rev. E **69**, 066133 (2004)
19. Raghavan, U.N., Albert, R., Kumara, S.: Near linear time algorithm to detect community structures in large-scale networks. Phys. Rev. E **76**, 036106 (2007)

A Novel Diversity Measure for Understanding Movie Ranks in Movie Collaboration Networks

Manqing Ma[1,2], Wei Pang[3], Lan Huang[1,2], and Zhe Wang[1,2(✉)]

[1] College of Computer Science and Technology, Jilin University,
Changchun 130012, China
rhythm_qing@hotmail.com
[2] Key Laboratory of Symbol Computation and Knowledge Engineering,
Ministry of Education, Jilin University, Changchun 130012, China
{huanglan,wz2000}@jlu.edu.cn
[3] School of Natural and Computing Sciences, University of Aberdeen,
Aberdeen AB24 3UE, UK
pang.wei@abdn.ac.uk

Abstract. We are interested in the relationship between the team composition and the outcome in the filmmaking process. We studied the "diversity" of the group of actors and directors and how it is related to the movie rank given by the audience. The "diversity" is considered as the representation of the degree of variety based on the possibilities of collaborations among its actors and directors. Their collaboration network for the movie was first generated from the "background" network of the collaborations from other works. Then a shortest-path method together with the Adamic/Adar method are used to form indirect links. Finally the "complete" collaboration network can be generated and the "diversity" measures are thus defined accordingly. We experimented on the France and Germany datasets and identified consistent patterns: the lower the "diversity" is, the lower the movie rank will be. We also demonstrated that a subset of our diversity measures were effective in the binary classification task for movie ranks, while the advantages are prone to *Precision/Recall* depending on the specific dataset. This further shows that the "diversity" measure is feasible and effective in distinguishing movie ranks.

Keywords: Collaboration network · Network analysis

1 Introduction

Movie collaboration networks are one of the earliest types of collaboration networks which have been studied [1]. However, due to the ambiguity of their link formation which is partially caused by involuntary collaborations among actors, this type of neworks is less studied in depth compared to other collaboration networks.

We are interested in the relationship between the team composition and the outcome in the filmmaking process. Therefore, in this research we studied the

© Springer International Publishing AG 2017
J. Kim et al. (Eds.): PAKDD 2017, Part I, LNAI 10234, pp. 750–761, 2017.
DOI: 10.1007/978-3-319-57454-7_58

"diversity" defined on the movie collaboration network composed of actors and directors. By diversity we basically mean the variety among the group of people - whether they have never co-operated, or are even not likely to co-operate. We added directors into the conventionally studied movie actors networks, with an aim to more effectively capture the nature of the collaboration. We also introduced the concept of indirect links (in contrast to the direct links which already exist between the network node pairs) to the collaboration network of a movie to compensate for the lack of information caused by the sparsity of direct links.

The movie collaboration networks we studied are weighted, undirected, and heterogeneous (different types of links and nodes). Thus the following challenges are imposed:

- How to deal with the complexity of link types induced by adding the director nodes?
- How to define indirect links?

We have used a shortest-path method along with the Adamic/Adar method [2] adapted from the link prediction techniques for the formation of indirect links and modify them to make them fit for our problems. After the collaboration network of a movie is fully generated, we used diversity measures adapted from common metrics in network analysis, and studied their relationships with movie ranks given by the audience.

The main contributions of our study are as follows:

1. We have proposed applicable diversity measures for weighted and heterogeneous collaboration networks.
2. We can observe a consistent pattern: the lower the diversity is, the lower the movie's rank will be. We have found a subset of our diversity measures indicative for the movie ranks.

The rest of the paper is structured as follows: In Sect. 2 we describe the generation of the networks. In Sect. 2 we introduce the movie collaboration network, the formation of indirect links, and the diversity measures. Section 4 reports the experimental results. Related work is presented in Sect. 5, and finally Sect. 6 concludes the paper and explores future work.

2 Preliminaries

2.1 Datasets and Network Generation

We extracted the original data from IMDb[1]. We then pre-processed them and divided them into several datasets by countries and regions *where the movies were filmed*. After this, we chose the France and Germany datasets due to the relatively mature film industry in both countries and the relative abundance of data.

---

[1] www.imdb.com/interfaces.

In the data preprocessing stage we removed duplicate titles (this happened in the circumstance of TV series), and we also discarded those movies with incomplete information.

According to the *means* of each attribute, we chose the movies in the France dataset released after 1981, which have ranks higher than 6.3 and votes in IMDb more than average. We chose the movies in the Germany dataset released after 1988, which have ranks higher than 6.4 and votes more than average. As for the choice of the time window, we'll present the feasibility later in this section.

From these movies we generated a list of directors and actors/actresses. This list includes anyone who worked in one of these chosen movies. For simplicity actresses and actors are all referred as actors in the rest of this paper.

For both France and Germany datasets we then generated various kinds of links which form the corresponding sub-graphs. The links have the following semantic meanings:

1. Actor-actor with the weight value w being n: Two actors have co-acted in n movies.
2. Actor-director with the weight value w being n: The actor has co-operated with n directors.
3. Director-director with the weight value w being n: Two directors have both co-operated with n actors.

An illustration of these links is presented in Fig. 1a.

Then we generated the subgraphs of the networks based on a subset of actors where we excluded the actors who worked in too few or too many movies. We chose those actors whose number of works among the *selected* movies is between 10% and 85% among all the previously selected actors. Namely, for the France dataset we chose the actors having the number of movies between 10 and 100 while for Germany dataset the number is between 12 and 120.

Among *these* actors, the mean length of the active periods is 66 years for the France dataset and 58 years for the Germany dataset, which easily cover the time windows we selected for the movies. Thus these actors can be considered to be contemporary with each other. Thus the feasibility of the choice of the time window is shown.

The experiments and analysis are all performed on the above-generated subgraphs. Overall, the summaries of the France and Germany networks are presented in Table 1 (the "Movies" actually include some TV series).

Note that the Director-Director networks are extremely dense. However, since the links are weighted (the number of actors that two directors have both co-operated with), they can still bring some variance to the result.

3   Diversity in Movie Collaboration Networks

3.1   Movie Collaboration Networks

In this research, we focus on two types of relationships in a collaboration network: actor-actor and actor-director.

Table 1. The statistics of the France and Germany networks

		Item count	Network density	
France	Movies	711	Actor-Actor Network	0.036
	Actors	7,874	Director-Director Network	0.541
	Directors	884	Actor-Director Network	0.016
Germany	Movies	394	Actor-Actor Network	0.098
	Actors	5,431	Director-Director Network	0.499
	Directors	440	Actor-Director Network	0.054

We consider that node pairs in the network can be either directly or indirectly connected. Direct links mean that the links already exist in the network. As for the indirect links, we basically assign a predictor for the formation of the connection between two nodes which are not directly linked in the network. We will get into the details of the indirect link predictor and discuss its applicability in Sect. 3.2.

For the actors/directors of a particular movie, we first generate the "background" network for their collaboration network. Let $M = \{m_1, m_2, ..., m_p\}$ be the movie set, and G be the whole network. For each movie $m_i \in M$ we generate a sub-network G_i whose nodes are $A_i = \{a_i^1, a_i^2, ..., a_i^{p_i}\}$, $D_i = \{d_i^1, d_i^2, ..., d_i^{q_i}\}$, where p_i and q_i are the numbers of actors and directors in movie m_i, respectively. Then the "background" collaboration network $\overline{G_i}$ of m_i is the sub-graph of G with links in G_i whose weights are reduced by 1. Thus the collaboration weight matrix for $\overline{G_i}$ is $\overline{W_i}$. An illustration of the "background" network generation is presented in Fig. 1b.

We then generate the direct and indirect links of the actors and directors in the movie from its "background" network to form a complete collaboration network.

3.2   Indirect Link Generation

Small World Test. Movie actor network is a well-known small world network since it has been studied in the primary work of "small-world-phenomenon" [1]. Small world network is by its name a network where there exist short paths for virtually every pair of nodes. This feature could be detrimental to the performance of potential connection searching - since the meaning of the potential connection will become trivial if all nodes are closely connected. Bearing this in mind, we preliminarily performed a small world test on the networks to be studied. The result has shown that for both France and Germany datasets the links manage to form a small world network, with node pairs getting reached from each other within about 4 hops (for the Germany network, this number is actually 3). Thus we believe that the searching for indirect links should be better kept within 1-2 hops.

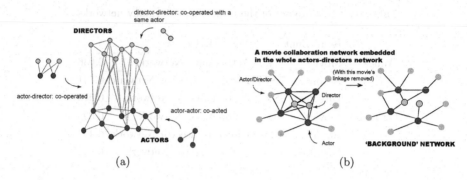

Fig. 1. (a) Actor-actor, actor-director, director-director links' semantic meanings and their formation. (b) A given movie's collaboration network embedded in the actors-directors network, and its "background" network (actors are represented by yellow nodes, and directors are blue nodes.) (Color figure online)

Indirect Link Formation. Using notations defined in Sect. 3.1, we can describe our link predictors for the problem of the indirect link formation. For demonstration, we will focus on a collaboration network for a single movie i, G_i (represented as G for simplicity thereafter). Correspondingly, the "background" collaboration network is $\overline{G_i}$ (represented as $\overline{G}$ thereafter), the weight matrix for the movie is $\overline{W_i}$ (represented as $\overline{W}$ thereafter). In our study, we use indirect links to compensate for the sparsity of links in movie collaboration networks.

Weighted Adamic/Adar in Heterogenous Networks. The simple Adamic/Adar [2] measure is originally used in link prediction tasks. Link prediction uses a predictor p that assigns a connection weight *score* to a pair of nodes $< x, y >$. Without loss of generality, we consider the link prediction task as predicting or discovering latent linkages. From this perspective link prediction could be smoothly adapted to our indirect link generation task.

The Adamic/Adar measure has proved to perform consistently well on collaboration networks [3] with appropriate modification. The original Adamic/Adar measure is only used on unweighted, homogenous networks [3]. In previous studies some researchers have managed to extend it to weighted [4] and heterogeneous networks [5].

Adding some modification to the measure proposed in [4] and [5], we can define a measure considering link weights meanwhile fit for the collaboration networks composed of several types of nodes and relationships (as shown in Fig. 2a):

$$
AA(node_1, node_2) =
$$
$$
\sum_{n \in N_{node_1} \cap N_{node_2}} \frac{\frac{2 \cdot \overline{W}_{node_1,n} \cdot \overline{W}_{n,node_2}}{\overline{W}_{node_1,n} + \overline{W}_{n,node_2}} \cdot}{log(\sum_{n' \in N_n, \sigma(n,n')=t_1} \overline{W}_{n,n'} + \sum_{n' \in N_n, \sigma(n,n')=t_2} \overline{W}_{n,n'})}, \tag{1}
$$

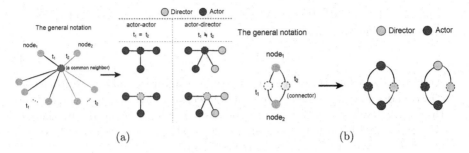

Fig. 2. Link types (t_1, t_2) in (a) Adamic/Adar and (b) One-hop shortest path measure for heterogenous movie collaboration network

where $\sigma(n', n)$ means the type of link (n', n), N_n means the neighborhood of n in the "background" network $\overline{G}$, $\overline{W}_{m,n}$ is the weight of link $m - n$ in the "background" collaboration network $\overline{G}$.

Simple One-hop Shortest Path. We also manually defined a simple graph-distance style measure which we term "one-hop shortest path", where distance is the inverse of link weight, as in the tradition of collaboration networks [6]. Namely, for a node pair (m, n) if there exists a link whose weight is 3, the distance along the link will be $1/3$. If there is a path with multiple nodes $(a_1, a_2, a_3, ...)$ on it, the path length will be the sum of all the distances between any two adjacent nodes.

Similar to the situation when we adapt heterogeneous links in the Adamic/Adar measure, for "one-hop shortest path" we consider indirect links under occasions in Fig. 2b.

The distance of a direct link is set to a global minimum value universally. The distances between pairs having no one-hop connection is set to a very large number (say 10000), and we will not include them in later computation. As for indirect links in Fig. 2b, we use the strategies as follows:

1. For node pairs which do not have direct links, if there exist multiple indirect links of any type, select the one with the shortest distance within each type.
2. Following the notations in Fig. 2b, we set the distance for the indirect connection $node_1 - node_2$ as the harmonic mean of the distance of the two types of links t_1 and t_2. Thus we define $dist(node_1, node_2)$ for indirectly connected nodes pair $(node_1, node_2)$ as follows:

$$dist(node_1, node_2) = \frac{2 \cdot dist(t_1) \cdot dist(t_2)}{dist(t_1) + dist(t_2)}, \tag{2}$$

where $dist(t_1)$ and $dist(t_2)$ are the distances of the indirect links t_1 and t_2. Suppose that the corresponding connectors on t_1 and t_2 are $connector_1$ and $connector_2$, $dist(t_1)$ and $dist(t_2)$ are then as follows:

$$\begin{aligned} dist(t_1) &= 1/\overline{W}_{node_1, connector_1} + 1/\overline{W}_{connector_1, node_2}, \\ dist(t_2) &= 1/\overline{W}_{node_1, connector_2} + 1/\overline{W}_{connector_2, node_2}, \end{aligned} \tag{3}$$

Since $w \geq 1, \forall w \in \overline{W}$, $\overline{W}$ is the link weight matrix of "background" network $\overline{G}$. Considering Formula (3) we have:

$$dist(t_1) \leq 2, dist(t_2) \leq 2 \qquad (4)$$

And consider when there is only one type of indirect links. Asserting that "Two nodes having different types of connections have stronger connections than those having only one single type of connections". Considering the fact that the global largest possible number for $dist(t_1)$ and $dist(t_2)$ is 2, we can assert $dist(node_1, node_2)$ under these occasions:

$$\frac{2 \cdot dist(t_n) \cdot d}{dist(t_n) + d} = dist(node_1, node_2), \forall d \leq 2, \qquad (5)$$

where there is only link t_n, $n \in \{1, 2\}$.

Indirect Link Formation. We use different strategies to form indirect links with weighted Adamic/Adar and one-hop shortest path:

1. For the weighted Adamic/Adar measure for heterogeneous relationships, the score of the Adamic/Adar measure is not on the same scale with the existing link weight which is in terms of the co-operation times between members. Thus we only decide whether the indirect link could be formed. By setting cutoff values from their score distribution, potential indirect links having score beyond the cutoff value could be formally formed.
2. For the one-hop shortest path measure, for each type of indirect links considered (as described in Fig. 2b) every potential indirect link is formally formed, with different distances assigned according to Eq. (2).

3.3 Diversity Measure: The D^- Family

D^- is essentially the inverse representation of a movie collaboration network's diversity. As links are segmented into the actor-actor and the actor-director, D^-s are thus computed separately on the subgraphs and the final D^- is the harmonic mean of the two. On each subgraph, the D^- is essentially defined as follows:

Using Adamic/Adar: The scores achieved from the Adamic/Adar measure can only be used to decide whether or not a potential link exists. We consider two types of D^-, namely, D^-_{cov} and D^-_{cc}.

D^-_{cov} is similar to the density measure used in network analysis. It merely consider the link coverage in all the possible links within the collaboration network of a movie. Use parameters to decide whether indirect links are added in and link weight considered.

D^-_{cc} is the average clustering coefficient [1] of the movie collaboration network, using parameters to decide whether indirect links are added in and link weight considered. For the actor-director subgraph, we only consider the local cluster coefficients of the director nodes.

Using One-hop Shortest Path: Different from the weighted Adamic/Adar measure, by definition, the weights computed for indirect links using "one-hop shortest path" is on the same scale with existing link weights. We also give two types of measures, namely, $D^-_{aver\_dist}$ and $D^-_{w\_cc}$, as detailed below. They can be seen as the distance-style modification of D^-_{cov} and D^-_{cc}, respectively.

$D^-_{aver\_dist}$ considers the average link weight in the network (with or without indirect links added), while normalized by a "direct link coverage" multiplier to eliminate the bias. Use parameters to decide whether indirect links are added in.

$D^-_{w\_cc}$ here is the weighted average clustering coefficient using a similar definition in [7] of a movie collaboration network G, using parameters to decide whether indirect links are added in. For the actor-director subgraph, we only consider the local cluster coefficients of the director nodes.

It is noted that we did not normalize the link weight $1/dist(n', n)$ (we did not extract the link weight from $\overline{W}$ because a formed indirect link does not naturally have its weight in $\overline{W}$) by the maximum weight in the network as in [7]. Because we were calculating the weighted average clustering coefficient for all the movie collaboration networks, and normalizing link weights by the maximum within each network will cause inconsistency in the result.

4   Experiments

4.1   The ROC Test of D^- Used as a Predictor in Binary Classification on Movie Rank, Using Incremental Cut-off Values

To see if our D^- measure can be used to predict whether a movie's rank is beyond a certain cutoff value, we use the ROC test for this challenge as ROC is commonly used in machine learning to evaluate features in binary classification. We directly extract the value from the CDF (Cumulative Distribution Function) of D^- as d^- used in the ROC test. The larger the value of AUC (area-under-curve) is, the more discriminative the predictor will be. The confusion matrix of our ROC test is presented in Table 2.

Table 2. The confusion matrix for ROC analysis of binary ranking and D^-

Predicted	Observed	
	$D^- > d^-$, $rank < cutoff$	$D^- > d^-$, $rank > cutoff$
$D^- > d^-$, $rank < cutoff$	TP	FP
$D^- > d^-$, $rank > cutoff$	FN	TN

The movie rank distributions in both France and Germany datasets are skewed as shown in Fig. 3, as we only select movies beyond the average rank in each dataset (see Sect. 2.1). We set the cutoff values of movie rank (*cutoff*) from 7 to 8.5 with the incremental step 0.1, since there are few movies whose ranks are beyond 8.5, and whatever result given by ROC for these movies is not convincing.

There are other details in the processing that are worth noting:

1. For the indirect link generation using the Adamic/Adar measure, we have experimented with the cutoff values from 0.3 to 0.8 of the score's distribution with the incremental step of 0.1. We found that at 0.6 the methods achieved consistently best performance in terms of the ROC analysis.
2. During the experiments, we only considered connected movie collaboration networks.

Compare with the Monte Carlo Simulation. We present the ROC analysis result in Fig. 3. We produced the heatmap (in Fig. 3) of the result obtained by Monte Carlo simulation on random scenarios(randomly shuffling the target movie rank list), upon which the real experimental result is plotted for comparison. Each cell in the heatmap of Monte Carlo has a value between $(0, 1)$, indicating for each cutoff value the probability of the AUC data point falling in the cell. We performed the Monte Carlo simulation for $1,000$ trials of random tests.

A valid predictor should perform significantly better on real world data than on random data, which means the AUC line of the real world result should be significant away from the random area (in Fig. 3 it is the area with darker color in the palette) and it should go towards the cells which the probability of random points falling in is less than 0.1 (those light yellow color cells).

4.2   Analysis

Given the evaluation standard above, we can see that the only consistently effective predictor for the France dataset and the Germany dataset is $D^-_{w\_cc}$ (with indirect links). For the France dataset there are another four D^-s which are indicative when the cutoff value is beyond 7.8: D^-_{cc} (weighted, with indirect links), D^-_{cc} (weighted, without indirect links), D^-_{cov} (weighted, with indirect links) and D^-_{cov} (weighted, without indirect links). Overall, we can observe a consistent pattern: the less diversity in a movie collaboration, the lower the movie rank will be. Our D^- is defined as the inverse of a movie collaboration network's diversity. Here "diversity" means the variety among the group of people based on their collaborations.

Diversity Measure Used in a Classification Task. We further applied our diversity measure D^-s to a movie rank binary classification task using cutoffs.

Due to the biased distribution of movie ranks, instead of computing the mean *Accuracy*, we only focused on the *Precision/Recall* of the predictions on ranks **beyond** the cutoff values. We used several classical machine learning classification algorithms [8] (Logistic Regression, Naive Bayes, SVC, Random

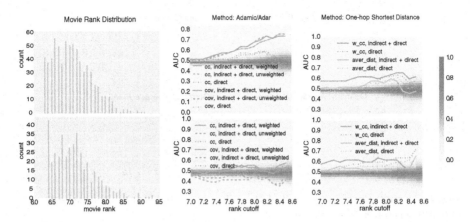

Fig. 3. AUCs for using D^-s as predictors of movie rank beyond cutoff values (France above and Germany below) (Color figure online)

Forest, Adaboost, and Nearst Neighbors) to perform a binary classification of movie ranks with the cutoff value ranging from 7 to 9. Among these models SVC and Logistic Regression tend to give high *Recall* and low *Precision* values, whereas Naive Bayes tends to give moderate *Precision* and *Recall* values.

There was little quantitative research on the factors affecting movie ranks, thus we empirically chose some of the possible factors in the preprocessing. Specifically, we set the "strong empirical" factors as { "average movie rank of the directors"}, and "moderate empirical" factors as { "genre", "count of the directors"}. These empirical factors are to be compared with the "selected":{$D^-_{w\_cc}$ (with indirect links)}, which is consistently effective on both dataset in the ROC test in Sect. 4.

Thus the experiments are performed on the datasets for 100 trials each with the datasets randomly separated (70/30) as training and test data. We then calculated the average *Precision* and *Recall* values on each cutoff value, ranging from 7 to 9 for all experiments.

The average *Precision/Recall* of all the classification models we used is presented in Fig. 4. Due to the "class imbalance" problem caused by the biased distribution of the movie ranks, when the cutoff value is beyond 7.3, the *Precision/Recalls* are not satisfactory (all below 0.5). Thus we only present results where the cutoff values are below 7.4 in Fig. 4.

The above results show that for the France dataset, the selected D^- has achieved highest average *Recall* value when the cutoff values are between 7.0 and 7.2, however it achieved only slightly higher *Precison* value than the random result. On the Germany dataset, the selected D^- has achieved average *Precision* value slightly less than the strong empirical factors, yet the average *Recall* value is lower than the random result. Overall, the selected D^- is enough for good predictions of movie ranks, however its advantage prones to different aspects of *Precision/Recall* for both datasets.

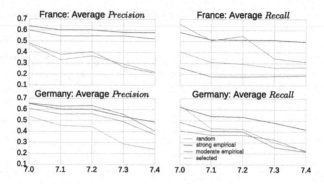

Fig. 4. Average *Precision/Recall* of the movie ranks binary classification using sets of features (France and Germany)

5 Related Work

Our study falls into the domain of collaboration networks. The movie actors network is less thoroughly studied due to its complex nature, even though some research has given a general profile of its structure [9–11]. Among the research on outcome aware collaboration networks [12,13], Kameshwaran et al. [12] studied the movie actors collaboration network. They used a modified eigenvector ranking method to describe how a collaboration outcome has influenced the intial nodes ranking in the network, however they assumed that the nodes did not determine the outcome directly.

6 Conclusions and Future Work

In this research we studied the diversity in movie collaboration networks, and its relationship with movie ranks. Experiments were performed on the movie collaboration networks composed of actors and directors, which are generated from the France and Germany datasets from IMDb. For indirect link generation we used adpted Adamic/Adar and "one-hop shortest path". We have observed consistent results in the France and Germany datasets: The less diversity in movie's collaboration network, the lower the movie rank will be. We also performed experiments on the selected diversity measures D^-s along with some empirically selected features in a movie ranks binary classification task and achieved satisfactory result on cutoff values below 7.3. Overall, we have proved the "diversity" measure is feasible enough in distinguishing movie ranks.

Our study serves as a first step towards further reasearch on the potential of using diversity measures for mining movie collaboration networks. The next step would include looking into the temporal evolving behavior of the network and relating it to the movie ranks.

Acknowledgements. This work is supported by the National Natural Science Foundation of China (Grant Nos. 61373051, 61472159, 61572227).

References

1. Watts, D.J., Strogatz, S.H.: Collective dynamics of small-world networks. Nature **393**(6684), 440–442 (1998)
2. Adamic, L.A., Adar, E.: Friends and neighbors on the web. Soc. Netw. **25**(3), 211–230 (2003)
3. Liben-Nowell, D., Kleinberg, J.: The link-prediction problem for social networks. J. Am. Soc. Inf. Sci. Technol. **58**(7), 1019–1031 (2007)
4. Murata, T., Moriyasu, S.: Link prediction of social networks based on weighted proximity measures. In: Web Intelligence, IEEE/WIC/ACM international conference on, pp. 85–88. IEEE (2007)
5. Davis, D., Lichtenwalter, R., Chawla, N.V.: Multi-relational link prediction in heterogeneous information networks. In: International Conference on Advances in Social Networks Analysis and Mining (ASONAM), pp. 281–288. IEEE (2011)
6. Newman, M.E.J.: Scientific collaboration networks. II. Shortest paths, weighted networks, centrality. Phys. Rev. E **64**(1), 016132 (2001)
7. Onnela, J.-P., Saramäki, J., Kertész, J., Kaski, K.: Intensity and coherence of motifs in weighted complex networks. Phys. Rev. E **71**(6), 065103 (2005)
8. Bishop, C.M.: Pattern recognition. Mach. Learn. 128 (2006)
9. Amaral, L.A.N., Scala, A., Barthelemy, M., Stanley, H.E.: Classes of small-world networks. Proc. Nat. Acad. Sci. **97**(21), 11149–11152 (2000)
10. Newman, M.E.J., Watts, D.J., Strogatz, S.H.: Random graph models of social networks. Proc. Nat. Acad. Sci. **99**(suppl 1), 2566–2572 (2002)
11. Ramasco, J.J., Dorogovtsev, S.N., Pastor-Satorras, R.: Self-organization of collaboration networks. Phys. Rev. E **70**(3), 036106 (2004)
12. Kameshwaran, S., Pandit, V., Mehta, S., Viswanadham, N., Dixit, K.: Outcome aware ranking in interaction networks. In: Proceedings of the 19th ACM International Conference on Information and Knowledge Management, pp. 229–238. ACM (2010)
13. Abbasi, A., Altmann, J.: On the correlation between research performance and social network analysis measures applied to research collaboration networks. In: 44th Hawaii International Conference on System Sciences (HICSS), pp. 1–10. IEEE (2011)

Local-to-Global Unsupervised Anomaly Detection from Temporal Data

Seif-Eddine Benkabou[1,2(✉)], Khalid Benabdeslem[1], and Bruno Canitia[2]

[1] University of Lyon1 - LIRIS, 43, Bd du 11 Novembre 1918, Villeurbanne, France
seif-eddine.benkabou@liris.cnrs.fr, kbenabde@univ-lyon1.fr
[2] Lizeo Online Media Group, 42 Quai Rambaud, Lyon, France

Abstract. Anomaly detection for temporal data has received much attention by many real-world applications. Most existing unsupervised methods dealing with this task are based on a sequential two-way approach (clustering and detection). Because of this, the clustering is less robust to anomalous series in data which distorts the detection step. Thus, to overcome this problem, we propose an embedded technique simultaneously dealing with both methods. We reformulate the task of anomaly detection as a *local*-weighting-instance clustering problem. The anomalous series are detected locally in each cluster as well as globally in the data, as a whole. Extensive experiments on benchmark datasets are carried out to validate our approach and compare it with other state-of-the-art methods of detection.

1 Introduction

In the last decade, the anomaly detection for temporal data has received much attention by the data mining community [8]. In fact, the rapid development of data acquisition tools has increased the accumulation of temporal data in several industries, such as financial time series, health care, astronomy, flight safety, traffic analysis, biology, environmental and industrial sensor data. In this work, we are interested in the global unsupervised anomaly detection for continuous time series data, i.e. *univariate real-value* time series (t_i is an ordered set of n real values, $t_i = \{t_{i1}, t_{i2}, \ldots, t_{in}\}$). Given a set of time series $T = \{t_1, t_2, \ldots, t_n\}$, we would like to detect the most anomalous time series in this set. In this task, it is assumed that most of the time series in the dataset are normal while a few are anomalous.

Similar to traditional anomaly detection, the usual recipe for solving such problems is to first learn a model based on all the time series sequences in the database, and then compute an anomaly score for each sequence with respect to the model. The model could be supervised or unsupervised depending on the availability of training data. Unsupervised global detection of an anomalous time series can be achieved by discriminative approaches based on a k-NN schema or by the two-way approach, which includes: clustering the time series and then computing the anomaly score of each time series as the distance to the centroid(medoid) of the closest cluster [3]. On the other hand, some parametric

© Springer International Publishing AG 2017
J. Kim et al. (Eds.): PAKDD 2017, Part I, LNAI 10234, pp. 762–772, 2017.
DOI: 10.1007/978-3-319-57454-7_59

approaches can be applied, for which anomalous series are not specified, and a summary model is constructed on the base data. A time series is then marked anomalous if the probability of generation of the sequence from the model is very low [7]. However, all these approaches deal with symbolic series (sequences) and applying them for continuous time series requires a discretization mechanism that could result in a loss of information. In contrast to the above methods, we propose an unsupervised approach which learns the anomaly score of time series by an embedded strategy where clustering and detection are performed simultaneously. In fact, we reformulate the task of anomaly detection as a *local* weighting-instance clustering problem based on dynamic time wrapping.

The remainder of this paper is organized as follows: In Sect. 2, we will review some related works on similarity measures for time series data and (weighted) clustering. We will describe our proposed algorithm in Sect. 3. Finally, in Sect. 4 we provide the experimental results for validating our proposal on some known benchmarks for time series data.

2   Related Works

Unsupervised direct detection of an anomalous time series can be achieved by a clustering mechanism. Once a similarity function between each pair of time series is defined, one can cluster them in different groups using this function. The anomaly score is then computed for each time series based on its distance to its closest centroid (or medoid). The main difference between these approaches is the choice of the similarity measure and the clustering process [3]. Time series data present challenges and opportunities for data mining, especially in clustering and representation learning. Such data exhibit noises, different lengths, and irregular sampling [1]. One can apply a discretization mechanism to transform time series to discrete sequences and then use some popular measures such as the simple match count based sequence similarity [12] or the normalized length of the longest common subsequence (LCS) [2]. A main concern is the loss of information resulting from the discretization mechanism. Other works transform the data into a more regular representation to which the simple Euclidian distance can be applied. However, it has only been recently shown in the functional data analysis literature that deformation-based metrics can be more robust to the curse of dimensionality than simple Euclidean distance [6]. For time series, Dynamic Time Warping (DTW) is a popular technique for measuring the distance between two time series with temporal deformations [19]. Unlike Euclidian distance, DTW can be used to compare time series, with different lengths, based on shape and permits distortions (e.g., shifting and stretching) along the temporal axis. This measure is known by its ability to return the best warping alignment between two time series despite its computational complexity in quadratic time using dynamic programming. Several variants are proposed to speed up this measure [17]. Given the distance metric, one can use k-means algorithm directly [15], or construct an affinity matrix and apply spectral clustering, or compute a dissimilarity matrix and build a taxonomy by a hierarchical clustering based on a linkage strategy [9]. Weighting-based clustering has been an important research topic in data analysis [10,13]. For example, the authors in [10,11] proposed the wk-means and ewk-means clustering algorithms that can automatically compute feature weights in the k-means clustering process. These algorithms extend the standard k-means algorithm with one additional step to compute feature weights at each iteration of the clustering process. As such, noise features can be identified and their affects on the clustering result are significantly reduced. These algorithms and their variants handle high-dimensional data and concern the weighting of *features* which is different than our work that considers time series (instance) weighting and especially anomaly detection.

3   Proposed Approach: L2GAD

In this section we present notations and definitions that we will use in the rest of the paper. Then, we describe in detail the theoretical aspects of our algorithm. We summarize the mathematical formulations in Algorithm 1 and discuss the convergence and the complexity of the proposed approach.

3.1   Notations and Definitions

In the paper, we use $T = \{t_1, t_2, \ldots, t_n\}$ to denote the set of n time series where each t_i is an ordered set of real values. The length of t_i is represented by N, which is equal or different to the lengths of other time series. The pairwise distance can be computed between each pair (t_i, t_j) by the DTW measure. **DTW** is a well-known technique to find an optimal alignment between two time series $t_i = (t_{i_1}, t_{i_2}, \ldots, t_{i_N})$ of length N and $t_j = (t_{j_1}, t_{j_2}, \ldots, t_{j_M})$ of length M. To compare two time points from t_i and t_j respectively, one needs a local cost measure, sometimes also referred to as local distance measure d. The goal is to find an alignment between t_i and t_j having minimal overall cost. The following definitions formalize the notion of an alignment.

Definition 1. An (N, M)-warping path is a sequence $s = (s_1, \ldots, s_C)$ with $s_c = (i_c, j_c) \in [1 : N] \times [1 : M]$ for $c \in [1 : C]$ satisfying the following three conditions.

- Boundary condition: $s_1 = (1, 1)$ and $s_C = (N, M)$
- Monotonicity condition: $i_1 \leq \ldots \leq i_C$ and $j_1 \leq \ldots \leq j_C$
- Step size condition: $s_{c+1} - s_c \in \{(1, 0); (0, 1); (1, 1)\}$ for $c \in [1 : C - 1]$

Definition 2. DTW is the warping path having minimal total cost among all possible warping paths. $\mathbf{dtw}(t_i, t_j) = min\{d_s(t_i, t_j)|s$ is an (N, M)-warping path$\}$; where $d_s(t_i, t_j) = \sum_{c=1}^{C} d(t_{i_c}, t_{j_c})$.

3.2   L2GAD Algorithm

In this section, we present our algorithm which detects a set of anomalous time series in an unsupervised way. We propose a local-to-global unsupervised anomaly detection approach for time series, named L2GAD, in short. The algorithm is an embedded technique dealing with clustering and detection, simultaneously. It detects the anomalous time series for each cluster (*local detection step*) and then aggregates their weights for the whole set of time series (*global detection step*). In doing this, we reformulate the task of anomaly detection as a *local-weighting-instance* clustering problem. For each cluster, we detect its anomalous time series by assigning smaller weights to the time series that increase its within-cluster distances. In fact, we believe that the weight of time series in some cluster represents the score of contribution of that time series in forming the cluster and computing its medoid.

Thus, such time series are considered anomalous for this cluster. By doing so, each time series would have as much weight as the number of clusters that represents its degree of abnormality within each cluster. Finally, the global detection step is done by assigning an *global* anomaly score to each time series by aggregating its weights with its distances to each medoid. The higher the score, the more anomalous the series.

In the following, we propose a new objective function, to be minimized, described in Eq. (1).

$$\min_{A,M,W} \Phi(A,M,W) = \sum_{l=1}^{k} \sum_{i=1}^{n} a_{il} w_{il}^{\alpha} \mathbf{dtw}(t_i, m_l) \tag{1}$$

subject to the following constraints:
$$\begin{cases} \sum_{l=1}^{k} a_{il} = 1; 1 \le i \le n \\ a_{il} \in \{0,1\}; 1 \le i \le n; 1 \le l \le k \\ \sum_{l=1}^{k} \sum_{i=1}^{n} w_{il} = 1; 0 \le w_i \le 1 \end{cases}$$

Where

- A is $n \times k$ partition matrix and a_{il} is a binary variable such as $a_{il} = 1$ indicates that the time series t_i is assigned to cluster l;
- $M = \{m_1, m_2, \ldots, m_k\}$ is a set of k time series representing the medoids of the k clusters;
- W is $n \times k$ weights matrix such that each time series has a k weights corresponding to the k clusters.
- α is parameter for weighting. Note that this parameter can neither be equal to zero nor to one. Indeed, if $\alpha = 0$, the weighting is removed and the detection cannot be performed. If $\alpha = 1$, w would disappear because of the bellow derivatives for solving the problem.

Minimization of Φ in Eq. (1) with the constraints forms a class of constrained nonlinear optimization problems whose solutions are unknown. Indeed, it is difficult to optimize three variables simultaneously. Thus, we adopt an alternating optimization to solve this problem, which works well for a number of practical optimization problems. To do this, we have to minimize the objective function with unknown variables A, M and W by iteratively solving the following three reduced minimization problems:

- **Problem1**: Minimizing Eq. (1) by fixing M and W for finding the solution for A. The optimization leads to:

$$a_{il} = \begin{cases} 1 \text{ if } \mathbf{dtw}(t_i, m_l) \leq \mathbf{dtw}(t_i, m_v); 1 \leq v \leq k \\ 0 \text{ } Otherwise \end{cases} \quad (2)$$

- **Problem2**: Minimizing Eq. (1) by fixing A and W for finding the solution of M. The medoid m_l is a representative time series of the cluster l whose distance to all time series in the same cluster is minimal.

$$m_l = \min_{t_j} \sum_{i|i \neq j, \; g_{il}=g_{jl}=1} w_{il}^\alpha \mathbf{dtw}(t_i, t_j). \quad (3)$$

- **Problem3**: Minimizing Eq. (1) by fixing A and M for finding weights of time series W. The optimization leads to (*see proof*):

$$w_{il} = \left(\sum_{l=1}^{k} \sum_{s=1}^{n} \left(\frac{a_{il}\mathbf{dtw}(t_i, m_l)}{a_{sl}\mathbf{dtw}(t_s, m_l)} \right)^{\frac{1}{\alpha-1}} \right)^{-1} \quad (4)$$

Then, we compute the global anomaly score s_i of each time series t_i as the scalar product of the weights vector $w_{(i,:)} = [w_{i1}, \ldots, w_{ik}]$ and the distance vector $d_{(i,:)} = [\mathbf{dtw}(t_i, m_1), \ldots, \mathbf{dtw}(t_i, m_k)]$

$$s_i = \langle w_{(i,:)}, d_{(i,:)} \rangle = [w_{i1}, \ldots, w_{ik}][dtw(t_i, m_1), \ldots, dtw(t_i, m_k)]^\top \quad (5)$$

Proof (Eq. (4)). We minimize the function by the Lagrangian multiplier. Let μ be the multiplier and $\Theta(W, \mu)$ be the Lagrangian.

$$\Theta(W, \mu) = \sum_{l=1}^{k} \sum_{i=1}^{n} w_{il}^\alpha a_{il} \mathbf{dtw}(t_i, m_l) - \mu \left(\sum_{l=1}^{k} \sum_{i=1}^{n} w_{il} - 1 \right) \quad (6)$$

To minimize $\Theta(W, \mu)$, the gradient of Θ on both W and μ must be equal to zero. Thus, we would have

$$\frac{\partial \Theta(W, \mu)}{\partial w_{il}} = (\alpha)w_{il}^{(\alpha-1)} (a_{il}\mathbf{dtw}(t_i, m_l)) - \mu = 0; 1 \leq i \leq n \quad (7)$$

and

$$\frac{\partial \Theta(W, \mu)}{\partial \mu} = \sum_{l=1}^{k} \sum_{i=1}^{n} w_{il} - 1 = 0 \quad (8)$$

From (7) we obtain:

$$w_{il} = \left(\frac{\mu}{(\alpha)a_{il}\mathbf{dtw}(t_i, m_l)} \right)^{\frac{1}{1-\alpha}} \quad (9)$$

Substituting (9) in (8) we have:

$$\sum_{l=1}^{k}\sum_{i=1}^{n}\left(\frac{\mu}{(\alpha)a_{il}\mathbf{dtw}(t_i,m_l)}\right)^{\frac{1}{1-\alpha}} = 1 \tag{10}$$

From 10 we derive:

$$\mu^{\frac{1}{1-\alpha}} = \left(\sum_{l=1}^{k}\sum_{i=1}^{n}\left(\frac{1}{(\alpha)a_{il}\mathbf{dtw}(t_i,m_l)}\right)^{\frac{1}{1-\alpha}}\right)^{-1} \tag{11}$$

Substituting (11) in (9):

$$w_{il} = \frac{\left(\sum_{l=1}^{k}\sum_{s=1}^{n}\left(a_{sl}\alpha\mathbf{dtw}(t_s,m_l)^{-1}\right)^{\frac{1}{1-\alpha}}\right)^{-1}}{\left(a_{il}\alpha\mathbf{dtw}(t_i,m_l)\right)^{\frac{1}{1-\alpha}}} \tag{12}$$

From (12) we obtain the final formula of weights:

$$w_{il} = \left(\sum_{l=1}^{k}\sum_{s=1}^{n}\left(\frac{a_{il}\mathbf{dtw}(t_i,m_l)}{a_{sl}\mathbf{dtw}(t_s,m_l)}\right)^{\frac{1}{\alpha-1}}\right)^{-1} \tag{13}$$

$\square$

Subsequently, we can summarize all the above mathematical developments in Algorithm 1.

Algorithm 1. L2GAD

1: **Input:** Set of time series $T = \{t_1, t_2, \ldots, t_n\}$, parameters k and α.
2: **Output:** Ranked time series
3: **Initialize:**
4: Randomly choose initial k time series from T as medoids $\{m_1, m_2, \ldots, m_k\}$
5: Randomly generate $n \times k$ weights of time series $w.r.t \sum_{l=1}^{k}\sum_{i=1}^{n} w_{il} = 1$
6: **repeat**
7:    calculate the cluster-memberships using Eq. (2)
8:    update the cluster-medoids by Eq. (3)
9:    compute the weights using Eq. (4)
10: **until** *Convergence* (no alteration in the weights)
11: Compute the global anomaly score s_i of each time series t_i using Eq. (5)
12: Rank the time series according to their anomaly score in descending order.

Lemma 1 (Convergence).
L2GAD converges to a local minimum solution in a finite number of iterations.

Proof (Lemma 1). Assume that $M^{p_1} = M^{p_2}$ where $p_1 \neq p_2$. Given M^p, we can compute the minimizer A^p, which is independent of W^p as shown in Eq. (2). For M^{p_1} and M^{p_2}, we have the minimizers A^{p_1} and A^{p_2}, respectively. It is clear that $A^{p_1} = A^{p_2}$ since $M^{p_1} = M^{p_2}$. Using M^{p_1} and A^{p_1}, M^{p_2} and A^{p_2}, we can compute the minimizers W^{p_1} and W^{p_2}, respectively according to Eq. (4). Again, it is clear that $W^{p_1} = W^{p_2}$. Thus, we get:

$$\Phi(A^{p_1}, M^{p_1}, W^{p_1}) = \Phi(A^{p_2}, M^{p_2}, W^{p_2}).$$

Note that the sequence $\Phi(A, M, W)$ generated by L2GAD is strictly decreasing, so the algorithm converges in a finite number of iterations.   □

Lemma 2 (Complexity).
L2GAD algorithm is computed $O((n(n-1)/2)L^2) + O(n \times \max(\log(n), pk))$ operations, where p is the number of iterations.

Proof (Lemma 2). Let L be the length of the longest time series in the whole dataset T. Computing the distance between two time series t_i and t_j using DTW, requires $O(max(|t_i|, |t_j|))^2)$ in general. Thus, computing the pairwise distance between each pair of time series, we need at most $O(L^2(\frac{n(n-1)}{2}))$ operations. Step 7 calculates the cluster-memberships values by $O(nk)$ operations. Step 8 also updates the medods by $O(nk)$ operations and the Step 9 provides the time series weights after $O(n \times 2k)$ operations. The two last steps compute the anomaly score of all time series in $O(n \times 2k)$ operations and rank them according to this score with $n \log(n)$ operations. Subsequently, L2GAD is computed in $O((n(n-1)/2)L^2) + O(n \times \max(\log(n), pk))$ where p is the total number of iterations.   □

4   Experiments

In this section, we present the used datasets and the compared algorithms, as well as the experimental setting used to conduct the empirical study. The results of the experiments are provided in Table 1.

4.1   Datasets and Methods

To evaluate the performance of L2GAD, we compare it with four anomaly detection algorithms: DTW+SPECTRAL [14], DTW+KMEDOID [2], DTW+HC [16] and FD-OCSVM [18]. The three first are two-way approaches based on clustering process, whereas DF-OCSVM computes a feature space by a frequency domain (FD) coding and then applies One-class SVM for anomaly detection in a set of time series [18]. The clustering based methods cluster the time series on k clusters based on the DTW measure. Then, they assign an anomaly score to a time series according to the distance with its closest (center or medoids). To assess the efficiency of our algorithm, an empirical study was carried out on 35 datasets to test different cases (length of series (shorter, longer), number of series, different number of classes). Theses datasets are available at the UCR Time Series Classification Archive [4]. The characteristics of these dataset are described in Table 1 (see the *Datasets description columns*).

4.2   Experimental Setting

For each dataset, experimental results are averaged over 20 runs. However, all these datasets are devoted to classification purpose and not for anomaly detection task. So, we propose to perturb 10% of time series in each dataset, randomly and consider them as anomalous. This perturbation is done by two parameters: *segment* and *snr*. The *segment* parameter represents the percentage of the values to perturb in a time series and the *snr* parameter represents the signal-noise-ratio to be added at each perturbed value in the time series. For the clustering based algorithms, the number of clusters k is set to the one indicated in Table 1 (*The # classes column*) and they will start with the same vector $m = m_1, \ldots, m_k$ that we generate randomly at each run. For L2GAD algorithm, the parameter α is taken randomly at each run such that $\alpha \in]1, 1000]$.

The performance is assessed via precision at top β which is the fraction of anomalous time series in the whole data. The β is known since we fix it in the perturbation process (*corresponding to the #anomalous column* in Table 1). For each algorithm we rank the set of all time series according to their anomaly scores (5) and we count the number of the true anomalous ones (*trueAnomalous*) in the top β portion. The detection accuracy is then computed as following:

$$DetectionAccuracy = \frac{trueAnomalous}{\beta} \qquad (14)$$

4.3   Results

For each dataset, the accuracies of detection of each algorithm are averaged over 20 runs and the results are reported in Table 1 (see *the Anomaly Detection Algorithm columns*). In order to better assess these results, we adopt the methodology proposed by [5] for comparison of several algorithms over multiple data sets. In this methodology, the nonparametric Friedman test is first used to evaluate the rejection of the hypothesis that all the algorithms perform equally well for a given risk level. It ranks the algorithms for each dataset separately, with the best performing algorithm obtaining the rank of 1, the second best rank 2, etc. In case of ties, it assigns average ranks. Then, the Friedman test compares the average ranks of the algorithms and calculates the Friedman statistic. If a statistically significant difference in performance is detected, we proceed with a post hoc test.

The Nemenyi test is used to compare all the algorithms with each other. In this procedure, the performance of two algorithms is significantly different if their average ranks differ more than some critical distance (CD). The critical distance depends on the number of algorithms, the number of data sets and the critical value (for a given significance level p-value) which is based on the studentized range statistic (see [5] for further details).

In this study, based on the values in Table 1, the Friedman test reveals statistically significant differences (p-value <0.05) between all compared algorithms. Furthermore, we present the result from the Nemenyi post hoc test with average rank diagrams as suggested by [5]. These are given in Fig. 1. The ranks are

Table 1. Comparison of Average Accuracy of Anomaly Detection on 35 data sets (Bold values are the best ones *(the higher score, the better performance)*

Data sets	Data sets description				Anomaly detection algorithm				
	# Time series	Length	# classes	# anomalous	L2GAD	HC$_{DTW}$	KM$_{DTW}$	Spectral$_{DTW}$	FD-OCSVM
1-RefrigerationDevices	375	720	3	37	**99.72**	91.89	97.83	97.29	37.83
2-ChronileConcentration	467	166	3	46	96.52	89.13	97.39	**97.82**	89.13
3-LargeKitchenAppliances	375	720	3	37	**97.56**	86.48	86.08	86.21	51.35
4-Adiac	390	176	37	39	**96.66**	7.69	90.12	91.15	64.10
5-Computers	250	720	2	25	**75.80**	56.00	69.00	56.00	36.00
6-CricketX	390	300	12	39	**86.02**	25.64	61.66	46.79	51.28
7-CricketY	390	300	12	39	**79.87**	20.51	62.05	64.87	58.97
8-CricketZ	390	300	12	39	**85.12**	30.76	55.51	52.05	51.28
9-DistalPhalanxOutlineAgeGroup	139	80	3	13	96.92	84.61	96.92	**100.00**	69.23
10-DistalPhalanxTW	139	80	6	13	**96.53**	61.53	**96.53**	91.92	69.23
11-FaceAll	560	131	14	56	**71.25**	26.78	57.23	48.66	55.35
12-FaceUCR	200	131	14	20	**82.00**	35.00	47.00	43.00	50.00
13-FordA	1320	500	2	132	**97.80**	44.69	79.43	71.21	52.27
14-FordB	810	500	2	81	**90.61**	53.08	78.20	69.13	51.85
15-PhalangesOutlinesCorrect	1800	80	2	180	96.52	**96.66**	**96.66**	**96.66**	67.77
16-InsectWingBeatSound	220	256	11	22	**82.27**	36.36	69.31	66.13	54.54
17-Lighting2	60	637	2	6	**77.50**	16.66	61.66	66.66	66.66
18-Lighting7	70	319	7	7	**75.71**	28.57	66.42	61.42	57.14
19-MedicalImage	381	99	10	38	**70.26**	42.10	58.55	66.97	57.89
20-NonInvasiveFatalECGThorax1	1800	750	42	180	93.77	77.22	97.72	**99.30**	65.55
21-NonInvasiveFatalECGThorax2	1800	750	42	180	98.00	77.22	97.69	**99.27**	65.55
22-OSULeaf	200	427	6	20	**78.25**	25.00	52.75	46.25	55.00
23-ProximalPhalanxT	205	80	6	20	**96.50**	75.00	**96.50**	94.50	65.00
24-ScreenType	375	720	3	37	**69.86**	40.54	50.81	45.67	35.13
25-SmallKitchenAppliances	375	720	3	37	**99.59**	94.59	98.24	**99.59**	67.56
26-SwedishLeaf	500	128	15	50	55.10	36.00	**75.60**	74.40	46.00
27-TwoPatterns	1000	128	4	100	**93.95**	39.00	77.30	72.10	57.00
28-uWaveGestureLibraryZ	896	315	8	89	**64.04**	23.59	55.67	45.16	49.43
29-uWaveGestureLibraryY	896	315	8	89	**69.71**	21.34	59.66	66.12	49.43
30-WaveGestureLibraryX	896	315	8	89	**64.66**	20.22	48.70	49.38	50.56
31-UWaveGestureLibraryAll	896	945	8	89	**76.23**	34.83	60.67	59.49	53.93
32-StarLightCurves	1000	1024	3	100	72.75	45.00	71.600	**92.75**	51.00
33-Wafer	1000	152	2	100	**73.15**	39.00	40.20	41.00	53.00
34-WordsSynonyms	267	270	25	26	**87.88**	11.53	65.76	73.46	57.69
35-50words	450	270	50	45	**91.00**	8.88	62.55	77.44	48.88

depicted on the axis, in such manner that the best ranking algorithms are to the right side of the diagram. Algorithms that do not differ significantly (at p-value $= 0.05$) are connected with a line. The critical distance CD is shown above the graph (here, CD $= 1.0311$). As can be seen, L2GAD tops the ranking, with no connections to other algorithms, meaning that it does statistically differ from others.

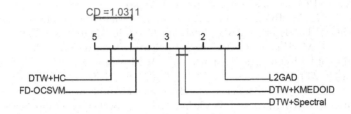

Fig. 1. Average ranks diagram comparing the anomalous time series detection algorithms in terms of accuracy Eq. (14).

5   Conclusion

In this paper, we proposed an unsupervised approach for anomaly detection from time series data. We presented L2GAD, a new algorithm that first detects anomalous series in each cluster and then in whole set of series. Unlike to the most existing methods, L2GAD deals with this task by doing clustering and detection simultaneously. Experiments on several datasets demonstrated superiority of L2GAD in comparison with state-of-the-art anomaly detection approaches. For future works, it would be interesting to extend L2GAD to deal with metric learning in which the alignment between time series could be learned instead of using DTW as a fixed measure. Another avenue would be to extend our approach to deal with semi-supervised learning on which the set of time series could be partially labeled.

References

1. Bahadori, M., Kale, D., Yingying, F., Yan, L.: Functional subspace clustering with application to time series. In: Proceedings of ICML, pp. 228–237 (2015)
2. Budalakoti, S., Srivastava, A., Otey, M.: Anomaly detection and diagnosis algorithms for discrete symbol sequences with applications to airline safety. IEEE Trans. Syst. Man Cybern. Part C: Appl. **39**(1), 101–113 (2009)
3. Chandola, V., Mithal, V., Kumar, V.: Comparative evaluation of anomaly detection techniques for sequence data. In: Proceedings of ICDM, pp. 743–748 (2008)
4. Chen, Y., Keogh, E., Hu, B., Begum, N., Bagnall, A., Mueen, A., Batista, G.: The UCR time series classification archive (2015). www.cs.ucr.edu/eamonn/time_series_data/
5. Demšar, J.: Statistical comparisons of classifiers over multiple data sets. J. Mach. Learn. Res. **7**, 1–30 (2006)

6. Ferraty, F., Vieu, P.: Nonparametric Functional Data Analysis: Theory and Practice. Springer, Heidelberg (2006)
7. Görnitz, N., Braun, L., Kloft, M.: Hidden Markov anomaly detection. In: Proceedings of ICML, pp. 1833–1842 (2015)
8. Gupta, M., Gao, J., Aggarwal, C., Han, J.: Outlier detection for temporal data: a survey. IEEE Trans. Knowl. Data Eng. 26(9), 2250–2267 (2014)
9. Hautamaki, T., Nykanen, P., Frant, P.: Time-series clustering by approximate prototypes. In: Proceedings of ICPR, pp. 1–4 (2008)
10. Huang, J., Ng, M., Rong, H., Li, Z.: Automated variable weighting in k-means type clustering. IEEE Trans. Pattern Anal. Mach. Intell. 27, 657–668 (2005)
11. Jing, L., Ng, M., Huang, Z.: An entropy weighting k-means algorithm for subspace clustering of high-dimensional sparse data. IEEE Trans. Knowl. Data Eng. 19(8), 1026–1041 (2007)
12. Lane, T., Brodley, C.: Sequence matching and learning in anomaly detection for computer security. In: AAAI Workshop: AI Approaches to Fraud Detection and Risk Management, pp. 43–49 (1997)
13. Modha, D., Spangler, S.: Feature weighting in k-means clustering. Mach. Learn. 52, 217–237 (2003)
14. Ng, A., Jordan, M., Weiss, Y.: Analysis and an algorithm. In: Proceedings of Neural Information Processing Systems (NIPS), pp. 849–856. MIT Press (2002)
15. Petitjean, F., Forestier, G., Webb, G., Nicholson, A., Chen, Y., Keogh, E.: Dynamic time warping averaging of time series allows faster, more accurate classification. In: Proceedings of ICDM, pp. 470–479 (2014)
16. Portnoy, L., Eskin, E., Stolfo, S.: Intrusion detection with unlabeled data using clustering. In: Proceedings of ACM CSS Workshop on Data Mining Applied to Security (DMSA), pp. 5–8 (2001)
17. Salvador, S., Chan, P.: Toward accurate dynamic time warping in linear time and space. Intell. Data Anal. 11(5), 561–580 (2007)
18. Schölkopf, B., Williamson, R., Smola, A., Shawe-Taylor, J., Platt, J.: Support vector method for novelty detection. In: Proceedings of Neural Information Processing Systems (NIPS), pp. 582–588 (1999)
19. Vintsyuk, T.: Speech discrimination by dynamic programming. Cybernetics 4(1), 52–57 (1968)

Mining Temporal Fluctuating Patterns

Shan-Yun Teng, Cheng-Kuan Ou, and Kun-Ta Chuang[✉]

Department of Computer Science and Information Engineering,
National Cheng Kung University, Tainan, Taiwan
{syteng,ckou}@netdb.csie.ncku.edu.tw, ktchuang@mail.ncku.edu.tw

Abstract. In this paper, we explore a new mining paradigm, called *Temporal Fluctuating Patterns* (abbreviated as *TFP*), to discover potentially fluctuating and useful feature sets from temporal data. These feature sets have some properties which are variant through time series. Once *TFPs* are discovered, we can find the turning points of patterns, which enables anomaly detection and transformation discovery over time. For example, the discovery of *TFPs* can possibly figure out the phenomenon of virus variation during the epidemic outbreak, further providing the government the clue for the epidemic control. However, previous work on mining temporal data computes frequent sets iteratively for different time periods, which is time-consuming. We, therefore, develop a union-based mining structure to speed up the mining process and dynamically compute the fluctuations of patterns through time series. As shown in our experimental studies, the proposed framework can efficiently discover *TFPs* on a real epidemic disease dataset, showing its prominent advantages to be utilized in real applications.

1 Introduction

In many applications, data are generated over time, such as location-based services, stock trading application, and disease reporting services, thus forming so-called temporal databases. Typically, temporal data consist of a sequence of data elements ordered by time. It is believed in the literature that temporal data usually behave with time-variant characteristics, leading to an interest in identifying hidden and evolving knowledge as time advances [6].

The observation on the fluctuation phenomenon is crucial to many applications. Analysis of fluctuation [11] is a required and practical strategy to catch up interesting trends of discovered patterns through time. The fluctuation, in this paper, refers to the difference between the recent property and the past property of a pattern. For example, as illustrated in Fig. 1, the support of pattern p_2 retrieved from an influenza dataset obviously changes temporally. Since the symptom behaves with dynamic fluctuation, it becomes a reliable indicator to sensitively detect the evolving events, such as virus variation. To the best of our knowledge, such phenomenon of pattern fluctuation is left unexplored thus far.

Traditionally, previous work usually attempt to capture the temporal-based frequent patterns [1,12], but the fluctuating factor of frequent patterns is not

© Springer International Publishing AG 2017
J. Kim et al. (Eds.): PAKDD 2017, Part I, LNAI 10234, pp. 773–785, 2017.
DOI: 10.1007/978-3-319-57454-7_60

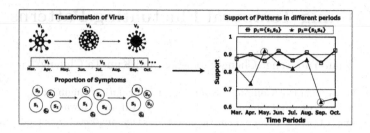

Fig. 1. An illustrative example of two symptom patterns.

seriously considered by these studies. The observation on the fluctuating factor of patterns enables us to discover the turning points of patterns. In this paper, we study a hospital-based dataset recording the physical symptom of patients during the influenza outbreak, attempting to timely capture the virus evolution and helping the government to deploy the intervention strategies, such as school closure or large-scale vaccination campaigns, based on the evolving evidence. We use Fig. 1 as an example to illustrate two symptom patterns $p_1 = \{$headaches (s_1), muscle aches $(s_3)\}$ and $p_2 = \{$runny nose (s_2), high fever $(s_4)\}$ from the example influenza data. The pattern p_2 has two significant fluctuations on May and September. It is reported to be likely to belong as the RNA virus variation, further affecting the first-revealed symptoms of patients. On the other hand, pattern p_1 has no significant fluctuations, which indicates that it includes stable features, headache and muscle aches, for disease implication. These applications of pattern fluctuations inspire us to consider the fluctuating factor into frequent pattern mining.

Specifically, we explore in this paper a practically interesting task, called mining *temporal fluctuating patterns* (abbreviated as *TFPs*), to identify frequent patterns which have significant or insignificant fluctuations in temporal databases. For example, the more significant *TFP* is $p_2 = \{$runny nose (s_2), high fever $(s_4)\}$ in Fig. 1. Alternatively, $p_1 = \{$headaches (s_1), muscle aches $(s_3)\}$ is the more insignificant *TFP*. The discovery of these *TFPs* enables anomaly detection and transformation discovery over time.

In fact, the framework to discover *TFPs* is highly challenging due to two major bottlenecks. First, the fluctuation can be variant, which cannot be represented by a static linear function. The adaptive fluctuation function in different cases should be designed. Second, the traditional frequent pattern mining algorithms applied on *TFPs* will lead to the inefficient situation. These algorithms cause multiple mining tasks to compute frequent sets iteratively for different time periods. Therefore, how to improve the time efficiency of this work is the most important issue. In this paper, we propose a weight based fluctuation function which would be dynamically changed by different fluctuations, and an *Unik* algorithm is devised with an union-based mining structure. As validated in real data, the performance and quality are comprehensively demonstrated the practicability of the proposed framework.

The remainder of this paper is organized as follows. In Sect. 2, we will give the problem definitions. We introduce our system framework in Sect. 3. The experimental results are conducted in Sect. 4. Sections 5 and 6 give the related works and conclusions.

2   Problem Definition

Before formally introducing our framework, we give the necessary definitions as follows.

Definition 1 (A Temporal Event). *A temporal event e_i is denoted by the 3-tuple (u_i, s_i, A_i), where s_i denotes the dispatch time when the temporal event appears, u_i stands for a user u_i, and A_i represents a set of user features $\{a_1, a_2, \ldots, a_j\}$, such as gender, age, and so on.*

For example, a temporal event e_i in a disease dataset can be $(u_1, 30\text{-}08\text{-}2015\ 08{:}12{:}00, \{male, 20, high\ fever, vomiting, headaches\})$. In this paper, we call temporal events as events for short.

Definition 2 (A Temporal Event Set). *Given a user defined unit r, the total data length of time T can be divided into different periods $\{t_1, t_2, \ldots, t_n\}$. Events took place in the j-th period form a temporal event set $E^{(t_j)} = \{e_1^{(t_j)}, e_2^{(t_j)}, \ldots, e_m^{(t_j)}\}$.*

This user defined unit r can be hours, days, weeks, months and years. For example, given a dataset which is collected starting from July and a user defined unit $r=$'week', the temporal event set $E^{(t_1)}$ includes the events appearing during the first week of July. By applying data mining algorithms, patterns can be retrieved from each temporal event set $E^{(t_j)}$, which can be defined as below.

Definition 3 (A Feature Pattern). *A feature pattern $p_i^{(t_j)}$ is defined as the form of user features $\{a_{i,1}^{(t_j)}, a_{i,2}^{(t_j)}, \ldots, a_{i,n}^{(t_j)}\}$, and is retrieved from temporal event set $E^{(t_j)}$ by using the proposed method mentioned in Sect. 3.2.*

For example, a feature pattern $p_1^{(t_1)}$ discovered from the temporal event set $E^{(t_1)}$ in a disease dataset can be $\{male, 20, high\ fever\}$. In this paper, we call feature patterns as patterns for short. In addition, the support value of a pattern $p_i^{(t_j)}$ is the proportion of occurrence count of this pattern p_i in period t_j. A top-k pattern set $K^{(t_j)}$, denoted by $\{p_1^{(t_j)}, p_2^{(t_j)}, \ldots, p_k^{(t_j)}\}$, consists of k patterns which have top-k support values during the j-th period. In this paper, we state the top-k pattern set as k-set.

Every pattern p_i has its support value set $Q_i = \{q_i^{(t_1)}, q_i^{(t_2)} \ldots, q_i^{(t_j)}\}$ which equals to the proportion of occurrence count in event sets during different periods. With these temporal support values, we give the definition of temporal fluctuation.

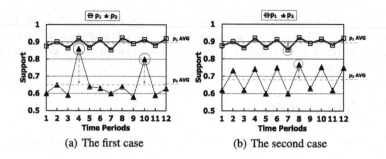

(a) The first case   (b) The second case

Fig. 2. The temporal fluctuation of different patterns.

Definition 4 (Temporal Fluctuation). *The fluctuation $f_i^{(t_j)}$ of pattern p_i between the j-th period and the $(j-1)$-th period is the difference of the support values $q_i^{(t_j)}$ and $q_i^{(t_{j-1})}$. The fluctuation set $F(p_i)$ of pattern p_i can be defined as*

$$F(p_i) = \{f_i^{(t_j)}|f_i^{(t_j)} = |q_i^{(t_j)} - q_i^{(t_{j-1})}|, 2 \le j \le n\}. \tag{1}$$

The temporal fluctuation $\mathcal{F}_{p_i}$ of pattern p_i can be defined as $\frac{1}{n-1}\sum_{j=2}^{n} f_i^{(t_j)}$, which is the average fluctuation of fluctuation set $F(p_i)$.

As shown in Fig. 2(a), the pattern p_2 has two huge peaks which are produced by significant fluctuations as illustrated as the red dash line. Such peaks should be highlighted since they do not appear as frequently as others. Therefore, we use the standard deviation $\sigma_{F(p_i)}$ of the fluctuation set $F(p_i)$ to define the weight of different fluctuations.

$$\sigma_{F(p_i)} = \sqrt{\frac{1}{n-1}\sum_{j=2}^{n}(f_i^{(t_j)} - \mu_{F(p_i)})^2}, \quad \mu_{F(p_i)} = \frac{1}{n-1}\sum_{j=2}^{n} f_i^{(t_j)}. \tag{2}$$

However, for both the patterns p_1 and p_2 shown in Fig. 2(b), their temporal fluctuations have no obvious peaks. It is obvious that the temporal fluctuation of pattern p_2 is bigger than pattern p_1, but deviation $\sigma_{F(p_i)}$ cannot draw such difference.

Lemma 1. *Suppose that all fluctuations of p_2 are the fluctuations of p_1 plus a difference Δx, the standard deviation $\sigma_{F(p_2)}$ of pattern p_2 is the same as the one $\sigma_{F(p_1)}$ of pattern p_1.*

Proof. The standard deviation $\sigma_{F(p_2)}$ of pattern p_2 is

$$\sigma_{F(p_2)} = \sigma_{F(p_1, \Delta x)} = \sqrt{\frac{1}{n-1}\sum_{j=2}^{n}((f_i^{(t_j)} + \Delta x) - (\mu_{F(p_i)} + \Delta x))^2} \tag{3}$$

$$= \sqrt{\frac{1}{n-1}\sum_{j=2}^{n}(f_i^{(t_j)} - \mu_{F(p_i)})^2} = \sigma_{F(p_1)}.$$

To deal with this problem, we need to make the function of temporal fluctuation strengthen the effect of huge fluctuations. Therefore, we redefine the temporal fluctuation $\mathcal{F}_{p_i}$ of pattern p_i as

$$\mathcal{F}_{p_i} = \frac{1}{n-1} \sum_{j=2}^{n} w_i^{(t_j)} \times f_i^{(t_j)}, \ where \ w_i^{(t_j)} = \frac{f_i^{(t_j)}}{\sqrt{\frac{1}{n-1} \sum_{j=2}^{n} (f_i^{(t_j)} - \mu_{F(p_i)})^2}}. \tag{4}$$

Problem Formulation (*Significant and Insignificant Temporal Fluctuating Patterns Discovery*): Suppose that a significant/insignificant temporal fluctuating pattern (abbreviated as *TFP* in the sequel) is defined as the form $p_i = \{a_{i,1}, a_{i,2}, \ldots, a_{i,j}\}$. Given a database D, desired number k, unit of time r, and the thresholds of temporal fluctuation, δ_a for significant one, and δ_b for insignificant one, the goal of our framework is to discover patterns which have significant/insignificant temporal fluctuation in the k-sets $K^{(t_j)}$, $1 \le j \le n$, where n is the number of periods. The discovery of significant and insignificant temporal fluctuating patterns over database D returns the result sets R_a and R_b:

$$\begin{cases} R_a(D) = \{p_i | \mathcal{F}_{p_i} \ge \delta_a \wedge p_i \in \bigcup_{j=1}^{n} K^{(t_j)}\}. \\ R_b(D) = \{p_i | \mathcal{F}_{p_i} \le \delta_b \wedge p_i \in \bigcup_{j=1}^{n} K^{(t_j)}\}. \end{cases} \tag{5}$$

3 Proposed Method

In this section, a naive generation (NG) is introduced as a straightforward way to discover *TFPs*. Then, the *Unik* approach is proposed as an improved solution to effectively and efficiently retrieve the significant and insignificant *TFPs*.

3.1 Naive Generation (NG)

The NG solution is devised based on the Apriori algorithm [2] to separately retrieve different k-sets from temporal event sets $\{E^{(t_1)}, E^{(t_2)}, \ldots, E^{(t_n)}\}$. After multiple mining tasks are executed, the obtained k-sets $\{K^{(t_1)}, K^{(t_2)}, \ldots, K^{(t_n)}\}$ are used to construct a union set $U = \bigcup_{j=1}^{n} K^{(t_j)}$, and the temporal fluctuation $\mathcal{F}_{p_i}$ of each pattern $p_i \in U$ is computed. However, the multiple mining tasks to compute k-sets iteratively for different time periods is time consuming. In addition, to independently retrieve different k-sets from temporal event sets causes a major problem: the algorithm cannot decide which generated pattern p_i can be discarded in the mining process. Since the algorithm cannot make sure if a pattern $p_i \in K^{(t_j)}$ may appear in another k-set $K^{(t_z)}$, where $z \ne j$, it needs to memorize all the generated patterns and the support of these patterns. To efficiently discover the *TFPs*, we propose the *Unik* generation.

3.2 Unik Generation

In this paper, the proposed *Unik* generation tries to produce n-item set of pattern candidates from the **Union** set U_{n-1} including **(n-1)**-items x_i in the **k**-set $K^{(t_j)}$ on the $(n-1)$-th iteration. This algorithm includes three major steps:

Step 1 (Generate): *Unik* generates n-item set X_n from the union set U_{n-1} produced on the $(n-1)$-th iteration, where

$$X_n = \{x_a \cup x_b | x_a, x_b \in U_{n-1} \wedge |x_a|, |x_b| = n-1\} - \{x_c | \{x_d \subseteq x_c \wedge |x_d| = n-1\} \nsubseteq U_{n-1}\},$$

(6)

and computes the support value set $Q_i = \{q_i^{(t_1)}, q_i^{(t_2)} \ldots, q_i^{(t_j)}\}$ of n-items $x_i \in X_n$.

Step 2 (Update): *Unik* updates all $K^{(t_j)}$ (for each $t_j \in T$) which include 1-items to $(n-1)$-items. If each n-item $x_i \in X_n$ can be inserted into these k-sets $K^{(t_j)}$ is independently checked in the algorithm.

$$K^{(t_j)} = \begin{cases} (K^{(t_j)} \cup x_i) - x_z, & if \ q_{x_i}^{(t_j)} > q_{x_z}^{(t_j)} \wedge x_z = \arg\min_{x_m} q_{x_m}^{(t_j)}, \ x_m \in K^{(t_j)}. \\ K^{(t_j)}, & otherwise. \end{cases}$$

(7)

Step 3 (Union): Finally, *Unik* unions the items in k-set $K^{(t_j)}$ on the n-th iteration of different periods $\{t_1, \ldots, t_m\}$ to form a new union set U_n.

$$U_n = \bigcup_{j=1}^{m} K^{(t_j)}.$$

(8)

The *Unik* generation iteratively executes these three steps until $X_n = \emptyset$. However, to apply the *Unik* generation, we need to prove that the patterns in k-sets can be derived from the union sets $\{U_1, \ldots, U_n\}$ on each iteration.

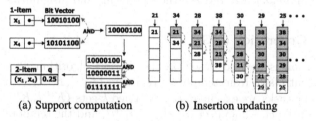

(a) Support computation   (b) Insertion updating

Fig. 3. Illustrative examples of support computation and insertion updating.

Lemma 2. *If an n-item x belongs to k-set $K^{(t_j)}$, it can be generated from the union set U_{n-1} on the $(n-1)$-th iteration and can be inserted into the k-set $K^{(t_j)}$ on the n-th iteration.*

Proof. Let $x = \{a_{i,1}^{(t_j)}, \ldots a_{i,2}^{(t_j)} \ldots, a_{i,n}^{(t_j)}\}$ and $x' \in x$ is an $(n-1)$-item. Given $x \in K^{(t_j)} = \{p_1^{(t_j)}, p_2^{(t_j)}, \ldots, p_k^{(t_j)}\}$, we have $q_x^{(t_j)} \geq q_{p_k}^{(t_j)}$ and $q_{x'}^{(t_j)} \geq q_x^{(t_j)}$. Therefore, $x' \in K^{(t_j)} \in U_{n-1}$. Similarly, given two $(n-1)$-items x', $x'' \in x$, we have x', $x'' \in K^{(t_j)} \in U_{n-1}$, so $x \in X_n$. Finally, as we know $q_x^{(t_j)} \geq q_{p_k}^{(t_j)}$, $x \in K^{(t_j)}$.

As proved in Lemma 2, it is applicable to retrieve the n-items $x_i \in K^{(t_j)}$ from the union set U_n on the n-th iteration.

Support Computation Using Bit Vectors: In order to speed up the support computation of the generated n-items $x_i \in X_n$, we use a bit vector $I_{x_i}^{(t_j)}$ associated with event IDs of the item candidate x_i in the j-th period. The length of $I_{x_i}^{(t_j)}$ is $|E^{(t_j)}|$, and the l-th bit of I_{x_i} is 1 if and only if the event $e_l^{(t_j)}$ consists of the item candidate x_i. Finally, the support value $q_{x_i}^{(t_j)}$ of an n-item x_i generated by two $(n-1)$-items, x_a and x_b, can be computed as

$$q_{x_i}^{(t_j)} = (\sum_{l=1}^{|I_{x_a}^{(t_j)}|} (i_l \ AND \ 1))/|I_{x_a}^{(t_j)}|, \ i_j \in (I_{x_a}^{(t_j)} \ AND \ I_{x_b}^{(t_j)}), \tag{9}$$

where AND is the operation of bitwise AND.

However, the intersection vector $I_n = (I_{x_a}^{(t_j)} \ AND \ I_{x_b}^{(t_j)})$ is actually sparse, and should be designed for running in time proportional to the number of 1 bits. Therefore, a mystical operator $(I_n \ AND \ (I_n - 1))$ is used to iteratively set the rightmost 1 bit in I_n to 0, which is shown in Fig. 3(a), and the support value $q_{x_i}^{(t_j)}$ of an n-item x_i is rewritten as

$$q_{x_i}^{(t_j)} = (\sum_{I_n=(I_n \ AND \ (I_n-1))}^{I_n=0} 1)/|I_n|. \tag{10}$$

Insertion Updating of k-sets: Given a sorted k-set $K^{(t_j)} = \{x_1, x_2 \ldots, x_k\}$ storing k items with high support values, the insertion updating only needs to check the support value $q_{x_i}^{(t_j)}$ of n-item $x_i \in X_n$ against **min** $q_{x_z}^{(t_j)}, 1 \le z \le k$ in each iteration. If support value $q_{x_i}^{(t_j)}$ is smaller than **min** $q_{x_z}^{(t_j)}$, n-item x_i can be discarded. Conversely, the insertion updating finds the correct order of n-item x_i within the sorted k-set $K^{(t_j)}$, discards **arg min** $q_{x_z}^{(t_j)}$, and then k-set $K^{(t_j)}$ is updated. An illustrative example is shown in Fig. 3(b). In this way, each n-item compares at least 1 time and at most k times, and count c_i of comparison for l n-items is $(l \times 1) \le \sum_{i=1}^{l} c_i \le (l \times k)$. Without the insertion updating on k-set, each n-item compares at least k times and at most l times, and the times of comparison would be $(l \times k) \le \sum_{i=1}^{l} c_i \le (l \times (l-1))$, where $(0 \le k \ll l)$.

3.3 TFPs Discovery

To discover significant and insignificant *TFPs*, our method retrieves the temporal fluctuations $\mathcal{F}_{p_i}$ of patterns $p_i \in U_n$ as defined in Definition 4, and returns the patterns with $\mathcal{F}_{p_i} \ge \delta_a$ or $\mathcal{F}_{p_i} \le \delta_b$. The computation of temporal fluctuation is straightforward. However, as can be expected, temporal data are continuously generated from applications in the future, and so that the temporal fluctuation of each pattern is not static. Therefore, it is not efficient to recompute new temporal fluctuations of patterns every time when new data come into the queue. Given m patterns and n time periods, the time complexity to update temporal fluctuations of patterns is $O(n^2)$. Therefore, in this section we derive how to efficiently update the old temporal fluctuation $\mathcal{F}_{p_i}$ of pattern p_i with added fluctuation $f_i^{(t_{n+1})}$. We start the derivation with the updating of the average $\mu_{F(p_i,f_i^{(t_{n+1})})}$, which can be derived as

$$
\begin{aligned}
\mu_{F(p_i,f_i^{(t_{n+1})})} &= \frac{1}{(n+1)-1} \sum_{j=2}^{n+1} f_i^{(t_j)} = \frac{1}{n}(n-1)\frac{1}{n-1}\left(\sum_{j=2}^{n}(f_i^{(t_j)}) + f_i^{(t_{n+1})}\right) \\
&= \frac{n-1}{n}\mu_{F(p_i)} + \frac{1}{n}f_i^{(t_{n+1})}.
\end{aligned}
\tag{11}
$$

The equation of standard deviation $\sigma_{F(p_i)}$ can be simplified as

$$
\sigma_{F(p_i)} = \sqrt{\frac{1}{n-1}\sum_{j=2}^{n}(f_i^{(t_j)} - \mu_{F(p_i)})^2} = \sqrt{\frac{\sum_{j=2}^{n} f_i^{(t_j)2}}{n-1} - \mu_{F(p_i)}^2},
\tag{12}
$$

and so that the new standard deviation $\sigma_{F(p_i,f_{n+1})}$ can be derived as

$$
\begin{aligned}
\sigma_{F(p_i,f_i^{(t_{n+1})})} &= \sqrt{\frac{1}{(n+1)-1}\sum_{j=2}^{n+1} f_i^{(t_j)2} - \mu_{F(p_i,f_i^{(t_{n+1})})}^2} \\
&= \sqrt{\frac{1}{n}(n-1)\frac{1}{n-1}\left(\sum_{j=2}^{n} f_i^{(t_j)2} + f_i^{(t_{n+1})2}\right) - \mu_{F(p_i,f_i^{(t_{n+1})})}^2} \\
&= \sqrt{\frac{n-1}{n}(\sigma_{F(p_i)}^2 + \mu_{F(p_i)}^2) + \frac{1}{n}f_i^{(t_{n+1})2} - \mu_{F(p_i,f_i^{(t_{n+1})})}^2}.
\end{aligned}
\tag{13}
$$

Finally, the temporal fluctuation $\mathcal{F}_{(p_i,f_{n+1})}$ can be written as

$$
\begin{aligned}
\mathcal{F}_{(p_i,f_i^{(t_{n+1})})} &= \frac{1}{(n+1)-1}\sum_{j=2}^{n+1}\frac{f_i^{(t_j)2}}{\sigma_{F(p_i,f_i^{(t_{n+1})})}} \\
&= \frac{1}{n} \times \frac{\sum_{j=2}^{n} f_i^{(t_j)2} + f_i^{(t_{n+1})2}}{\sigma_{F(p_i,f_i^{(t_{n+1})})}} = \frac{n-1}{n} \times \frac{\sigma_{F(p_i)}\mathcal{F}_{p_i} + f_i^{(t_{n+1})2}}{\sigma_{F(p_i,f_i^{(t_{n+1})})}}.
\end{aligned}
\tag{14}
$$

Therefore, the time complexity to update the temporal fluctuation of all patterns is reduced to $O(n)$, instead of $O(n^2)$.

4   Experimental Results

This section presents experimental studies. Our system framework is implemented in Python. All of the experiments are executed on a 3.40 GHz Core i7 machine with 4 GB of main memory, running on Windows 7 operating system.

4.1   Experimental Setup

Dataset Description: We exploit real data, which records the physical symptoms of all patients seeking for the medical cure during the epidemic outbreak.

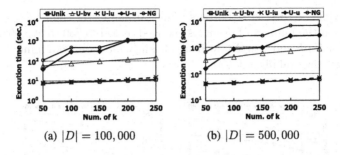

Fig. 4. The execution time analysis (when k is varied from 50 to 250).

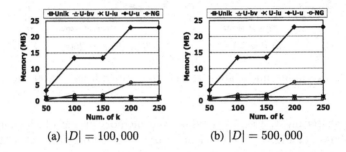

Fig. 5. The memory usage analysis (when k is varied from 50 to 250).

Specifically, the detailed information includes the home location (with the loose precision), age, gender, symptoms and onset time. More than 20000 patients are reported to be infected during five months (from July 2015 to Nov. 2015) in the urban area.

Other Methods: In our *Unik* algorithm, we use union (u) based generation, bit vector (bv) based support computation, and insertion updating (iu) of k-sets. In our experiment, we compare the effect of these three techniques with *Unik* by removing each of them from *Unik*, where these comparing methods are named U-u, U-bv, and U-iu, respectively.

4.2 Experiments on Real Dataset

In this section, the proposed *Unik*, U-u, U-bv, and U-iu methods are applied on real data recording physical symptoms of patients.

4.2.1 Effect of k: We first evaluate the execution time of different methods by the effect of parameter k of k-sets. We fix the number of events to 100,000 and 500,000, set the number of features to 40, and vary the number of k from 50 to 250. The results on the execution time of *TFPs* discovery are shown in Fig. 4. These figures are plotted in the log scale. As can be seen, trends of execution time all consistently increase when the number of k grows. Moreover, the efficiency ranking of these methods is *Unik*, U-iu, U-bv, U-u, and NG in sequence. The execution time of *Unik* is 1–2 orders of magnitude faster than that of U-u/NG. Finally, when the number of events is changed from 100,000 to 500,000, the time difference of *Unik* and U-u/NG will significantly increase from minutes to hours, which shows the stability of *Unik*.

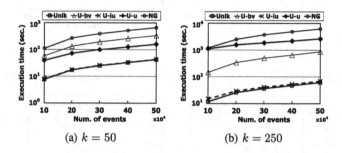

(a) $k = 50$ (b) $k = 250$

Fig. 6. The execution time analysis (when $|D|$ is varied from 100,000 to 500,000).

In this experiment, we also analyze the memory usage of these methods, which is shown in Fig. 5. The efficiency ranking is *Unik*, U-iu, U-bv, NG, and U-u in sequence. Since U-u uses the same generation process as NG, and applies bit vectors to save its execution time, its memory usage is more than that of NG. On the other hand, we can also observe that the memory usage of all methods is not affected by the number of events at all.

4.2.2 Effect of Event Quantity: We also evaluate the execution time of different methods by the effect of event quantity. We fix the number of k to 50 and 250, set the number of features to 40, and vary the number of events from 100,000 to 500,000. The results are shown in Fig. 6. Obviously, all trends of execution time increase when the number of events grows. Similar to the previous experiment, *Unik* is still the most efficient method, and it can significantly improve the execution time. However, the efficiency order of U-u and U-bv exchanges as k equals 50 and 250, which indicates that U-u is dramatically affected by factor k. Finally, though the efficiency of *Unik* and U-iu is close in the log scale, they differ from each other as the k is set as 250. In this experiment, the memory usage of *TFPs* generation is not analyzed, since it cannot be affected by the number of events as aforementioned.

4.2.3 Effect of Feature Quantity: In the epidemic disease dataset, four feature sets, symptoms (P_1), locations (P_2), age ranges (P_3), and gender (P_4), are obtained. Therefore, in this experiment, we evaluate the execution time and memory usage of different methods by the effect of feature quantity. We fix the number of k to 150 and 250, set the number of events to 500,000. The results are shown in Figs. 7 and 8. It is obvious that *Unik* algorithm is still the most efficient and saves lots of memory, comparing to other methods. All trends of execution time and memory increase when the number of features grows. Therefore, these

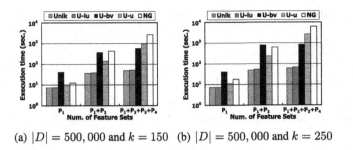

(a) $|D| = 500,000$ and $k = 150$ (b) $|D| = 500,000$ and $k = 250$

Fig. 7. The execution time analysis (when using different feature sets).

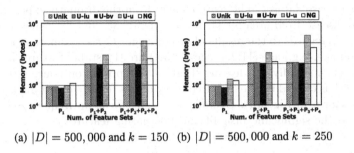

(a) $|D| = 500,000$ and $k = 150$ (b) $|D| = 500,000$ and $k = 250$

Fig. 8. The memory usage analysis (when using different feature sets).

experiments show that the *Unik* algorithm saves a huge amount of time and memory, which is more proper in our work.

4.2.4 Results of *TFPs* Mining: We aim to discuss different kinds of patterns from the real dataset. As shown in Table 1, we demonstrate the top-5 2-feature patterns for frequent patterns, significant *TFPs*, and insignificant *TFPs*. Obviously, the frequent patterns are usually not the most significant/insignificant *TFPs*, and so that it is necessary to develop a novel system to discover *TFPs*.

Firstly, let's take a look at the results of significant *TFPs*. From these patterns, we can observe that locations appear frequently on significant *TFPs* in

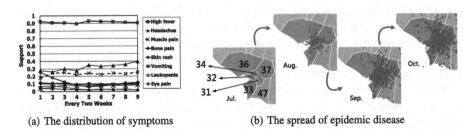

(a) The distribution of symptoms   (b) The spread of epidemic disease

Fig. 9. The distribution of symptoms through time series and the spread situation of epidemic disease.

Table 1. Top-5 2-feature patterns in epidemic disease dataset.

	Frequent patterns	Significant *TFPs*	Insignificant *TFPs*
1	{High fever, Female}	{Male, Area no.31}	{Muscle pain, Male}
2	{High fever, Male}	{High fever, Area no.31}	{Muscle pain, Female}
3	{High fever, Area no.34}	{High fever, Area no.33}	{High fever, Male}
4	{High fever, Muscle pain}	{Female, Area no.31}	{High fever, Age [11,20]}
5	{High fever, Headaches}	{High fever, Area no.37}	{High fever, Age [41,50]}

epidemic disease dataset. The reason comes from the dramatic expansion of this epidemic disease from area no.34 to areas no.31, 33, and 37, which is shown in Fig. 9(b). As regards to the results of insignificant *TFPs*, symptoms (e.g., muscle ache), and age (e.g., age [11,20], age [41,50]) are the stable features in epidemic disease events. Finally, the feature 'High fever' appears on many retrieved patterns, which is caused by the high frequency of this feature in our disease dataset, which is shown in Fig. 9(a). Overall, if we only focus on frequent patterns, we would miss the significant/insignificant fluctuating features.

5   Related Work

In this section, we review the literature related to the changes of patterns and then review the existing pattern discovery approaches.

At the beginning of patterns discovery, most researches try to extract frequent patterns on spatio-temporal data [7], streaming data [15], and non-identity data [3]. Currently, fluctuation becomes an important characteristic in temporal datasets. There exist many approaches for mining three kinds of fluctuating patterns. First of all, some work [4,5] focus on exploring interesting trend patterns. Their goal is to discover the most significant fluctuations in multidimensional spaces, so that the fluctuations in their paper are the gradient between two temporal series. Secondly, a few studies [9,13] discussed the fluctuating rules. They extracted specific rules of pattern changes over data streams. In these studies, structural changes are highlighted to provide better rules to developers. Finally, many researchers [8,10,14] pay attention to the sequential pattern changes mining. These researchers tried to detect the change by incremental mining of sequential patterns in a stream sliding window. Therefore, the fluctuations in these papers are represented as the difference between the current mining results and cumulative mining results.

However, none of these studies considers the changes of a single pattern between different time instants and takes the fluctuation factor into frequent pattern mining. These studies are orthogonal to our work.

6   Conclusions

In this paper, we propose a novel Temporal Fluctuating Patterns (*TFPs*) discovery on temporal databases and devise a system framework to retrieve the significant and insignificant *TFPs*. The union-based mining process, namely the *Unik* algorithm, bit vector based support computation, and insertion updating of k-sets, are proposed to improve the efficiency of our system framework. Moreover, we implement our method for the case studies on real disease datasets. The experimental results show some particular phenomenon and demonstrate that the proposed system framework is efficient and practical.

Ackowledgement. This paper was supported in part by Ministry of Science and Technology, R.O.C., under Contract 105-2221-E-006-140-MY2 and 105-2634-E-006-001.

References

1. Aggarwal, C.C., Han, J.: Frequent Pattern Mining. Springer, Heidelberg (2014)
2. Agrawal, R., Srikant, R.: Fast algorithms for mining association rules in large databases. In: VLDB (1994)
3. Almanie, T., Mirza, R., Lor, E.: Crime prediction based on crime types and using spatial and temporal criminal hotspots. CoRR, abs/1508.02050 (2015)
4. Alves, R., Belo, O., Ribeiro, J.: Mining top-K multidimensional gradients. In: Song, I.Y., Eder, J., Nguyen, T.M. (eds.) DaWaK 2007. LNCS, vol. 4654, pp. 375–384. Springer, Heidelberg (2007). doi:10.1007/978-3-540-74553-2_35
5. Alves, R., Ribeiro, J., Belo, O.: Mining significant change patterns in multidimensional spaces. Int. J. Bus. Intell. Data Min. **4**, 219–241 (2009)
6. Gaber, M.M., Zaslavsky, A., Krishnaswamy, S.: Mining data streams: a review. ACM Sigmod Rec. **34**, 18–26 (2005)
7. Giannotti, F., Nanni, M., Pinelli, F., Pedreschi, D.: Trajectory pattern mining. In: SIGKDD (2007)
8. Ho, C.-C., Li, H.-F., Kuo, F.-F., Lee, S.-Y.: Incremental mining of sequential patterns over a stream sliding window. In: Workshops Proceedings of ICDM (2006)
9. Hora, A.C., Anquetil, N., Ducasse, S., Valente, M.T.: Mining system specific rules from change patterns. In: WCRE (2013)
10. Li, I.-H., Huang, J.-Y., Liao, I.-E.: Mining sequential pattern changes. J. Inf. Sci. Eng. **30**, 973–990 (2014)
11. Liu, B., Hsu, W., Han, H.-S., Xia, Y.: Mining changes for real-life applications. In: Kambayashi, Y., Mohania, M., Tjoa, A.M. (eds.) DaWaK 2000. LNCS, vol. 1874, pp. 337–346. Springer, Heidelberg (2000). doi:10.1007/3-540-44466-1_34
12. Liu, Y., Zhao, Y., Chen, L., Pei, J., Han, J.: Mining frequent trajectory patterns for activity monitoring using radio frequency tag arrays. IEEE Trans. Parallel Distrib. Syst. **23**, 2138–2149 (2012)
13. Lo, D., Ramalingam, G., Ranganath, V.P., Vaswani, K.: Mining quantified temporal rules: formalism, algorithms, and evaluation. Sci. Comput. Program. **77**, 743–759 (2012)
14. Tsai, C.-Y., Shieh, Y.-C.: A change detection method for sequential patterns. Decis. Support Syst. **46**, 501–511 (2009)
15. Yang, D., Rundensteiner, E.A., Ward, M.O.: Shared execution strategy for neighbor-based pattern mining requests over streaming windows. ACM Trans. Database Syst. **37**, 5:1–5:44 (2012)

Marked Temporal Dynamics Modeling Based on Recurrent Neural Network

Yongqing Wang[✉], Shenghua Liu, Huawei Shen, Jinhua Gao,
and Xueqi Cheng

CAS Key Laboratory of Network Data Science and Technology,
Institute of Computing Technology, Chinese Academy of Sciences, Beijing, China
{wangyongqing,gaojinhua}@software.ict.ac.cn,
{liushenghua,shenhuawei,cxq}@ict.ac.cn

Abstract. We are now witnessing the increasing availability of event stream data, i.e., a sequence of events with each event typically being denoted by the time it occurs and its mark information (e.g., event type). A fundamental problem is to model and predict such kind of marked temporal dynamics, i.e., when the next event will take place and what its mark will be. Existing methods either predict only the mark or the time of the next event, or predict both of them, yet separately. Indeed, in marked temporal dynamics, the time and the mark of the next event are highly dependent on each other, requiring a method that could simultaneously predict both of them. To tackle this problem, in this paper, we propose to model marked temporal dynamics by using a *mark-specific* intensity function to explicitly capture the dependency between the mark and the time of the next event. Experiments on two datasets demonstrate that the proposed method outperforms the state-of-the-art methods at predicting marked temporal dynamics.

Keywords: Marked temporal dynamics · Recurrent neural network · Event stream data

1 Introduction

There is an increasing amount of event stream data, i.e. a sequence of events with each event being denoted by the time it occurs and its mark information (e.g. event type). Marked temporal dynamics offers us a way to describe this data and potentially predict events. For example, in microblogging platforms, marked temporal dynamics could be used to characterize a user's sequence of tweets containing the posting time and the topic as mark [9]; in location based social networks, the trajectory of a user gives rise to a marked temporal dynamics, reflecting the time and the location of each check-in [15]; in stock market, marked temporal dynamics corresponds to a sequence of investors' trading behaviors, i.e., bidding or asking orders, with the type of trading as mark [4]; An ability to predict marked temporal dynamics, i.e., predicting when the next event will take place and what its mark will be, is not only fundamental to understanding

© Springer International Publishing AG 2017
J. Kim et al. (Eds.): PAKDD 2017, Part I, LNAI 10234, pp. 786–798, 2017.
DOI: 10.1007/978-3-319-57454-7_61

the regularity or patterns of these underlying complex systems, but also has important implications in a wide range of applications, from viral marketing and traffic control to risk management and policy making.

Existing methods for this problem fall into three main paradigms, each with different assumptions and limitations. The first category of methods focuses on predicting the mark of the next event, formulating the problem as a discrete-time or continuous-time sequence prediction task [12,25]. These methods gained success at modeling the transition probability across marks of events. However, they lack the power at predicting when the next event will occur.

The second category of methods, on contrary, aims to predict when the next event will occur [10]. These methods either exploit temporal correlations for prediction [20,22] or conduct prediction by modeling the temporal dynamics using certain temporal process, such as self-exciting Hawkes process [2,6], various Poisson process [9,21], and other auto-regressive processes [8,16]. These methods have been successfully used in modeling and predicting temporal dynamics. However, these models are unable to predict the mark.

In recent years, researchers attempt to directly model the marked temporal dynamics [11]. A recent work [7] used recurrent neural network to automatically learn history embedding, and then predict both, yet separately, the time and the mark of the next event. This work assumes that time and mark are independent on each other given the historical information. Yet, such assumption fails to capture the dependency between the time and the mark of the next event. For example, when you have lunch is affected by your choice on restaurants, since different restaurants imply difference in geographic distance and quality of service. The separated prediction by maximizing the probability on mark and time does not imply the most likely event. In sum, we still lack a model that could capture the interdependency of mark and time when predicting the next event.

In this paper, we propose a novel model based on recurrent neural network (RNN), named RNN-TD, to capture the dependence between the mark of an event and its occurring time. The key idea is to use a mark-specific intensity function to model the occurring time for events with different marks. The benefits of our proposed model are three-fold: (1) It models the mark and the time of the next event simultaneously; (2) The mark-specific intensity function explicitly captures the dependency between the occurring time and the mark of an event; (3) The involvement of RNN simplifies the modeling of dependency on historical events.

We evaluate the proposed model by extensive experiments on large-scale real world datasets from Memetracker[1] and Dianping[2]. Compared with several state-of-the-art methods, RNN-TD outperforms them at prediction of marks and times. We also conduct case study to explore the capability of event prediction in RNN-TD. The experimental results indicate that it can better model marked temporal dynamics.

---

[1] http://www.memetracker.org.

[2] http://www.dianping.com.

2 Model

In this paper, we focus on the problem of modeling marked temporal dynamics. Before diving into the details of the proposed model, we first clarify two main motivations underlying our model.

2.1 Motivation

In real scenarios, mark and time of next event are highly dependent on each other. We use a case from Dianing to illustrate this phenomenon. We extract the trajectories starting from the same location (mark #6) and examine if the time interval between two consecutive events are discriminative to each other with respect to different marks. The distribution of time interval with different target marks are represented in Fig. 1(a). We can observe that large variance exists in the distributions when consumers make different choices. This motivates us to model mark-specific temporal dynamics.

Second, existing works [12] attempted to formulate marked temporal dynamics by Markov random processes with varying orders. However, the generation of next event requires strong prior knowledge on dependency of history. Besides, long dependency on history causes state-space explosion problem in practice. Therefore, we propose a RNN-based model which learns the dependency by deep structure. It embeds history information into vectorized representation when modeling sequences. The generation of next event is only dependent on history embedding.

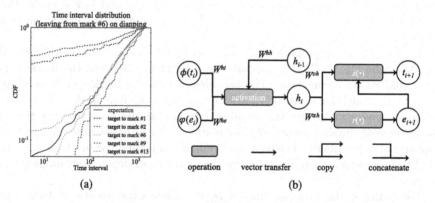

(a)   (b)

Fig. 1. (a) High variance existed in time interval distribution when targeting to different marks. (b) The architecture of RNN-TD. Given the event sequence $S = \{(t_i, e_i)\}_{i=1}$, the i-th event (t_i, e_i) is mapped through function $\phi(t)$ and $\varphi(e)$ into vector spaces as inputs in RNN. Then the inputs $\phi(t_i)$ and $\varphi(e_i)$ associated with the last embedding h_{i-1} are fed into hidden units in order to update h_i. Dependent on embedding h_i, RNN-TD outputs the next event type e_{i+1} and correspondent time t_{i+1}.

2.2   Problem Formulation

An event sequence $S = \{(t_i, e_i)\}$ is a set of events in ascending order of time. The tuple (t_i, e_i) records the i-th event in the sequence S, and the variables $t_i \in \mathcal{T}$ and $e_i \in \mathcal{E}$ denote the time and the mark respectively, where $\mathcal{E}$ is a countable state space including all possible marks and $\mathcal{T} \in \mathbb{R}^+$ is the time space in which observed marks take place. We could have various instantiation in different applications.

The likelihood of an observed sequence S can be written as

$$P(S) = \prod_{i=1}^{|S|} p(t_i, e_i | H_{t_i}),$$

where $H_{t_i} = \{(t_l, e_l) | t_l < t_i, e_l \in \mathcal{E}\}$ refers to all the historical events occurring before t_i. In practice, the joint probability of a pair of mark and time can be written by Bayesian rule as follows

$$p(t_i, e_i | H_{t_i}) = r(e_i | H_{t_i}) s(t_i | e_i, H_{t_i}), \tag{1}$$

where $r(e_i | H_{t_i})$ refers to the probability that the mark of next event is e_i and $s(t_i | e_i, H_{t_i})$ is the probability distribution function of time given a specific mark.

Next we propose a general model to parameterize $r(e_i | H_{t_i})$ and $s(t_i | e_i, H_{t_i})$ in marked temporal dynamics modeling, named RNN-TD. Recurrent neural network (RNN) is a feed-forward neural network for modeling sequential data. In RNN, the current inputs are fed into hidden units by nonlinear transformation, jointly with the outputs from the previous hidden units. The feed-forward architecture is replicative in both inputs and outputs so that the representation of hidden units is dependent on not only current inputs but also encoded historicial information. The adaptive size of hidden units and nonlinear activation function (e.g., sigmoid, tangent hyperbolic or rectifier function) make neural network capable of approximating arbitrary complex function [3].

The architecture of RNN-TD is depicted in Fig. 1(b). The inputs of an event (t_i, e_i) is vectorized by mapping function $\phi(\cdot)$ and $\varphi(\cdot)$. Then the i-th inputs associated with the last embedding h_{i-1} are fed into hidden units in order to update h_i. Given the i-th event (t_i, e_i), the embedding h_{i-1} and mapping function ϕ and φ, the representation of hidden units in RNN-TD can be calculated as

$$h_i = \sigma\left(W^{ht}\phi(t_i) + W^{he}\varphi(e_i) + W^{hh}h_{i-1}\right), \tag{2}$$

where σ is the activation function, and W^{ht}, W^{he} and W^{hh} are weight matrices in neural network. The procedure is iteratively executed until the end of sequence. Thus, the embedding h_i encodes the i-th inputs and the historical context h_{i-1}.

Based on the history embedding h_i, we can derive the probability of the $(i+1)$-th event in an approximative way,

$$p(t_{i+1}, e_{i+1} | H_{t_{i+1}}) \approx p(t_{i+1}, e_{i+1} | h_i) = r(e_{i+1} | h_i) s(t_{i+1} | e_{i+1}, h_i). \tag{3}$$

Firstly we formalize the conditional transition probability $r(e_{i+1}|h_i)$. The conditional transition probability can be derived by a softmax function which is commonly used in neural network for parameterizing categorical distribution, that is,

$$r(e_{i+1}|h_i) = \frac{\exp\left(W_k^{\alpha h} h_i\right)}{\sum_{j=1}^{K} \exp\left(W_j^{\alpha h} h_i\right)}, \qquad (4)$$

where row vector $W_k^{\alpha h}$ is k-th row of weight matrix indexed by the mark e_{i+1}.

Then we consider the probability distribution function $s(t_{i+1}|e_{i+1}, h_i)$. The probability distribution function describes *the observation that nothing but mark e_{i+1} occurred until time t_{i+1} since the last event.* We define a random variable T_e as the occuring time of next event with mark e, and the probability distribution function $s(t_{i+1}|e_{i+1}, h_i)$ can be formalized as

$$s(t_{i+1}|e_{i+1}, h_i) = P(T_{e_{i+1}} = t_{i+1}|e_{i+1}, h_i) \prod_{e \in \mathcal{E} \setminus e_{i+1}} P(T_e > t_{i+1}|e_{i+1}, h_i), \quad (5)$$

where the probability $P(T_e > t_{i+1}|e_{i+1}, h_i)$ depicts that the occuring time of event with mark e is out of the range $[0, t_{i+1}]$, and $P(T_{e_{i+1}} = t_{i+1}|e_{i+1}, h_i)$ is the conditional probability density function representing the fact that mark e_{i+1} occurs at t_{i+1}.

To formalize the Eq. (5), we define *mark-specific conditional intensity function* [1]

$$\lambda_e(t_{i+1}) = \frac{f_e(t_{i+1}|e_{i+1}, h_i)}{1 - F_e(t_{i+1}|e_{i+1}, h_i)}, \qquad (6)$$

where $F_e(t_{i+1}|e_{i+1}, h_i)$ is the cumulative distribution function of $f_e(t_{i+1}|e_{i+1}, h_i)$, referring to the probability that mark e_{i+1} will happen in $[0, t_{i+1}]$. According to Eq. (6), we can derive the cumulative distribution function

$$F_e(t_{i+1}|e_{i+1}, h_i) = 1 - \exp(-\int_{t_i}^{t_{i+1}} \lambda_e(\tau) d\tau). \qquad (7)$$

Thus, we have $P(T_e > t_{i+1}|e_{i+1}, h_i) = 1 - F_e(t_{i+1}|e_{i+1}, h_i)$. Then we can derive the mark-specific conditional probability density function by Eq. (7) as

$$P(T_e = t_{i+1}|e_{i+1}, h_i) = f_e(t_{i+1}|e_{i+1}, h_i) = \lambda_e(t_{i+1}) \exp(-\int_{t_i}^{t_{i+1}} \lambda_e(t) dt). \qquad (8)$$

Substituting Eqs. (7) and (8) into the likelihood of Eq. (5), we can get

$$s(t_{i+1}|e_{i+1}, h_i) = \lambda_{e_{i+1}}(t_{i+1}) \exp(-\int_{t_i}^{t_{i+1}} \lambda(t) dt), \qquad (9)$$

where $\lambda(\tau) = \sum_{e \in \mathcal{E}} \lambda_e(\tau)$ is the summation of all conditional intensity function.

The key to specify probability distribution function $s(t_{i+1}|e_{i+1}, h_i)$ is parameterization of mark-specific conditional intensity function λ_e. We parameterize λ_e conditioned on h_i as follows,

$$\lambda_e(t) = \nu_e \cdot \tau(t; t_i) = \exp\left(W_k^{\nu h} h_i\right) \tau(t; t_i), \tag{10}$$

where row vector $W_k^{\nu h}$ denotes to the k-th row of weight matrix corresponding to mark e. In Eq. (10), the mark-specific conditional intensity function is splited into two parts: $\nu_e = \exp(W_{j'}^{\nu h} h_i)$ is a nonnegative scalar as the constant part with respect to time t, and $\tau(t; t_i) \geq 0$ refers to an arbitrary time shaping function [10]. For simplicity, we consider two well-known parametric models for time shaping function: exponential and constant, i.e., $\exp(wt)$ and c.

Given a collection of event sequences $\mathcal{C} = \{S_m\}_{m=1}^N$, we suppose that each event sequence S_m is independent on each other. As a result, the logarithmic likelihood of a set of event sequences is the sum of the logarithmic likelihood of the individual sequence. Given the source of event sequence, the negative logarithmic likelihood of the set of event sequences $\mathcal{C}$ can be estimated as,

$$\mathcal{L}(\mathcal{C}) = -\sum_{m=1}^N \sum_{i=1}^{|S_m|-1} \left[W_k^{\alpha h} h_i - \log \sum_{j=1}^K \exp\left(W_j^{\alpha h} h_i\right) \right.$$
$$\left. + W_k^{\nu h} h_i + \log \tau(t; t_i) - \sum_{e \in \mathcal{E}} \exp\left(W_{j'}^{\nu h} h_i\right) \int_{t_i}^{t_{i+1}} \tau(t; t_i) dt \right].$$

In addition, we want to induce sparse structure in vector ν in order that not all event types are available to be activated based on h_i. For this purpose, we introduce lasso regularization on ν, i.e., $\|\nu\|_1$ [23]. Overall, we can learn parameters of RNN-TD by minimizing the negative logarithmic likelihood

$$\arg\min_W \mathcal{L}(\mathcal{C}) + \gamma \|\nu\|_1, \tag{11}$$

where γ is the trade-off parameter.

Finally, we estimate the next most likely events in two steps by RNN-TD: (1) estimate the time of each mark by expectation $t_{i+1} = \int_{t_i}^\infty t \cdot s(t|e_{i+1}, h_i) dt$; (2) calculate the likelihood of events according to the mark-specific expectation time, and then rank events in descending order of likelihood.

3   Optimization

In this section, we introduce the learning process of RNN-TD. We apply backpropagation through time (BPTT) [5] for parameter estimation. With BPTT method, we need to unfold the neural network in consideration of sequence size $|S_m|$ and update the parameters once after the completed forward process in sequence. We employ Adam [13], an efficient stochastic optimization algorithm, with mini-batch techniques to iteratively update all parameters. We also apply early stopping method to prevent overfitting in RNN-TD. The stopping criterion is achieved when the performance has no more improvement in validation set. The mapping function of $\phi(t)$ is defined by temporal features associated with t, e.g., logarithm time interval $\log(t_i - t_{i-1})$ and discretization of numerical attributes on year, month, day, week, hour, mininute, and second. Besides, we employ orthogonal initialization method for RNN-TD in order to speed up convergence in training process. The embedding learned by word2vec [18, 19] is used to initialize the parameter of mapping function $\varphi(e)$. The good initialization provided by the embedding can speed up convergence for RNN [17].

4   Experiments

Firstly, we introduce baselines, evaluation metrics and datasets of our experiments. Then we conduct experiments on real data to validate the performance of RNN-TD in comparison with baselines.

4.1   Baselines

Both mark prediction and time prediction are evaluated, and the following models are chosen for comparisons in the two prediction tasks.

(1) Mark sequence modeling.
  - **MC:** The markov chain model is a classic sequence modeling method. We compare with markov chain with order varying from one to three, denoted as MC1, MC2 and MC3.
  - **RNN:** RNN is a state-of-the-art model for discrete time sequence, successfully applied in language model. To fairly justify the performance between RNN and our proposed method, we use the same inputs in both RNN and RNN-TD.
(2) Temporal dynamics modeling. We choose point processes and mark-specific point processes with different characterizations as baselines.
  - **PP-poisson:** The intensity function related to mark is parameterized by a constant, depicting the leaving rate from last event.
  - **PP-hawkes:** The intensity function related to mark e is parameterzied by

$$\lambda(t; e) = \lambda(0; e) + \alpha \sum_{t_i < t} \exp\left(-\frac{t - t_i}{\sigma}\right), \tag{12}$$

  where $\sigma = 1$ and $\lambda(0; e)$ is a intrinsic rate defined on mark e when $t = 0$.
  - **MSPP-poisson:** We define the mark-specific intensity function by a parametric matrix, depicting the rate from one mark to another.
  - **MSPP-hawkes:** The mark-specific intensity function is parameterized by Eq. (12) where the constant rate is specialized according to mark pairs in parametric matrix.
    We also compare with the model that has the ability to generate both mark and temporal sequences.
  - **RMTPP:** Recurrent marked temporal point process (RMTPP) [7] is a method which independently models both mark and time information based on RNN.

4.2   Evaluation Metrics

Serveral evaluation metrics are used when measuring the performance in mark prediction and time prediction tasks. We regard the mark prediction task as a ranking problem with respect to transition probability. The prediction performance is evaluated by *Accuracy* on top k (Acc@k) and *Mean Reciprocal Rank* (MRR) [24]. On time prediction task, we define tolerance θ over the prediction error between estimated time and practical occuring time. The prediction accuracy on time prediction with respect to tolerance θ is formulated as,

$$\text{Acc@}\theta = \frac{\sum_{m=1}^{N} \sum_{i=1}^{|S_m|-1} \delta\left(|E(t; e_{i+1}, h_i) - t_{i+1}| < \theta\right)}{\sum_{m=1}^{N}(|S_m| - 1)},$$

where δ is an indicator function. Larger scores in Acc@k, MRR and Acc@θ indicate better predictions.

4.3   Datasets

We conduct experiments on two real datasets from two different scenarios to evaluate the performance of different methods:

Table 1. Performance of mark prediction on two datasets

		MRR	Acc@1	Acc@3	Acc@5	Acc@10	Acc@20
Memetracker	MC1	0.4634	0.2948	0.4595	0.6659	0.8253	0.9209
	MC2	0.4788	0.3155	0.4706	0.6773	0.8301	0.9186
	MC3	0.4670	0.3149	0.4583	0.6550	0.7891	0.8619
	RNN	0.4780	0.3202	0.4746	0.6825	0.8315	0.9201
	RMTPP	0.4833	0.3241	0.4834	0.6926	0.8386	0.9267
	RNN-TD(c)	0.4820	0.3220	0.4790	0.6895	0.8393	0.9270
	RNN-TD(exp)	0.4849	0.3266	0.4835	0.6929	0.8400	0.9273
	RNN-TD*(c)	0.4820	0.3220	0.4790	0.6895	0.8393	0.9270
	RNN-TD*(exp)	**0.4851**	**0.3266**	**0.4844**	**0.6937**	**0.8407**	**0.9274**
Dianping	MC1	0.6174	0.5231	0.6157	0.7212	0.7963	0.8787
	MC2	0.6260	0.5280	0.6396	0.7393	0.8007	0.8513
	MC3	0.5208	0.4462	0.5395	0.6035	0.6332	0.6569
	RNN	0.6355	0.5123	0.6135	0.7153	0.7905	0.8656
	RMTPP	0.6620	0.5482	0.6554	0.7578	0.8271	0.8935
	RNN-TD(c)	0.6663	0.5524	0.6601	0.7628	0.8346	0.8999
	RNN-TD(exp)	0.6635	0.5448	0.6560	0.7638	0.8345	0.8988
	RNN-TD*(c)	**0.6663**	**0.5524**	**0.6602**	0.7628	0.8346	**0.8999**
	RNN-TD*(exp)	0.6635	0.5452	0.6566	**0.7641**	**0.8351**	0.8990

p.s. the experimental results from * are dependent with given time.

- **Memetracker** [14]: Memetracker corpus contains articles from mainstream media and blogs from August 1 to October 31, 2008 with about 1 million documents per day. Contents in the corpus are organized according to topics by the proposed method in [14]. We use top 165 frequent topics and organize the posting sequence about posted blogs and post-time by users. The whole posting sequence of each user is splited into parts as follows, (1) get the statistics of time intervals between two consecutive posted blogs, (2) empirically estimate the period of user's posting behavior, (3) and divide the whole sequence into several parts according to the estimated period. We do not consider the sequences whose length are less than 3. The obtained dataset contains 1,481,491 posting sequences, and the time interval between two consecutive blogs is ranged from 2.77×10^{-4} to 99.68 h.
- **Dianping:** Dianping provides an online restaurant rating service in China, including coupon sales, bill payment, and reservation. We extract transaction coupon sales from top 256 popular stores located in Xidan bussiness district of Beijing from year 2011 to 2015. The consumption sequences of users are divided into segments as the same steps done in memetracker. Because of the existence of sparse shopping records in users, we also limit that time interval between two consecutive consumptions is two months. The processed dataset contains 221,893 event sequences, and the time interval between two consecutive consumptions is ranged from 2.77×10^{-4} to 1440 h.

On both datasets, we randomly pick up 80% of completed sequences in datasets as training, and the rest sequences are divided into two parts equally as validation set and test set respectively.

4.4 Performance of Mark Prediction

The performance of mark prediction is evaluated using metrics Acc@k and MRR. The experimental results are shown in Table 1. Comparing with MC1, MC2, MC3 and RNN, RNN-TD(c) and RNN-TD(exp) achieve significant improvements over all metrics in both datasets. In Memetracker, RNN-TD(exp) outperforms RMTPP in MRR at significance level of 0.1, and achieve a little improvements than RMTPP in Acc@1,3,5,10 and 20. However, the performance of RNN-TD(c) is worse than RMTPP. In Dianping, RNN-TD(c) achieves improvements than RMTPP in metrics of MRR and Acc@5 at significance level of 0.1 and metrics of Acc@10 and Acc@20 at significance level of 0.01. Besides, RNN-TD(exp) achieves improvements than RMTPP in metrics of Acc@20 at significance level of 0.1 and metrics of Acc@5 and Acc@10 at significance level of 0.01. The experimental results indicate that RNN-TD can better learn the mark generation by jointly optimizing mark-specific conditional intensity function with respect to different time shaping function applied in tasks.

We also conduct experiments according to event likelihood on RNN-TD with the given time, marked as RNN-TD*. The results of RNN-TD*(exp) performs little better than RNN-TD(exp) over all metrics in both datasets, However, the performance of RNN-TD*(c) is almost the same as RNN-TD(c). It demonstrates

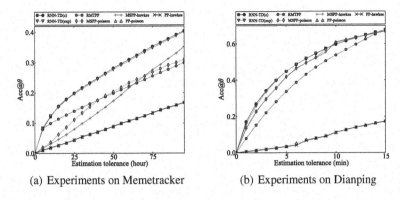

(a) Experiments on Memetracker   (b) Experiments on Dianping

Fig. 2. Performance of timing prediction on two datasets.

the robustness of RNN-TD on mark prediction whether or not given the occuring time. Besides, RNN-TD with exponential form of time shaping function has larger effects on given time than the constant form.

4.5   Performance of Time Prediction

We evaluate the performance of time prediction by Acc@θ. The predictions of RNN-TD and MSPP are based on true marks. Fig. 2(a) and (b) show the experimental results of RNN-TD and baselines on memetracker and dianping. As shown in Fig. 2, without considering any mark information, PP-poisson and PP-hawkes are unable to handle the temporal dynamics well on both Memetracker and Dianping. MPP can discriminate mark-specific time-cost, leading to better performance than PPs. In memetracker dataset, although RMTPP has better performance than PP, it does not overbeat MSPP-poisson and MSPP-hawkes. In dianping dataset, RMTPP(c) and RMTPP(exp) achieve better performance than MSPP-hawkes when tolerance $\theta \leq 65$ h, and also achieve better performance than MPP-poisson when tolerance $\theta \leq 15$ h. It is seen that RNN-TD(c) and RNN-TD(exp) achieve the best performance than all the baselines in the most cases on two datasets. The improvements achieved by RNN-TD indicate that our proposed method can well model marked temporal dynamics by learning mark-specific intensity functions, while RMTPP share the same intensity function for all the marks. Note that the variance of time distribution is quite larger in Dianping than Memetracker. Thus we need to give a smaller α in Dianping when training PP-hawkes and MSPP-hawkes model, leading to the similar performance than PP-poisson and MSPP-poisson shown in Fig. 2(b).

4.6   Case Study on Event Prediction

To explore the capability of event prediction of RNN-TD, we randomly choose one specific event sequence from memetracker and dianping respectively, and estimate the next events in the sequence. In RNN-TD, we select top 3 events in

Table 2. Case study on event prediction

(a) One specific event sequence prediction on memetraker

i-th event: mark,time (mins)		1th	2nd	3rd
RMTPP	c#1	Europe debt, 22.32	Europe debt, 12.29	Europe debt, 60.31
	c#2	LinkedIn IPO, 22.32	Dominique Strauss, 12.29	Amy Winehouse, 60.31
	c#3	Amy Winehouse, 22.32	LinkedIn IPO, 12.29	Dominique Strauss, 60.31
RNN-TD	c#1	Europe debt, 1.07	Dominique Strauss, 1.34	Dominique Strauss, 3.63
	c#2	Dominique Strauss, 0.45	Europe debt, 1.29	Europe debt, 3.12
	c#3	LinkedIn IPO, 0.44	LinkedIn IPO, 0.56	attack, 2.93
Ground truth		Dominique Strauss, 6.37	attack, 83.78	attack, 18.18

(b) One specific event sequence prediction on dianping

i-th event: mark,time (days)		1th	2nd	3rd
RMTPP	c#1	bibimbap, 2.34	bibimbap, 2.90	Sichuan cuisine, 3.02
	c#2	tea restaurnt, 2.34	cookies, 2.90	cookies, 3.02
	c#3	Yunnan cuisine, 2.34	Sushi, 2.90	tea restaurnt, 3.02
RNN-TD	c#1	bibimbap, 2.93	barbecue, 0.65	barbecue, 0.96
	c#2	Yunnan cuisine, 0.88	bibimbap, 0.85	Sichuan cuisine, 0.81
	c#3	bread, 0.92	Vietnamese cuisine, 0.51	bread, 0.48
Ground truth		barbecue,0.14	Sichuan cuisine,1.03	barbecue,1.06

descending order of event likelihood as candidates of next event, called c#1, c#2 and c#3. In RMTPP, we choose the most probable mark and expectation time independently and combine them as the candidates of next event. Table 2 lists the performance of RMTPP and RNN-TD. We can see that the predicted marks on RNN-TD are more accurate and relevant to ground truth than compared methods on both cases. Then, we categorize most relevant marks by empirical knowledge to evaluate the estimated time on mark-specific methods when marks are mismatched in all 3 candidates. For example, we consider bibimbap and barbecue belong to same regional cuisine, and Dominique Strauss is related to Europe debt. In this way, the average error of time prediction to ground truth for RNN-TD is 34.55 min, and the average error is up to 43.19 min for RMTPP in the case of Memetrack. In the case of Dianping, the average error of time prediction to ground truth for RNN-TD is 1.13 days, and the average error is nearly doubled to 2.04 days for RMTPP. Indeed, RNN-TD can provide more options according to possible event predictions which has more general applications, e.g., recommendation systems.

5   Conclusions

In this paper, we proposed a general model for marked temporal dynamics modeling. Based on RNN framework, the representation of hidden layer in RNN-TD learns the history embedding through a deep structure. The generation of marks and times is dependent on history embedding so that we can avoid strong prior knowledge on dependency of history. We observe that the generation processes of next event are significant different with respect to marks. To capture the dependence between marks and times, we unfolded the joint probability of mark and time and parameterized the mark transition probability and mark-specific conditional intensity function based on history embedding. We evaluate the effectiveness of our proposed model on two real-world datasets from memetracker and dianping. Experimental results demonstrate that our model consistently outperforms existing methods at mark prediction and time prediction tasks. Moreover, we conduct case study on event prediction demonstrating that our proposed model is well applicable in marked temporal dynamics modeling.

Acknowledgments. This work was funded by the National Basic Research Program of China (973 Program) under Grant Numbers 2013CB329602 and 2014CB340401, and the National Natural Science Foundation of China under Grant Numbers 61472400, 61572467, 61433014. H. W. Shen is also funded by Youth Innovation Promotion Association CAS and the CCF-Tencent RAGR (No. 20160107).

References

1. Arjas, E., Keiding, N., Borgan, O., Andersen, P.K., Natvig, B.: Survival models and martingale dynamics. Scand. J. Stat. **16**(3), 177–225 (1989)
2. Bao, P., Shen, H.W., Jin, X., Cheng, X.Q.: Modeling and predicting popularity dynamics of microblogs using self-excited hawkes processes. In: Proceedings of the 24th International Conference on World Wide Web, pp. 9–10 (2015)
3. Barzel, B., Liu, Y.Y., Barabási, A.L.: Constructing minimal models for complex system dynamics. Nat. Commun. **6**, Article no. 7186 (2015)
4. Cao, L., Ou, Y., Yu, P.S., Wei, G.: Detecting abnormal coupled sequences and sequence changes in group-based manipulative trading behaviors. In: Proceedings of the 16th ACM SIGKDD International Conference on Knowledge Discovery and Data Mining, pp. 85–94 (2010)
5. Chauvin, Y., Rumelhart, D.E.: Backpropagation: Theory, Architectures, and Applications. Psychology Press, Abingdon (1995)
6. Crane, R., Sornette, D.: Robust dynamic classes revealed by measuring the response function of a social system. Proc. Natl. Acad. Sci. **105**(41), 15649–15653 (2008)
7. Du, N., Dai, H., Trivedi, R., Upadhyay, U., Gomez-Rodriguez, M., Song, L.: Recurrent marked temporal point processes: embedding event history to vector. In: Proceedings of the 22nd ACM SIGKDD International Conference on Knowledge Discovery and Data Mining, pp. 1555–1564 (2016)
8. Engle, R.F., Russell, J.R.: Autoregressive conditional duration a new model for irregularly spaced transaction data. Econometrica **66**, 1127–1162 (1998)
9. Gao, S., Ma, J., Chen, Z.: Modeling and predicting retweeting dynamics on microblogging platforms. In: Proceedings of the 8th ACM International Conference on Web Search and Data Mining, pp. 107–116 (2015)
10. Gomez-Rodriguez, M., Leskovec, J., Schlkopf, B.: Modeling information propagation with survival theory. In: Proceedings of the 30th International Conference on Machine Learning, vol. 28, pp. 666–674 (2013)
11. Gunawardana, A., Meek, C.: Universal models of multivariate temporal point processes. In: Proceedings of the 19th International Conference on Artificial Intelligence and Statistics, pp. 556–563 (2016)
12. Isaacson, D.L., Madsen, R.W.: Markov Chains, Theory and Applications, vol. 4. Wiley, New York (1976)

13. Kingma, D.P., Adam, J.B.: Adam: A method for stochastic optimization. In: International Conference on Learning Representation (2015)
14. Leskovec, J., Backstrom, L., Kleinberg, J.: Meme-tracking and the dynamics of the news cycle. In: Proceedings of the 15th ACM SIGKDD International Conference on Knowledge Discovery and Data Mining, pp. 497–506 (2009)
15. Liu, Q., Wu, S., Wang, L., Tan, T.: Predicting the next location: a recurrent model with spatial and temporal contexts (2016)
16. Vaz de Melo, P.O.S., Faloutsos, C., Assunção, R., Loureiro, A.: The self-feeding process: a unifying model for communication dynamics in the web. In: Proceedings of the 22nd International Conference on World Wide Web, pp. 1319–1330 (2013)
17. Mikolov, T., Chen, K., Corrado, G., Dean, J.: Efficient estimation of word representations in vector space. arXiv preprint arXiv:1301.3781 (2013)
18. Mikolov, T., Sutskever, I., Chen, K., Corrado, G.S., Dean, J.: Distributed representations of words and phrases and their compositionality. Adv. Neural Inf. Process. Syst. **26**, 3111–3119 (2013)
19. Perozzi, B., Al-Rfou, R., Skiena, S.: Deepwalk: online learning of social representations. In: Proceedings of the 20th ACM SIGKDD International Conference on Knowledge Discovery and Data Mining, pp. 701–710 (2014)
20. Pinto, H., Almeida, J.M., Gonçalves, M.A.: Using early view patterns to predict the popularity of youtube videos. In: Proceedings of the 6th ACM International Conference on Web Search and Data Mining, pp. 365–374 (2013)
21. Shen, H., Wang, D., Song, C., Barabási, A.L.: Modeling and predicting popularity dynamics via reinforced Poisson processes. In: Proceedings of the 28th AAAI Conference on Artificial Intelligence, pp. 291–297 (2014)
22. Szabo, G., Huberman, B.A.: Predicting the popularity of online content. Commun. ACM **53**(8), 80–88 (2010)
23. Tibshirani, R.: Regression shrinkage and selection via the Lasso. J. R. Stat. Soc. Ser. B (Methodol.) **58**, 267–288 (1996)
24. Voorhees, E.M.: The TREC8 question answering track report. In: Text REtrieval Conference (1999)
25. Wang, Y., Shen, H., Liu, S., Cheng, X.: Learning user-specific latent influence and susceptibility from information cascades. In: Proceedings of the 29th AAAI Conference on Artificial Intelligence, pp. 477–483 (2015)

Mining of Location-Based Social Networks for Spatio-Temporal Social Influence

Yu-Ting Wen[1](✉), Yi Yuan Fan[2], and Wen-Chih Peng[1]

[1] National Chiao Tung University, Hsinchu, Taiwan
{ytwen,wcpeng}@cs.nctu.edu.tw
[2] EISTI, Cergy, France
fanyi@eisti.eu

Abstract. Following the advent of location-based social networks (LBSNs), location-aware services have attracted considerable attention among researchers. Research has shown that the social network is regarded as one of the strongest influences shaping individual attitudes and behaviors. This paper targets the mining of location-based social influences hidden in LBSNs. In other words, we sought to determine whether an individual's check-in behavior is influenced by friends' check-ins. Check-in data includes positional information; therefore, we refer to this type of influence as spatiotemporal social influences. This study proposes a framework for spatiotemporal social influence mining (*ST-SIM*) to identify users with the greatest influence on individuals (i.e., close friends and travel experts) from an LBSN and estimate the strength of these social connections. Explicitly, the proposed framework is able to infer a list of influential users of an individual under given conditions based on travel distance, visiting time or POI categories. We developed a diffusion-based mechanism for modeling the propagation of influence over time. Our experiment results demonstrate that the *ST-SIM* framework outperforms state-of-the-art methods in terms of accuracy and reliability, and is applicable in domains ranging from marketing to intelligence analysis.

Keywords: Influence propagation · Location-based social network

1 Introduction

Location-based social networks (LBSNs) enable users can share location-based information with their friends. Numerous studies have been conducted on the use of LBSNs for the discovery of popular attractions, travel planing and tour recommendations [2,4,6,12]. However, most of these studies have focused on mining movement patterns from crowds in LBSNs, and largely disregarding the potential impact of social influence hidden in LBSN. Social influence refers to situations in which a group of people influence individuals within the group in their decision making based on their interdependence or cohesion with the group.

© Springer International Publishing AG 2017
J. Kim et al. (Eds.): PAKDD 2017, Part I, LNAI 10234, pp. 799–810, 2017.
DOI: 10.1007/978-3-319-57454-7_62

Online social networks offer a rich forum for observing social interactions. Social influence analysis has considerable potential in fields such as marketing and recommendation system.

According to the Trust In Advertising report in 2015 from Nielsen[1], *Recommendations from people I know* have the greatest influence on consumers, with 83 percent of global average. In other words, recommendations from specific individuals, such as idols or friends with similar hobbies, may attract individuals to locations that are largely ignored by the general public. For example, many stores and restaurants now provide discounts to people who "like" them or check in on Yelp or Facebook. Also, when searching for travel tips, the opinion from a friend may be more convincing than rankings on an official website. However, recognizing members with the greatest influence on an individual can be a challenging problem for those users with a large friend base.

Many researchers have adopted social factors (the similarity of visited POIs between individuals and their friends) as weighted factors in recommendation systems [1,2,6,10,11]. They concluded that the social factors have far less impact than other factors such as geographic distance and user interest. In contrast, the researchers in [9] considered social influence from the viewpoint of the "users" rather than the "POIs", and found that (1) social relationships may differ in the degree of influence with regard to an individual's decisions; and (2) the influence is not necessarily generated directly by friends but may originate with the friends of friends. This is referred to as the *directionality* and *transition* of social influence.

This study developed an innovative framework for social influence mining on location-based social networks. We consider the fact that a person is influenced by her friends in her choices of where to visit, and weight the factors affecting this influence in term of space, time and POI categories. Using the information available through LBSNs, the proposed framework, *ST-SIM*, is used to mine the top-k influential users based on a user's query related to a specific geospatial region.

In summary, the contributions of this paper are four-fold:

- We propose a novel Spatio-temporal Social Influence Mining framework (*ST-SIM*) to identify influential users in an LBSN. The model captures the interaction among the social network, physical location and the effects of time to quantify the influence among user pairs.
- We define the spatio-temporal social follow relationship to formulate the spatio-temporal social influence on user behavior. Building on our empirical findings, *ST-SIM* use spatial and temporal features in order to quantify each connection between user pairs according to the probability of one user following the other's lead. We model the social influence over the network in terms of information propagation based on heat diffusion model.
- Considering the diversity of an individual's location interest as well as the impact of social effect, a dynamic weight tuning method is presented. Social effect and self effect are used in the computation of unified *followship* probability scores for top-k user recommendation.

---

[1] http://www.nielsen.com/.

– We conducted empirical experiments on real-world LBSN datasets to evaluate the effectiveness of *ST-SIM* framework.

2 Problem Formulation

Location-based social networks provide a platform on which the location of a particular user and the time of activities are recorded and shared. This means that a user is able to use online social network website/application to share her real-world mobility.

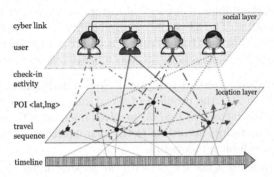

Fig. 1. An example of a heterogeneous graph, that captures user-user virtual community, user-POI mobility activities and time effects in an LBSN.

A location-based social network can be structured as a heterogeneous graph (*HG*) with multiple types of nodes, edges, static attributes and dynamic, interconnected activities (see Fig. 1). We characterize LBSNs according to three aspects: (1) a social layer S comprising nodes representing the members $u \in U$ of the service and edges showing their friendship links, (2) a location layer L containing all the POIs $p \in P$ that have been visited and (3) a set of check-in activities C which connects the social layer and the location layer; a check-in activity $c(u, p, t)$ represents a user u visits a location p at time t.

Definition 1. Spatio-temporal social influence: Social influence refers to the effect of implicit recommendations obtained on social network. The closer the relationship between two users is, the more effective the recommendation is in influencing the user. This study focused on the *spatio-temporal social influence* of LBSNs; i.e., if user u_i is influenced socially by u_j, then u_i will tend to visit a POI in accordance with the recommendation obtained from u_j. This reveals a relationship in which u_i checks in to the same POI after u_j shares her own check-in.

Definition 2. Spatio-temporal social follow relationship: A *spatio-temporal social follow relationship*, hereafter denoted as *followship*, represents a directed link from user u_i to her friend u_j iff u_i visits a location that previously visited by u_j.

Formally, a *followship* exists under the following conditions:

$$followship(c(u_i, l, t), u_j, \delta)$$
$$= \begin{cases} true & \exists t' : c(u_j, l, t') \bigwedge \delta = t - t' > 0 \\ false & otherwise. \end{cases} \tag{1}$$

where δ represents a valid time period. We can also define u_i as a *follower* and u_j as *influencer*.

Definition 3. ST-social strength: ST-social strength is defined as the quantitative measure of the influence of check-in histories, which is directed and varies with distance and time. The ST-social strength of how user u_j influence user u_i is abbreviated as s_{ij}. Note that friendship f_{ij} is undirected and ST-social strength s_{ij} is directed; and the members of the two sets are not necessarily equivalent.

Problem Definition. Spatial-temporal Social Influence Mining: Using heterogeneous graph HG, the problem of social influence mining on LBSN with spatial and temporal factors $(ST\text{-}SIM)$ involves inferring ST-social strength s_{ij} for any two users u_i and u_j according to the characteristics of their movements, i.e., whether they exhibit a followship.

Based on the inference, the k users with the greatest influence on each user u_i are identified. The result cam be personalized using optional queries associated with geospatial region or user preference.

3   Spatio-Temporal Social Follow Relationship and User Mobility

Since social influence has been verified in [9], in this section, we characterize the *spatio-temporal social follow relationship* by examining the influence of spatial and temporal features on user mobility.

3.1   Dataset Description

This study used the four real-world LBSN datasets listed in Table 1. The *FB* dataset is collected by the Facebook API[2]. We used the Facebook accounts of 96 volunteers as seeds (most of the users live in Taiwan). Once a user allows us to use the private information, we obtained details related to the location of all of the user's friends via check-ins and geo-tagged photos for the period of Jan. 2012 - Dec. 2014. For example, one user may have 300 friends. Then from this user we can create 301 user nodes and all the related locations as POI nodes. The *GWL* dataset [3][3], *FS* [5,8] and *FS-CA* [12] are check-in datasets within an undirected friendship network. Note that *GWL* and *FS* are larger but lack information related to POI categories.

---

[2] Facebook Developers. https://developers.facebook.com/.
[3] Stanford Network Analysis Project. http://snap.stanford.edu/.

Table 1. Details of the Heterogeneous Social Networks

	Property	Network			
		FB	GWL	FS	FS-CA
#records	check-in	869,317	6,442,890	2,201,511	483,813
#nodes	user	29,512	196,591	2,133,749	4,163
	POI	225,077	1,280,969	1,143,122	121,142
#edge	friend	39,513	950,327	27,098,472	32,512

3.2 Effects of Spatial and Temporal Features

We then sought to identify the factors that determine how much influence each *followship* has on the selected users. A number of assumptions were made prior to observation:

Assumption 1: The check-in behavior of users at times closer to the target time are more relevant, and thus more important with regard to their effectiveness as recommendations [12].

Assumption 2: Users tend to visit their nearby POIs [10].

Assumption 3: POI characteristics should be taken into consideration. Hot spots, such as train stations and shopping malls, are very popular and therefore more likely to result in *followship* [7].

To deal with Assumption 1, we measured the length of time that individuals maintain *followships*. Figure 2(a) plots the number of *followships* as a function of time for *FB*, *GWL*, *FS* and *FS-CA*. It was observed that the distribution corresponds to a power law with periodic peaks for each week. The distribution decays faster after the first week. Another interesting observation is that the larger the dataset (*GWL* > *FS* > *FB* > *FS-CA*), the flatter the distribution. Nonetheless, the periodic peaks are similar in all datasets.

Travel distance is considered to be the distance between the hometown of user and the target location (Assumption 2). One's hometown information is not

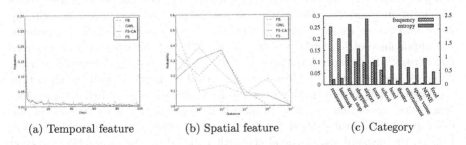

(a) Temporal feature   (b) Spatial feature   (c) Category

Fig. 2. Distribution of (a) time period, (b) distance and (c) POI characteristics of spatio-temporal social follow relationships.

explicitly given; therefore, we infer this as the location associated with the most frequent check-in events [3]. As shown in Fig. 2(b), we calculated the distribution of distances between the hometown of friends and where the *followship* events took place. However, the distribution was shown to vary greatly between the datasets, due differences in population cluster size among countries. We can find that the probability of *FB* approaches zero when the distance over 10^4 km, while the distance of other three datasets are farther (10^5 km).

Figure 2(c) illustrates the frequency of *followship* events by ratio and the entropy using the user frequency of POI categories, respectively. For example, the type of "restaurant" was shown to have the highest frequency, representing that the visiting activities at restaurants are socially influential. However, the "airport" category also has high *followship* frequency but with high user entropy. We deduce that the location is popular and the *followship* events may happen by coincidence.

Finally, we can make the following observed conclusions:

Observation 1: Individuals are more likely to visit the same place after friends with whom they have recent *followships*. This trend decays exponentially with time.

Observation 2: Most users tend to visit nearby POIs; however, in cases where an individual follows another user of a POI located at a long distance, then the leader may have stronger social influence.

Observation 3: POIs with high user entropy are considered hot spots. In other words, *followship* events associated with hot spots are considered less influential.

These three observations conclude three weighting features of the importance of each *followship* event, spatial, temporal and POI entropy factors, which will be applied in the computation of *ST-social strength* in our *ST-SIM* framework.

4   ST-SIM Model

This section describes the process of quantifying the social influence on LBSNs in terms of the *ST-SIM* model. A heterogeneous graph $HG = (S,C,L)$ was built using raw LBSN records in order to extract the interactions between user nodes and location nodes; i.e., *followship* events. In Sect. 4.1, we began by utilizing *followship* events as the main contribution to ST-social strength. To measure the importance of *followship* events, we modeled the background features into two classes: (1) **personal background** in the view of each individual user for different locations, and (2) **global background** in the view of all the users that has visited each locations. Moreover, we have already observed that the importance decays over time and may propagate from strangers. Thus, in Sect. 4.2, we developed a **diffusion-based model** to simulate the propagation of influence. Finally, the measure of ST-social strength is based on the interaction between the two users (**inter factor**) and the similarity of individual's preference on POI category (**intra factor**).

4.1   Background Featurization

Personal Background. The personal background models the individual's preference to be influenced. Users tend to frequent some locations more than others
based on the specific meaning they have for the user. Thus, it is important to
look into this user's location history in order to determine how different locations
affect the *followship* of users. Using the observation in Sect. 3, we extracted two
factors for the modeling of personal background.

The **temporal feature** considers the time difference Δt of the *followship*
event, which decays exponentially over time (Observation 1). $f_t = exp(-\Delta t)$.

The **spatial feature** considers the distance from user's hometown to the
location, and the probability of *followship* within the distance (Observation 2).
$f_s = \frac{1}{d(l_u,l)} \times P_d(d(l_u,l))$, where $d(l_u,l)$ represents the distance from user u's
hometown l_u to location l, and P_d is the probability of distance distribution as
shown in Fig. 2(b).

Global Background. It was also noted that the aggregation of location histories obtained from all of the users exhibited different characteristics. The global
background captures the popularity of specific locations, as inferred from all of
the users. *Followship* events in popular locations such as train stations are often
less indicative of the strength of mobility relationship. Conversely, two individuals could be expected to have a strong relationship in less popular locations
(Observation 3).

To model the popularity of a place, **POI entropy** is given by Shannon
entropy, as follows: $H_l = -\sum_{u,P_{u,l}\neq0} P_{u,l} log(P_{u,l})$, where $P_{u,l}$ is the probability
that user u has visited location l. A high value for POI entropy indicates that a
location is visited by many different users.

4.2   Diffusion-Based Influence Model

The process of exerting social influence can be seen as a specific type of information diffusion. By illustrating the physical diffusion of heat, a member in a
social network can be seen to act as a heat source diffusing influence to friends
via shared activities such as check-in events. Through these friends, the influence
gradually propagates. At a certain time point, influence is diffused to the margin
of the social network, whereupon complete strangers may be affected.

Spatio-Temporal Social Influence Propagation. As mentioned previously,
this study focused on *followship* events rather than simple friendships. Simple
social network is insufficient to capture the effects of social influence or its propagation among users. We have defined a novel **followship graph** G_F to represent
the possibility that an individual may visit a location because she is influenced
by her friends. $G_F = (U, E_F)$, where V is the set of users and E_F is the set of
spatio-temporal follow relationships among users in U.

Via the *followship*s in E_F, social influence may propagate among the users
within G_F. Formally, we define $p_{ij} = \frac{n_{ij}}{\sqrt{n_i}\sqrt{n_j}}$ as the probability of influence

moving from u_i to u_j; where n_{ij} denotes the *followship* from u_i to u_j and n_i denotes the total number of locations u_i has visited. Let us assume that user $u_i \in V$ is only influenced by herself initially, whereupon influence propagates to others in G_F.

The influence-based diffusion model two key parameters: (1) initial state probability for each *followship* event; (2) state transition probability from the *influencer* to the *follower*. During the process of propagation, users receive stimulation from their neighbors. Let vector $s(t)$ denote the proportion of the social influence score of users in V at time t. The change at u_i between time $t + \Delta t$ can be defined by applying the following equation to the diffusion model:

$$\frac{s(t + \Delta t) - s(t)}{\Delta t} = \alpha Inf s(t) \qquad (2)$$

where α is the propagation coefficient and Inf is a $N_{G_F} \times N_{G_F}$ matrix used to define the one-hop process of information diffusion (Fig. 3).

$$Inf_{ij} = \begin{cases} p_{ij} & (u_i, u_j) \in E_F \\ -\tau_i & i = j \\ 0 & otherwise. \end{cases} \qquad (3)$$

where τ_i denotes the amount of influence diffused from u_i via external links, such that $\tau_i = 0$ if u_i does not have any neighbors, otherwise, $\tau_i = \sum_{(u_i, u_j) \in E_F, i \neq j} p_{ij}$.

Using Eq. 2, we obtain the following differential equation when $\Delta t \to 0$:

$$\frac{ds(t)}{dt} = \alpha Inf s(t), s(t) = e^{\alpha t I} s(0) \qquad (4)$$

4.3   Spatio-Temporal Social Strength

Let s_{ij} denote the spatio-temporal social strength (*ST-social strength*) of user u_j for query user u_i in region r; i.e., the likelihood of u_i maintaining a *followship*

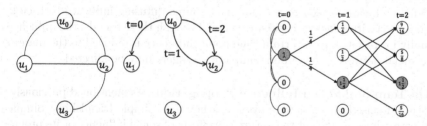

(a) Social Graph   (b) Followship Graph   (c) Influence propagation from t = 0 to
t = 2 with three followship events.

Fig. 3. An example of valid influence propagation among four users. The nodes with frame indicate the occurrence of spatio-temporal social follow relationships and the number in nodes indicate the followship weight.

with u_j proportional to the value of s_{ij}. We intuitively take s_{ij} as the sum of the influences of others and one's own interests (influenced by herself), which are denoted as s_{inter} and s_{intra} respectively. s_{inter} and s_{intra} are two weighting parameters ($0 \leq s_{inter} + s_{intra} \leq 1$). Here $s_{inter} = 1$ refers to the case where s_{ij} depends entirely on the prediction based on the social effect of u_j, while $s_{intra} = 1$ refers to the case where s_{ij} is based only on user interests. If we want to combine these two measures to produce an overall value for ST-social strength, it is necessary to determine the relative importance of each component-measure to ST-social strength.

Applying the above diffusion process to the follow graph, we obtain results that can be utilized in a dynamic weighting mechanism. s_{ij} represents the likelihood of a *followship* event by u_j to u_i, which fits the characteristic of social effect. In the case of user u_i, as the *inter factor* from any user $u_j, j \neq i$ represents the tendency of how u_i *follows* u_j, while the *intra factor* represents u_i's own interests, in other words, how u_i follows herself.

Further, while the s_{intra} represents u_i's own interests, *intra factor* should increase when u_j and u_i have similar preferences. The similarity is simply defined as the cosine similarity to weight the *intra factor* for different user pairs.

The unified geo-social strength can be revised as follows:

$$s_{ij} = \begin{cases} s_{inter} + s_{intra} \times (\sum_p^m w_{ij}^p) & i \neq j \\ 0 & i = j \end{cases} \qquad (5)$$

In the proposed *ST-SIM* framework, we consider *followship* events as new sources of influence in the follow graph. For each *followship* $< (u_i, l, t), u_j, \Delta t >$, $s_{ij}(0)$ is initialized to the *followship* weight based on the background features mentioned in Sect. 4.1, which jointly cover the three features:

$$\begin{aligned} s_{ij}(0) &= H_l \times f_s \times f_t \\ &= - \sum_{u, P_{u,l} \neq 0} P_{u,l} log(P_{u,l}) \times \frac{1}{d(l_i, l)} \times P_d(d(l_i, l)) \times exp(-\Delta t) \quad (6) \end{aligned}$$

where time period $\Delta t = current\ timestamp - t$.

5   Experimental Evaluation

In this section, we were particularly interested in the predictive performance of the *ST-SIM* framework; i.e., we sought to predict the set of users with the greatest influence on the travel behavior of an individual as accurately as possible.

5.1   Settings and Evaluation Methods

Experimental Setup. We employed the real-world *FS-CA* dataset described in Sect. 3.1. The data was ordered according to the creation time and then divided into two subsets, a training set and an evaluation set. The training set contained the first 70% of the check-in activities, whereas the evaluation set contained the remaining 30% of the data.

Performance Metrics. We use two popular measures to evaluate the performance of our techniques: average precision in overall results and MAP (Mean Average Precision) for ranked results. The definitions of the metrics are given as follows.

Precision@k is the fraction of the top-k users with influence over other users.

$$Precision@k = \frac{\# \text{ influential users in top-k results}}{k}$$

MAP stands for the mean of the AP values of all queries. AP is defined as the average of the precision values for all relevant results of a single query.

$$AP = \frac{\sum_{i=1}^{k} (Precision@i \times rel(i))}{\# \text{ influential users in top-k results}}$$

where Precision@i is the precision at cut-off i in the list, $rel(i)$ is an indicator function equal to 1 if the item at rank i is a relevant ranking and otherwise zero.

5.2   Comparison Methods

In addition to *ST-SIM*, the recommendation approaches under evaluation are listed below.

Baseline1 - Order by public frequency: this approach represents the public's trend by considering the top-k users with the most visiting counts in the query region.

Baseline2 - Order by following counts: this approach directly rates the users by the number of *geo-social following relations*. The result is confined to the friend circle.

Entropy-Based Model for Co-occurrence (EBM): this is one of the state-of-the-art model to infer social connection from LBSN [7]. EBM quantifies the strength of each social connection by considering the co-occurrences in the context of locations.

Only consider social effect (Inter): this is a special case of *ST-SIM* by setting the intra factor as zeros. In other words, only social effect from others is considered for recommendation.

Only consider self effect (Intra): this is also a special case of *ST-SIM* with the inter factor set to zeros. Only the user's interests are considered in the recommendation.

5.3   Performance Evaluation

Tuning Propagation Coefficient. Although the self-tuning technique of *ST-SIM* properly assigns the parameters for weighting inter factor and intra factor, the diffusion model of *ST-SIM* uses two parameters: α and t. Parameter α

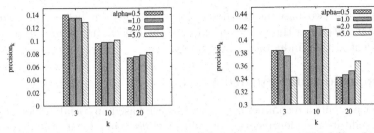

(a) Precision of top-200 check-in users (b) Precision of top-200 *followship* users

Fig. 4. Ranking results for different α value.

controls the diffusion rate of our model and time t varies from 0 to 1.0. As time $t = 0$, the influence score is centralized in query user vertex. When t increases, more and more people are influenced by their neighbors. Similarly, the magnitude of α represents how fast the influence diffuse. In this set of experiments, we want to examine how the propagation coefficient α controls the rate of influence diffusion and find the optimal value for α for the dataset.

We set $t = 1.0$ in all our experiments and Fig. 4 (a) and (b) shows the results of the query users with top-200 check-in counts and top-200 *followship* counts. Note that the value change has small influence on the final order when $\alpha t \leq 5.0$. But when αt increases more, the performance decreases because of most of the influence scores diffuse out and muti-degree friends may have similar scores to first-degree friends. Finally, we choose $\alpha = 1.0$ in the following experiments.

Goodness of Prediction with Baseline Heuristics. Our goal in this experiment is to evaluate how well the geo-social strength from training set fits the observed strength from evaluation set (ground truth).

Figure 5 depict the MAP and average of Precision@k results of the different recommendation methods at $k = 3, 10, 30$ under the following scenario: with the recommendation systems build from training set and given a member in LBSN, who we should choose as the top-k influential candidates and what is the performance according to the ground truth (stimulated by the individual's future behaviors in testing set). Each figure corresponds to an approach. Generally, *ST-SIM* and **Inter** performs the best in terms of all metrics, and **EBM** performs

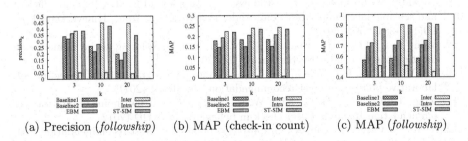

(a) Precision (*followship*)   (b) MAP (check-in count)   (c) MAP (*followship*)

Fig. 5. Ranking results for different recommendation methods

better than the two baseline methods. These all perform better than **Intra**. Specify that **Intra** has the worst hit value might reflect that *social influence* is more influential than individual's own preference.

6   Conclusion and Future Work

This paper presents a recommendation framework based on social influence (*ST-SIM*) to facilitate the identification of influential users in a location-based social network. We first built a heterogeneous graph to model the interaction between user-user pairs as well as user-category pairs. A diffusion-based influence model was also developed for the extraction of interactive features for user ranking. A dynamic weight tuning mechanism is included in the model to provide personalized recommendations for each user. We evaluated *ST-SIM* using real datasets of LBSN check-in logs. According to the experiment results, the proposed method provides recommendations that are more effective than many existing recommendation strategies.

Ackowledgement. Wen-Chih Peng was partially support by the TAIWAN MOST (104-2221-E-009 -138 -MY2 and 105-2634-E-009 -002) and Academic Sinica Theme project No. AS-105-TP-A07.

References

1. Bao, J., Zheng, Y., Mokbel, M. F.: Location-based and preference-aware recommendation using sparse geo-social networking data. In: SIGSPATIAL (2012)
2. Cao, X., Cong, G., Jensen, C.S.: Mining significant semantic locations from GPS data. VLDB Endowment **3**(1–2), 1009–1020 (2010)
3. Cho, E., Myers, S.A., Leskovec, J.: Friendship and mobility: user movement in location-based social networks. In: KDD (2011)
4. De Choudhury, M., Feldman, M., Amer-Yahia, S., Golbandi, N., Lempel, R., Yu,C.: Automatic construction of travel itineraries using social breadcrumbs. In: HT (2010)
5. Levandoski, J.J., Sarwat, M., Eldawy, A., Mokbel, M.F.: LARS: A location-aware recommender system. In: ICDE (2012)
6. Lu, X., Wang, C., Yang, J.-M., Pang, Y., Zhang, L.: Photo2Trip: generating travel routes from geo-tagged photos for trip planning. In: MM (2010)
7. Pham, H., Shahabi, C., Liu, Y.: EBM: an entropy-based model to infer social strength from spatiotemporal data. In: SIGMOD (2013)
8. Sarwat, M., Levandoski, J.J., Eldawy, A., Mokbel, M.F.: LARS*: an efficient and scalable location-aware recommender system. IEEE Trans. Knowl. Data Eng. **99**(1), 1384–1399 (2013)
9. Wen, Y.-T., Lei, P.-R., Peng, W.-C., Zhou, X.-F.: Exploring social influence on location-based social networks. In: ICDM (2014)
10. Ye, M., Yin, P., Lee, W.-C., Lee, D.-L.: Exploiting geographical influence for collaborative point-of-interest recommendation. In: SIGIR (2011)
11. Ying, J.J.-C., Lu, E.H.-C., Kuo, W.-N., Tseng, V.S.: Urban point-of-interest recommendation by mining user check-in behaviors. In: KDD (2012)
12. Yuan, Q., Cong, G., Sun, A.: Graph-based point-of-interest recommendation with geographical and temporal influences. In: CIKM (2014)

Multi-perspective Hierarchical Dirichlet Process for Geographical Topic Modeling

Yuan He$^{(\boxtimes)}$, Cheng Wang, and Changjun Jiang

Department of Computer Science, Tongji University, Shanghai, China
{chengwang,cjjiang}@tongji.edu.cn, yaronhe@outlook.com

Abstract. The pervasion of location acquisition technology has strongly propelled the popularity of geo-tagged user-generated content (UGC), which also raises new computational possibility for investigating geographical topics and users' spatial behaviors. This paper proposes a novel method for geographical topic modeling by combining text content with user information and spatial knowledge. Topics are estimated as the interests of users and features of locations. The joint modeling of the three heterogeneous sources (1) leads to high accuracy in predicting visit behaviors driven by personal interests, (2) discovers coherent topic representations for topic modeling, (3) enables the recommender system to suggest *interpretable* locations. Our framework is flexible to incorporate new dimensions of data such as temporal information without substantially changing the model structure. We also experimentally demonstrate the limitations of the traditional assumption that a topic is selected considerably dependent on the location. In many cases, the published topics are mainly affected by the user's interests rather than the current location. Our model discriminates these two scenarios. Through employing hierarchical Dirichlet process, we also need not predefine the number of topics like other mixture models. Experiments on three different datasets show that our model is effective in discovering spatial topics and significantly outperforms the state of the art.

Keywords: Hierarchical Dirichlet Process · Geographical topic modeling

1 Introduction

The pervasion of location acquisition technologies has strongly propelled the popularity of geo-tagged user-generated content (UGC), which offers great inspiration to the new research field, geographical topic modeling. Probabilistic graphical models [11,24,25] have been widely used in previous work, and proven promising in modeling such data. Usually, two latent variables, namely *topic* and *region*, are introduced to model the underlying semantics of the observed texts and locations. The conditional dependencies between all pairs of variables are comprehensively considered so that the algorithm is expressive enough to model the multi-dimensional data. However, the intricate variable dependencies

© Springer International Publishing AG 2017
J. Kim et al. (Eds.): PAKDD 2017, Part I, LNAI 10234, pp. 811–823, 2017.
DOI: 10.1007/978-3-319-57454-7_63

also lead to the complexity of modeling and difficulty in learning, which is the main reason that hinders the practice. Furthermore, when we try to incorporate a new dimension of data such as temporal information, both the model structure and learning algorithm have to make substantial changes.

To address these problems, we propose a flexible topic model, called Multi-perspective Hierarchical Dirichlet Process (MHDP). In our model, the text content is modeled from two perspectives, each of which corresponds to one kind of tag, namely the user id and location record. From the *user perspective*, historical postings of an individual are aggregated so that the per-document topic distribution tells the user's interests. From the *location perspective*, contents published at the same location are aggregated, thus the discovered topics represent the locations' features. Since the documents in the two perspectives are both reconstituted from the original corpus, the latent topics are assumed to be shared. This design decouples the complicate dependencies between variables, and makes our model extremely flexible. As depicted in Fig. 2(b), removing or adding a new dimension of data has very little impact on the model structure. Because different perspectives are inferred iteratively, the heterogeneous sources tightly interact with each other.

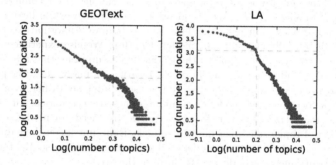

Fig. 1. The two latent categories of locations (*distracted locations* with few topics & *concentrated locations* with diverse topics) on two datasets, *GEOText* and *LA*

Two Categories of Locations: Previous work [1,7,12,26] regards the location as a vital factor affecting the generation of topics, and treats all locations the same. However, let's consider two kinds of places: one is locations like university or plaza, and the other is places like restaurant or bar. People at the first kind always publish diverse topics which present very little geographical relevancy. For example, a common user in a university may tweet about daily work, travelogues or even sudden feelings. The factor that affects the selection of topics is the user himself rather than the location. Conversely, users in a restaurant or bar mostly publish something about food or wine, therefore the location determines what he publishes. This property leads to the observation that topics at the first kind are diverse, and at the second kind are concentrated. In this paper, we distinguish the two scenarios, and define them as *distracted locations* and *concentrated* *locations* respectively. According to the definition of Dirichlet process [3,8,21], the expected number of clusters m for n observations is $E(m) \simeq \alpha \log(1 + \frac{n}{\alpha})$, which means that the number of topics (word clusters) can be approximated as the logarithm of word count. Thus we can plot Fig. 1 to estimate the statistics between *location* and *topic* on two different datasets, *GEOText* and *LA*, only observing the published words. The details of the datasets will be described in Sect. 3. Intuitively, it should follow a power law on each dataset, and present as a straight line. However, we observe an obvious cutoff around which there is an interesting *kink*. At both sides, there is a standard power law distribution. This observation verifies that locations can be inherently divided into two latent categories which are distinguished by the x-index of the kink (i.e., a threshold of topic count). We utilize this property to improve the modeling of topics from the location perspective.

The experimental evaluations are conducted on three different datasets. The results demonstrate both the preponderance of our model in discovering spatial topics and the capability in improving topic quality. The main contributions of this paper are as follows: **(1)** It proposes a flexible and modularized model MHDP (Multi-perspective Hierarchical Dirichlet Process) to model geo-tagged UGC. It is capable to incorporate new data dimensions without substantially modifying the model structure. **(2)** It proposes to classify locations into two categories, which are distracted places with diversity topics and concentrated ones with few topics. Experiments show that this hypothesis is more faithful to the underlying semantics of data. **(3)** It regards the latent topics as the intermediary between different perspectives, which enables our model to suggest *interpretable* locations for users.

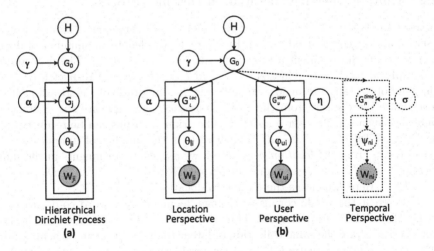

Fig. 2. Plate notations of HDP (Left) and MHDP (Right) model

2   The Proposed MHDP Model

In this section, we assume the reader is familiar with the mechanism of Dirichlet Process (DP), and give no introduction of its basic theories. For more details, you can refer to [16, 20, 22].

2.1   Hierarchical Dirichlet Process

One of the prototypical ingredients for nonparametric topic modeling is the Hierarchical Dirichlet Process (HDP) which is obtained by coupling draws from a Dirichlet process, and having the reference measure itself arise from another Dirichlet process. As depicted in Fig. 2(a), we have

$$G_0 \sim DP(\gamma, H)$$
$$G_j \sim DP(\alpha, G_0)$$

where γ and α are the *concentration parameters*, and H is the *base distribution*. This means that we first draw atoms from H to obtain G_0 which is used as the reference measure to obtain the set of measures G_j. All G_j's are discrete, and share atoms via G_0. In topic model, G_0 can be regarded as the global topic-word distribution shared by all documents. Each G_j corresponds to a document d_j, and the topic assignments of words in d_j are sampled from G_j. In our model, we generalize this structure, and let G_0 shared by all perspectives instead of documents.

2.2   Multi-perspective Hierarchical Dirichlet Process

In most cases, a piece of geo-tagged UGC consists of three parts, namely words, user id, and a geographical record. Formally, let M denote the number of documents, and each document d has N_d words $\{w_1, w_2, \ldots, w_{N_d}\}$. The user id and geographical record are denoted as u_d and $l_d = (lat_d, lon_d)$ respectively. Here the symbols *lat* and *lon* denote the latitude and longitude of the location. If we investigate the dataset from the *user perspective*, we aggregate the documents published by a specific user u as d_u^{user}. The original corpus can be reconstituted as $D^{user} = \{d_1^{user}, d_2^{user}, \ldots, d_U^{user}\}$ where U is the number of unique users in the dataset. From the *location perspective*, we aggregate documents in the same location l as d_l^{loc}, and create another corpus $D^{loc} = \{d_1^{loc}, d_2^{loc}, \ldots, d_L^{loc}\}$ of size $L = |D^{loc}|$. Here L denotes the number of locations.

If we treat the whole corpus as a singe document, the word co-occurrences in D^{user} and D^{loc} are the same. Thus it is reasonable to assume that the two perspectives share the same collection of latent topics, which can be achieved by sharing the global measure G_0. In other words, the latent topics are treated as the intermediary, which is affected by both perspectives, and also affects both perspectives. The basic prototype of our probabilistic model is illustrated as follows

$$G_0|\gamma, H \sim DP(\gamma, H)$$
$$G_u^{user}|\eta, G_0 \sim DP(\eta, G_0), \quad u \in [1, U]$$
$$G_l^{loc}|\alpha, G_0 \sim DP(\alpha, G_0), \quad l \in [1, L]$$

From the *user perspective*, the random measures G_u^{user} are conditionally independent given G_0 with distributions given by a Dirichlet process whose base probability measure is G_0. The same goes for G_l^{loc}. The distribution G_0 varies around the prior H with the amount of variability governed by γ. In our model, we set H a multinomial which denotes the topic-word distribution. Clearly, our model has a very flexible structure. By adding a new set of random measures G_n' which shares G_0 with other perspectives, we can easily incorporate a new dimension of data. Here we have

$$G_n'|\sigma, G_0 \sim DP(\sigma, G_0), \quad n \in [1, N]$$

which is depicted as the dashed part in Fig. 2(b). Taking temporal data as an example, the temporal perspective would capture the variation of topics in different intervals of a day. The shared global measure G_0 makes the three perspectives tightly affect each other.

2.3   Chinese Restaurant Franchises Metaphor

In this section, we give the details of a Chinese Restaurant Franchises (CRFs) representation for the MHDP model. This metaphor provides a concrete representation of draws from the MHDP, and it provides insights into the sharing of atoms across multiple perspectives. A CRFs representation is composed of five elements: a customer, a table, a dish, a restaurant and a franchise. The customer denotes a word in a document, the table denotes a latent variable, the dish denotes a topic, and the restaurant denotes a document. The franchise denotes a perspective.

In the Chinese Restaurant Process (CRP) metaphor [22] for Dirichlet process, each observed item is considered as a customer in a restaurant which has an infinite number of tables. Initially all tables are empty. A customer picks existing tables in proportion to their popularity, which is well known as "rich-get-richer" process. The probability for a customer i choosing a table t is

$$P(t_i = t) = \begin{cases} \dfrac{n_j}{n. + \alpha} & \text{for an existing table} \\ \dfrac{\alpha}{n. + \alpha} & \text{for a new table} \end{cases}$$

where t_i denotes the table chosen by customer i. n_j denotes the current number of customers sitting at table j, and $n.$ is the total number of customers so far. The dish chosen at table j is drawn i.i.d. from the base measure H. Specifically, we can map the notations in this metaphor perfectly to the variables in topic models. A dish chosen by a customer actually denotes the topic assigned to a word. Therefore, the posterior distribution of the dishes is the estimated topic distribution of the documents. In HDP, the Chinese restaurant is extended to a Chinese franchise which has M restaurants. The process defines a set of random probability measures G_j (one for each restaurant) and a global random probability measure G_0. The global measure G_0 is also distributed as a Dirichlet process, and shared by all G_j as the base probability measure. Since G_0 is discrete, the dishes in G_j would be repetitive. The shared measure G_0 is regarded as the corpus-level topic-word distribution.

In MHDP model, the one franchise is further extended to multiple franchises. In other words, the collection of documents in each perspective forms a new corpus, and is denoted as a franchise here. G_l^{loc} (or G_u^{user}) denotes the set of random probability measures from the perspective of location (or user), where the superscript (*loc* or *user*) indicates the franchise type. The global base probability measure G_0 is shared by not only restaurants but also franchises, which means that the topic-word distribution is a global variable shared by all perspectives. Based on this framework, we incorporate the two categories of locations, namely *distracted* and *concentrated* locations.

Location Perspective. In this setting, a restaurant l corresponds to a document d_l^{loc} with a document-level probability measure G_l^{loc} at location l. t_{li} denotes the table that customer θ_{li} chooses. Now we need a notation for counts. Let n_{lt} be the number of customers at table t at location l.

Distracted Category: Topics in such places tend to be diverse, and present very little geographical relevancy. Inspired by Pitman-Yor process [17] which encourages the model to have higher probability in choosing new tables, we alter the probability for a customer θ_{li} to choose a table t as

$$P(t_{li} = t) = \sum_{t=1}^{m_l} \frac{n_{lt} - \varepsilon}{n_{l.} + \alpha} \delta_t + \frac{\alpha + m_l \varepsilon}{n_{l.} + \alpha} G_0$$

where m_l is the number of tables already at location l, and $n_{l.}$ is the total number of customers at all tables. $\varepsilon \in [0, 1)$ is a discount parameter which encourages a customer to have higher probability to choose a new table and thereby sample a new dish from G_0. δ_t is a point mass centered at t. In other words, we encourage such locations to have more topics. When $\varepsilon = 0$, it reduces to a DP.

Concentrated Category: The probability equation is defined as

$$P(t_{li} = t) = \sum_{t=1}^{m_l} \frac{n_{lt}}{n_{l.} + \alpha} \delta_t + \frac{\alpha}{n_{l.} + \alpha} G_0$$

Since CRP well models Zipf's law, we keep this equation the same formula as CRP in choosing tables.

User Perspective. In this part, we define the notations similar to those in the previous section, and simply change the superscript of variables from *loc* to *user*. Since the locations' categories do not affect the user perspective in sampling tables, we define the probability equation as

$$P(t_{ui} = t) = \sum_{t=1}^{m_u} \frac{n_{ut}}{n_{u.} + \eta} \delta_t + \frac{\eta}{n_{u.} + \eta} G_0$$

where n_{ut} denotes the total number of customers choosing dish k with respect to the user u.

2.4   Classifying Locations

In this section, we give the details of how to calculate τ which plays as a threshold of topic count to classify locations. In our model, locations with less than τ topics are regarded as *concentrated* places, and those with more than τ topics are viewed as *distracted* ones. As depicted in Fig. 1, the power law distributions of the two categories present very different slopes, which is the major feature utilized by us. Here we maximize the difference between the two slopes by applying linear regression on both sides. In a two dimensional coordinate system, Least-Squares Regression is employed to calculate the slope degree. We first sort the numbers of topics at all locations as $\{k_1, \ldots, k_n\}$ ($k_1 < k_1 < \ldots < k_n$), and count the numbers of locations with different topic counts as $\{c_{k_1}, \ldots, c_{k_n}\}$. Specifically, c_{k_1} denotes the number of locations with k_1 topics. Now τ can be computed by applying Algorithm 1 .

Algorithm 1. Calculating threshold τ

Data:
The sorted topic counts: $k = (k_1, \ldots, k_n)$
The corresponding location counts: $c = (c_{k_1}, \ldots, c_{k_n})$

Logarithm of each element in the vector:
$k' \leftarrow log(k)$, $c' \leftarrow log(c)$
max $\leftarrow 0$ $\tau \leftarrow 0$
for $i = 1 \rightarrow n$ do
    $slop\_left$=**LeastSquares**($\{k'_1, \ldots, k'_i\}, \{c'_{k_1}, \ldots, c'_{k_i}\}$)
    $slop\_right$=**LeastSquares**($\{k'_i, \ldots, k'_n\}, \{c'_{k_i}, \ldots, c'_{k_n}\}$)
    if $|slope\_left - slope\_right| > max$ then
        $max \leftarrow |slope\_left - slope\_right|$
        $\tau \leftarrow i$
return Threshold τ

2.5   Learning and Inference

In the study of topic models, Gibbs sampling is a standard way to obtain a Markov chain over latent variables. It is used to produce a sample from a joint distribution when only conditional distributions of each variable can be efficiently computed. The framework of CRFs is also adapted to yielding a Gibbs sampling scheme for posterior sampling. In our model, we set the base measure H of G_0 having density $h(\cdot)$ which is a multinomial distribution with a Dirichlet prior parameterized by β. Let x_{ji} denote the i-th word in document j. The posterior probability of word x_{ji} under topic k given all data items except x_{ij} is

$$f_k^{-x_{ji}}(x_{ji}) = \frac{n_{k,x_{ji}}^{-x_{ji}} + \beta}{\sum_{v'=1}^{V}(n_{k,v'}^{-x_{ji}} + \beta)}$$

where $n_{k,x_{ji}}^{-x_{ji}}$ denotes the number of times that word x_{ji} appears in topic k at global level. Different from LDA [2], the state space consists of two variables, namely the table t and topic k in the CRFs metaphor. Thus we need to sample them sequentially.

Sampling Topic k: Let m_k denote the number of tables assigned to the topic k, and m. denote the total number of tables so far. $m_k^{-x_{ji}}$ represents the number of tables under the topic k when removing the word x_{ji}. $m_k^{-x_{ji}}$ changes only if the word x_{ji}'s corresponding table contains no words after removing x_{ji}. The conditional probability that a table t chooses a topic k is

$$p(k_t = k) \propto \begin{cases} m_k^{-x_{ji}} f_k^{-x_{ji}}(x_{ji}) & \text{for an existing topic} \\ \gamma f_{k^{new}}^{-x_{ji}}(x_{ji}) & \text{for a new topic} \end{cases}$$

where $f_{k^{new}}^{-x_{ji}}(x_{ji})$ equals to $1/V$, because the observation of word count $n_{k^{new},x_{ji}}^{-x_{ji}} = 0$. V is number of terms in the vocabulary. Note that the count of tables is the summation from both user and location perspectives.

Sampling Table t: When sampling an existing table, we can easily multiply the prior probability to the conditional density $f_k(\cdot)$ of topic k that the table serves. Since there are several branches discussed in the previous section, we do not repeat them. Here we only detail the probability from which a new table is sampled. Since a new table t can choose an existing topic or a totally new one, the probability should be calculated by integrating all the possible values:

$$P(t_{ji} = t_{new}) = \sum_{k=1}^{K} \frac{m_k^{-x_{ji}}}{m.^{-x_{ji}} + \gamma} f_k^{-x_{ji}}(x_{ji}) + \frac{\gamma}{m.^{-x_{ji}} + \gamma} f_{k^{new}}^{-x_{ji}}(x_{ji}).$$

Note that after each iteration, we update the parameter τ, and apply the new τ in the next iteration.

3   Experiments

In this section, we present both quantity and quality evaluations for our proposed model with some state-of-the-art baselines from two aspects: *topic modeling* and *spatial topic comparison*. *Topic modeling* evaluates the model's performance in extracting semantic coherent topics. Many efforts [5,6,18] have been devoted in this field. By incorporating geographical information, we expect the topic quality to achieve a significant improvement. Another important advantage of our model is its ability in bridging the geographical information with semantic topics. We evaluate this with *Geographical topic comparison*, which reveals the topic-specific distinctions at different locations. Four typical baselines are compared in the experiments:

- Geofolk: It is a geographical topic model proposed in [19]. It assumes that each *topic* corresponds to a latent *region*. It reports great performance in discovering location related topics for recommendation.
- LGTA: Latent Geographical Topic Analysis (LGTA) is proposed in [24] which incorporates both the spatial information and textual content. It is powerful in reducing the location and text perplexity.
- LDA: LDA [2] is the basic knowledge-free unsupervised topic model. It is impressive in discovering semantic coherent topics.
- HDP: Hierarchical Dirichlet Process [22] is a nonparametric Bayesian approach. It solves the intractable problem in LDA that the number of topics must be predefined.

Datasets: We apply our model on three public datasets: GEOText [7] and two check-in datasets [23] from Foursquare. GEOText contains geo-tagged microblog messages within the United States. The Foursquare dataset consists of the check-in data in New York City and Los Angeles, thus we briefly denote them as NYC and LA respectively. The texts in LA and NYC are tips commented by common users at different locations.

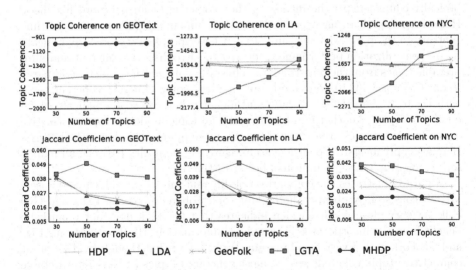

Fig. 3. Topic Coherence and Jaccard Coefficient of each model with different numbers of topics

Table 1. Three example topics of different models (*Good* words are marked in red and bold)

MHDP			LDA			Geofolk			LGTA		
Topic 1	Topic 2	Topic 3	Topic 1	Topic 2	Topic 3	Topic 1	Topic 2	Topic 3	Topic 1	Topic 2	Topic 3
chicken	wildlife	game	cheese	watch	good	case	lol	game	de	trail	taco
cheese	bird	earn	sweet	check	eat	chicken	yea	win	swag	green	de
delicious	plant	team	fries	inside	stuff	drink	kool	kobe	nom	de	swag
sandwich	nature	stadium	potato	game	cheap	sweet	aid	play	drink	dog	pee
burge	bald	soccer	sauce	west	ball	cheese	nature	team	taco	roll	kobe
salad	mammal	sport	mac	bear	cafe	hot	punk	fan	burge	lantern	nom
good	eagle	football	side	bird	beat	juice	bird	laker	game	nature	career
fries	watch	rugby	onion	video	quick	king	office	basketball	partner	cent	clip

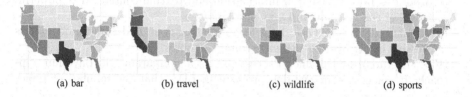

(a) bar   (b) travel   (c) wildlife   (d) sports

Fig. 4. Check-in heat map of four typical topics

3.1 Topic Modeling

Topic quality is an important metric in evaluating the performance of topic models. One of the most common ways is to compute the *perplexity*, which is algebraically equivalent to the inverse of the geometric mean per-word likelihood. However, researchers have pointed out that human judgments are sometimes contrary to the perplexity measure [4,6]. A lower perplexity score only indicates better generalization performance, but we expect to discover coherent and interpretable topics from the massive data. Thus we apply another widely used metric *Topic Coherence* [9,10,15]. Good topic models would generate coherent topics with accurate semantic clustering, and higher coherence scores indicate higher quality of topics. Another important metric for topic models is topic distinctiveness. Diversity topics are expected to offer more information beneath the texts. Here we apply Jaccard Coefficient to measure the similarity between all pairs of topics, and a lower score indicates better distinctiveness.

Figure 3 illustrates the comparison of different models on the three datasets. MHDP consistently obtains the highest topic coherence scores. LDA and GeoFolk get very close results, which indicates that GeoFolk does not remarkably benefit from the incorporated geographical information. HDP outperforms LDA and GeoFork with a very small margin. We also find that though LGTA obtains competitive topic coherent scores, most generated topics are less interpretable, and only contain some meaningless frequent words. Table 1 visually gives three corresponding example topics produced by MHDP, LDA, GeoFolk and LGTA. We can obviously find that our model generates topics with very accurate semantic clustering. LDA and GeoFolk produce some accurate topics such as topic 1 in LDA and topic 3 in GeoFolk. LGTA performs the worst, and we actually can not find better topic representations.

3.2 Spatial Topic Comparison

Another important purpose of geographical topic modeling is to discover the variation of topics at different regions, which is different from early work [13,14] that only focuses on modeling the spatial information. Figure 4 depicts a heat map which illustrates four typical topics appearing at different states of America. We can obviously distinguish the differences. The topic of *wildlife* is more popular in the midwestern with darker color, while *sports* gains its popularity all over the country. The *travel* topic covers all the states that have tourist attractions, such as beaches or mountains. This feature enables us to implement some interesting applications, such as the travel recommender system. If a user want to take photos of the wildlife, we would recommend him the popular states related to the *wildlife* topic. Meanwhile, we may also focus on another kind of applications which need fine-grained geographical topics. Our model divides places into two kinds: *distracted* and *concentrated* locations. The first kind always exists as public places such as universities or airports, while the second kind always provides some specific services such as restaurants or bars. From the Foursquare dataset, we extract the most frequent words appearing in the venues' names of the two categories, and show them in Table 2. The results are strongly consistent with our assumption. A point-of-interest (POI) recommender system employing our model can give higher weight to the locations which are labeled as *concentrated* to meet users' specific demands.

Table 2. The most frequent words appearing in the names of locations

Category	The most frequent words
Distracted locations	house, park, center, lounge, airport, school, university, museum, company, library, subway
Concentrated locations	cafe, restaurant, pizza, bar, grill, starbucks, club, salon, mexican, bakery, spa, home, hotel, mcdonald's, sephora

Table 3 lists some location-aware topics which are discovered by our model, but do not exist in the other four baselines. Our model clearly distinguishes the topic *noshery* and *restaurant*. Though LDA generates a very similar topic (Topic 1 in Table 1), it mixes them together. The topic *airport* is classified into the distracted category by MHDP. We find that the words in the topic *airport* are very diverse but strongly related to the services in a airport, which makes it easy to be distinguished from others. Thus it is reasonable to believe that our *two categories assumption* conforms to the underline semantics of geographical data, and makes the model perform better.

Table 3. Location-aware topics

Location	Topic words
Bar	beer, drink, bar, happy, music, party, night, sunday, open, enjoy
Restaurant	wine, food, menu, steak, brunch, delicious, cocktail, dinner, lobster
Noshery	chicken, cheese, sandwich, burger, salad, sauce, soup, fries, cream
Airport	line, wait, parking, card, coffee, airport, check, cash, flight, security
Planetarium	power, space, planetarium, NASA, science, energy, exhibit, air, solar

4   Conclusions

This paper proposes a novel geographical topic model MHDP which exploits the geo-tagged user-generated content from user and location perspectives. The flexible structure of MHDP makes it easy to incorporate new dimensions of data such as temporal information. We also classify locations into two categories, and experimentally demonstrate its reasonability on three public datasets. Experiments show that MHDP outperforms the baselines with significant improvements. With increasing studies focusing on location-based social networks, we believe that our proposed model is promising to advance the researches in this field.

References

1. Blei, D.M., Griffiths, T.L., Jordan, M.I.: The nested chinese restaurant process and Bayesian nonparametric inference of topic hierarchies. J. ACM (JACM) **57**, 7 (2010)
2. Blei, D.M., Ng, A.Y., Jordan, M.I.: Latent Dirichlet allocation. J. Mach. Learn. Res. **3**, 993–1022 (2003)
3. Buntine, W., Hutter, M.: A Bayesian view of the Poisson-Dirichlet process. arXiv preprint http://arxiv.org/abs/1007.0296 (2010)
4. Chang, J., Gerrish, S., Wang, C., Boyd-graber, J.L., Blei, D.M.: Reading tea leaves: how humans interpret topic models. In: Advances in Neural Information Processing Systems, pp. 288–296 (2009)
5. Chen, Z., Mukherjee, A., Liu, B., Hsu, M., Castellanos, M., Ghosh, R.: Discovering coherent topics using general knowledge. In: CIKM, pp. 209–218 (2013)
6. Chen, Z., Mukherjee, A., Liu, B., Hsu, M., Castellanos, M., Ghosh, R.: Leveraging multi-domain prior knowledge in topic models. In: IJCAI, pp. 2071–2077 (2013)
7. Eisenstein, J., O'Connor, B., Smith, N.A., Xing, E.P.: A latent variable model for geographic lexical variation. In: EMNLP, pp. 1277–1287 (2010)
8. Ferguson, T.S.: A Bayesian analysis of some nonparametric problems. Ann. Stat. **1**, 209–230 (1973)
9. He, Y., Wang, C., Jiang, C.: Discovering canonical correlations between topical and topological information in document networks. In: CIKM, pp. 1281–1290 (2015)
10. He, Y., Wang, C., Jiang, C.: Modeling document networks with tree-averaged copula regularization. In: WSDM, pp. 691–699 (2017)

11. Hong, L., Ahmed, A., Gurumurthy, S., Smola, A.J., Tsioutsiouliklis, K.: Discovering geographical topics in the twitter stream. In: WWW, pp. 769–778 (2012)
12. Hu, B., Ester, M.: Spatial topic modeling in online social media for location recommendation. In: RecSys, pp. 25–32 (2013)
13. Li, J., Adilmagambetovm, A., Jabbar, M.S.M., Zaïane, O.R., Osornio-Vargas, A., Wine, O.: On discovering co-location patterns in datasets: a case study of pollutants and child cancers. GeoInformatica **20**, 651–692 (2016)
14. Li, J., Zaïane, O.R., Osornio-Vargas, A.: Discovering statistically significant colocation rules in datasets with extended spatial objects. In: Data Warehousing and Knowledge Discovery, pp. 124–135 (2014)
15. Mimno, D., Wallach, H.M., Talley, E., Leenders, M., McCallum, A.: Optimizing semantic coherence in topic models. In: EMNLP, pp. 262–272 (2011)
16. Neal, R.M.: Markov chain sampling methods for Dirichlet process mixture models. J. Comput. Gr. Stat. **9**, 249–265 (2000)
17. Pitman, J., Yor, M.: The two-parameter Poisson-Dirichlet distribution derived from a stable subordinator. Ann. Probab. **25**, 855–900 (1997)
18. Ramage, D., Hall, D., Nallapati, R., Manning, C.D.: Labeled LDA: a supervised topic model for credit attribution in multi-labeled corpora. In: EMNLP, pp. 248–256 (2009)
19. Sizov, S.: Geofolk: Latent spatial semantics in web 2.0 social media. In: WSDM, pp. 281–290 (2010)
20. Teh, Y.W., Kurihara, K., Welling, M.: Collapsed variational inference for HDP. In: NIPS, pp. 1481–1488 (2007)
21. Teh, Y.W.: Dirichlet process. In: Encyclopedia of Machine Learning, pp. 280–287 (2010)
22. Teh, Y.W., Jordan, M.I., Beal, M.J., Blei, D.M.: Hierarchical Dirichlet processes. J. Am. Stat. Assoc. **101**, 1566–1581 (2006)
23. Wei, L.Y., Zheng, Y., Peng, W.C.: Constructing popular routes from uncertain trajectories. In: SIGKDD, pp. 195–203 (2012)
24. Yin, Z., Cao, L., Han, J., Zhai, C., Huang, T.: Geographical topic discovery and comparison. In: WWW, pp. 247–256 (2011)
25. Yuan, Q., Cong, G., Zhao, K., Ma, Z., Sun, A.: Who, where, when, and what: a nonparametric Bayesian approach to context-aware recommendation and search for twitter users. ACM Trans. Inf. Syst. **33**(1), 2 (2015)
26. Zheng, Y., Capra, L., Wolfson, O., Yang, H.: Urban computing: concepts, methodologies, and applications. TIST **5**, 38 (2014)

Enumerating Non-redundant Association Rules
Using Satisfiability

Abdelhamid Boudane[(✉)], Said Jabbour, Lakhdar Sais, and Yakoub Salhi

CRIL-CNRS, Université d'Artois, 62307 Lens Cedex, France
{boudane,jabbour,sais,salhi}@cril.fr

Abstract. Discovering association rules from transaction databases is a well studied data mining task. Many effective techniques have been proposed over the years. However, due to the huge size of the output, many works have tackled the problem of mining a smaller and relevant set of rules. In this paper, we address the problem of enumerating the minimal non-redundant association rules, widely considered as one of the most relevant variant. We first provide its encoding as a propositional formula whose models correspond to the minimal non redundant rules. Then we show that the set of minimal generators used for extracting non-redundant rules can also be encoded in this framework. Experiments on many datasets show that our approach achieves better performance with respect to the state-of-the-art specialized techniques.

1 Introduction

Extracting association rules from transactional databases have received intensive research since its introduction by Rakesh Agrawal et al. in [1]. Initially referring to data analysis, several new application domains have been identified, including among others, bioinformatics, medical diagnosis, networks intrusion detection, web mining, documents analysis, and scientific data analysis. This broad spectrum of applications enabled association analysis to be applied to a variety of datasets, including sequential, spatial, and graph-based data. Interestingly, association patterns are now considered as a building block of several other learning problems such as classification, regression, and clustering.

Most approaches have mentioned that the classical association rules mining task produces too many rules [2,5,6,14,18]. The huge size of such set of rules does not help the user to easily retrieve relevant informations. Such observation leads to various definitions of redundancy in order to limit the number of association rules. Thenceforth, many research have focused on eliminating redundant rules while maintaining the set of relevant ones called (minimal) non-redundant association rules. Different kinds of non-redundant rules have been introduced such as the Generic Basis [2], the Informative Basis [2], the Informative and Generic Basis [6], Minimum Condition Maximum Consequent Rules (MMR) [14] and the set of representative association rules [13] that cover all the association rules. To prune out redundant rules, almost approaches share the two following steps:

© Springer International Publishing AG 2017
J. Kim et al. (Eds.): PAKDD 2017, Part I, LNAI 10234, pp. 824–836, 2017.
DOI: 10.1007/978-3-319-57454-7_64

(1) find the set of minimal generators and closed itemsets, and (2) generate confident rules by considering the two sets already mined in step one.

Recently, declarative approaches have been proposed to tackle several data mining tasks through constraint programming (CP) and propositional satisfiability (SAT) [7,8,11,12,15]. In [3], the authors proposed a new framework for mining association rules in one step using propositional satisfiability leading to a competitive approach compared to specialized techniques. Encouraged by these results, we propose in this paper to extend this framework for extracting the minimal non-redundant rules. The redundancy is eliminated elegantly using new constraints combined to some others listed in [3]. We show that two kinds of non-redundant rules can be addressed. Furthermore, a restriction of our encoding can be used to extract the minimal generators.

2   Preliminaries

2.1   Propositional Logic and SAT Problem

We here define the syntax and the semantics of propositional logic. Let Prop be a countably set of propositional variables. We use the letters p, q, r, etc. to range over Prop. The set of *propositional formulas*, denoted Form, is defined inductively started from Prop, the constant $\bot$ denoting false, the constant $\top$ denoting true, and using the logical connectives $\neg$, $\wedge$, $\vee$, $\rightarrow$. We use $Var(\phi)$ to denote the set of propositional variables appearing in the formula ϕ. The equivalence connective $\leftrightarrow$ is defined by $\phi \leftrightarrow \psi \equiv (\phi \rightarrow \psi) \wedge (\psi \rightarrow \phi)$.

A formula ϕ in *conjunctive normal form (CNF)* is a conjunction of clauses, where a *clause* is a disjunction of literals. A *literal* is a positive (p) or negated ($\neg p$) propositional variable. The two literals p and $\neg p$ are called *complementary*. A CNF formula can also be seen as a set of clauses, and a clause as a set of literals.

An *interpretation* $\mathcal{I}$ of a propositional formula ϕ is a function which associates a value $\mathcal{I}(p) \in \{0, 1\}$ (0 corresponds to *false* and 1 to *true*) to the variables $p \in Var(\phi)$. A *model or an implicant* of a formula Φ is an interpretation $\mathcal{I}$ that satisfies the formula in the usual truth-functional way. *SAT problem* consists in deciding if a given CNF formula admits a model or not.

2.2   Association Rules

Let Ω be a finite non empty set of symbols, called *items*. From now on, we assume that this set is fixed. We use the letters a, b, c, etc. to range over the elements of Ω. An *itemset* I over Ω is defined as a subset of Ω, i.e., $I \subseteq \Omega$. We use 2^{Ω} to denote the set of itemsets over Ω and we use the capital letters I, J, K, etc. to range over the elements of 2^{Ω}.

A *transaction* is an ordered pair (i, I) where i is a natural number, called *transaction identifier*, and I an itemset, i.e., $(i, I) \in \mathbb{N} \times 2^{\Omega}$. A *transaction database* $\mathcal{D}$ is defined as a finite non empty set of transactions ($\mathcal{D} \subseteq \mathbb{N} \times 2^{\Omega}$) where each transaction identifier refers to a unique itemset.

Given a transaction database $\mathcal{D}$ and an itemset I, the *cover* of I in $\mathcal{D}$, denoted $\mathcal{C}(I, \mathcal{D})$, is defined as $\{i \in \mathbb{N} \mid (i, J) \in \mathcal{D} \text{ and } I \subseteq J\}$. The *support* of I in $\mathcal{D}$, denoted $Supp(I, \mathcal{D})$, corresponds to the cardinality of $\mathcal{C}(I, \mathcal{D})$, i.e., $Supp(I, \mathcal{D}) = |\mathcal{C}(I, \mathcal{D})|$. An itemset $I \subseteq \Omega$ such that $Supp(I, \mathcal{D}) \geqslant 1$ is a *closed itemset* if, for all itemsets J with $I \subset J$, $Supp(J, \mathcal{D}) < Supp(I, \mathcal{D})$.

Example 1. For instance, let us consider the transaction database $\mathcal{D}$ depicted in Table 1. We have $\mathcal{C}(\{c, d\}, \mathcal{D}) = \{1, 2, 3, 4, 5\}$ and $Supp(\{c, d\}, \mathcal{D}) = 5$ while $Supp(\{f\}, \mathcal{D}) = 3$. The itemset $\{c, d\}$ is closed, while $\{f\}$ is not since $Supp(\{f\}, \mathcal{D}) = Supp(\{c, d, f\}, \mathcal{D})$.

Table 1. A transaction database $\mathcal{D}$

tid	Transactions
1	$c\ d\ e\ f\ g$
2	$c\ d\ e\ f\ g$
3	$a\ b\ c\ d$
4	$a\ b\ c\ d\ \ f$
5	$a\ b\ c\ d$
6	$c\ \ \ \ e$

Table 2. Some association rules

Name	Asso. rules	Support	Confidence
r_1	$\{a\} \rightarrow \{b\}$	3/6	1
r_2	$\{a\} \rightarrow \{b, c, d\}$	3/6	1
r_3	$\{c\} \rightarrow \{d\}$	5/6	5/6
r_4	$\{c, d\} \rightarrow \{e, f, g\}$	2/6	2/5

In this work, we are interested in the problem of mining association rules (MAR). An *association rule* is a pattern of the form $X \rightarrow Y$ where X (called the antecedent) and Y (called the consequent) are two disjoint itemsets. In MAR, the interestingness predicate is defined using the notions of support and confidence. The *support of an association rule* $X \rightarrow Y$ in a transaction database $\mathcal{D}$, defined as $Supp(X \rightarrow Y, \mathcal{D}) = \frac{Supp(X \cup Y, \mathcal{D})}{|\mathcal{D}|}$, determines how often a rule is applicable to a given dataset, i.e., the occurrence frequency of the rule. The *confidence* of $X \rightarrow Y$ in $\mathcal{D}$, defined as $Conf(X \rightarrow Y, \mathcal{D}) = \frac{Supp(X \cup Y, \mathcal{D})}{Supp(X, \mathcal{D})}$, provides an estimate of the conditional probability of Y given X. When there is no ambiguity, we omit to mention the transaction database $\mathcal{D}$, and we simply note $Supp(X \rightarrow Y)$ and $Conf(X \rightarrow Y)$.

A *valid association rule* is an association rule with support and confidence greater than or equal to the minimum support threshold (minsupp) and minimum confidence threshold (minconf) respectively. More precisely, given a transaction database $\mathcal{D}$, a minimum support threshold minsupp and a minimum confidence threshold minconf, the problem of mining association rules consists in computing $MAR(\mathcal{D}, minsupp, minconf) = \{X \rightarrow Y \mid X, Y \subseteq \Omega, Supp(X \rightarrow Y, \mathcal{D}) \geqslant minsupp, Conf(X \rightarrow Y, \mathcal{D}) \geqslant minconf\}$.

Table 2 illustrates some association rules with their corresponding supports and confidences. For instance, $Supp(\{a\} \rightarrow \{b\}) = \frac{3}{6}$ and $Conf(\{a\} \rightarrow \{b\}) = 1$.

3   SAT-Based Association Rules Mining

In this section, we briefly review the recent approach proposed in [3] for mining association rules through Boolean satisfiability. The basic idea consists in modeling such mining task as a propositional formula whose models corresponds to the required association rules. In this encoding, two sets of Boolean variables are used to represent the items of an association rules $X \rightarrow Y$ and the transactions. Then, the support and the confidence of an association rule are captured through 0/1 linear inequalities over the Boolean variables associated to transactions. In order to define the SAT-based encoding, we fix, without loss of generality, a set Ω of n items, a transaction database $\mathcal{D} = \{(1, I_1), \ldots, (m, I_m)\}$ where $\forall i \in \{1, m\}, I_i \subseteq \Omega$, a minimum support threshold $minsupp$ and a minimum confidence threshold $minconf$.

In order to capture the two part of each association rule, we associate two Boolean variables to each item a, denoted x_a and y_a. The variables of the form x_a (resp. y_a) are used to represent the antecedent (resp. consequent) of each candidate rule. Then, to represent the cover of X and $X \cup Y$, each transaction identifier $i \in \{1, m\}$ is associated with two propositional variables p_i and q_i. The variables of the form p_i (resp. q_i) are used to represent the cover of X (resp. $X \cup Y$). More precisely, given a Boolean interpretation $\mathcal{B}$, the corresponding association rule, denoted $r_{\mathcal{I}}$, is $X = \{a \in \Omega \mid \mathcal{I}(x_a) = 1\} \rightarrow Y = \{b \in \Omega \mid \mathcal{I}(y_b) = 1\}$, the cover of X is $\{i \in \{1, m\} \mid \mathcal{I}(p_i) = 1\}$, and the cover of $X \cup Y$ is $\{i \in \{1, m\} \mid \mathcal{I}(q_i) = 1\}$. The SAT-based encoding of the problem of enumerating association rules consists in a set of constraints defined as follows.

$$(\bigvee_{a \in \Omega} x_a) \wedge (\bigvee_{a \in \Omega} y_a) \qquad (1)$$

$$\bigwedge_{i \in 1..m} \neg q_i \leftrightarrow \neg p_i \vee (\bigvee_{a \in \Omega \setminus I_i} y_a) \qquad (4)$$

$$\bigwedge_{a \in \Omega} (\neg x_a \vee \neg y_a) \qquad (2)$$

$$\sum_{i \in 1..m} q_i \geqslant m \times minsupp \qquad (5)$$

$$\bigwedge_{i \in 1..m} \neg p_i \leftrightarrow \bigvee_{a \in \Omega \setminus I_i} x_a \qquad (3)$$

$$\frac{\sum_{i \in 1..m} q_i}{\sum_{i \in 1..m} p_i} \geqslant minconf \qquad (6)$$

The two clauses of formula (1) express that X and Y are not empty sets. Formula (2) allows to express $X \cap Y = \emptyset$. It is simply defined by imposing that x_a and y_a are not both true for every item a. The third constraint is used to represent the cover of the itemset corresponding to the left part of the candidate association rule. Given an itemset X, we know that the transaction identifier i does not belong to $\mathcal{C}(X, \mathcal{D})$ if and only if there exists an item $a \in X$ such that $a \notin I_i$. This property is represented by constraint (3) expressing that p_i is $false$ if and only if X contains an item that does not belong to the transaction i. In the same way, the formula (4) allows to capture the cover of $X \cup Y$.

To specify that the support of the candidate rule has to be greater than or equal to the fixed threshold $minsupp$ (in percentage), and the confidence is greater than or equal to $minconf$, we use the constraints (5) and (6) expressed by 0/1 linear inequalities.

To extend the mining task to the closed association rules, the following constraint is added to express that $X \cup Y$ is a closed itemset [9]:

$$\bigwedge_{a \in \Omega} ((\bigwedge_{i \in 1..m} q_i \to a \in I_i) \to x_a \vee y_a) \tag{7}$$

This formula means that, for all item $a \in \Omega$, if we have $\mathcal{C}(X \cup Y, \mathcal{D}) = \mathcal{C}(X \cup Y \cup \{a\}, \mathcal{D})$, which is encoded with the formula $\bigwedge_{i \in \{1,m\}} q_i \to a \in I_i$, then we get $a \in X \cup Y$, which is encoded with $x_a \vee y_a$.

4   Minimal Non-redundant Association Rules

In this section, we present our encoding of the problem of extracting non-redundant rules into propositional satisfiability. First, we focus on the interesting representation that corresponds to the minimal non-redundant association rules (MNRs in short) [2,14].

Definition 1. *An association rule* $r : X \to Y$ *is a minimal non-redundant rule iff there is no association rule* $r' : X' \to Y'$ *different from* r *s.t. (i)* $Supp(r) = Supp(r')$, *(ii)* $Conf(r) = Conf(r')$ *and (iii)* $X' \subseteq X$ *and* $Y \subseteq Y'$.

Example 2. Consider again the association rules given in Table 2. In this set of rules, $r_2 : \{a\} \to \{b, c, d\}$ is a minimal non-redundant rule while $r_1 : \{a\} \to \{b\}$ is not.

In the following proposition, we point out that all the minimal non-redundant association rules are closed.

Proposition 1. *If* $r : X \to Y$ *is a minimal non-redundant association rule in a transaction database* $\mathcal{D}$ *then* $X \cup Y$ *is a closed itemset* $\mathcal{D}$.

Proof. Assume that $X \cup Y$ is not a closed itemset. Then, there exists an item $a \notin X \cup Y$ s.t. $Supp(X \cup Y, \mathcal{D}) = Supp(X \cup Y \cup \{a\}, \mathcal{D})$. Consider now the rule $r' : X \to Y \cup \{a\}$. Clearly, we get $Supp(r) = Supp(r')$ and $Conf(r) = Conf(r')$ since $Supp(X \cup Y, \mathcal{D}) = Supp(X \cup Y \cup \{a\}, \mathcal{D})$. Thus, r is not a minimal non-redundant association rule and we get a contradiction.

In other words, the minimal non-redundant association rules are the closed rules in which the antecedents are minimal w.r.t. set inclusion. Using this property, the authors of [2] provided a characterization of the antecedents of the minimal non-redundant rules, called minimal generators.

Definition 2 (Minimal Generator). *Given a closed itemset* X *in a transaction database* $\mathcal{D}$, *an itemset* $X' \subseteq X$ *is a minimal generator of* X *iff* $Supp(X', \mathcal{D}) = Supp(X, \mathcal{D})$ *and there is no* $X'' \subseteq X$ *s.t.* $X'' \subset X'$ *and* $Supp(X'', \mathcal{D}) = Supp(X, \mathcal{D})$.

Usual algorithms use the set of frequent closed itemsets together with minimal generators to extract the set of minimal non-redundant association rules. Then, most existing approaches to mine minimal association rules proceed in two steps. In our approach, we propose to extend the SAT-based encoding proposed in [3] to retrieve the minimal non-redundant association rules in one step.

In order to define a SAT-based encoding of the problem of generating the minimal non-redundant association rules, we only need to extend the encoding described in Sect. 3 with a formula that forces each antecedent to be a minimal generator. To this end, we use a formula that represents the fact that if $Supp(X \to Y, \mathcal{D}) = Supp(X \setminus \{a\} \to Y, \mathcal{D})$, then a has to be excluded from X, i.e., $a \notin X$. However, we write the contraposition of this property. Indeed, the following formula expresses that, for all item a, if a belongs to X then the support of X is smaller than the support of $X \setminus \{a\}$:

$$(\bigwedge_{a \in \Omega} x_a \to \bigvee_{(i \in 1..m,\ a \notin I_i)} (\bigwedge_{b \notin I_i \cup \{a\}} \neg x_b)) \vee (\sum_{b \in \Omega} x_b = 1) \tag{8}$$

We use $\mathcal{E}_{MNR}(\mathcal{D}, minsupp, minconf)$ to denote the encoding (1) $\wedge$ (2) $\wedge$ (3) $\wedge$ (4) $\wedge$ (5) $\wedge$ (6) $\wedge$ (7) $\wedge$ (8).

The soundness of $\mathcal{E}_{MNR}(\mathcal{D}, minsupp, minconf)$ comes directly from the following proposition:

Proposition 2. *The association rule* $r : X \to Y$ *is a minimal non-redundant rule iff* r *is a closed association rule, and* $|X| = 1$ *or, for all item* $a \in X$, $Supp(X, \mathcal{D}) > Supp(X \setminus \{a\}, \mathcal{D})$.

Proof.
Part $\Rightarrow$. Using Proposition 1, we know that r is a closed association rule. Assume now that there exists an item $a \in X$ s.t. $Supp(X, \mathcal{D}) = Supp(X \setminus \{a\}, \mathcal{D})$. Then, $r' : X \setminus \{a\} \to Y \cup \{a\}$ is a closed association rule s.t. $Supp(r, \mathcal{D}) = Supp(r', \mathcal{D})$ and $Conf(r, \mathcal{D}) = Conf(r', \mathcal{D})$. Thus, we get a contradiction since r is a minimal non-redundant association rule.

Part $\Leftarrow$. Using the fact that r is a closed association rule, we know that there is no association rule $r' : X' \to Y'$ s.t. $X \cup Y \subset X' \cup Y'$ and $Supp(r, \mathcal{D}) = Supp(r', \mathcal{D})$. Moreover, knowing that $Supp(X, \mathcal{D}) > Supp(X \setminus \{a\}, \mathcal{D})$ for every $a \in X$, we get $Conf(X \setminus \{a\} \to Y \cup \{a\}, \mathcal{D}) < Conf(r, \mathcal{D})$ for every $a \in X$. As a consequence, r is a minimal non-redundant association rule.

The soundness of our encoding means that a Boolean interpretation $\mathcal{I}$ is a model of $\mathcal{E}_{MNR}(\mathcal{D}, minsupp, minconf)$ if and only if $X = \{a \in \Omega \mid \mathcal{I}(x_a) = 1\} \to Y = \{b \in \Omega \mid \mathcal{I}(y_b) = 1\}$ is a minimal non-redundant association rule.

Proposition 3. *The encoding* $\mathcal{E}_{MNR}(\mathcal{D}, minsupp, minconf)$ *is sound.*

Proof. It come from the soundness of the encoding (1) $\wedge$ (2) $\wedge$ (3) $\wedge$ (4) $\wedge$ (5) $\wedge$ (6) $\wedge$ (7) w.r.t. the problem of generating closed association rules, Proposition 2 and the fact that (8) expresses that $Supp(X, \mathcal{D}) > Supp(X \setminus \{a\}, \mathcal{D})$ for every $a \in X$.

Let us note that the constraint (8) is not a CNF formula. In order to avoid the blow up in terms of the number of clauses resulting from the transformation of (8) into CNF, new additional variables can be added to present the subformulas of the form $\bigwedge_{b \notin I_i \cup \{a\}} \neg x_b$ i.e., $z_i \leftrightarrow \bigwedge_{b \notin I_i \cup \{a\}} \neg x_b$. Nonetheless, using this transformation, the number of resulting clauses from constraint (8) is in $O(m \times |\Omega|^2)$ which may make the model enumeration much more harder. To limit the number of clauses, we propose the following transformation which is equivalent to the property captured by (8).

$$ (\bigwedge_{a \in \Omega} (x_a \rightarrow \bigvee_{(i \in 1..m, \, a \notin I_i)} \neg z_i)) \wedge (\bigwedge_{i \in 1..m} (\neg z_i \rightarrow \sum_{b \notin I_i} x_b \leq 1)) \vee (\sum_{b \in \Omega} x_b = 1) \ (9) $$

In fact, this transformation comes from the fact that $(\bigwedge_{b \notin I_i \cup \{a\}} \neg x_b)$ is equivalent to $(\sum_{b \notin I_i} x_b \leq 1)$ in the case where I_i does not contain a. As a consequence, (9) expresses exactly the requirements of (8). The additional variables z_i allow to obtain an efficient encoding.

Note that (9) can be encoded in $O(m \times |\Omega|)$ rather than $O(m \times |\Omega|^2)$ of the previous formulation. A linear constraint of the form $\sum_{i=1}^{n} x_i \leq 1$, commonly called AtMostOne constraint, can be encoded in a linear way [16] using additional variables as follows.

$$ (\neg x_1 \vee s_1) \wedge (\neg x_n \vee \neg s_{n-1}) \wedge \bigwedge_{1 < i < n} (\neg x_i \vee s_i) \wedge (\neg s_{i-1} \vee s_i) \wedge (\neg x_i \vee \neg s_{i-1}) \quad (10) $$

Thus, the constraint $(\neg y \rightarrow \sum_{i=1}^{n} x_i \leq 1)$ can be obtained by adding y to each clause of (10). However, this can slow down the unit propagation process. In fact, when more than one x_i is assigned to true, y is not deduced to be true directly by unit propagation. To increase the power of unit propagation, one need to add y only on negatives binary clauses of (10) as shown in (11).

$$ (\neg x_1 \vee s_1) \wedge (y \vee \neg x_n \vee \neg s_{n-1}) \wedge \bigwedge_{1 < i < n} (\neg x_i \vee s_i) \wedge (\neg s_{i-1} \vee s_i) \wedge (y \vee \neg x_i \vee \neg s_{i-1}) \ (11) $$

It is worth noting that one can use some of the constraints above to enumerate all the minimal generators. As mentioned before, the minimal generators are the antecedents of the minimal non-redundant rules. As a consequence, the encoding (3) ∧ (5) ∧ (9) (restricted to X) allows us to get all the minimal generators.

Another notion of non-redundant rules has been defined in the work of Zaki [18]. It is slightly different from representative rules defined in [13]. It consists in mining association rules, called the most general rules (MGR in short), that have the shortest antecedent and consequent (in terms of inclusion) in an equivalent class of rules (with the same confidence and support).

Definition 3. [18] *An association rule* $r : X \rightarrow Y$ *is a non-redundant rule iff there is no association rule* $r' : X' \rightarrow Y'$ *different from* r *s.t. (i)* $Supp(r) = Supp(r')$, *(ii)* $Conf(r) = Conf(r')$ *and (iii)* $X' \subseteq X$ *and* $Y' \subseteq Y$.

Unlike the non-redundant notion in Definition 1, the closure constraint on $X \cup Y$ in Zaki's notion is obviously omitted.

Example 3. Considering again the association rules of Table 2. The rule r_1 : $\{a\} \to \{b\}$ is non-redundant while $r_2 : \{a\} \to \{b, c, d\}$ is not.

Proposition 4 provides a characterization of Zaki's non-redundant association rules.

Proposition 4. *Given an association rule $r : X \to Y$ in a transaction database $\mathcal{D}$, r is a non-redundant rule iff (i) $|X| = 1$ or $\forall a \in X$, $Supp(X \setminus \{a\}, \mathcal{D}) > Supp(X, \mathcal{D})$; and (ii) $|Y| = 1$ or $\forall b \in Y$, $Supp(X \cup Y) < Supp(X \cup Y \setminus \{b\})$.*

Proof.
Part $\Rightarrow$. Assume that $|X| > 1$ and there exists $a \in X$ such that $Supp(X \setminus \{a\}, \mathcal{D}) = Supp(X, \mathcal{D})$. Then, $Supp(X \setminus \{a\} \to Y, \mathcal{D}) = Supp(r, \mathcal{D})$ holds. Moreover, we have $Supp(X \cup Y \setminus \{a\}, \mathcal{D}) = Supp(X \cup Y, \mathcal{D})$. Thus, we have $Conf(X \setminus \{a\} \to Y, \mathcal{D}) = Conf(r, \mathcal{D})$. As a consequence, we get a contradiction since r is non-redundant rule, and we obtain the property (i).

Assume now that there exists $b \in Y$ such that $|Y| > 1$ and $Supp(X \cup Y \setminus \{b\}, \mathcal{D}) = Supp(X \cup Y, \mathcal{D})$. Then, $Conf(X \to Y \setminus \{b\}, \mathcal{D}) = Conf(X \to Y, \mathcal{D})$ holds. Moreover, we have $Supp(X \to Y \setminus \{b\}, \mathcal{D}) = Supp(X \to Y, \mathcal{D})$. Thus, using the fact that r is a non-redundant rule, we get a contradiction, and then we obtain the property (ii).

Part $\Leftarrow$. Assume that r is a redundant rule. Then, there exists $a \in X \cup Y$ s.t. $Supp(X \setminus \{a\} \to Y, \mathcal{D}) = Supp(r, \mathcal{D})$ if $a \in x$, and $Conf(X \to Y \setminus \{a\}, \mathcal{D}) = conf(r, \mathcal{D})$ otherwise. Thus, we get $Supp(X \setminus \{a\}, \mathcal{D}) = Supp(X, \mathcal{D})$ if $a \in X$, and $Supp(X \cup Y) = Supp(X \cup Y \setminus \{b\})$ otherwise. As a consequence, using the properties (i) and (ii) we get a contradiction. Therefore, r is non-redundant.

Using the characterization provided in Proposition 4, we only need to add to the encoding $\mathcal{E}_{MNR}(\mathcal{D}, minsupp, minconf)$ without the closeness constraint a new constraint representing the property (ii) to get an encoding for mining Zaki's non-redundant rules. Our definition of such constraint is as follows:

$$\bigwedge_{a \in \Omega} y_a \to (\bigvee_{(i \in 1..m, \, a \notin I_i)} (p_i \wedge \bigwedge_{b \notin I_i \cup \{a\}} \neg y_b)) \vee (\sum_{b \in \Omega} y_b = 1) \qquad (12)$$

It is worth noting that the constraint (12) is very similar to (8). Indeed, the difference is in the fact that we use the variables p_i to reason about the cover of $X \cup Y$ and not only Y. Furthermore, one can easily see that (12) can be encoded into a CNF formula in the same way as (8).

5   Experiments

In this section, we present a comparative experimental evaluation of our proposed approach with specialized association rules mining algorithms. We consider the minimal non redundant (MNR) association rules mining task.

To enumerate the set of models of the resulting CNF formula, we follow the approach of [3]. The proposed model enumeration algorithm is based on a backtrack search DPLL-like procedure. In our experiments, the variables ordering heuristic, focus in priority on the variables of respectively X and Y to select the one to assign next. The main power of this approach consists in using watched literals structure to perform accurately the unit propagation. Let us also note that the constraint (5) and (6) dedicated to frequency and confidence are managed without translation into CNF form, leading to an hybrid SAT-CSP model enumeration algorithm. Indeed, the linear inequalities (5) and (6) are managed and propagated on the fly as usually done in constraint programming. Each model of the propositional formula encoding the association rules mining task, corresponds to an association rule obtained by considering the truth values of the propositional variables encoding the antecedent (X) and the consequent (Y) of this rule.

In the experiments, *SAT4MNR* indicates our SAT based solver for mining the minimal non redundant association rules. In addition we consider *SAT4MNR-D* that partition the search as in [10]. This is done as follows: Let $\Omega = \{a_1, \ldots, a_n\}$, we transform the problem into n mining problem where each one encodes rules $X \rightarrow Y$ s.t. $\{a_1 \ldots, a_{i-1}\} \not\subset X$ and $a_i \in X$. Moreover, we denote by *SAT4MGR* our SAT based solver for mining most general rules (Definition 3).

To assess the performance of our constraint based encoding for minimal non-redundant rules, we compare our solver to two specialized association rules mining solvers namely *CORON*[1] and *SPMF*[2] [4]. *CORON* and *SPMF* are two multi-purpose data mining toolkits, implemented in Java, and which incorporate a rich collection of data mining algorithms. For *minimal non redundant* association rules, we compare our approach to the *ZART* algorithm implemented in *CORON* and *SPMF* toolkits, which is one of the recent and the most efficient state-of-the-art algorithms for enumerating minimal non redundant association rules [17]. Let us recall that *ZART* finds the minimal non redundant associations rules in two steps. Firstly, the set of all frequent closed itemsets and the minimal generators are extracted rapidly. Second, the identification of non-redundant rules is then performed. This two steps-based procedure is more time consuming.

To compare the performances of our proposed approach, for each data we proceed by varying the support from 5% to 100% with an interval of size of 5%. The confidence is varied in the same way. Then, for each data, a set of 400 configurations is generated. All the experiments were done on Intel Xeon quad-core machines with 32 GB of RAM running at 2.66 Ghz. For each instance, we fix the timeout to 15 min of CPU time.

Results: Table 3 describes our comparative results. We report in column 1 the name of the data and its characteristics in parenthesis: number of items (#items), number of transactions (#trans) and density. For each algorithm, we report the number of solved configurations (#S), and the average solving time (*avg.time* in seconds). For each unsolved configuration, the time is set to 900 s (time out). In the last row of Table 3, we provide the total number of solved configurations and the global average CPU time in seconds.

[1] Coron: http://coron.loria.fr/site/system.php.
[2] SPMF: http://www.philippe-fournier-viger.com/spmf/.

Table 3. Non-redundant associations rules: $SAT4MNR$ vs $CORON$ vs $SPMF$

data (#items, #trans, density)	SAT4MNR-D		SAT4MNR		CORON		SPMF		SAT4MGR	
	#S	avg. time (s)	#S	avg. time (s)	#S	avg. time (s)	#S	avg. time (s)	#S	avg. time (s)
Audiology (148, 216, 45%)	21	854,82	21	854.87	20	855.01	20	855.00	20	855.00
Zoo-1 (36, 101, 44%)	400	0.23	400	0.27	400	1.35	373	108.60	400	0.71
Tic-tac-toe (27, 958, 33%)	400	0.34	400	0.14	400	0.24	400	0.20	400	0.61
Anneal (93, 812, 45%)	279	337.25	248	405.82	160	591.39	80	724.46	221	461.05
Australian-c (125, 653, 41%)	298	265.74	278	309.32	251	352.01	220	417.94	263	358.40
German-c (112, 1000, 34%)	354	149.03	328	212.58	321	206.34	278	294.45	304	272.88
H-cleveland (95, 296, 47%)	331	200.28	317	235.79	271	307.57	240	368.21	286	289.28
Hepatitis (68, 137, 50%)	360	140.69	343	170.89	286	284.09	260	331.57	315	228.13
Hypothyroid (88, 3247, 49%)	150	615.13	126	649.22	104	681.52	80	751.23	109	676.03
kr-vs-kp (73, 3196, 49%)	198	504.62	172	556.85	168	552.04	140	627.64	158	583.25
Lymph (68, 148, 40%)	400	6.78	400	19.21	357	131.07	280	316.78	395	37.15
Mushroom (119, 8124, 18%)	400	146.87	389	77.02	400	3.81	360	97.25	354	181.89
P-tumor (31, 336, 48%)	400	2.08	400	4.61	400	4.15	379	87.66	400	8.11
Soybean (50, 650, 32%)	400	0.36	400	0.20	400	0.61	380	48.51	400	2.26
Vote (48, 435, 33%)	400	5.43	400	30.46	364	87.56	380	84.82	372	111.06
Total	**4790**	**215.31**	**4622**	**235.15**	4302	270.58	3870	340.94	4397	271.05

According to such results, $SAT4MNR$ outperforms the two specialized solvers $CORON$ and $SPMF$. It solves 488 configurations more than $CORON$ and 920 more than $SPMF$. $SAT4MNR$-D is the best on all the data in terms of the number of solved configurations and average CPU time, Except for *mushroom* data where $CORON$ is better in term of time but $SAT4MNR$-D solves all the configurations. Let us remark that for *mushroom* data, the number of minimal non redundant association rules is very limited. This explains why $SAT4MNR$ is worse than $CORON$ on this data. For instance, on *anneal* data, $SAT4MNR$ is remarkably efficient. It solves about 100 configurations more than $CORON$ and about 200 configurations more than $SPMF$. We can also remark that for *Lymph* data $SAT4MNR$-D solves all the configurations in an average time of 7 s where $CORON$ and $SPMF$ cannot solve all the configurations and they take a lot of time compared to $SAT4MNR$-D. More generally, the higher the density of the data, the better are the performances of $SAT4MNR$. Interestingly enough, partitioning the mining, allows to push further the performances of $SAT4MNR$. In fact, $SAT4MNR$-D allows us to obtain better performances i.e., 168 more solved instances and the average time solving is improved from 235.15 to 215.31. Unsurprisingly, $SAT4MGR$, solves less configurations than $SAT4MNR$. In fact, the set of minimal non-redundant rules is known to be reduced related to most general non-redundant ones.

Figure 1 depicts the behavior of the considered association rules mining app-roach on two representative data, *Anneal* and *kr-vs-kp*. The results are obtained by varying one parameter, while maintaining the others fixed. When the mini-mum support decreases, the time needed to find all the rules increases. Let us remark that for $CORON$ and $SPMF$ the time increases rapidly compared to $SAT4MNR$-D. For *anneal* data $SPMF$ (resp. $CORON$) is not able to provide all non redundant rules when the minimum support is lower than 85%(resp. 65%). In contrast, with $SAT4MNR$ and $SAT4MNR$-D it is possible to obtain all rules

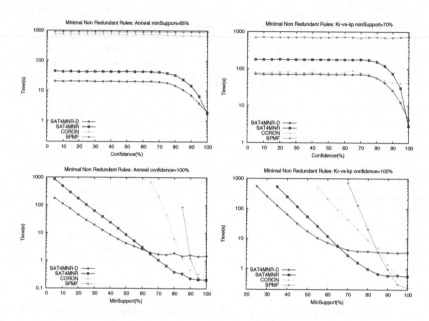

Fig. 1. Results highlights: *Anneal* and *kr-vs-kp*

for all values in the minimum support range. For *kr-vs-kp* it is important to note that the time needed to extract rules increases drastically for *SPMF* and *CORON* even if the confidence is higher. For instance, when the minimum support goes from 100% to 80% the time is multiplied by at least 10. Such increasing is very limited for *SAT4MNR* and *SAT4MNR-D*.

Finally, in Table 4, we provide the variation of the ratio between the number of classical (pure) rules, closed, generalized non redundant rules, and the minimal non-redundant rules for *kr-vs-kp* data. As we can observe, the number of minimal non-redundant association rules is smaller than those of generalized ones. The latter is smaller than closed association rules that is itself smaller than pure ones especially. For instance, when minimum support is equal to 40, the minimal non-redundant association rules presents 2.85% from all the classical association rules where the generalized ones is about 3.90%.

Table 4. *kr-vs-kp* : Pure vs Closed vs MNR vs MGR

Minimum support (%)	40	45	50	55	60	65	70
#Pures/#Closed	7.67	5.68	3.64	2.99	2.46	1.95	1.67
#Closed/#MGR	2.40	2.16	1.95	1.78	1.61	1.46	1.35
#MGR/#MNR	1.94	1.83	1.73	1.63	1.54	1.45	1.38

6   Conclusion and Perspectives

In this paper we proposed a novel approach for discovering non-redundant association rules. We show that non-redundant rules with minimum antecedent and maximum consequences can be captured by modeling this problem into propositional satisfiability. We demonstrated that our approach is highly declarative and flexible. Indeed, we have shown that minimal generators can be extracted using similar kind of constraints. We have also shown how to catch the non-redundant rules with minimum antecedent and minimum consequences. The experimental evaluation shows that our proposed approach achieves better performance than specialized mining techniques.

As a future work, we plan to address the question of mining most general rules having adjacent itemsets [18] using satisfiability to have a compact representation of the set of most general non-redundant rules.

References

1. Agrawal, R., Imielinski, T., Swami, A.: Mining association rules between sets of items in large databases. In: Proceedings of SIGMOD 1993, pp. 207–216 (1993)
2. Bastide, Y., Pasquier, N., Taouil, R., Stumme, G., Lakhal, L.: Mining minimal non-redundant association rules using frequent closed itemsets. In: Lloyd, J., Dahl, V., Furbach, U., Kerber, M., Lau, K.-K., Palamidessi, C., Pereira, L.M., Sagiv, Y., Stuckey, P.J. (eds.) CL 2000. LNCS (LNAI), vol. 1861, pp. 972–986. Springer, Heidelberg (2000). doi:10.1007/3-540-44957-4_65
3. Boudane, A., Jabbour, S., Sais, L., Salhi, Y.: A sat-based approach for mining association rules. In: Proceedings of IJCAI 2016, pp. 2472–2478 (2016)
4. Fournier-Viger, P., Gomaric, A., Gueniche, T., Soltani, A., Wu, C.-W., Tseng, V.S.: SPMF: a java open-source pattern mining library. J. Mach. Learn. Res. **15**(1), 3389–3393 (2014)
5. Fournier-Viger, P., Tseng, V.S.: TNS: mining top-k non-redundant sequential rules. In: Proceedings of SAC 2013, pp 164–166 (2013)
6. Gasmi, G., Yahia, S.B., Nguifo, E.M., Slimani, Y.: IGB: a new informative generic base of association rules. In: Proceedings of PAKDD 2005, pp. 81–90 (2005)
7. Guns, T., Nijssen, S., Raedt, L.D.: Itemset mining: a constraint programming perspective. Artif. Intell. **175**(12–13), 1951–1983 (2011)
8. Guns, T., Nijssen, S., Raedt, L.D.: k-pattern set mining under constraints. IEEE Trans. Knowl. Data Eng. **25**, 402–418 (2013)
9. Jabbour, S., Sais, L., Salhi, Y.: The top-k frequent closed itemset mining using top-k sat problem. In: Proceedings of ECML/PKDD 2013, pp. 403–418 (2013)
10. Jabbour, S., Sais, L., Salhi, Y.: Decomposition based SAT encodings for itemset mining problems. In: Cao, T., Lim, E.-P., Zhou, Z.-H., Ho, T.-B., Cheung, D., Motoda, H. (eds.) PAKDD 2015. LNCS (LNAI), vol. 9078, pp. 662–674. Springer, Cham (2015). doi:10.1007/978-3-319-18032-8_52
11. Järvisalo, M.: Itemset mining as a challenge application for answer set enumeration. In: Delgrande, J.P., Faber, W. (eds.) LPNMR 2011. LNCS (LNAI), vol. 6645, pp. 304–310. Springer, Heidelberg (2011). doi:10.1007/978-3-642-20895-9_35
12. Khiari, M., Boizumault, P., Crémilleux, B.: Constraint programming for mining n-ary patterns. In: Cohen, D. (ed.) CP 2010. LNCS, vol. 6308, pp. 552–567. Springer, Heidelberg (2010). doi:10.1007/978-3-642-15396-9_44

13. Kryszkiewicz, M.: Representative association rules. In: Wu, X., Kotagiri, R., Korb, K.B. (eds.) PAKDD 1998. LNCS, vol. 1394, pp. 198–209. Springer, Heidelberg (1998). doi:10.1007/3-540-64383-4_17
14. Kryszkiewicz, M.: Representative association rules and minimum condition maximum consequence association rules. In: Żytkow, J.M., Quafafou, M. (eds.) PKDD 1998. LNCS, vol. 1510, pp. 361–369. Springer, Heidelberg (1998). doi:10.1007/BFb0094839
15. Métivier, J., Boizumault, P., Crémilleux, B., Khiari, M., Loudni, S.: A constraint language for declarative pattern discovery. In: Proceedings of SAC 2012, pp. 119–125 (2012)
16. Sinz, C.: Towards an optimal CNF encoding of boolean cardinality constraints. In: Beek, P. (ed.) CP 2005. LNCS, vol. 3709, pp. 827–831. Springer, Heidelberg (2005). doi:10.1007/11564751_73
17. Szathmary, L., Napoli, A., Kuznetsov, S.O.: ZART: a multifunctional itemset mining algorithm. In: Proceedings of ICCLTA 2007 (2007)
18. Zaki, M.J.: Mining non-redundant association rules. Data Min. Knowl. Discov. **9**, 223–248 (2004)

Author Index